Biochemistry and Molecular Biology

Second Edition

William H. Elliott

Department of Molecular Biosciences, University of Adelaide

Daphne C. Elliott

School of Biological Sciences, Flinders University of South Australia

OXFORD

UNIVERSITY PRESS

UNIVERSITY PRESS

Great Clarendon Street, Oxford OX2 6DP

Oxford University Press is a department of the University of Oxford.
If furthers the University s objective of excellence in research, scholarship,
and education by publishing worldwide in

Oxford New York

Athens Auckland Bangkok Bogotá Buenos Aires
Cape Town Chennai Dar es Salaam Delhi Florence Hong Kong Istanbul
Karachi Kolkata Kuala Lumpur Madrid Melbourne Mexico City Mumbai
Nairobi Paris São Paulo Singapore Taipei Tokyo Toronto Warsaw

with associated companies in Berlin Ibadan

Oxford is a registered trade mark of Oxford University Press
in the UK and in certain other countries

Published in the United States
by Oxford University Press Inc., New York

A catalogue record for this book is available from the British Library

Library of Congress Cataloging in Publication Data
(Data applied for)

ISBN   0 19 870045 8

Typeset by Best-set Typesetter Ltd., Hong Kong
Printed in Italy
on acid-free paper by Giunti Industrie Grafiche, Florence

It may metaphorically be said that natural selection is daily and hourly scrutinising, throughout the world, the slightest variations; rejecting those that are bad and adding up all that are good; silently and insensibly working, whenever and wherever opportunity offers, at the improvement of each organic being in relation to its organic and inorganic conditions of life.

Charles Darwin

Evolution is a tinkerer.

Francois Jacob

Preface

Biochemistry has been in an ever-increasing state of excitement for decades with its seemingly never-ending stream of almost incredible discoveries; this has brought with it the problem of how teachers and students can cope with the avalanche of new information. Our approach in this book has been to put the emphasis on the understanding of the molecular strategies that constitute life, with the presentation stripped of detail not essential to that understanding. The policy has permitted the inclusion of more explanation of concepts than perhaps is usual and has allowed extra emphasis to be given to crucially important or difficult concepts. It also has allowed the subject to be dealt with in a moderate-sized text without sacrificing coverage of important areas or at the price of superficial treatment; more biological background than is commonly found in biochemistry books is also included.

We have coupled this with what we hope is a readable style of presentation which students new to the subject will find easy to handle. Presentation of material is layered so that the reader is always given the overall picture before going into each topic more deeply. Our aim is that the book should bring students to a level of understanding from which they could move on to many of the 'Trends in Biochemical Sciences' type of review, available in a number of journals, without too much difficulty, and make some keen to do so.

The more traditional areas of biochemistry are dealt with in the earlier parts of the book and the molecular biology in the later parts, but an extensive system of cross-referencing between the chapters has been included so that they can be studied in any order. So far as metabolism goes, we have been selective in what to include. Metabolic pathways of central importance are fully covered but others are given more limited treatment. To illustrate this policy with two examples, in the amino acid metabolism area, the important concepts are explained, but pathways of the metabolism of individual amino acids have been confined to those of special interest such as medical relevance. In cholesterol synthesis, the initial steps that are important in understanding control of cholesterol synthesis are dealt with, but the subsequent steps in the pathway are omitted. By contrast cholesterol homeostasis and its transport around the body are dealt with at some length because of their medical importance.

To keep metabolic biochemistry to a reasonable size we have largely concentrated on pathways found in animals, since for the most part differences in other organisms tend to be in detail only. Photosynthesis is the exception in this respect; it is dealt with because of its central importance. In the molecular biology area, prokaryotic systems are often described as well as eukaryotic ones, because the two may differ in significant ways, for example in gene transcription, and in many cases the information derived from prokaryotes is essential to the story.

This second edition has, in addition to general updating, been extensively rewritten and expanded, especially in areas of molecular biology that have advanced dramatically since the first edition of this book was published.

To give some detail of the changes made to the new edition, starting with the more traditional areas, a new chapter on enzymes and their properties is included and the chapter on proteins expanded. Mechanisms of catalysis at enzyme active centres using proteases as the illustrative examples are dealt with. The current increased interest in connective tissue proteins has been recognized by an extended coverage of this area. A section on the main protein technologies has been added, including a simple brief account of the use of protein and DNA databases which, together with the application of mass spectrometry to protein studies, is having a tremendous impact on the subject. Students are introduced to the terms bioinformatics, proteomics and genomics.

In metabolic control, which is covered in a separate integrated chapter, the recent elucidation of the mechanism of insulin regulation of glycogen synthase and of other cellular events has been included. Attention has also been given to the area of metabolic control analysis with a description of what this is about and of its potential importance in biotechnology. In the membrane chapter, gated ion channels have been dealt with in the context of nerve impulse generation and transmission. In other chapters, the rotary mechanism of ATP synthase, in which much fascinating progress has been made in the past few years, has been dealt with fully and the swinging lever mechanism of muscle contraction explained. The realization of the importance of protein breakdown in fundamental cellular processes has been reflected in a rewritten and expanded chapter on this topic which includes coverage of those astonishing organelles, the proteasomes.

Turning to the molecular biology area, protein synthesis and

folding mechanisms have been updated. Eukaryotic gene transcription and its regulation have been greatly expanded in a separate new chapter in the light of the major advances. This includes an extended coverage of self-splicing and of ribozymes, together with the medical potential of the latter. The cell-signalling chapter has been completely rewritten to give a comprehensive coverage of the dramatic advances in this vital field during the past few years. Recent progress in protein targeting has been described in a new separate chapter, which includes the remarkable research achievements in understanding nuclear–cytoplasmic traffic. A new chapter on cancer molecular biology deals with cell cycle control and its abnormalities. In the recombinant DNA chapter, knockout mice and stem cell technology are now included, both topic of great interest in fundamental and medical research.

The book is aimed at students embarking on their first university course in the subject. Our hope is that it gives a coherent and up-to-date picture of the state of biochemistry and molecular biology today.

Adelaide
January 2001

W.H.E.
D.C.E.

Acknowledgements to the second edition

We are greatly indebted to the following colleagues, who have reviewed sections or given valuable advice:

Dr M. Abbey, Division of Human Nutrition, CSIRO, Adelaide, South Australia

Professor B. A. Askonas, Imperial College, London

Professor P. Atwood, Department of Biochemistry, University of Western Australia

Dr David A. Bender, Royal Free and University College Medical School, Department of Biochemistry and Molecular Biology, University College London, UK

Dr A. J. Berden, Swammerdam Institute of Life Sciences, Amsterdam, The Netherlands

Dr G. W. Booker, Department of Molecular Biosciences, University of Adelaide, South Australia

Dr M. A. Bogoyevitch, Department of Biochemistry, University of Western Australia

Professor John Bowie, Department of Chemistry, University of Adelaide, South Australia

Dr T. K. Bradshaw, Senior Lecturer in Biological Chemistry, School of Biological and Molecular Sciences, Oxford Brookes University, UK

Professor L. A. Burgoyne, Biological Sciences, Flinders University of South Australia

Professor D. Catcheside, Biological Sciences, Flinders University, South Australia

Professor M. C. Clark, Department of Biochemistry, University of Tasmania

Professor M. James C. Crabbe, Division of Cell and Molecular Biology, School of Animal and Microbial Sciences, The University of Reading, UK

Dr S. Dalton, Department of Molecular Biosciences, University of Adelaide, South Australia

Professor Alan P. Dawson, School of Biological Sciences, University of East Anglia, Norwich, UK

Professor Keith Dyke, Microbiology Unit, Department of Biochemistry, University of Oxford, UK

Dr Malcolm East, School of Biological Sciences, Biochemistry and Molecular Biology Division, University of Southampton, UK

Professor J. B. Egan, Department of Molecular Biosciences, University of Adelaide, South Australia

Professor David Fell, School of Biological and Molecular Sciences, Oxford Brookes University, UK

Dr G. Findlay, Biological Sciences, Flinders University, South Australia

Professor F. Gibson, John Curtin School of Medical Research, Australian National University

Dr G. Goodall, Hanson Cancer Institute, IMVS, South Australia

Dr T. Gonda, Hanson Cancer Institute, IMVS, South Australia

Dr G. Gream, Department of Chemistry, University of Adelaide, South Australia

Professor N. Hoogenraad, Department of Biochemistry, La Trobe University, Victoria

Dr D. Jans, Division of Biochemistry and Molecular Biology, John Curtin School of Medical Research, Australian National University

Dr J. Kelly, Department of Genetics, University of Adelaide, South Australia

Professor P. Kuchel, Department of Biochemistry, Sydney University, New South Wales

Liisa Laakkonen, Institute of Biomedicine, University of Helsinki, Finland

Professor B. K. May, Department of Molecular Biosciences, University of Adelaide, South Australia

Dr T. Miles, Department of Physiology, University of Adelaide, South Australia

Professor W. O'Sullivan, School of Biochemistry and Molecular Genetics, University of New South Wales

Dr Ralph Rapley, Department of Biosciences, University of Hertfordshire, UK

Professor P. Rathjen, Department of Molecular Biosciences, University of Adelaide, South Australia

Mrs R. Rogers, Department of Molecular Biosciences, University of Adelaide, South Australia

Professor R. H. Symons, Waite Agricultural Research Institute, University of Adelaide, South Australia

Professor John M. Walker, Head of Department of Biosciences, University of Hertfordshire, UK

Professor J. C. Wallace, Department of Molecular Biosciences, University of Adelaide, South Australia

Professor J. Wheldrake, Biological Sciences, Flinders University, South Australia

Our especial thanks are again due to Professor L. A. Burgoyne, Biological Sciences, Flinders University of South Australia, who patiently read the entire typescript of new material in draft form, and made many valuable comments.

We also wish to thank the staff at Oxford University Press for their help and co-operation in the development and production of this edition.

Brief contents

Contents

Part 3 Metabolism

Part 4 Information storage and utilization

Part 5 Transport of oxygen and CO₂

Part 6 Mechanical work by cells

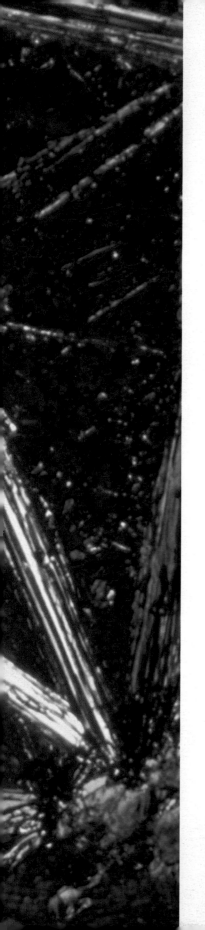

Part 1

Introduction to the chemical reactions of the cell

Crystals of the enzyme lactate dehydrogenase.

Chapter summary

Chemistry, energy, and metabolism

Life is a chemical process involving thousands of different reactions occurring in an organized manner. These are called **metabolic reactions** and, collectively, **metabolism**.

In studying biochemistry you will therefore be confronted with a fairly large number of chemical equations describing the essentials of cell chemistry (though not with all of the thousands of reactions). The chemical strategies perfected by millions of years of evolution are elegant and fascinating but to appreciate these, or for that matter to easily understand and enjoy biochemistry, you first need to appreciate the problems that faced the establishment of life.

The consideration underlying everything in life is **energy**. A printed chemical equation on its own lacks a vital piece of information and that is the energy change involved. It may not be immediately obvious what is meant by this, for chemical energy change in reactions occurring in solution is not something of which we are necessarily conscious. Energy considerations determine whether a reaction is possible on a significant scale and whether the reverse reaction can occur to a significant degree.

To give an example of how this aspect applies directly to living processes, during vigorous exercise, glycogen (a polymer of glucose) in muscles is converted to lactic acid by a series of chemical reactions. During the ensuing rest period, lactic acid is converted to glycogen but not by the reverse of exactly the same chemical reactions by which it was formed. With a knowledge of simple energy considerations it is possible to see why the cell does what it does. Without such a knowledge, the chemistry can seem pointlessly complicated.

The level of knowledge of thermodynamics needed to understand biochemistry, in general, is little more than common sense. While there are specialist areas of the subject where a sophisticated knowledge of thermodynamics is needed, it is probably true that the majority of biochemists do not often use a thermodynamic equation, but nevertheless understand the simple principles involved. The aim of this section is to give you this simple understanding.

What determines whether a chemical reaction is possible?

As already implied, energy considerations permit certain chemical reactions to occur while others are not allowed. Note that in biochemistry we are concerned with reactions occurring to a *significant* extent. In principle, energy considerations do not completely prevent a reaction occurring (unless reactants and products are precisely at equilibrium concentrations; see below), but if the reaction ceases after a minute amount of conversion has taken place, then, for the practical purposes of life, it has not occurred.

First, what do we mean by energy change in a chemical system (the latter being an assemblage of molecules in which chemical reactions can take place, such as occurs inside a cell)? This is not self-evident as is the case with, say, gravitational energy change in a weight falling. A chemical system involves a huge number of individual molecules each of which contains a certain amount of energy dependent on its structure. This energy can be described as the heat content or **enthalpy** of the molecule. When a molecule is converted to a different structure in a chemical reaction, its energy content may change; the change in the enthalpy is written as ΔH (delta H). The ΔH may be negative (heat is lost from molecules and released, so raising the temperature of the surroundings) or positive (heat is taken up from the surroundings, which, correspondingly, cool).

At first sight it may seem surprising that reactions with a positive ΔH can occur since it might seem analogous to a weight raising itself from the floor. This is the point at which physical analogies such as weights falling become inadequate as models for chemical reactions. In the latter, a negative ΔH favours the reaction and a positive ΔH has the opposite effect. But, in a chemical reaction, ΔH is not the final arbiter as is gravitational energy with a weight system, for entropy change, or ΔS, also has a say in the matter.

Entropy can be defined as the degree of randomness of a system. In a chemical system, this can take three forms: first, a molecule is not usually rigid or fixed—it can vibrate, twist around bonds, and rotate. The greater the freedom to do these things—really, to indulge in any molecular movement—the greater the randomness or entropy. Secondly, in a chemical system, vast numbers of individual molecules are involved, which may be randomly scattered or be in some sort of orderly arrangement as occurs, for example, to a great degree in living cells. Thirdly, the number of individual molecules or ions may change as a result of a chemical change. The greater the number of individual molecules, the greater the randomness; the greater the randomness, the greater the entropy. Thus a chemical reaction may change the entropy of the system. Increasing entropy lowers the energy level of the system; decreased entropy (increased order) increases the energy level.

Both ΔH and ΔS have a say in determining whether a chemical reaction may occur. A negative ΔH and a positive ΔS both reinforce the 'yes' decision; a positive ΔH and negative ΔS both reinforce the 'no' decision. A negative ΔH and negative ΔS disagree on the decision, as do a positive ΔH and a positive ΔS, and whether the outcome is yes or no depends on which is quantitatively the larger.

The driving force of increasing entropy is often illustrated by the example of the melting of a block of ice in warm water. Heat is taken up as the ice melts (the ΔH is positive) but the scattering of the organized molecules of the ice crystal, as it dissolves, increases the entropy so that the process proceeds. It must be admitted, however, that it is not necessarily easy to visualize how entropy drives a chemical reaction. A usual statement is that, if energies are similar, the disordered state occurs more readily than the ordered state. If you find it difficult to conceptualize entropy as a driving force, rather than have a mental block it would be best to accept the concept for the moment; it will become more familiar later.

The situation we have described between ΔS and ΔH is not convenient—we have two terms of variable sizes that may reinforce or oppose each other in determining whether a reaction is possible. Moreover, in biological systems it is difficult or impossible to measure the ΔS term directly. The situation was greatly ameliorated by the **Gibbs free energy** concept that combined the two terms in a single one. The change in free energy (ΔG after Gibbs) is given by the famous equation

$$\Delta G = \Delta H - T\Delta S$$

where T is the absolute temperature. This equation applies to systems where the temperature and pressure remain constant during a process, which is the case for biochemical systems.

The term 'free' in free energy means free in the sense of being *available* to do useful work, not free as in something for nothing. ΔG represents the *maximum* amount of energy available from a reaction to do useful work. It is somewhat like available cash being free for you to make purchases with. Useful work includes muscle contraction, chemical synthesis in the cell, and osmotic and electric work. ΔG values are expressed in terms of calories or joules per mole (1 calorie = 4.19 joules); the former term is used in some books, but the latter is now the official version. Since the values are large, the terms kilocalories or kilojoules per mole are used. The latter is usually abbreviated to $kJ\,mol^{-1}$. The ΔG of a reaction is the all-important thermodynamic term; in its application to chemical reactions, the rule is as follows.

A chemical reaction can occur only if the ΔG is negative—that is, that at the *prevailing conditions* the products have less free energy than the reactants have. That's all.

Reversible and irreversible reactions and ΔG values

You may have learned in chemistry that, strictly speaking, *all* chemical reactions are reversible. This may puzzle you because it might imply that the ΔG value must be negative in both directions—remember that a reaction cannot occur unless the ΔG value is negative. The answer to this apparent paradox is that the ΔG of a reaction is not a fixed constant but varies with the reactant and product concentrations. The relationship is given later in this chapter. Thus in the reaction A $\rightleftharpoons$ B, if A is at a high concentration and B at a low one, the ΔG may be negative in the direction A $\rightarrow$ B and, of course, positive from B $\rightarrow$ A. Reverse the concentrations and the ΔG can be negative for the reverse direction. A reaction will proceed to the point at which A and B are in concentrations at which the ΔG is zero in both directions and then no further net reaction can occur. This is the **chemical equilibrium point**.

If the ΔG of a biochemical reaction A $\rightleftharpoons$ B is small, significant reversibility may be possible in the cell because changes in reactant concentrations in the cell may be sufficient to reverse the sign of the ΔG of the reaction. If it is large, for all practical purposes the reaction is irreversible. In cellular reactions there is relatively little scope for concentration change in the **metabo-**

lites (as reactants and products are called)—the concentrations are always relatively low: 10^{-3}–10^{-4} would be typical of many. The net result is that reactions with large negative ΔG values are irreversible because concentration changes are insufficient to reverse the sign of those values. (We later explain why, in certain reactions, this may appear to be contradicted; see page 149.)

As a general guide, hydrolytic reactions in the cell—reactions in which a bond is split by water—are irreversible, in the sense that synthesis of substances in the cell does not occur by reverse of the hydrolytic reactions. As you proceed through the chapters on metabolism you will become familiar with which reactions are reversible and which are irreversible.

To summarize, a reaction with a small ΔG is likely to be reversible in the cell, the direction being determined by small changes in metabolite concentrations. A reaction with a large ΔG value will, in cellular terms, proceed in one direction only and, moreover, will proceed to virtual completion because the equilibrium point is so far to the side of the reaction. Put in another way, in the latter case, the ΔG of the reaction does not become zero until virtually all of the reactant(s) have been converted to product(s).

The importance of irreversible reactions in the strategy of metabolism

In the preceding section we have emphasized the effect of the ΔG of a reaction in determining whether that reaction can proceed in the reverse direction. Why is there such emphasis on reversibility? We will now explain this. The major physiological chemical processes of the cell usually involve, not single reactions, but series of reactions organized into metabolic pathways in which the products of the first reaction are the reactants of the next and so on. In the example of the glycogen → lactic acid conversion in muscle mentioned earlier, a dozen successive reactions are involved.

An important general characteristic of metabolic pathways is that they are, as a whole, irreversible. Many of the individual reactions in a pathway may be freely reversible but it virtually always contains one or more reactions that cannot be directly reversed in the cell. Such irreversible reactions act as one-way valves and ensure that from a thermodynamic viewpoint the pathway can proceed to completion. Metabolic pathways are decisive—they go with a bang, as it were.

All this is not the same as saying that overall physiological chemical processes are irreversible. Lactic acid *is* converted back to glycogen in the body and, while many steps in the process are the simple reversal of those in the forward process, there are steps that cannot be reversed directly and alternative reactions

are necessary. These involve the input of energy, which makes the alternative reactions also irreversible, but in the opposite direction. So the forward pathway (glycogen → lactic acid) is directly irreversible as in the reverse pathway (lactic acid → glycogen). A typical metabolic situation is

$$A \rightleftharpoons B \; \circlearrowright \; C \rightleftharpoons D \rightleftharpoons E \; \circlearrowright \; F \rightleftharpoons Products .$$

The red arrows represent irreversible reactions with large negative ΔG values. You will want to know how energy is provided to achieve irreversibility in both directions. This will become clear later.

A general biochemical principle emerges. *Whenever the overall chemical process of a metabolic pathway has to be reversed, the reverse pathway is not exactly the same as the forward pathway—some of the reactions are different in the two directions.*

Why is this metabolic strategy used in the cell?

There are basically two reasons. The alternative to the strategy we have outlined is that *all* the reactions of a metabolic pathway are reversible, that is,

$$A \rightleftharpoons B \rightleftharpoons C \rightleftharpoons D \rightleftharpoons E \rightleftharpoons F \rightleftharpoons products.$$

The major drawback of this arrangement is that the whole process is subject to mass action. If the concentration of A increases (perhaps due to something you ate), the reaction would swing to the right and more products would be formed. If substance A decreased in amount, some of the products would revert to A to maintain the equilibrium. Imagine that the pathway is for the synthesis of molecules such as the DNA of genes or of vital proteins, etc. and you can see how impossible this scenario is. It would be rather like building the walls of a house with the laying of bricks being a reversible process. The walls would rise and fall to maintain a constant equilibrium with the number of bricks lying on the ground. In the cell, if the concentration of A were to rise, more DNA and protein would be synthesized; if it were to fall, the DNA or proteins would break down—a totally unworkable situation for a living cell.

There is a second, important but related reason for having dissimilar reactions in the forward and reverse direction of metabolic pathways. Metabolism must be controlled. As already stated, during violent exercise muscles convert glycogen to lactic acid. During rest, the lactic acid is converted to glycogen and the forward conversion switched off. To independently control the two directions (that is, to switch one on and the other off) there must be separate reactions to control; otherwise it would be possible only to switch both directions on and off together. Thus, the irreversible reactions are usually the control

points. Metabolic control is a major subject that we will deal with in Chapter 13.

How are ΔG values obtained?

The ΔG value of a reaction, as explained is *not* a fixed constant but the ΔG value of a reaction under specified standard conditions is a fixed constant. The standard conditions are with reactants and products at 1.0 M, 25°C, and pH 7.0, that is, the free energy difference between separated one-molar solutions of reactants and products. The ΔG value, *under these conditions*, is called the **standard free energy change of a reaction**. It is denoted as $\Delta G^{0\prime}$, the prime being used in biochemical systems to indicate that the pH is 7.0 rather than the value at pH 0, which is used in physical sciences.

The $\Delta G^{0\prime}$ value may often be calculated from physical chemistry tables. In these, the standard free energies of formation of large numbers of compounds are listed. For many reactions, the $\Delta G^{0\prime}$ value can be calculated by adding up the free energies of formation of the reactants and (separately) those of the products. The difference is the $\Delta G^{0\prime}$ value. An alternative is to determine experimentally the equilibrium constant for the reaction and from this the $\Delta G^{0\prime}$ value is easily calculated.

There is a simple direct relationship between $\Delta G^{0\prime}$ values and ΔG values at given reactant and product concentrations. If, therefore, we know the relevant metabolite concentrations in the cell, the ΔG value for the reaction in the cell is readily determined (see page 12 for an illustrative calculation). This has been done for a good many biochemical reactions so that such values are often quoted.

There is a snag—determining the actual concentrations of the thousands of metabolites in a cell is a nontrivial matter. They are present at low concentrations and are changing anyway. So, for many biochemical reactions, we do not have this data and therefore do not have the ΔG values. However, it is found that $\Delta G^{0\prime}$ values usually correlate very well with known cellular happenings so that they are a useful guide in understanding metabolic reactions. Thus, although such values are not directly applicable to cells since metabolite concentrations are never 1.0 M, they are frequently quoted to explain why certain reactions behave as they do. It is a compromise but a very useful one.

Standard free energy values and equilibrium constants

A particularly useful aspect of knowing the $\Delta G^{0\prime}$ value of a reaction is that, from it, the **equilibrium constant** of that reaction is readily determined. The equilibrium constant K'_{eq} of a reaction represents the ratio at equilibrium of the products to reactants. (The prime indicates that it is the K_{eq} at pH 7.0.) Thus, if one

Table 1.1 Relationship of the equilibrium constant (K_{eq}) of a reaction* to the $\Delta G^{0\prime}$ value of that reaction

Approximate $\Delta G^{0\prime}$ (kj mol^{-1})	K'_{eq}
+17.1	0.001
+11.4	0.01
+5.7	0.1
0	1.0
−5.7	10.0
−11.4	100
−17.1	1000

*For a reaction $A + B \leftrightarrow C + D$, the equilibrium constant is the molar concentration of $C \times D$ divided by that of $A \times B$.
$\left(K_{eq} = \dfrac{[C][D]}{[A][B]} \right)$; K'_{eq} is the K_{eq} at pH 7.0.)

considers the reaction $A + B \rightleftharpoons C + D$, the K'_{eq} is calculated from the concentrations of A, B, C, and D present after the reaction has come to equilibrium—that is, when there is not net change in their concentrations,

$$K'_{eq} = \frac{[C][D]}{[A][B]}.$$

The relationship between the value of the $\Delta G^{0\prime}$ for a reaction and the K'_{eq} for that reaction is

$$\Delta G^{0\prime} = -RT \ln K'_{eq} = -RT\,2.303 \log_{10} K'_{eq}.$$

R is the gas constant (8.315 J mol^{-1} K^{-1}) and T is the temperature in degrees Kelvin (298 K = 25°C). At 25°C, $RT = 2.478$ kJ mol^{-1}. Thus, if the K'_{eq} is determined, the $\Delta G^{0\prime}$ for the reaction can be calculated and vice versa. Table 1.1 shows the relationship between $\Delta G^{0\prime}$ value and the K'_{eq} value for chemical reactions.

Given that a reaction has a negative ΔG value, what determines whether it actually takes places at a perceptible rate in the cell?

The change in free energy of a chemical reaction must, as described above, be negative for that reaction to occur, but, given that condition, it does not follow automatically that such a reaction actually proceeds at a perceptible rate. Thermodynamic considerations of sugar reacting with oxygen (burning) to form CO_2 and H_2O are highly favourable, but sugar is quite stable in air. There is a barrier to the occurrence of chemical reactions, even if they involve large negative free energy changes. If this weren't so, all combustible material on Earth would burst into flame—or, rather, it would have done so long ago. However, sugar molecules, so stable in the bowl, can be rapidly oxidized when they enter a living cell. The question then is, what causes reactions to occur in the cell that doe not occur outside the cell (or do so at an imperceptible rate)?

The answer lies in enzyme catalysis. An **enzyme**, in its simplest form, is a **catalyst** that brigs about one chemical reaction. There are many thousands of biochemical reactions, each catalysed by a separate enzyme. However, it is more convenient to deal with enzymes in the next chapter, so for the time being please just keep in mind that reactions in the cell are brought about by specific enzymes.

How is food breakdown in cells coupled to drive energy-requiring reactions in the cell?

To introduce this topic we will start with a very simple basic but important concept. Energy cannot be created or destroyed; the total amount of energy in the universe remains constant. This is known as the **law of conservation of energy** expressed in the first law of thermodynamics. However, energy can be transformed from one form to another; kinetic energy (movement) can be converted into heat; heat can be converted into electricity and vice versa; light can be converted into the chemical energy of food by plants, food energy can be converted into body constituents by animals, and so on. Put colloquially, you can't get energy for nothing; there always must be a pre-existing source. The source of energy in animals is obviously the chemical energy contained in food which ultimately comes from nuclear reactions in the sun conveyed as light energy and converted into chemical energy by photosynthesis. A major biochemical task is to release that energy in a form which can be used for performing useful work. Heat cannot be used to perform useful work in a living cell so energy release from food and its utilization clearly cannot be the simple oxidation (burning) of food. What we want to do in the next few pages is to explain how the cell utilizes the chemical energy of food for its own purposes.

The conversion of simple precursor molecules to larger cellular molecules (such as DNA and proteins) involves large increases in free energy and therefore cannot occur spontaneously. 'Spontaneous' in a thermodynamic context has a special limited meaning, namely *energetically* unassisted. (Enzymes are still needed to catalyse biochemical reactions even if spontaneous in this sense.) The required energetic assistance comes ultimately from food oxidation. Chemical conversions involving positive free energy changes—the synthetic or 'building up' processes—are collectively called **anabolism** or anabolic reactions. (The anabolic steroids of sporting ill-repute promote increase in body mass, hence their name.) The other half of metabolism consists of the 'breaking down' reactions with negative free energy changes—the catabolic reactions, collectively, **catabolism**. Metabolism is comprised of catabolism and anabolism. Catabolism of food liberates free energy, which is used to drive the energy-requiring processes of anabolism. A chemical *conversion* involving a positive free energy change can, as stated, occur only if energetically assisted but the ΔG of the *actual reaction(s)* by which the conversion is accomplished must be negative—as is the case with all reactions. Catabolic reactions are often referred to as exergonic since they give out energy and those requiring energetic assistance are sometimes referred to as endergonic reactions. A distinction needs to be made between a chemical reaction and a chemical conversion or process occurring by a number of separate reactions. Thus, compound A may be converted in the cell to compound X even if X has a much higher standard free energy level than A, and this necessitates energetic assistance. But, the conversion must, by definition, be achieved by chemical reactions, each of which has a negative ΔG value. We will deal with how energetic assistance is given shortly.

We can summarize the overall situation as in Fig. 1.1. Food oxidation releases energy, which is used to drive energetically unfavourable processes. To keep the system going on a global scale, CO_2 and H_2O are reconverted during photosynthesis to food molecules (glucose) using light energy. Glucose is converted by organisms to other food molecules such as fat. Although the assembly of large cellular structures involves a decrease in entropy (unfavourable), the oxidation of food molecules involves a greater entropy increase (favourable). The entropy change of the total system (cell and surroundings) is positive, and so the **second law of thermodynamics** is obeyed. (This law states that all processes must result in an increase in the total entropy of the universe.)

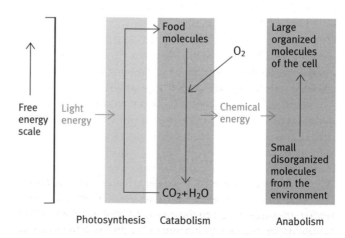

Fig. 1.1 The energy cycle in life.

An analogy can be made of petroleum oxidation driving a car uphill, but, if you placed a bucket of petroleum under the bonnet and set it alight, while energy would be liberated, it wouldn't be useful in driving the car—the negative ΔG of petroleum oxidation must be appropriately coupled to the wheels of the car. Similarly, oxidation of food in the cell without appropriate coupling to the energy-requiring reactions would simply liberate heat and this cannot be used to do chemical or other work in the cell. Instead, the free energy change involved in food oxidation must somehow be coupled to the energy-requiring processes.

How is this done? It cannot be done by heat liberation. Heat at body temperature is waste energy—it cannot, as indicated, be used to perform work. With reference also to this problem, energy release from different foods occurs by a variety of reactions and this energy release has to drive many forms of energy-requiring processes in the cell. These include mechanical (muscular) work, osmotic work, and electrical work, in addition to chemical work. It would be impracticable to *directly* and inflexibly link up the different energy-producing reactions to each of the many energy-requiring processes. Instead, there is a common energetic intermediary that accepts the energy from *all* types of food oxidation and that can deliver it to *all* types of processes involving work and requiring energy. This is a more flexible arrangement. There is an analogy in electrical energy generation. The free energy released by burning coal, oil, wood, or gas may be converted to electrical energy, which can be transmitted to where it is needed and used for a wide variety of applications. (The analogy isn't quite right, because electricity generation involves heat as an intermediary which is *not* true of the cell.) In the case of the cell, the energy intermediary has to be chemical in nature. Which leads to the next important question—what is this universal intermediary?

The high-energy phosphate compound

The answer to the above central question is remarkably simple and applies universally throughout life (to be strictly correct there are exceptions, but they are so rare that assertion of a single universal energy intermediary is justified).

The concept can be stated very simply. As a result of catabolic reactions of food molecules, inorganic phosphate ions are converted into phosphoryl groups in molecules of a **high-energy phosphate compound** (defined below). The high-energy phosphate compound is transported to wherever work is to be performed in the cell, where the phosphoryl groups are converted back to inorganic phosphate ions, with the liberation of the free energy that went into the formation of the phosphate compound. It is very important to note, however, that this does not

mean that *direct* hydrolysis of the phosphate compound occurs, as this could only liberate the energy as useless heat. The mechanisms by which the energy is harnessed for work will be described shortly.

What is a 'high-energy phosphate compound'?

If we consider cellular compounds containing a phosphoryl group, they can be divided into two categories—low-energy phosphate compounds, whose hydrolysis to liberate inorganic phosphate (P_i) is associated with negative $\Delta G^{0'}$ values in the range of about 9–20 kJ mol^{-1} and high-energy phosphate compounds with corresponding negative $\Delta G^{0'}$ values larger than about 30 kJ mol^{-1}. The concept of a high-energy phosphate compound used to be described in terms of the 'high-energy phosphate bond', but this usage has been abandoned because a high-energy bond in chemical terms refers to a bond whose breakage requires a large input of energy, the reverse of the intended biochemical concept. For clarity of presentation we will occasionally use the term 'high-energy phosphoryl group' as a shorthand way of referring to a phosphoryl group in a high-energy phosphate compound. (The energy resides in the molecule as a whole and not in the actual phosphoryl group.) With that important qualification, the **high-energy phosphoryl group** can be usefully regarded as the universal energy currency of the living cell. This concept is illustrated in Fig. 1.2 which is a refinement of Fig. 1.1.

What are the structural features of high-energy phosphate compounds?

Phosphoric acid, H_3PO_4, is an oxyacid of phosphorus with three dissociable protons as shown below.

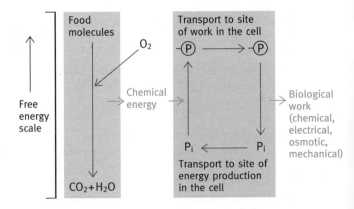

Fig. 1.2 The role of the phosphoryl group in the energy economy of a cell. —Ⓟ represents the phosphoryl group in a molecule whose hydrolysis to liberate P_i is associated with a $\Delta G^{0'} > 30$ kJ mol^{-1}.

$$HO-\overset{\overset{\displaystyle O}{\|}}{\underset{\underset{\displaystyle OH}{|}}{P}}-OH \quad \underset{pK_{a_1}=2.2}{\rightleftharpoons} \quad HO-\overset{\overset{\displaystyle O}{\|}}{\underset{\underset{\displaystyle OH}{|}}{P}}-O^- + H^+$$

$$pK_{a_2}=7.2 \;\updownarrow$$

$$HO-\overset{\overset{\displaystyle O}{\|}}{\underset{\underset{\displaystyle O^-}{|}}{P}}-O^- + H^+$$

$$pK_{a_3}=12.3 \;\updownarrow$$

$$^-O-\overset{\overset{\displaystyle O}{\|}}{\underset{\underset{\displaystyle O^-}{|}}{P}}-O^- + H^+$$

At normal cell pH, a mixture of the singly negatively and doubly negatively charged phosphate ions predominates in the solution. In biochemistry, this mixture of phosphate ions is symbolized as P_i, the i indicating 'inorganic'; it represents the lowest energy form of phosphate in the cell and can be regarded as the ground state of phosphate when considering its energetics. The degree of ionization of the molecule is a function of the three dissociation constants of phosphoric acid. These are known as **pK_a values**; a pK_a is the pH at which there is 50 per cent dissociation of the group. An explanation of pK_a values and buffers is given in the appendix to this chapter; this should be studied and learned now if you are not already familiar with the subject. At physiological pH (say 7.4) the group with a pK_a of 2.2 will be fully ionized and that with a pK_a of 12.3 will be undissociated. The group with a pK_a value of 7.2 will be partially dissociated. The actual degree of the latter can be calculated from the **Henderson-Hasselbalch equation**,

$$pH = pK_{a_2} + \log\frac{[Salt]}{[Acid]};$$

$$7.4 = 7.2 + \log\frac{[HPO_4^{2-}]}{[HPO_4^-]}.$$

Therefore,

$$\log\frac{[HPO_4^{2-}]}{[H_2PO_4^-]} = 0.2$$

so that

$$\frac{[HPO_4^{2-}]}{[HPO_4^-]} = 1.58$$

Inorganic phosphate, esterified with an alcohol, is a phosphate ester.

$$H_2O + R-O-\overset{\overset{\displaystyle O}{\|}}{\underset{\underset{\displaystyle O^-}{|}}{P}}-O^- \longrightarrow ROH + HO-\overset{\overset{\displaystyle O}{\|}}{\underset{\underset{\displaystyle O^-}{|}}{P}}-O^-$$

Phosphate ester   Alcohol Inorganic phosphate (P_i)

The $\Delta G^{o'}$ for the hydrolysis of this ester is roughly of the order of $-12.5\,kJ\,mol^{-1}$, resulting in the equilibrium for the hydrolytic reaction being strongly to the side of hydrolysis. In the cell the reverse reaction does not occur.

As well as forming phosphate esters with alcohols, inorganic phosphate can form a phosphoric anhydride called **inorganic pyrophosphate** (*pyro* means fire; pyrophosphate can be made by driving off water from P_i at high temperatures) and in biochemistry it is often written as PP_i. The $\Delta G^{o'}$ for the hydrolysis of this compound is $-33.5\,kJ\,mol^{-1}$, which is much higher than the value for hydrolysis of the ester phosphate.

$$HO-\overset{\overset{\displaystyle O}{\|}}{\underset{\underset{\displaystyle O^-}{|}}{P}}-O-\overset{\overset{\displaystyle O}{\|}}{\underset{\underset{\displaystyle O^-}{|}}{P}}-O^- \quad \underset{\Delta G^{o'}=-33.5\,kJ\,mol^{-1}}{\xrightarrow{\quad +\ H_2O \quad}} \quad 2\ HO-\overset{\overset{\displaystyle O}{\|}}{\underset{\underset{\displaystyle O^-}{|}}{P}}-O^- + H^+$$

Inorganic pyrophosphate (PP_i)   Inorganic phosphate (P_i)

The very high free-energy release associated with the hydrolysis of the phosphoric anhydride group is due to several factors. When the pyrophosphoryl group is hydrolysed, several things result.

1 The destabilization of the structure caused by electrostatic repulsion of the two negatively charged phosphoryl groups is relieved. This factor is made clear by the reflection that, in the reverse reaction to synthesize a pyrophosphoryl bond, the electrostatic repulsion of the two ions has to be overcome in bringing them together.

2 The products of the reaction (in this case, two P_i ions) are **resonance stabilized**—they have a greater number of possible resonance structures than has the pyrophosphate structure. This is because the two phosphoryl groups in the latter have to compete for electrons in the bridging oxygen atom; these electrons are essential for the resonating structure. Therefore, in the pyrophosphate molecule one or other group can resonate—they compete in this. The two P_i ions can resonate simultaneously. In an inorganic phosphate ion,

all of the P—O bonds are partially double-bonded in character rather than the proton being associated with any one oxygen, resulting in an increase in entropy. The main resonating forms of the phosphate ion are shown below:

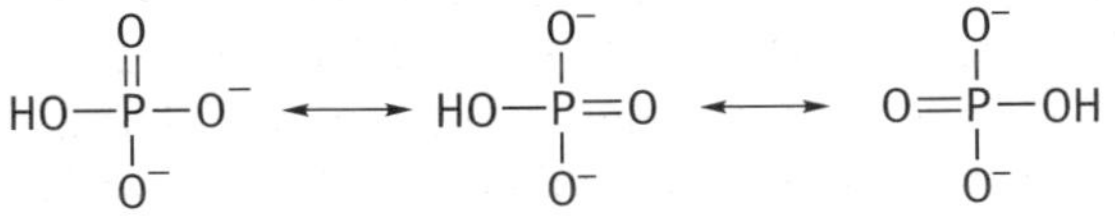

It should be noted that the ↔ symbol has a special meaning in chemistry; it is not the same as ⇌. It does not imply that the different ionized forms are interconverting but that the structure which exists is not any of the forms shown but is an intermediate one in which all oxygens have a partial negative charge and the proton is not associated with any one form. The same comments apply to the structures of the resonating forms of a carboxylic acid and guanidino compounds described below.

These factors all favour the hydrolysis reaction, which has an equilibrium constant decisively in favour of P_i formation (equals a large negative $\Delta G^{0'}$ value). These considerations apply not only to inorganic pyrophosphate but to any phosphoric anhydride group. Other 'high-energy' phosphoryl groups will also be referred to later. Points 1 and 2 above are not applicable to hydrolysis of ester phosphates, which explains why the energy release is so much less.

The phosphoric anhydride structure discussed above is not the only biological high-energy phosphate compound (though it is the predominant one). Three other types are found.

The first is the anhydride between phosphoric acid and a carboxyl group, hydrolysis of which by the reaction shown has a very large negative $\Delta G^{0'}$ value ($-49.3\,\text{kJ mol}^{-1}$ in a typical case). The large free-energy change is associated with the resonance stabilization of both products, namely, P_i and the carboxy acid (the latter illustrated below).

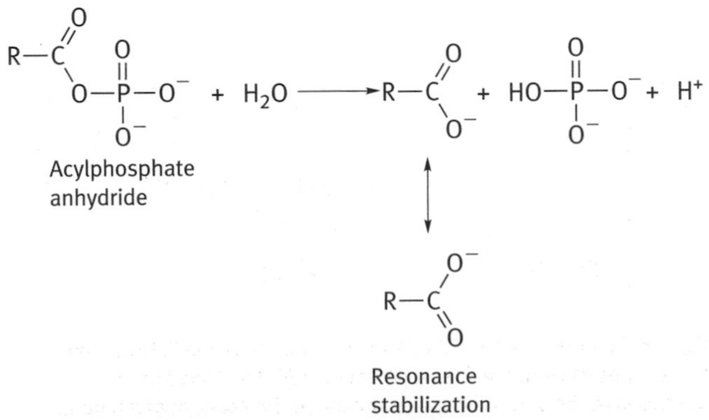

Acylphosphate anhydride

Resonance stabilization

The second structure is that of guanidino-phosphate, whose hydrolysis to produce P_i also has a very large negative $\Delta G^{0'}$ value ($-43.0\,\text{kJ mol}^{-1}$ in a typical case). Again this is due to resonance stabilization of both products.

Guanadino-phosphate

Resonance stabilization

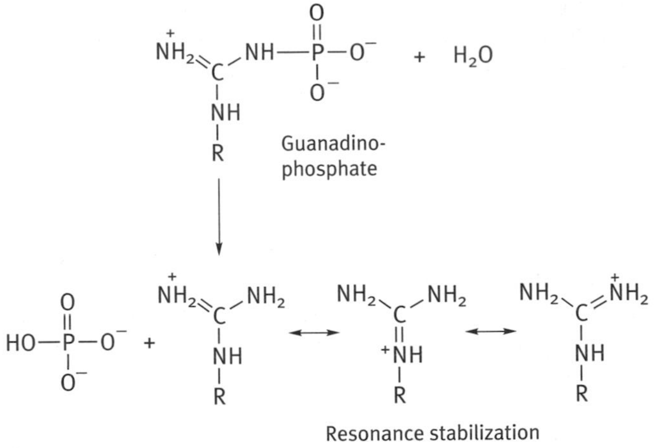

A third type is in a different category—an enol-phosphate structure. This is found in the metabolite, phosphoenolpyruvate. This looks an unlikely candidate for high-energy status, but, on removal of the phosphate group by hydrolysis, the enol structure so formed spontaneously tautomerizes into keto form of pyruvate, the equilibrium of this being far to the right.

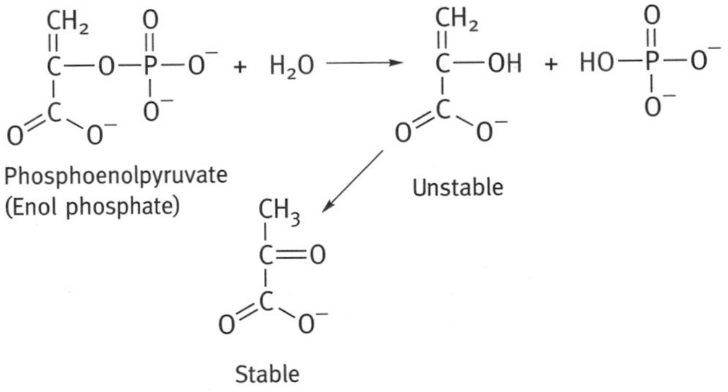

Phosphoenolpyruvate (Enol phosphate)

Unstable

Stable

The overall conversion of phosphoenolpyruvate to P_i and the keto form of pyruvate has a $\Delta G^{0'}$ value of $-61.9\,\text{kJ mol}^{-1}$.

With that description of the 'high-energy phosphoryl group', we can refine the situation in Fig. 1.1 to that in Fig. 1.2 which, in the overall sense, is a complete description of energy generation and utilization.

As you progress through the book you will meet a good number of phosphate compounds and we will give their individual structures and $\Delta G^{0'}$ values for their hydrolysis as we come to

them. However, there is one of absolutely central importance that we will deal with now.

We have so far spoken of the phosphoryl groups present in high-energy phosphate compounds in general terms. Thus, in Fig. 1.2, P_i is shown as being elevated to a 'high-energy phosphoryl group' and then transported around the cell to where it is needed to supply the energy for work. Clearly, the phosphoryl group (—Ⓟ) must be covalently attached to another molecule that acts as its carrier within the cell. Which brings us to the next question.

What transports the —Ⓟ around the cell?

The general carrier is AMP or **adenosine monophosphate**, whose structure is given in Fig. 1.3. AMP carrying a single —Ⓟ group is called ADP or **adenosine diphosphate**. AMP carrying two —Ⓟ groups is ATP or **adenosine triphosphate**. ATP therefore can be written as AMP—Ⓟ—Ⓟ. The two terminal phosphates are attached to AMP by phosphoric anhydride linkages of the same type as already met in PP_i, and the reasons given earlier for the large $\Delta G^{0'}$ for hydrolysis off of these two groups apply equally here. Note that AMP itself is a low-energy ester phosphate and does not itself directly participate in the energy cycle except as carrier for the two —Ⓟ groups. You will see from Fig. 1.3 that the $\Delta G^{0'}$ of hydrolysis of each of the two terminal groups is $-30.5\,kJ\,mol^{-1}$; that is, of the high-energy type. When food is oxidized, the energy released, as explained, is coupled to the conversion of ADP (adenosine diphosphate or AMP—Ⓟ) to ATP,

$$ADP + P_i \xrightarrow[+energy]{} ATP + H_2O.$$

Each cell has only a very small quantity of ATP in it at any one time. The amount would last only a very short time—and it can't get any from the outside; neither ATP, ADP, or AMP can spontaneously diffuse through the cell membrane because they are highly charged (see Chapter 4). Each cell has to synthesize the entire molecule itself. ATP thus 'turns over' or cycles very rapidly, by which we mean it breaks down to ADP and P_i and is resynthesized to ATP. We can modify the diagram given in Fig. 1.2 to include this fact (Fig. 1.4). The poison, cyanide, stops most of the resynthesis of ATP and death results in a very short time because insufficient energy can be supplied for cellular processes once the small amount of pre-existing ATP is broken down.

How does ATP perform chemical work?

Suppose the cell needs to synthesize X—Y from the two reactants, X—OH + Y—H, and the $G^{0'}$ change involved in the conversion is $12.5\,kJ\,mol^{-1}$. The simple reaction XOH + YH → XY + H_2O cannot occur to any significant extent because the $\Delta G^{0'}$ is positive and the equilibrium is far to the left. The solution is to use **coupled reactions** involving ATP breakdown, as shown below. Coupled reactions are two or more reactions in which the product of one becomes the reactant for the next. No physical coupling is necessarily involved—they simply have to be present in the same chemical system. Note that all reactions in the cell involving ATP must be enzymically catalysed—the fact that hydrolysis of each of the two phosphoric anhydride groups is strongly exergonic does not mean that ATP is an unstable or

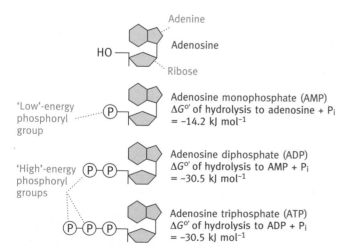

Fig. 1.3 Diagrammatic representation of adenosine and its phosphorylated derivatives. Adenosine is a nucleoside with ribose and adenine as its components. Details of their structures are irrelevant here but are dealt with in later chapters.

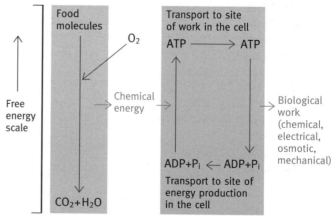

Fig. 1.4 The role of ATP in the energy economy of a cell. Note that some types of work involve breakdown of ATP to AMP, but, as described in the text, this does not change the concept given here.

highly reactive molecule. (It can be bought in bottles as a white powder and can be stored indefinitely in the freezer.)

Reaction 1:   XOH + AMP—Ⓟ—Ⓟ → X—Ⓟ + AMP—Ⓟ

Reaction 2:   X—Ⓟ + YH → X—Y + P_i

Sum of 1 + 2: XOH + YH + AMP—Ⓟ—Ⓟ
$\quad$ → X—Y + AMP—Ⓟ + P_i

The overall $\Delta G^{0'}$ for coupled reactions is the arithmetic sum of the $\Delta G^{0'}$ values of the component reactions. The $\Delta G^{0'}$ for the XOH + YH → X—Y + H_2O we take to be 12.5 kJ mol^{-1}; the $\Delta G^{0'}$ for the reaction ATP + H_2O → ADP + P_i is −30.5 kJ mol^{-1}. Therefore, the $\Delta G^{0'}$ of the overall process is −18.0 kJ mol^{-1}, a strongly exergonic process that will proceed essentially to completion—the equilibrium constant will be about 10^3—which means that the reactions can proceed to approximately 99.9% to the side of X—Y formation (see Table 1.1). It is to be noted that the use of ATP here involves phosphoryl group transfer from ATP to one of the reactants and only then is P_i liberated. *Direct* hydrolysis of ATP to ADP and P_i is not occurring. In the cell, the actual ΔG value of the ATP breakdown to ADP and P_i is considerably larger than the $\Delta G^{0'}$ value for this, because ΔG values are affected by reactant concentrations and cellular levels of ATP, ADP, and P_i are, of course, nowhere near 1.0 M (the concentrations specified in $\Delta G^{0'}$ determinations). If the actual cellular concentrations of these components are known, the ΔG value for the hydrolysis of ATP to ADP + P_i can be calculated from the equation

$$\Delta G = \Delta G^{0'} + RT2.303 \log_{10} \frac{[\text{Products}]}{[\text{Reactants}]}$$

where R is the gas constant (8.315 J mol^{-1} K^{-1}); and T is the temperature in degrees Kelvin (298 K = 25°C).

The concentrations of ATP, ADP, and P_i will vary from cell to cell. ATP 'levels' generally fall in the range 2–8 mM, ADP levels are about a tenth of this, and P_i values are similar to those of ATP. However, for simplicity let us assume a situation in which all three are at 10^{-3} M. (For reactions in dilute aqueous solution, water has the value of 1.) Substituting these values into the equation

$$\Delta G = \Delta G^{0'} + RT2.303 \log_{10} \frac{[\text{ADP}][P_i]}{[\text{ATP}]},$$

we obtain

$$\Delta G = -30.5 \text{ kJ mol}^{-1} + (831.5 \times 10^{-3})(298)$$
$$\times 2.303 \log_{10} \frac{[10^{-3}][10^{-3}]}{[10^{-3}]}$$
$$= -30.5 - 17.1 = -47.6 \text{ kJ mol}^{-1}.$$

The reaction sequence given above is used for many biochemical reactions coupled to ATP breakdown. But, more commonly, for the synthesis of molecules such as nucleic acids and proteins, the cell uses an even more energetically effective trick. Instead of breaking off one —Ⓟ of ATP to P_i it releases two. The negative $\Delta G^{0'}$ value that results from this is so large that the equilibrium is such that the reaction is completely irreversible in the cell. The way this is done is as follows, again taking XOH + YH → X—Y + H_2O as the example.

Reaction 1:   XOH + AMP—Ⓟ—Ⓟ → X—AMP + PP$_i$

Reaction 2:   X—AMP + YH → X—Y + AMP

Reaction 3:   PP$_i$ + H_2O → 2P_i

Sum: XOH + YH + AMP—Ⓟ—Ⓟ + H_2O → X—Y + AMP + 2P_i

Assume that the $\Delta G^{0'}$ for the reaction from XOH + YH → X—Y + H_2O is 12.5 kJ mol^{-1}: the $\Delta G^{0'}$ for the reaction ATP + H_2O → AMP + PP$_i$ is −32.2 kJ mol^{-1}; that for PP$_i$ hydrolysis is −33.5 kJ mol^{-1}.

The $\Delta G^{0'}$ for the reaction ATP + H_2O → AMP + 2P_i is −65.7 kJ mol^{-1}. The overall process of synthesizing X—Y at the expense of ATP breakdown thus has a negative $\Delta G^{0'}$ of 65.7 − 12.5 = 53.2 kJ mol^{-1}, a very large value indeed.

The mechanism depends on the fact that inorganic pyrophosphate, PP$_i$, is rapidly broken down in the cell. It explains why cells contain an enzyme, called **inorganic pyrophosphatase** that catalyses the reaction (already shown above)

Inorganic pyrophosphate (PP$_i$)   Inorganic phosphate (P_i)

This enzyme, once regarded as an unimportant one, is a driving force in biochemical syntheses and is widely distributed.

How does ATP drive other types of work?

As well as performing chemical work, ATP breakdown powers muscle contraction, the generation of electrical signals, and pumping of ions and other molecules against concentration gradients. The mechanisms are, in principle, the same. Whatever the process, provided ATP breakdown is coupled into the mechanism the resultant free-energy liberation will drive it. Many of these processes will be dealt with in subsequent chapters.

A note on the relationship between AMP, ADP, and ATP

There might seem to be a problem. As explained, many ATP-requiring chemical syntheses in the cell produce AMP + 2P_i,

not ADP + P$_i$. AMP cannot, itself, be converted to ATP by the foodstuff oxidation system; only ADP is accepted as shown in Fig. 1.4. However, AMP is easily rescued by an enzyme that transfers —℗ from ATP to AMP. The reaction is

AMP + AMP—℗—℗ ⇌ 2AMP—℗

or, as written more conventionally,

AMP + ATP ⇌ 2ADP.

The enzyme is called AMP-kinase. **Kinase** is the term for carrying a phosphoryl group from ATP; AMP-kinase means it transfers the group to AMP (AMP is also called adenylic acid; hence AMP-kinase may also be called adenylate kinase). Note that the —℗ group is directly transferred from one molecule to another without hydrolysis or significant release of energy. You will find as you go through the book that such freely reversible 'shuffling' at the 'high-energy level' occurs frequently. The ADP (AMP—℗) so produced is now accepted by the food oxidation system and converted to ATP.

To give perspective to the subject, we have not in this account dealt with the mechanisms by which ATP is synthesized from ADP and P$_i$ at the expense of food catabolism. This is a very large topic that forms a substantial part of the chapters on metabolism later in this book. Also, we have dealt with the utilization of ATP energy to perform work only in general terms so far—as you progress through the book you will encounter example after example of this in later chapters, for ATP utilization is involved in virtually all biochemical systems where work is performed.

We come now to a change of topic though it is still concerned with free-energy changes in chemical processes.

Weak bonds and free-energy changes

The chemistry we have been discussing so far in this chapter involves covalent bonds; now we come to a different type of molecular interaction involving **noncovalent bonds**, also referred to as secondary or **weak bonds**. The latter two terms belie their importance in life.

In enzymatically catalysed reactions of the type discussed so far, atoms or groups of atoms are rearranged so that new compounds result and covalent bonds are broken and made. However, although the covalent bonds may have different atoms attached to them, at the end of the reaction there is the same total number of covalent bonds.

There's a good reason why, in biochemical reactions, there is no net destruction of covalent bonds. When two atoms become

joined to form a chemical bond, a certain amount of energy is released (which is why the bonds form at all),

X + X → X − X + energy.

(X here represents a free *atom*, two of which are becoming joined by a bond *that did not previously exist*. We are *not* talking of ordinary chemical reactions in which there is no increase in the number of covalent bonds.)

To reverse this process, to break the chemical bond, exactly the same amount of energy has to be supplied as was released in its formation,

X − X + energy → X + X.

The formation of a typical covalent bond releases a very large amount of energy and, therefore, its destruction also requires a very large amount of energy. Such bond destruction does not occur in biochemical systems. To illustrate the point, oxygen exists as O$_2$—the $\Delta G^{0'}$ value of its formation from atomic oxygen is so large (about 460 kJ mol^{-1}), and negative, that the equilibrium is such that the number of single atoms in oxygen gas is an infinitely small proportion of the total existing as O$_2$.

In chemical reactions, the breakage of a covalent bond is accompanied by the simultaneous formation of another covalent bond via the transition state (page 22) so that the net energy change involved is far less than the very large value required to *destroy* a covalent bond. However, noncovalent, weak or secondary bonds between atoms and groups of atoms in solution are continually being formed and broken. They are sufficiently strong to form structures and yet sufficiently weak for those structures to be readily disassembled—they allow very great structural flexibility. This general statement may have little reality to you at this stage but their central role will become clear when we've dealt with membranes, proteins, and DNA. They are weak or secondary, not in importance, but in energy of formation, involving $\Delta G^{0'}$ values in the range of only 4–30 kJ mol^{-1} so that correspondingly small amounts of energy are required to break them.

There are three types of weak bonds: ionic bonds, hydrogen bonds, and **van der Waals** attractions. **Ionic bonds** arise from the electrostatic attraction between two permanently charged groups that are close to each other. A typical example would be that between a negatively charged carboxyl group and a positively charged amino group,

—COO$^-$---H$_3^+$N—.

Next we have **hydrogen bonds**. These also are due to electrostatic attractions between atoms of polar molecules, but the attractions are due not to ionized groups but to a much weaker

form of positive and negative charge separation, the electric dipole moment. Consider a water molecule as shown below.

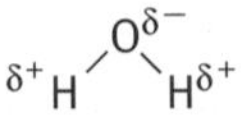

$$\delta^+ H \overset{O^{\delta-}}{\diagup \diagdown} H^{\delta+}$$

The oxygen atom is electronegative—it has a greater attraction for the bond electrons than has a hydrogen atom—it is greedy for electrons and takes more than its half share. The molecule has a partial negative charge on the oxygen atom and corresponding partial positive charges on the hydrogen atoms. This results in weak attractions between the O and H atoms of adjacent water molecules known as hydrogen bonds. Each water molecule is bonded to four other molecules because the oxygen atom can form two hydrogen bonds and the hydrogen atoms one each. A hydrogen atom of any molecule can participate in hydrogen bond formation provided it is attached to an electronegative atom. The main relevant electronegative atoms in biochemistry are nitrogen and oxygen. The bond must involve also an electronegative acceptor atom—also usually nitrogen or oxygen.

The types of hydrogen bonds common in biochemistry are illustrated here.

—O—H---O—

—O—H---N—

—N—H---O—

—N—H---N—

The hydrogen bond is highly directional and of maximal strength when all of the atoms are in a straight line. Van der Waals forces is the collective term used to describe a group of weak interactions between closely positioned atoms; fluctuations in electron density of the atoms cause induced dipoles. They are nonspecific and are the only type of weak bond possible between nonpolar molecules that cannot form hydrogen or ionic bonds. Van der Waals forces can operate between *any* two atoms, which must be positioned close together so that their electron shells are almost touching. The attraction between them is inversely proportional to the sixth power of the distance, so close positioning is essential. If atoms are forced to become too close together, their electron shells overlap and a repulsive force is generated. In short, precise positioning is a prerequisite for van der Waals attraction to arise.

What are the characteristics of the three types of weak bonds? First, while they vary in length, *they are all short-range attractions*; they vary somewhat in their energy of bond formation. The $\Delta G^{0'}$ values are, on average, about $20\,kJ\,mol^{-1}$ for

ionic bonds, $12–29\,kJ\,mol^{-1}$ for hydrogen bonds, and about $4–8\,kJ\,mol^{-1}$ for the average **van der Waals** bond. Compare these values with bond energy of covalent bonds—several hundred $kJ\,mol^{-1}$.

The importance of weak bonds is that they promote associations between and within molecules. In this context we must add another phenomenon that also promotes this but is not an actual bonding—it is called **hydrophobic force** or sometimes **hydrophobic attraction**.

A hydrophilic (polar) molecule such as sugar that can form hydrogen bonds is soluble in water because, although it disrupts hydrogen bonding between water molecules, it can itself form bonds with the latter, making the process energetically feasible. Salts such as NaCl are soluble because the Na^+ and Cl^- ions become surrounded by hydration shells in which the ion–water attractions exceed those between the ions themselves and, when separated, there is a large increase in entropy from the crystalline state. If an attempt is made to dissolve a nonpolar substance such as olive oil in water, the oil molecules get in the way of hydrogen bonding between water molecules. Since hydrogen bonding is highly directional, the water molecules around the oil molecules must arrange themselves so that they can still be fully bonded—that is, with none of their bonding sites aimed at the oil molecules. This more highly ordered arrangement is at a lower entropy level than that of randomly arranged water molecules (that is, a higher energy level) and so the solubilization of the oil is opposed. The nonpolar molecules are forced to associate together so as to present the minimum oil/water interface area. The olive oil forms droplets and then a separate layer which is the minimum free energy state. This repelling effect, pushing together nonpolar molecules by water molecules, is called a hydrophobic force. Hydrophobic groups occur in proteins and DNA and other cellular molecules. Hydrophobic forces on these play a crucial role in the structure of these molecules as will be seen later in this book; it is remarkable how the necessity of 'hiding' hydrophobic groups from water determines so much in living cells.

What causes weak bond formation and breakage?

Formation of weak bonds involves a fall in free energy and the energies of activation involved in their formation are very low. They therefore form spontaneously between appropriate atoms that are sufficiently close, *without the need for enzyme catalysis*. It has already been mentioned that there is a direct relationship between the $\Delta G^{0'}$ of a reaction and its equilibrium constant. Table 1.1 gives the equilibrium constants for given $\Delta G^{0'}$ values. Although those for weak bond formation are small, nonetheless

from Table 1.1 you can see that a $\Delta G^{0'}$ as small as $-5.7\,\mathrm{kJ\,mol^{-1}}$ gives an equilibrium well to the side of bond formation. When you get up to a $\Delta G^{0'}$ value of around $-11.4\,\mathrm{kJ\,mol^{-1}}$, the equilibrium is 99% in favour of bonding. In short, weak bond formation is favoured by energy considerations.

However, such a bonding equilibrium is a dynamic one—individual bonds are continually being broken and reformed. What causes the breakage? All that is required to break a weak bond is to supply the same amount of energy as was released when the bond was formed. This energy comes from the kinetic energy of molecules in motion (thermal motion) when they collide with the atoms forming the weak bonds; thermal motion of molecules at the higher end of the energy distribution spectrum, even at physiological temperatures, has sufficient kinetic energy to disrupt weak bonds.

The vital role of weak bonds in molecular recognition

Given that, say, a van der Waals attraction has a $\Delta G^{0'}$ of formation only slightly above the average thermal energy of molecules in solution and can be made or broken spontaneously, it might be thought that such ephemeral bonds can't be of much importance. To a lesser degree the same could be thought about the other weak bonds.

Paradoxically, it is their very weakness that makes them of central importance to life. The essence of almost everything in life involves biological specificity, which depends on one molecule binding to another with such exquisite precision that only the particular molecules designed by evolution to interact do so. To emphasize the point again, the precisely specific binding of molecules one to another is the basis of virtually all biological processes. To give a few examples at random, an enzyme is specific for one reaction because its active site will combine only with its own substrate(s). A hormone has specific effects because it combines only with its own receptor protein on cells. A gene is controlled because controlling proteins recognize and bind to a particular piece of DNA. An antibody combines only with its relevant antigen. The list is very long. You'll come across one case after another as you proceed through the subject.

How are these precise molecular interactions achieved? The binding processes in the examples referred to depend on weak bonds and the specificity arises from: (1) the fact that individual bonds are so weak that several must be formed for the attachment to occur; and (2) the weak bonds, being short-range ones, can form only if the relevant atoms are precisely positioned very close to each other. The combination of (1) and (2) means that only molecules shaped to fit their designated partner(s) can bind in this way. The relationship can be so precise that, if a single atom is changed in an enzyme substrate, or in the active centre of an enzyme, the interactions do not occur.

Another important consideration is that biological processes often demand that specific bonding between molecules be transitory. Binding of substrates to an enzyme must not be so tight that the products of the reaction can't diffuse away; gene action and reversible gene control demand that interactions are easily reversible. This is where $\Delta G^{0'}$ values for bond energies of the weak bonds are exactly right. They are sufficiently large to give equilibrium constants that favour bond formation but sufficiently small that bonds can be easily broken and the attachments freely reversible. In the case of antibody–antigen interactions, a sufficiently large number of weak bonds results in bonding that is essentially irreversible. In summary, weak bonds permit precise binding between molecules and give the basis for biological specificity. Small numbers of bonds permit easy reversibility of the binding while large numbers can produce stable inter- and intramolecular interactions. The role of weak bonds in biochemistry will be more apparent when we come to the chapters on membranes, protein structure, gene structure, and gene action.

With that introduction we will deal in the next two chapters with two important topics. These are protein structure and membranes. Both topics are necessary for understanding metabolism which is the major area covered in the first half or so of the book.

Appendix: Buffers and pK_a values

It is very important in biochemistry to understand what buffers and pK_a values are. We have placed this is an appendix to avoid disrupting the text and because many will have dealt with this in their chemistry studies. If not, this material *should be learned now*.

The pH of a living cell is maintained in the range 7.2–7.4. Special situations occur, such as in the stomach where HCl is secreted and in lysosomes into which protons are pumped to maintain an acid pH, but, otherwise, the pH of cells and of circulating fluids is maintained within narrow limits. This is despite the fact that metabolic processes, such a lactic acid and acetoacetic acid formation and CO_2 conversion to H_2CO_3 in the blood, occur on a large scale.

This pH stability is largely due to the buffering effect of weak acids. An acid in this context is defined as a molecule that can release a hydrogen ion.

A carboxylic acid dissociates, liberating a proton

R COOH $\leftrightarrow$ R COO$^-$ + H$^+$.

The proton donor and proton acceptor in the above equation is called a conjugate acid and base pair. Written in a more general form, the equation for acids is

HA $\leftrightarrow$ H$^+$ + A$^-$.

Acids vary in their tendency to dissociate. Stronger acids do so more readily than weaker ones—this is why, say, a 0.1 M solution of formic acid (HCOOH) has a lower pH than a 0.1 M solution of acetic acid (CH$_3$COOH). The tendency to dissociate is quantitated as a dissociation constant, K_a, for each acid: the larger the value of K_a, the greater the tendency to dissociate and the stronger the acid,

$$K_a = \frac{[H^+][A^-]}{[HA]}.$$

For acetic acid $K_a = 1.74 \times 10^{-5}$. This value is not much used, as such, by biochemists, for there is another way of expressing the strength of the acid by a much more convenient term—the pK_a value. The two are related by the equation

pK_a = $-\log K_a$.

Thus, acetic acid has a pK_a of 4.76 and formic acid a pK_a of 3.75. These values represent the pH at which an acid is 50% dissociated. As the pH increases, the acid becomes more dissociated; as it decreases, the reverse. The HA $\rightleftharpoons$ H$^+$ + A$^-$ equilibrium is affected by the H$^+$ concentration as would be expected (see Fig. 1.5).

An amine base also has a pK_a value because the ionized form can dissociate to liberate a proton and, in this sense, is an acid

R NH$_3^+$ $\rightleftharpoons$ R NH$_2$ + H$^+$.

In this case, as the pH increases, the dissociation decreases; with both carboxylic acids and amine bases, increase in H$^+$ concentration causes increased protonation. The difference, of course, is that protonation of a carboxylic acid reduces the amount of ionized form while, with amine bases, the proportion in the ionized form increases (Fig. 1.5).

pK_a values and their relationship to buffers

If you take a 0.1 M solution of acetic acid and gradually add 0.1 M NaOH measuring the pH at each step, the curve shown in Fig. 1.6 is obtained.

At the beginning of the titration, the added OH$^-$ neutralizes existing H$^+$ (the acetic acid is slightly dissociated) and the pH rises rapidly. However, as the pH begins to approach the pK_a

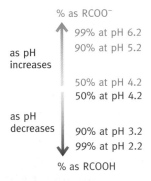

(a) for the molecule RCOOH with a pK_a of 4.2:

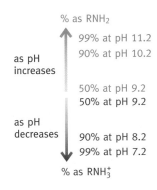

(b) for the molecule RNH$_2$ with a pK_a of 9.2:

Fig. 1.5 Effect of pH on the ionization of **(a)** —COOH and **(b)** —NH$_2$ groups. Kindly provided by R. Rogers.

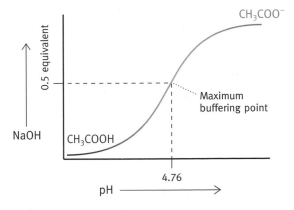

Fig. 1.6 Titration curve for acetic acid (pK_a = 4.76).

value of acetic acid, as more NaOH is added, the acetic acid dissociates into the acetate ion liberating hydrogen ions, which neutralize the hydroxide ions so that the pH change is relatively small. The reaction is

$$CH_3COOH + OH^- \rightarrow CH_3COO^- + H_2O.$$

(Note that sodium acetate is fully dissociated.)

Similarly, from the other side of the pH scale, additions of H^+ cause little change to the pH because of the reaction

$$CH_3COO^- + H^+ \rightarrow CH_3COOH.$$

This is the pH buffering effect of acetic acid. It is maximal at the pK_a so that an equimolar mixture of acetate and acetic acid gives its maximum buffering effect at that pH. As a rule of thumb, in biochemistry, a useful buffer covers a range of 1 pH unit centred at the pK_a. The pH of a solution containing an acid–base conjugate pair can be calculated from the Henderson–Hasselbalch equation

$$pH + pK_a + \log \frac{[\text{Proton acceptor}]}{[\text{Proton donor}]}$$

or, as often expressed,

$$pH = pK_a + \log \frac{[\text{Salt}]}{[\text{Acid}]}.$$

The pH of any mixture of acetic acid and sodium acetate can be calculated from this equation so that the composition of a buffer of any desired pH can be determined.

As an example for a solution containing 0.1 M acetic acid and 0.1 M sodium acetate,

$$pH = 4.76 + \log \frac{[0.1]}{[0.1]}$$
$$= 4.76 + 0 = 4.76.$$

If the mixture contains 0.1 M acetic acide and 0.2 M sodium acetate

$$pH = 4.76 + \log \frac{[0.2]}{[0.1]}$$
$$= 4.76 + \log 2 = 4.76 + 0.30 = 5.06.$$

Suppose that to the buffer containing 0.1 M acetic acid and 0.1 M sodium acetate you added NaOH to a concentration of 0.05 M; the pH of the resultant solution would be

$$4.76 + \log \frac{[0.15]}{[0.05]}$$

(since half of the acetic acid would be converted to sodium acetate)

$$= 4.76 + \log 3 = 4.76 + 0.48 = 5.24.$$

Acetic acid/acetate mixture has no significant buffering power at physiological pHs, but compounds with pK_as close to pH 7 exist in the body, and these are effective buffers. Among the most important are the phosphate ion and its derivatives. Phosphoric acid has three dissociable groups (page 9), the second one ($H_2PO_4 \rightleftharpoons HPO_4 + H^+$) has a pK_a value of 6.86 so that phosphate is an excellent buffer in this region.

Another buffering structure is the imidazole nitrogen of the histidine residue in proteins with a pK_a around 6. The dissociation involved is shown below.

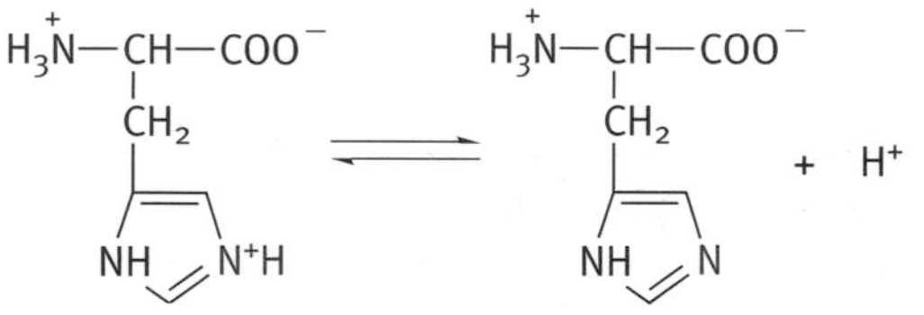

The phenomenon of buffering can be illustrated by a very simple observation. If you take a test tube containing distilled water and adjust the pH to 7.0 (dissolved CO_2 makes it slightly acid) with NaOH and then add a few drops of 0.1 M HCl there will be a precipitous drop in pH. If you take a test tube containing 0.1 M sodium phosphate, also adjusted to pH 7.0, and add the same amount of HCl, the pH will hardly change, due to the buffering action of the phosphate ion.

The buffering action of compounds with pK_a values near 7 thus protects the cell and body fluids against large pH changes.

FURTHER READING

Thermodynamics

Newsholme, E. A. and Leach, A. R. (1983). *Biochemistry for medical sciences*, pp 10–45. Wiley.

Gives an account of the laws of thermodynamics with special emphasis on their application to metabolism.

Nelsestuen, G. L. (1989). A partial remedy for the nonrelationship between reversibility of a reaction in the cell and the value of $\Delta G^{0'}$. *Biochemical Education*, **17**, 190–2.

Discusses why reactions such as aldolase with a large $\Delta G^{0'}$ value are freely reversible.

The high energy phosphate compound concept

Lipmann, F. (1941). Metabolic generation and utilisation of phosphate bond energy. *Adv. Enzymol.*, 1, 99–162.

The famous review which initiated the concept of ATP driving cellular processes. Very old, but the concepts still apply.

PROBLEMS FOR CHAPTER 1

1   Suppose that a 70-kg man, moderately active, has a food intake for his energy needs of 10 000 kJ per day. Assume that the free energy available in his diet is utilized to form ATP from ADP and P_i with an efficiency of 50%. In the cell, the ΔG for the conversion of ADP + P_i is approximately 55 kJ mol^{-1}. Calculate the total weight of ATP the man synthesizes per day, in terms of the disodium salt (molecular weight = 551).

2   The $\Delta G^{0'}$ for the hydrolysis of ATP to ADP and P_i is 30.5 kJ mol^{-1}. Explain why, in Problem 1, a value of 55 kJ mol^{-1} was suggested as the amount of free energy required to synthesize a mole of ATP from ADP and P_i in the cell.

3   The hydrolysis of ATP to ADP + P_i and that of ADP to AMP + P_i have $\Delta G^{0'}$ values of −30.5 kJ mol^{-1}, while the hydrolysis of AMP to adenosine and P_i has a value of −14.2 kJ mol^{-1}. What are the reasons for the large difference?

4   Explain
    (a)   why benzene will not dissolve in water to any significant extent;
    (b)   why a polar molecule such as glucose is soluble in water;
    (c)   why NaCl is soluble in water.

5   In the cell, ADP is converted to ATP using the energy derived from food catabolism, but AMP is not utilized by the ATP synthesizing system; many synthetic processes convert ATP to AMP. How is AMP brought back into the system?

6   An enzyme catalyses a reaction that synthesizes the compound XY from XOH and YH, coupled with the breakdown of ATP to AMP and PP_i. The $\Delta G^{0'}$ of the reaction XOH + YH → XY + H_2O is 10 kJ mol^{-1}. Determine the $\Delta G^{0'}$ of the reaction: (a) in the cell; and (b) using a completely pure preparation of the enzyme. Explain your answer. (You are told that the $\Delta G^{0'}$ values for ATP hydrolysis to AMP and PP_i and for PP_i hydrolysis are −32.2 and −33.4 kJ mol^{-1}, respectively.)

7   What are the different types of weak bonds of importance in biological systems and their approximate energies. Why is it that their formation does not require enzymic catalysis? If they are so weak, why are they of importance?

8   The pK_a values for phosphoric acid are 2.2, 7.2, and 12.3.
    (a)   Write down the predominant ionic forms at pH 0, pH 4, pH 9, and pH 14.
    (b)   What would be the pH of an equimolar mixture of NaH_2PO_4 and Na_2HPO_4?
    (c)   Suppose that you wanted a buffer fairly close to physiological pH and all you had available were the 20 amino acids found in proteins; which one would be the most suitable?

9   Virtually all biological processes involve specific interactions between proteins and other molecules (which may also be proteins). Explain how the specific interactions are achieved.

10   The use of the term 'high energy phosphate bond' (indicated by a squiggly bond) by Lipmann in 1940 has been of great importance in development of the concept of biological energy. Although sometimes still used, because it is a convenient shorthand notation, it has fallen out of favour because it is not chemically correct. Discuss this.

Chapter summary

Enzymes

Given that a reaction has a negative ΔG value, what determines whether it actually takes place at a perceptible rate in the cell? This question follows on immediately from the previous chapter where we explained that on the question of whether a reaction is possible (*may* occur) energy considerations have absolute authority. If the free energy change involved in the reaction under the prevailing conditions is negative it may occur; if it is zero or positive it cannot occur and nothing in the universe can alter that. Note however that as mentioned in the previous chapter, chemical *conversions* (such as the synthesis of compound X–Y from reactants XOH and YH), involving positive ΔG values do occur in the cell by incorporating ATP breakdown into the process. These are not exceptions to the universal rule, because they involve reactions each of which must have a negative free energy change. The sole qualification of the rule that a reaction must have a negative ΔG is that it refers to *net* chemical change in the reaction. At equilibrium, where the free energy change of the reaction is zero there cannot be any net chemical change but it can be shown that this is because the reaction in one direction is exactly balanced by that in the reverse direction. The qualification has no relevance to the practical business of life which is based on net chemical changes.

We have briefly touched on what determines whether a reaction occurs in Chapter 1, but we will elaborate on this here, for it is of basic importance. Energy considerations, as stated, determine whether a reaction *may* occur, not whether it *does* occur. This is just as well, because energy considerations say that everything combustible around you may burst into flame or rather would have done so long ago. But petrol does not ignite by itself, and sugar in the bowl on the table does not burn. Nonetheless petrol in a car cylinder burns beautifully and sugar inside the body participates in a cauldron of chemical activities.

From these very simple considerations you can see that there is a barrier to the occurrence of chemical reactions. Even though a reaction has a strong negative ΔG something restrains the reaction from happening. In the case of petrol in a car cylinder, a spark from the spark plug overcomes that barrier. This brings us to one of the most fundamental problems that had to be solved before life could exist. How can chemical reactions be caused to occur in the cell? The answer to the problem lies in **enzyme catalysis**. It is perhaps worth reflecting on what a formidable obstacle this problem posed for the development of life, for it leads to an appreciation of what an astonishing phenomenon enzyme catalysis is. It was necessary to develop a means whereby at low temperatures, at almost neutral pH, in aqueous solutions, and at low reactant concentrations, otherwise stable molecules undergo the rapid chemical conversions needed for life.

Enzyme catalysis

An enzyme is a catalyst which brings about one particular chemical reaction but itself remains unchanged at the end of the reaction (as required in the definition of a catalyst). There are thousands of biochemical reactions, each catalysed by a separate enzyme. Multi-functional enzymes with several catalytic activities on the same molecule and multi-enzyme complexes exist for special situations but the general principle remains. It automatically follows from what we have said about energy considerations that an enzyme cannot affect the equilibrium or direction of a reaction; it can only promote a reaction subject to the limitations of energy considerations. This

is little different from saying that you can't get energy for nothing.

An enzyme is a protein molecule (with rare exceptions in which RNA, a nucleic acid, catalyses a reaction; see page 372). The structure of proteins is dealt with in Chapter 3, but a brief description will be useful here. Proteins are built up of 20 different species of amino acids that are linked together to form one or more long chains. Enzymes are therefore large molecules— a molecular weight of 10 000 corresponds to a small enzyme, and they range in size up to hundreds of thousands in molecular weight. One of the reasons why they are so large is that the long chain(s) of which they are constituted must be folded such that there is an **active site** on the surface so shaped that it is a three-dimensional pocket or cleft into which the compounds attacked by the enzyme (known as **substrates**), fit with exquisite precision. This site occupies a very small part of the protein in most cases. As with all such specific protein–ligand binding (ligand = any combining molecule) the substrate attaches reversibly by noncovalent, or weak bonds. As explained earlier, the specificity of all such attachments arises from the fact that weak bonds are very short range and directional and because they are weak, several are needed. It follows that unless there is a precise fit between the interacting groups on the substrate and enzyme, the attachment will not occur. Almost everything in life depends on this simple principle by which specific interactions of proteins with other molecules (which may also be proteins) are achieved. The active site is called an **active centre** or **catalytic site**.

The nature of enzyme catalysis

To explain this, we must look at the nature of chemical reactions. A reaction occurs in two stages. Consider a reaction converting substrate to product, S → P. In this reaction S must first be converted to the **transition state**, $S^{\ddagger}$, which might be thought of as a 'halfway house' in which the molecule is distorted to an electronic conformation which readily converts to P. The transition state has an exceedingly brief existence of 10^{-14}–10^{-13} of a second. Appreciation of the energy considerations involved is needed.

The overall reaction S → P must have a negative free energy change; otherwise it couldn't occur. An important thermodynamic principle is that the free energy change of a reaction is determined solely by the free energy difference between the starting and final products. The 'energy pathway' or energy profile, the route by which the reaction takes place, does not affect the free energy change of the reaction. Hence it is in no way a contradiction that the transition state $(S^{\ddagger})$ is at a higher free energy level than S. The free energy change for S → $S^{\ddagger}$ is posi-

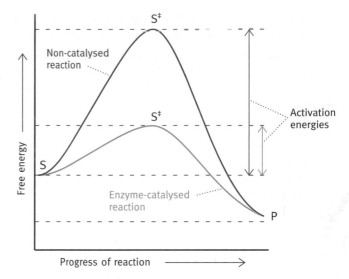

Fig. 2.1 Energy profiles of noncatalysed and enzyme-catalysed reactions. The inverse relationship between the rate constant and the activation energy of the reaction is exponential so that the rate of the reaction is extremely sensitive to changes in the activation energy. S, substrate; $S^{\ddagger}$ transition state; P, products.

tive. It is called the **energy of activation** for that reaction (Fig. 2.1).

This energy hump constitutes a barrier to chemical reactions occurring. If it weren't present and the energy profile S → P were a straight downward slope then, as stated earlier, everything which could react would do so.

It follows that energy of activation must be supplied to permit a reaction to occur. In a car cylinder, the spark causes a few molecules of petrol to be activated to the transition state which then react and in oxidising, produce enough heat to activate further molecules which results in the explosion. In a noncatalysed reaction in solution, the energy is supplied by collisions between molecules. Provided that colliding molecules are appropriately oriented, and of sufficient kinetic energy, reactant molecule(s) can be distorted into the appropriate higher energy transition state which enables the reaction to occur. The rate of formation of the transition state therefore determines the reaction rate. High temperatures, which increase molecular motion and increase collision frequency between molecules, facilitate this. Hence the organic chemist usually employs high temperatures to promote reactions. At physiological temperatures however, around 37°C, most biochemical reactions, uncatalysed, proceed at imperceptible rates.

As stated, each enzyme has one or more combining sites within the active sites, to which the substrate(s) bind. The binding has several effects.

- First, it positions substrate molecules in the most favourable relative orientations for the reaction to occur.

- Secondly, the active site is perfectly complementary, not to the substrate in its ground state (as the unactivated substrate molecule is referred to), but to the **transition state**, which is intermediate between the reactant molecules and products.

A substrate in its transition state, binds to the enzyme more tightly than does the substrate in its ground state. This is due to the structure of the active centre, which has evolved to do this. A transition state tightly bound to the enzyme is at a lower energy level than the same transition state in free solution as occurs in a noncatalysed reaction. The transition state has too ephemeral an existence to measure just how tightly it binds to the enzyme, but **transition state analogues** have been synthesized. These are stable molecules similar in structure to the transition state. It has been found that these bind to enzymes with remarkably high affinity—in one case, binding to the enzyme thousands of times more tightly than the substrate. As described later, transition state analogues have considerable medical interest.

Another way of looking at this is that formation of the transition state involves a partial redistribution of electrons. The amino acid groups in the structure of the active centre of the enzyme are of such a nature, and so positioned, that they stabilize the electron distribution of the transition state. (You will find this explained in greater detail when you come to the mechanism of the enzyme chymotrypsin, but we defer a description of this to the chapter on protein structure (page 47).) The fact that the active centre is a less perfect fit to the substrate than it is to the transition state results in the substrate being strained on binding to the active centre, and this favours transition state formation. The net effect is to lower the activation energy of the reaction; the energy hump barrier to the reaction is reduced (Fig. 2.1). This is the central 'secret' of enzyme catalysis. Once formed, the transition state rapidly converts to products. The products bind less tightly to the enzyme and diffuse away. The catalysis rate is very sensitive to changes in the activation energy, there being an inverse exponential relationship. A very small reduction the activation energy, equivalent to the small amount of energy released on formation of a single average hydrogen bond (page 3), can increase the rate of the reaction by a factor of 10^6. Enzymes increase the rate of chemical reactions sometimes by a factor of 10^7–10^{14}. The enzyme, urease, which destroys urea, reduces the energy of activation by $84 \, kJ \, mol^{-1}$ and increases the reaction rate 10^{14} times. Most molecules are capable of participating in many different chemical reactions, each with its own transition state. In an uncatalysed reaction, promoted by high temperatures, molecules collide unpredictably and different transition states are formed, resulting in a whole variety of side reactions. By contrast, an enzyme catalyses a specific reaction for which there are only a limited number of well-defined products.

From what has been said, it might be deduced that if one could produce a protein with a high affinity for the transition state of a given chemical reaction, it would catalyse that reaction. Such has been found to be the case. Antibodies (described in Chapter 27) are proteins which bind tightly to specific structures in molecules, and moreover they can be produced to bind to selected molecules. It was found that an antibody against the transition state involved in the hydrolysis of a synthetic ester behaved as a hydrolytic enzyme towards that ester. The term **abzyme** has been coined for such proteins, ab denoting 'antibody'. (The antibody was raized against a stable analogue of the transition state.)

The induced fit mechanism of enzyme catalysis

As implied already, the first step in an enzyme-catalysed reaction is for the substrate to combine with the active site of the enzyme to form an enzyme–substrate complex. This implies that the relationship between an enzyme and its substrate(s) is simply a 'lock and key' model in which the active site is a envisaged to be a rigid structure in an unchanging form and the substrate (the key) fits to it (Fig. 2.2(a)). A more recent concept, however, is the **'induced fit'** mechanism which is based on the view that the enzyme is not a rigid structure analogous to a lock, but rather is a flexible structure capable of changing its conformation slightly in an interactive way when its substrate binds. ('Conformation' refers to the particular arrangement of the

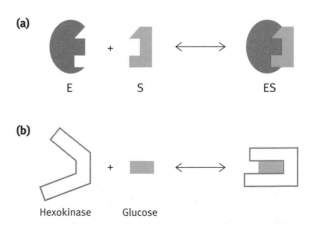

Fig. 2.2 (a) Lock and key model of enzyme mechanism. E, enzyme; S, substrate. (b) Induced fit model of the hexokinase mechanism.

protein chain(s) in three-dimensional space.) In other words, it changes its shape slightly, which has the important effect of altering the spatial arrangements of groups on the molecule. Conformational changes in proteins, as will become clear later in the book, are of central importance to the existence of life in many ways. This induced fit mechanism has been fully established for the enzyme **hexokinase** where the postulated conformational change has been proved to occur by X-ray crystallographic studies. The enzyme catalyses the transfer of a phosphoryl group from ATP to glucose (a hexose). The enzyme has two 'wings' to its structure. In the absence of glucose, these have an 'open' conformation, but on binding of glucose the wings close in a jaw-like movement which results in the creation of the catalytic site (Fig. 2.2(b)). The conformational change explains why the enzyme is specific for glucose. A structure which binds this molecule would be expected to be capable to some extent of binding other polyhydric molecules or even water. On the lock and key model this might be expected to result in some activity with other molecules. In the case of water being bound, this might result in hexokinase hydrolysing ATP—an undesirable reaction, which does not, in fact, occur. Since the catalytic site forms only on binding of the correct substrate, glucose, the enzyme phosphorylates only that substrate. Other molecules may bind, but unless they are capable of bringing about the conformational change, no reaction occurs.

Enzyme kinetics

This term refers to the study of enzymes by determining their reaction rates. In a typical case, an enzyme rate is measured by incubating it with substrate(s) and any required activators at a defined temperature and pH and following the production of a reaction product with time. As seen in Fig. 2.3 the reaction gradually diminishes in rate with time; this is due to accumulation of product, which inhibits the enzyme by binding to the enzyme (E), forming EP and /or depletion of substrate. The initial rate of reaction is denoted by V_0. For meaningful quantitative assays of enzyme activity, it is necessary to ensure that initial velocities are measured, which typically means using short time periods before there is a significant amount of product formed. In this situation the reaction velocity is linearly proportional to the amount of enzyme added.

Hyperbolic kinetics of a 'classical' enzyme

Michaelis and Menten proposed a model of the way in which enzymes act to fit the observed kinetics of enzyme catalysis for a single substrate enzyme

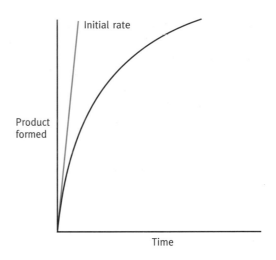

Fig. 2.3 Time course of a typical enzyme reaction.

$$E + S \rightleftharpoons ES \rightarrow E + P$$

What the equation says, is that substrate binds reversibly at the active site of the enzyme, the reaction occurs and the products (P) diffuse away into the solution. At low [S] the rate of an enzyme-catalysed reaction is determined by the rate of formation of ES. In this situation ES formation is not favoured and hence the rate of reaction is relatively low. As the concentration of S increases so will the proportion of enzyme in the form of ES, and therefore the rate of reaction increases. However, as [S] further increases, the point is reached at which the rate of formation of ES ceases to be rate limiting so that the catalytic activity of the enzyme or the rate of product release then becomes limiting. The catalytic rate of the enzyme is an inherent property and varies from enzyme to enzyme. Some enzymes work extremely rapidly, and others more slowly. At high levels of [S], the enzyme is said to be **saturated** with substrate. The rate of the reaction is then called V_{max} (for **maximum velocity**), and further increases in the substrate have no effect on the rate of catalysis. V_{max} is a function of the amount of enzyme present in a given experiment and the rate at which each molecule of enzyme catalyses the reaction. If the molar concentration of the enzyme in an experiment is known, the **turnover number** or K_{cat} can be calculated. This is the number of molecules of substrate converted to product by a molecule of enzyme at saturating levels of substrate per second, values ranging up to 4×10^7/second for catalase (page 283). In the case of many or most enzymes, when the velocity of enzyme activity is plotted against substrate concentration a **hyperbolic curve** is obtained, as shown in Fig. 2.4. An enzyme displaying these kinetics is referred to as **Michaelis–Menten enzyme** and the kinetics as Michaelis–Menten kinetics or **hyperbolic kinetics**. Enzymes of

this type are sometimes referred to as 'classical' enzymes because for a long time they were the only ones known. (Other types are referred to below.)

In the derivation of the equation which describes the relationship between the velocity of an enzyme reaction and the substrate concentration certain assumptions are made, namely that only a single substrate is involved, and that initial velocities (V_0) are measured so that the concentration of product is negligible compared with that of the substrate. This eliminates the complication of product binding significantly to the enzyme. It is also assumed that the system is in a steady state in which the rate of formation of ES is exactly balanced by the rate of its removal and that the substrate is in vast molar excess over the enzyme. The latter two conditions are usually met because steady state kinetics are established almost instantly and the molar amount of enzyme is usually negligible. The equation which describes the relationship between the velocity of an enzyme reaction and the substrate concentration which results in the hyperbolic curve shown in Fig. 2.4 is known as the Michaelis–Menten equation:

$$V = \frac{[S]V_{max}}{[S] + K_m}$$

The K_m value is a very useful constant. What might, in molar terms, be a low concentration of S for one enzyme, may be a saturating concentration of substrate for another. It depends on how tightly the substrate binds to the enzyme, or as it is usually expressed, what the **affinity** of the enzyme is for its substrate.

This depends on the precise relationship of the substrate to its enzyme—the nature and number of the weak bonds established between them. This affinity is given a numerical value as the **Michaelis constant** (K_m) of the enzyme. It is defined as that concentration of S at which the enzyme is working at half maximal velocity, as shown in Fig. 2.4, and is therefore independent of the total amount of E. At the K_m value of substrate concentration, half the total number of enzyme active sites are occupied and half vacant. Other methods of plotting graphs are used to determine K_m values more precisely; if instead of plotting reaction velocity (V_0) against [S] as is done in Fig. 2.4, $1/V$ is plotted against $1/S$ this gives a straight line which, if extrapolated back (Fig. 2.5) intercepts the horizontal axis at the reciprocal of the K_m. This is known as a **double reciprocal plot**, or a Lineweaver–Burke plot after its authors (Fig. 2.5). The intercept of the line

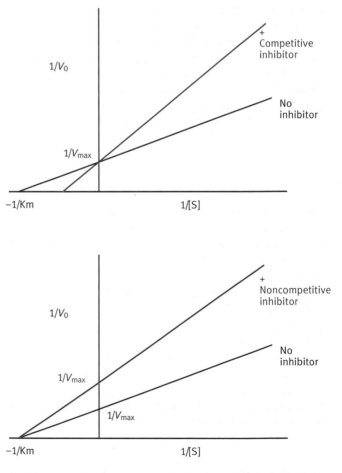

Fig. 2.5 Double reciprocal plots of enzyme reactions in the presence of (a) competitive and (b) noncompetitive inhibitors respectively. See text for explanation of inhibitors (page 27).

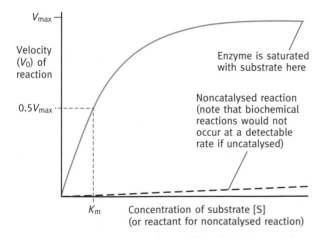

Fig. 2.4 Effect of substrate concentration on the reaction velocity catalysed by a classical Michaelis–Menten type of enzyme. K_m, Michaelis constant. The dashed line shows, for comparison, the effect of reactant concentration on a noncatalysed chemical reaction. Note that the two lines are drawn to illustrate their shapes, not their relative rates.

with the vertical axis also gives a value for $1/V_{max}$. The higher the K_m, the lower is the affinity of the enzyme for its substrate. K_m values can be used to compare the affinities of different enzymes for their substrates. They are of interest from the viewpoint of metabolism for they tell us how an enzyme will respond to changes in the concentration of substrate in the cell. The affinity of a protein for a ligand is a function of the equilibrium between the protein–ligand complex and the free components. In the case of many enzymes, the K_m represents a true affinity constant (the reciprocal of the dissociation constant) but only for those in which the rate of dissociation of ES back to E + S is much faster than the catalytic step of ES to E + P. The reason is that since an affinity constant represents the position of the equilibrium E + S $\rightleftharpoons$ ES, if ES is rapidly removed by conversion to E + P, a true equilibrium is not established. The K_m is a true affinity constant in those cases where this requirement is met which is true with many enzymes. However, even where it is not, the *apparent* affinity of the enzyme for its substrate, which is in all cases reflected in the K_m value, is a useful way of comparing how different enzymes will respond to substrate concentration changes. As a general statement, most enzymes have K_m values such that they are working in the cell at sub-saturating substrate levels. For many enzymes the K_m is in the range of 10^{-4}–10^{-6}.

Allosteric enzymes

There is another very important class of enzymes which do not have hyperbolic kinetics. These are regulatory enzymes whose activities are controlled by chemical signals in the cell and which in turn control metabolism. Metabolic control is a major subject described in Chapter 13 and a description of the 'special' kinetics of these enzymes is deferred until then (page 212).

General properties of enzymes

Nomenclature of enzymes

Usually enzyme names end in '-ase', preceded by a term which most often indicates the nature of the reaction it catalyses and/or indicates the substrate. Thus *amylase* attacks *amylose*. A *dehydrogenase* removes hydrogen atoms from a substrate: lactate dehydrogenase is an example. This is not always the case, for proteolytic enzymes often end in '-in'; *pepsin, chymotrypsin, plasmin*, and *thrombin*, are examples. We will explain names more fully when the enzymes are considered in biochemical systems. An international committee has systematized all enzyme names, but as the systematic names are necessarily rather long there are also shorter recommended names for everyday use. The systematic names are usually given once in published research papers and reference works to avoid any ambiguity.

Isozymes

It is quite common to find that the same reaction is catalysed by a number of distinguishable different enzymes, called isozymes or isoenzymes. There is considerable similarity in the amino acid composition of the different isozymes catalysing a given reaction, suggesting that they all have the same evolutionary ancestor but the genes coding for them have diverged somewhat to suit particular roles in the body. Isozymes are usually found in different tissues or in different locations in cells. The reason for multiple versions of the same enzyme is to tailor them to the specific needs of the cell. Thus in some cases the isozymes differ in substrate affinities, or have different regulatory mechanisms or other properties. This is not too surprising because different tissues have quite different roles. The enzymes often have a number of different subunits, or protein molecules which join together to form the complete enzyme. Thus an enzyme which you will meet later (page 138), **lactate dehydrogenase**, has four subunits of two different types, one known as H because it is the main one in the heart enzyme and the other M because of its association with muscle. Various combinations of H and M subunits produce the different isozymes in different tissues. The isozymes can often be separated because of their different electrical charge. Placed in a gel with a voltage gradient across it they migrate at different rates, a process known as **electrophoresis** (see Chapter 3). Another enzyme known as **creatine kinase**, which again you will meet later, has a form characteristic of heart muscle and another characteristic of skeletal muscle. If a person has a heart attack or myocardial infarction, some of the heart's creatine kinase leaks out. By measuring the amount of the heart-specific isozyme, the severity of the attack can be assessed since there is no confusion with any enzyme that may have originated from skeletal muscle.

Enzyme cofactors and activators

In the simplest case, the enzyme protein combines with a single substrate, reaction occurs, and the products leave the active site to make way for another molecule. Frequently the enzyme requires a **cofactor** for activity. This may be a metal ion such as Mg^{2+} or Zn^{2+} which participates in the reaction mechanism (or it may be an organic molecule attached to the enzyme in which case it is known as a **prosthetic group**. The protein part is called an **apoenzyme** (*apo* = detached or separate) and the complete enzyme with the prosthetic group attached, a **holoenzyme**. The

prosthetic group is sometimes a vitamin derivative (page ••). Since each vitamin molecule in this case activates an enzyme which can bring about reactions in vast numbers of substrates it explains why such small amounts of vitamins can have such a huge effect on the body. **Coenzymes** also have vitamins as components and behave somewhat similarly but are not firmly attached to the enzyme (details will be given at appropriate places later in the book).

Effect of pH on enzymes

The activity of an enzyme is influenced by several factors. The stability of the protein is influenced by the state of its ionizable groups and the function of the active centre may be likewise dependent on this. The ionization of the substrate itself may also be affected. The rate of catalysis is therefore dependent on the pH. Enzyme pH activity profiles vary from one to another but the optimum is generally around neutral pH; a typical plot is shown in Fig. 2.6(a). Exceptions occur such as the case of the digestive enzyme, pepsin, which functions in the acidic stomach contents; its pH optimum is near 2.0.

Effect of temperature on enzymes

Temperature also affects enzyme activity rates. As the temperature increases, the rate of most chemical reactions increases (approximately twofold for each 10°C) but because of the inherent instability of most protein molecules (page 35), the enzyme is inactivated at higher temperatures. Thus a typical enzyme optimum temperature plot would appear as shown in Fig. 2.6(b) but, although perhaps useful in a practical sense, this has little absolute significance since the optimum will depend on the experimental time period used in measuring the rates; the shorter the time the less will be the destructive effect of higher temperatures. A few enzymes are stable to high temperatures but, in general, temperatures over 50°C are destructive and some enzymes are more labile still.

Effect of inhibitors on enzymes

Reversible inhibitors

The activity of an enzyme can also be affected by inhibitory compounds. Inhibitors may be reversible or irreversible. In the former case the enzyme and inhibitor exist in a reversible equilibrium (E + I $\rightleftharpoons$ EI). Irreversible inhibitors bind to the enzyme and do not dissociate from it to an appreciable extent; the extreme case of this is where the inhibitor becomes covalently attached to the enzyme. The effect of aspirin on an enzyme, described below, is one such case.

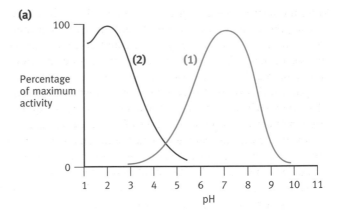

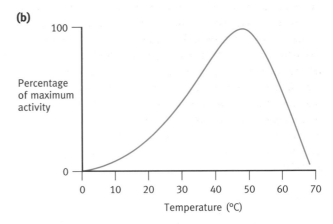

Fig. 2.6 Effect of pH and temperature on enzyme activity. (a) Effect of pH on enzyme activity. Curve (1) is typical of the majority of enzymes with maximal activity near physiological pH. Curve (2) represents pepsin, an exceptional case, since the enzyme functions in the acid stomach contents. (b) Effect of temperature on a typical enzyme. The precipitous drop at high temperatures is due to enzyme destruction (though a number of heat-stable enzymes are known). Note that, as described in the text, the temperature 'optimum' of an enzyme has little significance.

Competitive and noncompetitive inhibitors

Reversible inhibition of an enzyme may be of different types. One class, called **competitive inhibitors**, simply mimic the substrate and compete with the latter for binding to the active site. The degree to which the latter is occupied by the inhibitor will determine the degree of inhibition. The inhibition will be a function of the relative affinities of substrate and inhibitor and their relative concentrations. It is possible to distinguish between competitive and noncompetitive inhibition using the double reciprocal plot already described and shown in Fig. 2.5. A competitive inhibitor will have no effect at infinite substrate

concentration since the substrate will completely win in the competition to bind to the active site. The intersection of the reciprocal plot with the vertical axis (which represents infinite [S]) will be the same whether inhibitor is present or not. It will, however, change the K_m so that the intersection with the horizontal axis which gives the reciprocal of the K_m value is changed by the inhibitor.

A third type of inhibition is known as **uncompetitive.** In this, the inhibitor binds to the enzyme in a noncompetitive way but binds only to the ES form of the enzyme.

Transition state analogues can be very effective enzyme inhibitors because they may bind to the active centre of an enzyme very much more tightly than does the substrate. This is the basis of the so-called 'statin' drugs, which are used to treat people with high cholesterol levels. Drugs such as simvastatin are transitional state analogue inhibitors of an enzyme which controls the synthesis of cholesterol (page 130). These compounds originated from fungi.

A **noncompetitive inhibitor** binds to the enzyme at a position separate from the active site so that there is no competition with the substrate; this is readily seen in a double reciprocal plot as shown in Fig. 2.5(b) in which it is seen that the V_{max} at infinite substrate concentration is reduced while the K_m is unchanged.

Noncompetitive inhibitors work in various ways. For example, a heavy metal such as mercury may covalently react with a thiol group essential for catalytic activity. Removing the metal with a thiol compound or a metal-chelating agent may reverse this. In other cases, an inhibitor may covalently acylate the enzyme in an irreversible manner. Thus aspirin inactivates an enzyme (cyclooxygenase) involved in prostaglandin synthesis (page 195) as follows:

ENZ—OH   +   Aspirin (acetylsalicylic acid)

Cyclooxygenase   Aspirin

↓

ENZ—O—C(=O)—CH₃   +   2-Hydroxybenzoic acid

Acylated enzyme   2-Hydroxybenzoic acid

The acetylated group is a serine residue which is part of the active centre of the enzyme.

In some cases the enzyme covalently participates as a transitory reactant, though without itself being altered at the end of the process. As an example, an important class of enzymes are the serine proteases, so called because they have at their active centres the amino acid serine (page 48) which participates covalently in the catalytic process. We discuss the mechanism by which the digestive enzyme **chymotrypsin** brings about catalysis in terms of the structure of the active site after we have dealt with protein structure (page 47).

FURTHER READING

Steitz, T. A., Shoham, M., and Bennett, Jr., W. S. (1981). Structural dynamics of yeast hexokinase during catalysis. *Phil. Trans. Roy. Soc. Series B, London*, **293**, 43–52.

Discusses the conformational change of the enzyme on binding of glucose.

Srere, P. A. (1984). Why are enzymes so big? *Trends Biochem. Sci.*, **9**, 387–90.

A necessarily speculative, but interesting article.

Koshland, Jr., D. E. (1987). Evolution of catalytic function. *Cold Spring Harbor Symp. Quant. Biol.*, **LII**, 1–7.

A concise clear summary of the basic roles of active sites and of enzyme conformational change in catalysis.

Kraut, J. (1988). How do enzymes work? *Science*, **242**, 533–40.

Enzyme catalysis explained by the principle of transition state stabilization. Examples from the zinc protease mechanism and the generation of catalytic antibodies specific for transition state analogues.

Schramm, V. L. (1998). Enzymatic transition state analog design. *Ann. Rev. Biochem.*, 67, 693–720.

PROBLEMS FOR CHAPTER 2

1   What information can be drawn from a plot of the rate of activity of an enzyme against temperature?

2   The oxidation of glucose to CO_2 and water has a very large negative $\Delta G^{0'}$ value, and yet glucose is quite stable in the presence of oxygen. Why is this so?

3   Explain how an enzyme catalyses reactions.

4   Describe how you would determine whether an inhibitor of an enzyme reaction is a competitive or noncompetitive one. Draw a graph to illustrate your answer.

5   In the case of some enzymes a K_m value is a true measure of the affinity of the enzyme for the substrate. In other cases it is not. Explain this.

6   Some therapeutic drugs work by inhibiting a specific enzyme by blocking its active site with an analogue of the substrate. Transition state analogues have been found to be very effective in some cases. Why would you expect such a molecule to be more effective than a competive analogue of the substrate of the same enzyme?

7   If you wish to study the properties of an enzyme and how its rate of catalysis responds to various factors, what precautions do you need to take in making measurements of reaction velocities?

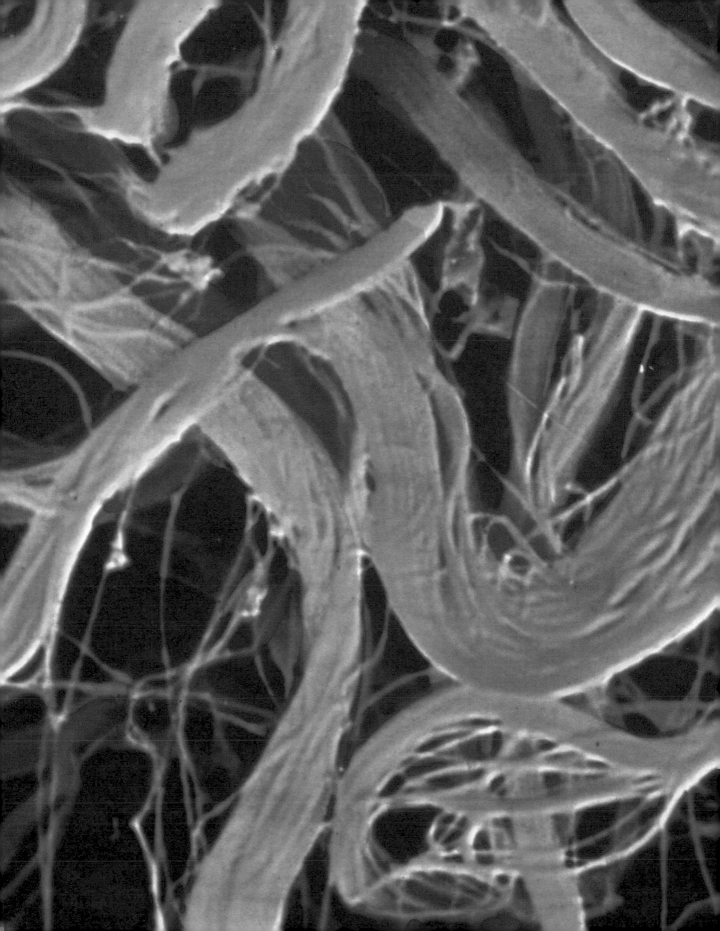

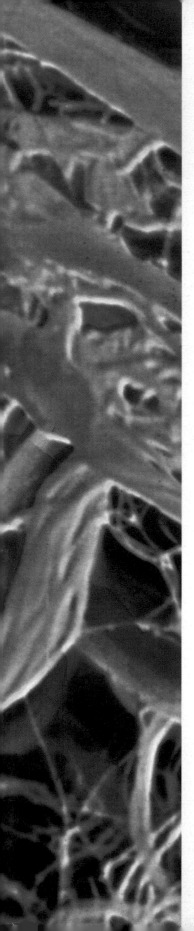

Structure of proteins and membranes

Colorized scanning electron micrograph of collagen fibres ($\times$ 500) in connective tissue reticulum.

Chapter summary

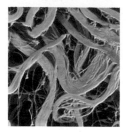

The structure of proteins

It has been said, with considerable justification, that it is improbable that there exists in the entire universe any type of molecule with properties more remarkable than those of proteins. To understand biochemistry, it is essential to appreciate the overwhelmingly dominant role of proteins in life. If a specific job is to be done, it is almost always a protein that does it. As stated in the previous chapter, life depends on thousands of different proteins whose structures are fashioned so that individual protein molecules combine, with exquisite precision, with other molecules. Chemical reactions in the cell depend on enzymes combining with substrates and these are often controlled by other molecules combining with specific sites on the proteins (page 212). Structures such as muscle depend on protein–protein interactions, gene control depends on protein–DNA combinations, hormone control on hormone interactions with protein receptors. Transport across membranes involves protein–solute interactions, nerve activity requires transmitter substance–protein interactions and immune protection requires antibody–antigen interactions. All are specific interactions; in detail, the list is almost endless. The strategy for achieving specificity of interactions is explained on page 15.

This chapter deals with the chemical structure of these giant molecules which carry out such a range of activities. As you might imagine, the study of these large delicate molecules requires special methodologies and the subject of protein chemistry is a big one in its own right. Central to this topic are the methods of separating the hundreds or thousands of proteins which may be present in a crude cell extract and the methods of sequencing proteins. We give a brief description of some of these methods at the end of this chapter.

The primary structure of proteins

Proteins are polypeptide chains. A protein molecule may have more than one such chain, but for the moment we will refer to single-chain proteins which are quite common. A polypeptide chain consists of a large number of amino acids linked together. Twenty different amino acid structures are used (with the qualification that one of them, proline, is an imino acid—explained shortly); if we regard these different amino acids as an alphabet of 20 letters, a polypeptide chain is a word usually hundreds of letters in length. The biggest polypeptide chain has about 5000 amino acids but most have less than 2000. There can, in principle, be an almost infinite number of different 'words' or proteins. Evolution is not limited by the number of different primary structures that could, in principle, exist. This is presumably one major reason why proteins were selected for the vast number of different roles that they have. A limited number of different primary structures to select from wouldn't satisfy the needs of evolution for, as you will see, the chance of a random series of amino acids linked together forming a functional protein is small.

An α-amino acid, written in the nonionized form has the structure (a), but in aqueous neutral solution it exists as the zwitterionic form (b).

(a)
$$H_2N-\overset{\alpha}{\underset{R}{C}H}-COOH$$

(b)
$$^+H_3N-\overset{\alpha}{\underset{R}{C}H}-COO^-$$

Every amino acid, with the exception of proline (see below), has the same H_2N—CH—COOH part—only the R group attached

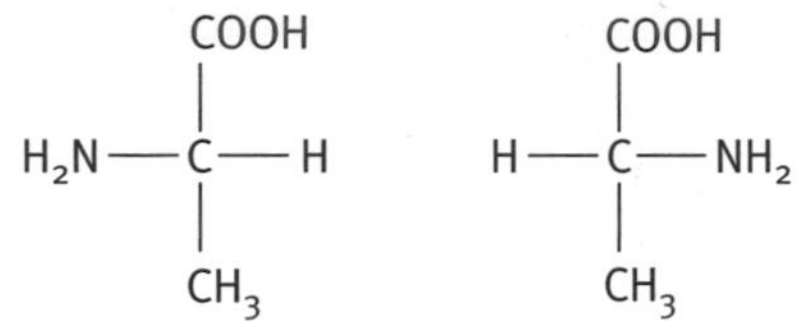

Fig. 3.1 Stereo-isomers of L- and D-alanine drawn on the Fischer projection. In this, vertical lines represent bonds which project below the plane of the paper and horizontal lines bonds which project outward from the paper. Note that the two enantiomorphs are mirror images of one another.

Table 3.1 Single-letter and three-letter symbols for amino acids

Amino acid	One letter	Three letter
Alanine	A	Ala
Arginine	R	Arg
Asparagine	N	Asn
Aspartic acid	D	Asp
Cysteine	C	Cys
Glutamine	Q	Gln
Glutamic acid	E	Glu
Glycine	G	Gly
Histidine	H	His
Isoleucine	I	Ile
Leucine	L	Leu
Lysine	K	Lys
Methionine	M	Met
Phenylalanine	F	Phe
Proline	P	Pro
Serine	S	Ser
Threonine	T	Thr
Tryptophan	W	Trp
Tyrosine	Y	Tyr
Valine	V	Val
Unspecified or unknown	X	Xaa

to the α carbon atom varies. With the exception of glycine, which has no asymmetric carbon atom, amino acids in proteins are of the L-configuration. Using alanine as an example, Fig. 3.1 gives the conventional structures of L and D amino acids, the two being mirror images of one another.

Two amino acids can be linked together by the notional removal of H_2O, but note carefully that this is only a notional reaction; it doesn't happen by such a direct process in the cell. In an aqueous medium, formation of a peptide bond requires chemical energy. The result would be a **dipeptide**, as shown below (structures written in the nonionized form for clarity). The —CO—NH— bond is the **peptide bond** or link.

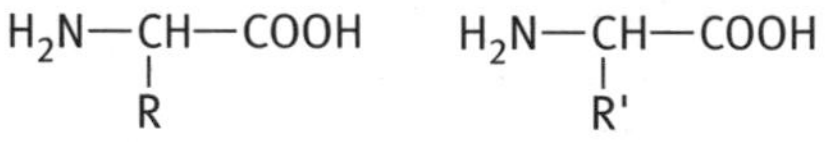

Two amino acids

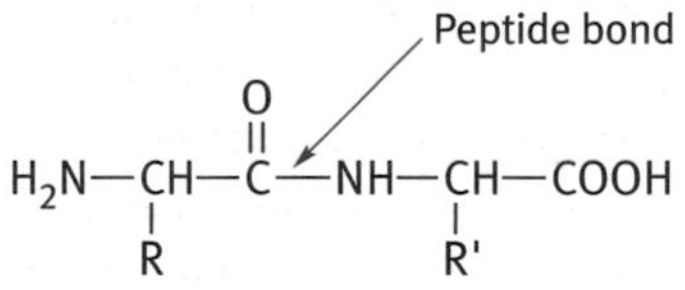

A dipeptide

The dipeptide still has an amino group at one end and a carboxyl group at the other, so that more and more amino acids can be added, giving the **polypeptide** structure shown.

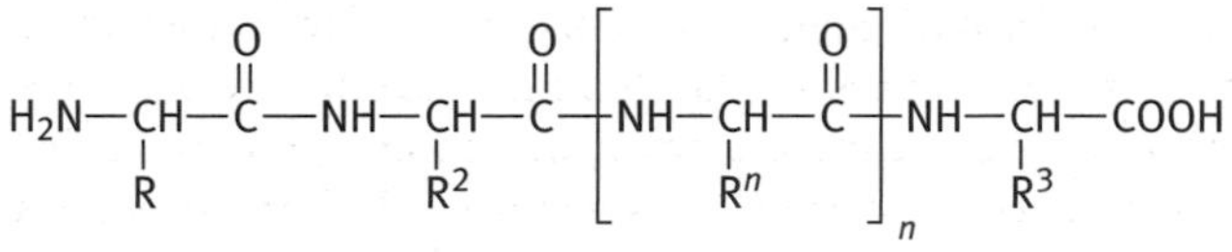

A polypeptide

The naming of peptides of different lengths is an arbitrary matter but as a rough guide a short chain—say up to 20 amino acids—is called a peptide or oligopeptide (*oligo* = few), a polypeptide has more than about 80 amino acids, and anything in between might be called a large peptide. A protein may have several polypeptide chains which may be discrete subunits which assemble together by weak bonds into a multimeric or multi-subunit protein, or they may be covalently linked.

A few terms first: the —NH_3^+ end of a protein is referred to as the **amino terminal** or **N-terminal** end and the other as the **carboxy terminal** or **C-terminal** end. The central chain, without the R groups, (—CH—CO—NH—CH—CO—NH—CH—, etc.) is called the **polypeptide backbone**. An amino acid in a protein is referred to as an **amino acid residue**; the R groups are called amino acid side chains, protein side chains, or quite often simply **side chains**.

The order in which the 20 different amino acids are arranged in the polypeptide is the **amino acid sequence**. The sequence is the **primary structure** of a protein. Determining the sequence is referred to as **protein sequencing**, or amino acid sequencing.

Some methods of protein sequencing are given at the end of this chapter.

What is a native protein?

The primary structure of a protein gives the impression that it resembles an extended piece of string in shape. This is misleading, as can be illustrated by reference to egg white which is largely composed of the protein egg albumin, a water-soluble colourless protein. This can be purified and crystallized as sharp, shiny crystals. The crystallizability depends on the molecules being defined in their compact three-dimensional structure so that they can pack together in the ordered array of a crystal, for crystallization normally demands that all the molecules are identical in shape. The polypeptide chain is, in its natural state, in globular proteins, folded up in a complex, specific way into a compact structure. There is no exact definition of a globular protein except that it has a small axial ratio in contrast to a fibrous protein which has an elongated shape. The three-dimensional structure of globular proteins—the precise folding of the polypeptide chain—is usually destroyed by heat. The folded form is called the **native** protein—the unfolded form, a randomly arranged polypeptide chain, is called the **denatured** protein and the conversion from the native to the denatured state is known as **denaturation.**

When an egg is cooked, the transparent egg white, as it is subjected to mild heat, turns opaque and becomes an insoluble white mass. The discrete folded chains of the native protein molecules unfold with heat and become irreversibly tangled up together to form an insoluble mass (Fig. 3.2). The phenomenon illustrates that the folding of the native protein is held together completely or largely by weak noncovalent bonds (page 13), for covalent bonds in proteins are not disrupted by heat at neutral pH. The polypeptide chain itself remains intact with this treatment. The biochemical function of proteins, such as enzyme catalysis, is almost always destroyed during the heating process (though a few heat-stable proteins do exist). This is why heat is lethal to cells.

Having discussed what the primary structure of a protein is, we must now look at the folded three-dimensional arrangement of that polypeptide chain.

What are the basic considerations which determine the three-dimensional structure of a protein?

As already indicated, most proteins are globular, compact molecules and, except for fibrous proteins are not elongated chains. From Chapter 1 (page 14) you will realize that for a molecule to achieve a stable globular shape in an aqueous medium, polar

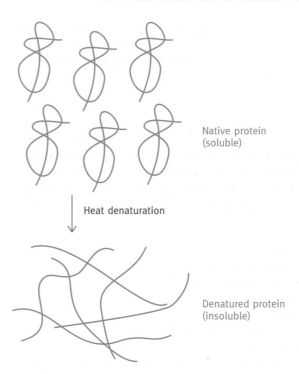

Native protein
(soluble)

Heat denaturation

Denatured protein
(insoluble)

Fig. 3.2 Hypothetical representation of egg albumin denaturation. (The folded configuration is drawn arbitrarily.)

groups must be on the outside in contact with water, whereas the hydrophobic groups must be, to the maximum possible extent, inside, out of contact with water. Any groups inside the molecule that have the potential to form hydrogen bonds, or ionic bonds, but are not in fact involved in appropriate bonding, will tend to destabilize the molecule. Thus, for example, an ionized group inside the hydrophobic interior of a protein would tend to destabilize it. (If you would like to be reminded of the meaning of the terms hydrophobic and hydrophilic please look at page 14.) As described later (page 77), many proteins are located in cell membranes; in this case the requirements are different in that hydrophobic residues are exposed on the outside of the molecule where they contact the hydrophobic membrane structures, whereas hydrophilic groups are hidden away from them.

The problems of making a functional globular protein might well appear to verge on the impossible. There is a chain of perhaps hundreds of amino acid residues all of which must fold up to form a tightly packed solid globular structure with polar side chains on the outside and hydrophobic side chains on the inside, shielded from water. Also on the outside there are pockets shaped to fit to other molecules such as substrates and other ligands with exquisite precision.

Evolution, in essence, involves the development of new proteins, and with the above considerations in mind it is not surprising that it is a slow business, being measured in millions of years; starting from scratch, the chance of accidentally producing a polypeptide which will fold up into a protein capable of efficiently performing a specific task is exceedingly small. This helps to explain why, repeatedly, it is found that proteins with different functions are related in structure as if they have evolved one from another. It seems that once evolution has found a polypeptide structure that folds nicely and performs a function, it may often have been preferable to tinker with it and adapt it to other uses rather than start all over to find totally different structures *de novo*.

Before we can deal with the way proteins are folded we must first look at the building blocks from which they are constructed—the 20 different amino acids.

Structures of the 20 amino acids

Learning to recognize 20 different structures of amino acids may be less daunting if you keep in mind what the different side groups are for. They are the building blocks with which evolution juggles to produce polypeptide chains that will fold up to satisfy the almost impossible requirements listed above. Clearly, there must be a variety of amino acids with different shapes, sizes, and polarity characteristics to permit the evolution of proteins. You can almost imagine the process—a small hydrophobic group needed here to fill a hole adjacent to a large group, a strongly polar group here and a weaker one there, and so on. Having 20 different units to play with gives evolution flexibility in designing protein structures.

More than 20 amino acids are found in different organisms, but only what Francis Crick has called the 'magic 20' are used for protein synthesis—they are the only ones specified in the genetic code. If the latter statement means nothing to you, don't worry; it is explained in later chapters. We will start with aliphatic side chains presented in order of increasing size and hydrophobicity. First the smallest:

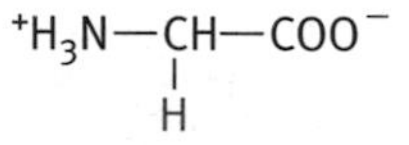

Glycine

Glycine with H as its side group is the smallest amino acid and has no marked hydrophilic or hydrophobic properties. The rest of the aliphatic nonpolar amino acids, from alanine to isoleucine, have side chains of increasing hydrophobicity.

Hydrophobic amino acids

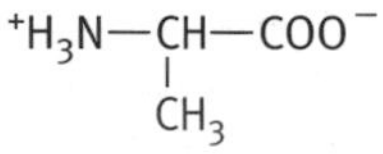

Alanine 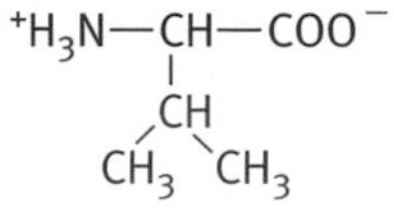 Valine

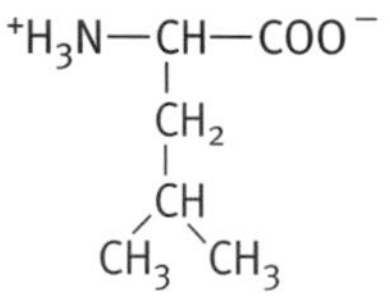

Leucine 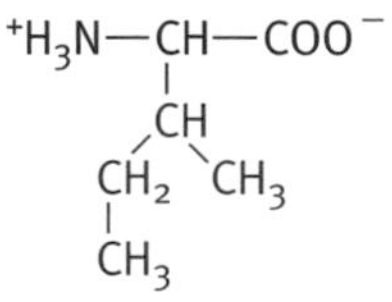 Isoleucine

The latter three are known as the branched chain aliphatics. Methionine is a rather special hydrophobic aliphatic amino acid (special more because of its role in methyl group metabolism than its role in protein structure, page ••).

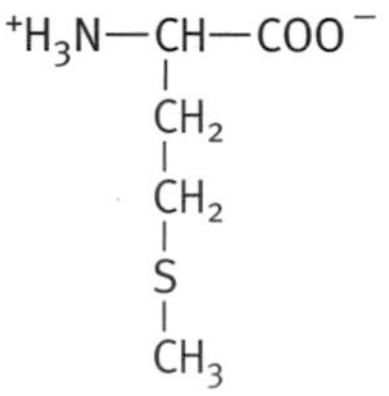

Methionine

Then we come to the really large hydrophobic side chains—aromatic ones.

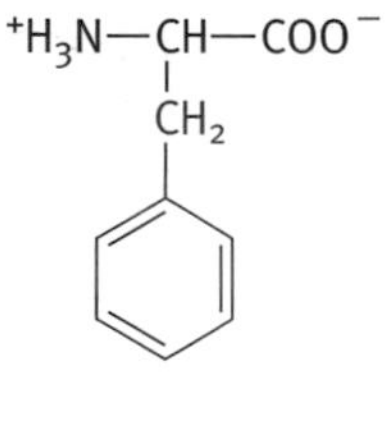

Phenylalanine 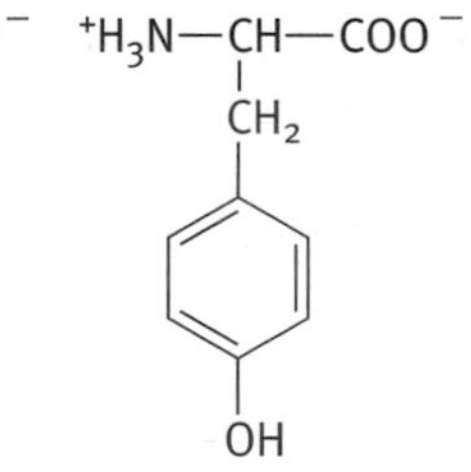 Tyrosine 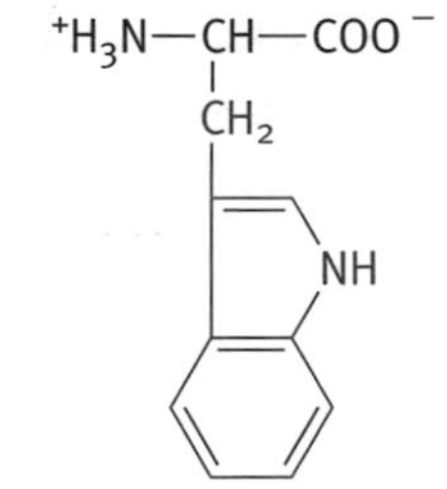 Tryptophan

Tyrosine with its —OH group can form a hydrogen bond, but its aromatic ring is large and hydrophobic so it is of somewhat mixed classification. Tryptophan can also form a hydrogen bond.

Hydrophilic amino acids

An ionized group is very hydrophilic. Acidic amino acids (with an extra —COOH on the side chain) are negatively

charged at the pH of the cell, and thus hydrophilic, and basic amino acids (extra —NH_2 groups on the side chains) are positively charged and also hydrophilic. Aspartic and glutamic acids both have acidic side chains, lysine and arginine have strongly basic side chains and histidine is a weakly basic amino acid. The ring structure of histidine is known as the **imidazole ring**.

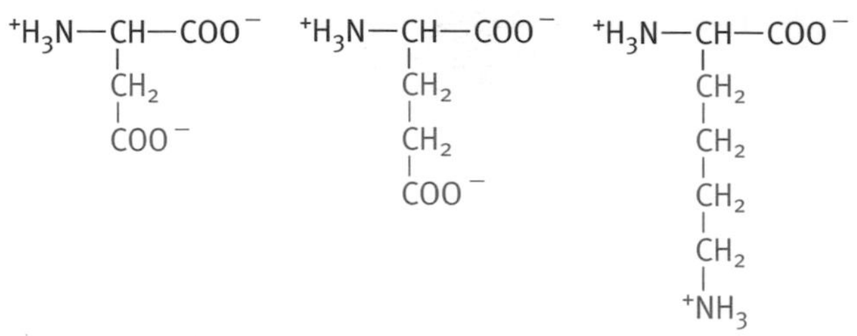

$$^+H_3N-CH-COO^-$$
$$|$$
$$CH_2$$
$$|$$
$$COO^-$$

Aspartic acid

$$^+H_3N-CH-COO^-$$
$$|$$
$$CH_2$$
$$|$$
$$CH_2$$
$$|$$
$$COO^-$$

Glutamic acid

$$^+H_3N-CH-COO^-$$
$$|$$
$$CH_2$$
$$|$$
$$CH_2$$
$$|$$
$$CH_2$$
$$|$$
$$CH_2$$
$$|$$
$$^+NH_3$$

Lysine

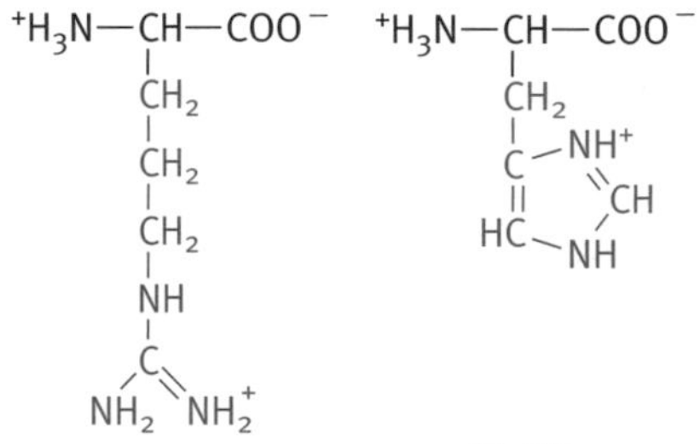

$$^+H_3N-CH-COO^-$$
$$|$$
$$CH_2$$
$$|$$
$$CH_2$$
$$|$$
$$CH_2$$
$$|$$
$$NH$$
$$|$$
$$C$$
$$NH_2 \quad NH_2^+$$

Arginine

$$^+H_3N-CH-COO^-$$
$$|$$
$$CH_2$$
$$|$$
$$C-NH^+$$
$$||\qquad CH$$
$$HC-NH$$

Histidine

The two acidic amino acids, aspartic acid and glutamic acid, exist also as the amides, asparagine and glutamine, respectively. Since the amide group does not ionize, these two amides are less hydrophilic than the parent compounds.

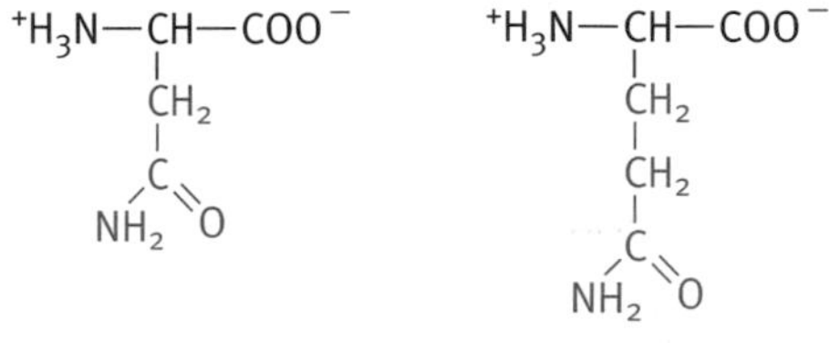

$$^+H_3N-CH-COO^-$$
$$|$$
$$CH_2$$
$$|$$
$$C$$
$$NH_2 \quad O$$

Asparagine

$$^+H_3N-CH-COO^-$$
$$|$$
$$CH_2$$
$$|$$
$$CH_2$$
$$|$$
$$C$$
$$NH_2 \quad O$$

Glutamine

Serine and threonine are hydrophilic.

$$^+H_3N-CH-COO^-$$
$$|$$
$$CH_2OH$$

Serine

$$^+H_3N-CH-COO^-$$
$$|$$
$$CHOH$$
$$|$$
$$CH_3$$

Threonine

Amino acids for special purposes

Cysteine is similar to serine but with —SH instead of —OH. It plays two special roles in protein structure—supplying external —SH groups such as in active centres of enzymes, and forming covalent —S—S— bonds or disulfide bonds internally—but more of this later.

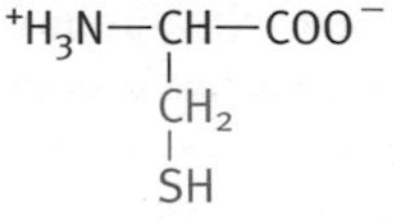

$$^+H_3N-CH-COO^-$$
$$|$$
$$CH_2$$
$$|$$
$$SH$$

Cysteine

Proline is really the oddity—literally a kinky amino acid in that it can put a kink into the conformation of polypeptide chains (described later). It has an imino (—NH) group rather than an amino group. (If you have trouble in memorizing proline—which can happen—remember that it is an ordinary amino acid except that the three-carbon side chain, shown in red, is linked to the nitrogen atom making it an —NH_2^+ imino acid rather than an —NH_3^+ amino acid.

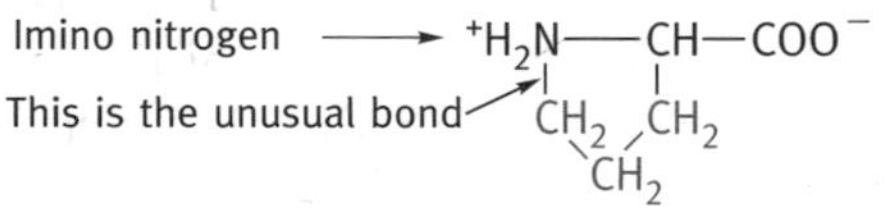

Imino nitrogen $\longrightarrow$ $^+H_2N-CH-COO^-$
This is the unusual bond $CH_2 \quad CH_2$
CH_2

Ionization of amino acids

As already mentioned, free amino acids in aqueous solution have the zwitterionic structures shown above, in which the α-amino and α-carboxyl groups are ionized. Their pK_a values are around 8–10 and 2 respectively; (see page 15 if you wish to be reminded about such values). However, when incorporated into a polypeptide chain, these groups are no longer ionizable, except for the terminal amino and carboxyl groups. The ionized state of a protein is therefore almost entirely dependent on the amino acid residues of aspartic and glutamic acid, lysine, arginine, and histidine since their ionizable side chain groups are not blocked by peptide bond formation, as shown for glutamic acid and lysine in the illustration below.

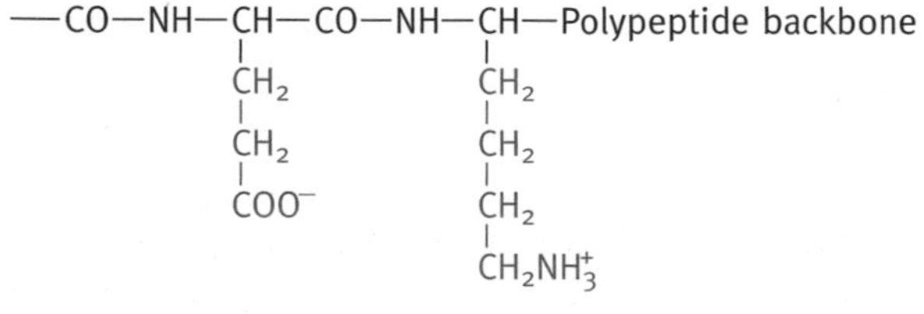

$$—CO-NH-CH-CO-NH-CH-\text{Polypeptide backbone}$$
$$|\qquad\qquad\qquad |$$
$$CH_2\qquad\qquad\quad CH_2$$
$$|\qquad\qquad\qquad |$$
$$CH_2\qquad\qquad\quad CH_2$$
$$|\qquad\qquad\qquad |$$
$$COO^-\qquad\qquad CH_2$$
$$|$$
$$CH_2NH_3^+$$

Glutamic acid side chain   Lysine side chain

The side chain —COO⁻ groups of aspartic and glutamic acids have pK_a values around 4, so they are virtually fully dissociated at pH 7.4—these provide the negatively charged side groups of proteins. The basic amino acids lysine and arginine, with pK_a values for their —NH_3^+ side chain groups of 10.5 and 12.5 respectively, are fully ionized at physiological pH (page 16). The third basic amino acid, histidine, has an imidazole ring as the side chain group whose pK_a value is near neutrality (around 6 in proteins), so that histidine is often found in active sites of enzymes where movement of a proton is involved in the reaction catalysed; the imidazole group can accept or donate a proton at a pH near that existing in the cell.

The distribution of charged amino acid residues in a protein has an important effect on the conformation (the arrangement of the three-dimensional structure in space) that a polypeptide chain can adopt. Charges of the same sign close to each other repel each other, i.e. negative repels negative and positive repels positive. Closely positioned positive and negative charges will attract each other. Phosphorylation of proteins, so commonly used to control enzyme activity (page 215), introduces a strong negative charge which can affect conformation of the molecule.

Symbols for amino acids

There are two types of abbreviations for amino acids as shown in Table 3.1. The three letter system is often (but not always) used when short sequences are represented because they have the advantage of being self-evident. For longer sequences the single letter abbreviations are less cumbersome.

The different levels of protein structure—primary, secondary, tertiary, and quaternary

We first give a brief overview of this topic, to be followed by a more detailed treatment.

The sequence of amino acids which are linked together covalently into a polypeptide chain is, as stated, its primary structure (Fig. 3.3(a)). It says nothing about how that polypeptide is arranged in a three-dimensional space. The determination of the three-dimensional structure of proteins is a technically sophisticated area. In brief outline there are two main methods.

- The one that has determined most of the known protein structures is **X-ray diffraction** of protein crystals. For this it is essential to purify a protein completely so that large crystals can be obtained for study and this may be the most dif-

ficult part of the work; the protein also has to be 'labelled' with at least two atoms of heavy metal (which do not affect the structure of the molecule) to provide 'reference points' for the mathematical processing of results. Diffraction studies are made on crystals with and without this labelling. Proteins such as hemoglobin and digestive enzymes were the earliest to be tackled because they exist in large amounts and are relatively easily purified. The crystal is mounted in an apparatus which bombards it with X-rays whose wavelength is sufficiently short to be diffracted by the atoms in

(a) Primary

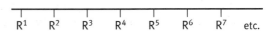

R¹   R²   R³   R⁴   R⁵   R⁶   R⁷   etc.

This structure will be represented below as a simple line.

(b) Secondary

The polypeptide **backbone** exists in different sections of a protein either as an α helix, a β-pleated sheet, or random coil. (These have yet to be explained.)

α helix   β-pleated sheet   Random coil or loop region

(c) Tertiary

The secondary structures above are folded into the compact globular protein.

 This protein will be represented below as:

(d) Quaternary

Protein molecules known as subunits assemble into a multimeric protein held together by weak forces.

Fig. 3.3 Diagrammatic illustration of what is meant by primary, secondary, tertiary, and quaternary structures of proteins. Note that the structures referred to will be described shortly in the text.

the molecule. The reflections are collected as spots on a photographic plate or more recently by photomultipliers linked to a computer. The atomic reflections are processed mathematically to give the positions of individual atoms within the protein molecule from which the entire structure is determined.

- A second method of increasing importance is **nuclear magnetic resonance (NMR)** which gives structural information on proteins in concentrated solutions, the latter a major advantage given the difficulty often in obtaining crystals of some proteins. It also studies the proteins in an environment more closely resembling that in the cell and permits changes to be observed.

The polypeptide **backbone** itself is arranged in a particular conformation known as the **secondary structure** (Fig. 3.3(b)). The secondary structure is folded yet again to give the **tertiary structure** (Fig. 3.3(c)). The complete molecule so formed by the primary, secondary, and tertiary structures may be the final functional protein or it may be a protein monomer or subunit so that to form a functional protein it must associate with other protein monomers (which may be the same or different). This association is known as the **quaternary structure** (Fig. 3.3(d)), the resultant multi-component molecules being known as **polymeric** or **multi-subunit proteins**.

With this overview, we will now deal in more detail with the different levels of protein structure; the primary structure we have already covered.

Secondary structure of proteins

This is concerned primarily with the arrangement of the polypeptide backbone rather than of the side chains. We have said that in a water-soluble globular protein, as far as possible, polar side chains are arranged (due to tertiary structure) on the outside and the hydrophobic ones on the inside, but there is a separate problem with the polypeptide backbone. As it crosses back and forth to fold the whole molecule into a compact shape, it cannot avoid being exposed to the hydrophobic interior of the protein molecule. The problem is that the backbone has polar groups with the capability of hydrogen bonding—in fact, two bonds per amino acid unit, since the C=O and N—H of the peptide bond are capable of hydrogen bond formation. Unless this bonding potentiality is satisfied by actual bond formation, with a resultant liberation of free energy, the structure will be destabilized. The nonpolar side groups of amino acids in the interior of the molecule cannot hydrogen bond to the polypeptide backbone groups. So what can the latter hydrogen bond with? The answer is with groups on the same, or an adjacent, polypeptide backbone.

It should be emphasized again that we are not (yet) talking about side chain packing, which comes into play in tertiary structure. We are still dealing with the problem of how the hydrogen bonding requirements of the C=O and N—H groups of the polypeptide backbone are satisfied. There are two main classes of structures which solve the problem—the **α helix** in which the backbone is arranged in a spiral-like coil and the **β-pleated sheet** in which extended polypeptide backbones are side by side. These structures are very stable; they can occur at the exterior of proteins with appropriate hydrophilic side chains or in the hydrophobic interior of proteins, with appropriate hydrophobic side chains.

The α helix

In this, the polypeptide backbone is twisted into a right-handed helix (Fig. 3.4(a)). For L-amino acids, a right-handed helix is more stable than a left-handed one. You can visualize the direction of twist of the right-handed helix by imagining that you are a right-handed person driving home a screw with a screwdriver. Your fingers then follows the right-handed twist of the chain. Alternatively, in case you find this simpler, when you look down the axis either way the helix turns clockwise. Its shape is shown in Fig. 3.4(a).

It is almost natural to assume that, in such a helix, the number of amino acids constituting one turn should be a whole number, and this idea bedevilled the first attempts to produce a model of the α helix since the workers tried to build it with an integral number of amino acids per turn and indeed the X-ray data seemed to support that idea (but see below). Linus Pauling was the first to discard the assumption and discovered the α helix structure with a pitch of 3.6 amino acid units per turn. This results in the C=O of each peptide bond being aligned to form a hydrogen bond with the peptide bond N—H of the fourth distant amino acid residue. The C=O groups point in the direction of the axis of the helix and are nicely aimed at the N—H groups with which they hydrogen bond, giving maximum bond strength and making the a helix a very stable structure. Thus, every C=O and N—H group of the polypeptide backbone are hydrogen bonded in pairs, forming a stable, cylindrical, rod-like structure (Fig. 3.4(a)). In cross section an α helix is a virtual solid cylinder with all the R groups projecting outwards (Fig. 3.4(b)). The van der Waals atomic radii are such that there is almost no space along the axis of the α helix, again making it a stable structure.

As mentioned earlier, the X-ray data indicated that the number of amino acids in one pitch of the α helix was a whole number rather than the correct 3.6. The reason is that α helices do not lie exactly parallel but a pair makes a long slow coil around one another which distorts the reflections. Pauling made his

(a)

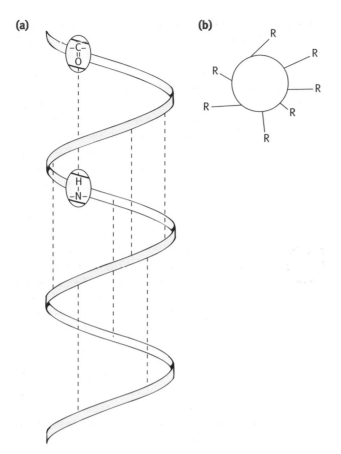

(b)

which rotation might be expected), in fact the peptide bond is a hybrid between two structures: (1) in which the bond between the carbon and oxygen atoms is a pure double bond and (2) in which it is a single bond.

The actual electron density is between the two giving the C—N bond about 40% double-bonded character. This is sufficient to prevent rotation about it, making the polypeptide chain more rigid. With proline in peptide linkage, there is no hydrogen atom on the nitrogen available for hydrogen bonding and the structure of the residue restricts rotation, so that it cannot assume the conformation needed to fit into an α helix. Some amino acids are excellent helix formers; a few, like proline, tend to be helix breakers.

Imino nitrogen in peptide linkage not able to hydrogen bond

We will return shortly to how runs of amino acids in α helices fit into protein structures but, before that, we must deal with the alternative to the α helix, namely the β-pleated sheet. Proteins are often made up of mixtures of the two with each constituting different sections of the polypeptide chain.

Fig. 3.4 The α helix form of a polypeptide chain. (a) Hydrogen bonding between C=O and N—H groups of the polypeptide backbone (side chains not shown). The hydrogen bonds (broken lines) are shown in approximate positions only. (b) Looking down the axis of an α helix, with the amino acid side chains projecting from the cylindrical structure (each at a different distance below the plane of the paper). The R groups are not drawn in their exact orientation from the axis. Since there are 3.6 residues per turn, each residue occurs every 100° around a circle (360°/3.6 = 100°).

important discovery of the structure while in bed with a cold, using folded sheets of paper as models.

Not all of the polypeptide chain of a globular protein is in the α helix form. The sections that are, average 10 residues in length, but the range of lengths varies a great deal from this average in different proteins.

Amino acids vary in their tendency to form α helices. Proline in particular is an α helix breaker or terminator—a proline residue forces a bend in the structure. With the other amino acids, the polypeptide chain can rotate around two bonds per residue, as shown below. Note that although the CO—NH peptide bond is written as an ordinary single bond (about

The β-pleated sheet

This also forms a stable structure in which the polar groups of the polypeptide backbone are hydrogen bonded to one another, thus enabling the structure to be stable, even in the hydrophobic interior of a globular protein.

The principle is simple. The polypeptide chain lies in an extended or β form with the C=O and N—H groups hydrogen bonded to those of a neighbouring chain (which may be formed by the same chain folding back on itself, or may be a separate chain lying alongside, Fig. 3.5). Several chains can thus form a sheet of polypeptide. It is pleated because successive α-carbon atoms of the amino acid residues lie slightly above and below the plane of the β sheet alternately. The adjacent polypeptide chains bonded together can run in the same direction (parallel) or opposite directions (antiparallel). In the latter case, the polypeptide makes tight 'β turns' to fold the chain back on itself.

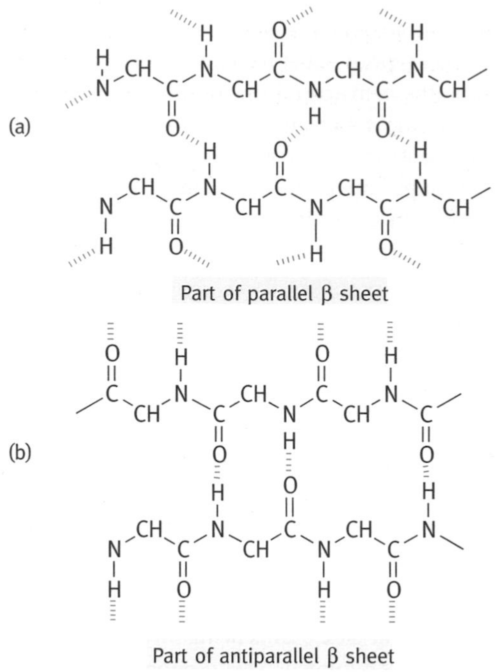

Fig. 3.5 (a) Hydrogen bonding between two polypeptide chains running in the same direction forming a parallel β sheet. The R groups attached to the —CH— groups are above and below the plane of the paper. (b) Adjacent polypeptide chains running in opposite directions can also mutually hydrogen bond forming an antiparallel β sheet. This enables a single chain to form a β sheet by folding back on itself.

Connecting loops

In a protein the α helices and β sheet sections are connected together by unstructured polypeptide sometimes known as random coils—a confusing name, because they are neither random nor coils, and connecting loop is more descriptive. The term indicates a section of polypeptide in a protein whose conformation is not recognizable as one of the defined structures described above. The structure of a loop region in a protein is mainly determined by side chain interactions and, within a given protein, is fixed rather than varying in a random way. The connecting loop may be in any conformation as determined by the various group interactions in the protein structure. Since the structure of a loop may not satisfy the hydrogen bonding potentials of the C=O and N—H groups of the backbone, or those of the side groups, such sections are often found at the exterior of proteins, in contact with water.

Tertiary structure of proteins

To sum up so far: evolution has 20 'units', the amino acids, to play with in designing proteins. These are assembled into a polypeptide chain, their sequence being the primary structure, and this is folded into secondary structures, namely the α helix and the β-pleated sheet linked together by unstructured loop regions. The former two structures internally satisfy the hydrogen bonding potential of the polypeptide backbone groups. A single polypeptide chain in a protein can be arranged as a mixture of these various structures in different parts of the chain and these secondary structures are folded up and packed together to form the protein molecule. This arrangement of the various secondary structures into the compact structure of a globular protein is referred to as the **tertiary structure.**

To simplify structural diagrams of proteins, conventions have been adopted to depict secondary structures. An α helix is represented either as a solid cylinder, sometimes with α helix inside it, or alternatively as α helical ribbon (Fig. 3.6(a)). The individual sections of polypeptide chains or β strands which participate in β-pleated sheet formation are represented as broad arrows (Fig. 3.6(b); an arrow, let it be clear, is not a sheet but a polypeptide in a β-pleated sheet). An extended polypeptide chain in the β configuration has been shown always to twist slightly to the right so that the arrows are usually drawn in protein structures with this twist. Connecting loops are shown as any sort of line.

How are proteins made up from the three motifs—the α helix, the β-pleated sheet, and connecting loops?

In principle, the answer is that any combination of the three can be used to assemble a globular protein provided that the side chains can be packed together in ways appropriate to their various affinities and repulsions, the unstructured sections having their backbone polar groups satisfactorily bonded—usually to water—and exposure of hydrophobic groups to water or other polar groups minimized. These are formidable provisions, which probably only one in countless numbers of random amino acid sequences can fulfil. Quite large numbers of proteins have had their three-dimensional structure determined by X-ray diffraction studies. The picture is emerging that certain patterns of structure are preferred, so that the number of basic 'recipes' for protein structures may be limited. The three-dimensional structures of proteins are not things a nonspecialist biochemist usually memorizes but they are very useful, for example, in discussing biochemical mechanisms. In Fig. 3.7, representations of a few illustrative protein structures are given. Myoglobin, described later (page 513), has only α helices connected by loop sections with its heme group inserted into a cleft. The staphylococcal nuclease (an enzyme hydrolysing nucleic acid) has a mixture of an antiparallel β sheet and three α helices connected by unstructured polypeptide (Fig. 3.7(b)). A

(a)

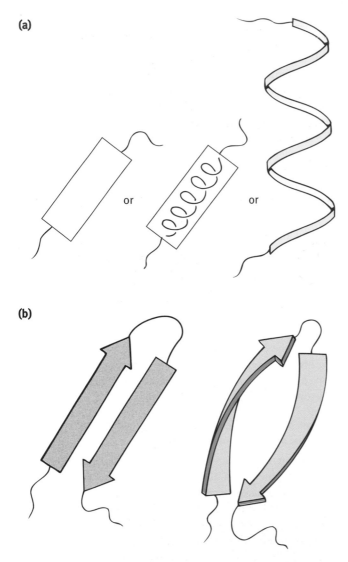

(b)

Fig. 3.6 Symbols used to indicate (a) α helix structures; or (b) β-pleated sheets. In (b) a pair of polypeptide stretches are shown making an antiparallel sheet. Because of the slight right-handed twist the arrows are represented as on the right. The single lines connecting structures represent random coil or loop sections.

common arrangement is the so-called α/β barrel. In this, there is a central core of β strands arranged like the staves of a wooden barrel except that they are twisted. The barrel encloses tightly packed hydrophobic side chains. Surrounding this barrel are α helices. The diagram of the triosephosphate isomerase in Fig. 3.7(c) illustrates this. Note that although the arrows representing the eight β strands appear to have insufficient contact for the mutual hydrogen bonding involved in forming a β sheet, this is the effect of representing the structure as viewed from the top of

the barrel. The enzyme pyruvate kinase, shown in Fig. 3.7(d), shows another type of structure with three such sections.

Useful as these structural diagrams ar, there is the danger that you get the totally incorrect impression that proteins are openwork structures of strings and ribbons. In fact, when the side groups are represented, they are solid molecules with no space in them. If you turn to page 271 you will see (Fig. 17.3) a protein, represented as a model, which makes the point clear.

What forces hold the tertiary structure in position?

We have seen that the secondary structures of proteins (α helices and β sheets) depend on hydrogen bonding and, as explained, these secondary structures are arranged in specific ways to form the tertiary structure of the protein. Something must hold these sections in their tertiary structure. Hydrogen bonding and ionic interactions, between side chains, influence the folded structure, and the hydrophobic force (page 14) is a major contributor to the drive which folds a protein and results in the hydrophobic side chains being forced into the centre of the molecule, thus minimizing hydrophobic–water interactions and maximizing van der Waals bonding of hydrophobic groups. Most globular proteins depend entirely on the weak forces described above for their structure—but now see below for a qualification of this.

Where do the disulfide or S—S covalent bonds come into protein structure?

It was mentioned earlier that cysteine has special functions in proteins. One is that the thiol group often functions in specific reaction centres on proteins. (The glycolytic enzyme glyceraldehyde 3-phosphate dehydrogenase on page 150 is a good example of this.) However, cysteine has another function in proteins. Although their tertiary structure is largely the result of weak forces between the side chains, many extracellular enzymes are stabilized by disulfide (S—S) bridges which, being covalent, are very strong. These include proteins liberated into the blood (insulin is one example), or the intestine (digestive enzymes). This is achieved by pairs of thiol groups of the cysteine side chains, brought together by polypeptide folding, forming a covalent disulfide bond. This provides a very strong interlocking bond—more or less like a steel rivet in the structure. A few of these makes the folded shape more stable. Insulin has three disulfide or S—S bridges. Proteins with disulfide bonds are often less easily denatured by heat. Few intracellular proteins have disulfide bonds in their structure, probably because the interior of cells are strongly reducing and this might be sufficient to disrupt them.

A simple observation may help to explain how these bonds

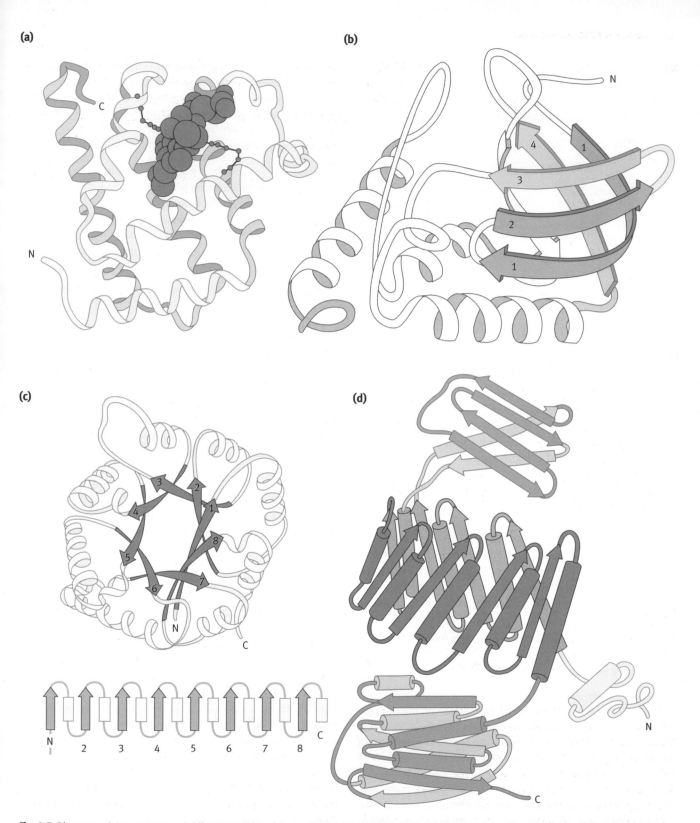

Fig. 3.7 Diagrams of the structures of different proteins: (a) myoglobin; (b) staphylococcal nuclease; (c) triosephosphate isomerase (the lower diagram shows the arrangement of the numbered β sheets and α helices); (d) pyruvate kinase. (See text for description.)

form. If a water-clear solution of the amino acid cysteine is left at neutral pH in an open beaker at room temperature, within hours the surface of the liquid becomes covered with a layer of white insoluble crystals of cystine. The chemical reaction is:

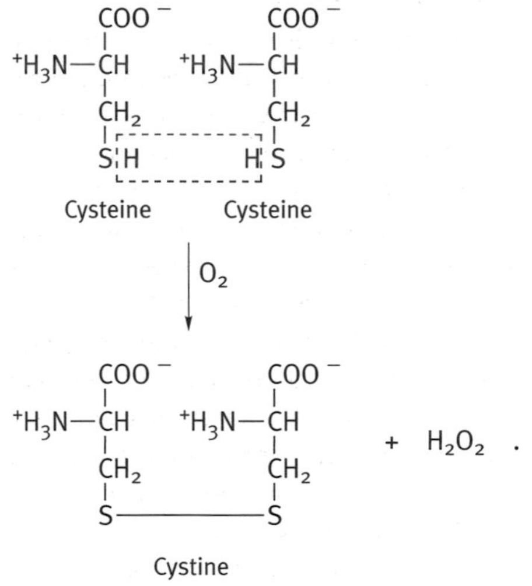

Cystine

The same formation of S—S bonds occurs between appropriate thiols in proteins but in the cell it is an enzymic process; the oxidation there involves electrons being transported to oxygen by electron transport proteins (page 162) with the formation of water rather than hydrogen peroxide.

An extreme example of stabilization of a protein by disulfide bridges is the **keratin** protein of hair. The long polypeptides of this (described below) are interlinked by many disulfide bonds which are important in determining the configuration of the hair. In permanent waving, these are broken by reduction followed by setting the hair into a new configuration. This is made permanent by the 'neutralizer' which reoxidises cysteine —SH groups to reform disulfide bonds. However, these bonds are new ones between —SH groups which have been brought together in the new curled configuration of the hair.

Quaternary structure of proteins

With the tertiary structure, we now have a protein molecule, and for many proteins that is the end. However, many functional proteins have more than one such protein molecule in them which remain as discrete molecules held together in a single complex by secondary bonds. The molecules have to be structured so as to fit via complementary surface patches so that only the correct subunits complex together. The individual molecules in such a structure are called **subunits**, protein subunits, protomers, or **monomers**. The whole protein is called an **oligomeric**, **multimeric**, or **multi-subunit protein**. Allosteri-

cally regulated enzymes are mostly of this type (see page 213), and so is hemoglobin (see page 513). This arrangement of subunits into a single functional complex is referred to as the quaternary structure of a protein (Fig. 3.3(d)).

Membrane proteins

We have so far described globular proteins as being water soluble, but many membrane proteins have to have an external section that is water soluble at each end and a hydrophobic section in the form of an α helix in the middle (see page 78). Polar groups in contact with the hydrocarbon layer in the centre of a membrane would be an unstable situation. The structure of membrane proteins will be dealt with in the next chapter.

Conjugated proteins

Many proteins are simply as already described—they need nothing but the folded polypeptide chain(s) for their function. However, many enzymes require a metal ion to be complexed with the protein for activity, as described on page 26. Some have more complex structures attached to them which participate in the reaction. When the protein is an enzyme and an additional molecule is firmly attached to form the active enzyme, the attachment is called a **prosthetic group**, the active complex a **holoenzyme**, and the protein alone, the **apoenzyme;** ($apo =$ detached or separate). The cytochromes (page 164) with heme prosthetic groups and the dehydrogenases with bound FAD (page 137) are examples. Other proteins have carbohydrate attachments and are called **glycoproteins**. Most membrane proteins have oligosaccharides attached on the outside surface (page 78) to the —OH of serine or threonine side chains (*O*-linked) or to an asparagine side chain (*N*-linked). The latter attachment method is illustrated below.

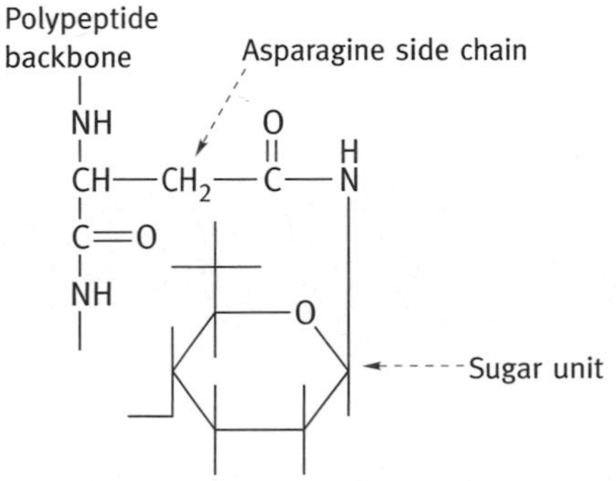

Many secreted proteins such as blood proteins are glycoproteins. The function(s) of the carbohydrate attachments are not always, or perhaps not usually, understood. They may in some

cases be related to the stability of the protein to which they are attached, or to the longevity of the structure of which the glycoprotein is a component. For example, the degradation of carbohydrate attachments to serum proteins and to erythrocyte membrane proteins appear to mark them for uptake and destruction by liver cells—the carbohydrates are acting as indicators of age of the components. Conversely, glycosylation of some proteins may protect them from proteolytic attack. In some cases they are involved in recognition, for many combinations of different sugars can exist and thus might constitute unique 'labels'. In sorting of proteins by the Golgi apparatus they play this role in some cases (discussed in Chapter 25).

In the case of the mucin glycoproteins which protect mucus membranes, the O-linked carbohydrate attachments cause the polypeptide to adopt an extended configuration which facilitates the formation, even in very dilute solutions, of networks trapping much water. This ability of O-linked carbohydrate attachment to cause the polypeptide to adopt an extended form probably has other applications. In LDL receptors (page 127) the membrane protein is anchored into the membrane while the functional receptor domain is extended well clear of the membrane surface, readily accessible to its binding protein, by a rigid 'stalk' formed by O-glycosylation of the polypeptide in this region (Fig. 3.8). A whole family of proteoglycans exist in which the carbohydrates are rich in amino sugars, sulfated sugars, and carboxylic acid sugars; these are dealt with in a later section in this chapter.

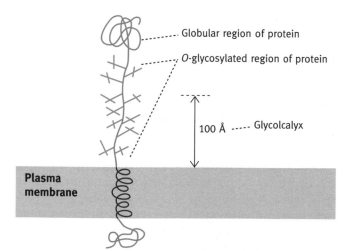

Fig. 3.8 The low-density lipoprotein (LDL) receptor showing, the O-glycosylation region as a stiff extended structure projecting >100 Å above the plasma membrane, with a globular protein domain (the functional receptor) beyond this.

What are protein modules or domains?

If you consider a protein molecule consisting of a single polypeptide chain, its folded structure may be a single compact entity, no part of which could exist on its own, because the folding of the protein would be disrupted.

However, especially in proteins which are larger than about 200 amino acids in length, this may not be the case. In such proteins when the three-dimensional structure is determined it is often seen that there are two or more regions which form compact 'islands' of folded structure usually linked together by unstructured polypeptide. Inspection of the structure leads to the impression that if these folded sections were to be obtained separately they would remain neatly folded in their native state. Indeed, in certain cases, they *have* been obtained separately and their integrity of structure confirmed. A definition of a **protein module** or **domain**, as these sections are called, is that it is a sub-region of the polypeptide which possesses the characteristics of a folded globular protein. A domain must be formed from a discrete section of polypeptide chain—it can't involve the folded structure being formed by the chain leaving the domain and then doubling back into it. It must be self-contained. A (somewhat fanciful) analogy is to liken domains to individual movements of a symphony; each is self-contained with its own structural characteristics, and could reasonably be listened to as a separate piece of music, but nonetheless is part of a single musical entity, linked to the others to form the whole and each with its purpose within the whole. The structure of the enzyme pyruvate kinase illustrated in Fig. 3.7(d) clearly illustrates three domains joined together to form a single protein.

Why should domains be of interest?

Protein domains are often associated with different partial activities of a protein that are involved in the function of the protein. The case of mammalian fatty acid synthase with seven catalytic activities needed for fat synthesis (page 188) being accommodated on a single polypeptide chain is an extreme example. In this case, since the same activities reside on separate proteins in bacteria, it is reasonable to speculate that the separate domains or modules of the mammalian enzyme arose from gene fusions. Many enzymes with a single catalytic function combine with at least two substrates. The NAD^+ dehydrogenases (to be described later, page 136, so don't be concerned with what NAD^+ is at this stage) are a typical case. They bind NAD^+ and also the substrate to be oxidized. Different NAD^+ dehydrogenases all bind NAD^+, but each binds a different oxidizable substrate and all catalyse the reaction:

$$AH_2 + NAD^+ \rightleftharpoons A + NADH + H^+$$

It is found that enzymes catalysing such reactions have separate domains for binding NAD$^+$ and the substrate (AH$_2$). However, the NAD$^+$ binding domains of the different enzymes examined have a similar basic structure, whereas the AH$_2$ binding domains are different in the different enzymes. There are at least two possible interpretations of this. One, that it represents convergent evolution—that the NAD$^+$ binding structure is the only one capable of binding NAD$^+$ and therefore it is inevitable that all would be similar. The other is that once evolution had successfully developed an NAD$^+$ binding module, it was used over and over again. This gives rise to the concept of **domain shuffling** in evolution.

The concept envisages that modular construction of enzymes and other proteins permits rapid evolution of new functional proteins. It is somewhat like the modular assembly of electronic instruments. If you have 'boards' (made up of many components) with different broad functions, they can be assembled into different instruments by using various combinations of 'plug-in' boards. Similarly, linking together different domains from various pre-existing proteins might make many different functional proteins. Admittedly it wouldn't do to construct musical symphonies by movement shuffling, to return to our previous analogy, but natural selection ruthlessly discards the disharmonies and preserves the successes

All this might seem to be pure speculation based on similarities between proteins of similar functions, but as you'll see later (Chapter 23), domains are often coded for by specific separate gene sections, called exons (don't be concerned with what these are just yet). This strongly suggests that 'exon shuffling' and the consequent domain shuffling may well have been an evolutionary mechanism of importance. Developing new proteins by using different combinations of existing domains (with appropriate 'tinkering' for limited modified function) could be a more rapid process than relying on random changing of individual amino acids in entire proteins. We will return to other examples of protein domains when we deal with antibodies (Chapter 27).

The problem of protein folding

When a protein is synthesized, the polypeptide emerges from the synthetic machinery (the subject of Chapter 24) as a linear structure whereas most native proteins are compact structures (fibrous proteins have secondary structure but little tertiary folding). A water-soluble protein has its hydrophilic groups largely on the outside and the hydrophobic groups inside, shielded from water. The polypeptide has to fold up, making the correct quite elaborate interactions to achieve this result. The folding pattern is entirely determined by the amino acid se-

quence. However, while the primary structure of a polypeptide chain dictates what folded protein structure is *possible*, the question of *how* the three-dimensional folding is actually achieved is a major problem. An 'incorrect' amino acid sequence may prevent the correct structure being achievable, but how does a 'correct' structure become folded, in the correct way? When synthesized in the cell, a typical polypeptide chain of a protein folds up in a couple of minutes or so. If it had to do this by randomly trying every theoretically possible folded structure, it would take countless millions of years. In other words, the solution of trying every possible folded conformation until reaching the lowest free energy state achievable is not feasible. Viewed in this light, this looks to be one of the greatest problems to be overcome for life to be established. It is believed that certain secondary structural features known as 'molten globules' in a protein are rapidly formed and this somehow facilitates the entire folding process, but the process is far from completely understood.

There is, however, a related but separate major problem. In Chapter 24 we describe how polypeptides are synthesized on ribosomes, from which the newly synthesized polypeptides emerge in an extended form known as 'nascent polypeptides'. In these, the hydrophobic groups, which will ultimately be sequestered away from water inside the protein, are exposed to the aqueous environment of the cytosol. Hydrophobic groups in this situation automatically aggregate together either within the molecule and/or between other molecules so there is the danger of improper associations being made which, unless something is done about it, could prevent the polypeptide ever being able to fold up correctly. Inside the cell the concentration of proteins is very high so that hydrophobic aggregation between nascent polypeptides could result in a denatured mass somewhat akin to a fried egg. On the face of it, the problem is of such magnitude that one might have thought it would preclude the possibility of life in its known form. However, an experiment by Christian Anfinsen, in which denatured (unfolded) ribonuclease folded up satisfactorily on its own, shows that proteins can fold up if given appropriate conditions. In the experiment referred to the ribonuclease was in very low concentration, and its small size is relevant also. So what happens in the cell to proteins in general?

There is a family of proteins known as **chaperones**, because like human chaperones in Victorian times they prevent improper associations—in this case between hydrophobic groups which shouldn't interact. These hydrophobic associations could occur before the polypeptide is completely synthesized and prevent proper associations in the completed chain. The chaperones are proteins which bind to hydrophobic sections of polypeptide chains; by preventing improper associations and then detaching, they preserve the possibility of correct folding.

If they don't succeed and the protein fails to fold the first time, they repeat the operation until it does. Most proteins, as they are synthesized, need only this type of chaperone. Even large proteins can fold with this assistance because they usually consist of multiple domains each of which can fold sequentially as a small protein as the polypeptide emerges from the ribosome, on which it is synthesized (Chapter 24). However, for reasons not clearly pinpointed certain proteins need a different type of helper protein, known as a **chaperonin**. The protein to be folded is enclosed in a cavity so that it is separated from the other components of the cell which might interfere with the folding process. The folded protein is then released. In a real sense it is mimicking the ribonuclease experiment carried out by Anfinsen, so that the chaperonin box is sometimes known as an 'Anfinsen cage' in that the molecule is given ideal conditions in which to have a go at folding properly. If it fails it is allowed back for further attempts. It is very important to be clear that the chaperones and chaperonins do not *direct* the folding; they have no structural information but simply provide conditions under which the primary structure can direct proper folding. It also should be noted that while these mechanisms are essential to folding they do not explain how the folding is actually arrived at in such a short time; this remains one of the big problems in biology.

There is another aspect to this topic. A nascent polypeptide emerging from the ribosome as it is synthesized is essentially a denatured protein—as already explained, denaturation unfolds proteins. The chaperones were first discovered when it was noticed that in cells suddenly exposed to a temperature rise, a new group of proteins rapidly appeared. Their role is to facilitate refolding of proteins denatured by the heat shock. For this reason the chaperones are known as **heat shock proteins** with names such as **Hsp 70** (a ubiquitous example of the widely used type which simply binds to nascent chains). The best-known chaperonin type is **Hsp 60,** known as **Groel** in *E. coli.*

The heat shock proteins and the way in which they work are described in more detail in the chapter on protein synthesis (page 394).

What are the structural features of enzyme proteins that confer catalytic activity on them?

In Chapter 1 we briefly discussed the factors that enabled enzymes to be such effective catalysts, but no specific example of how this is achieved in terms of actual protein structure was given since it would have been premature there. Now that you know what a protein is, we can illustrate how one class of enzymes work in structural terms, using the enzyme **chymotrypsin** as an example. It is worth understanding; enzymes typically increase reaction rates by a factor of 10^{10}–10^{15}. A 10^{15}-fold increase means that an enzyme catalyses in 1 second an amount of chemical reaction that without the enzyme would require 30 000 000 years. It is one of the many fascinating aspects of the living process.

A special note

Some courses include this topic at this introductory stage whereas others may treat it in more advanced courses. If it is not appropriate to your requirements, skipping it will not impair understanding of the rest of the chapter. If in doubt we suggest you try it, for it is one of the most satisfying illustrations of the remarkable abilities of proteins to perform specific chemical tasks at incredible speeds and by mechanisms, which are in concept very simple. The chemistry of the chymotrypsin reaction which follows involves little more than a proton jumping from one group to another and back again.

Mechanism of the chymotrypsin reaction

Chymotrypsin is a digestive enzyme produced by the pancreas; it hydrolyses certain peptide bonds of proteins in the diet. Its digestive role is described on page 100.

Every enzyme has an **active centre**, a region of the protein into which fits the substrate to be attacked. The site is so constructed that the specific substrate(s) can make sufficient noncovalent bonds with it to form an enzyme–substrate complex (see pages 15 and 23). In the case of chymotrypsin, the natural substrate is a polypeptide and it hydrolyses those peptide bonds whose carbonyl group is part of a large hydrophobic amino acid residue (mainly the aromatic ones phenylalanine, tyrosine, and tryptophan but also methionine). The active centre has a large hydrophobic pocket to accommodate the bulky hydrophobic group. The enzyme is an **endopeptidase**—it hydrolyses internal peptide bonds of proteins, in contrast to an **exopeptidase** which hydrolyses terminal peptide bonds (*endo* = within; *exo* = without). The reaction given here in nonionized structures is as follows:

$$\text{R-CONH-R}' + \text{H}_2\text{O} \longrightarrow \text{RCOOH} + \text{R}'\text{N}\overset{\text{H}}{\underset{\text{H}}{\diagup}}$$

R represents a section of the polypeptide substrate and the amino acid residue containing the carbonyl group of the

peptide bond is one of the large hydrophobic amino acids. R′ represents the rest of the peptide.

Chymotrypsin is one of a group of proteases known as **serine proteases** because the active centre contains a serine residue, which during the reaction becomes an intermediate acyl ester. Hydrolysis of the peptide bond takes place in two stages. (Non-ionized structures are used for clarity; R and R′ are as defined in the equation already given above.)

Stage 1:

RCONHR′ + Enz-OH ⟶ RCOO-Enz + R′N⟨H H

| Peptide substrate | Enzyme with serine–OH group | Intermediate acyl enzyme |

First product

Stage 2:
The ester bond in the intermediate acyl enzyme is hydrolysed by water releasing the second product (RCOOH) and restoring the enzyme to its original state.

RCOO-Enz + HOH ⟶ RCOOH + Enz-OH

Second product | Enzyme in original state

There are two main questions to be considered.

- Why is the —OH of the serine residue so reactive in stage 1 of the enzymic reaction? Serine itself is a very stable molecule and its hydroxyl group unreactive at neutral pH; other serine residues in the chymotrypsin molecule are inert. The peptide bond of the substrate to be attacked is likewise quite stable at neutral pH and on its own does not react at a perceptible rate.

- Why does water so readily hydrolyse the ester bond in stage 2 of the reaction? A carboxylic ester is relatively stable at neutral pH in aqueous solution. To answer these questions we must look at the structure of the catalytic centre of the enzyme.

The catalytic triad of the active centre

Projecting into the active site of the enzyme are the side chains of three amino acid residues which are part of the polypeptide chain comprising the enzyme—aspartate, histidine, and serine, known as the **catalytic triad**. Although quite widely separated in the polypeptide chain of the enzyme, folding of the chain brings them together in the active centre as diagrammatically shown in Fig. 3.9. This is an important role of enzyme proteins—to bring together and fix in optimal relationships the reactive groups involved in catalysis. In solution the

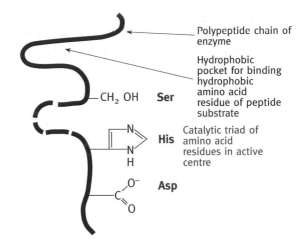

Fig. 3.9 The active site of chymotrypsin. Folding of the polypeptide brings together the three amino acid residues of serine, histidine, and aspartate (residues 195, 57, and 102 of the polypeptide chain respectively—in the unfolded chain they are quite widely separated). The substrate specificity of chymotrypsin for peptides whose carbonyl group is donated by a hydrophobic amino acid derives from the specific hydrophobic binding site of the active centre.

formation of such a relationship by three free amino acids would virtually never occur.

A few points about these groups first:

- The side chain carboxyl group of aspartate (the ionized form of aspartic acid) has a pK_a of about 4 and is therefore dissociated at physiological pH.

- The serine —OH group with a pK_a of about 14 is not significantly dissociated.

- The imidazole side chain of histidine is the interesting one with a pK_a in its protonated form of about 6 (see below). This means that at physiological pH there is a rapid equilibrium between the protonated and unprotonated states and on the Bronsted–Lowry definition of an acid being a proton donor and a base being a proton acceptor, in its protonated form it is an acid and in its unprotonated form a base.

Thus the histidine side chain in its protonated form can readily function as a proton donor in which case it is acting as a **general acid** or in its unprotonated form, it can accept a proton and function as a **general base** and thus can promote **general acid-base catalysis**, as will shortly be explained

It might be helpful if we remind you of some chemical principles. (*Those with a good knowledge of chemistry might want to skip this part.*) A covalent bond involves two atoms sharing a pair of electrons between them. Each electron is attracted to both atoms and this mutually holds the two atoms together.

Formation of the bond releases energy which is why it forms. A simple example is the formation of a covalent bond between two hydrogen atoms to form a hydrogen molecule (H_2):

$$H\cdot \ + \ H\cdot \ \longrightarrow \ H{:}H$$

Each hydrogen atom has a single electron so that the two atoms contribute equally to formation of the bond. In some molecules or ions an atom has an unshared pair of electrons which can be donated to another atom to form a covalent bond. The chemical entities containing such donor atoms are called **nucleophiles**. The acceptor of the pair of electrons is called an **electrophile**. The process of forming such bonds is known as a **nucleophilic attack**. (In the one specific case of a proton (H^+), the chemical entity supplying the electrons is called a base and not a nucleophile.)

$$X{:} \ + \ Y \ \longrightarrow \ X^+{:}Y^-$$

Nucleophile Electrophile

To get back to the chymotrypsin mechanism, one of the nitrogen atoms of the imidazole side chain of histidine has an unshared pair of electrons which can interact with a proton. The dissociation of the imidazole group of histidine occurs as follows:

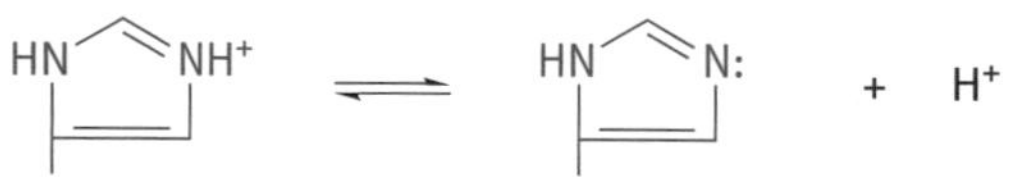

In donating two electrons to H^+ to form the covalent bond in the reverse reaction, the nitrogen atom acquires a positive charge. Note that the imidazole ring of histidine can exist in two tautomeric forms. In the unprotonated form, the single hydrogen atom can be on either of the two nitrogen atoms. With that background we can proceed to the mechanism of chymotrypsin catalysis.

The reactions at the catalytic centre of chymotrypsin

All three amino acid residues—aspartate, histidine, and serine—are essential for the enzyme to function properly but actual chemical changes (reactions) occur only between the histidine and serine so we will deal with these first to keep it simple. The role of aspartate will be explained later.

A peptide substrate molecule attaches to the catalytic site of chymotrypsin by the binding of its hydrophobic group to a specific nonpolar pocket such that the carbonyl carbon ($C{=}O$) atom of the peptide bond to be attacked is close to the serine —OH. Simultaneously the hydrogen atom of the serine —OH transfers to the histidine nitrogen atom and the oxygen atom forms a bond with the carbonyl carbon atom of the substrate as shown in step 1 of Fig. 3.10. This produces the first tetrahedral intermediate shown in the yellow box in the figure. (The term

tetrahedral refers to the organization of bonds of the carbon atom of the bond to be broken; a tetrahedral carbon has its four bonds pointing to the vertices of a tetrahedron.)

That is what happens, but how can serine react in this way? The serine —OH group is normally unreactive at neutral pH. For the oxygen atom to form the bond with the carbon atom of the substrate, it has to lose its hydrogen atom as a proton, but, with a pK_a of 14, the —OH group does not significantly dissociate except in strongly alkaline solutions whose pH is near the pK_a of the group. The answer lies in the ability of the histidine ε N atom to abstract the hydrogen—it is acting as a **general base** (defined, we remind you, as a group that accepts a proton). It does so because the serine and histidine of the catalytic triad are oriented so that the hydrogen of the —OH group is perfectly positioned to interact with the nitrogen atom of the histidine.

The histidine with its acquired proton is now a **general acid**—it can donate a proton. (You can now see why the process is referred to as **general acid—base catalysis**.) The proton is transferred to the tetrahedral intermediate causing it to break down as shown in step 2 of Fig. 3.10, to liberate the first product ($R'NH_2$) and form the acyl enzyme intermediate.

We are halfway there; the next step is to hydrolyse the ester bond of the acyl enzyme intermediate and thus liberate the second product of peptide hydrolysis (RCOOH) and restore the enzyme to its original state ready for reaction with the next substrate molecule. It is basically a repeat of the strategy used in the first stage. What is required is for the oxygen atom of water to make a nucleophilic attack on the carbonyl carbon atom of the ester bond of the acyl enzyme intermediate. When a molecule of water enters the active site (Fig. 3.10, step 3) the histidine group, acting as a general base abstracts a proton (just as it did from serine in the previous stage of the reaction). The water oxygen atom makes a nucleophilic attack forming the second tetrahedral intermediate (Fig. 3.10, step 4). Exactly as before, the protonated histidine is now a general acid; it donates its acquired proton back to the tetrahedral intermediate (Fig. 3.10, step 5) resulting in completion of the reaction and liberation of the second product (RCOOH). The serine —OH is restored to its original state ready for reaction with the next substrate molecule.

The actual mechanisms by which the chemical changes are brought about are thus remarkably simple, involving little more than histidine acquiring a proton and giving it back at each of the two stages.

What is the function of the aspartate residue of the catalytic triad?

If the aspartate residue is converted to an asparagine residue in which the carboxyl group becomes an amide group that

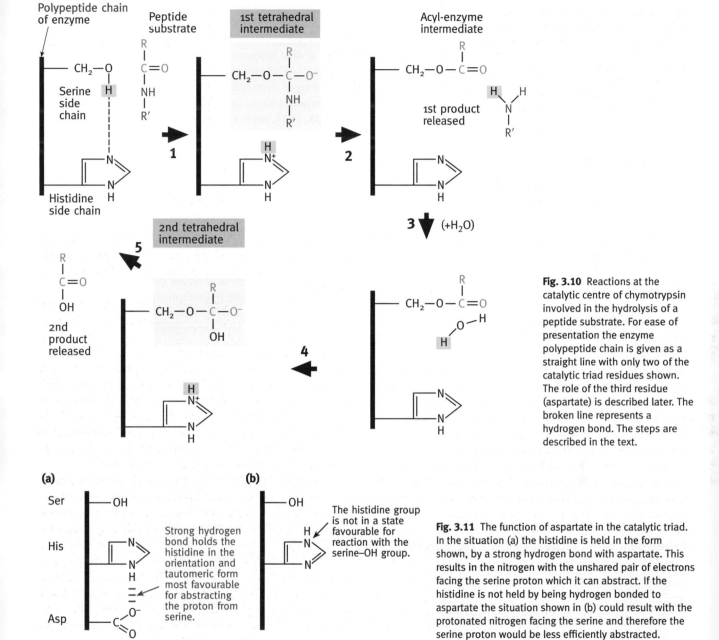

Fig. 3.10 Reactions at the catalytic centre of chymotrypsin involved in the hydrolysis of a peptide substrate. For ease of presentation the enzyme polypeptide chain is given as a straight line with only two of the catalytic triad residues shown. The role of the third residue (aspartate) is described later. The broken line represents a hydrogen bond. The steps are described in the text.

Fig. 3.11 The function of aspartate in the catalytic triad. In the situation (a) the histidine is held in the form shown, by a strong hydrogen bond with aspartate. This results in the nitrogen with the unshared pair of electrons facing the serine proton which it can abstract. If the histidine is not held by being hydrogen bonded to aspartate the situation shown in (b) could result with the protonated nitrogen facing the serine and therefore the serine proton would be less efficiently abstracted.

does not significantly dissociate, the catalytic activity of chymotrypsin falls by a factor of 10 000. The aspartate carboxylate anion is thus essential but nevertheless it does not undergo any chemical reaction during the catalytic process. Why then is it needed? The aspartate carboxylate anion forms a strong hydrogen bond with the histidine side chain as shown in Fig. 3.11. Its main function is to hold the histidine residue in the orientation and tautomeric form shown in the figure, so that the nitrogen atom which accepts the proton from the serine residue is always facing the latter, optimally positioned to abstract the proton from the —OH group. If the aspartate is converted to asparagine, the hydrogen bonding potentiality is very much weaker and this immobilizing effect on the histidine residue is missing. It is historically interesting that initially it was believed that this residue in the catalytic triad is in fact asparagine. When it was appreciated that aspartate would make more sense mechanistically, the structure determination was re-checked revealing that this residue is indeed aspartate. The reference

Fig. 3.12 Simplified diagram of the pockets in the active centres of chymotrypsin, trypsin, and elastase into which fit the amino acid side chains of their respective substrates. That of chymotrypsin accommodates a bulky hydrophobic group such as the side chains of phenylalanine or tryptophan, that of trypsin accommodates the positively charged side chain of lysine or arginine which bind to the negatively charged aspartic acid residue present in the binding site. The elastase pocket accepts only smaller amino acid side chains of substrate molecules, the entrance to the binding site being restricted by the side chains of valine and threonine residues.

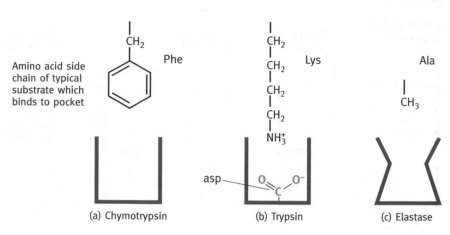

(a) Chymotrypsin   (b) Trypsin   (c) Elastase

by D. M. Blow describes the fascinating discovery of the mechanism of chymotrypsin action.

Other serine proteases

The catalytic triad mechanism has been adopted by a variety of hydrolytic enzymes. The serine proteases chymotrypsin, trypsin, and elastase have the same mechanism all with the aspartate, histidine, and serine residues. They all hydrolyse peptides but have different specificities for the component amino acid residues forming the peptide bond. The active centres differ in the pockets in the active sites required for the binding of the substrates; they will accept only the particular amino acid side chains of their specific substrates. As stated, the pocket in chymotrypsin is hydrophobic (Fig. 3.12(a)) whereas that of trypsin has an aspartate residue (different from that in the catalytic triad) to which binds the basic side chains of trypsin-specific substrates (Fig. 3.12(b)). That of elastase is smaller and access to it restricted by threonine and valine residues so that the enzyme is specific for peptide bonds whose carbonyl group is contributed by amino acid residues with small side chains (Fig. 3.12(c)). The three enzymes described above are structurally closely related to one another and clearly have an evolutionary relationship. The bacterial proteolytic enzyme, subtilisin (from *B. subtilis*) is a totally different protein but still has the identical catalytic triad; the independent evolution of this emphasizes its basic importance.

A brief description of other types of proteases

As well as the serine proteases there are three other classes of proteases in terms of the structures of their catalytic sites. These are the thiol, aspartic, and zinc proteases. The **thiol proteases** are very similar to the serine types but instead of an activated serine hydroxyl group as in chymotrypsin, there is an activated thiol group of cysteine (page 37). An intermediate thioester

(RCO—S—Enz) is formed instead of the carboxylic ester RCO—O—Enz) as in chymotrypsin. The plant proteolytic enzyme papain (prevalent in the latex juice of the papaya) is an example.

Pepsin, whose digestive role is described on page 99, belongs to the **aspartic protease** class. It has a catalytic diad of two aspartate residues: one is unprotonated and can accept a proton, the other is undissociated and can donate one. The two act in turn as a general acid and a general base respectively and reverse their roles with each round of reaction. Several enzymes of this class are known, such as the kidney enzyme **renin** which is involved in blood pressure control. The finding that HIV (the AIDS virus) has an aspartic protease required for its replication has heightened interest in the group. Inhibition of this enzyme is a potential site for therapeutic attack on the disease.

An example of the class of **zinc proteases** is **carboxypeptidase A**, a digestive enzyme which hydrolyses off the terminal carboxyl residue from peptides with a preference for hydrophobic amino acid residues. It exhibits a remarkably large conformational change on substrate binding (see page 23 for the induced fit theory). Mechanistically the catalytic process has strong resemblances to that of chymotrypsin in that general acid base catalysis is involved—a proton is removed from water by transfer in this case to a glutamate residue. Activation of the water molecule in this case is promoted by its binding to a zinc atom, which is bound to the enzyme active site.

Extracellular matrix proteins

In structural terms we now come to a different class of proteins. The proteins we have discussed so far have been globular proteins defined loosely as having a small axial ratio—in other words globular in shape. Proteins of the extracellular matrix that we will be dealing with are elongated fibrous proteins.

Not so very long ago, extracellular matrix proteins were generally regarded as a specialist and rather mundane area of biochemistry. This has undergone a remarkable change with the realization that they play extremely important roles not previously appreciated. There is great medical interest in the area.

Some general description of the extracellular matrix may be helpful here to give perspective on its role. All cells are bathed by it, but in tissues such as liver the cells are in close contact and the layer through which molecules diffuse to reach cell surfaces is very thin. Between tissues there are often large spaces and these are filled with **connective tissue** which, as the name implies, connects tissues together.

The dense connective tissues include bone and tendons, the latter linking muscles to bone and transmit the tension of contraction. Both are predominantly **collagen**, the most plentiful protein in the mammalian body; in the case of bone, the tissue is calcified. At the other end of the spectrum are the loose connective tissues found under all epithelial layers; wherever there is a bodily cavity such as the intestine and the blood vessels, there is a layer of epithelial cells lining the cavity. This has no mechanical strength but is supported by a layer of protein fibres known as the **basal lamina**. Underneath this is a layer of loose connective tissue as illustrated in Fig. 3.13 where skin is used as an illustrative example. The connective tissue joins up the epithelial layer to the underlying tissue. It is flexible and resists compression. The background substance is a soft, highly hydrated gel formed by proteoglycans (described below) which on its own has no mechanical strength but this is reinforced with **collagen** and **elastin** fibres. Some fibres link the basal lamina to the epithelial cells above it and to the connective tissue below. Further links join the underlying tissue cells to components of the connective tissue so that the whole is a stable structure. There are also adhesive proteins that help to interconnect everything, the best known of which is **fibronectin**.

The components of connective tissue are secreted by cells known as **fibroblasts** dotted around in the background matrix and occupying little of the volume of connective tissue. In bone and cartilage are special fibroblasts known as osteoblasts and chondroblasts respectively.

With that general introduction we now describe the structures of the reinforcing proteins, collagen and elastin; the structures of proteoglycans which form the jelly like ground substance of connective tissue; and then the adhesion proteins which bind components together. Finally we will mention the proteins which connect the extracellular matrix to intracellular components; these are currently viewed with intense interest.

Structure of collagens

Collagen occurs outside cells. The protein from which it is assembled is secreted by cells in the form of procollagen which is subjected to a variety of chemical changes catalysed by enzymes, resulting in the mature collagen. **Procollagen** consists of a triple superhelix—three helical polypeptides twisted around each other (see Fig. 3.14(a)). At the ends are extra peptides which, after secretion of the procollagen, are cleaved to give tropocollagen molecules from which collagen is assembled. Each of the polypeptides in the triple superhelix of tropocollagen is an unusual left-handed helix (not the common right-handed α helix of globular proteins). About one in three amino acid residues is proline and every third residue is glycine.

After synthesis of the molecule, many of the proline and also lysine residues are hydroxylated to form **hydroxyproline** and **hydroxylysine** in the polypeptide. These hydroxylated amino acids are not in the 'magic 20' used to synthesize proteins and are formed after the parent amino acids are in polypeptide form. Hydroxylation of proline in the polypeptide requires ascorbic acid or vitamin C, which keeps an essential Fe^{2+} atom in the enzyme prolyl hydroxylase in the reduced form. In defi-

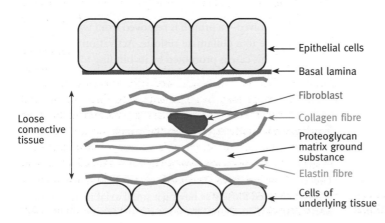

Fig. 3.13 Components of loose connective tissue that underlies epithelial cell layers, e.g. of skin or intestinal lining. The structures of the collagen and elastin fibres and of the proteoglycans are described later in the text.

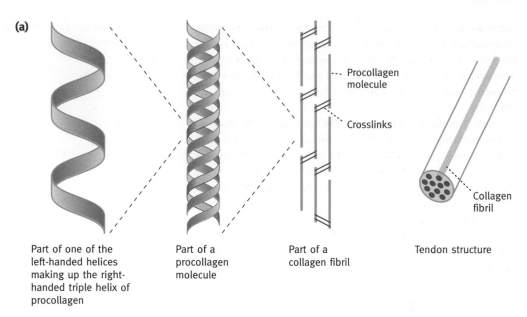

Fig. 3.14 (a) Arrangement of collagen fibrils in collagen fibres. The bonds in red are covalent links formed between lysine residues. They exist also within the triple helix. (b) One type of cross-link formed between two adjacent lysine residues. Note that several specific types of collagen exist for individual functions.

(a)

Part of one of the left-handed helices making up the right-handed triple helix of procollagen

Part of a procollagen molecule

Part of a collagen fibril

Tendon structure

Procollagen molecule

Crosslinks

Collagen fibril

(b)   Polypeptide chains

$$ \overset{|}{C}=O \qquad\qquad \overset{|}{C}=O $$
$$ | \qquad\qquad | $$
$$ NH \qquad\qquad NH $$
$$ | \qquad\qquad | $$
$$ CH-(CH_2)_4-NH-(CH_2)_4-CH $$
$$ | \qquad\qquad | $$
$$ C=O \qquad\qquad C=O $$
$$ | \qquad\qquad | $$

Crosslink derived from two lysine side chains

ciency of this vitamin, connective tissue fails to be properly formed, resulting in the painful consequences of scurvy.

$$ \overset{O}{\underset{}{\overset{\|}{}}} \qquad\qquad \overset{O}{\underset{}{\overset{\|}{}}} $$

—N—CH—C— $\xrightarrow[\substack{\text{Ascorbic acid}\\ \text{Prolyl}\\ \text{hydroxylase}}]{O_2}$ —N—CH—C—
| | | |
CH_2 CH_2 CH_2 CH_2
CH_2 CH
                                     OH

Proline residue in polypeptide   Hydroxyproline residue in polypeptide

(Note that the hydroxylation reaction is more complex than shown here and also involves α-ketoglutarate in a reaction which disposes of the second oxygen atom).

As stated, the peculiar amino acid sequence of collagen results in each polypeptide being a left-handed helix with three residues per turn, three strands of this helix being twisted together into a triple superhelix. The design of the collagen superhelix illustrates rather beautifully what can be done by the protein 'engineering' that occurs in evolution. The three strands of the triple superhelix are in very close association with one another, forming a very strong structure. The bulky side chains of polypeptide chains would normally prevent such close association. In collagen, the helical structure of each polypeptide is more extended than in an α helix with three amino acid residues per turn. Every third residue is glycine and, in the triple superhelix, the contacts between the chains occurs always at glycine residues whose side chain is a hydrogen atom that does not get in the way of close contact. The hydroxylated lysines and prolines form hydrogen bonds between the three chains, thus stabilizing the super helix.

The procollagen molecule so far described has about 1000

amino acid residues. These assemble into collagen fibrils by staggered head to tail arrangement as shown in Fig. 3.14(a). The 'holes' in this structure are believed to be sites where crystals of hydroxyapatite ($Ca_{10}(PO_3)_6OH)_2$) are laid down to initiate bone mineralization. The structure described so far would not have the required strength for a tendon. This is achieved by the formation of unusual covalent links between ends of the tropocollagen units; two adjacent lysine side chains are modified to form the link, one type of which is shown in Fig. 3.14(b). The collagen fibrils so formed aggregate to form tendons by a parallel arrangement. In skin, it is more of a two-dimensional network. Different subclasses of collagen exist dependent on the precise structure of the three polypeptides in the triple superhelix. Variations between the different types of chains include the content of hydroxylysine and hydroxyproline and the degree of glycosylation of these residues.

Genetic diseases of collagen

The functions of collagens are always to provide tough reinforcing filaments in connective tissues, but there are slightly different needs in different situations. This is reflected by the existence of almost 20 types of collagens (identified by roman numerals); there are about 30 different genes coding for the constituent polypeptides of these collagen types. All contain the triple helix but this can vary from occupying the entire length of the collagen molecule to chains where it is quite short. Not surprisingly, with so many genes for the collagen chains and for the enzymes carrying out posttranslational modification there are many associated genetic diseases, with a wide variety of results: for example, weakened blood vessels leading to aortic rupture, to hyperelastic skin and hypermobility of joints. In another disease the small filaments which attach the basal lamina to the underlying collagen fibres in the connective tissue of skin is deficient. In people with this condition the skin forms blisters as a result of the slightest provocation as it becomes detached from the connective tissue.

Structure of elastin

The elastin molecule has a unique structure, different from that of collagen. It contains α helices, which can stretch reversibly. Unusual covalent cross-links are formed between polypeptide chains, forming an elastic network that can stretch in any direction. The cross-links are derived from four lysine residues forming a loose network which, together with extendible helices, produces a structure more elastic than rubber. The structure of the four-way cross-linkage (desmosine) is shown in Fig. 3.15. Reduction in the amount of elastin contributes to ageing of the skin.

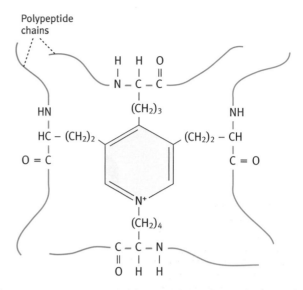

Fig. 3.15 Desmosine cross-link between four polypeptide chains of elastin. The structure is formed by enzyme modification of four lysine residues.

Structure of proteoglycans

As stated, the proteoglycans provide the background jelly-like substance of the connective tissues. Hydrated jellies in nature are based on charged carbohydrate polymers. The molecular strategy is clear; the chains of polar sugars are highly hydrophilic and the mutual repulsion of the polar groups ensure that the chains are fully extended and occupy a large volume, thus entrapping a lot of water. A proteoglycan consists of chains of charged sugars attached to the serine side groups of core protein molecules (Fig. 3.16). The protein chain is fully extended, as are the carbohydrate chains. The negative charges attract a cloud of cations which contribute to the osmotic pressure of the matrix.

The carbohydrate chains are all made of repeating disaccharide units each of which has either **N-acetylglucosamine** or **N-acetylgalactosamine** (Fig. 3.17) as one component so that the polysaccharides are known as **glycosaminoglycans** (GAGs). The second sugar usually has a carboxyl group and a sulfate group (though exceptions with only one of these groups occur). The general pattern of the repeating disaccharide is shown in Fig. 3.18.

There are many different types of proteoglycan; the basic design is extremely flexible. In different proteoglycans the length of the core protein varies from about 1000 to 5000 amino acid residues, and the number of polysaccharide chains attached to the core protein varies up to about 100; the length of the polysaccharide chain varies but typically is about 80–100 sugar residues long. Finally, the number and type of charged groups

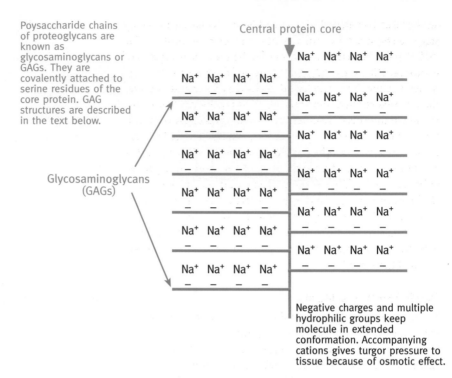

Poysaccharide chains of proteoglycans are known as glycosaminoglycans or GAGs. They are covalently attached to serine residues of the core protein. GAG structures are described in the text below.

Central protein core

Glycosaminoglycans (GAGs)

Negative charges and multiple hydrophilic groups keep molecule in extended conformation. Accompanying cations gives turgor pressure to tissue because of osmotic effect.

Fig. 3.16 The design of proteoglycans. Note that there can be many variations on the common theme: the size of the central protein core can vary; the number of GAGs attached to it can vary; the GAGs can vary in length, in the number, position, and nature of the charged groups, and in the details of their chemistry. We show the accompanying cloud of cations as Na^+ but other cations could be involved. The structures of the GAGs are described in the text.

N-Acetylglucosamine   *N*-Acetylgalactosamine

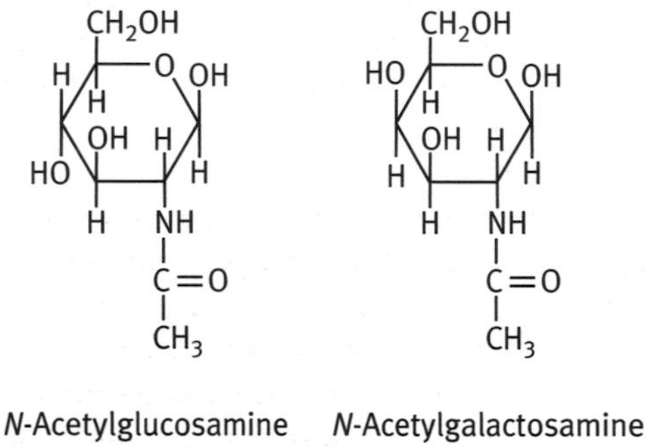

Fig. 3.17 N-Acetylglucosamine and N-acetylgalactosamine.

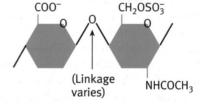

(Linkage varies)

Fig. 3.18 A disaccharide unit of glycosaminoglycan (GAG). For simplicity all bonds and substituents except those characteristic of GAGs have been omitted. The polysaccharide portion of proteoglycans are made of long unbranched chains of these disaccharides. Different GAGs vary in the number and positions of sulfate and carboxyl groups and in other details such as the nature of the glycosidic link between the sugars.

can vary. The main GAGs are known as **chondroitin sulfate**, **dermatan sulfate**, **heparan sulfate**, and **keratan sulfate**, which differ in the ways described above. (**Heparin** is structurally similar to heparan sulfate but has more sulfate groups. It has however a different role in blood vessels where it exists as free GAG and is important in controlling blood clotting; see page 279.) The GAGs are used in the assembly of different types of proteoglycan for different functions. The chief proteoglycan of cartilage provides an example to illustrate this. Cartilage has to withstand very large compressive forces and be very tough. In knee joints, for example, it is needed to prevent direct contact of leg bones and has to withstand enormous pressures. The pro-

teoglycan in such cartilage made up of a protein core with two different GAGs attached to the serine side groups, and large numbers of these proteoglycan molecules are complexed non-covalently to yet another long GAG called hyaluronate forming a gigantic molecule (Fig. 3.19) resembling a bottle brush as seen in the electron microscope. This matrix is heavily reinforced with collagen fibres so that the cartilage resists both compression and tearing. The GAG hyaluronate is widely found in soft extracellular matrices and synovial fluid, where it exists free rather than associated with protein.

Aggrecan is a proteoglycan
consisting of many copies of two
GAGs attached to a long core
protein via serine residues

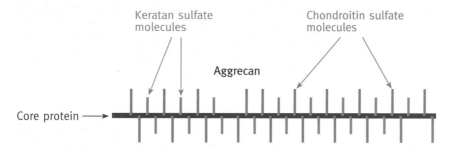

In cartilage many molecules of aggrecan are attached noncovalently to a third
GAG (hyaluronan) via link protein molecules to form a huge complex

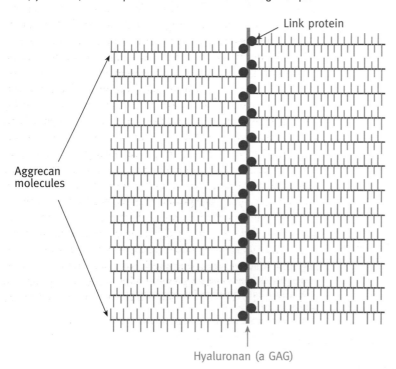

Fig. 3.19 The main proteoglycan found in cartilage. Note that hyaluran is simply a GAG; It forms a huge noncovalent complex with multiple copies of the proteoglycan aggrecan, the attachment being via link protein molecules. Hyaluronate is also widely found in soft extracellular matrices where it exists free, not linked to proteins.

Adhesion proteins of the extracellular matrix

The best known of the adhesion proteins is **fibronectin**. It consists of two flexible elongated protein chains joined at one end by two disulfide bridges. It has several different sites on it which bind to proteoglycans, collagen, and cell surfaces respectively which serves to bind the entire structure together (Fig. 3.20). The binding sites at cell surfaces are a whole family of transmembrane proteins called **integrins** (Fig. 3.21). On the connective tissue side these bind to fibronectin and also directly to the other components of the connective tissue such as collagen and proteoglycans. On the cytoplasmic side of the cell membrane the integrins bind to the cytoskeleton; the latter has not yet been mentioned in this book (and is dealt with in Chapter 33) but for the moment it is enough to know that cells are pervaded by a complex net of protein filaments called the cytoskeleton. It is believed that integrins transmit signals to the cytoskeleton. Integrin-associated signalling from matrix to the cell may be involved in cell growth control, cell migration, and differentiation, though understanding of this is far from complete. As an example, different cells have different integrins; these may change at different stages of differentiation and

determine their adhesion characteristics. As already stressed, relatively little is understood of the regulatory effects of integrin signalling across cell membranes but it is a field of intense research interest.

Experimental handling and separation of proteins

An extract of cells will contain many different proteins and it is often necessary to purify or identity a particular protein amongst the many. Proteins may be separated mainly because of their differing sizes, or their different electrical charges. It may

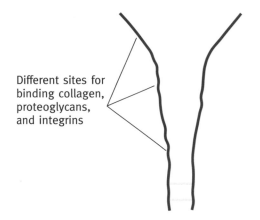

Different sites for binding collagen, proteoglycans, and integrins

Fig. 3.20 A fibronectin molecule consisting of two polypeptides linked by two disulfide bridges (yellow). The integrins are transmembrane receptors on cells. The binding of fibronectin to collagen fibres, proteoglycans, and integrins links together the extracellular matrix structure.

be that the aim is to obtain in purified form a significant amount of the protein, perhaps to crystallize it or determine its chemistry. Column chromatography is commonly used for this purpose. Alternatively it may simply be necessary to be able to examine the amount of a given protein in samples, or to obtain its molecular weight or assess its purity or other characteristics but not to obtain samples of the protein. Gel electrophoresis is commonly used for this, as described below.

Column chromatography

The term **chromatography** is widely used for analytical and preparative techniques in which components of a mixture are separated. The word chromatography implies something to do with colour, which probably originated from its use in separating pigments but now is a term widely used for separating procedures. In column chromatography, the material which is used to effect the separation is packed into a column and the material to be separated flows through it. The separation can be based on a variety of characteristics. A common method is to separate on the basis of molecular size using **gel filtration.** A gel filtration column is packed with beads of a gel with known pore sizes (readily obtainable commercially) and the protein mixture applied and then washed through it. Protein molecules too large to enter the pores of the beads flow unimpeded through the interstices between the beads but those that are small enough to enter the beads are retarded in their rate of movement through the column (see Fig. 3.22). The proteins are washed through with appropriate buffers and samples are collected in a fraction collector so that you end up with a series of separated proteins in tubes. When quantitated and plotted on a graph, the individual components are seen to emerge as a series of 'peaks'. Alternatively the column packing may be a polymer with ionized groups on it so that different proteins bind with different affinities and can be

Fig. 3.21 The role of integrins in connecting components of the extracellular matrix to the cytoskeleton. A family of fibronectins exist. The fibronectin is represented as a single line rather than its dimer shape (Fig. 3.20) for simplicity. The cytoskeleton is described in Chapter 33.

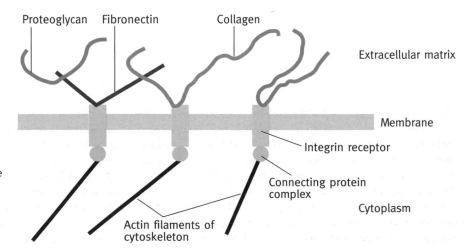

Proteoglycan   Fibronectin   Collagen

Extracellular matrix

Membrane

Integrin receptor

Connecting protein complex

Cytoplasm

Actin filaments of cytoskeleton

Sample is
applied

Sample is washed
through

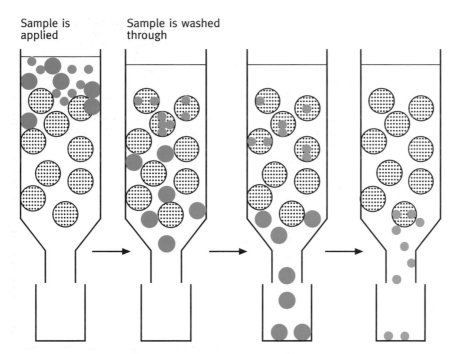

Fig. 3.22 Protein separation by gel filtration. The column is packed with beads of a gel which has pores of defined size. A mixture of large and small protein molecules is allowed to enter the column. The green molecules are too large to enter the beads but the small ones can do so. The column is washed with a suitable buffer to move the molecules down the column. The blue molecules are retarded behind the green ones because they enter the beads and so emerge from the column later than the green ones. Pure samples of each can be collected as separate fractions.

eluted sequentially with appropriate buffers. This is known as **ion exchange chromatography.** Another variation is to use **affinity chromatography.** Suppose your protein of interest is known to specifically bind strongly to compound X; if an inert column packing material has X attached to it by chemical means, then the protein you want will be retarded by binding to the column while the rest are washed through. The protein can then be eluted, perhaps with a solution of different ionic strength or by a solution of component X. Usually a variety of methods are needed to purify a protein completely and it can be a long job. The speed of column chromatography separations may be increased by using **high pressure liquid chromatography (HPLC)**, also sometimes known as high performance liquid chromatography; in this the column material is packed in a steel tube and the liquids forced through at high pressure.

SDS polyacrylamide gel electrophoresis

For analytical separation only small amounts of protein, in the microgram ranges, are involved. Gel electrophoresis is often used. The principle is that charged molecules migrate in an electrical field. A polyacrylamide gel is used as the stationary phase because a solid gel is more convenient to handle than solutions, and also mixing due to heating effects is minimized in the gel. A very common procedure is to use a **denaturing gel**. In this, a porous polyacrylamide gel is cast between two glass plates, the polymerization occurring *in situ*. The gel is made to a recipe

giving suitable porosity so that as the proteins migrate in the electric field, they are separated by **molecular sieving**; small ones move fastest through the gel and the largest ones may not move at all while others migrate at rates in between according to size. The gel contains a detergent, **sodium dodecylsulfate (SDS)** which has a hydrophobic tail and a negatively charged sulfate group ($CH_3 (CH_2)_{10} CH_2 OSO_3^- Na^+$) and the protein sample is also dissolved in a solution of this which denatures it. Disulfide bonds are disrupted by including a reducing agent in the solution. The SDS inserts itself by its hydrophobic tail into the proteins, which are thus covered, on the surface, with negative charges that swamp whatever charge the native protein had. The detergent also solubilizes water-insoluble hydrophobic membrane proteins so that these can also be studied, a major advantage of the technique. Large amounts of the SDS attach, roughly one molecule per two amino acid residues. When the SDS–proteins are electrophoresed on the SDS gel they all move towards the anode and, and since they all have a large negative charge, separation is by molecular sieving according to size of the protein molecules. The proteins appear as bands in the gel when visualized by staining with a dye, Coomassie blue. The resolving power of this technique is very great. Such gels have many uses. Several samples can be run as separate 'tracks' on a single gel and if one of the tracks has a mixture of pure proteins of known molecular weight, the gel is 'calibrated' so that the molecular weight of proteins in the samples can be estimated. Or the variation of the amount of a given protein in cells may be

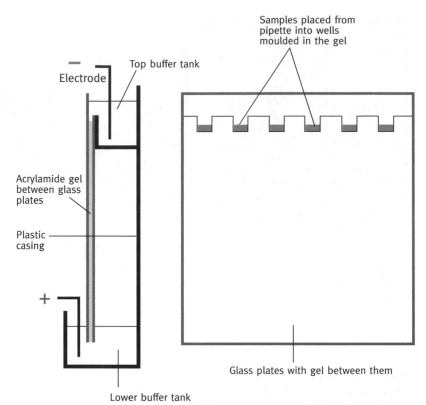

Fig. 3.23 View of polyacrylamide gel electrophoresis apparatus (a) and a front view of the gel between the plates (b). The samples are injected into the wells through the buffer solution with a syringe or pipette. To prevent mixing of the sample in the wells with the buffer, the samples contain glycerol to make them dense. A blue dye makes it easy to see what is happening in the loading.

studied. In practical terms the apparatus is quite simple (Fig. 3.23). The plates with the gel sandwiched in between is held vertically with the top and bottom edges of the gel exposed to tanks of the SDS–buffer solution and a voltage applied across the two tanks. When the gel is cast between the glass plates, before polymerization, a plastic 'comb' with teeth about 1 cm wide, is inserted into the top edge so that, when this is removed following solidification, the gel has a series of separate wells into which different samples can be introduced. The gel is set up in the apparatus with buffer in the tanks and the samples are applied with a pipette into the wells under the buffer. The solution containing the samples contains glycerol to make it dense so that it settles into the wells without mixing with the buffer; a blue dye in the sample makes it easy to see that the loading process is satisfactory. One gel can run multiple analyses in separate tracks, as they are called. A typical SDS gel is shown in Fig. 3.24. This figure also shows the purification of a protein achieved by selective elution from an ion exchange column.

Nondenaturing polyacrylamide gels can also be run. In these the gel is prepared and run so that proteins are not denatured. In this case the separation is based partly on the charge and varying electrophoretic mobilities of the proteins. Another variant is **isoelectric focusing**. Here, the gel has a pH gradient established

electrophoretically with commercially available mixtures of small polymers, known as **ampholines**, with varying isoelectric points. Native proteins migrate electrophoretically until they reach a point in the gel at which the net charge on the protein is zero (its **isoelectric point**) because the positive and negative charges are exactly balanced. The molecule then remains stationary. Nondenaturing gels have the advantage that they can be tested for a biological activity such as enzyme activity.

There are many variations in protein electrophoresis techniques. For maximum resolving power, two-dimensional gel electrophoresis can be used. In this, a single sample is separated by isoelectric focusing on a pH gradient gel strip (available commercially) and the 'strip' is physically transplanted to an SDS gel and electrophoresed at right angles. A crude cell extract so treated can give rise to hundreds of separate protein spots on a single gel. Electrophoretic analysis may also be rapidly performed using cellulose acetate strips or even paper may be used as the solid support, particularly for resolution of simple mixtures (and in class experiments).

As well as these fairly sophisticated techniques of protein separation, there are still methods applicable to large scale preparations, which have been used for decades and have achieved crystallization of great many proteins. These include

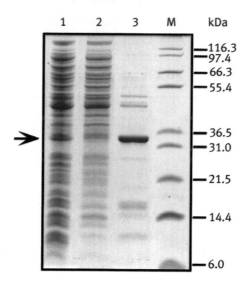

Fig. 3.24 SDS polyacrylamide gel electrophoresis. The figure also shows the purification of a protein from *E. coli*. Samples from each stage were reduced and denatured, then electrophoresed on a 10% polyacrylamide gel in the presence of SDS. Separated proteins were visualized by staining with Coomassie brilliant blue. Lane 1 shows proteins extracted from *E .coli*; the arrow indicates the protein of interest in this experiment. The mixture was chromatographed on a column of cation exchange resin (see pages 57–58). Lane 2 shows the proteins that did not bind to the resin and were recovered in the flow-through. Lane 3 shows the protein of interest in a fraction eluted from the column. Lane M (for markers) shows proteins of known molecular weights. The protein of interest had a molecular weight of 35.3 kilodaltons.

fractionation by adding large amounts of a salt, ammonium sulfate being the usual one, which precipitates proteins selectively. If your protein of interest is more heat stable than others, selective denaturation may achieve purification by precipitating unwanted proteins. There are many variants.

Methods of protein sequencing

Sequencing is very important in protein chemistry. **Fred Sanger** in Cambridge was awarded a Nobel Prize in 1958 for determining the sequence of insulin, the first time this had ever been done for a protein. **Pehr Edman**, working in Melbourne, later developed a machine determining peptide sequences automatically, using different chemistry. In the latter, briefly, the N-terminal amino acid of the protein is labelled with a reagent which under suitable conditions causes the amino acid, still with its label, to detach. This exposes the next N-terminal

amino acid and the same cycle is repeated over and over, working along the polypeptide chain. As each labelled amino acid is detached it is passed through a separating column which identifies it (the column is pre-calibrated with known amino acid derivatives) and a printout of the sequence is automatically recorded. Nowadays machines will produce sequences from as little as $0.01 \mu g$. However, the sequenator, as the machine is called, produces sequences limited to about 30 amino acids from a single sample. To determine complete protein sequences the protein has be fragmented into suitably sized peptides and isolated separately before sequencing them. There are several ways of fragmenting a polypeptide at defined places. Specific enzymes may be used, and treatment of the protein with cyanogen bromide (CNBr) breaks the chain wherever there is a methionine residue. When all of the peptides so obtained have been sequenced, the next step is to determine their order in the protein; to do this it is necessary to obtain overlapping peptides by breaking the chain by an alternative method such as enzymic hydrolysis. From this the complete polypeptide sequence of the protein can be deduced, but it is a very laborious task to sequence a complete protein in this way.

Mass spectrometry is increasingly becoming the choice for structure determination of peptides and proteins when only nanogram quantities of material are available. This is particularly so when analysis of mixtures is required. Initially, mass spectrometry was applicable only for volatile peptides, but today, modern ionization techniques such as fast atom bombardment (FABMS) and electrospray mass spectrometry (ESMS) will produce protonated molecules (MH^+) from non-volatile peptides and proteins with molecular weights as high as 500 000. The mass spectrometric fragmentations of the MH^+ species provide sequencing information, particularly for the lower molecular weight peptides. To give reality to the topic we show in Fig. 3.25 the spectrum obtained from a small neuropeptide. The information from single analysis in the mass spectrometer is sufficient to determine the structure of the peptide as Glu–Gln–Asp–Tyr–Thr–Gly–Ala–Gly–Ala–His–Met–Asp–Phe, the last residue being amidated. (The area is a specialist one and it is not appropriate here to attempt a description of how the spectrum is interpreted.)

A protein is generally too large to be sequenced directly by mass spectrometry. Enzymic cleavage of the protein with trypsin or some other suitable enzyme is carried out to form smaller molecular weight peptides, each of which is sequenced by mass spectrometry. A combination of the sequencing data from the different peptides may provide much of the amino acid sequence of the protein. Mass spectrometry often cannot distinguish between the isomers leucine and isoleucine or

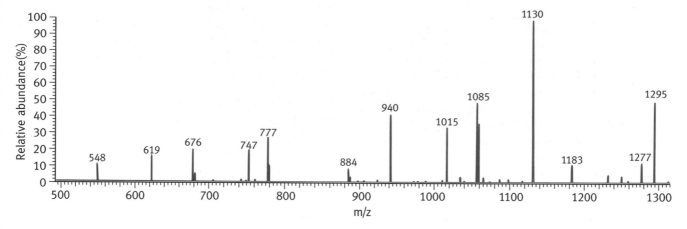

Fig. 3.25 Mass spectrum obtained from a peptide of 11 amino acid residues. As described in the text, from this the amino acid sequence of the peptide can be deduced. The values represent masses of generated fragments.

lysine and glutamine, the latter pair having the same molecular weight but different numbers of atoms. These residues may be distinguished by the Edman procedure. Mass spectrometry also provides a very rapid and accurate method of determining the molecular weights of small amounts of proteins. Even single spots of proteins on two-dimensional gels can be used for this.

Deduction of a protein sequence from the nucleotide structure of its gene

Today a frequently used procedure for protein sequencing is, where applicable, to isolate the gene coding for the protein and to determine its nucleotide sequence by a method developed by the same Fred Sanger who obtained the Nobel prize for insulin sequencing; he received a second Nobel prize for this in 1980. The procedure involved in this is the subject of Chapter 29 (page 483). There may be even simpler ways of obtaining information about protein structures, thanks to the Internet, computer technology and international cooperation in science, which we will now mention briefly.

Protein databases and proteomics

As the techniques for sequencing proteins have improved, the number of proteins for which full or partial primary sequences are known has increased at an explosive rate. Even more rapid has been the number of protein sequences derived from gene

sequencing. This is area is dealt with extensively in Chapter 29 but for the present purposes, the DNA of genes codes for the amino acid sequences of proteins. Each amino acid is coded for by a triplet of nucleotides in the DNA. Since the genetic code (page 381) is fully known, DNA nucleotide sequences can be readily converted to amino acid sequences of the proteins coded for by given genes. There is the limitation that this will not reveal post-translational modifications to the amino acids of a protein; an example is the hydroxylation of proline and lysine in collagen, described on page 52. The proportion of proteins undergoing such modification after synthesis is, however, small.

There is the major problem of how the vast amount of information that is accumulating can be utilized. To this end a number of databases have been established by international cooperation. Protein information is added to these as it is reported. The branch of science, which deals with the utilisation of this data is referred to as **proteomics**. The partner branch which deals with corresponding databases of DNA sequences is called **genomics** (page 485) and the two are collectively referred to as **bioinformatics** which has no precise definition but refers to computer assisted handling and analysis of biological data.

Bioinformatics is an area requiring multiple skills of protein and DNA biochemistry together with computer technology and is an area mainly used by research workers. It promises to be of revolutionary importance and is becoming a major subject. There is a wealth of information available in databases and here we will give only a very few illustrative examples of their uses.

In one databank, SWISS-PROT, there are over a quarter of a million protein sequences entries. The databases are generally

in the public domain and linked to web sites that contain interactive software which allows sequence analysis. This can be used for example to match a newly determined sequence of a protein, or a short section of a sequence, to see if it, or a closely related sequence already exists in the database. If so a large amount of information on the protein may be at once available. Especially if a similar protein already exists for which secondary structure is known, predictions of the secondary structure of a new sequence can be made. Data obtained from mass spectrometric analysis of a minute amount of protein obtainable, say, from a two-dimensional gel separation of a mixture of proteins can be enough to identify the protein in a databank and thus make available much information. Other specialized databanks permit the identification of individual proteins on 2-D electrophoretic gels. The likely role of a protein may be sometimes be inferred from the use of databases. To give two examples, amino acid sequences may be analysed to see whether there is a characteristic domain such as that found in transmembrane proteins, which span the membrane. Or some domains are characteristic of specific classes of proteins. Certain types of gene control proteins are an example of this.

Another database is that which holds the coordinates for 3-dimensional structures of proteins, or domains of proteins, usually obtained by X-ray diffraction or NMR. It contains over 12 000 structures with about 40 new ones added weekly. Coordinates of the structures are given and software is available to visualize these structures which can also be used to make computer models of polypeptide structures with sequences similar to those in the database. The three-dimensional structures have immense importance; since drugs work by combining with specific sites on proteins, protein structure determination is important in the design of new medicines. A sample of web site addresses is given below to give reality to the topic and in case it particularly interests you.

Database of protein sequences
http://www.ebi.ac.uk/swissprot/Information/information.html
Tools for protein analysis
http://expasy.proteome.org.au/
Database of three-dimensional structures
http://www.rscb.org/pdb/

A note on where we are at this point in the book

In this chapter we have described the structure of proteins from a general viewpoint. You will now understand what proteins are, some of the structural problems involved and how these are solved. The best-understood examples of the relationship of protein structure to function are the **hemoglobins** involved in oxygen transport and the **antibodies** in immune protection. However, instead of dealing with these now, it will be more meaningful to do so in the chapters describing their biological functions (Chapters 31 and 27 respectively).

The general plan of the book is to deal with metabolism before the molecular biology of the genome. The prerequisites for understanding metabolism include as well as the first two chapters, a knowledge of membrane structure, for reasons that will become clear. We will therefore now go on to membranes and their protein components in the next chapter.

FURTHER READING

Branden, C. and Tooze, J. (1999). *Introduction to protein structure*, 2nd Edn. Garland Publishing.

A comprehensive account of all aspects of protein structure, including the basic aspects and the structure of membrane proteins dealt with in this chapter. In addition it covers specific classes such as DNA-binding proteins, antibodies and receptors relevant to later chapters. Superbly illustrated with protein structures.

Protein size and structure

Chothia, C. (1984). Principles that determine the structure of proteins. *Ann. Rev. Biochem.*, **53**, 537–72.

Excellent for basics of protein structure.

Goodsell, D. S. (1991). Inside a living cell. *Trends Biochem. Sci.*, **16**, 203–6.

An attempt to present a picture of the distribution of molecules on a proper scale in a cell of *E. coli*.

Goodsell, D. S. (1993). Soluble proteins: size, shape and function. *Trends Biochem. Sci.*, **18**, 65–8.

A survey of protein structures—their size and shape. Why are they so big? Why are they oligomeric?

Protein modules or domains

Doolittle, R. F. (1995). The multiplicity of domains in proteins. *Ann. Rev. Biochem.*, **64**, 287–314.

A discussion of domains, domain shuffling, exons, and introns. (The latter two topics are dealt with in Chapter 21 of this book).

Proteins of hair and connective tissue

Francis, M. J. O. and Duksin, D. (1983). Heritable disorders of collagen metabolism. *Trends Biochem. Sci.*, **8**, 231–4.

A relatively simple account of this very complex and medically important field.

Hulmes, D. J. S. (1992). The collagen superfamily—diverse structures and assemblies. *Essays in Biochem.*, **27**, 49–67.

An account of the very complex structures found in different collagens, together with medical aspects.

Catalytic mechanisms

Blow, D. M. (1997). The tortuous story of Asp . . . His . . . Ser: structural analysis of α-chymotrypsin. *Trends Biochem. Sci.*, **22**, 405–8.

A very personal story of how the classical elucidation of the mechanism of this enzyme took place.

Dodson, G. and Wlodawer, A. (1998). Catalytic triads and their relatives. *Trends Biochem. Sci.*, **23**, 347–52.

A review of the mechanism of chymotrypsin catalysis together with a discussion of other enzymes using the 'serine' catalytic triad and their evolutionary relationships.

Mass spectrometry

Bieman, K. and Martin, S. (1987). Mass spectrometric determination of the amino acid sequence of peptides and proteins. *Mass Spectrom. Rev.*, **6**, 1–78.

Reference given in case of special interest in the area.

Zhu, X. and Desidero, M. (1996). Peptide qualification by tandem mass spectrometry. *Mass Spectrom. Rev.*, **15**, 213–41.

Reference given in case of special interest in the area.

PROBLEMS FOR CHAPTER 3

1   What is the primary structure of a protein?

2   What is meant by denaturation of a protein?

3   Write down the structure and name of an amino acid with each of the following side chains:
    (a) H;
    (b) aliphatic hydrophobic;
    (c) aromatic hydrophobic;
    (d) acidic;
    (e) basic.

4   Give the approximate pK_a values of:
    (a) acidic amino acid side chains;
    (b) basic amino acid side chains;
    (c) the histidine side chain.

5   Which amino acids are the major determinants of the charge of a polypeptide chain containing all 20 amino acids?

6   What are the four levels of protein structure?

7   Give an account of the secondary structures of proteins, together with their apparent purpose.

8   What is the peculiar structural feature of elastin that gives it its elastic properties?

9   In collagen, sections of the polypeptide chains have glycine as every third amino acid residue. What is the significance of this?

10   The activities of proteins in general are readily destroyed by mild heat. The peptide bond is quite heat stable.
   (a) Why are proteins so inactivated by mild heat?
   (b) A few proteins, particularly extracellular ones are more stable than usual. What structural feature is probably responsible for this?

11   What is a protein domain and why are they of interest?

12   In a globular protein where would you statistically expect to find most of the residues of (a) phenylalanine?; (b) aspartic acid?; (c) arginine?; (d) isoleucine?

13   Describe, without chemical detail, three methods for determining the primary structure of a protein.

14   In the active centre of chymotrypsin there is a serine residue which makes a nucleophilic attack on the carbonyl carbon atom of the peptide substrate forming a covalent bond to it. For this, the proton of the serine —OH has to be removed at the same time. Other serine residues in the protein are totally inert in this respect as is free serine. Explain how this reaction is caused to occur in the catalytic centre of the enzyme.

15   The active centre of chymotrypsin contains an aspartate residue; although this does not participate in the chemical reaction involved in catalysis, its conversion to asparagine lowers the activity of the enzyme ten thousand fold. Why is this so?

16   Explain why the molecular structure of proteoglycans is very suitable for creating mucins and gels with a very high water content.

17   The α helix and β-sheet structures are prevalent in proteins. What is the common feature that they have which makes them suitable for this role?

Chapter summary

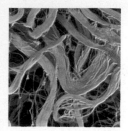

The cell membrane: a structure depending only on weak forces

In Chapter 1 we discussed, in general, the importance of three types of weak, noncovalent bonds as well as the hydrophobic force resulting from water molecules rejecting nonpolar molecules which interfere with hydrogen bonding between those water molecules. Formation of weak bonds involves a decrease in free energy so that only when formation of such bonds is maximized do you have the preferred most stable structure. This forms a link to the topic now to be dealt with—the cell membrane and its functions.

The plan of the next section of the book, on metabolism, is to start with food and then look at how it is digested, absorbed, and carried around the body, and its fate in the various tissues. After this, the metabolic pathways by which the processes occur will be dealt with.

However, before food can enter the bloodstream it must pass through intestinal cell membranes. It is therefore necessary to deal with the biochemistry of membranes before discussing food utilization. Also, the constituents of cell membranes are involved in transport of fats around the body and it will help to first understand what these are.

An overview of cellular membranes

Inside most eukaryotic cells there is a variety of membrane structures. All have the same basic structure but have their own retinue of proteins for their various functions. The internal membranous structures separate the cell into separate compartments. The main components of a typical animal cell are illustrated in Fig. 4.1. The term **eukaryote** arises from the fact that such cells have a nucleus surrounded by a double membrane which contains all of the DNA (apart from that in mitochondria and chloroplasts) so that the genetic apparatus is separated from the cytoplasm (*eu* as a prefix derives from the Greek and implies 'true' or 'good'; *caryote* means nucleus). Eukaryotic cells are referred to as 'higher' cells since they include all animal and plant cells; prokaryotes are the bacteria. With the exception of cyanobacteria (which used to be called blue-green algae) prokaryotes have no internal membranes, the entire cytoplasm being a single compartment not separated from the DNA of the cell. The internal membranes of cyanobacteria are responsible for photosynthesis. Bacteria are surrounded by a very strong cell wall which resists the very high osmotic pressure found in such cells and prevents the cell membrane from bursting.

We will now summarize the functions of the various organelles of cells very briefly since they are all much more fully described in later chapters, so that it doesn't matter if a few terms are unfamiliar to you.

- The **nucleus** contains the DNA and the genetic machinery; it is the site of DNA replication and RNA synthesis. Communication with the cytoplasm is via nuclear pores. In prokaryotes the genetic machinery is in intimate contact with the cytoplasm.

- The **endoplasmic reticulum** (ER) is an extensive network of tubular membranes and flattened sacs enclosing one continuous space which in some cells amounts to 10% of the volume of the cell. The outer membrane of the nucleus is continuous with the ER membrane. The amount of ER membrane varies greatly between cells, being minimal in mammalian red blood cells (which have no nucleus) and very extensive in protein-secreting cells such as liver and pancreas. It is concerned with the synthesis of new proteins

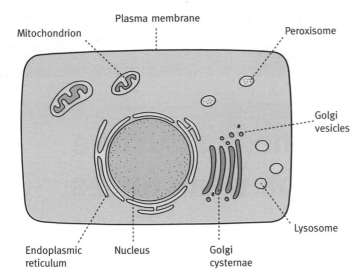

Fig. 4.1 Membranous structures in a typical animal cell.

destined for secretion or incorporation into **lysosomes**. The membrane itself is also the site of much membrane lipid synthesis.

- The **Golgi apparatus** is a stack of flattened membrane sacs involved in sorting out newly synthesized proteins which arrive from the ER and despatching them to their correct cellular destinations in transport vesicles.

- **Mitochondria** are the site of the oxidative metabolism which generates most of the ATP in aerobic cells. They have double membranes, the inner one being the site of ATP generation. Mitochondria are about the same size as an *E. coli* cell and they represent, in evolutionary terms, prokaryotic cells which were captured by cells and became essential to them. Mitochondria have some DNA content. They replicate inside the cell, but are not genetically independent of the cell nucleus whose DNA codes for most of the mitochondrial proteins which are transported in. In prokaryotes the plasma membrane is the site of oxidative generation of ATP which in this context is equivalent to the inner mitochondrial membrane.

- **Lysosomes** contain hydrolytic enzymes whose task is to destroy unwanted material.

- **Peroxisomes** are membrane-bounded organelles which carry out oxidative reactions producing hydrogen peroxide. The latter is rapidly destroyed by the enzyme catalase. The oxidation reactions are not coupled to ATP production but rather seem to have the purpose of carrying out reactions not performed elsewhere in the cell.

Plant cells are eukaryotes and possess the same internal membranes as animal cells but in addition may contain chloroplasts (the latter being the site of photosynthesis described in Chapter 15) and water-filled vacuoles. Glyoxysomes are found only in plant cells. They are the site of the glyoxylate cycle (page 206), a variant of the citric acid cycle (page 154), not found in animal cells.

Since the functions of intracellular membranes are covered in appropriate later chapters, in the rest of this chapter we will deal only with plasma (external) membranes.

Why are cell membranes needed?

The obvious answer to the question is that they are necesary to hold together the contents of the cell. It is not certain how life originated, though knowledge has reached the point where possible mechanisms can be formulated. But whatever the mechanism, one of the prerequisites must, at some stage, have been to contain the primitive self-replicating molecular system. If it wasn't contained, presumably the molecular constituents would have been dispersed and the process aborted.

This requirement seems to pose a difficult problem, for a cell membrane is a complex structure. How could a primitive self-replicating system produce a containment membrane? The early containment structure may have been different from modern ones, but it may be no coincidence that, given the right type of molecule, structures resembling the permeability barrier of cells can form with ease—in fact, hollow vesicles are produced just by agitation of the appropriate molecules in water. If the primitive replicating system happened to have available the right type of amphiphilic substance (see below), agitation could have enclosed minute drops of solution, containing the replicative molecular system, inside vesicles enclosed by structures similar to the membranes of a modern living cell. In other words, the primitive cell membrane may have been self-assembling. The modern cell, of course, synthesizes its membranes, as described later.

What type of substance has such properties? The material extractable by organic solvents from egg yolk is one. The egg yolk extract contains **polar lipids**. A polar lipid (Fig. 4.2) is called an **amphipathic** or **amphiphilic molecule** (*amphi* = both kinds of) because it has a polar part (the so-called **head group**) and a nonpolar hydrophobic part (usually called a **hydrophobic tail**). The amphipathic molecules which might have formed the (speculative) primitive membrane need not have been the modern polar lipids; as long as their amphipathic properties

were similar, the same forces would operate on them with the same result.

In the presence of water, such molecules are forced into structures in which the hydrophilic head groups have maximum contact with water, to maximize weak bond formation. Conversely the hydrophobic tails are forced into minimum contact with water because such contact forces water into a higher energy arrangement (page 14) and therefore is resisted. Also, the contact between hydrophobic tails maximizes van der Waals interactions (page 14), again minimizing free energy of the structure and therefore maximizing stability.

When polar lipids are agitated in an aqueous medium, one of the structures formed is called a **liposome**—a hollow spherical lipid bilayer. A lipid bilayer has two layers of polar lipids with their hydrophobic tails pointing inwards and their hydrophilic heads outwards in contact with water and each other as shown in Fig. 4.3(a).

If a synthetic liposome is sectioned and stained with a heavy metal that attaches to the polar heads, the lipid bilayer appears in the electron microscope as a pair of dark 'railway lines' owing to absorption of electrons by the metal stain. A living cell membrane treated in the same way has an identical appearance. All biological membranes have the lipid bilayer, as their basic structure.

Another type of structure taken up by amphipathic molecules in water is the solid cylindrical micelle (Fig. 4.3(b)); but

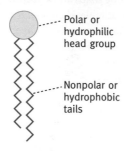

Polar or hydrophilic head group

Nonpolar or hydrophobic tails

Fig. 4.2 An amphipathic molecule of the type found in cell membranes.

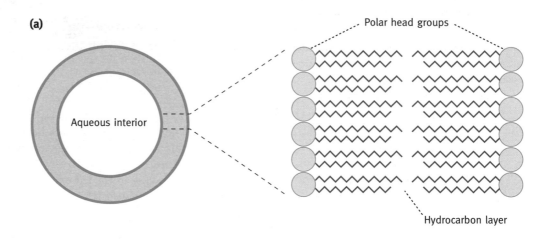

(a)

Aqueous interior

Polar head groups

Hydrocarbon layer

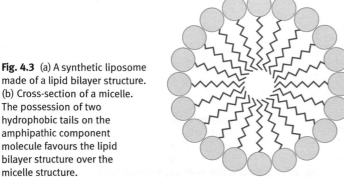

(b)

Fig. 4.3 (a) A synthetic liposome made of a lipid bilayer structure. (b) Cross-section of a micelle. The possession of two hydrophobic tails on the amphipathic component molecule favours the lipid bilayer structure over the micelle structure.

the dual hydrophobic tails on membrane polar lipids favour the bilayer structure, rather than the micelle structure.

The lipid bilayer

What are the polar lipid constituents of cell membranes?

There are a number of membrane polar lipids structures that look quite different from one another on paper, but in space-filling models they all have the same basic shape as the molecule shown in Fig. 4.2 with a polar head and two hydrophobic tails. We will now deal with the structures of the various polar lipids, but the nonpolar lipid structure will also be discussed for comparison.

A lipid is fat. A neutral fat is derived from **fatty acids**; neutral fats never occur in membranes, but it is helpful to discuss them first. A fatty acid has the structure RCOOH where R is a long hydrocarbon chain. The commonest lengths are C_{16} and C_{18}. They can be called hexadecanoic and octadecanoic acids respectively, or by their common names, palmitic and stearic acids. They can also be represented as $16:0$ and $18:0$, indicating the number of carbon atoms and the number of unsaturated bonds.

Stearic acid (C_{18}) has the structure

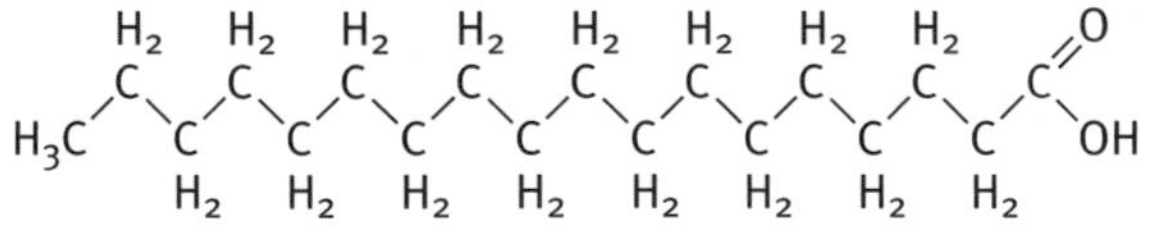

For convenience, fatty acids are often represented simply as

or $CH_3(CH_2)_{16}COOH$ (where the aim is simply to indicate a long hydrocarbon chain we usually don't bother to count the number of zigzags.) The corresponding C_{16} is *palmitic acid*. At neutral pH, as say, the sodium salt, a fatty acid is a soap—what you wash yourself with:

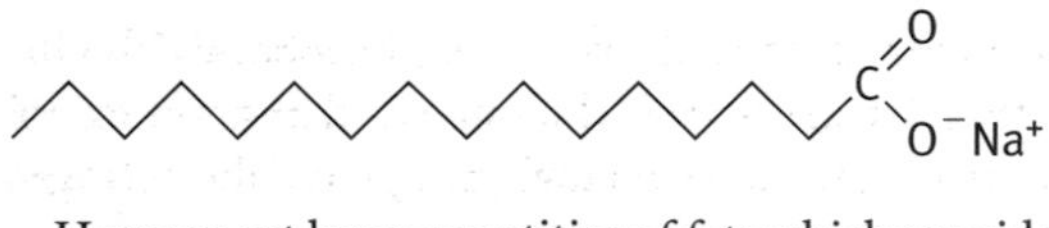

Humans eat large quantities of fats which provide a substantial proportion of our food, but soaps are not good to eat. They don't taste nice and their detergent action at high concentrations could disrupt cell membranes. Instead we eat mainly neu-

tral fat, more properly called **triacyglycerols (TAGs)** in which three molecules of fatty acids are esterified to the three hydroxyl groups of glycerol as shown in Fig. 4.4. A carboxylic acid attached in ester linkage is an acyl group, and there are three acyl groups attached to one glycerol molecule, hence the name triacylglycerol. The term triglyceride is sometimes used but is chemically incorrect.

If neutral fat is boiled with NaOH or KOH, the ester bonds are hydrolysed, forming soaps and glycerol, which is how soap is made. Neutral fat is suitable as a food; it is not a detergent. As stated, *neutral fat does not occur in membranes*. We have discussed it to help in understanding what a polar lipid is by contrast. A TAG molecule has no polar groups and therefore forms oily droplets or particles in water—it could never produce a lipid bilayer structure.

Membrane lipids based on glycerol are the most abundant form and are known as **glycerophospholipids**. They are based on glycerol-3-phosphate.

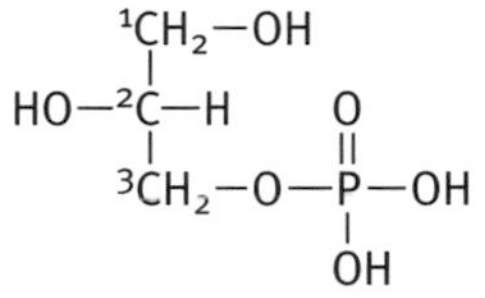

sn-Glycerol-3-phosphate

The prefix sn, for stereospecific numbering, refers to the nomenclature system used here . The central carbon atom of this compound is asymmetric; in glycerol-3-phosphate found in the cell, the secondary hydroxyl is represented on the Fischer projection to the left and the carbon atoms numbered from the top. If two fatty acids are esterified to the primary and

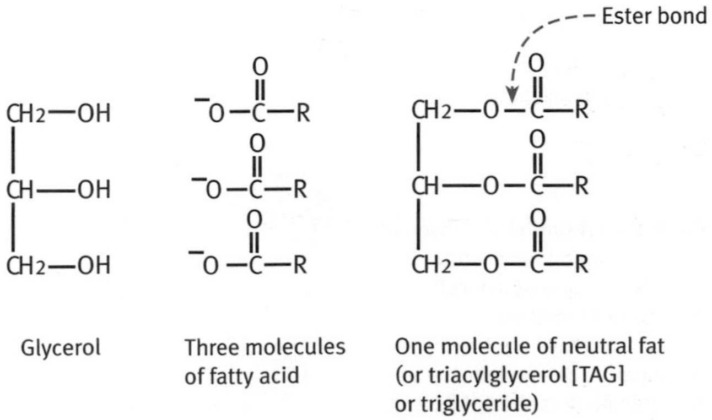

Fig. 4.4 A triacylglycerol and its component parts.

secondary hydroxyl groups of glycerol-3-phosphate, the product is **phosphatidic acid**.

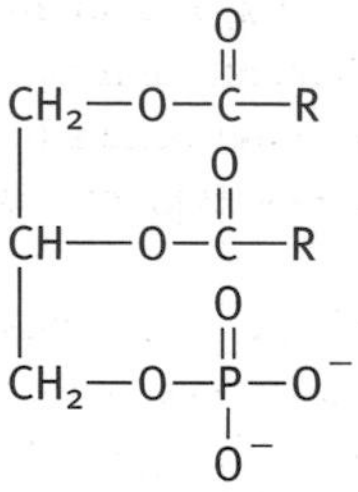

Phosphatidic acid

Fix the name phosphatidic acid in your mind for it makes the nomenclature of the rest simple. We can attach other polar molecules to the phosphoryl group. If phosphatidic acid is attached to something else (e.g. 'X') it becomes a **phosphatidyl group** (e.g. phosphatidyl-X) with the structure:

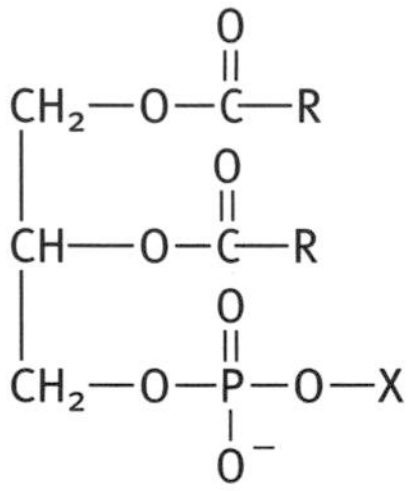

If X is also a highly polar molecule we now have an extremely polar head group.

The structure above gives a misleading impression of the shape of the molecule. A space-filling model would look somewhat like this:

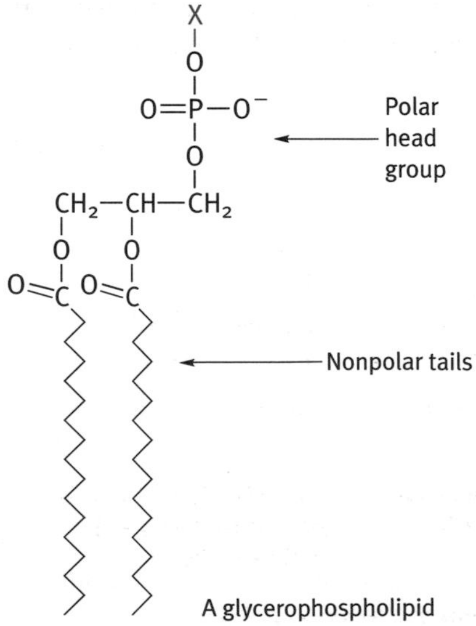

A glycerophospholipid

As emphasized in the introductory statement, it is the amphipathic membrane lipid shown in Fig. 4.2 in diagrammatic form which, as shown earlier, assembles into lipid bilayer—a highly polar head group and two hydrophobic tails.

What are the polar groups attached to the phosphatidic acid?

In principle, only one species of polar lipid molecule is needed to form a lipid bilayer. However, living cells use a variety of structures to form their membranes, some quite complex, but all structures used have *the same overall shape and amphipathic properties*. We therefore have several different polar groups attached to phosphatidic acid molecules.

The first polar substituent is **ethanolamine** ($HOCH_2CH_2 NH_3^+$) giving the phospholipid **phosphatidylethanolamine** (PE for short, or the trivial name **cephalin**). Its structure is

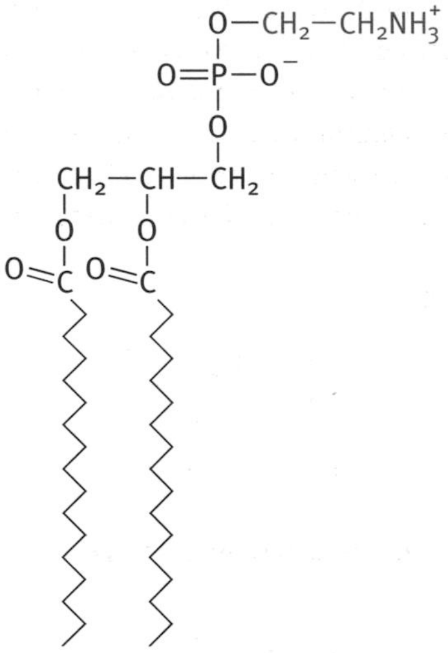

If the nitrogen of ethanolamine is trimethylated, it is **choline**

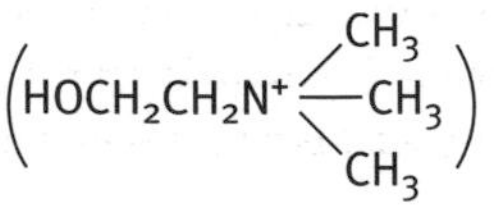

and the phospholipid derived from it is **phosphatidylcholine** (PC) or **lecithin**. Its structure is the same as that given for PE above, apart from the three methyl groups on the nitrogen atom.

If the ethanolamine is carboxylated we have a **serine** substituent giving **phosphatidylserine** or PS (serine is $HOCH_2CHNH_2COOH$). The attachment is like this:

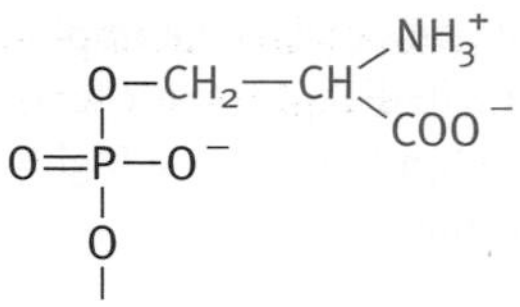

$$O-CH_2-CH \begin{smallmatrix} NH_3^+ \\ COO^- \end{smallmatrix}$$
$$O=P-O^-$$
$$O$$

Another quite different polar substituent is the hexahydric alcohol, **inositol**:

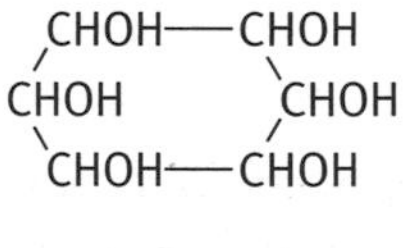

$$\text{Inositol}$$

attached like this:

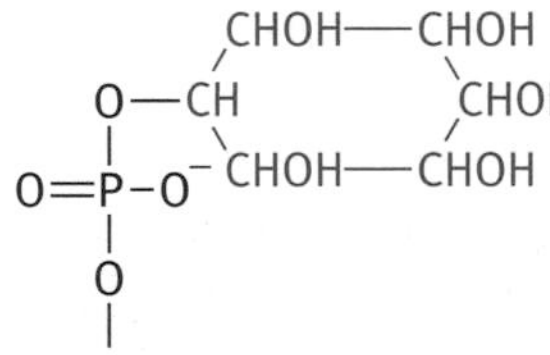

giving **phosphatidylinositol**, usually abbreviated to PI.

There are many different ways to ring the changes on glycerol-based polar lipids to form components of lipid bilayers. Here's yet another one—**cardiolipin** or **diphosphatidylglycerol**—in which two phosphatidic acids are linked by a third glycerol unit.

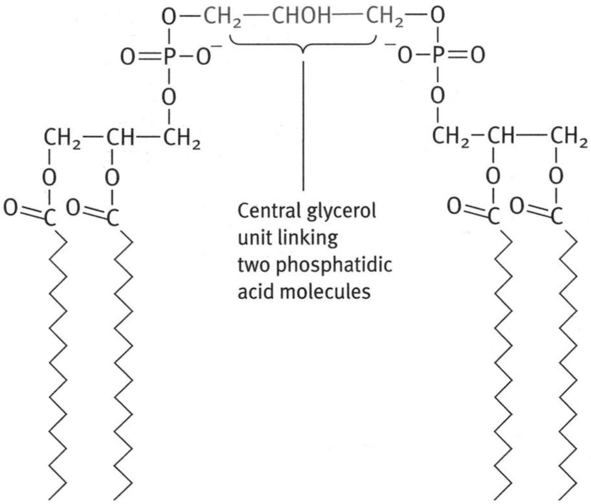

$$O-CH_2-CHOH-CH_2-O$$

Central glycerol unit linking two phosphatidic acid molecules

The resultant structure still has the amphipathic shape to fit into lipid bilayers. It occurs in the inner mitochondrial membrane and in bacterial cell membranes.

All of the above polar lipids are based on glycerol. However, the process of evolution has, with fascinating ingenuity, produced molecules of almost identical overall shape derived from a different structure called **sphingosine**.

The basic molecule isn't all that different from glycerol; the middle (C-2) hydroxyl group of glycerol is replaced by an —NH$_2$ group and a hydrogen on carbon atom 1 is replaced by a C$_{15}$ hydrocarbon group—more or less a permanently 'built-in' hydrocarbon tail. One tail isn't enough but another can be added by attaching a fatty acid to the central —NH$_2$ group by means of a —CO—NH— linkage.

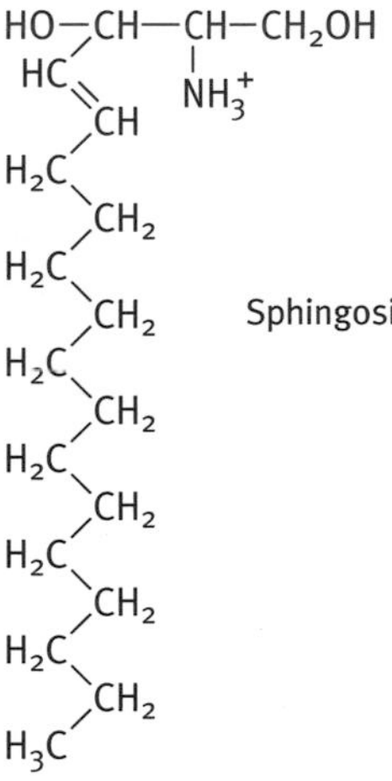

$$HO-CH-CH-CH_2OH$$

Sphingosine

This gives a molecule called a **ceramide**, which is very similar in shape to a diacylglycerol.

$$HO-CH-CH-CH_2$$
$$OH$$
$$NH$$
$$O=C$$

A ceramide

If now we add a phosphorylcholine group, as in lecithin, the product is **sphingomyelin**

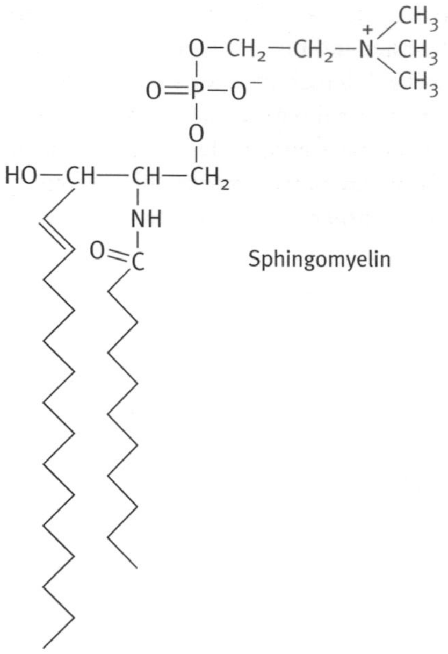

$$O-CH_2-CH_2-\overset{+}{N}\big\langle\begin{matrix}CH_3\\CH_3\\CH_3\end{matrix}$$

$$O=\overset{|}{\underset{|}{P}}-O^-$$

$$O$$

$$HO-CH-CH-CH_2$$

$$\overset{|}{NH}$$

$$O=C$$

Sphingomyelin

which is similar in shape to lecithin. It is prevalent in the myelin sheath of nerve axons.

Evolution has been described as a tinkerer—it goes on modifying things and if the change is beneficial it is preserved. It has tinkered with the sphingosine-based polar molecules by adding different polar groups to the free —OH of the ceramide molecule. Sugars are highly polar and are used for this purpose.

Using a single sugar as the polar group we get a **cerebroside**, important in brain cell membranes. Cerebrosides contain either glucose or galactose. The latter is a stereoisomer of glucose in which the C-4 hydroxyl is inverted.

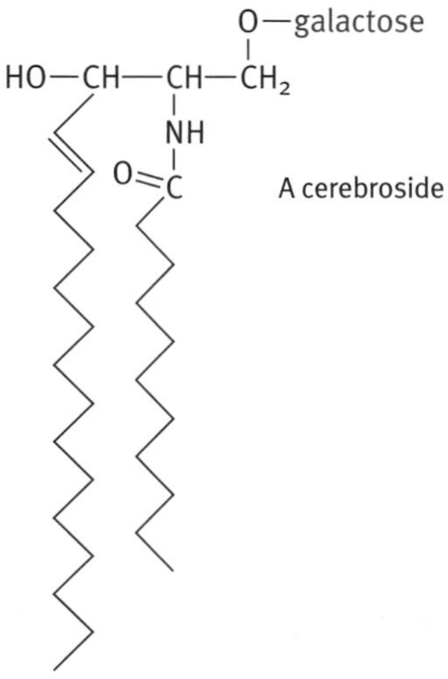

$$O-galactose$$

$$HO-CH-CH-CH_2$$

$$\overset{|}{NH}$$

$$O=C$$

A cerebroside

There is a wide variety of sugars in nature. Some are **amino sugars** such as glucosamine (whose structure is given below) or derivatives of them.

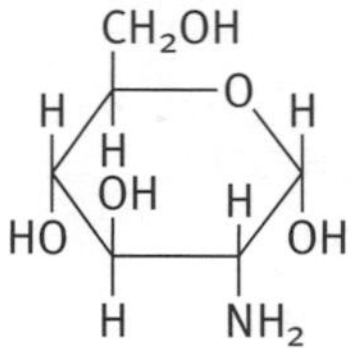

$$CH_2OH$$

H   O   H

H

OH   H

HO   OH

H NH_2

Combinations of the different sugars (**oligosaccharides**) can produce a variety of molecules in which a small number of different sugars are linked together in a branched oligosaccharide structure. (*Oligo* = few, so an oligosaccharide is a small polymer of sugars—perhaps 3–20 sugars rather than the hundreds or thousands found in polysaccharides.) Such an oligosaccharide is highly polar. When one of these is attached to a ceramide we have a **ganglioside**. This gives the molecule a very large highly polar head.

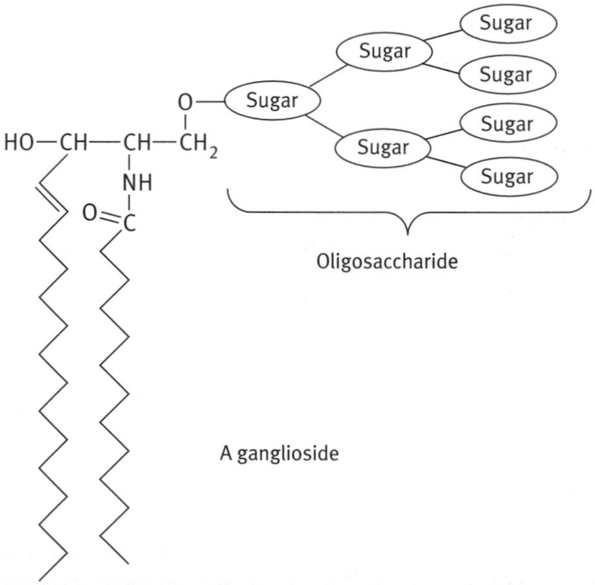

$$O-Sugar$$

$$HO-CH-CH-CH_2$$

$$\overset{|}{NH}$$

$$O=C$$

Sugar
Sugar
Sugar
Sugar
Sugar
Sugar

Oligosaccharide

A ganglioside

Not many biochemists would memorise the carbohydrate structures in these unless they had a special interest in them. However, you should be familiar with structures such as *N*-acetyl glucosamine and sialic acid.

CH_2OH

OH

H

OH

HO

H

NH

$C=O$

CH_3

H_3C-C-N

O

R

H

COO$^-$

OH

OH

H

R =

CHOH

CHOH

CH_2OH

N-Acetylglucosamine   Sialic acid
(also called N-acetylneuraminic acid)

Both these compounds are constituents of gangliosides and of carbohydrate attachments to membrane proteins (see page 44). Sialic acid is particularly interesting since it is involved in infection of cells by the influenza virus (page 469). Gangliosides, based on sphingosine, are what distinguishes the human blood groups O, A, and B.

A note on membrane lipid nomenclature

The names of individual polar lipids have been given above, but alternative collective terms are sometimes used. The terms **membrane lipids** and **polar lipids** include any lipid found in cell membranes. A **phospholipid** is any lipid containing phosphorus. Those based on glycerol can be called **glycerophospholipids,** as opposed to sphingomyelin which is a single specific sphingosine-based phospholipid. The ceramide-based membrane lipids with carbohydrate polar groups and lacking any phosphoryl group are called **glycolipids** or **glycosphingolipids** to indicate that they are based on sphingosine. A plasmalogen is a glycerophospholipid in which one of the hydrophobic tails is linked to glycerol by an ether bond. (Cholesterol, another membrane component described below, is also classified as a lipid.)

Why are there so many different types of membrane lipids?

The answer to this question is not known with certainty. The different membrane lipids confer different properties on the membrane surface. The choline substituent of lecithin (PC) has a positive charge, the serine of PS is a zwitterion, and the carbohydrate of a cerebroside has no charge. Different cells have quite different membrane lipid compositions. Cerebrosides and gangliosides are common in brain cell membranes, and the cell membranes of the myelin sheath of nerve axons are rich in glycosphingolipids. As well as different cells having different lipid compositions, the outer and inner halves of the bilayer of the one membrane may be different from one another and different membranes within the cell have different compositions. For example, glycolipids are always on the outer side of the bilayer so that their sugars point outwards from the cell into the external aqueous environment. This asymmetry is preserved by the fact that transverse movement of lipids, known colloquially as 'flip-flop' (from one side of the membrane to another), is severely restricted; such movement would involve pushing the polar heads through the central hydrocarbon layer to get to the other side. This is energetically unfavourable and so the asymmetry is preserved. Proteins catalysing energy-dependent membrane flip-flop are known and may be involved in the creation and maintenance of asymmetry.

The very limited flip-flop movement of membrane lipids contrasts with their potential for rapid lateral movement within the plane of the bilayer. The assembly of lipids into the bilayer structure does not involve covalent bonds and, in general, they move around freely—a molecule can cover the length of the cell very rapidly.

In the case of one particular membrane lipid, phosphatidylinositol, we know of a very special role as the source of two chemical signals which have profound effects on control of cellular events, dealt with later (pages 433, 439) but, in general, we have little idea why membranes contain so many different polar lipid components or why the composition varies from cell to cell and between inner and outer halves of a given bilayer. Some membrane lipids are of special interest. For example, there is a clinical interest in gangliosides, because of children's genetic diseases called **glycosphingolipidoses** (Tay–Sachs disease and Gaucher's disease). In these, the glycosphingolipids are not broken down properly and residues of them accumulate causing severe brain disorders. However, although these diseases show that turnover of such molecules normally occurs, they do not indicate their specific functions. The essential message is that cells use a variety of polar lipids in forming lipid bilayers, and a precise reason for this cannot be given. There may be many different reasons, but one is probable. As will become clear shortly, the other major component of membranes is proteins. The association of different proteins with different characteristics and different functions is likely to require different lipid components, and indeed the association and dissociation of proteins from membranes may well play a vital part in the regulation of the activities of cells. Given the wide variety of different membrane proteins and their functions, it is not surprising that different types of membrane lipids are needed.

What are the fatty acid components of membrane lipids?

The fatty acyl 'tail' components of a membrane lipid such as lecithin vary. In length they may range from C_{14} to C_{24} (almost always an even number, for reasons explained in Chapter 9), but C_{14}–C_{18} are the commonest. The two fatty acyl residues in a single phospholipid molecule may be the same or different, but usually in a glycerophospholipid, the fatty acid attached to the C-1 —OH group is saturated and that attached to the C-2 —OH group is unsaturated. The degree of saturation of fatty acid tails is of great importance because the central hydrocarbon core of the bilayer must be fluid rather than solid and unsaturated hydrocarbon tails lower the temperature at which a bilayer loses its essential fluidity. Membrane fluidity is essential to allow lateral movement of transmembrane proteins so that they can interact with one another. In addition, such proteins undergo conformational changes needed for ion channels, transporters, and receptors to function. Without a fluid bilayer such changes would be difficult to accommodate. (The proteins referred to are described later.) A saturated fatty acid tail is straight, but a double bond introduces a kink into it as illustrated. Natural unsaturated fatty acids are almost always in the *cis* configuration.

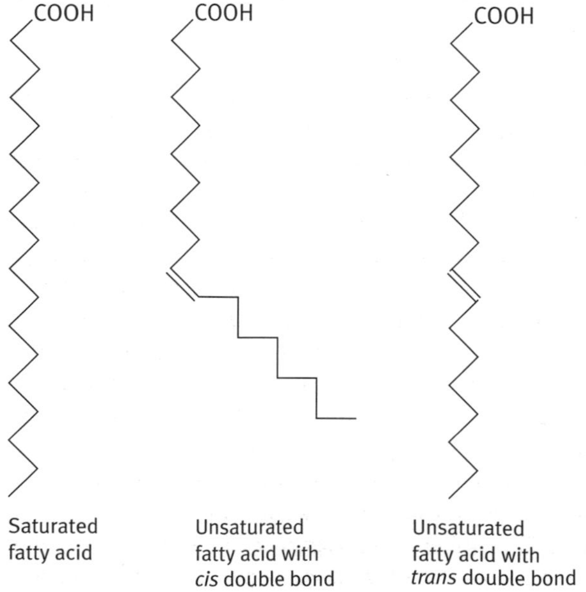

| Saturated fatty acid | Unsaturated fatty acid with *cis* double bond | Unsaturated fatty acid with *trans* double bond |

The saturated chains comfortably pack together and interact, but the kinked ones cannot do so. The physical effect of unsaturation can be seen by comparing hard mutton fat with olive oil. Both are TAGs but the olive oil is rich in unsaturated

fatty acyl tails. (But note again that such TAGs never occur in membranes.)

The importance of maintaining membrane fluidity is illustrated by the fact that bacteria adjust the degree of unsaturation of the fatty acid components of their membranes bilayers according to growth temperature. In special situations, such as during hibernation, animals modulate the degree of saturation of cell membrane components to cope with lower body temperature.

A variety of fatty acids, often with more than one double bond, are present in membrane lipids. The usual ones are C_{18} and C_{16} with one double bond in the middle (oleic and palmitoleic acids, respectively), linoleic (C_{18} with two double bonds), and arachidonic acid (C_{20} with four double bonds). The nomenclature of such acids is dealt with in more detail in a later chapter (page 191).

What is cholesterol doing in membranes?

From its structure, as conventionally drawn (Fig. 4.5(a)), cholesterol seems an improbable membrane constituent. However, the actual conformation of the ring system is more like that shown in Fig. 4.5(b). The molecule is elongated, the steroid nucleus being rigid and the hydrocarbon chain flexible. It is an amphipathic molecule, the —OH group being weakly polar.

Cholesterol in membranes acts as a 'fluidity buffer'. It is shorter than the hydrocarbon tails and inserted between the membrane lipids (with the —OH level with the polar heads); it acts as a wedge between the chains near the surface of the membrane and thereby probably makes the bilayer less permeable

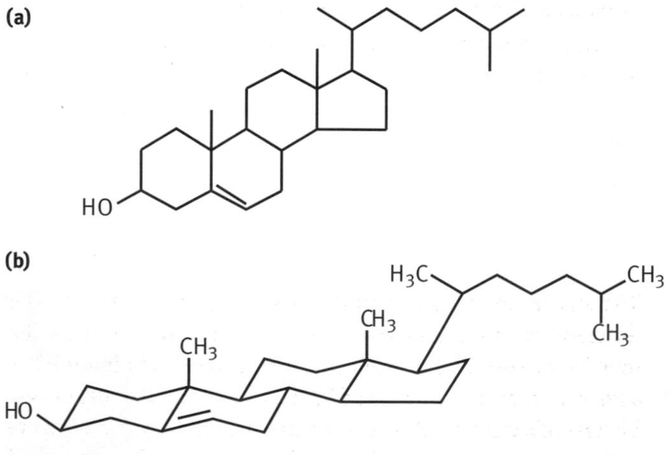

Fig. 4.5 (a) Structure of cholesterol as conventionally drawn. (b) Structure drawn to give a better indication of the actual conformation of cholesterol.

but also prevents close packing of the ends of the hydrocarbon tails and thus lowers the melting point of the bilayer. The observed effect is that cholesterol 'blurs' the melting point of a lipid bilayer. Without cholesterol, the transition from solid to liquid is sharper than when cholesterol is present. It is reminiscent of impurities in a crystal, giving a diffuse melting point instead of the sharp one of a pure substance. A red blood cell membrane may be about 25% cholesterol. On the other hand, bacterial membranes have no cholesterol and animal cell mitochondria have very little. Plants contain other sterols, known as phytosterols.

The self-sealing character of the lipid bilayer

The lipid bilayer is effectively a two-dimensional fluid. If you had a small enough probe you could poke it through without difficulty and the lipids would seal around it. The bilayer gives cells flexibility and a self-sealing potential, the latter being essential when a cell divides (Fig. 4.6(a)).

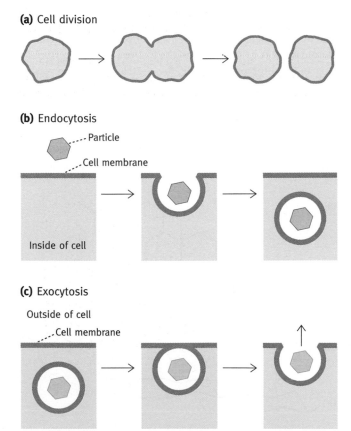

(a) Cell division

(b) Endocytosis

(c) Exocytosis

Fig. 4.6 (a) Cell division; (b) endocytosis; (c) exocytosis for the release of, for example, a digestive enzyme from cells.

This versatility is also essential in **endocytosis**—the process by which cells can engulf structures much too large to pass through the bilayer. The cell engulfs the object by forming an invagination which is nipped off to form a vacuole containing that object inside the cell (Fig. 4.6(b)). When macromolecular particles are to be endocytosed they accumulate in depressions on the cell surface called **clathrin-coated pits** which are very numerous on animal cell membranes. Clathrin is a protein molecule with three arms and is capable of assembling into a basketlike network. The internal surface of a pit is coated with clathrin molecules. The pit invaginates trapping the material to be engulfed in a vesicle completely covered in a network of clathrin. The final stage is that a protein called **dynamin** assembles around the neck of the vesicle which becomes pinched off, energy being required in the process. Possibly an extra enzyme colloquially referred to as a 'pinchase' does the final job guided by the noose of dynamin. The internalized object in the vacuole can be processed in ways that will be described later (page 268).

The reverse can also happen. If the cell needs to secrete a substance such as a digestive enzyme it is synthesized inside the cell and enclosed within a membrane-bounded vesicle which migrates to the plasma membrane, fuses with it, and releases its contents to the outside. The process is called **exocytosis**. A good example of this is the secretion of digestive enzymes into the intestine by pancreatic cells (Fig. 4.6(c)).

Permeability characteristics of the lipid bilayer

The ability of a substance to diffuse through a lipid bilayer is largely related to its fat solubility. Small fat-soluble molecules penetrate easily. Strongly polar molecules such as ions traverse the bilayer extremely slowly if at all. Large, more weakly polar molecules such as glucose penetrate very slowly, although smaller ones such as ethanol or glycerol diffuse across more readily. Ionized groups of polar molecules and inorganic ions are surrounded by a shell of water molecules which must be stripped off for the solute to pass through the lipid bilayer hydrocarbon centre, and this is energetically unfavourable. The lipid bilayer is therefore almost impermeable to such molecules and to ions, but a slow leak inevitably occurs.

The majority (but not all) of molecules of biochemical interest are polar in nature and cannot pass through the lipid bilayer at rates commensurate with cellular needs. This means that although the bilayer structure is ideal for holding in the contents of a cell, special arrangements must be made to permit rapid movement of molecules across the membrane as needed.

Surprisingly, water molecules, despite being polar, pass through the lipid bilayer with sufficient ease for the needs of the cell. Presumably this is due to the small size of the molecule and

its lack of charge, but there is some uncertainty as to how it traverses the bilayer with such apparent ease. Cells involved in water transport, such as those of renal tubules and secretory epithelia, have a transmembrane protein, **aquaporin**, which allows free movement of water. The lipid bilayer also readily allows gases in solution such as oxygen to diffuse through it.

To sum up so far: cells must be bounded by a barrier which prevents leakage of cell constituents which are mainly polar. This is achieved by the lipid bilayer structure dependent on weak noncovalent forces acting on a variety of amphipathic or polar lipids. The two halves of the lipid bilayer have different polar lipid compositions, the difference established at the time of synthesis being preserved by the fact that flip-flop movement of lipids across the bilayer is an energetically unfavourable process. Lateral movement within the bilayer halves occurs readily. Different cells have different lipid compositions in their membranes. It is not understood why so many different lipids are used though one component, phosphatidylinositol, has recently emerged as a key compound in cellular control mechanisms. The hydrocarbon layer of the lipid bilayer must be in a liquid, 'noncrystalline' form. This is achieved in animal cells by a proportion of the hydrocarbon tails of the lipids containing double bonds which confer a kink in the hydrocarbon chains thus preventing close packing and lowering the solidification temperature. Branched chains achieve the same result in bacteria. Cholesterol also plays a role in maintaining the correct fluidity of the hydrocarbon layer. The lipid bilayer is effectively impermeable to most polar substances, except weakly polar ones of small molecular size, a condition obviously essential for retaining cellular constituents against leakage. The lipid bilayer allows for fusion of membranes and for nipping off of vacuoles—it is self-sealing. These characteristics are essential for cell division, endocytosis, and exocytosis.

Where do we go from here? The lipid bilayer alone isn't enough for a cell membrane. One vital need is to allow essential traffic across the bilayer and the traffic has to be specific to selected molecules. Membrane proteins are responsible for this.

Membrane proteins and membrane design

A beautiful aspect of cell membrane structure is that it provides a totally flexible system which can be modified in the evolutionary sense. Different cells need different membrane functions and hence different cells have quite different protein molecules in their membranes. If a new functional protein is evolved to act

in a membrane it can be inserted into the membrane to function there. (We explain in Chapter 25 how this is achieved.) Proteins can be inserted into the bilayer provided that they have the requisite properties (explained below)—the bilayer self-seals around the proteins. To emphasize this, functional proteins experimentally isolated from membranes have been incorporated into synthetic phospholipid liposomes where they function exactly as in a cell membrane. It is the ultimate in flexible design. The membrane structure is called a **lipid fluid mosaic** (Fig. 4.7). In this, 'icebergs' of protein are floating in a two-dimensional sea of lipid. Such proteins are called **integral proteins** (labelled I) because they are integrated into the membrane. Other proteins are called **peripheral proteins** (labelled P) because they associate with the periphery of the membrane (see below).

What holds integral proteins in the lipid bilayer?

Integral membrane proteins are so structured that they are held in position in the lipid bilayer by the weak forces with which you are already familiar.

As stated in the previous chapter, proteins are, in primary structure, long strings of amino acids covalently joined together by peptide bonds. If the side chains exposed on the outside of

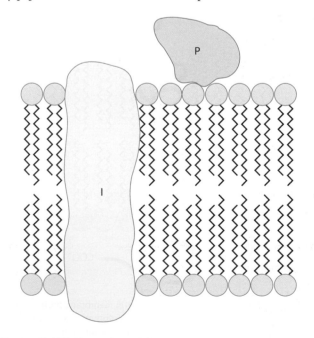

Fig. 4.7 Lipid fluid mosaic model for membrane structure. I, intergral protein; P, peripheral protein.

the polypeptide are hydrophobic, the protein in that section will be extremely hydrophobic; if a section of the protein predominantly has exposed hydrophilic groups, that section will be hydrophilic. Integral membrane proteins have ends made of hydrophilic amino acids and the middle section of hydrophobic ones as shown diagrammatically in Fig. 4.8(a).

In Fig. 4.8(b) is a diagram of **glycophorin**, a major component of the erythrocyte membrane. On the external surface of the membrane is a large N-terminal section of polypeptide, rich in hydrophilic amino acids; attached to serine, threonine and asparagine residues are carbohydrates (see page 44), making this part of the protein extremely hydrophilic. On the inside face of the bilayer is a shorter C-terminal section devoid of carbohydrate attachments but containing hydrophilic amino acids. (Carbohydrate attachments to membrane proteins are always external.) Connecting the two external sections and spanning the lipid bilayer is a stretch of 19 amino acid residues which, in the form of an α-helix, are sufficient to just span the hydrocarbon central layer of the membrane which is about 30 nm wide. In this section of the chain, isoleucine, leucine, valine, methionine, and phenylalanine, all strongly hydrophobic and good α helix formers, predominate. There are no strongly hydrophilic residues (except in those cases where the protein is a transmembrane aqueous pore lined with hydrophilic residues).

The essential feature of an integral membrane protein is that the hydrophobic section corresponds in length with the hydrocarbon middle zone of the membrane lipid bilayer. The parts of the protein projecting from the membrane that are in contact with water, or the polar head groups of membrane lipids, have hydrophilic residues as illustrated in Fig. 4.8(b). Such a protein, with sections of different polarity is called an **amphipathic or amphiphilic protein**, indicating that the molecule has *both* polar parts and hydrophobic parts.

In some proteins the polypeptide chain loops back and crosses the bilayer several times—for this, alternating hydrophilic and hydrophobic sections are required, each of the latter being about 20 amino acids long and forming α-helices which nicely span the bilayer. One of the best-studied examples is **bacteriorhodopsin**, a protein present in the membrane of the purple bacterium *Halobactium halobium* found in salty ponds. The protein criss-crosses the membrane seven times, forming a cluster of seven α-helices spanning the membrane and connected by hydrophilic loops (Fig. 4.9). The cluster has a light-absorbing pigment at its centre to capture light energy which drives the pumping of protons from the cell to the outside. The energy of the proton gradient so formed is used to drive the synthesis of ATP by a mechanism to be described in Chapter 9.

The arrangement of a membrane protein described maximizes hydrophilic bonding and hydrophobic bonding. If the protein tended to move out of the membrane, the hydrophobic groups of the protein would come into contact with water and hydrophilic groups into contact with the hydrocarbon layer. Both are energetically resisted and the protein is fixed in the transmembrane sense. Unless otherwise restrained it could move laterally in the bilayer. The mechanism which puts the proteins into the membrane with the correct inside-outside orientation is described in a later chapter (page 410).

Anchoring of peripheral membrane proteins to membranes

A peripheral water-soluble membrane protein may be associated with a membrane by hydrogen bonding and ionic attrac-

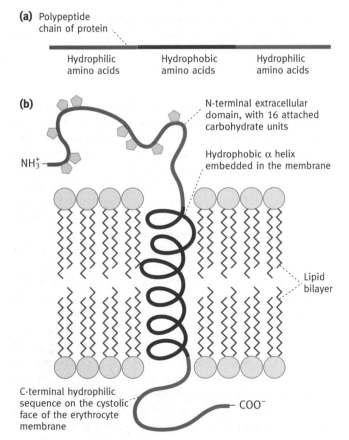

(a) Polypeptide chain of protein

Hydrophilic amino acids | Hydrophobic amino acids | Hydrophilic amino acids

(b)

N-terminal extracellular domain, with 16 attached carbohydrate units

NH_3^+

Hydrophobic α helix embedded in the membrane

Lipid bilayer

C-terminal hydrophilic sequence on the cystolic face of the erythrocyte membrane

COO^-

Fig. 4.8 (a) The structural plan of an integral membrane protein. (b) Glycophorin, a protein of the erythrocyte membrane. The α helix, containing about 19 hydrophobic amino acid residues, is approximately 30 Å in length which is sufficient to span the nonpolar interior of the lipid bilayer.

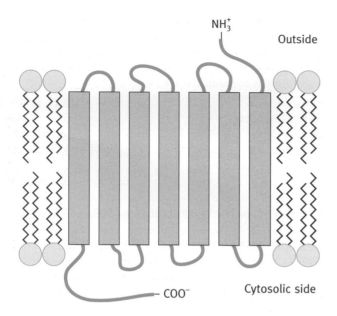

Fig. 4.9 Topological representation of bacteriorhodopsin, with seven α helices spanning the lipid bilayer. The α helices are not actually arranged linearly, as shown here for convenience, but are clustered compactly together.

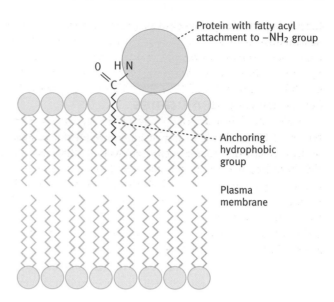

Fig. 4.10 A fatty acid anchoring molecule joined by CO—NH linkage. to the amino group of a protein. Alternative linkages such as ester and thiol ester can occur with appropriate amino side groups of the protein.

tions. However, an alternative method exists in the case of certain proteins. In these, the proteins have attached to them a fatty acid whose hydrocarbon chain is inserted into the lipid bilayer thus anchoring the protein to the membrane (Fig. 4.10). As so often is the case, evolutionary tinkering has produced variations on the one theme. *N*-**Myristoylation** of the protein is one in which myristic acid (C_{14}) is linked to an N-terminal glycine of the protein:

$$R—CO—NH—CH_2—CO—Polypeptide$$

In addition C_{14}, C_{16}, and C_{18} fatty acids may anchor proteins in a similar fashion, but attached to serine or threonine groups by ester linkage or to cysteine by thioester linkage. More complex acids are also found in ether linkage for the same purpose. A much more elaborate anchoring method is for phosphatidylinositol (PI) to be linked via an oligosaccharide to a protein; the PI structure itself is inserted into the bilayer exactly as a normal membrane phospholipid.

Glycoproteins or membrane proteins with sugars attached on the exterior surface

We have already seen that glycolipids occur in cell membranes (page 76). In these, single sugars or small-branched complexes of sugars (oligosaccharides) form the polar groups of mem-

brane lipids—the cerebrosides and gangliosides are on the external surface of the cell. Many membrane proteins also have attached to them branched oligosaccharides, again on the external surface. The carbohydrates are covalently attached to the side chains of asparagine or serine (page 44).

The sugars which make up the oligosaccharides include fucose, galactose, mannose, *N*-acetylgalactosamine, *N*-acetylglucosamine, and sialic acid. The last two you have already met (page 74). Fucose and mannose are isomers of glucose.

In glycophorin of the red blood cell, the glycoprotein described earlier, half the weight of the protein molecule is carbohydrate. The exact function of these carbohydrate attachments is an area of uncertainty but may, in some cases, have recognition roles on cell surfaces (see also page 74).

To summarize the topic of membrane proteins: since membranes have many functions there are many different membrane proteins. Integral proteins are embedded in the membrane, peripheral proteins are superficially attached (by secondary bonds). Integral proteins have an amphipathic structure with a hydrophobic 'middle section' in contact with the nonpolar centre of the bilayer and hydrophilic ends in contact with the aqueous phase on either side of the membrane. Membrane proteins are often glycosylated (have sugars or sugar complexes covalently attached) on the exterior surface of the cell. An evolutionary advantage of a membrane with functions determined by its inserted proteins, is the great flexibility of the

system. Any number of different proteins can be put into the lipid bilayer.

Functions of membranes

Membranes have a wide variety of functions; for the moment we'll deal with the plasma membrane surrounding the cell. We'll come to the internal membranes in later chapters.

The main functions of such membranes, apart from retaining cell contents, are:

- transport of substances in and out of the cell
- ion transport and nerve impulse conductance
- signal transduction (transmission of signals across membrane from outside the cell to the inside—explained later)
- maintaining the shape of the cell
- cell–cell interactions.

Let us now look at these more closely.

Transport of substances in and out of the cell

As mentioned, many, or most, of the substances which must be taken up by the cell cannot diffuse across the lipid bilayer at anywhere near the rate required. (There are exceptions to this statement for a small number of membranes have quite large pores in them which indiscriminately permit the passage of molecules smaller than about 600 Da. This is true of the outer mitochondrial membrane and the outer chloroplast membrane.) Polar structures such as sugars, amino acids, inorganic ions, etc. require specific proteins to allow passage of such substances through the membrane. Simple holes in the membrane will not do—they would allow nonspecific leakage in and out and would be lethal. Each transport system has to handle only specific molecules so there are many different transport systems and therefore many different **transport proteins.**

Active transport

We can separate transport systems into active and passive types. The former is transport that requires performance of work which, as you will know from Chapter 1, involves (with exceptions in bacteria, see page 174) the hydrolysis of ATP. It requires work because the movement of substances occurs against a concentration gradient.

The amount of energy required to transport a solute into a cell, against a gradient, can be calculated from the equation for chemical reactions given in Chapter 1.

$$\Delta G = \Delta G^{o'} + RT\,2.303\log_{10}\frac{[\text{Products}]}{[\text{Reactants}]}$$

For transport of a solute whose structure does not change, $\Delta G^{o'}$ is zero and the equation becomes:

$$\Delta G = RT\,2.303\log_{10}\frac{[\text{C2}]}{[\text{C1}]}$$

where C_1 is the concentration outside and C_2 is the concentration inside, taken in this example to be in the ratio of 10/1 so that

$$\Delta G = (8.315\ \text{J mol}^{-1}\text{K}^{-1})(298\,\text{K})2.303\ \log_{10}10/1 = 5706\ \text{J mol}^{-1}$$

Thus the transport of 1 mol, under these conditions (at 25°C) requires 5706 J. A cell such as a nerve cell, which is very active in Na^+ and K^+ pumping, uses a large proportion of its total ATP production in this activity. Note that the above calculation applies to the transport of an uncharged solute. With a charged solute, generation of an electrical potential requires an additional term to correct for this.

A good example of active transport is the Na/K$^+$ pump present in animal cells. Almost all animal cells have a high internal K^{I} concentration (140 mM) and a low Na^+ concentration (12 mM) as compared with those of the blood (4 mM and 145 mM respectively). This is energetically expensive—it consumes perhaps a third of the total energy usage in a resting animal. The interior of the cell requires high K^+ and low Na^+ levels to function properly. The ion gradients are necessary for electrical conduction in excitable membranes and for driving solute transport across membranes.

Mechanism of the Na$^+$/K$^+$ pump

The Na$^+$/K$^+$ pump is also called the Na$^+$/K$^+$ ATPase because ATP is hydrolysed to ADP + P$_i$ as Na$^+$ is pumped out and K$^+$ in. (ATP is described on page 11). The protein is a complex of four polypeptides or subunits. In the Na/K$^+$ pump there are two identical so-called α subunits and two β subunits. Some proteins have the property of slightly changing their shape when specific ligands bind to them (page 23). (A **ligand** is any molecule which binds to a specific site.) This may be due to a change in the conformation of a simple protein, or to a relative change in position of subunits of a protein complex (see hemoglobin on page 514). The covalent attachment of a phosphoryl group to the Na$^+$/K$^+$ ATPase by transference from ATP causes a **conformational** (shape) **change.** Such a change can alter the ability of a protein to bind a given ligand (if it has been designed to do so). Thus, in one form, the protein of the pump is believed

to bind Na⁺ but not K⁺ and another form K⁺ but not Na⁺. After that preamble, let's look at the pump mechanism which has been postulated.

In the model shown in Fig. 4.11, the Na⁺/K⁺ ATPase protein in effect exists in two conformations. Form (a) is open to the interior of the cell and binds Na⁺ but not K⁺. ATP now phosphorylates the protein, yielding ADP + phosphorylated protein. In this form it is open to the outside but no longer binds Na⁺ (which diffuses away) but does bind K⁺ (which therefore attaches). This gives form (b) (Fig. 4.11). The —P group is now hydrolysed from the protein giving P_i and the protein reverts to form (a) to which the K⁺ no longer binds. The latter therefore enters the cell and more Na⁺ attaches to the pump (Fig. 4.11(c)).

The net result is that hydrolysis of ATP pumps Na⁺ out and K⁺. The ratio is 3 Na⁺ out and 2 K⁺ in.
The overall equation is:

$$3\,Na^+(in) + 2K^+(out) + ATP + H_2O \rightarrow$$
$$3\,Na^+(out) + 2K^+(in) + ADP + P_i$$

There is an interesting medical aspect. **Cardiac glycosides** are a group of compounds found in the digitalis species of plants (foxglove). They are steroids or steroid glycosides; steroids resemble cholesterol in structure; glycosides are molecules with sugars attached. Such compounds inhibit the Na⁺/K⁺ ATPase by preventing the removal of the phosphoryl group from the transport protein. This 'freezes' the pump in one form (Fig. 4.9(b)) and stops the ion transport. The cardiac glycosides have long been used clinically as treatment for congestive heart failure. The compounds are lethal in sufficient amounts, but in appropriate doses, the partial inhibition of the Na⁺/K⁺ ATPase increases the Na⁺ concentration inside heart muscle cells and thus lowers the Na⁺ gradient from outside to inside. This has the effect of raising the cytoplasmic Ca^{2+} level because there is another system which transports Na⁺ into the cell and Ca^{2+} out of the cell. The ejection of Ca^{2+} is driven by the Na⁺ gradient (see below); if the latter is lowered by cardiac glycosides then ejection of Ca^{2+} is reduced and the internal level of Ca^{2+} rises. The raised Ca^{2+} level stimulates heart muscle contraction; the role of Ca^{2+} in contraction is dealt with in Chapter 32 (page 530). Ouabain, the African arrow tip poison, has similar effects. There is also a Ca^{2+} ATPase which ejects calcium ions from cells.

(a)   Outside of cell

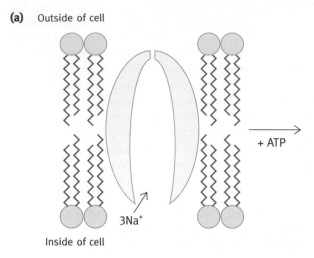

Inside of cell

(b)   Outside of cell

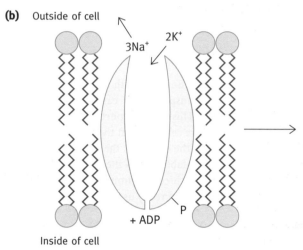

Inside of cell

(c)   Outside of cell

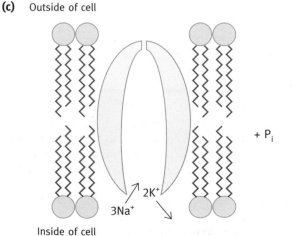

Inside of cell

Fig. 4.11 A possible mechanism for the Na⁺/K⁺ pump: (a) Na⁺/K⁺ ATPase in conformation (a); (b) Na⁺/K⁺ ATPase in conformation (b); (c) Na⁺/K⁺ ATPase after the phosphoryl group is hydrolysed off the protein, returning the pump to conformation (a).

Symport systems

The Na$^+$/K$^+$ ATPase, described above, produces a steep Na$^+$ gradient across the cell membrane. Any gradient has potential energy and, given a chance, the accumulated exterior Na$^+$ will flow back into the cell. What happens in Na$^+$ cotransport is that the transport protein conducts Na$^+$ into the cell, driven by the Na$^+$ gradient and another substance hitches a thermodynamic ride on the back of this flow. Such a transport system is called a **symport** (it simultaneously transports two components), or **cotransport** system. The glucose symport protein or symporter system permits movement of Na$^+$ and glucose across the membrane *when both are present* but not when only one is present. It can transport glucose against a concentration gradient at the expense of the Na$^+$ gradient. The Na$^+$ entering the cell is pumped out by the Na/K$^+$ ATPase to restore the Na$^+$ gradient so that it is ATP hydrolysis which indirectly supplies the energy for glucose uptake. Absorption from the gut of glucose and amino acids occurs by this mechanism (Fig. 4.12), separate symport proteins being required for the different substances transported.

Antiport systems

As mentioned above, in connection with cardiac glycoside action on the heart, the cotransport of Na$^+$ can be used in another way—to pump out Ca^{2+}. This is an **antiport system** (Fig. 4.13). Na$^+$ is transported into the cell only if at the same time as Ca^{2+} is transported out. Again the energy in the Na$^+$ gradient is the driving force and this is established by the Na$^+$/K$^+$ ATPase system. When we come to muscle contraction you will find a different type of Ca^{2+} pump which directly uses ATP (page 530).

Uniport systems

In this, a transport protein transfers a single specific molecule across the bilayer. The Ca^{2+} ATPase of sarcoplasmic reticulum just referred to is such a case . You will later meet ATP-

driven proton pumps which cause acidification of lysosomal vesicles.

To summarize transport across plasma membranes: transmembrane transport proteins designed to transport specific solutes are required and therefore there are many different transporting systems. Transport can be passive, the systems simply allowing passage either way depending on relative concentrations inside and outside, the anion channel of erythrocytes and the glucose transporter of animal cells being important examples. There are many situations where solutes must be transported against a concentration gradient. Animal cells possess a Na/K$^+$ ATPase which pumps Na$^+$ out and K$^+$ in to maintain steep gradients. This is inhibited by the cardiac glycosides which are used clinically. The Na$^+$ gradient outside a cell can be used to drive uptake of, for example, glucose or amino acids by a cotransport mechanism of the symport type. Such systems are involved in absorption from the gut. It can also be used in antiport systems where Na$^+$ goes in and Ca^{2+} goes out, again the drive being supplied by the Na$^+$ gradient, the latter continuously maintained by the Na/K$^+$ ATPase. A different Ca^{2+} ATPase is used in muscle contraction and will be discussed later.

Passive transport or facilitated diffusion

In passive transport, a protein permits the movement of a substance across the membrane so that the movement will be down whatever concentration gradient exists across the membrane, no energy being involved in the actual transport process. This is **facilitated diffusion**. A good example is the anion transport protein in red blood cells which lets HCO$_3^-$ and Cl$^-$ ions pass through the membrane in either direction (Fig. 4.14). The purpose of this is discussed in Chapter 31 (page 517). Another important example of facilitated diffusion is the glucose transporter possessed by many animal cells. It allows glucose to diffuse passively across the membrane. When we come to

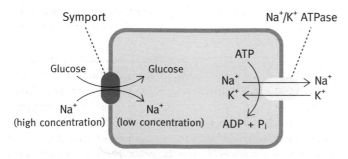

Fig. 4.12 The Na$^+$/glucose cotransport system—a symport.

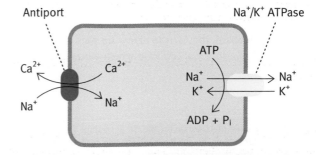

Fig. 4.13 The Na$^+$/Ca^{2+} cotransport system—an antiport.

Symport systems

The Na$^+$/K$^+$ ATPase, described above, produces a steep Na$^+$ gradient across the cell membrane. Any gradient has potential energy and, given a chance, the accumulated exterior Na$^+$ will flow back into the cell. What happens in Na$^+$ cotransport is that the transport protein conducts Na$^+$ into the cell, driven by the Na$^+$ gradient and another substance hitches a thermodynamic ride on the back of this flow. Such a transport system is called a **symport** (it simultaneously transports two components), or **cotransport** system. The glucose symport protein or symporter system permits movement of Na$^+$ and glucose across the membrane *when both are present* but not when only one is present. It can transport glucose against a concentration gradient at the expense of the Na$^+$ gradient. The Na$^+$ entering the cell is pumped out by the Na/K$^+$ ATPase to restore the Na$^+$ gradient so that it is ATP hydrolysis which indirectly supplies the energy for glucose uptake. Absorption from the gut of glucose and amino acids occurs by this mechanism (Fig. 4.12), separate symport proteins being required for the different substances transported.

Antiport systems

As mentioned above, in connection with cardiac glycoside action on the heart, the cotransport of Na$^+$ can be used in another way—to pump out Ca^{2+}. This is an **antiport system** (Fig. 4.13). Na$^+$ is transported into the cell only if at the same time as Ca^{2+} is transported out. Again the energy in the Na$^+$ gradient is the driving force and this is established by the Na$^+$/K$^+$ ATPase system. When we come to muscle contraction you will find a different type of Ca^{2+} pump which directly uses ATP (page 530).

Uniport systems

In this, a transport protein transfers a single specific molecule across the bilayer. The Ca^{2+} ATPase of sarcoplasmic reticulum just referred to is such a case. You will later meet ATP-driven proton pumps which cause acidification of lysosomal vesicles.

To summarize transport across plasma membranes: transmembrane transport proteins designed to transport specific solutes are required and therefore there are many different transporting systems. Transport can be passive, the systems simply allowing passage either way depending on relative concentrations inside and outside, the anion channel of erythrocytes and the glucose transporter of animal cells being important examples. There are many situations where solutes must be transported against a concentration gradient. Animal cells possess a Na/K$^+$ ATPase which pumps Na$^+$ out and K$^+$ in to maintain steep gradients. This is inhibited by the cardiac glycosides which are used clinically. The Na$^+$ gradient outside a cell can be used to drive uptake of, for example, glucose or amino acids by a cotransport mechanism of the symport type. Such systems are involved in absorption from the gut. It can also be used in antiport systems where Na$^+$ goes in and Ca^{2+} goes out, again the drive being supplied by the Na$^+$ gradient, the latter continuously maintained by the Na/K$^+$ ATPase. A different Ca^{2+} ATPase is used in muscle contraction and will be discussed later.

Passive transport or facilitated diffusion

In passive transport, a protein permits the movement of a substance across the membrane so that the movement will be down whatever concentration gradient exists across the membrane, no energy being involved in the actual transport process. This is **facilitated diffusion**. A good example is the anion transport protein in red blood cells which lets HCO$_3$$^-$ and Cl$^-$ ions pass through the membrane in either direction (Fig. 4.14). The purpose of this is discussed in Chapter 31 (page 517). Another important example of facilitated diffusion is the glucose transporter possessed by many animal cells. It allows glucose to diffuse passively across the membrane. When we come to

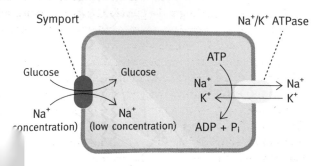

...sport system—a symport.

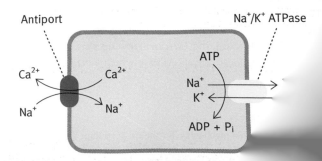

Fig. 4.13 The Na$^+$/Ca^{2+} cotransport system—an a...

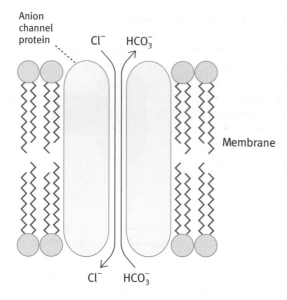

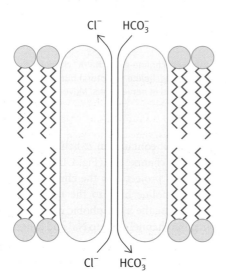

Fig. 4.14 An anion channel of red blood cells. Cl^- and HCO_3^- may move in either direction, according to concentration gradients across the membrane. The counterflow of the ions prevents any electrical potential developing across the membrane.

discuss the utilization of blood glucose by cells this becomes very important.

Gated ion channels

An extremely important class of passive transport systems are the gated pores or channels. These are aqueous pores, highly selective for specific ions, that open and close on receipt of a signal. The most important gated pores are those for Na^+, K^+, and Ca^{2+} (each selective for one ion) and the Na^+/K^+ channel. The passage of ions through these channels, when open, is the result of concentration gradients across the membrane so that the flow proceeds in either direction as determined by these. Since the gradients are steep in respect of ions such as Na^+, K^+, and Ca^{2+} the rate of movement of these ions through open channels is very much faster than the rate of transport of Na and K ions brought about by the Na^+/K^+ ATPase pump.

The control of entry and exit of ions into cells is central to cellular function; lipid bilayer membranes are relatively impermeable to them because the ions are hydrated. The transmembrane traffic of ions is carried out by specialized transmembrane proteins. We have already described the universal Na^+/K^+ pump which establishes the gradients of these two ions across cell membranes. As you will learn in later chapters, Ca^{2+} is an important signalling agent so the strategy is always to keep its level in the cytoplasm very low by means of Ca^{2+} pumps thus producing a steep concentration gradient across cell membranes since the extracellular level is high.

These ionic gradients are made good use of by cells which have gated ion channels in their membranes as appropriate to the function of the cells. When channels open, rapid flows (fluxes) of ions occur across the cell membrane resulting in biochemical responses. The most spectacular use of this principle is in nerve conduction which we will use as an example. We will first describe the nature of gated ion channels and then their role in nerve impulse generation and propagation along the nerve fibre.

Nerve impulse transmission

A gated ion channel is one which opens and closes as instructed by a signal which may be the binding of a ligand (acetylcholine or other chemical signal) or changes in the membrane potential (explained below), known as **ligand-gated** and **voltage-gated** channels respectively. The concept of a gated channel is illustrated in Fig. 4.15. In the open state they permit the movement of specific ions according to the prevailing concentration gradients. The ones involved in nerve conduction are those for Na^+/K^+, Na^+ alone, K^+ alone, and Ca^{2+}.

A nerve impulse is transmitted by a series of neurons or nerve cells which have a central body, a thin axon which may be very long, even metres in length and terminating in branches. The gaps between one neuron and another are called **nerve synapses** (Fig. 4.16(a)); the signal is transmitted chemically across synapses by **acetylcholine** (or another neurotransmitter substance) which is released from the neuron ending to stimulate the next neuron by combining with acetylcholine receptors on

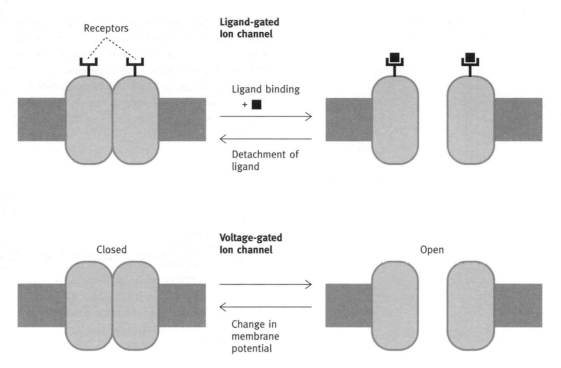

Fig. 4.15 Ligand-gated and voltage-gated ion channels. Examples of ligand-gated channels are the acetylcholine-gated Na$^+$/K$^+$ channels found on postsynaptic membranes and on muscle membranes at neuromuscular junctions (also known as the acetylcholine receptors) but the principle applies to other neurotransmitters. Examples of voltage-gated channels are the Na$^+$ and K$^+$ channels found in nerve axons. Movement of ions is passive, driven solely by concentration gradients.

the postsynaptic membrane. (We confine our discussion here to acetylcholine as the transmitter substance). The acetylcholine in the synapse is rapidly hydrolysed by acetylcholinesterase bound to the synaptic membrane, the bound acetylcholine molecules dissociate from the receptor and the neuron becomes ready to accept a new signal. If the hydrolysis of the acetylcholine is inhibited, transmission across these synapses is blocked. This is the basis of the organophosphate nerve gases and insecticides. Snake venoms such as cobra toxin achieve the same result by attaching to the channels in the postsynaptic membrane and inactivating them.

The same receptors occur at neuromuscular junctions (Fig. 4.16(b)); acetylcholine liberated by motor nerves binds to the receptors and triggers muscle contraction. Curare, the arrow poison, blocks the action of acetylcholine and is used by surgeons during operations to relax voluntary muscles by blocking the signal from a motor nerve to a muscle.

Figure 4.17 gives a simplified overview of how a neuron conducts a signal.

The acetylcholine-gated Na$^+$/K$^+$ channel or acetylcholine receptor

The channel consists of a pore created by five protein subunits arranged in a circle. Two of the subunits have a binding site for

acetylcholine. Each subunit contains an α-helix kinked in the middle thus restricting the channel size (Fig. 4.18). In the closed form hydrophobic groups project into the channel but when two molecules of acetylcholine bind to the receptor the α-helices tilt slightly and cause the hydrophobic groups to swing out of the way leaving the channel open to Na$^+$ and K$^+$ ions. The gate remains open only for about a millisecond even if acetylcholine is still bound because the gate rapidly becomes desensitized and closes. The channel is highly selective for Na$^+$ and K$^+$ ions; anions are repelled by strategically placed negatively charged carboxyl groups of amino acid side chains.

How does acetylcholine binding to a membrane receptor result in a nerve impulse?

Consider a nerve synapse at which acetylcholine is liberated by the presynaptic neuron (the membrane at the end of the neuron carrying the impulse to the synapse). It diffuses the short distance to the postsynaptic membrane of the next neuron where it can trigger a nerve impulse in that neuron. As with other cells, neurons have a high level of K$^+$ and a low level of Na$^+$ inside relative to the levels outside, the result of the Na$^+$K$^+$ pump.

In the resting stage (where nothing is happening in terms of nerve impulses) the cell membrane is more permeable to K$^+$

(a)

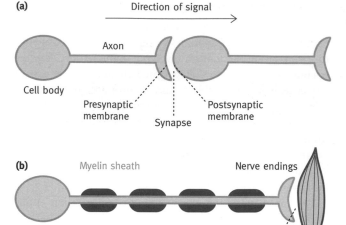

than to Na⁺ because of special K⁺ 'leak' channels. K⁺, high in concentration inside, leaks out creating a negative charge inside and a positive one outside since the membrane is impermeable to anions. The K⁺ leakage is self-limiting because the internal negative charge so created holds them back and an equilibrium is established in which the resting cell potential across the membrane is about $-60\,mV$ (more negative inside). This results in a separation of electric charges at the membrane (which is

(b)

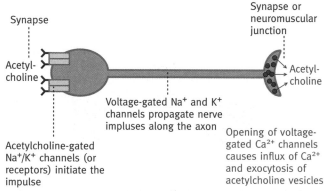

Fig. 4.16 (a) A synapse between two neurons. A nerve impulse in the first one causes the liberation of a neurotransmitter such as acetylcholine into the cleft. This stimulates the succeeding neuron. (b) A myelinated neuron. The example shown is of a motor neuron to skeletal muscle. An impulse causes the release of acetylcholine from the nerve ending into the neuromuscular junction which stimulates the muscle to contract (Chapter 32) The myelin sheath insulates the nerve axon from signal leakage and greatly speeds up the transmission of Impulses along axons (see text).

Fig. 4.17 Simplified diagram of the events in the transmission of a nerve impulse along a neuron, starting with the acetylcholine stimulation of the postsynaptic membrane and ending with the release of acetylcholine into a nerve synapse or into a neuromuscular junction. The acetylcholine-gated channel conducts both Na⁺ and K⁺ ions; there are two separate voltage-gated channels for Na⁺ and K⁺ respectively for the propagation of the impulse along the axon.

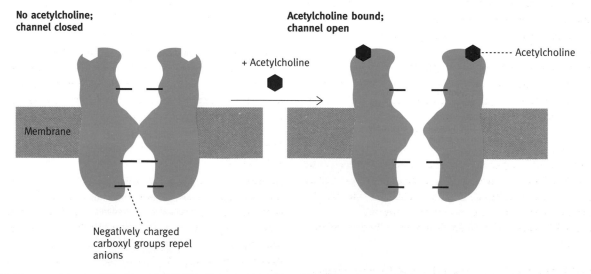

Fig. 4.18 Diagram of the acetylcholine-gated Na⁺/K⁺ channel of postsynaptic membrane and the receptor at neuromuscular junctions. The channel is composed of five subunits arranged to form a pore but only two are shown for simplicity; two are identical and each has an acetylcholine binding site which induces the allosteric change to open the channel on ligand-binding. It opens for only a fleeting period.

an electrical insulator) with a surplus of negative charges inside and positive outside. They attract one another other across the lipid bilayer as shown in Fig. 4.19(a), creating an electrical or **membrane potential**. The membrane is then said to be **polarized**. If you placed an electrode inside the cell and another outside, a **voltage** would be recorded by a meter placed in the circuit connecting the two.

When Na^+/K^+ channels in the postsynaptic membrane open in response to acetylcholine binding, the resulting Na^+ influx is much greater than the efflux of K^+ because the negative charge inside the membrane opposes the latter. While the ion fluxes are large, they only occur for about a millisecond and there is negligible effect on the overall Na^+ and K^+ gradients between the inside and the outside of the cell. The membrane polarization in the postsynaptic membrane in the vicinity of the channels is reversed—it changes from 60 mV to +65 mV and the section is then said to be **depolarized** (Fig. 4.19(b)). Again, this could be measured as a **voltage change** across the membrane by suitably placed electrodes. We are almost finished with the acetylcholine-gated channels; they have opened, the membrane is locally depolarized and the channels close again (Fig. 4.20). It is the initial stimulus for the transmission of a nerve impulse which may have to travel metres or so along the axon (in the giraffe, many metres).

How is the initial signal propagated along nerve axons?

This is an ingenious process that utilizes ion gradients which exist across all cell membranes to supply the required energy, the propagation mechanism involving nothing more than the opening and closing of voltage-gated channels in the correct sequence. The remarkable feature is that nothing physically moves along the length of the axon in the sense of a flow of molecules or ions. It is not like an electric current in which electrons flow along the wire. All that is need is that at the end of the neuron, the presynaptic membrane has to become locally depolarized just as the signal started out with a similar local depolarization at the postsynaptic membrane.

The mechanism of nerve impulse propagation along the axon was elucidated by the classical work of Hodgkin and Huxley in Cambridge, England, in 1952. The acetylcholine-gated channel as described, creates a small patch of depolarized membrane at the postsynaptic membrane (Fig. 4.21(a)). This, in turn, partially depolarizes the adjacent section of membrane due to the movement of ions along the axon for a short distance (Fig. 4.21(b)). In the axon membrane there are **voltage-gated Na^+ channels** and **K^+ channels**. When the membrane in which they are located is depolarized to the extent of 20 mV to give a membrane potential of −40 mV (the threshold value for triggering channel opening), a few of the Na^+ channels in that area open and Na^+ flows in. This flow further increases the depolar-

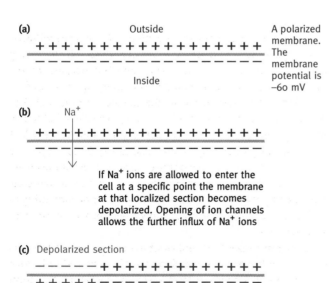

Fig. 4.19 Diagram to illustrate the meaning of the term 'polarized membrane' and its depolarization. In talking about depolarization in the context of neuronal function it is important to remember that only localized sections of membrane are referred to, not the whole cell membrane, and also that the depolarization is transitory as explained in the text.

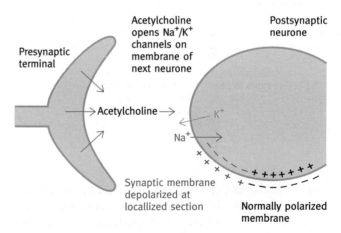

Fig. 4.20 Transmission of a nerve impulse across a nerve synapse. The presynaptic membrane liberates acetylcholine (or other neurotransmitter) on arrival of a nerve impulse. The acetylcholine binds to the receptors (the Na^+/K^+-gated channels) on the postsynaptic membrane of the next neuron. The resultant channel opening causes an influx of Na^+ ions and a smaller efflux of K^+ ions. This causes a local depolarization of the postsynaptic membrane. This depolarization triggers the propagation of the nerve impulse along the axon of the neuron as described in the text.

(a)

Direction of signal →

Depolarized section resulting from acetylcholine binding to receptors on the synaptic membrane

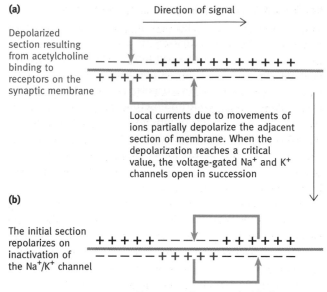

Local currents due to movements of ions partially depolarize the adjacent section of membrane. When the depolarization reaches a critical value, the voltage-gated Na⁺ and K⁺ channels open in succession

(b)

The initial section repolarizes on inactivation of the Na⁺/K⁺ channel

The pattern is repeated with the next small section of membrane. Section by section the depolarization progresses along the axon. Note that the depolarization is transient due to the rapid repolarization of each section

Fig. 4.21 How a nerve impulse is propagated along the axon: (a) Starting with the small depolarized section caused by acetylcholine at the synaptic membrane as shown in the previous figure, (b) when a small section of membrane is depolarized, local currents due to ion movements (brown arrows) cause partial depolarization of the adjacent section of membrane. When the membrane potential in the latter has fallen to the threshold value of −40 mV, opening of voltage-gated Na⁺ channels causes rapid further depolarization of that section. Closing of the Na⁺ channels and opening of the K⁺ channels restores the polarization but meanwhile the next section has become sufficiently depolarized to trigger channel opening in that section. The same cycle is repeated until the end of the neuron is reached. The impulse is prevented from going backwards by a mechanism described in the text.

izization and even more Na⁺ channels are opened as a result. The influx of Na⁺ rapidly depolarizes the local section of membrane up to a value of +40 mV. The Na⁺ channels are inactivated after about a millisecond. (This will be returned to shortly.) At this point, the depolarization peaks and the membrane potential reverts to that of the resting stage (it actually transiently overshoots a little). The repolarization is due to opening of voltage-gated K⁺ channels which open slightly later than the Na⁺ channels and allow efflux of K⁺ ions. The K⁺ channels also are rapidly inactivated so that the resting potential is restored. The whole thing takes about 3 milliseconds.

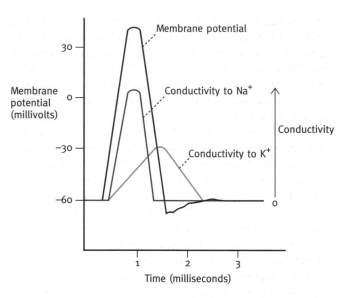

Fig. 4.22 Events in an action potential. The membrane potential (black line) is a measurement of the depolarization and repolarization of the membrane section. The conductivity is a measurement of the degree of opening of the voltage-gated Na⁺ channels (red line) and of the K⁺ channels (blue line) in the axon membrane which cause the polarization changes. These curves represent experimental measurements of the depolarization and repolarization events at the membrane. It is the latter which propagate the nerve impulse along the axon.

The depolarization of a small section of membrane followed by its repolarization can be measured experimentally by having electrodes placed across the membrane but because of the rapidity of the changes a cathode ray oscillograph was used by Hodgkin and Huxley to record the changes in potential. The result is shown in Fig. 4.22 (black curve). The electrical 'spike' of voltage change is known as an **action potential**. At the risk of overexplaining what may be obvious, this spike is simply the experimental measurement of the changes in membrane polarization. It is the changes in membrane polarization that conduct the impulse; the action potential spike is not something in addition to this but, as stated, simply a demonstration or measurement of it.

At the same time, the Cambridge workers measured the conductivity of the patch of membrane to Na⁺ ions and to K⁺ ions over the 3 milliseconds or so involved in generating the action potential. This gives an accurate picture of the state of opening of the relevant channels (red and blue curves in Fig. 4.22). You can see from these curves that the depolarization is due to the opening of Na⁺ channels and the reversal due to their inactivation coupled with the opening of the K⁺ channels. (It will be clear that when the K⁺ channels open in the depolarized section,

the surplus of positive charges is lost as the K⁺ flows out so that the original polarization is restored.) Following the repolarization of the membrane, the Na⁺ and K⁺ channels close. What we have achieved so far in nerve impulse propagation along the axon is the depolarization of a short section of membrane adjacent to the synapse. This depolarization has to travel right along the axon, one small section at a time, until finally the distal synaptic membrane is depolarized causing the release of the transmitter substance to carry the impulse across the synaptic cleft. How does this travel along the axon occur? The answer is remarkably simple; exactly the same is repeated so that the next small patch of membrane is depolarized by spread of the Na⁺ ions along the inside face of the membrane so there is another action potential generated; this spreads to the next and so on producing an action potential at one small section of membrane after another very rapidly until at the end of the neuron the synaptic membrane is likewise depolarized.

It may help if we summarize the events. A small section of membrane becomes depolarized by the influx of Na⁺ ions due to opening of voltage-gated channels. The Na⁺ ions spread so that enough depolarization of the next section occurs to trigger the opening of more Na⁺ channels in that section so that there is a rapid depolarization there, followed by restoration of the original state due to inactivation of the Na⁺ channels and opening of K⁺ channels. But meanwhile the next section has been depolarized due to spread of Na⁺ ions so that Na⁺ channels open in that section and so on along the whole length of the axon, one section at a time.

Perhaps you can see a potentially disastrous problem with the scheme described so far. The Na⁺ ions from the depolarized section can spread in both directions. What is to stop the depolarization spreading in the in the backwards direction as well? The Na⁺ and K⁺ channels have a special property which copes with this. There is a **refractory period** during which the channels which had opened and closed cannot be re-opened until after a slight delay. By the time they are capable of re-opening, the depolarization has moved too far along the axon for it to affect the channels; they will open only when the next impulse arrives. The impulse can only go forwards.

Mechanism of control of the voltage-gated Na⁺ and K⁺ channels

The Na⁺ and K⁺ channels show considerable structural homology, suggesting that they have a common ancestor. The Na⁺ channel is a protein with four transmembrane domains linked on the cytoplasmic side by extensive peptide loops and smaller loops on the outside. The K⁺ channel has four subunits. The four domains or subunits are arranged in the membrane to form the channel. Each of the four domains or subunits is comprised of six transmembrane α-helices, so the Na⁺ channel is a

very large protein. In each of the transmembrane domains, one of the transmembrane helices is a **voltage sensor** rich in basic residues which slightly changes its position in response to membrane potential changes and this is what causes channel opening and closing. However, we have pointed out above that the channels, opened by a depolarization of the membrane, are inactivated after about 1 millisecond. Although in this state they do not allow passage of ions, this does not apparently involve restoration of the original closed state until after the refractory period mentioned above. How does this inactivation occur?

In the case of the K⁺ channel, the evidence points to an ingenious 'ball and chain' mechanism as illustrated in Fig. 4.23. In this an N-terminal peptide (the ball) is tethered to the channel by an unstructured section of polypeptide such that it is free to move. It is believed that when the channel opens, the positively charged ball, probably attracted by a negative charge, fits into the channel and blocks it. From experiments with site directed mutagenesis by which the protein can be experimentally altered (described in Chapter 29) it seems that the shorter the tether the more quickly the channel is closed—the ball finds its place more rapidly. There is evidence that the Na⁺ channel may in principle have a similar inactivation mechanism, but in this case it is one of the cytoplasmic loops connecting two of the domains that is believed to be involved rather than a terminal peptide 'ball'.

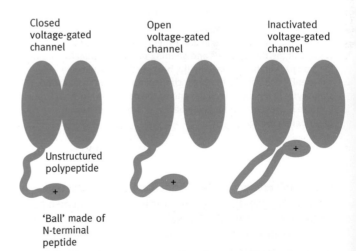

Closed voltage-gated channel

Open voltage-gated channel

Inactivated voltage-gated channel

Unstructured polypeptide

'Ball' made of N-terminal peptide

Fig. 4.23 The general principle of the 'ball and chain' inactivation of the voltage-gated K⁺ channels. Only two of the subunits forming the pore are shown. Restoration of the original closed state occurs after the inactivated state. The Na⁺ channels are believed to be inactivated by a mechanism similar in principle but the blocking peptide is a cytoplasmic loop of the protein connecting two transmembrane domains rather than an N-terminal 'ball'.

At the next synapse (or at neuromuscular junctions) the membrane depolarization activates voltage-gated Ca^{2+} channels. The resultant inflow of Ca^{2+} causes acetylcholine-filled vesicles to discharge the neurotransmitter from the synaptic membrane (see Fig. 4.17).

A central feature of the nerve impulse propagation mechanism is that the signal strength, the size of the action potentials, is maintained right along the lengths of axons. There is no modulation of the strength of a nerve impulse: it is all or nothing. If the initial signal is sufficient to trigger the initial depolarization, the impulse will travel. What can be modulated is the frequency of the impulses. The ingenious feature to note is that the signal strength is maintained by the electrochemical gradient due to the asymmetric distribution of ions across the neuronal membrane, generated by the Na^+/K^+ ATPase. In effect the signal is boosted every time an action potential is generated. Without this a signal would be dissipated in a very short distance. An undersea telephone cable has an analogous problem in that over the great distances an electrical signal gets progressively weaker; to cope with this, amplifiers are placed in the line to boost the signal at regular intervals. In axons the signal is boosted by the ion gradient of Na^+ and K^+ with the effective amplifiers located at molecular distances from each other. The analogy is imperfect, however, in that an electrical current flows through the length of the cable which is not true of the axon, as explained above.

In motor and other neurons there is yet another refinement. The axon, instead of being bare, is insulated by a myelin sheath (Fig. 4.24) which prevents ion movement across it. The myelin sheath is interrupted every couple of millimetres at places called the **nodes of Ranvier** and it is here that the voltage-gated channels are located. This means that the depolarization of the membrane occurs in defined places; the nerve impulse thus leaps rapidly from one node to the next resulting in much faster nerve conduction. In **multiple sclerosis** there is breakdown of the myelin sheath (see page 73 for structure of a main myelin component) which has serious effects on the propagation of the impulses and therefore effects the target voluntary muscles.

Why doesn't the Na^+/K^+ pump conflict with the propagation of action potentials?

We have explained (page 80) that the high internal concentration of K^+ and low Na^+ inside in cells is maintained by the Na^+/K^+ ATP-dependent pump. Since the action potential propagation depends on migration of these ions in and out of the neuron, it might be thought that the pump in continually restoring the balance would be confusing the mechanism. This does not occur because the ion movements in nerve conduction are very much faster than those caused by the pump—50 or more times faster. The Na^+/K^+ pump acts as a slow 'trickle charger', keeping the ion gradient topped up. The membrane potential derives from the relatively small number of ions near the membrane surfaces and the movements of ions in generating action potentials involve only a minute proportion of the ions in the bulk of the cells and extracellular medium. There is little concentration change involved. This can be illustrated by the fact that if you take a squid giant axon and inactivate the Na^+/K^+ ATPase by inhibiting ATP production, it is still possible to elicit 10^6 action potentials before the ion gradients are depleted.

Now for the next membrane function:

Signal transduction—a brief preliminary note

The dictionary defines transduction as 'carrying' or 'leading across'. The cells of an animal must work cooperatively according to physiological needs and instructions must therefore be sent to cells. Life is a chemical process and signals are sent in chemical form. Hormones released by glands and neurotransmitter substances released by nerve cell endings are examples.

One class of chemical signals, such as steroid hormones, are lipid soluble, and enter the cell directly by diffusion through the lipid bilayer (because of their nonpolar nature) so the membrane has no relevance to their action. The second type are water-soluble and do not directly enter the cell but bind to specific receptor proteins on the external membrane surface and in doing so deliver an 'instruction' to the cell. The combination of the chemical signal with the exterior of the membrane receptor results in molecular events inside the cell. This is the **signal**

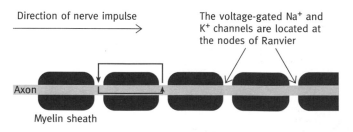

Direction of nerve impulse

The voltage-gated Na^+ and K^+ channels are located at the nodes of Ranvier

Axon

Myelin sheath

Fig. 4.24 Myelinated nerve found in motor and many other neurons. Breakdown of the insulating myelin sheath occurs in multiple sclerosis, and the resulting defect in transmission of action potentials causes the symptoms of this disease. The red arrows represent the local currents caused by depolarization at the nodes from one node to the next so that depolarization and triggering of action potentials leaps along the axon in a saltatory (jumping) fashion: this greatly speeds up the rate of transmission.

transduction role of the cell membrane. It is a very important area of biochemistry and is the subject of Chapter 26.

Role of the cell membrane in maintaining the shape of the cell and in cell mobility

Eukaryotic cells have an internal scaffolding which maintains the shape of the cell and is involved in amoeboid motility. It is called the **cytoskeleton** (the subject of Chapter 33). This is not a rigid inflexible structure like the bony skeleton of an animal but rather is composed of protein microfilaments which pervade the cytoplasm and, at various places, are attached to proteins of the cell membrane. Such membrane proteins are not then free to move laterally in the lipid bilayer as are other proteins.

An extreme example of the attachment of membrane proteins to the cytoskeleton is in the red blood cell which has special 'cell shape' proteins. The cell is a biconcave disc, which has the advantage of presenting a large surface area for gaseous exchange, but it is always on the move and therefore subject to shearing forces as it squeezes through capillaries, demanding a robust but flexible cell membrane. Underneath the membrane is a dense scaffolding of fibres of the protein, **spectrin**, anchored to the anion channel protein by a protein appropriately called **ankyrin**. The name spectrin comes from the fact that you can release the red cell contents but retain the empty membrane with its cytoskeleton, producing a red cell ghost (or spectre). The spectrin is linked by ankyrin to the anion channel protein described earlier. The latter is a large protein which projects into the cytoplasm providing a cytoskeleton attachment point (Fig. 4.25). Spectrin is also linked to glycophorin by other specific linking proteins.

In a small number of people the cytoskeleton of red blood cells is deficient because of faulty spectrin or ankyrin due to a genetic defect. The cells are abnormally shaped and tend to be destroyed by the spleen. The diseases are called **hereditary spherocytosis** or **hereditary elliptocytosis**.

Cell–cell interactions—tight junctions and gap junctions and cellular adhesive proteins

In an epithelial tissue such as the lining of the intestine, the products of food digestion are selectively taken up by the cells

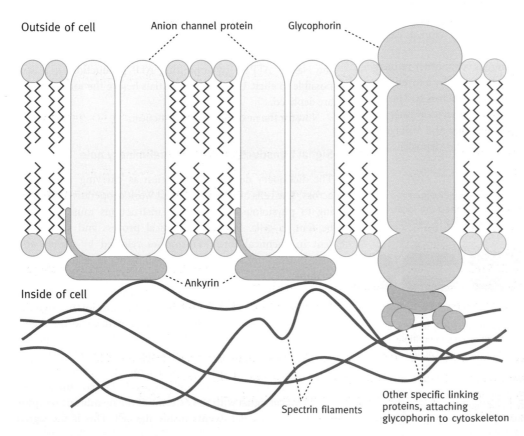

Outside of cell

Anion channel protein

Glycophorin

Ankyrin

Inside of cell

Spectrin filaments

Other specific linking proteins, attaching glycophorin to cytoskeleton

Fig. 4.25 Diagram to illustrate attachment of anion channel protein and glycophorin to the cytoskeleton.

and then transported from the cell by appropriate systems in the membrane of the opposite side of the cell facing the blood vessels. (More of this in the next chapter.) Special bands of membrane proteins encircle the cells and these bind to corresponding proteins of the neighbouring cells, creating 'tight' junctions—i.e. nonleaky cell–cell contacts. On the other hand, it can be desirable for adjacent cells to exchange molecules so as to coordinate the chemical activities throughout a tissue. This is achieved by special proteins forming a tunnel or hole between cells, known as **gap junctions**. Gap junctions will not allow large molecules such as proteins to pass, but molecules as large as ATP can do so. The coordination of heart cell contractions depends on gap junctions.

Another function of membrane proteins is to promote adhesion between cells of a tissue to form that tissue. If different types of embryo cells are mixed together they will re-associate with cells of their own type—kidney cells to kidney cells and so on. This is a function of tissue-specific cell–cell adhesion proteins called **cadherins**. Another family of proteins, the **N-CAMS** (for **n**erve **c**ell **a**dhesion **m**olecules), are important in nervous tissue formation.

FURTHER READING

Membrane lipids and movement

Higgins, C. F. (1994). Flip-flop: the transmembrane translocation of lipids. *Cell*, 79, 393–5.

Raises the general question of how lipids get to where they should be and discusses particularly how asymmetry of the lipid bilayer is achieved.

Rooney, S. A., Young, S. L., and Mendelson, C. R. (1994). Molecular and cellular processing of lung surfactant. *FASEB J.*, 8, 957–67.

Reviews a more physiological role of phospholipid—that of lining the alveoli.

Membrane proteins

Macdonald, C. (1985). Gap junctions and cell-cell communication. *Essays in Biochem.*, 21, 86–118.

A comprehensive review.

Von Heijne, G. (1994). Membrane proteins: from sequence to structure. *Ann. Rev. Biophys. Biomolec. Structure*, 23, 167–92.

A review tracing the steps from sequence to structure of integral membrane proteins.

Von Heijne, G. (1995). Membrane protein assembly: rules of the game. *BioEssays*, 17, 25–30.

Interesting discussion of how proteins can be 'stitched' into membranes with multiple transmembrane α helices.

Avery, J. (1999). Synaptic vesicle proteins. *Current Biology*, 9, R624.

A single-page quick guide.

Choe, S., Kreusch, A., and Pfoffinger, P. J. P. (1999). Towards the three-dimensional structure of voltage-gated potassium channels. *Trends Biochem. Sci.*, 24, 345–9.

Describes work on the structure and mechanism of action of the channels.

Ion channels

Neher, E. and Sakmann, B. (1992). The patch clamp technique. *Sci. Amer.*, 266(3), 44–51.

Enables studies to be made on ligand-gated and voltage-gated ion channels.

Changeux, J.-P. (1993). Chemical signalling in the brain. *Sci. Amer.*, 269(5), 30–7.

The acetylcholine receptor and how it works.

Catterall, W. A. (1995). Structure and function of voltage-gated ion channels. *Ann. Rev. Biochem.*, **64**, 493–531.

Covers sodium, calcium, and potassium channels.

PROBLEMS FOR CHAPTER 4

1   What structural characteristics of amphipathic molecules favours liposome (lipid bilayer) formation, rather than micelle formation?

2   What is the role of cholesterol in eukaryotic membranes?

3   Give the structure of phosphatidic acid.

4   Name three glycerophospholipids based on phosphatidic acid. Name the substituent in each case attached to the latter.

5   What is the significance of *cis*-unsaturated fatty acyl tails in membrane phospholipids?

6   Why don't inorganic ions readily pass through membranes?

7   What is meant by facilitated diffusion through a membrane? Give an example.

8   Compare the structure of a triacylglycerol and a polar lipid. Why can the former never be used in lipid bilayers?

9   Explain the rationale behind the structure of sphingosine.

10   Explain how an ion gradient across a cell membrane can be harnessed to provide the energy for the active transport of an unrelated organic molecule.

11   What is a gated ion channel? Name two different types.

12   Digitalis is used to strengthen the heartbeat in patients with congestive heart failure. What is the mechanism of this?

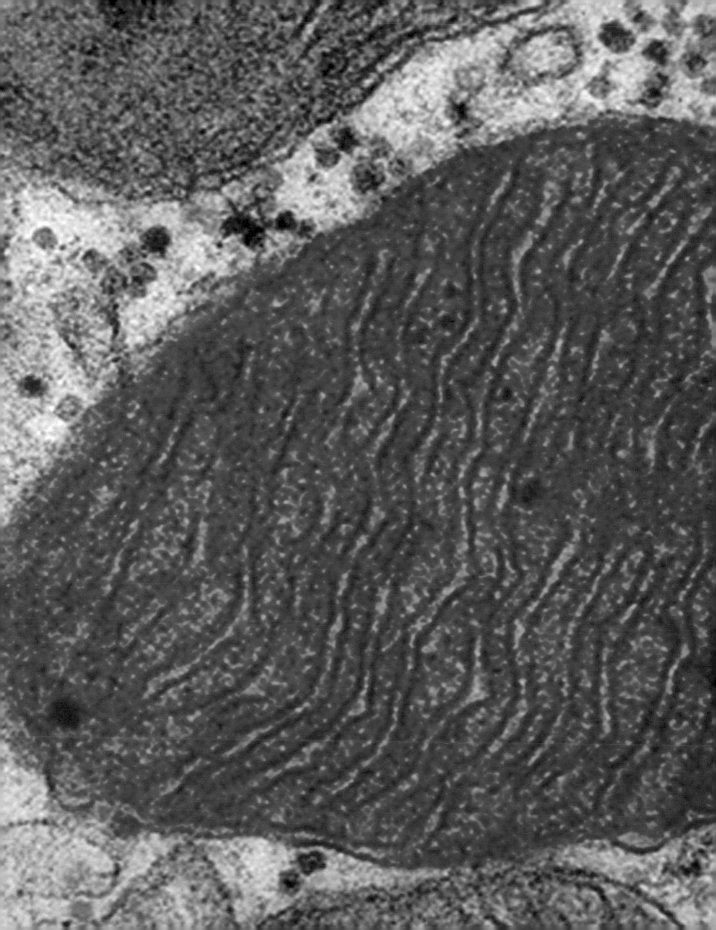

Metabolism

Chapter summary

Previous pages:
Colorized electron micrograph of a heart muscle mitochondrion, showing cristae.

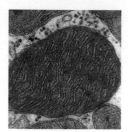

Digestion and absorption of food

First, where does this topic fit in the overall arrangement of the book? In Chapter 1 we looked at what is necessary for chemical reactions to occur and at the chemistry of weak bonds. Next, we dealt with the properties of enzymes which bring about the chemical reactions (Chapter 2), then in Chapter 3, the structure of protein molecules laid the basis for the key reactions and processes on which life depends. Chapter 4 dealt with the cell membranes because this information is essential for understanding digestion and absorption. We now progress to the major area of metabolism. Food digestion and absorption is a logical place at which to start, for that's where metabolism begins. At a gross level, the subject may seem mundane; at the molecular level it is as elegant and interesting as are the topics with a more exciting image.

In this chapter we will discuss what food is in chemical terms, how it is prepared for absorption into the bloodstream, and how it reaches the bloodstream. What happens to the food once it gets there will be taken up in the next chapter.

Chemistry of foodstuffs

There are three main classes of food—proteins, carbohydrates, and fats.

- **Proteins** are, as described in Chapter 3, large polymers of 20 species of amino acids linked together to form polypeptide chains.
- **Carbohydrates** are the sugars and their derivatives; the name comes from their empirical formula with carbon atoms and the elements of water in the ratio of 1:1 (CH_2O).

Although simple monosaccharide sugars such as glucose occur in food, most of the carbohydrate is in the form of disaccharides such as sucrose or polysaccharides such as starch.

- **Fats** in the diet are mainly in the form of neutral fat or triacylglycerols (page 70). Butter, olive oil, and the visible fat of meat are typical examples. Polar lipids are also present.

Digestion and absorption

With the exception of simple free sugars (monosaccharides) such as glucose, all of the foodstuffs mentioned above must be digested by hydrolysis into their constituent parts in the intestine. To be absorbed, substances must enter the epithelial cells lining the gut and this involves crossing the cell membranes. Triacylglycerols cannot do this, nor can proteins, and amongst the carbohydrates only monosaccharides can be absorbed. Thus, digestion consists mainly of the conversions

Proteins → amino acids;

Carbohydrates → sugar monomers (monosaccharides);

Neutral fats → fatty acids and monoacylglycerol.

Anatomy of the digestive tract

The following regions are involved.

- In the **mouth** food is masticated and lubricated for swallowing; limited starch digestion occurs.
- The **stomach** contains HCl which 'sterilizes' food and denatures proteins; partial digestion of proteins occurs.

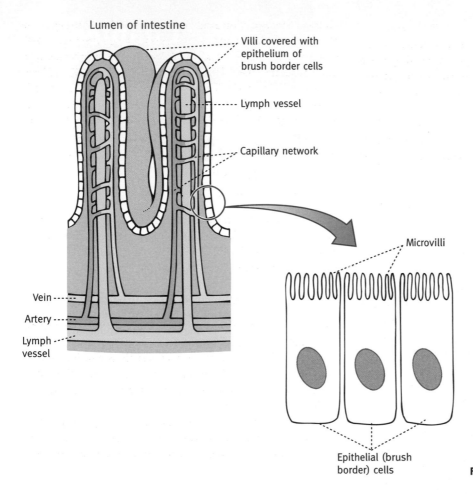

Fig. 5.1 The surface of the small intestine.

- The **small intestine** is the major site of digestion and absorption of all classes of food. It is lined by fine finger-like processes, the villi which are covered by epithelial cells—known as brush border cells because the microvilli of the epithelial cells resemble bristles on a brush (Fig. 5.1). The microvilli are on the external membrane of the epithelial cells of the villi and such an arrangement, the combination of the villi and the microvilli, gives the enormous surface area needed for absorption.

- The **large intestine** is mainly involved in the removal of water.

What are the energy considerations in digestion and absorption?

So far as digestion goes, there are no thermodynamic problems. Hydrolytic reactions such as hydrolysis of proteins to amino acids, of di- and polysaccharides to monosaccharides, and of fats to fatty acids and monoacylglycerols are all exergonic processes—they have negative ΔG values sufficient to push the equilibrium entirely to the side of hydrolysis. Hydrolytic reactions (splitting or lysis by water) in biochemistry are invariably of this type. Absorption is a different matter since it is often an active process in which molecules are absorbed against a concentration gradient and energy is needed.

A major problem in digestion—why doesn't the body digest itself?

Food is, chemically, little different from the tissues of the animal that eats it. A fearsome array of enzymes is produced by the digestive system to completely digest the food into its smaller components, and those enzymes have to be produced inside living cells, which, if exposed to their action, would be destroyed. There are two major types of defence against this.

Zymogen or proenzyme production

This refers to the production of enzymes as inactive **proenzymes** or **zymogens** which are activated only when they reach the intestine. Thus, glands producing digestive enzymes secrete most of them as inactive proteins and are never themselves exposed to the destructive processes. In the case of enzymes that pose no threat (amylase, the enzyme that hydrolyses starch, is one), proenzymes are not involved. The question also arises as to how cells selectively secrete digestive enzymes, but this is best left until we deal with protein targeting (Chapter 25) for the mechanism is complex and not directly related to digestion. We will deal with the mechanisms of zymogen activation when we come to particular enzymes.

Protection of intestinal epithelial cells by mucus

The cells lining the intestinal tract are protected from the action of activated digestive enzymes, the main line of defence being the layer of mucus that covers the entire epithelial lining of the gut. The essential components of mucus are the **mucins**. These are extremely large glycoprotein molecules—proteins with large amounts of carbohydrates attached to their polypeptide chains in the form of oligosaccharides that contain a mixture of sugars you've already met in membrane glycoproteins (pages 55 and 79)—glucosamine, fucose, sialic acid, etc. The mucins form a network of fibres, interacting by noncovalent bonds and resulting in a gel containing more than 90% water that protects intestinal cells. The carbohydrates may protect the mucin proteins themselves from digestion. The mucins are synthesized and secreted by special goblet cells in the epithelial lining of the gut. The amount of mucin secreted is controlled.

Digestion of proteins

In a normal or 'native' protein, the polypeptide chain is folded up, the shape being largely determined by weak bond interactions. In this compact folded form, many of the peptide bonds are hidden away inside the molecule where they are not accessible to hydrolytic enzymes. An important step is to **denature** the native proteins. This is done by acid in the stomach where, due to HCl secretion, the pH is about 2.0. This disrupts many weak bonds and the polypeptide folding is disrupted, making the polypeptide chain very susceptible to **proteolysis**, as the hydrolytic process is called. The HCl also kills microorganisms.

HCl production in the stomach

The stomach epithelial lining contains **parietal** or **oxyntic cells** that secrete acid. In essence, the process consists of ejecting H⁺

against a concentration gradient—this is similar to the ejection of Na^+ from cells against a concentration gradient (page 80) which is achieved by a Na^+/K^+ ATPase in the membrane. In a similar manner, the acid-secreting cells have a H^+/K^+ ATPase. Using energy from ATP hydrolysis, they eject H^+ and import K^+, the latter recycling back to the exterior of the cell. The process is shown in Fig. 5.2. Where do the protons come from? CO_2 enters the cell from the blood. An enzyme, **carbonic anhydrase**, converts this to carbonic acid, which dissociates as shown.

$$CO_2 + H_2O \rightleftharpoons H_2CO_3 \rightleftharpoons H^+ + HCO_3^-$$
$$\text{Carbonic anhydrase}$$

The bicarbonate ions exchange via an anion transport protein with Cl^- in the blood (Fig. 5.2), the latter moving to the stomach lumen to form HCl. We have already met this type of exchange in the anion channel of the red blood cell.

Pepsin, the proteolytic enzyme of the stomach

A note on nomenclature: enzymes have names usually ending in 'ase', but many digestive enzymes were amongst the earliest to be discovered and were given names such as pepsin, trypsin, chymotrypsin. To indicate an inactive precursor, the suffix 'ogen' is used; for example, pepsinogen is capable of generating, or giving rise to, **pepsin**.

Cells of the stomach epithelium secrete pepsinogen. The secretion is stimulated by the hormone gastrin released by stomach cells into the blood in response to food entering the

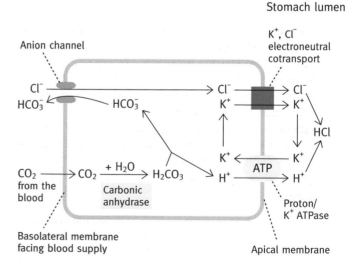

Fig. 5.2 Mechanism of gastric HCl secretion.

stomach. Pepsinogen is pepsin with an extra stretch of 44 amino acids added to the polypeptide chain. This additional segment covers, and blocks, the active site of the enzyme. When pepsinogen meets HCl in the stomach, a slight change in conformation exposes the catalytic site which self-cleaves—it cuts off the extra peptide from itself to form active pepsin. As soon as a small amount of active pepsin is produced in this way, it rapidly converts the rest of the secreted pepsinogen to pepsin. The activation is therefore autocatalytic.

Pepsin is unusual for an enzyme in that it works optimally at acid pH (Fig. 2.6(a)), most enzymes requiring a pH near neutrality. It hydrolyses peptide bonds of proteins within the molecule producing a mixture of peptides. It is therefore an **endoenzyme** or **endopeptidase**; that is, it does not attack terminal peptide bonds at the end of molecules, but only those within the molecule. The derivation is Greek, not English; *endo* = within.

Proteolytic enzymes (proteases) are usually relatively specific in their action—they hydrolyse only peptide bonds adjacent to certain amino acids. Pepsin has this characteristic so that only partial digestion of proteins can occur in the stomach. It hydrolyses only those peptide bonds in which one of the aromatic amino acids—tyrosine, phenylalanine, or tryptophan (page 36)—supplies the NH of the peptide bond.

Infants take most of their protein in the form of milk. Another stomach enzyme, rennin, acts on the casein of milk, causing clotting, so that it doesn't pass too rapidly through the stomach and escape gastric digestion.

Completion of protein digestion in the small intestine

The chyme, as the partially digested stomach contents are called, enters the duodenum at the start of the small intestine. The acid stimulates the duodenum to release hormones (secretin and cholecystokinin) into the blood, which stimulate the pancreas to release pancreatic juice. This is alkaline and (together with bile juice) neutralizes the HCl giving a slightly alkaline pH suitable for pancreatic enzyme action and terminating pepsin activity.

A battery of pancreatic proteases is produced from clusters of cells in the pancreas and secreted into the intestine as pancreatic juice via the main pancreatic duct, which is formed by the joining together of smaller ducts draining the cell clusters. There are three endopeptidases—**trypsin**, **chymotrypsin**, and **elastase**—all entering the intestine in the form of the inactive proenzymes, trypsinogen, chymotrypsinogen, and proelastase, respectively. An exopeptidase, **carboxypeptidase**, secreted as an inactive proenzyme, chops off terminal amino acids from the C-terminal end of peptides.

$$H_2N-CH-CO-NH-CH-CO-NH-----NH-CH-COOH$$

$$\quad\quad\;\; R \quad\quad\quad\quad\;\; R' \quad\quad\quad\quad\quad\quad\;\; R''$$

Amino terminal end or N-terminal end

Carboxy terminal end or C-terminal end

The details of how these enzymes work has been given in Chapter 3.

Activation of the pancreatic proenzymes

As with pepsinogen activation, pancreatic proenzymes are activated by proteolytic cleavage of the proenzymes. The activation process is autocatalytic and is triggered off by an enzyme produced by the cells of the small intestine, an active enzyme (not a proenzyme in this case) called **enteropeptidase**, which hydrolyses a single peptide bond of trypsinogen thus activating it to trypsin. The initial amount of active trypsin so produced now activates all proenzymes (including trypsinogen itself) so that all are rapidly activated in an autocatalytic cascade (Fig. 5.3). The activation details differ for different enzymes. This is an elegant mechanism for ensuring that the fearsome array of *active* enzymes is present only in the intestinal lumen. Their premature activation in the pancreas or even the pancreatic duct is deleterious—if it occurs, the disease pancreatitis ensues. Blockage of the pancreatic duct, or damage to the gland, can trigger this disease. After synthesis, the proenzymes are stored in the cells producing them in membrane-bound secretory vesicles. On hormonal or neurological stimulation these fuse with the cell membrane and release their contents by exocytosis (this is described in detail later). Presumably as a safeguard against

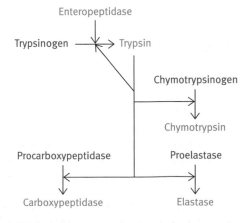

Fig. 5.3 Activation of the pancreatic proteolytic enzymes. Proenzymes (zymogens) are shown in red; activated proteolytic enzymes are shown in green.

accidental leakage of the vesicles before secretion, the cells contain a trypsin-inhibitor protein capable of inactivating any trypsin that might accidentally come in contact with the cytoplasm. The inhibitor protein fits the trypsin active site so perfectly that a sufficient number of weak bonds are formed to make the combination of the two almost completely irreversible. The mechanism by which enzymes to be secreted are produced and enveloped in the secretory vesicles is more suitably dealt with later (page 409).

An additional enzyme, called **aminopeptidase** (an **exopeptidase**), attached to the outside (lumen side) of intestinal cells hydrolyses off terminal amino acids from the amino ends of peptides to be digested. Thus the three endopeptidases (trypsin, chymotrypsin, and elastase), each with a different preference for the peptide bonds they attack, chop away in the middle of polypeptides, while carboxypeptidase and aminopeptidase chop away at each end, and proteins are completely converted to free amino acids in the lumen of the intestine.

Absorption of amino acids into the bloodstream

Amino acids are transported from the intestine across the cell membrane of epithelial cells (the brush border cells; Fig. 5.1) and into the cell. The brush border cells actively concentrate amino acids inside themselves from where they diffuse into the blood capillaries inside the villi. Because there are so many different amino acids, there are several different transport pumps in the membrane. The cotransport mechanism has already been described in Chapter 4 in which the Na^+ gradient is used to drive the uptake of some of the amino acids. Figure 4.12 shows glucose being transported by this mechanism but it applies equally well to amino acids using a different transport protein.

We will take up the subject of what happens to the amino acids in the blood in the next chapter.

Digestion of carbohydrates

The main carbohydrates in the diet are starch and other polysaccharides, and the disaccharides sucrose and lactose (the latter from milk). Free glucose and fructose also occur. As stated, only free sugars or monosaccharides are absorbed from the intestine so the aim of digestion is to hydrolyse the dietary components to such monosaccharides.

Sugars are linked together by **glycosidic bonds** to form glycosides. This is an important bond from several viewpoints and needs to be described. Glucose has the following structure.

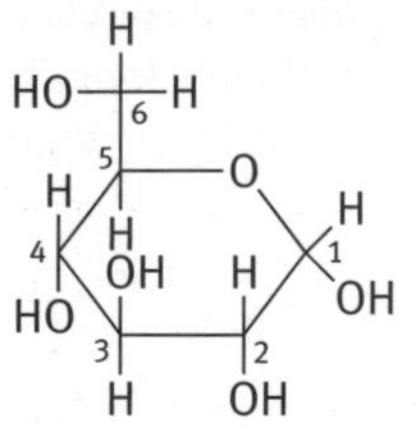

α-D-glucose 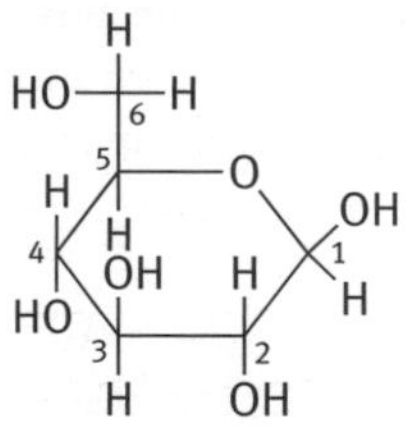   β-D-glucose

In α-D-**glucose** the —OH on carbon atom 1 points below the plane of the ring, and in β-D-**glucose** upwards. In free sugars, the two are in equilibrium in solution by a **mutarotation** (via an open-chain structure). Suppose we have two glucose molecules.

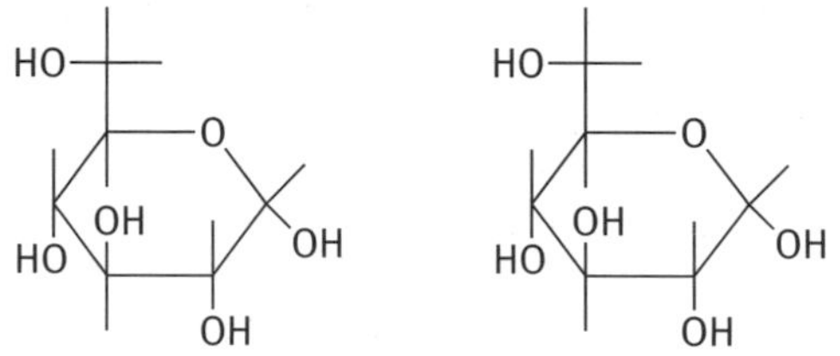

They can (by appropriate chemistry) be joined together by a glycosidic bond.

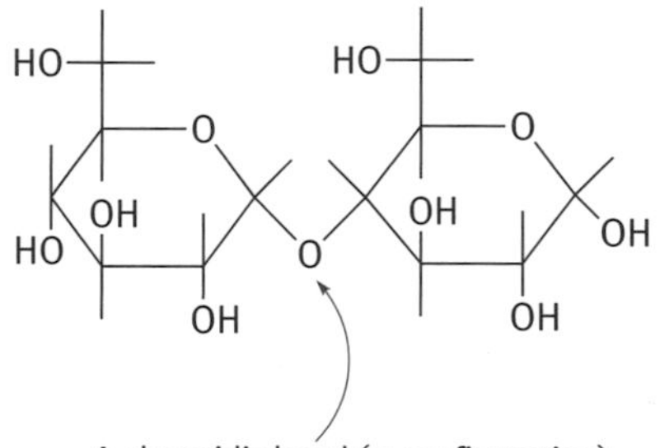

A glycosidic bond (α-configuration)

The glycosidic bond fixes the configuration on carbon atom 1 of the first sugar into one form—it no longer mutarotates. In the example shown, the glycosidic bond is between carbon atoms 1 and 4 of the two units and is in α-configuration. The compound is therefore glucose-$\alpha(1 \rightarrow 4)$-glucose. It is a disaccharide plentiful in malted barley and has the trivial name **maltose**. As will be described shortly, disaccharides with β-glycosidic bonds also exist.

Digestion of starch

From the structure of maltose above it can be seen that glucose units can be joined together indefinitely to form a huge

polysaccharide molecule. The **amylose** component of starch, a molecule hundreds of glucose units (glucosyl units) long, is precisely this. If we represent an α-glucosyl unit as ⌲ then amylose is

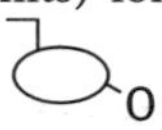

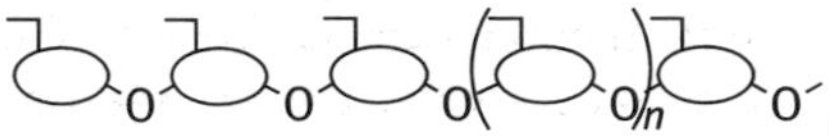

⌲ , etc.

where n is a large number.

The second component of starch is **amylopectin**. This molecule also is a huge polymer of glucose but instead of it being one long chain it consists of many short chains (each about 30 glucosyl units in length) crosslinked together. The glucosyl units of the chains are $1 \rightarrow 4$-α-glycosidic bonds but the crosslinking between chains is by $1 \rightarrow 6$-α-glycosidic bonds, as shown below. (We omit bonds and groups not relevant to the topic in hand.)

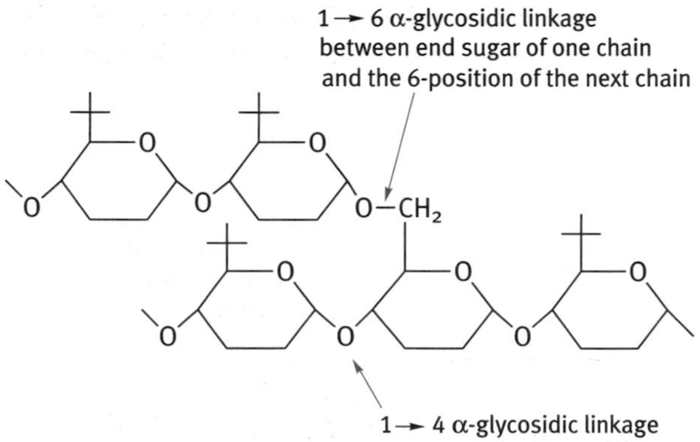

$1 \rightarrow 6$ α-glycosidic linkage between end sugar of one chain and the 6-position of the next chain

O–CH$_2$

$1 \rightarrow 4$ α-glycosidic linkage

Many chains are linked together in this way. Amylopectin is therefore like this.

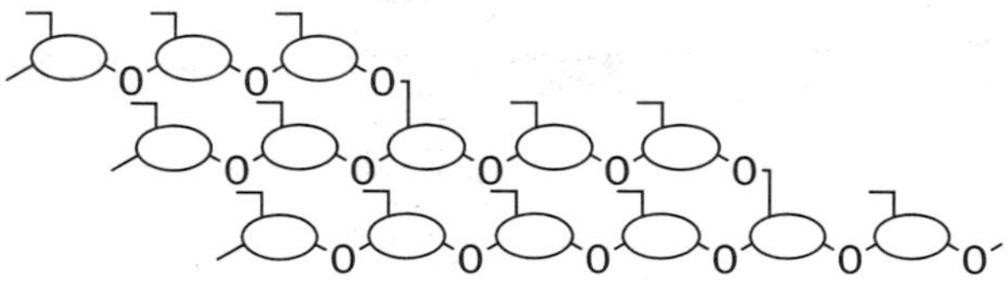

etc.

Digestion of starch is done by α-**amylase**, which is present in the saliva and pancreatic juice. It is an **endoenzyme**, hydrolysing glycosidic bonds anywhere inside the molecule except that, in amylopectin, it can't attack a bond near the $1 \rightarrow 6$ linkage. The amylase thus trims off the molecule but leaves a core containing the $1 \rightarrow 6$ bonds plus the nearby glucosyl units. This limit dextrin, as it is called (the limit of hydrolysis), is hydrolysed by an intestinal enzyme (an amylo α-$1 \rightarrow 6$ glucosidase) that hydrolyses the $1 \rightarrow 6$ linkages. The 2 and 3 unit sugars produced

by amylase digestion is attacked by intestinal α-**glucosidase** (**maltase**) which produces free glucose. Salivary amylase has only a brief time to attack starch while food is in the mouth, for stomach acid destroys the enzyme. Most of the starch digestion therefore occurs in the intestine. The digestion of the other two main dietary carbohydrates, lactose and sucrose, also occurs in the small intestine.

Lactose is galactose-$\beta\,(1 \rightarrow 4)$ glucose, the principal sugar in milk. Galactose has the carbon 4-OH of glucose inverted.

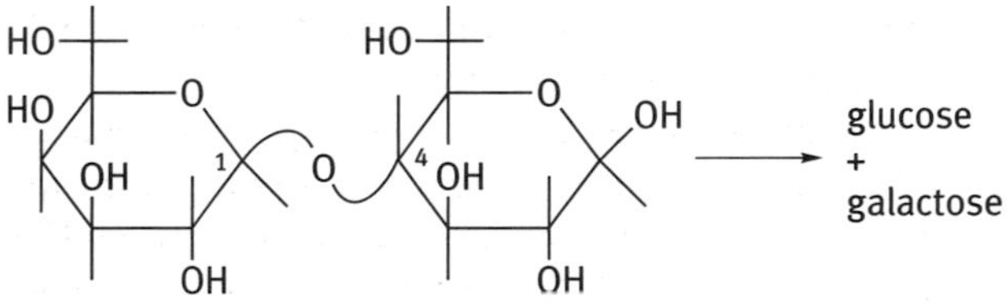

HO, HO, OH, OH, OH $\longrightarrow$ glucose + galactose

The curly bond shown is normally used to simplify the presentation of the structure. An intestinal enzyme attached to the external membrane of the epithelial cells, **lactase**, hydrolyses lactose to the free sugars. Because lactose is a β-galactoside, lactase is also known as β-**galactosidase**. Many adults, particularly Asians, lose the ability to produce lactase and develop milk intolerance as they leave childhood. They are unable to hydrolyse the sugar and, since the disaccharide is not absorbed, it passes to the large intestine where it is fermented by bacteria giving rise to severe discomfort, water intake, and diarrhoea. The other major disaccharide of food, sucrose, is digested by the intestinal enzyme **sucrase**, liberating glucose and fructose which are absorbed.

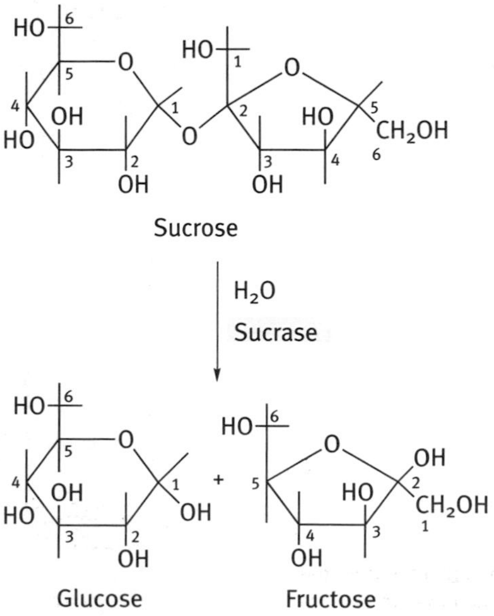

Sucrose

H_2O
Sucrase

Glucose   Fructose

Absorption of glucose into epithelial cells occurs by the same Na^+ cotransport mechanism described earlier (page 82) as shown in Fig. 5.4. The Na^+ is continually pumped to the outside by the Na^+/K^+ ATPase so this maintains the Na^+ concentration difference and drives the cotransport. The glucose has to exit the cells on the side opposite that of the lumen so as to enter the blood capillaries for transport to the liver by the hepatic portal vein. The glucose transporter for this movement is the facilitated diffusion type; the sugar moves down the cell/blood concentration gradient created by active uptake from the intestine (Fig. 5.4). Fructose is absorbed from the gut by a Na^+-independent passive transport system.

Amino acids and sugars and other non-lipid molecules absorbed from the intestine are collected by the portal blood system which delivers digestion products directly to the liver. This arrangement means that these digestion products pass through the liver before being released into general circulation—an advantageous arrangement since the liver is responsible for removing most toxic foreign compounds entering the body via the intestine, and for processing some of the absorbed nutrients.

Digestion and absorption of fat

In Chapter 4 we described what fat is—neutral fat, or triacylglycerol (TAG). Since there are no polar groups, liquid fat in water forms droplets with minimum contact between the lipid and water. The TAG molecules cannot be absorbed as such.

The main digestion of fat occurs in the small intestine by the action of the pancreatic enzyme, **lipase**, which mainly attacks *primary* ester bonds; the middle ester bond is a secondary ester bond and is not significantly attacked.

Simple as this digestion reaction is, there are problems. Fat is physically unwieldy in an aqueous medium. Lipase can attack only at the oil/water interface and the area of this in a simple fat/water mixture is insufficient for the required rate of digestion. The available surface area is increased by emulsifying the fat. The monoacylglycerol and free fatty acids, produced by lipase action together with bile salts (see below), help to disperse the oily liquid into droplets (**emulsification**) and

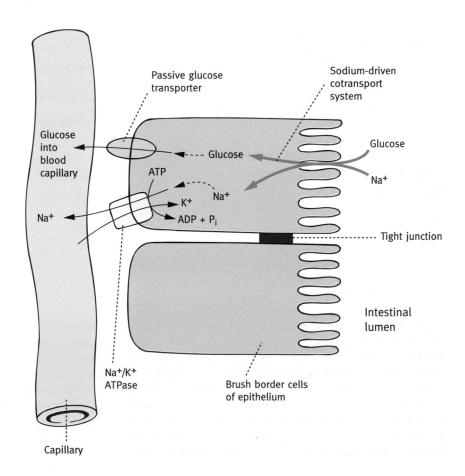

Fig. 5.4 Absorption of glucose from the intestinal lumen by cotransport with Na^+.

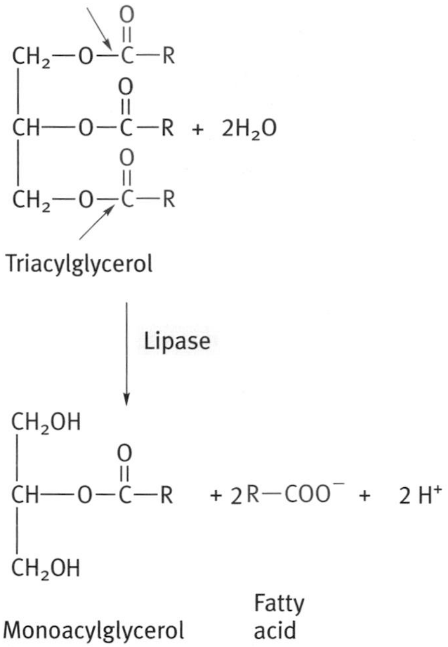

Triacylglycerol

$\downarrow$ Lipase

Monoacylglycerol   Fatty acid

phospholipids in the food also help this process. The products of lipase action are relatively insoluble in water but they must be moved from the emulsion to the cells lining the intestine to be absorbed. This movement is facilitated by bile acids, or bile salts as they are interchangeably called.

Bile acids are produced in the liver and stored in the gall bladder until discharged into the duodenum. They are produced from cholesterol, whose structure they resemble. The main change is that the hydrophobic side chain of the cholesterol molecule is converted to a carboxyl group, and —OH groups may be added.

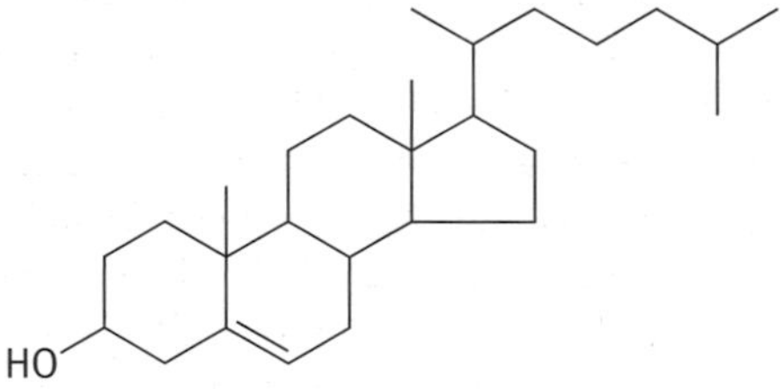

Cholesterol

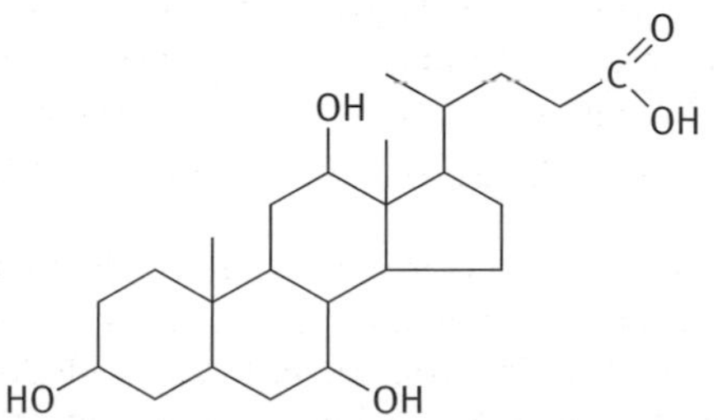

Cholic acid

The main bile acid is cholic acid; others varying in the number and position of hydroxyl groups exist. The cholic acid mostly has attached to it either glycine or the sulfonic acid, taurine. Glycine is $NH_3^+CH_2COO^-$. Taurine is $NH_3^+CH_2CH_2SO_3^-$ (it does *not* occur in proteins). If cholic acid is represented as $RCOO^-$ then glycocholic acid is $RCONHCH_2COO^-$ and taurocholic acid $RCONHCH_2SO_3^-$. These conjugated acids have lower pK_a values (approximately 3.7 and 1.5, respectively) than the parent cholic acid (approximately 5.0). Possibly the reason for the conjugation is to ensure full ionization in the intestinal contents.

The monoacylglycerols and fatty acids, in the presence of bile salts, form **mixed micelles**. These are disc-like particles in which bile salts are arranged around the edge of the disc surrounding a more hydrophobic core containing digestion products of fat, together with cholesterol and phospholipids. The particles are much smaller than emulsion droplets, giving a clear solution, and carry higher concentrations of lipid digestion products than is possible in simple solution. In this form, the lipid digestion products diffuse to enter the epithelial cells; possibly the micelle breaks down at the cell surface and the lipids diffuse in. Bile salts are also partially reabsorbed and transported back to the liver.

What happens to the fatty acids and monoacylglycerol absorbed into the intestinal brush border cells?

Resynthesis of neutral fat in the absorptive cell

The products of digestion of fat absorbed into the intestinal cell are resynthesized into fat, the fatty acids being re-esterified producing triacylglycerols.

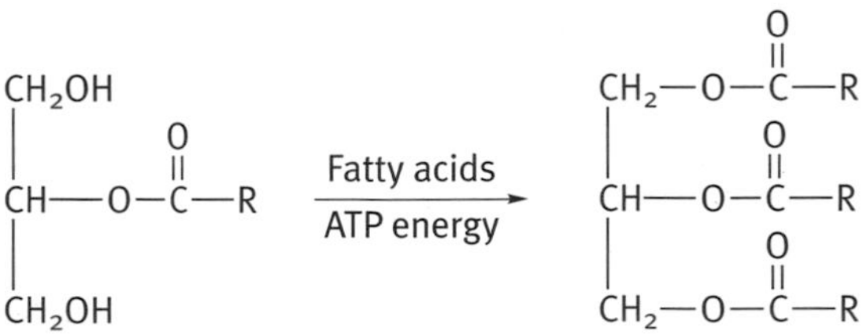

The mechanism of this resynthesis is not given here because it would divert too much from the topic in hand—however, in Chapter 11 (page 192) on fat metabolism, the matter is dealt with. The neutral fat and cholesterol must now be transported from the intestinal cell to the tissues of the body. Neutral fat cannot diffuse out of the cell through membranes and, in any case, it would be insoluble in the blood—it cannot simply be ejected into the circulation.

The solution is that neutral fat and cholesterol are arranged inside the epithelial cells into fine particles, called **chylomi-**

crons. They are enclosed in membrane vesicles and released from the cell by exocytosis, since they are too large to traverse the membrane in any other way.

What are chylomicrons?

Chylomicrons (Fig. 5.5) are spherical particles, in the middle of which are hydrophobic molecules—neutral fat and cholesterol esters. Cholesterol has a hydrophilic —OH group and a hydrophobic part; a **cholesterol ester** has a fatty acid attached to that —OH group.

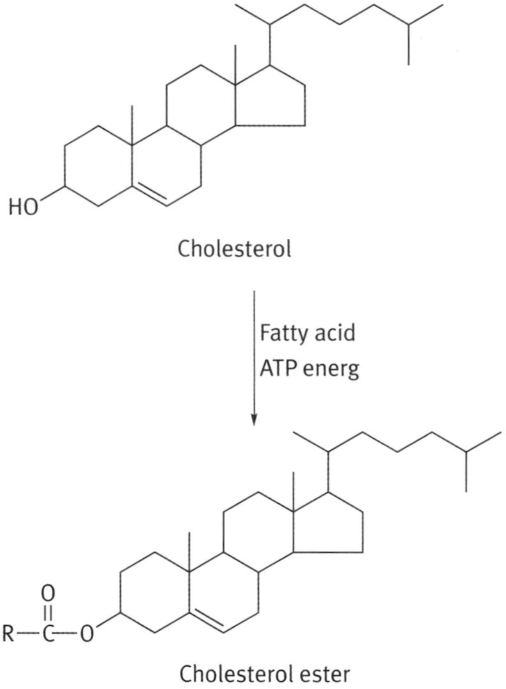

Cholesterol

Fatty acid
ATP energ

Cholesterol ester

This is a simplified diagram of esterification of cholesterol. The enzyme (called ACAT, or acyl CoA: cholesterol acyl transferase) requires coenzyme A (not yet described).

In making a cholesterol ester from cholesterol, the polar —OH group, which would interfere with packing cholesterol into the hydrophobic centre of chylomicrons, is eliminated. It illustrates the importance that the polarity-hydrophobicity characteristics of biological molecules play in life. Hydrophobic particles of neutral fat and cholesterol ester would coalesce into an insoluble mass unless they were stabilized. For this, there is a 'shell' containing phospholipids, some free cholesterol which is weakly amphipathic, and, very importantly, some special proteins.

There are several different proteins involved, a major one being **apolipoprotein B (apoB)**, necessary for chylomicron

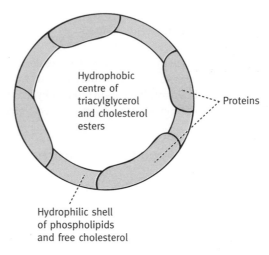

F g. 5.5 Cross-section of a chylomicron.

synthesis. ApoB is a glycoprotein, the carbohydrate attachment providing a highly polar group. The prefix *apo* mean 'detached' or 'separate'. Thus an apolipoprotein is a protein normally in lipoproteins but now is detached or separate from the lipoprotein. The whole chylomicron tructure is called a lipoprotein. The hydrophilic shell stabilizes the chylomicron so that it remains as a suspended particle in the lymph chyle (see below) and blood. The overall composition of chylomicrons is about 90% or more neutral fat, giving them a low density.

The chylomicrons are not released directly into the blood (unlike absorbed amino acids and sugars) but into the lymph vessels, called lacteals, that drain the villi. The suspension of chylomicrons in lymph fluid is called chyle. A few words on lymph might be useful here. As blood circulates through the capillaries, a clear lymph fluid containing protein, electrolytes, and other solutes filters out to bathe cells in an interstitial fluid. All tissues have a fine network of lymph capillaries closed at the fine ends into which lymph drains. The lymph capillaries join into lymphatic ducts and discharge the lymph back into the major veins of the neck via the thoracic duct. It is not actively pumped but movements of the body propel it along. After a fatty meal, chylomicrons, entering the bloodstream via lymph, give the blood a milky appearance. The circulate in the blood and their contents are used by tissues, the subject of the next chapter.

We want to temporarily end the story here—we have the food components, amino acids, sugars, and fat, in the blood where they are now utilized by cells. It is convenient to deal with the next phase in the next chapter. Meanwhile, we would like to finish the topic of digestion with a few notes.

Digestion of other components of food

There are components in food other than the ones dealt with so far and digestive enzymes exist to hydrolyse them into their constituent parts. Thus phospholipids in food are hydrolysed by phospholipases; nucleic acids (which you'll meet later) are hydrolysed by ribonuclease (RNase) and deoxyribonuclease (DNase) also in the pancreatic juice. Plant material contains carbohydrate molecules such as cellulose, for which animals do not produce digestive enzymes (herbivores have microorganisms in the rumen which do produce them). In humans, these components constitute the fibre of the diet.

FURTHER READING

Digestion and absorption of fats

Riddihough, G. (1993). Picture of an enzyme at work. *Nature*, **362**, 793.

A news and views summary of the fascinating way pancreatic lipase is activated in the digestion of fats.

PROBLEMS FOR CHAPTER 5

1   Which digestive enzymes are produced in an inactive zymogen form? Why should this be? Why do you think amylase is not produced as an inactive zymogen?

2   Explain how the inactive proteases are activated in the intestine.

3   If pancreatic proenzymes are prematurely activated due to physical and chemical damage or pancreatic duct blockage, what is the result?

4   Why is digestion of foods necessary?

5   Many people, especially Asians, suffer severe intestinal distress on consuming milk or milk products. Why is this so?

6   Write down the structure of a neutral fat. What is the alternative name for this? Indicate which groups in the molecule are primary esters.

7   Fat in water forms large globules with little fat/water interface for lipase to attack. Explain how the digestive system copes with the physical intractability of lipid.

8   Amino acids and sugars absorbed into the intestinal cells move into the portal bloodstream and are carried via the liver and thence to the rest of the body. What happens to the absorbed digestion products of fat?

Chapter summary

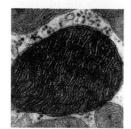

Preliminary outline of fuel distribution and utilization by the different tissues of the body

In the previous chapter we dealt with food digestion and absorption from the intestine and left the various products of digestion just having reached the blood—fat and cholesterol into the general circulation via the lymph system as chylomicrons, and everything else headed for the liver in the portal vein and thence into the general circulation. With this chapter we start on the subject of metabolism; if you have already heard of metabolic pathways such as glycolysis, the citric acid cycle, etc. we suggest you put them out of your mind for the time being.

We will, first, deal with the traffic of food molecules between the blood and tissues and between different tissues, and with how this is regulated to satisfy different physiological needs—in essence, the logistics of fuel movements in the body. In this chapter we do not give any details of the metabolites involved—the chapter is designed to give a broad picture of the situations. The details of metabolic pathways covered in subsequent chapters will make more sense through this arrangement.

Purposes of metabolism

The purposes of metabolism can be summed up in these terms.

- The major role of metabolism is to oxidize food to provide energy in the form of ATP.
- Food molecules are converted to new cellular material and essential components.
- Waste products are processed to facilitate their excretion in the urine.

- Some specialized cells in human babies oxidize food to generate heat. This is exceptional since heat is, in general, a byproduct of metabolism.

Storage of food in the body

Animals take in food at periodic intervals; they do not eat continuously. The body is, in effect, continually exposed to cyclical feast and famine conditions. The periods between meals can for humans vary from short intervals during the day to longer periods during sleep and very long periods during starvation. The biochemical machinery in the body copes with these situations.

After a meal, the blood becomes loaded with food absorbed from the intestine. To state the obvious, this food isn't held in the blood until it is all used up by metabolism, but rather it is rapidly cleared from the blood by uptake into the tissues so that blood levels quickly return to normal. After a fatty meal, the lipemia (presence of milky blood plasma) is cleared within a few hours. Similarly, blood glucose may increase after a meal from its normal $90\,mg\,dl^{-1}$ to $140\,mg\,dl^{-1}$ in an hour (5 and 7.8 mM, respectively) but it reverts back to normal, in a non-diabetic person, in 2 hours. The tissues are not using all of this food at once; most of it is stored.

How are the different foods stored in cells?

Glucose storage as glycogen

It would not be practicable for cells to store glucose as the free sugar—the osmotic pressure of the sugar at high

concentrations would be too high. The osmotic pressure of a solution is proportional to the number of particles of solute in the solution. If, therefore, large numbers of glucose molecules are joined together to form a single macromolecule the osmotic pressure, exerted by the store of glucose residues, is accordingly reduced. The resultant large polymerized molecule may fall out of solution as a granule. In animals, glucose is polymerized into a highly branched so-called 'animal starch', or glycogen, with the same chemical bondings as in amylopectin (page 102), but more highly branched. Its synthesis requires energy. When needed, glycogen is broken down again into glucose. A very important fact is that *glycogen storage in animals is extremely limited*. In humans, the glycogen reserves of liver, which are used to supply glucose to the blood for utilization by other tissues, are exhausted after about 24 hours of starvation. This fact has a profound effect on the biochemistry of an animal, as will become apparent later in the chapter.

In all of the above, we have talked about glucose. What about other sugars such as galactose and fructose, present in lactose and sucrose respectively? Glucose is the sugar of central metabolic importance; other sugars are converted to glucose (or glycogen), or else to compounds on the main glucose-metabolizing pathways.

Storage of fat in the body

Fat is stored by cells as neutral fat or triacylglycerols (page 70). The bulk of it is stored in adipose tissue or fat cells, which are distributed in many areas of the body and which specialize in fat storage. A loaded fat cell can, under the microscope, look like droplets of oil surrounded by a thin layer of cytoplasm and membrane as shown in Fig. 6.1.

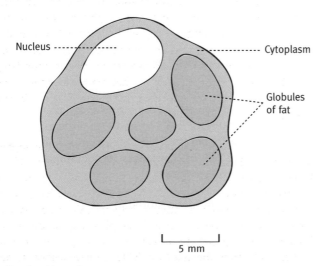

Fig. 6.1 A fat cell.

Unlike glycogen storage, that of fat is essentially unlimited and large reserves of energy can be stored in the body as neutral fat in the adipose cells. In a typical human the reserves of fat, in terms of stored energy, are about 50 times those of the glycogen reserves.

Since this limited glycogen storage causes the metabolic problems described later, it may be asked why the body stores so much fat, and so little glucose, when the latter is essential to life in animals? Fat is more highly reduced (less oxidized) than is carbohydrate and hence contains more energy per unit mass, particularly since glycogen in the cell is hydrated while fat is not. This means that the latter occupies much less volume per unit of stored potential energy than does glycogen. If the energy equivalent of fat stored in our bodies was in the form of glycogen we'd need to be much larger than our existing sizes.

Since glucose storage is so limited and sugar intake in the diet potentially almost unlimited, it follows that glucose in excess of that used for glycogen synthesis must be stored in another way—in fact as fat. The position then is as shown in Fig. 6.2.

An important fact is that, while glucose is readily converted to fat, *in the human body there is no net conversion of fatty acids to glucose*. (The glycerol moiety of neutral fat *can* be converted to glucose but this represents a small fraction of the energy available in the three long-chain fatty acids.)

Are amino acids stored by the body?

Digestion of the third main class of food, protein, results in amino acids being absorbed and taken into the blood, but there is no dedicated storage form of amino acids in animals if we exclude egg and milk proteins which don't negate the general concept. Plant seeds have storage proteins whose only function appears to be storage of amino acids in a convenient form to supply the developing embryo. In animals then, dietary amino acids in the blood are taken up by tissues as needed for the synthesis of cellular proteins, neurotransmitters, etc., and then those in excess of immediate needs are destroyed, the amino group ($-NH_2$) being removed and, in mammals, converted into urea and excreted in the urine. Urea is highly water-soluble,

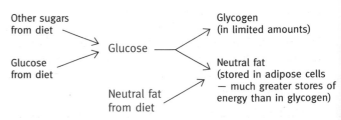

Fig. 6.2 Storage of sugars and fat from the diet in the post-absorptive period.

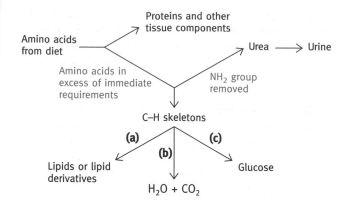

Fig. 6.3 Summary of the fate of amino acids in the diet. Which of the metabolic routes **(a)**, **(b)**, or **(c)** is followed depends on the particular amino acid, the physiological state, and biochemical control mechanisms.

neutral, and nontoxic, even in relatively high concentration. This leaves the carbon-hydrogen 'skeletons' of the amino acids, which contain chemical energy. There are converted to glycogen or fat, depending on the particular amino acid, or they can be oxidized (Fig. 6.3), according to the physiological needs at the time.

In one sense, there is a storage of amino acids in the body—in the proteins of all cells. Muscle proteins are quantitatively of greatest importance. However, they are all functional proteins mainly involved in contraction and are not dedicated to storage. When amino acids must be made available to the body in general, these proteins are broken down and the result is muscle wasting. You will shortly see in what situation this becomes essential.

Characteristics of different tissues in terms of energy metabolism

The different tissues in the body have special biochemical characteristics. However, in our present context of looking at overall food traffic in the body, several organs are of overriding importance. These are the liver, the skeletal muscles, the brain, the adipose cells, and the red blood cells. (We are excluding regulatory organs such as the pancreas and adrenal glands from the list though their roles are crucial.)

● The **liver** has a central role in maintaining the blood glucose level; it is the glucostat of the body. When blood glucose is high, such as after a meal, the liver takes it up and stores it as glycogen.

When blood glucose is low, it breaks down the glycogen and releases glucose into the blood.

In starvation, after about 24 hours, the reserves of glycogen in the liver are exhausted by this activity; the concentration of glucose in the blood would decline to lethal levels if nothing were done about it. During starvation, there will, for a prolonged period, be reserves of fat that are released into the blood by the fat cells of adipose tissue. This will keep the muscles and other tissues happy, but not the brain and red blood cells, which must have glucose for they cannot use fatty acids. The liver saves the situation by converting amino acids to glucose, a process called **gluconeogenesis**. The main sources of amino acids in this situation are muscle proteins, which are broken down to provide them. This means destruction of functional muscle proteins, causing muscle wasting, but this is preferable to death resulting from low blood glucose.

The liver also plays an important role in fat metabolism. In starvation, the fat cells release fatty acids into the blood that can be directly used by most tissues—brain and red blood cells excepted, as stated. However, in such conditions, where massive fat utilization is occurring, the liver converts some of the fatty acids to small compounds called ketone bodies and releases them into the blood for use by other tissues (red blood cells again excepted). Of special importance, the brain can adapt to utilize ketone bodies which, during starvation, can supply about half of that organ's energy needs. The rest must come from glucose. For the moment, ketone bodies might be regarded as fatty acids partially metabolized by the liver. The liver is also the major site of fatty acid synthesis, which is exported to adipose cells and other tissues (as TAG).

So, energy-wise, the liver stores glucose as glycogen and releases glucose when needed, as long as there are glycogen reserves left. In starvation it produces glucose from amino acids and ketone bodies from fat, which are turned out into the blood for use by other tissues (Fig. 6.4). It synthesizes fatty acids from glucose and exports them. It preferentially oxidizes fatty acids for energy supplies. The liver has many other functions but for now we are concentrating on energy supplies in the body.

● The **brain** must, as already stated, have a continuous supply of glucose in the blood. It has no significant fuel reserves. If the blood becomes hypoglycemic (low glucose level), the rate of glucose entry into the brain is reduced since it involves passive transport and neuron function is impaired. Convulsions and coma result. However, as already emphasized, although the brain *must* have glucose, in starvation it can adapt to use ketone bodies produced by the liver from fat for perhaps half its energy needs. This helps to economize on scarce glucose.

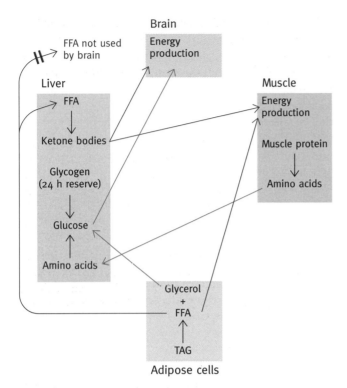

Fig. 6.4 Main fuel movements in the body during starvation. TAG, triacylglycerols; FFA, free fatty acids.

• **Skeletal muscles** require large-scale energy production for muscle contraction. They can utilize most types of energy sources. They take up glucose from the blood and store it as glycogen but *in contrast to the liver, they do not release any glucose back into the blood.* They can oxidize fatty acids, ketone bodies, and amino acids as well as glucose. In the presence of a high level of free fatty acids and ketone bodies in the blood, these are preferentially oxidized. In starvation, muscles sacrifice their protein to supply the liver with amino acids for glucose synthesis.

• The role of the **adipose cells** can be stated simply. After a meal, they take up fatty acids, which they store as neutral fat, and take up glucose and convert it also to stored neutral fat. As soon as the blood glucose retreats as in mild fasting (or in starvation) they release reserves of fat into the blood as fatty acids for the rest of the body to use. Adipose cells and liver are the major sites of triacylglycerol formation, the fatty acids coming mainly from the diet and from synthesis in the liver. Glucose is needed to provide the glycerol moiety.

• **Red blood cells** can utilize only glucose, producing lactate. They are terminally differentiated cells (that is, fully formed and

don't divide) without mitochondria and cannot oxidize foodstuffs as can other cells. But their Na^+/K^+ ATPase pumps must run and there are other energy-requiring processes.

Overall control of the logistics of food distribution in the body by hormones

With that description of the major organs involved in food logistics in the body, let us now see in overview how food distribution is directed—how the various organs are coordinated in their activities to cope with different physiological situations. These situations are as follows.

• Excess food available—the postprandial condition after a good meal.

• Mild fasting—this refers to the condition between meals—for example, before breakfast after the night's fasting.

• Starvation—fasting that goes on for longer than about 1 or 2 days.

• Abnormal conditions—for example, in diabetes, glucose cannot be adequately utilized so that, irrespective of its nutritional state, the body is biochemically starved of carbohydrate.

• Emergency responses—this relates to conditions, for example, where violent muscular action is needed to avoid a threat.

The overall picture of food traffic in the body in the various nutritional situations is summarized below.

Post-absorptive phase

All food components are at a high level in blood. Glucose is taken up by the liver, muscle, and adipose cells and used to replenish glycogen stores. Beyond this, excess glucose is taken up by the adipose tissue and by the liver and other tissues and converted into neutral fat. The liver does not retain this in large amounts as do the adipose cells but transfers it to other tissues. A 'fatty liver' is pathological.

Amino acids are taken up by all tissues and used to synthesize protein and other cell components. Fat is taken from chylomicrons and absorbed by tissues including adipose tissue, which stores it, and by the lactating mammary glands, which utilize it in milk production. While active storage of glycogen and fat is going on, the breakdown of these components in the cells is stopped.

The main 'signal' for all of this storage work to take place is

the pancreatic hormone, **insulin**, whose release from the pancreas into the blood is signalled by high blood sugar levels. The high insulin level is the signal to the tissues that times are good and that food should be stored both as glycogen and fat. Correlated with this is the *low* level of **glucagon**, another pancreatic hormone, whose release is *inhibited* by high blood sugar levels. Glucagon has the opposite effect to insulin. It signals that the tissues are short of fuel and that storage organs should release some into the blood. Note that the brain is not affected by insulin; it goes on using glucose from the blood all the time.

Fasting condition

The insulin-stimulated storage activity in the post-absorptive period lowers the blood glucose and amino acids to normal levels and clears the chylomicrons. With time, the blood glucose level begins to fall and, with it, insulin secretion. The hormone is rapidly destroyed after release, so its blood level rapidly falls when secretion stops. In concert with this, the pancreas releases glucagon, whose level rises in response to low blood glucose. This stimulates the liver to break down glycogen and release glucose into the blood. It also stimulates adipose tissue to hydrolyse neutral fat and release free fatty acids into the blood. Muscles and liver, and other tissues, use the fatty acids to provide energy and this conserves glucose for the brain. Glycogen synthesis and fat synthesis from glucose cease due to the low insulin/high glucagon ratio.

Prolonged fasting or starvation

After about 24 hours the liver has no glycogen left to provide blood sugar and no other tissue, other than the quantitatively unimportant kidney, is capable of releasing glucose. The insulin level is low and that of glucagon high. In this situation the adipose cells pour out free fatty acids into the blood, for use by muscle and other tissues; the fat reserves in adipose cells are sufficient perhaps for weeks. Muscles break down their proteins to amino acids, which the liver uses to synthesize glucose, a process activated by glucagon. As starvation proceeds, the high glucagon causes the adipose tissue to turn out more and more free fatty acids into the blood and the liver metabolizes some of this into **ketone bodies** which it turns out into the blood. Ketone bodies can be regarded in general terms as 'predigested' or partly metabolized fatty acids. There are two components, acetoacetate and β-hydroxybutyrate.

$CH_3COCH_2COO^-$ Acetoacetate

$CH_3CHOHCH_2COO^-$ β-Hydroxybutyrate

The name 'ketone bodies' is a misnomer; they are not 'bodies', nor is β-hydroxybutyrate a ketone, but the term is an old one. Once regarded as occurring only in pathological situations, production of ketone bodies is recognized to be an important normal process.

The situation in diabetes

If a patient lacks insulin, even in the presence of abundant glucose, the sugar is not utilized by muscles. The uptake of glucose by the brain and red blood cells is not dependent on the presence of insulin and so can survive in the diabetic state. Thus, the diabetic state resembles that of extreme starvation resulting in the adipose cells turning out large amounts of fatty acids into the blood because the insulin/glucagon ratio is obviously very low. The liver, in this situation, produces ketone bodies in large amounts. Acetoacetate, a component of ketone bodies, spontaneously decarboxylates to acetone, which is not metabolized, giving the characteristic sickly sweet smell of the breath of untreated juvenile-onset diabetics.

$$CH_3COCH_2COO^- \rightarrow CH_3COCH_3 + CO_2$$
acetoacetate   acetone

There are two types of diabetes. Type 1, the juvenile onset form of the disease, is due to the failure of the pancreas to secrete insulin. This type responds to insulin injection. In type 2, also known as maturity onset diabetes, there is adequate insulin, which may be present at greater than normal levels but the insulin responsive cells (adipocytes, muscle, and liver) do not respond in the normal way, which may be due to lack of receptors on the cell membrane. (Hormone receptors are dealt with on page 428). This disease is not usually characterized by ketoacidosis. The condition is described more fully later—see page 229.

The emergency situation—fight or flight

When an animal such as a human is presented with a dangerous situation, the adrenal glands release **epinephrine (adrenalin)** into the blood in response to a neurological signal from the brain. Adipose tissue is innervated and can also be stimulated by epinephrine and norepinephrine released from nerve endings. Epinephrine, as it were, presses the biochemical panic button. It over-rides normal control and stimulates the liver to pour out glucose into the blood and the adipose cells to release free fatty acids so that muscles have no shortage of fuel. It also stimulates glycogen breakdown in skeletal muscle so that the cells produce ATP, at the maximal rate. The pattern has an overall logic of ensuring that muscles can react violently and instantly to escape the threat.

PROBLEMS FOR CHAPTER 6

1   Why do cells of the body store glucose as glycogen? Why not save trouble and store it as glucose?

2   Triacyglycerol (TAG) is stored in much greater quantities in the body than is glycogen. why is this so?

3   Can glucose be converted to fat? Can fatty acids be converted in a net sense to glucose? Do amino acids have a special dedicated long-term storage form in the body? Give brief explanations.

4   In terms of food logistics, describe the chief metabolic characteristics of the liver.

5   Can the brain use fatty acids for energy generation?

6   What are the chief metabolic characteristics of fat cells?

7   What is the main energy source used by red blood cells?

8   What are the main hormonal controls on the logistics of food movement in the body in different nutritive states?

9   At one time ketone bodies were regarded as entirely abnormal. Discuss this briefly.

Chapter summary

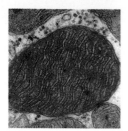

Biochemical mechanisms involved in food transport, storage, and mobilization

In this chapter we deal with the biochemical mechanisms involved in fuel transport, storage, and release processes described in Chapter 6.

There are two aspects to the biochemistry of metabolic pathways; first there are the pathways themselves and secondly the regulation of the pathways. In this book we have collected the regulation of the main pathways of fat and carbohydrate metabolism into a separate chapter (Chapter 13) for several reasons. The pathways themselves are complicated enough on their own and understanding metabolic regulation requires a considerable amount of prior information which we can give in the separate chapter. Finally the regulation of different pathways are integrated together and there are advantages in dealing with them together. However, if it is preferred to study regulation along with the pathways, the appropriate sections can easily be found in Chapter 13.

Glucose traffic in the body

Mechanism of glycogen synthesis

The synthesis of glycogen from glucose is an **endergonic process**—it requires energy. Hydrolysis of the glycosidic bond joining glucose units in glycogen has a $\Delta G^{0'}$ value of about $16 \, \text{kJ mol}^{-1}$ which would give an equilibrium totally to the side of glucose for the simple reaction: glycogen plus $H_2O \rightarrow$ glucose. Therefore energy must be supplied from high-energy phosphoryl groups to form the glycosidic bond.

Glycogen synthesis occurs by enlarging pre-existing glycogen molecules (called 'primers') by the sequential addition of glucose units. The synthesis of a glycogen granule is initially primed by the protein, **glycogenin**, which transfers eight glucosyl residues to a tyrosine–OH on itself. The protein remains inside the glycogen granule. The new units are always added to the nonreducing end of the polysaccharide. Glucose is an aldose sugar. This is seen clearly in the open-chain form which is in spontaneous equilibrium in solution with the ring form, the latter predominating.

Glucose (ring form) ⇌ Glucose (open-chain form)

An aldehyde is a reducing agent, so that the carbon 1 end of the ring sugar is the reducing end and carbon 4 the nonreducing end. The glycogen chain thus always has a nonreducing end.

Non-reducing end of chain

(The glucose unit is represented as ⬡.)

The process of glycogen synthesis in essence therefore is as follows.

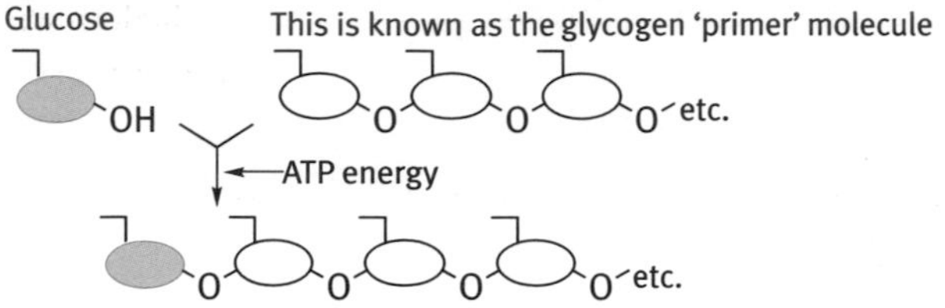

Glucose   This is known as the glycogen 'primer' molecule

—ATP energy

The process is repeated over and over again, elongating the polysaccharide chain.

How is energy injected into the process?

When glucose enters the cell it is phosphorylated by ATP. The reaction is catalysed in brain and muscle by the enzyme hexokinase. A **kinase** transfers a phosphoryl group from ATP to something else—in this case, glucose, which is a hexose sugar—hence the name **hexokinase**. In liver, a different enzyme catalysing the same reaction is called **glucokinase**.

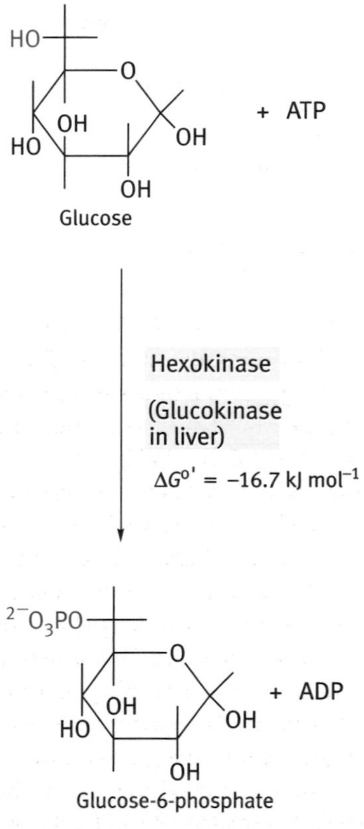

Glucose

+ ATP

Hexokinase

(Glucokinase in liver)

$\Delta G^{o'} = -16.7$ kJ mol^{-1}

+ ADP

Glucose-6-phosphate

The ΔG of this reaction is strongly negative, making the reaction irreversible. The highly charged glucose-6-phosphate molecule is unable to traverse the cell membrane so that glucose phosophorylation has the effect of trapping the sugar inside the cell and facilitating entry of more glucose from the blood, since in facilitated diffusion (by which glucose enters) it can move only down a concentration gradient. The phosphoryl group is now switched around to the 1 position by a second enzyme, **phosphoglucomutase**, a freely reversible reaction.

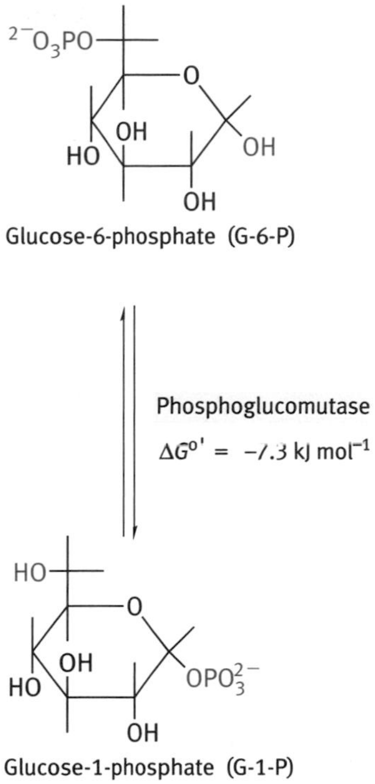

Glucose-6-phosphate (G-6-P)

Phosphoglucomutase

$\Delta G^{o'} = -7.3$ kJ mol^{-1}

Glucose-1-phosphate (G-1-P)

The enzyme is called phosphoglucomutase because it mutates (changes) phosphoglucose.

The free energy of the phosphoryl ester group of glucose-1-phosphate (G-1-P; $\Delta G^{o'}$ for its hydrolysis $= -21.0$ kJ mol^{-1}) is about the same as that of the glycosidic bond in glycogen ($\Delta G^{o'}$ for hydrolysis of the same $\alpha\,(1 \rightarrow 4)$-bond of maltose is -15.5 kJ mol^{-1}). You might expect that the glucose unit would be transferred from G-1-P to a glycogen primer and the synthesis would be done. It was thought years ago that this was what happened. But it's not so. We explained (page 5) that biochemical pathways are usually arranged to be thermodynamically irreversible—if necessary, energy is pumped in at the expense of—Ⓟ groups and the process given such a large negative ΔG value that the pathway is one-way only. If glycogen synthesis occurred directly from G-1-P, the process would be freely reversible and hence uncontrollable.

An extra step is inserted that renders glycogen synthesis thermodynamically irreversible. This extra step is somewhat curious. To understand it, first go back to the structure of ATP (Fig. 1.2). A similar compound, uridine triphosphate (UTP), exists in which adenine is replaced by uracil (U) whose structure doesn't matter in this context (but is given on page 289 where it is relevant). (Uridine is uracil-ribose, a structure analogous to adenosine.) The cell synthesizes UTP (the energy coming from

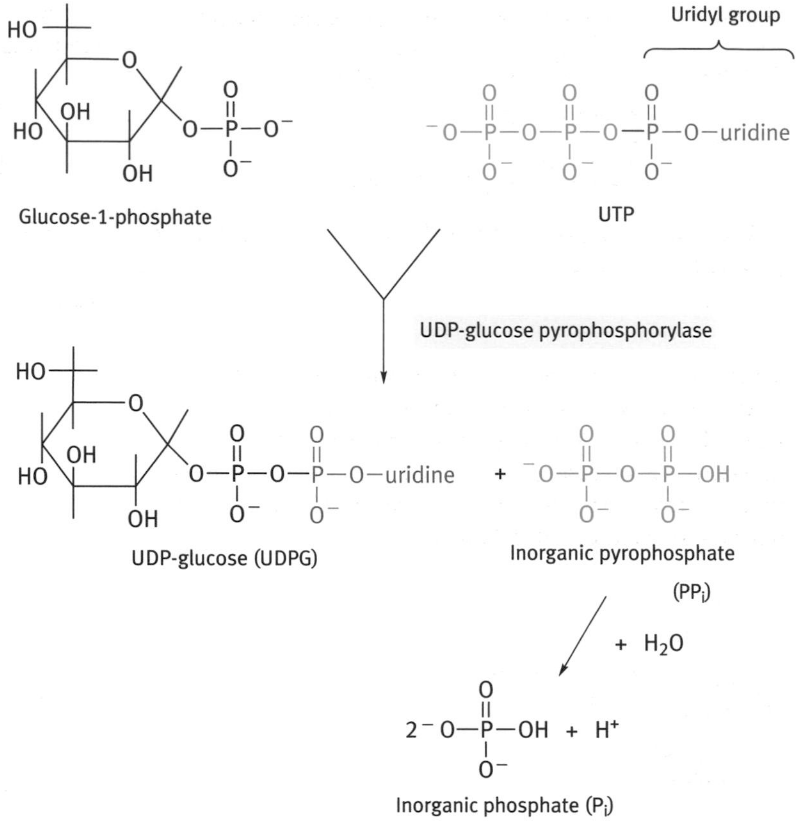

Uridyl group

Glucose-1-phosphate

UTP

UDP-glucose pyrophosphorylase

UDP-glucose (UDPG)

+ O—P—O—P—OH

Inorganic pyrophosphate
(PP$_i$)

+ H$_2$O

$2\ ^-O—P—OH\ +\ H^+$

Inorganic phosphate (P$_i$)

Fig. 7.1 Formation of UDP-glucose by a reaction between UTP and glucose-1-phosphate catalysed by UDP-glucose pyrophosphorylase. See text for an explanation of the name of this enzyme.

ATP). In glycogen synthesis, uridine diphosphoglucose (UDP-glucose) is made by a reaction between UTP and G-1-P as shown in Fig. 7.1.

This enzyme, for systematic nomenclature reasons, is named for the reverse reaction (which doesn't occur in the cell). This reaction (were it to occur) would cleave UDPG with pyrophosphate. The enzyme is therefore a pyrophosphorylase and is called **UDP-glucose pyrophosphorylase**; inorganic pyrophosphate is not available in the cell because it is rapidly destroyed and therefore the reverse reaction is inoperative.

UDPG is the 'activated' or reactive glucose compound that donates its glucosyl group to glycogen. Wherever in animals an 'activated' sugar has to be produced for a chemical synthesis, as a general rule it is a uridine diphosphate derivative. Whether there is a fundamental reason for this or whether it is an accidental quirk that evolution was locked into isn't known. (Plants use the corresponding adenine compound for starch synthesis.) Synthesis of glycogen using UDPG as glucosyl donor occurs as shown in Fig. 7.2. The enzyme that does this is called **glycogen synthase** ('synthase' is used wherever a synthesizing enzyme does *not* directly use ATP and 'synthetase' where ATP is involved). The UDPG pyrophosphorylase produces inorganic pyrophosphate. This is hydrolysed by inorganic pyrophosphatase,

which, as explained on page 9, has a $\Delta G^{o\prime}$ value of $-33.5\,\mathrm{kJ\,mol^{-1}}$, and thus pulls the process of UDPG synthesis from G-1-P and UTP completely to the right. Glycogen synthesis in the cell is not therefore directly reversible by the route of its synthesis. The synthesis is summarized in the scheme shown in Fig. 7.3.

There's one more point about glycogen synthesis. If more and more glucosyl groups were added to primer molecules, the result would be very long polysaccharide chains—which is precisely what plants synthesize as the amylose starch component. Glycogen is different—instead of consisting of long chains of glucosyl units, it is a highly branched molecule. This structure is achieved by another enzyme, called the **branching enzyme**. When the glycogen synthase has made a 'straight chain' extension more than 11 units in length, the branching enzyme transfers a block of about seven terminal glucosyl units from the end of an $\alpha(1 \rightarrow 4)$-linked chain to the C(6)—OH of a glucose unit on the same or another chain (Fig. 7.4). The energy in the $\alpha(1 \rightarrow 6)$ link is about the same as in the $(1 \rightarrow 4)$ link so the reaction is a simple transfer one. It creates more and more ends both for glycogen synthesis and breakdown (see page 22).

Thus, to return to the starting point, blood glucose after a meal is taken up by tissues and converted into glycogen by the reactions outlined.

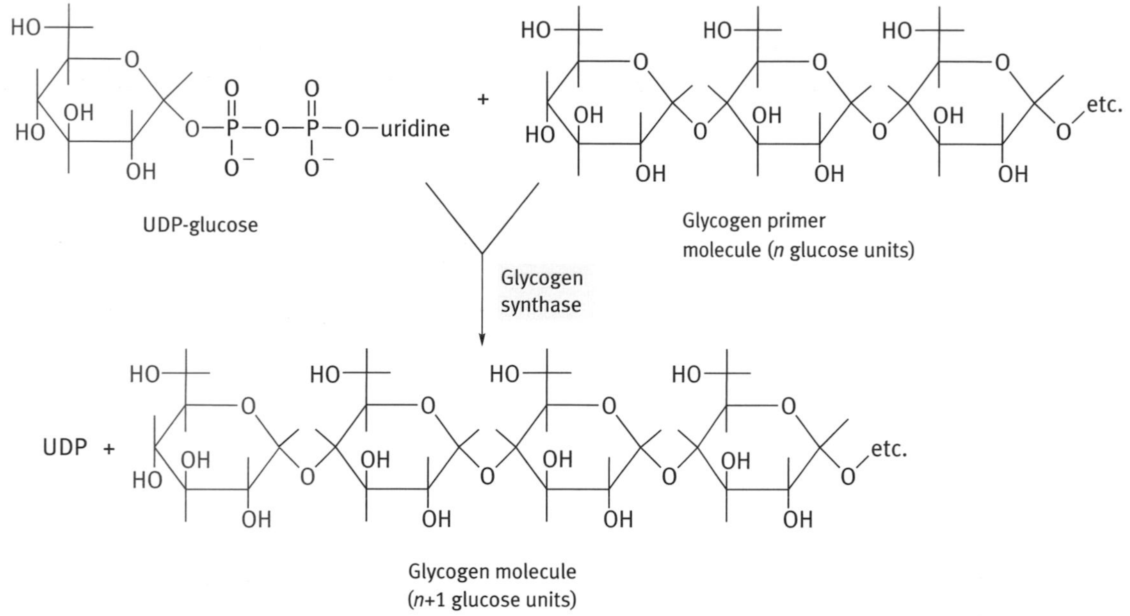

Fig. 7.2 Synthesis of glycogen by the elongation of a glycogen primer molecule using UDP-glucose (UDPG) as glucosyl donor.

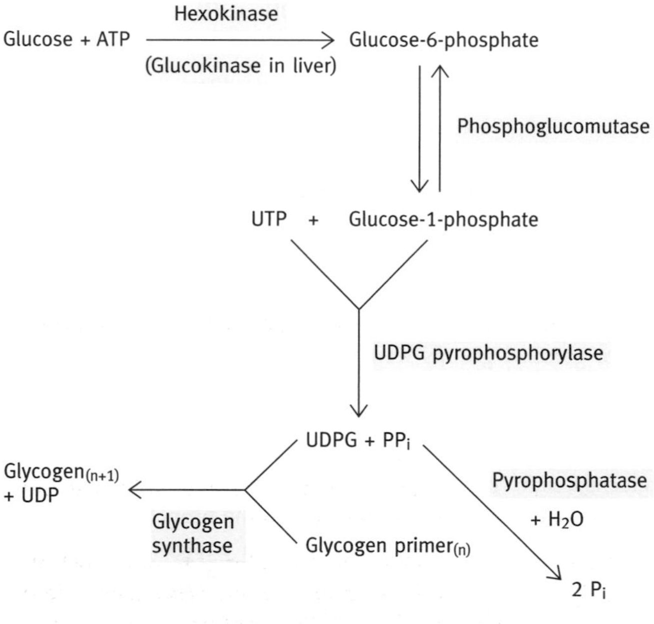

Fig. 7.3 Summary of glycogen synthesis from glucose.

How does the liver release glucose?

The liver, you'll remember, stores glycogen, not so much for it-self but for other tissues and especially the brain and red blood cells. In fasting, such as between meals, the liver breaks down glycogen (high glucagon/low insulin are the signals for this) and releases glucose to keep up the blood glucose level, for otherwise the brain would cease to function normally.

As with synthesis, **glycogen breakdown** occurs at the non-reducing end. Instead of splitting off (lysing) glucose by water (*hydro*lysis) is splits it off with phosphate (*phosphoro*lysis). The enzyme is called **glycogen phosphorylase**.

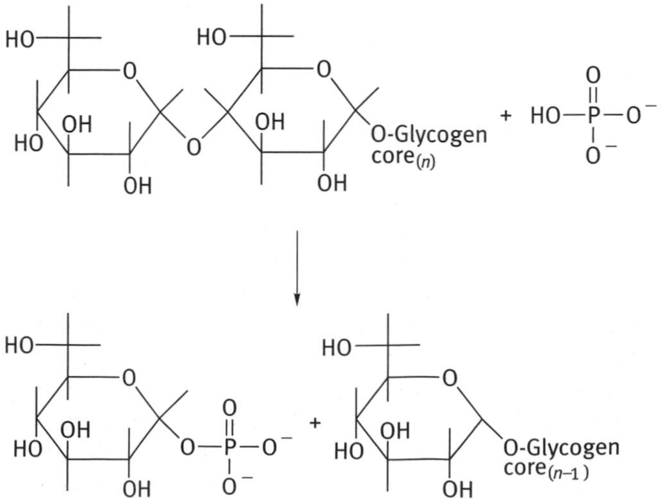

The G-1-P so formed, is converted by phosphoglucomutase working in the reverse direction, to glucose-6-phosphate and

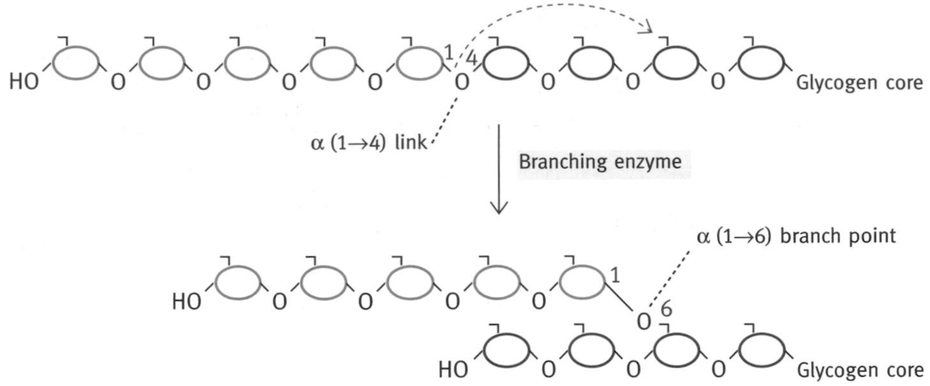

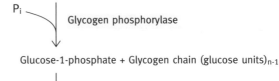

Fig. 7.4 Action of the branching enzyme in glycogen synthesis.

this is hydrolysed (in liver and kidney only) by the enzyme **glucose-6-phosphatase** to give free glucose which is released into the blood.

Glucose-1-phosphate ⇌ Glucose-6-phosphate,

Glucose-6-phosphate + H_2O → Glucose + P_i

A summary of the production of glucose from glycogen is shown in Fig. 7.5.

Glucose-6-phosphatase is located in the membrane of the endoplasmic reticulum (page 67). It is a remarkable enzyme in that its active site is exposed to the lumen of the endoplasmic reticulum (ER), not the cytoplasm. Glucose-6-phosphate is transported through the ER membrane (by a transport protein) where it is hydrolysed. The products, glucose and P_i, are transported back into cytoplasm by separate transport systems.

The branch points of glycogen create a problem, for phosphorylase cannot function within four glucose units of such points—the enzyme is big and presumably the branch gets in the way of the enzyme attaching to the site of the glucosyl bond to be attacked. A pair of enzymes takes care of this problem, so that the release of glucose can go on. The first of these, the **debranching enzyme**, transfers three of the glycosidic units of the branch to the 4-OH of another chain. This makes them part of a chain long enough for phosphorylase to work on. The last unit, in 1 → 6 linkage, is hydrolysed off by an $\alpha\,(1 \rightarrow 6)$-glucosidase activity of *the same enzyme* (which has two activities) and this opens up the chain for further attack by phosphorylase (Fig. 7.6). Glucose uptake, storage, and release in the liver is summarized in Fig. 7.7.

There are some important points to note, especially the following.

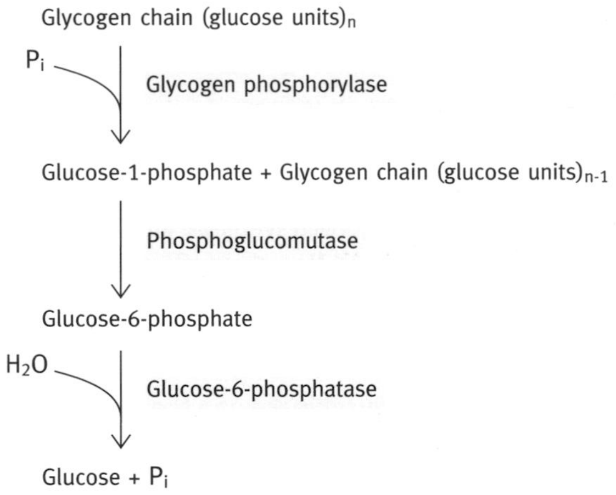

Glycogen chain (glucose units)$_n$

P_i → Glycogen phosphorylase

Glucose-1-phosphate + Glycogen chain (glucose units)$_{n-1}$

Phosphoglucomutase

Glucose-6-phosphate

H_2O → Glucose-6-phosphatase

Glucose + P_i

Fig. 7.5 Degradation of glycogen by phosphorolysis and the ultimate release of free glucose into the blood.

- The synthesis and degradation pathways are different and therefore independently controllable.

- Glucose-6-phosphatase is found in liver but not muscle or adipose tiss ues, which therefore cannot release glucose into the blood. Only the kidney, among other tissues, has this enzyme.

- In extrahepatic cells (as well as in liver) glucose-6-phosphate is on the path to glucose oxidation (Chapter 9).

- Most importantly, the effect of insulin on cells is to activate glycogen synthesis and inactivate glycogen breakdown. The effect of glucagon on cells is the reverse.

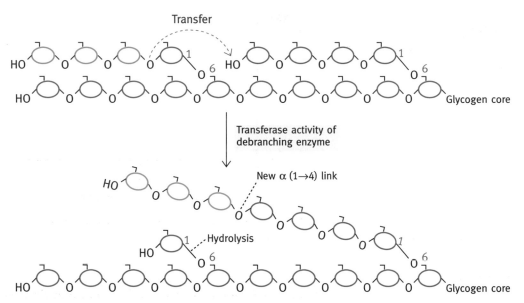

Fig. 7.6 The debranching process. Before debranching, the structure above cannot be attacked by glycogen phosphorylase. After the transferase and hydrolysis actions of the debranching enzyme, both chains are now open to phosphorylase attack.

Thus, insulin promotes glycogen synthesis; glucagon promotes release of glucose from liver. How these effects are achieved is the subject of later chapters when we come to deal with molecular signals to cells.

Why does liver have glucokinase and the other tissues hexokinase?

You will recall that once glucose enters a cell it is phosphorylated to glucose-6-phosphate. Glucokinase in liver and hexokinase in brain and other tissues catalyse the same reaction. Why is this so? In starvation, when glucose supply to the blood is the all-important problem, the entry of glucose into muscle and other cells is restricted because of the lack of insulin (the mechanism of this control is described on page 218), while the entry of glucose into brain, liver, and red blood cells is *not insulin-dependent*. A potentially illogical situation exists. The liver synthesizes glucose from amino acids supplied by the muscle so that it can keep the blood glucose level up to permit normal brain function and it would not make sense for the liver to take up glucose in competition with the brain. The liver glucokinase has a much lower affinity for glucose than has brain hexokinase (see Fig. 7.8). This means that the liver takes up glucose less readily when its concentration in blood is low since its passive entry depends on phosphorylation within the cell. After a meal,

when blood glucose levels are high and when insulin signals the liver to make glycogen, glucokinase works maximally. Also hexokinase is inhibited by glucose-6-phosphate while glucokinase is not. Thus glycogen synthesis can proceed even at high glucose-6-phosphate levels in the cells. (In addition to this automatic control, the glucose transporter of liver has a higher Michaelis constant, K_m (see page 25) than that of many other cells which has the same effect.

What happens to other sugars absorbed from the intestine?

The diet of animals usually results in large amounts of other sugars being absorbed into the portal bloodstream. For example, the sugar in milk is lactose (page 102), which on hydrolysis in the gut yields galactose and glucose. Sucrose yields fructose and glucose. So far as energy metabolism goes, the policy of the body is simple—convert sugars such as galactose and fructose either to glucose or to compounds on the main glucose metabolic pathways in the liver. Thus, these sugars can be converted to anything to which glucose is converted—glycogen, fats, or to CO_2 and H_2O. What happens to galactose has special interest.

To convert galactose to glucose, the H and OH on carbon atom 4 must be inverted or **epimerized**.

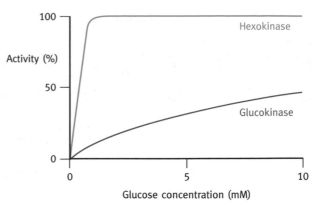

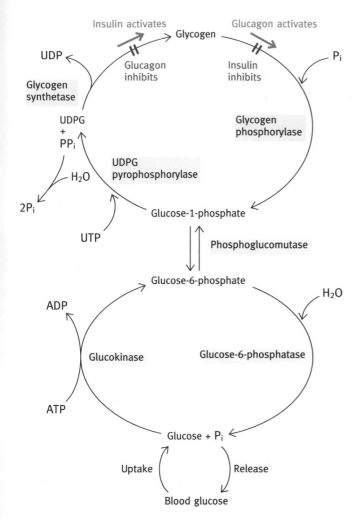

Fig. 7.8 The response of hexokinase and glucokinase to glucose concentration.

Fig. 7.7 Summary of glucose uptake, storage, and release in the liver. Note that glucagon and insulin do not act directly on glycogen metabolism enzymes (see Chapter 13).

This is not done directly but by an epimerase that epimerizes not galactose itself, but UDP-galactose to UDP-

glucose. UDP-glucose, you will recall, is formed in glycogen synthesis.

First, galactose is phosphorylated by galactokinase to galactose-1-phosphate.

You might have expected this to react with UTP by analogy with UDP-glucose formation but that is not what happens. Instead, galactose-1-phosphate displaces glucose-1-phosphate from UDP-glucose forming UDP-galactose and glucose-1-phosphate as follows.

UDP-glucose

Galactose-1-phosphate

Uridyl transferase

UDP-galactose

Glucose-1-phosphate

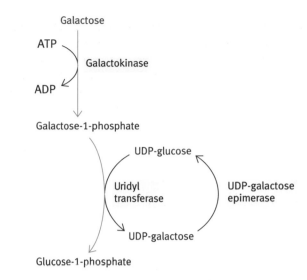

Fig. 7.9 How galactose enters the main metabolic pathways by conversion of galactose to glucose-1-phosphate. The green arrows represent the net effect of the reactions.

You will see that what happens in this reaction is that the—P— uridine group (uridyl) is transferred from UDPG to galactose-1-phosphate and the enzyme is therefore a uridyl transferase. Now epimerization of UDP-galactose occurs.

$$\text{Galactose} + \text{ATP} \xrightarrow{\text{Galactokinase}} \text{Galactose-1-P} + \text{ADP}$$

$$\text{Galactose-1-P} + \text{UDP-glucose} \xrightleftharpoons{\text{Uridyl transferase}} \text{Glucose-1-P} + \text{UDP-galactose}$$

$$\text{UDP-galactose} \xrightleftharpoons{\text{UDP-galactose epimerase}} \text{UDP-glucose}$$

The net effect is to convert galactose to glucose-1-phosphate (as summarized in Fig. 7.9).

In the infant genetic disease, galactosemia, uridyl transferase is missing and galactose cannot be converted as shown. Accumulation of galactose and its metabolic products leads to impairment of brain development and blindness. Elimination of galactose from the diet avoids this. Since the epimerase is reversible, UDP-glucose can be converted to UDP-galactose so the patient can survive without any intake of galactose in the diet for essential syntheses of glycolipids (page 73) and glycoproteins (page 44). These require a donation of galactose residues (from UDP-galactose).

UDP-galactose

UDP-galactose epimerase

UDP-glucose

Amino acid traffic in the body (in terms of fuel logistics)

Since amino acids are not *deliberately* stored, there is no whole body story comparable to that of glycogen and fat traffic, but amino acid traffic does occur. The movement of greatest impor-

Putting the three reactions together, the following sequence occurs.

tance is that in starvation, when muscle proteins break down to produce amino acids that are transported to the liver to provide substrates for glucose synthesis. It will be more convenient to deal with this in detail when we come to the mechanisms of metabolic pathways. We will therefore move on to the traffic of fat in the body—a very important subject.

Fat and cholesterol traffic in the body

We need to explain why cholesterol comes into this story, since the chapter is concerned with the transport of energy-yielding fuels, and cholesterol is certainly not one of these. However, cholesterol is absorbed from the diet and transported by the same lipoproteins as is fat, no doubt because they both present the same problem of water insolubility. It is therefore necessary to discuss the two together.

Uptake of fat from chylomicrons into cells

After a fatty meal, the blood is loaded with **chylomicrons**. The chylomicron (page 105) has a shell of phospholipids, cholesterol, and proteins surrounding a core of neutral fat (TAG) and cholesterol esters.

Let's consider the fate of TAG first. It is taken up (but not as such) by the adipose cells for storage and the lactating mammary gland for secretion, by muscle and other tissues for energy production and also by liver.

Free fatty acids (FFA) readily pass through cell membranes; TAG cannot do so. The TAG in chylomicrons is hydrolysed in the capillaries by a lipase attached to the *outside* cells lining the blood capillaries to produce glycerol and FFA which immediately enter the adjacent cells.

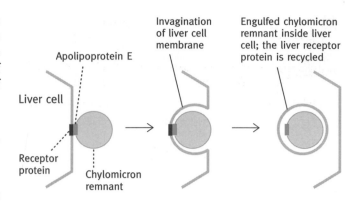

Fig. 7.10 Uptake of chylomicron remnants into a liver cell by receptor-mediated endocytosis. The receptor on the liver cell is specific for apolipoprotein E present on the chylomicron remnant. See Fig. 17.2 for further details of the fate of the engulfed particle and of the mechanism of endocytosis.

capillaries. The amount of lipase is varied according to physiological need; for example, a high-insulin, low-glucagon ratio causes increases in the amount of the enzyme. The chylomicron is a lipoprotein—a complex of protein and lipid. Hence the capillary lipase is called a **lipoprotein lipase** to distinguish it from other lipases. There is evidence that lipoprotein lipase cause the chylomicron to transitorily bind to the cell surface, presumably to facilitate TAG removal.

The progressive removal of triglyceride from chylomicrons reduces these in size to become **chylomicron remnants**, containing all the cholesterol and its esters and about 10% of the triglyceride of the original chylomicron. The remnants are taken up by the liver by receptor-mediated endocytosis (Fig. 7.10) to be destroyed inside the liver cell, thus delivering fat and cholesterol from the intestine to the liver.

A picture of chylomicron usage is shown in Fig. 7.11(a), (b).

Logistics of fat and cholesterol movement in the body

An overview

The movement of lipid and cholesterol around the body is a subject of great importance, particularly from a medical viewpoint.

The liver is the major source of lipid and cholesterol circulating in the blood in the form of lipoproteins, except for chylomicrons, which, as already described, carry absorbed fat and cholesterol from the intestine. The liver is active in synthesizing

$$CH_2-O-\overset{\displaystyle O}{\overset{\displaystyle \|}{C}}-R_1$$
$$CH-O-\overset{\displaystyle O}{\overset{\displaystyle \|}{C}}-R_2 \;+\; 3H_2O \longrightarrow$$
$$CH_2-O-\overset{\displaystyle O}{\overset{\displaystyle \|}{C}}-R_3$$

$$CH_2OH$$
$$CHOH \;+$$
$$CH_2OH$$

$$R_1-C\overset{\displaystyle O}{\underset{\displaystyle O^-}{\diagdown}}$$
$$R_2-C\overset{\displaystyle O}{\underset{\displaystyle O^-}{\diagdown}} \;+\; 3H^+$$
$$R_3-C\overset{\displaystyle O}{\underset{\displaystyle O^-}{\diagdown}}$$

TAG   Glycerol   Free fatty acids

The amount of lipase activity present in the capillaries of a particular tissue determines the amount of FFA release in that tissue and hence the amount of FFA uptake by the tissue. Adipose tissue is rich in lipase, as is lactating mammary gland, while tissues with less need for fat have less of the enzyme in the

(a)

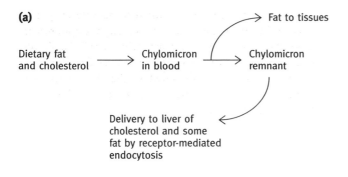

(b)

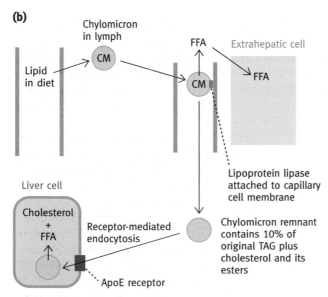

Fig. 7.11 **(a)** Transport of fat and cholesterol by chylomicron route. **(b)** Mechanism for the transport of fat and cholesterol from the intestine to the tissues by chylomicrons. CM, chylomicron; FFA; free fatty acids; HDL, high-density lipoprotein.

TAG from glucose and other metabolites and it receives about 10% of the absorbed fat due to uptake of chylomicron remnants. Nonetheless, it is not a storage organ for lipid—a 'fatty liver' in which there are extensive depositions of fat is pathological. The liver exports its TAG to peripheral tissues in the form of VLDL (very low-density lipoprotein), a lipoprotein similar in structure to a chylomicron. The rationale appears to be that liver, as a major source of fat, supplies this to major fat users such as muscle, which oxidize it for energy production, and to adipose cells for storage.

The liver is the major site of cholesterol synthesis in the body, and also receives dietary cholesterol via the chylomicron remnants. This also is exported to peripheral tissues in the VLDL resulting in an outward flow, from liver to peripheral tissues, of lipid and cholesterol.

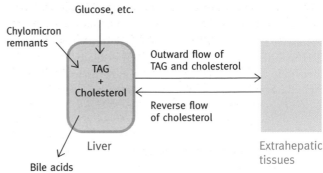

Fig. 7.12 Overview of the major movements of fat (TAG) and cholesterol to and from the liver. The liver synthesizes both components from metabolites and receives them from chylomicron remnants. It also converts cholesterol to bile acids.

There is also a reverse flow of cholesterol from peripheral tissues to the liver (Fig. 7.12). The physiological rationale for this apparent 'equilibration' of cholesterol between the liver and the rest of the body is not self-evident. It might be a way of ensuring that all cells have adequate cholesterol supplies with any excess returned to the liver, which excretes part of it into the intestine in the form of bile acids. Most cells are, however, capable of cholesterol synthesis.

A somewhat different speculative viewpoint has been put forward—that the outward flow of cholesterol may not be primarily a means of transporting this substance, but rather that cholesterol is needed for the lipoprotein structure used in lipid transport. On this view cholesterol transport is not the main aim of the outward flow, but is, as the authors of the view express it, the carrier bag that should be distinguished from the groceries (fat), the reverse flow of cholesterol being regarded as the return of carrier bag material to the liver. They suggest that perhaps this concept has been obscured by the intense interest in cholesterol movement because of its very great medical interest in atherosclerosis (see below).

With that introduction we shall proceed to an account of the mechanisms involved, but it should be kept in mind that in detail it is an extraordinarily complex business, is incompletely understood, and is the subject of a massive continuing research effort.

Utilization of cholesterol in the body

Cholesterol is an important constituent of animal cell membranes (page 75). Because of the effects of its oversupply in causing cardiovascular disease, mechanisms for its removal from the body are of great interest.

The main route for disposing of cholesterol is bile acid formation by the liver. (The structures of cholesterol and of a bile acid are shown on page 104.) About 0.5 g per day is disposed of in humans by this route. Bile acids are reabsorbed from the gut and re-used. One therapeutic approach to lowering cholesterol levels in patients is to administer a compound that complexes bile acids in the intestine and prevents their reabsorption; a more recent approach is to inhibit its production, as described on page 130. Cholesterol is also used in the adrenal glands and gonads for the synthesis of steroid hormones. (Examples of the structures of the latter are given page 354.)

Cholesterol ester, whose structure and synthesis are given on page 105, is the storage form of cholesterol in cells and is the form in which much of the cholesterol in lipoproteins is carried.

Lipoproteins involved in fat and cholesterol movement in the body

There are two lipoproteins produced by the liver, **VLDL** (**very-low-density lipoprotein**) and **HDL** (**high-density lipoprotein**). The production of lipoproteins involves the endoplasmic reticulum and the Golgi apparatus (see page 405), which packages them into membrane-bound vesicles for release by exocytosis. In the rat, about 80% of lipoprotein (other than chylomicrons) is made by the liver and the rest by intestinal cells. The VLDL is converted by the mechanism described below to **IDL** (**intermediate-density lipoprotein**) and then **LDL** (**low-density lipoprotein**) so that, including chylomicrons, five different lipoproteins are found in circulation. Figure 7.13 shows electron micrographs of lipoproteins.

Apolipoproteins

Each type of lipoprotein has its own particular associated set of **apolipoproteins**. A dozen or more are already known. Their functions are not established in every case but they have several roles.

- Some are required as components for the production of lipoproteins. In the case of chylomicrons the main one is $apoB_{48}$ and, in that of VLDL, it is $apoB_{100}$.

- Some are required as destination-targeting signals—specific apoproteins designed to bind to specific receptors on the surface of cells. Binding leads to uptake of the bound lipoprotein by receptor-mediated endocytosis (see Fig. 7.10). In this way individual lipoproteins are taken up only by designated cells. Example of this targeting function are $apoB_{100}$ on LDL which binds to LDL receptors, and apoE on chylomicron remnants which binds to liver receptors.

- Some apolipoproteins are required to activate enzymes. Thus apoCII on chylomicrons is necessary for lipoprotein lipase activity, which removes fatty acids from TAG. Not all of the apoproteins are present when secreted from the liver or intestine; some are delivered in the circulation. For example, apoCII produced by the liver is delivered to chylomicrons after release.

Mechanism of TAG and cholesterol transport from the liver and the reverse cholesterol transport in the body

Let us start with VLDL and the outward flow of TAG and cholesterol from the liver. VLDL secretion is inhibited during the high insulin postabsorptive phase because the hormone reduces the amount of the required $apoB_{100}$ available for VLDL synthesis. In people who are deficient in the synthesis of this apolipoprotein, accumulation of TAG in the liver occurs due to the failure of its export via VLDL.

Once VLDL is released, the TAG is progressively removed by lipoprotein lipase action in the way already described for chylomicrons. As the amount of fat diminishes, the percentage of cholesterol and its esters rises (see Table 7.1), causing an increase in density and the lipoprotein structures shrink in size. The end product of the process is LDL, formed via the intermediate state, IDL (intermediate density lipoprotein). LDL is taken up by peripheral tissue cells via receptor-mediated endocytosis so that the remaining TAG and cholesterol are delivered to extrahepatic tissues. The number of LDL receptors on cells is regulated to control LDL uptake. This outward flow of cholesterol from the liver is counterbalanced by a reverse flow; how does this occur?

How do cells get rid of cholesterol?

The question brings us to the last of the lipoproteins, **HDL** (**high-density lipoprotein**). HDL is produced mainly by the liver in an 'immature' disc-like form, known as nascent HDL,

Table 7.1 Approximate content of fat and cholesterol as percentage dry weight of different lipoproteins

	Percentage dry weight of				
	Chylomicron	VLDL	IDL	LDL	HDL
Neutral fat (TAG)	80	50	30	10	8
Cholesterol + cholesterol ester	8	20	30	50	30

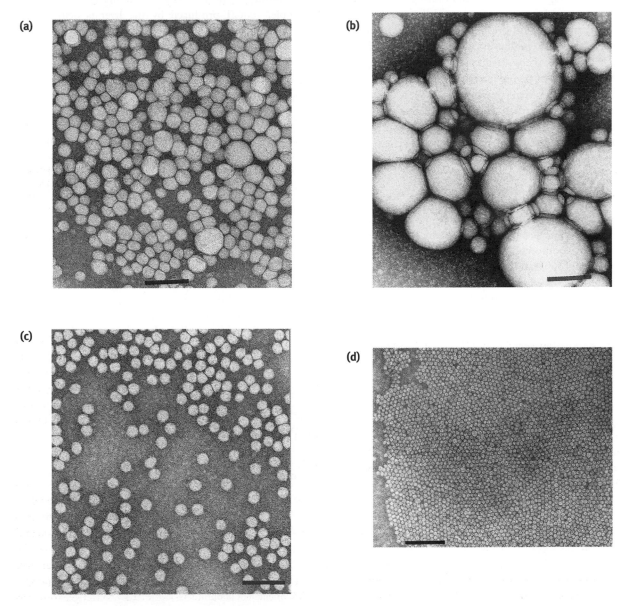

Fig. 7.13 Electron micrographs of: **(a)** chylomicrons (Chylo); **(b)** very low-density lipoproteins (VLDL); **(c)** low-density lipoproteins (LDL); and **(d)** high-density lipoproteins (HDL). Scale bar, 1000 Å. Kindly provided by Dr Trudy Forte of the Lawrence Berkeley Laboratory, University of California.

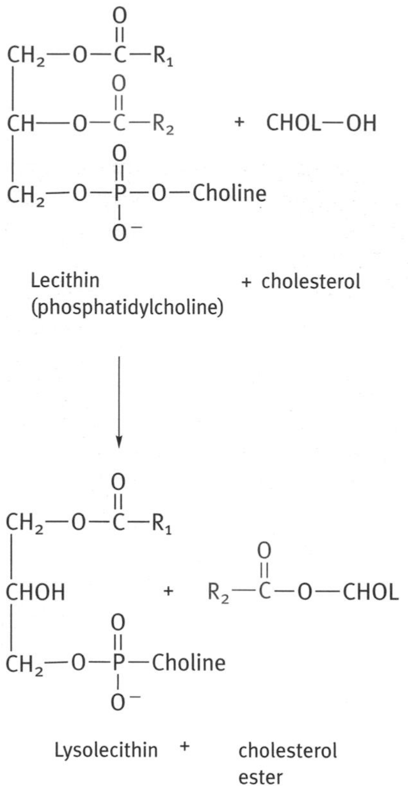

$$
\begin{array}{l}
CH_2-O-\overset{\displaystyle O}{\overset{\displaystyle \|}{C}}-R_1\\[4pt]
CH-O-\overset{\displaystyle O}{\overset{\displaystyle \|}{C}}-R_2 \qquad + \quad CHOL-OH\\[4pt]
CH_2-O-\overset{\displaystyle O}{\overset{\displaystyle \|}{P}}-O-Choline\\
\qquad\qquad\quad\underset{\displaystyle O^-}{\big|}
\end{array}
$$

Lecithin   + cholesterol
(phosphatidylcholine)

$$
\begin{array}{l}
CH_2-O-\overset{\displaystyle O}{\overset{\displaystyle \|}{C}}-R_1\\[6pt]
CHOH \qquad + \quad R_2-\overset{\displaystyle O}{\overset{\displaystyle \|}{C}}-O-CHOL\\[6pt]
CH_2-O-\overset{\displaystyle O}{\overset{\displaystyle \|}{P}}-Choline\\
\qquad\qquad\quad\underset{\displaystyle O^-}{\big|}
\end{array}
$$

Lysolecithin   +   cholesterol
                         ester

Fig. 7.14 Reaction catalysed by lecithin: cholesterol acytransferase where R_1 and R_2 are fatty acyl groups; CHOL—OH is cholesterol.

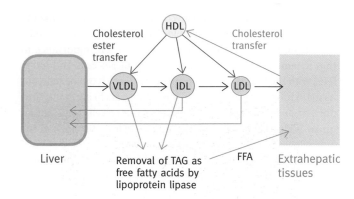

Fig. 7.15 The role of lipoproteins in the outward and reverse flow of cholesterol from, and to, the liver. The heavy red lines represent transfer of cholesterol esters. The transfer of cholesterol esters between lipoproteins is reversible and more complex than shown. In addition, TAG interchange occurs. The situation shown is that in the fasting condition in which chylomicrons are not present. After a meal containing fat, chylomicrons participate in the cholesterol ester transfer from HDL and, in delivering cholesterol to the liver via chylomicron remnants, also contribute to the reverse cholesterol flow. VLDL, very-low-density lipoproteins; IDL, intermediate-density lipoproteins; LDL, low-density lipoproteins; HDL, high-density lipoproteins.

made of phospholipid and apoproteins but containing little TAG or cholesterol. The nascent HDL picks up cholesterol from peripheral tissue cells. Free cholesterol picked up by HDL as such would remain in the peripheral layer of the lipoprotein with its -OH group in contact with water, but, once in the HDL, it is converted to cholesterol ester. Hydrophobic force (page 14) causes the latter to migrate to the centre of the HDL, away from contact with water. The accumulation of the ester causes the HDL to fatten out into a spherical particle. Cholesterol esterification *in cells* requires ATP but HDL has none of this available and an alternative strategy is used. In this, a fatty acyl group of lecithin, which is present in the hydrophilic shell, is transferred to cholesterol in an energy-neutral reaction by an enzyme associated with HDL, called **lecithin:cholesterol acyltransferase (LCAT)** (Fig. 7.14). (The ATP requiring synthesis which occurs in cells is catalysed by an enzyme called **ACAT** (page 105).

The HDL transfers its cholesterol ester to VLDL, IDL and LDL and also to chylomicrons (present after a fat-containing meal), by means of **a cholesterol ester transfer protein (CETP)**. Although IDL and LDL deliver their contents to extrahepatic tissues as described above, a proportion return to the liver (they both use the LDL receptor). This is the explanation of the seemingly pointless recycling of IDL and LDL now enriched by cholesterol from HDL; it is the reverse flow of cholesterol. This is illustrated in Fig. 7.15.

How does cholesterol exit cells to be picked up by HDL?

The HDL must transitorily associate with peripheral cells to accept the cholesterol, but how does the latter get out of the cell to be picked up? It has been known for a long time how it gets in—this occurs by LDL receptors receiving LDL from the blood and endocytosing the complex—but only very recently has the mechanism by which it gets out been discovered. The answer came from an isolated island in Chesapeake Bay called Tangier Island. Some of the inhabitants exhibited a condition known as **Tangier disease** in which high concentrations of cholesterol accumulate in lymphatic organs such as tonsils and spleen, the latter becoming enlarged. Genetic studies on people on the island enabled the responsible gene defect to be pinpointed. It codes for an enzyme that was already known, called ABC1 transporter ATP cassette-binding protein 1. This protein (further described

on page 281) is a membrane glycoprotein which was known to transport a variety of pharmacological drugs out of cells resulting in multidrug resistance which has clinical importance in that it removes therapeutic drugs from target cells and interferes with the therapy. It also transports steroids out of cells, but its main function was in doubt. It now appears likely that an important one is to transport cholesterol out of cells to be picked up by HDL. This results in cholesterol being sent back to the liver where it can be disposed of as bile salts.

It has long been known that high levels of HDL cholesterol (colloquially known as 'good cholesterol') correlates with low risk of coronary disease and clinical syndromes are known in which low levels of HDL in the blood lead to increased risk of coronary occlusions. It is therefore perplexing that people with Tangier disease have low levels of HDL but not increased heart disease risk. The new discovery has considerable medical potential for investigating the causes of vascular disease.

Cholesterol homeostasis in cells

The control of cholesterol levels is very important, and several mechanisms have been identified. Cholesterol levels in liver increase with high levels of fat in the diet and the cells respond by downregulating (decreasing the number) of the LDL receptors, which reduces the uptake of LDL and IDL from the blood. This means that blood LDL cholesterol rises since uptake of LDL and IDL is the route by which cholesterol in the reverse flow re-enters the liver.

LDL cholesterol is colloquially known as 'bad cholesterol' because high levels are associated with increased atherosclerosis risk. The high LDL level lead to the development of plaques in blood vessels, a complex process in which cholesterol deposition is involved. These block blood vessels and if present in coronary arteries may cause heart attacks. An extreme case of this is found in genetic disease **familial hypercholesterolemia**, in which people have defective LDL receptors leading to very high levels of LDL cholesterol and early vascular disease.

In addition to control of intake, the rate of synthesis of cholesterol is controlled. If the cellular level is low, synthesis increases; if the level is high, synthesis is inhibited so that cholesterol homeostasis is achieved. How do these control mechanisms work? HMG-CoA is a metabolite you haven't met yet in this book; it is described on page 196 where we briefly deal with cholesterol synthesis, but at this point it is sufficient that it is needed for cholesterol synthesis and is produced by the enzyme HMG-CoA synthase. At low cholesterol levels in the cell, the genes coding for this enzyme, and for LDL receptor protein,

are activated so that cholesterol synthesis and intake of cholesterol into cells are both stimulated. At high cellular cholesterol levels the genes are not activated and, since there is a turnover of the two proteins, cholesterol synthesis diminishes and intake is reduced. A novel mechanism of this control has recently been discovered (described in the Osborne reference and illustrated in Fig. 7.16). It involves a gene-regulatory protein normally tethered to the endoplasmic reticular membrane inside the cell; at low cholesterol levels, a proteolytic mechanism is activated to release it. The released protein then migrates to the nucleus where it attaches to the DNA at sites known as **sterol-regulatory elements** (SREs) which leads to activation of genes coding for LDL receptors and HMG-CoA synthase. This leads to an increase in the level of the latter two proteins involved in cholesterol uptake into cells and in cholesterol synthesis respectively. In the presence of cholesterol, the gene regulatory proteins remain tethered to the membrane of the endoplasmic reticulum and so cannot enter the nucleus. The relevant genes are therefore not activated. There is a further control. HMG-CoA reductase, a regulatory enzyme in cholesterol synthesis (page 197), is inhibited at high cholesterol levels; the enzyme becomes phosphorylated, in which form it is inactive. Yet a further control is that at high cholesterol levels the enzyme is also more rapidly destroyed, but the mechanism of this is unknown.

The above account emphasizes the reverse cholesterol flow, but the transfer of cholesterol ester between lipoproteins occurs reversibly and a more complex traffic occurs between all of them; in addition, transfer of TAG takes place between the various components, catalysed by CETP which actually exchanges cholesterol ester for TAG. The precise reasons for this traffic complexity are not known.

Inhibitors of cholesterol synthesis

Therapeutically, drugs have been developed to lower cholesterol levels. These resemble mevalonic acid (page 197) in structure, an intermediate in cholesterol synthesis formed by **HMG-CoA reductase.** In the body they are converted to analogues of the transition state (page 27) involved in the HMG-CoA reductase reaction. They inhibit this very effectively by combining with the active site of the enzyme. One of the drugs, simvastatin, was isolated from fungi.

How is fat released from adipose cells?

A lipase releases FFA from stored neutral fat. But there is a very important difference here. The lipase inside the adipose cell is

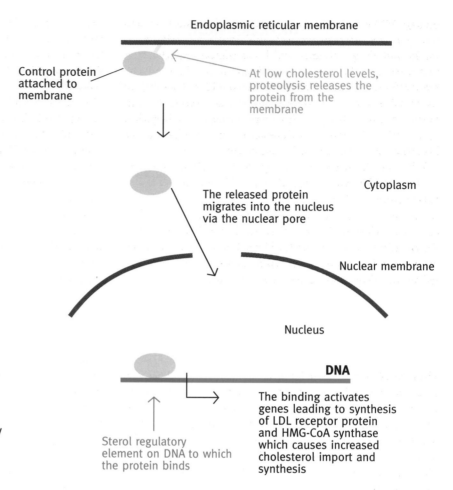

Fig. 7.16 Simplified diagram giving the principle of the mechanism by which mammalian cells respond to low cholesterol levels by activating genes for the synthesis of LDL receptor protein and HMG-CoA synthase. For simplicity the diagram does not show that, for reasons not fully understood, there are two separate activating proteins, or that two proteolytic steps are involved in the release of the regulatory protein.

not the same (lipoprotein) lipase as in the capillaries. It is a *hormone-sensitive lipase, activated by a glucagon signal to the cells and inhabited by an insulin signal to the cells.* The mechanism of how this occurs is described in Chapter 13. You can see how it all fits in. After a meal, insulin is high and glucagon low. This instructs adipose cells to take up glucose from plasma and FFA from chylomicrons and convert them into neutral fat; release of FFA is inhibited. In fasting, the reverse occurs and tissues are supplied with FFA in the blood. In juvenile-onset diabetes (type 1), the low insulin/glucagon ratio me ans that the adipose cell hormone-sensitive lipase is activated and FFA pours out into the blood (In type 2, mature-onset diabetes, insulin is produced but the target cells respond inadequately—see page 229 for further details.). The liver partially converts these to ketone bodies and this results in severe ketonemia (that is, ketone bodies in the blood).

A further point: the adipose cell hormone-sensitive lipase is also activated by epinephrine (adrenaline). Thus, in response to an epinephrine-liberating alarm, the adipose cells pour out FFA to supply the muscles with FFA to support any required violent muscular activity. How epinephrine does this is an important topic, but it is more suitable to deal with it in Chapter 13 . *Nor-epinephrine* liberated from the sympathetic nervous system that innervates fat cells has the same effect.

How are free fatty acids carried in the blood?

It would be unsatifactory for relatively high levels of FFA to be carried in the blood since, at neutral pH, it would literally be a soap solution. Instead they are carried adsorbed to the surface of the blood protein, serum albumin. This protein has hydrophobic patches on its surface and is the carrier for many molecules with hydrophobic parts. The adsorbed FFA are in equilibrium with FFA in free solution so that, as cells take them up, more dissociate from the protein carrier into the serum where they are available to cells.

FURTHER READING

Glycogen synthesis

Alonso, M. D., Lomako, J., Lomako, W. M., and Whelan, W. J. (1995). A new look at the biogenesis of glycogen. *FASEB J.*, **9**, 1126–37.

A beautiful article describing the role of glycogenin. Summarizes new concepts in a field many thought was a closed book.

Van Schaftingen, E., Detheux, M., and Da Cunha, M. V. (1994). Short-term control of glucokinase activity: role of a regulatory protein. *FASEB J.*, **8**, 414–19.

Describes the occurrence, properties, and control of this enzyme.

Lipid transport and lipoproteins

Nilsson-Ehle, P., Garfinkel, A. S., and Schotz, M. C. (1980). Lipolytic enzymes and plasma lipoprotein metabolism. *Ann. Rev. Biochem.*, **49**, 667–93.

Gives general summary of lipoproteins, including useful information at the physiological level.

Gannon, B. and Nedergaard, J. (1985). The biochemistry of an inefficient tissue. *Essays in Biochem.*, **20**, 110–64.

Deals with heat production by brown fat cells.

Angelin, B. and Gibbons, G. (1993). Lipid metabolism. *Curr. Opin. Lipidology*, **4**, 171–6.

An excellent overview of lipoproteins.

Tall, A. (1995). Plasma lipid transfer proteins. *Ann. Rev. Biochem.*, **64**, 235–57.

Excellent review of the medically important area of the cholesterol ester transport protein (CETP), lipid transport, forward and revese cholesterol transport between liver and extrahepatic tissues.

Cholesterol homeostasis and transport

Osborne, T. F. (1997). Cholesterol homeostasis: Clipping out a slippery regulator. *Current Biology*, **7**, R172–4.

Describes the unusual control in which membrane-bound sterol regulatory element-binding proteins are released to enter the nucleus and activate genes for HMG-CoA synthetase and LDL receptor protein among others. Low levels of cholesterol are the trigger for this.

Gura, T. (1999). Gene linked to faulty cholesterol transport. *Science*, **285**, 814–15.

A short informative report on the new discovery of the way in which cholesterol is removed from cells. The work relates to the genetic disease (Tangier disease) found on the Tangier Island of Chesapeake Bay.

PROBLEMS FOR CHAPTER 7

1 Starting with glucose-1-phosphate, explain how the synthesis of glycogen is made thermodynamically irreversible?

2 Have you any comments on the name of enzyme UDP-glucose pyrophosphorylase?

3 Which tissues release glucose into the blood as a result of glycogen breakdown? Why is this so?

4 What reaction do the two enzymes glucokinase and hexokinase catalyse? Why should the liver have glucokinase while brain and other tissues have hexokinase?

5   In the genetic disease galactosemia, infants are unable to metabolize galactose correctly. Removal of galactose from the diet can prevent the deleterious effects of the disease. UDP-galactose epimerase is reversible; why is this important in connection with the above statement?

6   How is TAG removed from chylomicrons and utilized by tissues?

7   Explain what VLDL is and its likely role.

8   What is meant by reverse cholesterol flow?

9   What is the main route of cholesterol removal from the body? Why is this of medical interest?

10 Cholesterol is esterified in HDL. Conversion of cholesterol to its ester is an energy-requiring process but HDL has no access to ATP. How is the esterification achieved?

11 How is fat released from adipose cells and under what conditions? How is it transported in the blood for transportation to other tissues? Compare this to the transport of fat from the liver to other tissues.

12 A genetic disease exists in which cholesterol levels in the blood are very high. Explain the cause of this.

13 How is fat from the adipose tissue transported and utilized by cells? How is fat from the liver transported and utilized by cells?

Chapter summary

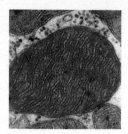

Energy release from foodstuffs— a preliminary overview

The generation of energy in the form of ATP from glucose, fats, and amino acids involves fairly long and somewhat involved metabolic pathways. There is the possibility that, if we go at once into these pathways, the overall strategy of energy generation by cells may be lost amongst the detail. We will therefore, at this stage, deal with major phases of the processes, picking out landmark metabolites only, and view the strategies on a broad basis. When this has been done, subsequent chapters will deal in more detail with the pathways; in short, we want you to see the wood before the trees. For convenience we will start with energy production from glucose.

Energy release from glucose

The main phases of glucose oxidation

Overall, glucose is oxidized as follows

$$C_6H_{12}O_6 + 6O_2 \rightarrow 6CO_2 + 6H_2O.$$

The $\Delta G^{0'}$ for this reaction is $2820\,\text{kJ}\,\text{mol}^{-1}$.

In the cell, this oxidation process is accompanied by the synthesis of more than 30 molecules of ATP from ADP and P_i. The entire process of glucose oxidation to CO_2 and H_2O can be divided up into three phases. If you cannot fully understand the summaries below don't be concerned—they will become clear shortly.

- **Glycolysis**—which means lysis or splitting of glucose into two C_3 fragments in the form of pyruvic acid, accompanied by reduction of an electron carrier. This occurs in the **cytoplasm** of cells.

- The **Krebs, citric acid, or tricarboxylic acid (TCA) cycle** (these are alternative names) in which carbon atoms two and three of pyruvate are converted to CO_2. Electrons are transferred to electron carriers. No oxygen is involved in this phase. Carbon atoms are released as CO_2. The cycle is located inside **mitochondria** in eukaryotes.

- The **electron transport system** in which electrons are transported from the electron carriers to oxygen where, with protons from solution, water is formed. It is in stage 3 that most of the ATP is generated. This occurs in the inner mitochondrial membrane in eukaryotes.

Befroe we go into these stages we need to say a little about biological oxidation.

Biological oxidation and hydrogen transfer systems

Oxidation does not necessarily involve oxygen; it involves the removal of electrons. It may involve *only* electron removal such as in the ferrous/ferric system

$$Fe^{2+} \rightarrow Fe^{3+} + 2e^-$$

or it may be the removal of electrons accompanied by protons from a hydrogenated molecule

$$AH_2 \rightarrow A + 2e^- + 2H^+$$

In such chemical oxidation systems, the electrons must be transferred from the electron donor to an electron acceptor. Depending on the particular electron acceptor, the electron transferred may be accompanied by the proton in which case it is equivalent to a hydrogen atom being transferred or the proton may be liberated into solution.

The ultimate electron acceptor in the aerobic cell is oxygen. Oxygen is **electrophilic**—it has an avidity for electrons. When it accepts electrons, protons from solution join up to produce water

$$O_2 + 4e^- + 4H^+ \rightarrow 2H_2O.$$

However, oxygen is only the *ultimate* electron acceptor in the cell—there are other electron acceptors, which form a chain, carrying electrons from metabolites to oxygen. They accept electrons and hand them on to the next acceptor and thus are electron carriers. This is the electron transport chain, which plays a predominant part in ATP generation. The essential concept is given in Fig. 8.1. Two of the electron carriers involved in energy production are of such central importance in metabolism that we must now describe them. It is essential for you to be completely familiar which these.

NAD⁺—an important electron carrier

The first electron carrier involved in the oxidation of most metabolites is NAD⁺, which stands for **nicotinamide adenine dinucleotide.** You have already met a nucleotide in the form of AMP (Fig. 1.6); a **nucleotide** has the general structure, base–sugar–phosphate, the base in AMP being adenine. (The structures of adenine and other bases are dealt with later in Chapter 19 and need not concern us now.) In the case of NAD⁺, a **dinucleotide** is formed by linking the two phosphate groups of two nucleotides (which is unusual and not how nucleotides are linked together in nucleic acids, as will be evident in later chapters). The general structure of NAD⁺ is

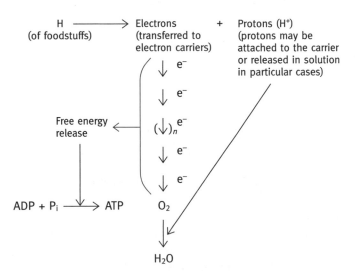

Fig. 8.1 The concept of electron transfer to oxygen and ATP generation.

Base–sugar–phosphate–phosphate–sugar–base.

The two bases are adenine (as in ATP) and nicotinamide, respectively—giving the structure of NAD⁺ below.

Adenine	Nicotinamide
|	|
Ribose	Ribose
|	|
Phosphate ——	Phosphate

NAD⁺ is a **coenzyme**—a small organic molecule that participates in enzymic reactions. It differs from an ordinary enzyme-substrate only in that the reduced compound leaves the enzyme and attaches to a second enzyme where it donates its reducing equivalents to a second substrate. NAD⁺ thus acts catalytically by being continually reduced and re-oxidized, and in doing so transfers electrons from one molecule to another.

The 'business end' of the molecule is the nicotinamide group. It is derived from the vitamin, nicotinic acid or niacin. Niacin is an essential vitamin for animals, but humans can make it from the amino acid tryptophan. The rest of the molecule fits the coenzyme to appropriate enzymes but does not undergo any chemical change.

Nicotinamide has the structure

When linked in NAD⁺ the structure is

where R is the rest of the NAD molecule.

NAD⁺ can be reduced by accepting two electrons plus one proton as a hydride ion (H:⁻), to give the following structure (the second proton being liberated into solution).

Reduced NADH + H⁺

NAD⁺ is the coenzyme for several **dehydrogenases** that catalyse this type of reaction:

$AH_2 + NAD^+ \rightleftharpoons A + NADH + H^+$.

The reduced NAD^+ can then diffuse to a second enzyme and participate in a reaction such as

$B + NADH + H^+ \leftrightarrow BH_2 + NAD^+$.

In this way NAD^+ acts as the carrier for the transfer of a pair of hydrogen atoms from A to B

$AH_2 + B \rightarrow A + BH_2$.

In biochemistry such reactions are often presented in the form of whirligigs.

$$AH_2 \searrow \quad NAD^+ \searrow \quad BH_2$$
$$A \nearrow \quad NADH \nearrow \quad B$$
$$+ H^+$$

In case it requires extra emphasis, $NADH + H^+$ can add *two* H atoms to a substrate since the second electron it transfers to an acceptor molecule may be joined by a proton from solution as shown above. In equations, reduced NAD^+ is written as $NADH + H^+$. In the text, the term NADH is used, but it always implies $NADH + H^+$.

FAD—another important electron carrier

Another electron (hydrogen) carrier is **FAD** or **flavin adenine dinucleotide**. It is derived from vitamin B_2 or riboflavin. The important feature of this molecule is that it can (in combination with appropriate proteins) accept two H atoms to become $FADH_2$. FAD is a prosthetic group—it is a permanent attachment to its apoenzyme, unlike NAD^+ which moves from one dehydrogenase to another. Its role will become clear; for the present it is sufficient that FAD can be reduced. The chemistry of this reduction is given below, where a related carrier, FMN or flavin mononucleotide, is also described.

FAD, FMN, and their reduction

FAD has the structure:

Isoalloxazine ring system—ribitol—phosphate—phosphate—
    ribose—adenine.

It is very unusual in having the linear ribitol molecule instead of ribose. The left-hand portion of the molecule is not, therefore, in the strict sense, a nucleotide, but nonetheless is always designated as such from long usage.

Oxidation–reduction reactions occur in the isoalloxazine structure.

FAD (Oxidized form)   FADH₂ (Reduced form)

Another carrier, FMN or flavin mononucleotide, has the structure isoalloxazine–ribitol–phosphate. It is reduced in the same way. The vitamin, riboflavin, lacks the phosphate. The flavins can exist in the semiquinone form and thus can participate in single-electron transfer reactions as well as in two-electron transfer ones.

Stages in the release of energy from glucose—first glycolysis

As stated earlier, glucose oxidation occurs in three main stages, the first of which is called **glycolysis**. It does not involve oxygen and very little energy is produced—in fact only two ATP molecules per molecule of glucose lysed. The end-products are pyruvate and NADH as shown in Fig. 8.2. In aerobic glycolysis, when the oxygen supply is plentiful, the NADH is re-oxidized to NAD^+ by mitochondria in eukaryotic cells; the pyruvate is taken up by the mitochondria where it is oxidized to CO_2 and H_2O.

However, in the body, oxygen isn't necessarily always plentiful. In muscle, especially in the initial phase of emergency flight reactions before the heart has increased its pumping rate, the rate of glycolysis is dramatically increased with the aim of rapidly producing the ATP needed for muscle contraction.

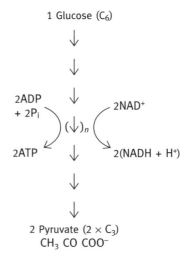

Fig. 8.2 The net result of aerobic glycolysis of glucose.

Since NAD$^+$ acts catalytically and is present in cells in small amounts, for glycolysis to proceed, the NADH must be recycled back to NAD$^+$; no NAD$^+$, no glycolysis. If the capacity of mitochondria to do this is inadequate at high glycolytic rates, glycolysis and the muscle's ability to maximally generate ATP, needed for contraction, would be impaired unless another way were found to regenerate NAD$^+$ from NADH. An 'emergency' system comes into play in this situation. The NADH is re-oxidized by reducing pyruvate to produce lactate. The reaction is

$$CH_3COCOO^- + NADH + H^+ \rightleftharpoons CH_3CHOHCOO^- + NAD^+.$$
Pyruvate   Lactate
              Lactate
              dehydrogenase

The enzyme catalysing this reaction is **lactate dehydrogenase**. The production of lactate from glucose is known as **anaerobic glycolysis**, as opposed to **aerobic glycolysis** which forms pyruvate, the NADH being oxidized by mitochondria. The object of anaerobic glycolysis is not to produce lactate but to re-oxidize NADH and thus permit continued ATP production from glycolysis (Fig. 8.3). Since there is a lot of lactate dehydrogenase in muscle, the NADH can be rapidly re-oxidized in this way and this in turn permits glycolysis to proceed at a very fast rate. The advantage of this is that, although only two ATP molecules are generated per molecule of glucose, relatively vast amounts of glucose can be broken down. The ATP so produced might make the difference between being eaten by a tiger and not being eaten if you happen to be chased by one. After a short time, the increased heart rate and dilatation of blood vessels supplying muscle increase the oxygen supply and the oxidative regenera-

tion of NAD$^+$ increases. The lactate formed leaks out into the blood but is not wasted; it is used mainly by the liver as described later (page 204).

As a parenthetic note, yeast can live entirely on anaerobic glycolysis using an analogous trick to re-oxidize NADH. The pyruvate is converted to acetaldehyde + CO$_2$; a second enzyme, alcohol dehydrogenase, reduces the acetaldehyde to ethanol.

$$CH_3COCOO^- + H^+ \longrightarrow CH_3CHO + CO_2$$
              Pyruvate decarboxylase

$$CH_3OH + NADH + H^+ \longrightarrow CH_3CH_2OH + NAD^+$$
              Alcohol dehydrogenase

Because of the low yield of ATP, vast quantities of glucose are broken down, producing alcohol and CO$_2$. The latter causes the explosion of bottles if, during ginger beer making, it is bottled too early.

An explanatory note

In the above account, for simplicity, we have talked about glucose being the carbohydrate that is glycolysed. It can indeed be free glucose, but in muscle it is mainly glycogen, the storage form of glucose, that is broken down. This does not affect the overall account given above except that, as described in later chapters, the initial steps of metabolism of glycogen and glucose are slightly different and the ATP yield from glycogen is three per glucosyl unit not two as from glucose. We also again remind you that in the next chapter the detailed mechanisms of glycolysis, the citric acid cycle, and the electron transport system will be given. At this stage the main thing is to get an overview.

Stage 2 of glucose oxidation—the citric acid cycle

The **mitochondria** are small organelles located in the cytoplasm of the cell. They are the main energy-generating units of aerobic cells where most of the ATP is produced. They are bounded by two membranes. The outer membrane is permeable and plays little role in energy generation. The inner membrane is highly impermeable to most substances, except where specific transport systems exist. Its surface area is increased by folds or cristae: the more energetic the tissue, the more cristae there are (Fig. 8.4). The mitochondrion is filled with a concentrated solution of enzymes—it is called the **matrix**. It is here that stage two of glucose metabolism mainly occurs, only one reaction being located in the inner membrane.

Aerobic glycolysis, as stated, produces pyruvate and NADH in the cytoplasm. To be further oxidized the pyruvate must enter the mitochondria. A transport system in the inner mito-

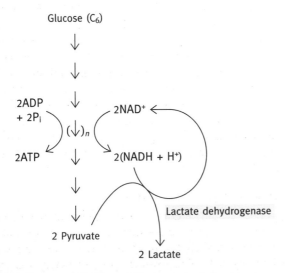

Fig. 8.3 The net result of anaerobic glycolysis of glucose.

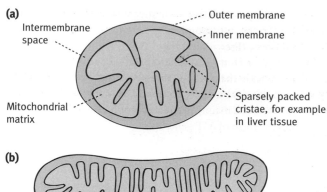

(a)

Intermembrane space

Outer membrane

Inner membrane

Mitochondrial matrix

Sparsely packed cristae, for example in liver tissue

(b)

Densely packed cristae for example in heart muscle or insect flight muscle

Fig. 8.4 Mitochondria from **(a)** liver and **(b)** heart tissue. The number of cristae reflects the ATP requirements of the cell.

chondrial membrane takes the pyruvate from the cytoplasm into the mitochondrion, but the cytoplasmic NADH cannot *itself* enter the mitochondrion for reoxidation. Instead, it hands over its reducing equivalent to either NAD$^+$ or an FAD enzyme *inside* the mitochondrion by what are known as shuttles, whose mechanisms are given in the next chapter. The end result is that NADH produced by glycolysis is re-oxidized by mitochondria leaving NAD$^+$ in the cytoplasm to participate further in glycolysis. The NADH and FADH$_2$ inside the mitochondrion are oxidized by the electron transport system (stage 3), but we must now deal with stage 2, the citric acid cycle, which metabolizes pyruvate, before we come to this. Some cells such as erythrocytes lack mitochondria and so must generate ATP by glycolysis only; they are glucose-dependent.

How is pyruvate fed into the citric acid cycle?

We now come to an enzyme reaction of major importance in which pyruvate, transported into mitochondria, is converted to a compound which is at the crossroads of metabolism. This is acetyl-CoA, a compound not previously mentioned in this book.

What is coenzyme A?

Coenzyme A is usually referred to as **CoA** for short, but is written in equations as CoA—SH because its thiol group is the reactive part of the molecule.

Unlike NAD$^+$ and FAD, coenzyme A is *not* an electron carrier, but an acyl group carrier (A for acyl). Like NAD$^+$ and FAD it is a

dinucleotide and, as so often happens with cofactors, incorporates a water-soluble vitamin in its structure, in this case pantothenic acid. Pantothenic acid has an odd sort of structure (which, we suggest, it is not necessary to learn).

$$HO—CH_2—\overset{\overset{\displaystyle CH_3}{|}}{\underset{\underset{\displaystyle CH_3}{|}}{C}}—\overset{\overset{\displaystyle H}{|}}{\underset{\underset{\displaystyle OH}{|}}{C}}—\overset{\overset{\displaystyle O}{\|}}{C}—NH—CH_2—CH_2—\overset{\overset{\displaystyle O}{\diagup}}{\underset{\underset{\displaystyle OH}{\diagdown}}{C}}$$

Pantothenic acid

You would normally expect the vitamin moiety of a coenzyme to play a part in the reaction for which the coenzyme is required (as with, for example, riboflavin in FAD and nicotinamide in NAD$^+$), but the pantothenic acid moiety is just 'there' and apparently quite inert. It presumably provides a recognition group to help bind the CoA to appropriate enzymes but why this particular structure is used is not obvious. It may be a quirk of evolution that was used by chance early in evolution and the cell became locked into its use.

The structure of CoA is as follows.

$$\begin{array}{ccc} \text{Adenine} & & \beta\text{-mercaptoethylamine} \\ | & & | \\ \text{Phosphate—Ribose} & & \text{Pantothenic acid} \\ | & & | \\ \text{Phosphate——Phosphate} \end{array}$$

The β-mercaptoethylamine part is the business end of the molecule.

RCO—NHCH$_2$CH$_2$—SH

β-Mercaptoethylamine moiety

The CoA molecule carries acyl groups as thiol esters. For example, acetyl-CoA can be written as CH$_3$CO—S—CoA. *The thiol ester is a high-energy compound* (unlike a carboxylic ester). It has a $\Delta G^{0'}$ of hydrolysis of approximately -31 kJ mol^{-1} as compared with about -20 kJ mol^{-1} for a carboxylic ester. The difference is due to the fact that the latter compound is resonance stabilized and so has a lower free-energy content than a thiol ester, which is not resonance stabilized. This is important in several areas of biochemistry. With that description of CoA we can now return to the question of what happens to pyruvate transported into mitochondria.

Oxidative decarboxylation of pyruvate

The pyruvate is subjected to an irreversible **oxidative decarboxylation** in which CO$_2$ is released (decarboxylation), a pair of electrons is transferred to NAD$^+$ (oxidation), and an acetyl group is transferred to CoA. Note that this is different from the nonoxidative decarboxylation carried out by yeast pyruvate

decarboxylase, described earlier. In the latter there is, as implied, no oxidation, NAD^+ is not involved, and the product is acetaldehyde, not an acetyl group. The large negative free-energy change means that the reaction is irreversible. The reaction catalysed by pyruvate dehydrogenase is as follows.

Pyruvate + CoA—SH + $NAD^+ \longrightarrow$ Acetyl—S—CoA
+ NADH + H^+ + CO_2 $\Delta G^{o\prime}$ = −33.5 kj mol^{-1}

The acetyl group of acetyl-CoA is now fed into the citric acid cycle. As indicated earlier, there are alternative names for this; it is known as the tricarboxylic acid (TCA) cycle since citric acid has three carboxyl groups, and as the Krebs cycle after its discoverer. The name citric acid cycle will be used here, but at this stage we will not be concerned with the reactions in it. The essential point is that the carbon atoms of the acetyl group of acetyl-CoA produced from pyruvate are converted to CO_2 while three molecules of NAD^+ are reduced to NADH. In addition, a molecule of FAD is reduced to $FADH_2$, the reducing equivalents coming partly from water. Almost as a sideline, the cycle generates one 'high-energy' phosphoryl group from P_i for each acetyl group fed in. This stage is summarized in Fig. 8.5.

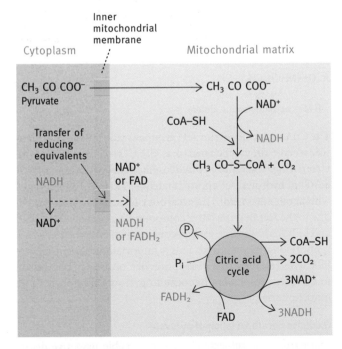

Fig. 8.5 What happens to pyruvate and NADH generated in glycolysis and the generation of further NADH and $FADH_2$ inside mitochondria. The FAD is always attached to an enzyme. The source of the reducing equivalents is dealt with in the next chapter. The NADH and $FADH_2$ are oxidized by the electron transport pathway described shortly.

In summary, a molecule of pyruvate from the cytoplasm is converted in the mitochondria to three molecules of CO_2 and, in the process, three molecules of NAD^+ and one molecule of FAD are reduced but we have hardly started to make ATP—only two molecules in the glycolysis and two in the cycle (per starting glucose molecule) but there are almost 30 still to be made.

So far, it's been mainly preparation of fuel—now for the big return in the form of ATP generation.

Stage 3 of glucose oxidation—electron transport to oxygen

The oxidation of NADH and $FADH_2$ takes place in the inner mitochondrial membrane which contains a chain of electron carriers.

The electron transport chain—a hierarchy of electron carriers

The problem we are now concerned with, we remind you, is the transference of electrons from NADH and $FADH_2$ to oxygen with the formation of water

NADH + H^+ + $\frac{1}{2}O_2 \rightarrow NAD^+$ + H_2O.

The $\Delta G^{o\prime}$ of this reaction is −220 kJ mol^{-1}. To understand how the process occurs, we need to discuss the **redox potential** of compounds.

Compounds capable of being oxidized are, by definition, electron donors; in any oxido-reduction reaction there must be an electron acceptor (oxidant) and a donor (reductant). The reaction

$X^- + Y \rightleftharpoons X + Y^-$

can be considered to occur by two theoretical half-reactions, as follows,

(a) $X^- \rightleftharpoons X + e^-$;
(b) $Y + e^- \rightleftharpoons Y^-$.

Each of these involves the reduced and oxidized forms of each reactant which are called **conjugate pairs** or **reduction-oxidation couples** or, the most convenient term, **redox couples**. X and X^- are such a couple and Y and Y^-, another couple. Clearly, any real-life oxidation–reduction reaction must involve a pair of redox couples, because one must donate electrons and one must accept them.

Different redox couples have different affinities for electrons. One with a lesser affinity will tend to donate electrons to another of higher affinity. The **redox potential value** (E_0' is a measurement of the electron affinity or electron-donating potential

of a redox couple). This is of importance in biochemistry for it is an indicator of the direction in which electrons will tend to flow between reactants. Equally important E'_0 values are directly related to free-energy changes (see below).

E'_0 values are expressed in volts—the more negative, the lower the electron affinity, the greater the tendency to hand on electrons, the greater the reducing potential, and the higher the energy of the electrons.

The reason why this chemical property is expressed in volts is due to the method of redox potential determination. As stated above, two redox couples must be involved in an oxido-reduction reaction but, since electron transfer is involved, they can be in two separate vessels (half-cells) if they are connected by a copper wire to conduct electron flow. The method is to use the $2H^+ + 2e^- \rightleftharpoons H_2$ equilibrium (catalysed by platinum black) in one half-cell as the reference redox couple, and compare the unknown sample in the second half-cell with it. Electrons will flow through the wire according to the relative electron affinities of the two systems. The positive ions in each half-cell (H^+ in the hydrogen electrode half-cell and, for example, ferrous and ferric ions in the sample half-cell) are accompanied by anions. Change in the positive ions due to loss or gain of electrons in the half-cells necessitates a compensating anion migration from one half-cell to the other. The agar–salt bridge shown in Fig. 8.6 provides the route for this. The electrical potential between the half-cells is measured by a voltmeter inserted into the copper wire connecting electrodes in each half-cell. The reference (hydrogen electrode) half-cell is arbitrarily assigned the value of zero and the relative value of the sample cell is the redox potential of the redox pair in it. In physics, the convention is that electrical current flows in the opposite direction to electron flow and, therefore, the half-cell that is donating electrons has the more negative voltage. Thus, if the sample half-cell is more reducing than the reference half-cell, its redox potential value is more negative.

The E_0 values (written without a prime) are standard values measured at 1.0 M concentrations of components or H_2 gas at 1 atmosphere pressure. In biochemistry, values are adjusted to pH 7.0 instead of pH 0 and this brings the redox potential of the reference half-cell to a value of -0.42 V. Redox potentials so corrected are written as E'_0 values. The half-reaction $NAD^+ + 2H^+ + 2e^- \rightarrow NADH + H^+$ has an E'_0 value of -0.32 V, and that of the half-reaction $\frac{1}{2}O_2 + 2H^+ + 2e^- \rightarrow H_2O$, an E'_0 value of $+0.82$ V. The very large difference indicates that NADH has the potential to reduce oxygen to water but the reverse will not occur.

There is a direct relationship between the $\Delta G^{0'}$ value and the E'_0 value of an oxido-reduction reaction, quantified by the Nernst equation,

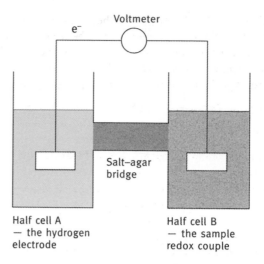

Fig. 8.6 Apparatus for measurement of redox potentials. The reference hydrogen electrode in A contains the redox couple $H_2/2H^+$ catalysed by platinum black on the electrode. The sample half-cell B contains the redox couple whose redox potential is being determined. If B is more reducing than A, electrons will flow from B to A and reduce $2H^+$ to H_2 if the half-cells are connected by a connecting wire. The E'_0 value is therefore more negative than that of the hydrogen electrode whose E'_0 is assigned the value of -0.42 V. As protons in A are reduced to hydrogen atoms, anions will flow from A to B via the agar–salt bridge to preserve charge neutrality. If the sample in B is less reducing than the reference hydrogen electrode, a reverse series of events will occur and the sample redox couple will have an E'_0 value more positive than -0.42 V.

$$\Delta G^{0'} = -nF\Delta E'_0,$$

where n equals the number of electrons transferred in the reaction, F is the Faraday constant ($96.5\,kJ\,V^{-1}\,mol^{-1}$), and $\Delta E'_0$ is the difference in redox potential between the electron donor and electron acceptor.

If we consider the oxidation of NADH, the half reactions are

(1) $NAD^+ + 2H^+ + 2e^- \rightarrow NADH + H^+$ $E'_0 = -0.32$ V,

(2) $\frac{1}{2}O_2 + 2H + 2e^- \rightarrow H_2$ $E'_0 = 0.816$ V.

For the overall reaction, subtracting (1) from (2), we get

$NADH + H^+ + \frac{1}{2}O_2 \rightarrow NAD^+ + H_2O$,
$\Delta E'_0 = +0.816 - (-0.32) = +1.136$ V.

Therefore,

$$\Delta G^{0'} = -2(96.5\,kJ\,V^{-1}\,mol^{-1})\,(-1.136\,V) = -219.25\,kJ\,mol^{-1}.$$

The transport of electrons to oxygen does not happen in a single step. In the electron transport system of the

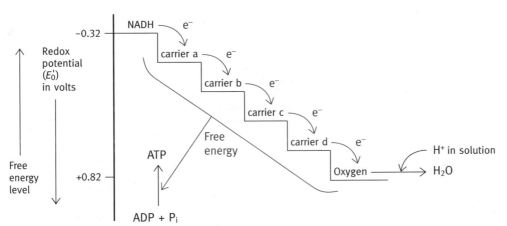

Fig. 8.7 Principle of the electron transport chain. (The number of carriers shown is arbitrary; each is a different carrier.) With some carriers only electrons are accepted, protons being liberated into solution, while, in the case of other carriers, protons accompany the electrons. The final reaction with oxygen involves protons from solution. $FADH_2$ donates electrons to the chain at a point lower down than does NADH.

mitochondria there is a chain of electron carriers of ever-increasing redox values (decreasing reducing potentials) terminating in the ultimate electron acceptor, oxygen. In effect, electrons from NADH and $FADH_2$ bump down a staircase, each step being a carrier of the appropriate redox potential and each fall releasing free energy (Fig. 8.7). The free energy is thus liberated in manageable parcels—manageable in the sense that it can be harnessed into mechanisms that (indirectly) result in ATP generation from ADP and P_i rather than being wasted as heat, as occurs in simple burning of glucose. The oxidation of NADH and $FADH_2$ thus drives the conversion of ADP and P_i to ATP. Hence the complete process is called **oxidative phosphorylation**. Thirty or more ATP molecules are synthesized from the oxidation of one molecule or glucose (the exact number is discussed in the next chapter).

In Fig. 8.8, all three phases are put together in the one scheme.

Energy generation from oxidation of fat and amino acids

In addition to glucose and glycogen oxidation to supply the energy for ATP generation, the body oxidizes fat and excess amino acids for the same purpose. To start with fat; in terms of energy production, it is the fatty acid components of neutral fat (TAG) that are quantitatively important, the glycerol portion being less significant. Chemically, fatty acids are quite different from glucose and you might well expect their oxidation to be correspondingly different. It is at this point that you might get your first glimpse of how, despite complexity in detail, the metabolism of different foodstuffs dovetails together with majestic

simplicity. Glucose, as described, is manipulated so that acetyl-CoA is formed and fed into the citric acid cycle. Fatty acids are also manipulated so that carbon atoms are detached two at a time as acetyl-CoA which is fed into the citric acid cycle. Fatty acid oxidation, therefore, differs from glucose oxidation only in this preliminary formation of acetyl-CoA. Moreover, in this preliminary manipulation, NAD^+ and FAD are reduced and the electrons they carry are fed into the electron transport system as in glucose oxidation. Figure 8.9 gives the relationship between glucose and fat oxidation.

The situation with regard to amino acid oxidation is more complex in detail but similar in concept. There are 20 different amino acids and, as explained earlier, if these are present in excess of immediate requirements they are deaminated and the carbon–hydrogen skeletons used as fuel. Once again, the metabolism of the latter shows a simplicity of concept. The carbon-hydrogen skeletons are converted either to pyruvate or to acetyl-CoA or to intermediates in the citric acid cycle, so that they also join the same metabolic path (as shown in Fig. 8.10). The citric acid cycle thus plays a central role in metabolism.

The interconvertibility of fuels

Although this chapter is concerned primarily with energy generation by food oxidation, a brief note here can throw light on the fuel logistics already described in earlier chapters. We described how glucose, in excess, can be stored as fat. This is because fatty acids can be synthesized from acetyl-CoA,

Glucose → pyruvate → acetyl-CoA → fatty acids.

However, in animals, fatty acids *cannot* be converted, in a net sense, to glucose, because to synthesize the latter (the topic of

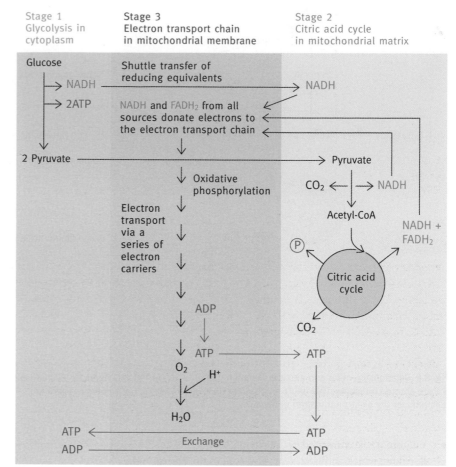

Stage 1
Glycolysis in
cytoplasm

Stage 3
Electron transport chain
in mitochondrial membrane

Stage 2
Citric acid cycle
in mitochondrial matrix

Glucose

Shuttle transfer of
reducing equivalents

NADH → NADH

2ATP

NADH and FADH$_2$ from all
sources donate electrons to
the electron transport chain

2 Pyruvate → Pyruvate

Oxidative
phosphorylation

CO_2 ← | → NADH

Electron
transport
via a
series of
electron
carriers

Acetyl-CoA

NADH +
FADH$_2$

(P)

Citric acid
cycle

CO_2

ADP

ATP → ATP

O$_2$

H$^+$

H$_2$O

ATP ←

ATP

Exchange

ADP →

ADP

Fig. 8.8 Summary of the oxidation of glucose. For ease of presentation only products are shown; obviously, reduced NAD is formed from NAD$^+$, etc. The same scheme applies to glycogen except that three ATP molecules are produced per glucose unit. We indicate the production of one —(P) from the citric acid cycle rather than as one ATP because, in this case, GDP rather than ADP is the acceptor. Energetically, it amounts to the same thing and is explained in the next chapter. The ATP generated is transported to the cytoplasm by a mechanism involving ADP intake. An important point to note is that NAD$^+$ and FAD collect electrons from different metabolic systems including fatty acid oxidation and not just from glucose (see Fig. 8.10). NADH and FADH$_2$ donate electrons to different points in the electron transport chain. The glycerophosphate shuttle transfer to FAD in the mitochondrial membrane is not shown. See Figs 9.8 and 9.9 for details of shuttles.

Chapter 12) pyruvate is needed. In animals, acetyl-CoA cannot be converted to pyruvate because the pyruvate dehydrogenase reaction is irreversible and, therefore, fatty acids cannot be converted to glucose but glucose can be converted to fatty acids and thence to neutral fat (Fig. 8.11). Those amino acids that give rise to pyruvate or a citric acid cycle acid can be converted to glucose (glucogenic amino acids). This is why, in starvation, muscle proteins are destroyed and the amino acids so produced transported to the liver, which converts them to glucose. There are organisms such as bacteria and plants that do have the capacity to convert fat and C$_2$ compounds such as acetate to glucose, but this involves a special cycle known as the glyoxylate cycle (a modified citric acid cycle) that is not present in animals. This is also described later.

The directions of metabolism, whether fat and glucose are oxidized or are synthesized, are the result of control mechanisms to be described in Chapter 13.

A survey of vitamins

A range of vitamins is essential in the diet and they have widely different structures and functions. We have already in this chapter described the roles of nicotinic acid, B$_2$ or riboflavin, and pantothenic acid. Rather than compiling the structures and functions of all the vitamins together in one section we will give a brief summary of them here together with page references to where they are dealt with in the appropriate places. Their functions are not easy to understand if presented prematurely.

The vitamins are divided into fat-soluble and water-soluble ones. First the fat-soluble components:

- Vitamin A or retinol is involved in the visual process (page 442).

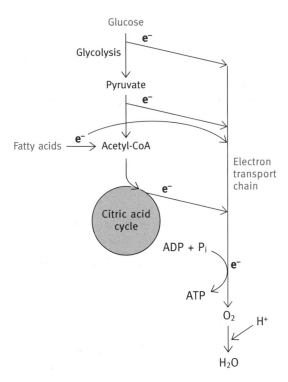

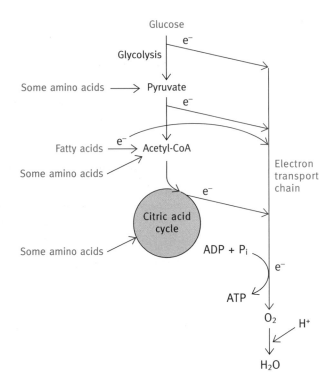

Fig. 8.9 Relationship of fat and glucose oxidation. Details given in Fig. 8.7 are omitted for clarity. This figure emphasizes the collection of electrons for the electron transport pathway.

Fig. 8.10 Relationship of glucose, fat, and amino acid oxidation for energy generation.

- Vitamin D is converted to calcitriol; this is a signalling hormone which activates genes involved in the absorption of calcium from the intestine (page 426).
- Vitamin E is an antioxidant (page 283).
- Vitamin K (for *koagulation*) is involved in blood clotting (page 279).

To turn to the water-soluble vitamins:

- Ascorbic acid (vitamin C) is involved in hydroxylation reactions, of especial importance in collagen formation (page 53).
- Biotin is involved in carboxylation reactions (page 161).
- Folic acid is involved in 1-carbon transfer reactions (page 290).
- Thiamine (B_1) is involved in pyruvate metabolism (page 155).
- Pyridoxin group (B_6) is involved in amino acid metabolism (page 256).
- Vitamin B_{12} (cobalamin) is involved in a number of special reactions (pages 183, 298).
- Riboflavin (B_2), niacin (B_3), pantothenic acid (B_5), are covered earlier in this chapter.

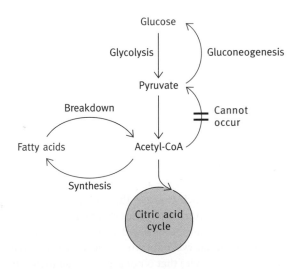

Fig. 8.11 Why glucose can be converted into fats but fats cannot be converted into glucose in animals. The reverse pathways in red are not completely the same as the forward pathways and are described in later chapters. In plants and bacteria, fat can be converted into glucose but not by the reversal of the pyruvate dehydrogenase reaction.

PROBLEMS FOR CHAPTER 8

1   What are the three major phases involved in the oxidation of glucose and where do they occur?

2   Write down the overall structure of NAD^+ in words; show the structure of the electron-accepting group in the oxidized and reduced form. Explain how NAD^+ acts as a hydrogen carrier between substrates.

3   What is FAD and its role?

4   Explain the difference between aerobic and anaerobic glycolysis in muscle and the circumstances in which they occur. What is the point of anaerobic glycolysis?

5   Write down the structure of coenzyme A in words; also give the structure of its acyl-accepting group. What is the $\Delta G^{0'}$ of hydrolysis of a thiol ester? How does this compare with that of a carboxylic ester?

6   The pyruvate dehydrogenase reaction is of central importance. Write down the reaction and give its $\Delta G^{0'}$ value.

7   What normally happens to the acetyl-CoA generated in the pyruvate dehydrogenase reaction?

8   Glycolysis and the citric acid cycle produce NADH and $FADH_2$. What happens to these?

9   The redox couple: $FAD + 2H^+ + 2e^- \rightarrow FADH_2$ has an E'_0 value of $-0.219\,V$. That of $\frac{1}{2}O_2 + 2H^+ + 2e^- =$ $0.816\,V$. Calculate the $\Delta G^{0'}$ value for the oxidation of $FADH_2$ by oxygen to water. The Nernst equation is $\Delta G^{0'} = -nF\Delta'_0$ where $F = 96.5\,kJ\,V^{-1}\,mol^{-1}$.

10   What is the major source of acetyl-CoA other than the pyruvate dehydrogenase reaction?

11   (a)   Can glucose be converted to fat? Explain your answer.
     (b)   Can fatty acids be converted to glucose in animals. Explain your answer.

Chapter summary

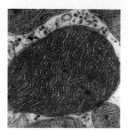

Glycolysis, the citric acid cycle, and the electron transport system: reactions involved in these pathways

In the previous chapter you saw the overall pattern of the way in which various foodstuffs are metabolized. We now want to fill in the mechanisms of the metabolic pathways, starting with carbohydrate oxidation. In subsequent chapters we will deal with how fatty acids are metabolized to join up to the glucose oxidation pathway at acetyl-CoA and then do the same with amino acid metabolism. A potential problem in studying these pathways is to forget their purposes—to get lost in the detail. If necessary, keep on going back to the previous chapter to refresh your memory on where pathways are heading.

The note at the beginning of Chapter 7 applies here also. Regulation of these pathways is dealt with in Chapter 13, but if it is preferred to deal with them now the appropriate sections can be found in that chapter.

Stage 1—glycolysis

This, we remind you, is the splitting of glucose or a glucosyl unit of glycogen into two molecules of pyruvate.

Glucose or glycogen?

So far, for simplicity, we have talked mainly of glucose metabolism. However, in normal circumstances most tissues will have glycogen stores, and glycolysis may be proceeding from this rather than from free glucose. There is a difference between the two. When glycogen is broken down (page 120), glucose-6-phosphate is produced via glucose-1-phosphate. In the liver this can be hydrolysed to release free glucose into the blood. However, glucose-6-phosphate is on the glycolytic pathway also and in all tissues is broken down to pyruvate. Free glucose is also

converted to glucose-6-phosphate by phosphorylation using ATP. You have met this reaction already (page 118) because it is the same as that involved in glycogen synthesis. Whether glucose-6-phosphate goes to glycogen or to pyruvate or to blood glucose in the case of liver depends on how the metabolic control switches are set, according to physiological needs, and this is a major later topic. The relationship between glycolysis, glucose, and glycogen is shown in Fig. 9.1. It costs a molecule of ATP to produce glucose-6-phosphate from glucose, but no ATP is used if we start with glycogen, because the energy of the glucosyl group in the latter (that is, the $\Delta G^{0'}$ of its hydrolysis) is similar to that of the phosphate ester and is preserved by phosphorolysis.

Why use ATP here at the beginning of glycolysis?

It may seem odd that a pathway designed to produce ATP should start by using up ATP. Why use it here? The reason is that glycolysis involves phosphorylated compounds and ATP must be used to phosphorylate glucose—it has the necessary energy potential. Glucose-6-phosphate is a low-energy phosphoryl compound so certainly we have lost a high-energy phosphoryl group in using ATP. Think of it as an investment for, as you'll see, a 100% profit is made on the ATP used in glycolysis of glucose.

Why is glucose-6-phosphate converted to fructose-6-phosphate?

The next step is to convert glucose-6-phosphate, an aldose sugar, to fructose-6-phosphate, its ketose isomer. The aim, you'll not forget, is to split glucose into two C_3 compounds. To (apparently) digress for a moment, organic chemistry

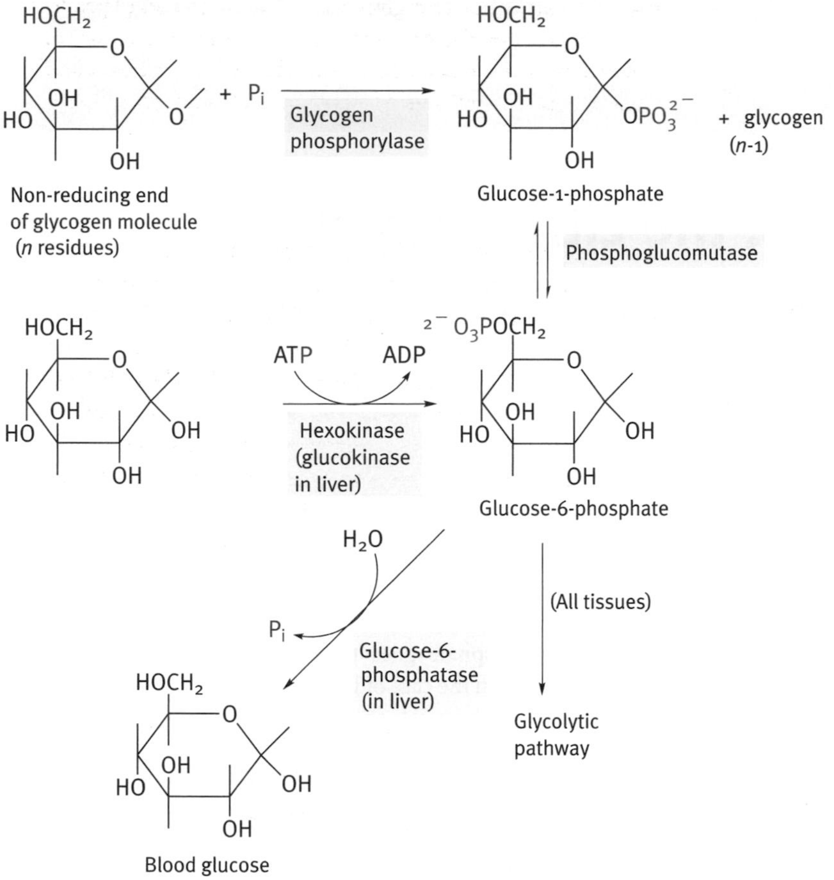

Fig. 9.1 Production of glucose-6-phosphate from glycogen or free glucose and its fate. Which routes are operative depends on control mechanisms described in Chapter 13.

textbooks describe a test-tube reaction called the **aldol condensation**. In this, an aldehyde and a ketone (or aldehyde) condense together as shown in Fig. 9.2. The reaction is freely reversible and therefore can be used to split an aldol into two parts—an aldol being the β-hydroxycarbonyl compound shown. Now turn back to glucose-6-phosphate; this does not have the correct aldol structure for an aldol split, but fructose-6-phosphate does, as is easily seen in the straight-chain formulae in Fig. 9.3. By forming the fructose isomer, the sugar phosphate can be split into two. The glucose-6-phosphate is isomerized into fructose-6-phosphate by the enzyme **phosphohexose isomerase**. Before splitting, another phosphoryl group from ATP is transferred to the fructose-6-phosphate by the enzyme **phosphofructokinase** (PFK), yielding fructose-1:6-bisphosphate. The fructose-1:6-bisphosphate is now split by the enzyme **aldolase** catalysing the aldol reaction; the second phosphate means that each of the two C_3 products is phosphorylated to give glyceraldehyde-3-phosphate and dihydroxyacetone phosphate (Fig. 9.4). In Fig. 9.5, the same reactions are presented with the more commonly used ring structures for the sugars; the fructose-6-

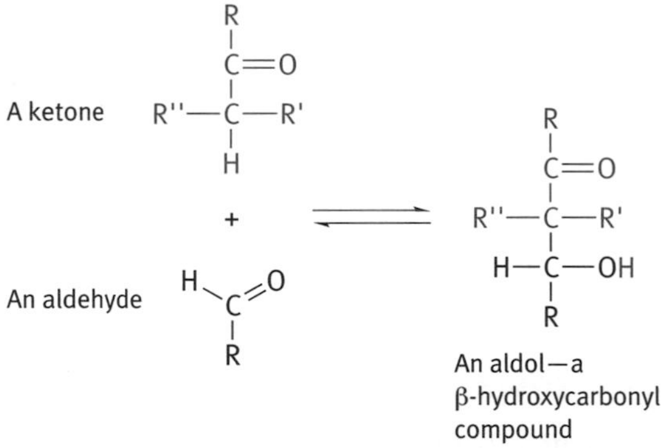

Fig. 9.2 The aldol condensation. A chemical reaction between an aldehyde and a ketone (or aldehyde).

phosphate is in the five-membered ring (furanose) configuration. The $\Delta G^{0'}$ for the aldolase reaction is $+24.3\,\mathrm{kJ\,mol^{-1}}$, which would seem to preclude its ready occurrence. There are, however, special considerations applying to this reaction because

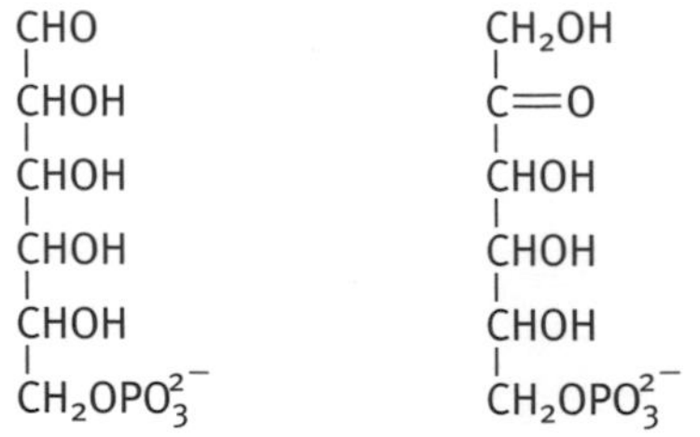

CHO	CH$_2$OH
CHOH	C=O
CHOH	CHOH
CHOH	CHOH
CHOH	CHOH
CH$_2$OPO$_3^{2-}$	CH$_2$OPO$_3^{2-}$

Glucose-6-phosphate,
an aldose sugar.
This is not an aldol

Fructose-6-phosphate,
the ketose isomer.
This *is* an aldol

Fig. 9.3 The straight-chain formulae of an aldose sugar and its ketose isomer. The glucose-6-phosphate is in equilibrium with the six-membered ring (pyranose) form and the fructose-6-phosphate with the five-membered ring (furanose) form. The ring structures are shown in Fig. 9.5.

one molecule of reactant gives rise to two molecules of product which means that the ΔG of the reaction is influenced by concentration to an unusual degree (see below). In cellular conditions, the ΔG is small and the reaction freely reversible.

A note on the $\Delta G^{0\prime}$ and ΔG values for the aldolase reaction

The reaction catalysed by aldolase has a $\Delta G^{0\prime}$ value of +24.3 kJ mol^{-1} and is freely reversible while reactions with smaller $\Delta G^{0\prime}$ values will not proceed in the cell. $\Delta G^{0\prime}$ values are determined at 1 M concentrations of reactants and products; since concentrations in the cell are more likely to be at 10^{-3}–10^{-4} M, ΔG values are always different from $\Delta G^{0\prime}$ values. Nonetheless, the latter are usually useful guides to metabolic events. This is not true in the case of aldolase where the correlation between $\Delta G^{0\prime}$ values and ΔG values in the cell is very poor.

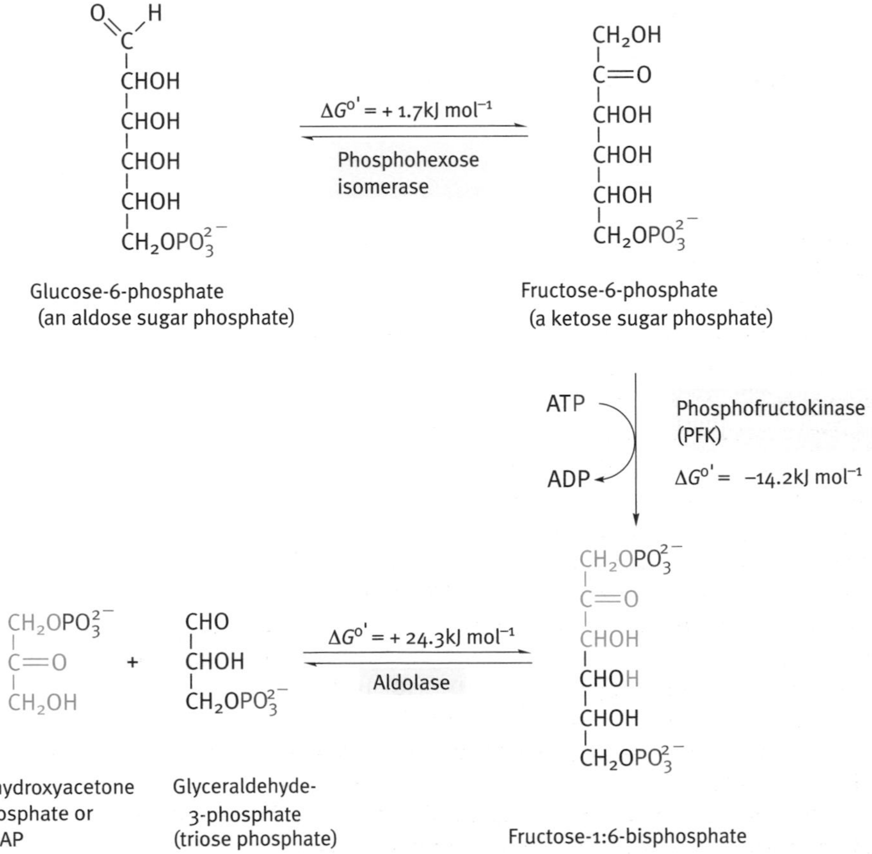

Fig. 9.4 Conversation of glucose-6-phosphate into two C$_3$ compounds. The $\Delta G^{0\prime}$ value for the aldolase reaction in the forward direction would appear to preclude its occurrence, but see text for explanation of this. Straight-chain formulae for the sugars are used here for clarity; the reactions are commonly presented as in Fig. 9.5.

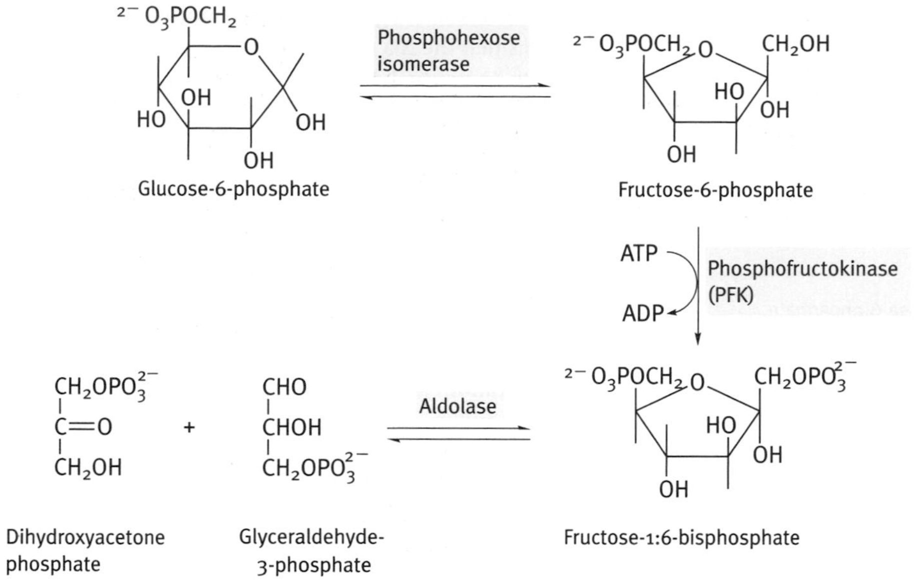

Fig. 9.5 This is the same scheme as in Fig. 9.4 but presented in the more usual form with sugars as ring structures.

The reason is that in the aldolase reaction, one molecule of reactant (fructose-1:6-bisphosphate) gives rise to two molecules of produce (glyceraldehyde-3-phosphate and dihydroxyacetone phosphate). The relationship between $\Delta G^{0\prime}$ and ΔG values is given in the equation

$$\Delta G = \Delta G^{0\prime} + RT \ln \frac{[products]}{[reactants]},$$

that is,

$$\Delta G = \Delta G^{0\prime} + RT$$
$$\times \ln \frac{[glyceraldehyde\text{-}3\text{-}phosphate] \times [dihydroxyacetone\ phosphate]}{[fructose\text{-}6\text{-}phosphate]}.$$

Because there are two products of low concentration, the $RT \ln\{[products]/[reactants]\}$ moiety has a large negative value, giving a ΔG value compatible with ready reversibility in the cell. To illustrate this, assume that all reactants are present in the cell at 10^{-4} M. Then

$$\Delta G = +24.3\ \text{kJ mol}^{-1}$$
$$+ (8.315 \times 10^{-3})(298)2.303 \log_{10} \frac{(10^{-4})(10^{-4})}{(10^{-4})}$$
$$= +24.3 - 22.8 = +1.5\ \text{kJ mol}^{-1}.$$

It can be seen that the actual ΔG value at these assumed concentrations is compatible with free reversibility.

Interconversion of dihydroxyacetone phosphate and glyceraldehyde-3-phosphate

Glyceraldehyde-3-phosphate and dihydroxyacetone phosphate are isomeric molecules. An enzyme, **triose phosphate isomerase**, interconverts these.

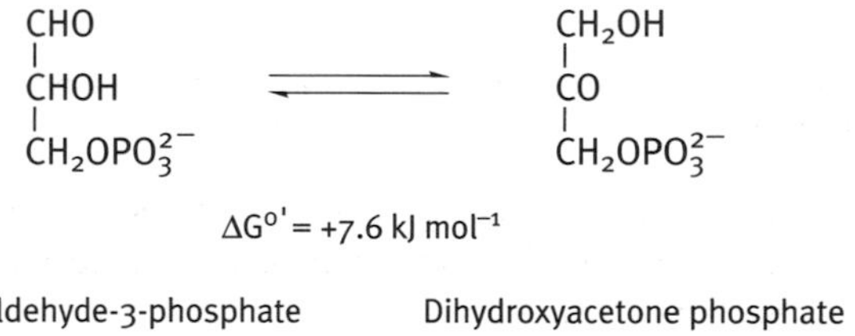

$$\Delta G^{0\prime} = +7.6\ \text{kJ mol}^{-1}$$

Glyceraldehyde-3-phosphate   Dihydroxyacetone phosphate

The two compounds are in equilibrium but, since glyceraldehyde-3-phosphate is continually removed by the next step in glycolysis, all of the dihydroxyacetone phosphate is progressively converted to glyceraldehyde-3-phosphate.

Glyceraldehyde-3-phosphate dehydrogenase—a high-energy-phosphoryl-compound-generating step

The aldehyde group of glyceraldehyde-3-phosphate is oxidized by NAD^+. You would expect this to produce a carboxyl group (and so it does, ultimately) but oxidation of a —CHO group to —COO$^-$ has a large negative ΔG value, sufficient in fact to

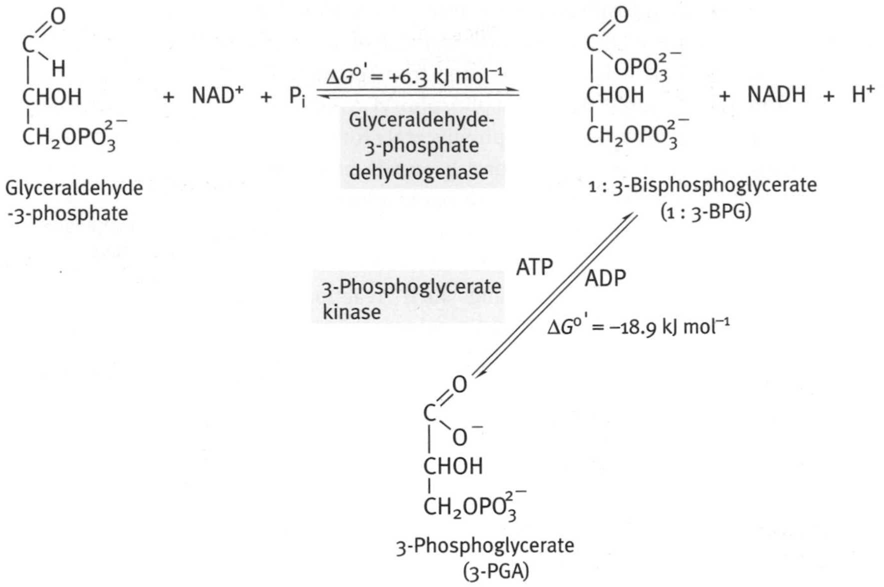

Fig. 9.6 Conversion of glyceraldehyde-3-phosphate to 3-phosphoglycerate.

generate a high-energy phosphate compound from P_i on the way. The mechanism by which this is achieved is as follows: at the active site of the enzyme there is the amino acid cysteine which has a thiol or sulfhydryl group (—SH) on its side chain.

The aldehyde glyceraldehyde-3-phosphate condenses with the thiol to form a thiohemiacetal.

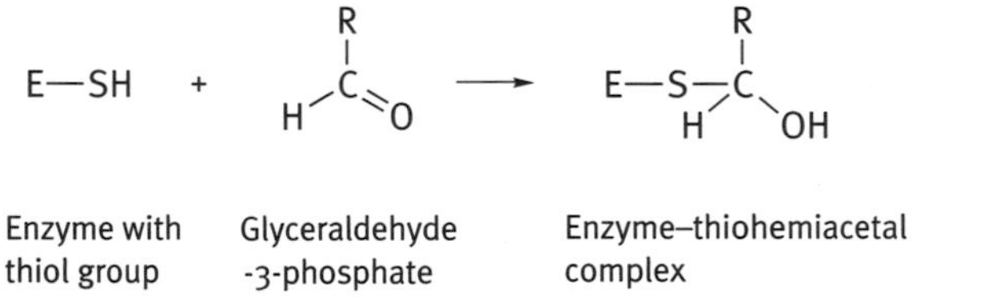

Enzyme with thiol group   Glyceraldehyde-3-phosphate   Enzyme–thiohemiacetal complex

The enzyme substrate complex is now oxidized, the electrons being accepted by NAD^+, and a thiol ester is formed with the enzyme sulfhydryl group.

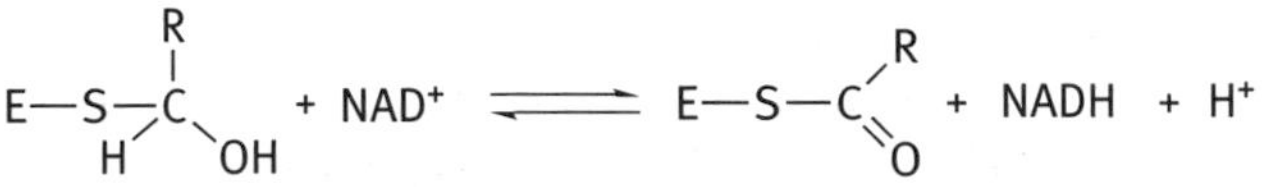

A thiol ester (R—CO—S—), as explained earlier (page 139), is a high-energy compound—of the same order as that of a high-energy phosphate compound. It is therefore thermodynamically feasible for P_i to react as follows.

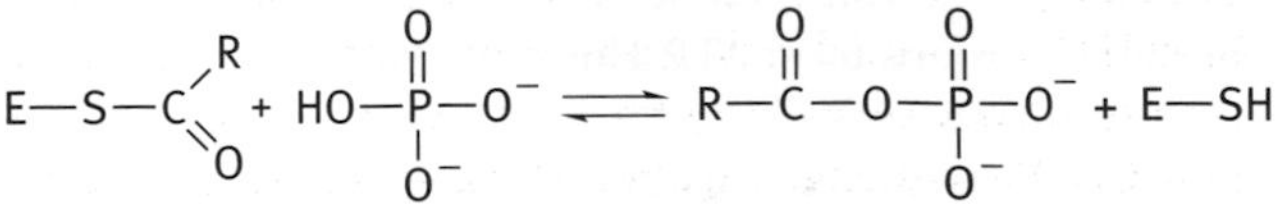

The $\overset{O}{\underset{}{R\overset{||}{C}}}{-}PO_3^{2-}$ group is also high energy and so the phosphoryl group can be transferred to ADP forming ATP.

The responsible enzyme is **phosphoglycerate kinase** because, in the reverse direction, it transfers a phosphoryl group from ATP to 3-phosphoglycerate (3-PGA). Kinases are always named from the ATP side of the reaction (Fig. 9.6).

The generated phosphoryl group in this process is formed attached to the actual substrate of an enzyme. For this reason it is called **substrate-level phosphorylation**, a point to which we'll refer later. 3-Phosphoglycerate is a low-energy phosphate compound and cannot phosphorylate ADP. However, the next steps in glycolysis manipulate the molecule so that this low-energy phosphate ester becomes a high-energy phosphoryl group, transferable to ATP. This is not energy for nothing—you'll see how it is done within the law.

The final steps in glycolysis

First, the phosphoryl group of 3-phosphoglycerate is transferred from the 3 to the 2 position as shown.

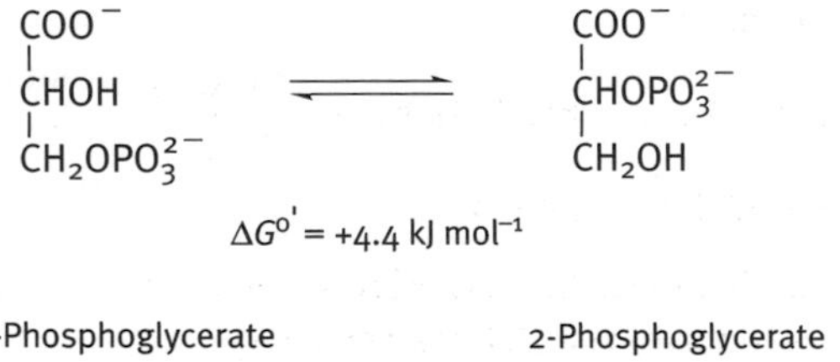

$$\Delta G^{o'} = +4.4 \text{ kJ mol}^{-1}$$

3-Phosphoglycerate   2-Phosphoglycerate

This is called the **phosphoglycerate mutase reaction**. The reaction is not really an *intra*molecular transfer of the phosphoryl group (though it is in the enzyme from plants). The enzyme from rabbit muscle contains a phosphoryl group that it donates to the 2-OH group of 3-phosphoglycerate forming 2:3-bisphosphoglycerate. The 3-phosphoryl group is now transferred to the enzyme to replace the donated phosphate, so that the net effect is the reaction shown above. The next step in glycolysis is that a water molecule is removed from the 2-phosphoglycerate. Enzymes catalysing such reactions are usually called dehydratases but in this particular case in glycolysis, the old established name is **enolase** (because it forms a substituted enol).

$$\begin{array}{l} COO^- \\ | \\ CHOPO_3^{2-} \\ | \\ CH_2OH \end{array}$$

2-Phosphoglycerate

Enolase
$-H_2O$

$$\begin{array}{l} COO^- \\ | \\ C-O-PO_3^{2-} \\ \| \\ CH_2 \end{array} \xrightarrow[\text{ADP}]{\substack{\text{Pyruvate}\\\text{kinase}}} \left[\begin{array}{l} COO^- \\ | \\ C-OH \\ \| \\ CH_2 \end{array}\right] \longrightarrow \begin{array}{l} COO^- \\ | \\ C=O \\ | \\ CH_3 \end{array}$$

$+$ ATP

Phosphoenolpyruvate   Enolpyruvate   Pyruvate
(PEP)

The enolase reaction has a $\Delta G^{0\prime}$ of only $+1.8\,\text{kJ}\,\text{mol}^{-1}$, but the enolphosphate compound is of the 'high-energy' type, with a $\Delta G^{0\prime}$ of hydrolysis of $-62.2\,\text{kJ}\,\text{mol}^{-1}$; a reason for this is that the immediate product of the reaction, the enol form of pyruvate, spontaneously converts to the keto form, a reaction with a large negative $\Delta G^{0\prime}$ value. The conversion of a low-energy phosphate to a high-energy one by enolase is not a case of energy for nothing but rather that the energy of the molecule is rearranged. The phosphoryl group is transferred to ADP by the enzyme **pyruvate kinase**; this name might misleadingly imply that pyruvate can be phosphorylated by ATP by reversal of the reaction; the name of the enzyme derives from the convention mentioned earlier that a kinase is named from the reaction involving ATP even though, in this case, that reaction never occurs. The irreversibility of the conversion of PEP to pyruvate has important metabolic repercussions as you'll see when we come later to gluconeogenesis. (There is a potential source of

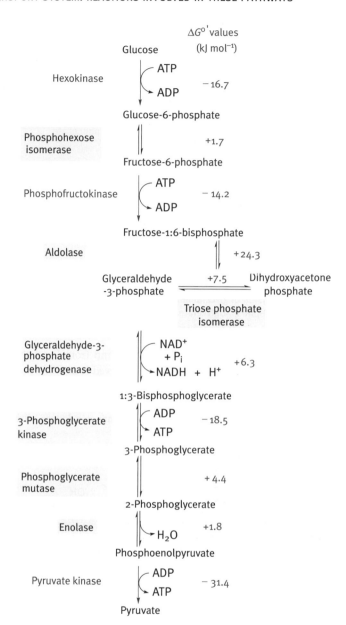

Fig. 9.7 The glycolytic pathway. Irreversible reaction are indicated in red. The free reversibility of the aldolase reaction would appear to be inconsistent with such a large $\Delta G^{0\prime}$ value (see text for explanation).

confusion arising from the face that in certain plants and microorganisms, pyruvate *is* directly converted to PEP by a quite different enzyme (described on page 248) that utilizes two phosphoryl groups from ATP. However, this reaction does *not* occur in animals.)

The complete glycolytic pathway is shown in Fig. 9.7.

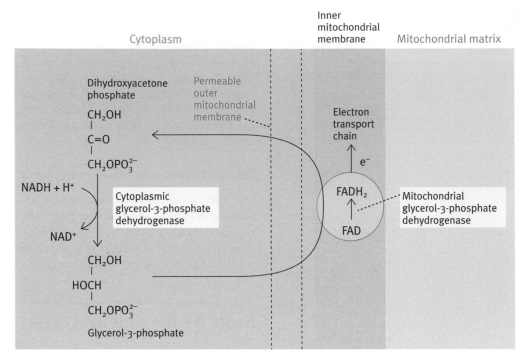

Fig. 9.8 The glycerophosphate shuttle that transfers electrons from cytoplasmic NADH to the electron transport chain of mitochondria.

The ATP balance sheet from glycolysis

Starting with glucose, two molecules of ATP were used to form fructose-1:6-bisphosphate. The phosphoglycerate kinase produced two ATP molecules per original glucose and the pyruvate kinase two—a total of four and a net gain of two. Starting from glycogen, only one molecule of ATP was consumed so the net gain is three molecules of ATP formed from ADP + P_i per glucosyl unit.

Reoxidation of cytoplasmic NADH by electron shuttle systems

In the aerobic situation, the NADH generated in glycolysis is re-oxidized by transferring its electrons into mitochondria. As already explained (page 139) NADH itself cannot enter the mitochondrion; there are two systems for transferring its electrons into mitochondria. The first involves dihydroxyacetone phosphate (DHAP, generated by the aldolase reaction). An enzyme in the cytoplasm transfers electrons from NADH to DHAP giving glycerol-3-phosphate (Fig. 9.8). The enzyme is called **glycerol-3-phosphate dehydrogenase**, working in reverse in the above reaction.

Glycerol-3-phosphate can reach the inner mitochondrial membrane (the outer one being highly permeable) where a different type of glycerol-3-phosphate dehydrogenase, built into the membrane—this time with an FAD prosthetic group—transfers electrons from glycerol-3-phosphate to the mitochondrial electron transport chain. The DHAP so produced cycles (shuttles) back into the cytoplasm to pick up more electrons (Fig. 9.8). Note that in this, the electron carrier (glycerol-3-phosphate) does not have to enter the mitochondrial matrix but its pair of electrons gain access to the electron chain carrying electrons to oxygen, located in the inner mitochondrial membrane. The net effect is to transfer electrons from cytoplasmic NADH to the mitochondrial electron transport chain.

Another shuttle, the **malate–aspartate shuttle**, transfers electrons from cytoplasmic NADH to mitochondrial NAD⁺. The mitochondrial NADH thus generated is then oxidized by the electron transport chain. This shuttle system involves transfer of electrons from NADH to oxaloacetate to form malate in the cytoplasm which is transported into the mitochondrion, by a specific carrier, where it is re-oxidized to oxaloacetate, mitochondrial NAD⁺ being reduced (Fig. 9.9). The net effect is that NADH outside reduces NAD⁺ inside. This shuttle is a little more complicated in that the oxaloacetate, so formed, can't traverse the mitochondrial membrane to get back to the cytoplasm. It is

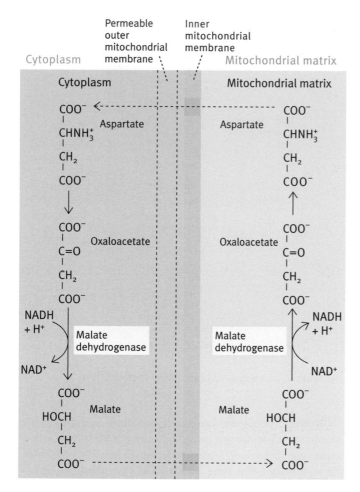

Fig. 9.9 The malate–aspartate shuttle for transferring electrons from cytoplasmic NADH to mitochondrial NAD$^+$. The mechanism of the interconversion of oxaloacetate and aspartate is dealt with on page 200. This shuttle, unlike the glycerophosphate shuttle, is reversible, and can operate as shown, bringing NADH into the mitochondrion, only if the NADH/NAD$^+$ ratio is higher in he cytoplasm than in the mitochondrial matrix.

converted to aspartate, which is transported, again by a specific carrier, to the cytoplasm and reconverted to oxaloacetate there; hence the name, the malate-aspartate shuttle. At this stage we'll not give the mechanism of aspartate $\rightleftharpoons$ oxaloacetate interconversions since it will be more convenient to do this later when we deal with amino acid metabolism (page 256). Different tissues probably use the two shuttles to different extents. The two shuttles differ in a significant way—the glycerophosphate shuttle results in cytoplasmic NADH reducing mitochondrial FAD in the form of the prosthetic group of the mitochondrial flavo-protein, glycerol-3-phosphate dehydrogenase. The FADH$_2$ has a

higher redox potential then NADH (lower energy); it hands on its electrons to the electron transport chain at a point that is further along the chain from that at which NADH hands on its electrons. The net ATP generation from the oxidation of a *cytoplasmic* molecule NADH by this FADH$_2$ route is 1.5 molecules (see below). The aspartate–malate shuttle starts with one molecule of cytoplasmic NADH and ends up with one molecule of mitochondrial NADH whose oxidation generates 2.5 molecules of ATP (explained later).

Transport of pyruvate into the mitochondria

The other product of glycolysis besides NADH is pyruvate. Unless it is reduced to lactate (see page 138) the pyruvate, as explained earlier, is transported into the mitochondrial matrix by an antiport type of membrane transport protein (page 82), which exchanges it for OH$^-$ inside the matrix.

Stage 2—the citric acid cycle

Before we get to the cycle itself we will deal with the preparation of pyruvate to enter the cycle, by which we mean its conversion to acetyl-CoA.

Conversion of pyruvate to acetyl-CoA—a preliminary step before the cycle

As outlined earlier (page 139), pyruvate in the mitochondrial matrix is converted to acetyl-CoA, which feeds the acetyl group into the citric acid cycle (the structure of CoA is described on page 139).

The overall reaction catalysed by pyruvate dehydrogenase is repeated here

Pyruvate + NAD$^+$ + CoA—SH → Acetyl—S—CoA
     + NADH + H$^+$ + CO$_2$.

Pyruvate dehydrogenase, the responsible enzyme, is a very large complex composed of many polypeptides. It essentially consists of three different enzyme activities aggregated together for efficiency, each catalysing one of the intermediate steps in the process (Fig. 9.10). The first step is decarboxylation of pyruvate to produce CO$_2$ and a hydroxyethyl group $CH_3\overset{\displaystyle H}{\underset{\displaystyle OH}{C}}-$ attached to the cofactor **thiamin pyrophosphate (TPP)**. TPP is derived from thiamin, or vitamin B$_1$, deficiency of which impairs the ability to metabolize carbohydrate. The hydroxyethyl group, in a series of steps, is converted to the acetyl group of acetyl-CoA

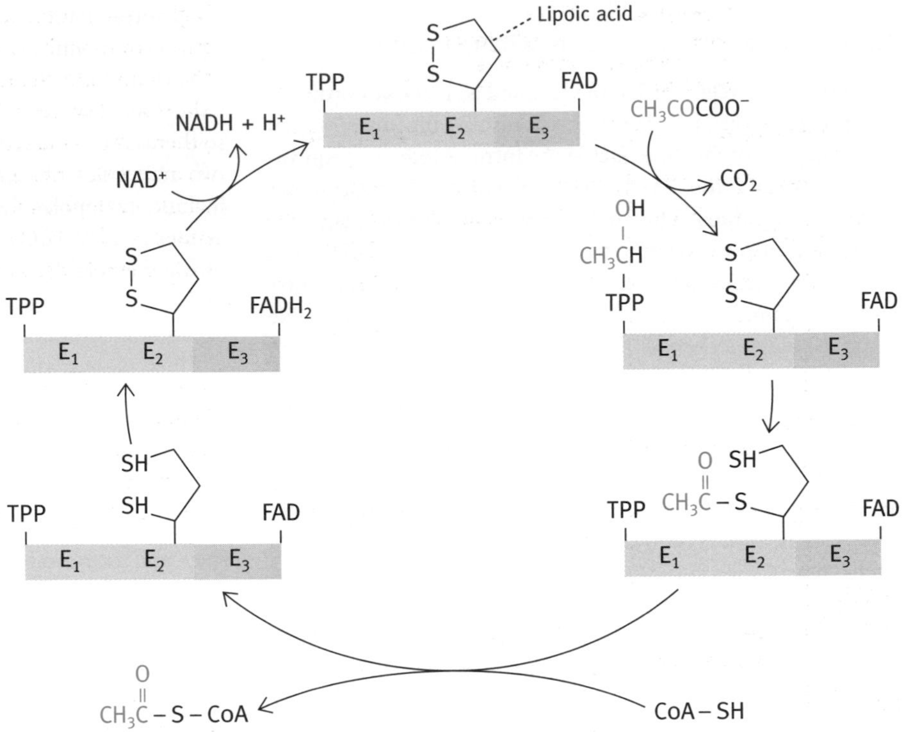

Fig. 9.10 Mechanism of the pyruvate dehydrogenase reaction. TPP, thiamin pyrophosphate; E_1–E_3, enzyme regions of complex.

with the reduction of NAD^+. The process is known as an **oxidative decarboxylation** for obvious reasons. The structure of components of the reaction given below (the cofactor TPP, and the lipoic acid that takes part in the reaction as part of the enzyme) are worth a look, but it is suggested that you need not learn these structures.

This conversion of pyruvate to acetyl-CoA is irreversible; the $\Delta G^{0\prime}$ of the reaction is $-33.5\,\mathrm{kJ\,mol^{-1}}$. As you will see later, this is of profound significance for irreversibility means that fatty acids can never be converted in a net sense to glucose in the animal body though, as mentioned, bacteria and plants have a special mechanism for achieving this. The acetyl-CoA now enters the citric acid cycle.

Components involved in the pyruvate dehydrogenase reaction

TPP has the structure

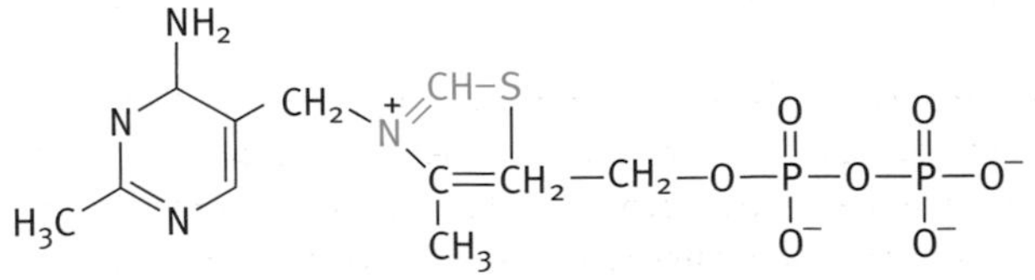

TPP-hydroxyethyl has the structure

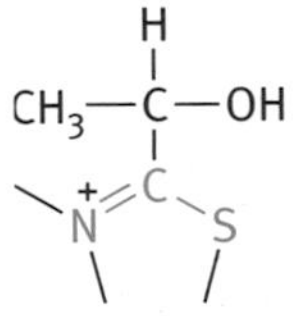

Lipoic acid in its reduced form has the structure

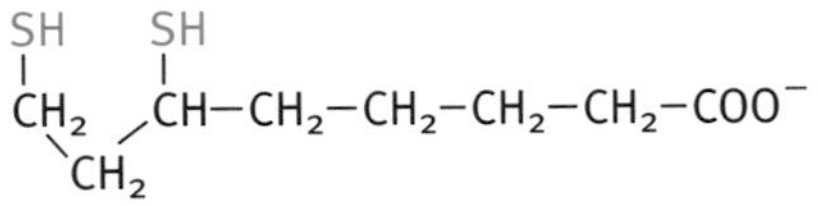

and in its oxidized form the structure is

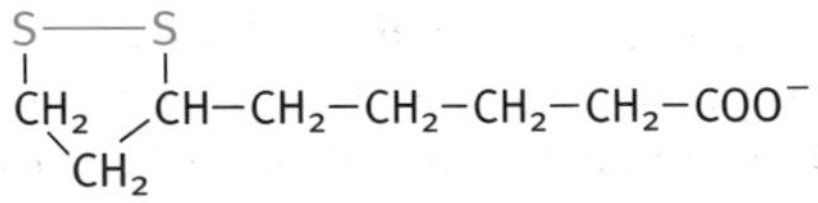

Lipoyllysine consists of lipoic acid attached to a lysine side chain of the enzyme by —CO—NH— linkage. In Fig. 9.10 lipoyllysine is represented by the disulfide structure. The three enzymes represented by E_1, E_2, and E_3 are part of a very large protein complex.

The citric acid cycle as a water-splitting machine

The citric acid cycle, which looks like, and is, a series of ordinary enzymic reactions, is a device of astonishing ingenuity in that it produces a combustible fuel in the form of reducing equivalents of NADH and $FADH_2$, part of which, in effect, comes from splitting water. This fuel is burned in the next stage, the electron transport system, to produce ATP from ADP and P_i. Unlike schemes to run cars on water, the cycle is thermodynamically sound because it uses the free energy made available from the destruction of the acetyl group of acetyl-CoA to drive the process. It must be noted that the splitting of water is not a direct 'head on' process as occurs in photosynthesis (page 244) and oxygen is not liberated as such, but as CO_2. Rather it is an indirect splitting that can easily be overlooked and is seldom referred to in texts. We need to explain this aspect more clearly for it is central to appreciating what the cycle is all about.

Acetyl-CoA enters the cycle by a reaction with oxaloacetate to produce citrate. Do not bother at this stage with how this occurs—it is described later. The reaction involves the input of a molecule of water. With one complete 'turn' of the cycle (again explained later) oxaloacetate is re-formed and the acetyl group of acetyl-CoA has disappeared. As a result of the cycle reactions, the products are as follows: two molecules of CO_2; the reduction of three molecules of NAD^+ to NADH; and that of one molecule of FAD to $FADH_2$. In addition, the CoA—S— of acetyl-CoA becomes CoASH (Fig. 9.11). (A single high-energy phosphoryl group, as GTP, is produced from P_i.) If you add up all the reducing equivalents of the three NADH and one $FADH_2$, there are eight. (Remember that NAD^+ accepts two electrons.) The formation of the thiol group of CoASH (from CoA—S—) requires

one more—a total of nine reducing equivalents (effectively equivalent to nine H atoms).

The CH_3CO—S—CoA supplies three of these so that there is a shortfall of six. There is no involvement of oxygen in the cycle so there is also a shortfall of three oxygen atoms to produce the two molecules of CO_2. The source of these 'missing' atoms includes two molecules of H_2O. You will see below that, as well as the input of H_2O into citrate synthesis, a second molecule of water enters the cycle. But, this still leaves a shortfall of two hydrogen and one oxygen atoms. It will be more convenient to explain the source of these later (page 159).

Thus, we see the remarkable feat—electrons of H_2O are raised up the energy scale to become reducing equivalents of NADH and $FADH_2$ and, of course, electrons from the acetyl group also are utilized for this purpose. Again, it is emphasized that this does not mean that the components of H_2O go *directly* to these products but they are required for the chemical balance sheet and the net effect is that H_2O reduces NAD^+ and FAD and the oxygen is taken up by CO_2 production. As stated, the chemical reactions involved in converting the acetyl group to its products provide the free energy to convert H_2O into reducing equivalents. In fact, the whole cycle has a negative ΔG value and so proceeds thermodynamically 'downhill'.

With that preamble we can now turn to the reactions of the cycle.

A simplified version of the citric acid cycle

Possibly one obstacle to learning the cycle is that the progression of reactions doesn't make much sense until you have completed it, so we will first look at a simplified version devoid of detail.

Be sure you know the structure of oxaloacetic acid for that's where it all starts and finishes. Acetate is CH_3COO^-, the oxalo group is $^-OOC—C{\overset{O}{\diagup}}$ so oxaloacetate is

$$\overset{O}{\underset{H_2C—COO^-}{\overset{\diagdown}{C}—COO^-}}.$$

The acetyl group of acetyl-CoA is joined to oxaloacetate to form citric acid; it is easy to see how citrate can be derived from oxaloacetate.

$$\underset{H_2C—COO^-}{\overset{CH_2COO^-}{\underset{|}{\overset{|}{HO—C—COO^-}}}}$$

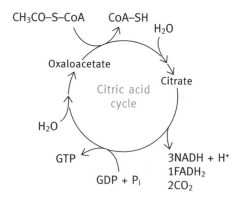

Fig. 9.11 The inputs and outputs of the citric acid cycle. Individual cycle reactions are not indicated.

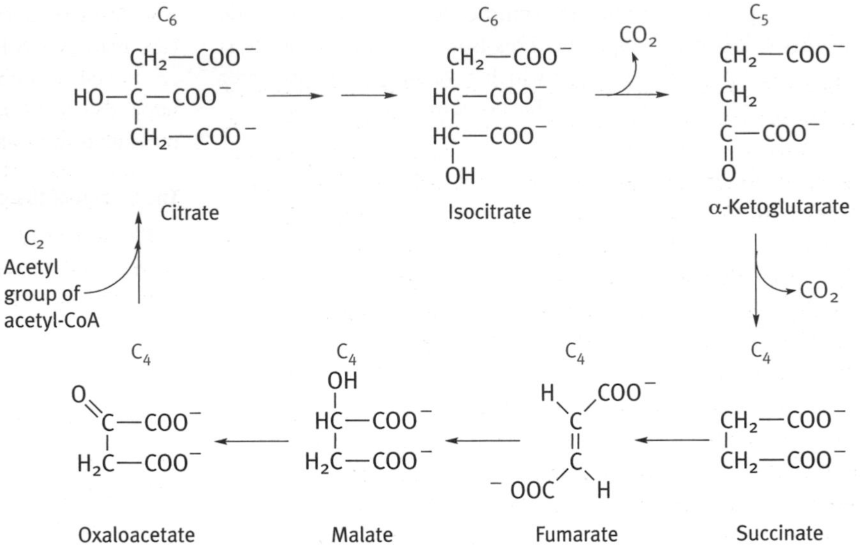

Fig. 9.12 Simplified citric acid cycle showing the component acids and their sequence. The purpose of this diagram is for you to learn the structures of the main cycle acids and their interrelationships without complicating details. When you are familiar with these it will be easier to appreciate the full cycle.

It is helpful to remember that citrate is a C_6 *symmetric tricarboxylic* compound. This is converted to its asymmetric isomer, isocitrate, which, in the cycle, is progressively converted to α-ketoglutarate (C_5), succinate (C_4), fumarate (C_4), malate (C_4), and oxaloacetate (Fig. 9.12). This means that one turn of the cycle eliminates the acetyl group fed into it.

We suggest that you make yourself completely familiar with these acids of the cycle. With that preparation, a more detailed consideration of this superhighway of carbon compound metabolism can be given.

Mechanisms of the citric acid cycle reactions

We might, for convenience, divide the reactions into three groups. (1) the synthesis of citrate—this is the reaction feeding acetyl groups into the cycle; (2) the 'top part' of the cycle—involving conversion of C_6 citrate to C_5 α-ketoglutarate; and (3) the 'carbon 4' part-conversion of succinate to oxaloacetate.

The synthesis of citrate

The name of the enzyme involved here is **citrate synthase**. The enzyme catalyses the condensation of acetyl-CoA with oxaloacetate to give citryl-CoA. This is unstable and hydrolyses to citrate. Citrate formation has a large negative ΔG value ($-32.2\,\text{kJ}$ mol^{-1}) and hence the reaction is irreversible.

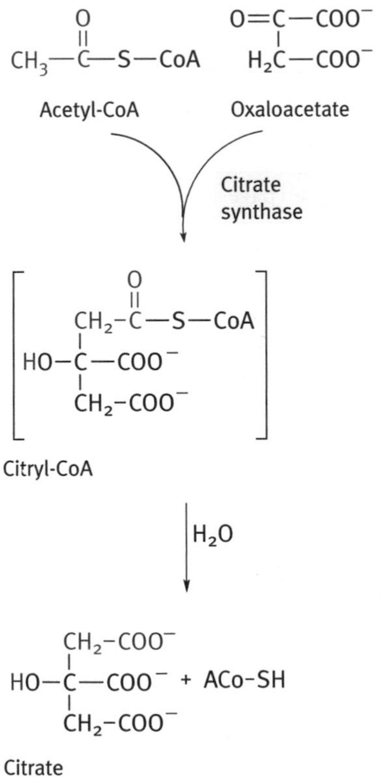

Conversion of citrate to α-ketoglutarate

Citrate → Isocitrate
The strategy of this reaction is to switch the hydroxyl group of citrate (a symmetric molecule) from the 3 position to the 2

position, giving isocitrate (an asymmetric molecule). The logic of this will become apparent. This isomerization of citrate is catalysed by a single enzyme which reversibly removes water and adds it back across the double bond in either direction.

$$H-\underset{|}{\overset{|}{C}}-OH \quad \underset{+H_2O}{\overset{-H_2O}{\rightleftarrows}} \quad H-\overset{|}{C}$$
$$H-\underset{|}{\overset{|}{C}}-H \qquad\qquad H-\overset{||}{C}$$

Enzymes catalysing such reactions are usually called dehydratases but in this case, due to long tradition, its name is **aconitase** because the unsaturated product is *cis*-aconitate (found first in the plant genus *Aconitum*).

$$\underset{\text{Citrate}}{\begin{array}{c}CH_2-COO^-\\|\\HO-C-COO^-\\|\\CH_2-COO^-\end{array}} \underset{+H_2O}{\overset{-H_2O}{\rightleftarrows}} \underset{\text{cis-Aconitate}}{\begin{array}{c}CH_2-COO^-\\|\\CH\\||\\CH-COO^-\end{array}} \underset{-H_2O}{\overset{+H_2O}{\rightleftarrows}} \underset{\text{Isocitrate}}{\begin{array}{c}CH_2-COO^-\\|\\HC-COO^-\\|\\CHOH-COO^-\end{array}}$$

Since isocitrate is now metabolized further in the cycle the net effect is that aconitase catalyses the reaction sequence

Citrate → *cis*-Aconitate → Isocitrate

Isocitrate dehydrogenase
You have already met the class of NAD⁺-requiring dehydrogenases in lactate dehydrogenase (page 138). This again is a common reaction of the type

$$H-\underset{|}{\overset{|}{C}}-H \atop H-\underset{|}{\overset{|}{C}}-OH \quad +\ NAD^+ \longrightarrow \quad H-\underset{|}{\overset{|}{C}}-H \atop \underset{|}{C}=O \quad +\ NADH\ +\ H^+ \ .$$

In the cycle **isocitrate dehydrogenase** catalyses the reaction

$$\underset{\text{Isocitrate}}{\begin{array}{c}CH_2-COO^-\\|\\CH-COO^-\\|\\CHOH-COO^-\end{array}}$$

$$\Big\downarrow \ +\ NAD^+$$

$$\left[\underset{\text{Oxalosuccinate}}{\begin{array}{c}CH_2-COO^-\\|\\\alpha CH-COO^-\\|\\\beta C-COO^-\\||\\O\end{array}}\right] \longrightarrow \underset{\text{α-Ketoglutarate}}{\begin{array}{c}CH_2-COO^-\\|\\CH_2\\|\\C-COO^-\\||\\O\end{array}} \ +\ CO_2\ .$$

$$+\ NADH\ +\ H^+$$

The immediate product is oxalosuccinate which is a β-keto acid (the keto group is β to the centre-COOH group). Such acids are unstable and readily lose the carboxyl group as CO_2. This happens on the surface of the isocitrate dehydrogenase so the product is the C_5 acid α-ketoglutarate as shown.

The C₄ part of the cycle

α-Ketoglutarate is an analogue of pyruvate. We can write both as

$$\begin{array}{c}R\\|\\C-COO^-\\||\\O\end{array}$$

For pyruvate R=—CH₃; for α-ketoglutarate R=—CH₂CH₂COO⁻. We have already seen that pyruvate dehydrogenase converts pyruvate to acetyl-CoA and CO_2. The reaction is repeated here.

$$\begin{array}{c}R\\|\\C-COO^-\\||\\O\end{array} \ +\ NAD^+\ +\ CoA-SH$$

$$\Big\downarrow$$

$$\begin{array}{c}R\\|\\C-S-CoA\\||\\O\end{array} \ +\ CO_2\ +\ NADH\ +\ H^+$$

An equivalent enzyme complex exists for α-ketoglutarate and precisely the same equation applies as above except that R=—CH₂CH₂COO⁻; the enzyme complex attacks α-ketoglutarate rather than pyruvate. The product of **α-ketoglutarate dehydrogenase** is therefore succinyl-CoA, analogous to acetyl-CoA.

$$\underset{\text{Succinyl-CoA}}{\begin{array}{c}CH_2-COO^-\\|\\CH_2\\|\\C-S-CoA\\||\\O\end{array}}$$

However, in the cycle, whereas acetyl-CoA is used to form citrate, succinyl-CoA is broken down to free succinate plus CoA—SH. In principle, the simplest way to do this would be to hydrolyse succinyl-CoA. However, the $\Delta G^{0\prime}$ of this hydrolysis of the thiol ester is $-35.5\ \text{kJ mol}^{-1}$—enough energy to raise P_i to a high-energy phosphate compound. So, why waste this energy? Why not trap it instead? This is precisely what happens.

Generation of GTP coupled to splitting of succinyl-CoA

The reaction is

Succinyl-CoA + GDP + P_i ⇌ Succinate + GTP + CoASH;
  $\Delta G^{0'} = -2.9\,kJ\,mol^{-1}$.

The enzyme is named for the back reaction—hence **succinyl-CoA synthetase**—it synthesizes succinyl-CoA from succinate and GTP but, in the citric acid cycle, it works in the other direction, of course. In plants, ADP, not GDP, is used in this reaction; it is not known why GDP should be used in animals.

The mechanism of the reaction is as follows: P_i displaces CoA producing succinyl phosphate.

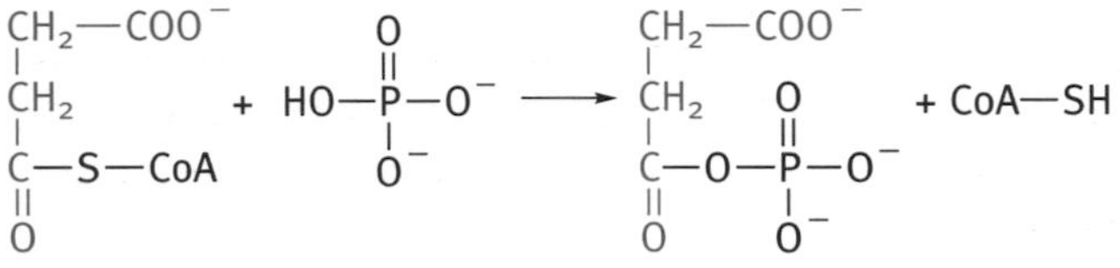

Succinyl-CoA P_i Succinyl phosphate

The phosphoryl group is now transferred to GDP, giving succinate and GTP.

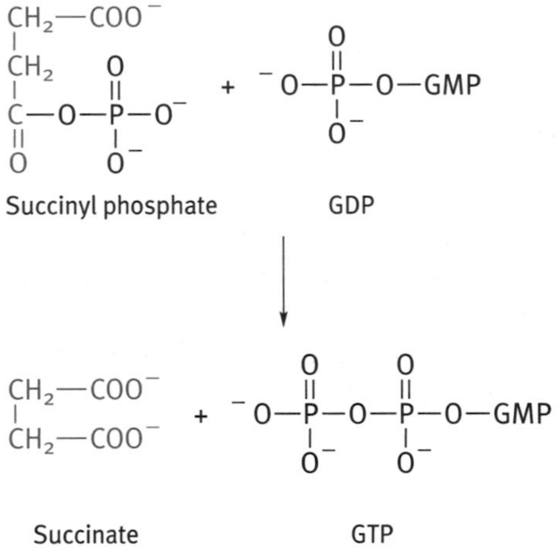

Succinyl phosphate   GDP

Succinate   GTP

Earlier (page 156), we pointed out that, to balance the output of CO_2 and reducing equivalents from the cycle with the input components, in addition to the two H_2O molecules entering the cycle we need two more hydrogen atoms and one oxygen. These arise from the involvement of inorganic phosphate in the breakdown of succinyl-CoA as described above. In case this is not clear, the *actual* reactions given above can be notionally regarded for balance sheet purposes as being *equivalent* to the two reactions

GDP + P_i → GTP + H_2O;

Succinyl-CoA + H_2O → Succinate + CoASH.

We emphasize that the reaction *does not* proceed in that way but it illustrates how the 'missing' elements of H_2O are supplied to the cycle to put the balance sheet in order.

Conversion of succinate to oxaloacetate

First, succinate is dehydrogenated but by **succinate dehydrogenase** whose electron acceptor is FAD (page 137), firmly bound to the enzyme and which can reversibly accept a pair of hydrogen atoms. Why do we use NAD^+ for the other dehydrogenation reactions of the cycle and FAD here? It's a question of redox potentials. On page 140, we described how electrons will, for thermodyamic reasons, flow from a lower redox potential (higher reducing potential or energy level) to electron acceptors of higher redox potential (lower reducing potential or energy level). In the case of succinate dehydrogenase the reaction is of the type

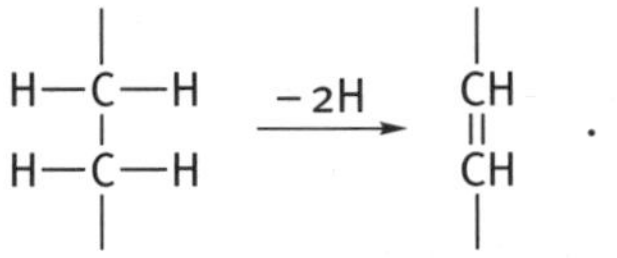

The redox potential or reducing potential of this system is such that it cannot reduce NAD^+ but can reduce FAD (which is more strongly oxidizing than NAD^+). The reaction is therefore

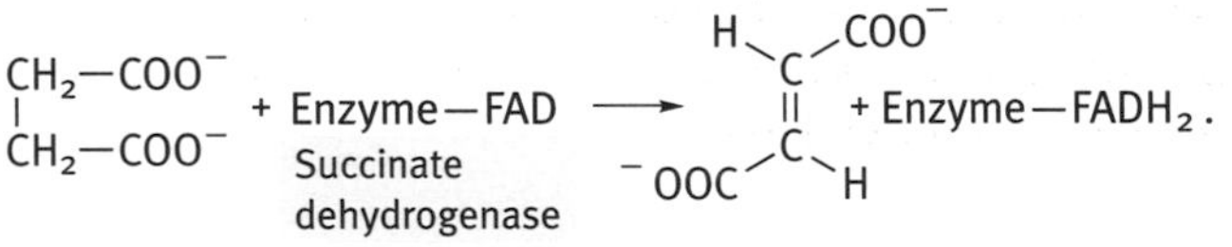

Succinate   Fumarate

The rest of the cycle, conversion of fumarate to oxaloacetate is plain sailing because you have already met the reaction types involved. A molecule of water is added to fumarate (cf. aconitase, above). The enzyme should logically be called fumarate hydratase but, by long usage, is called **fumarase**.

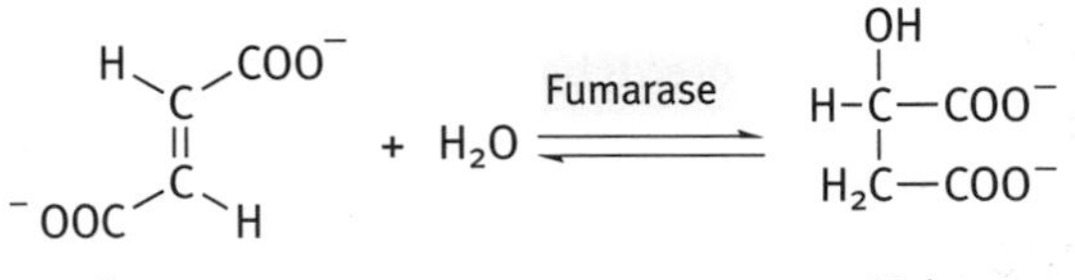

Fumarate   Malate

The malate so produced is dehydrogenated by **malate dehydrogenase**, an NAD^+ enzyme, and so the cycle is back to the starting point, oxaloacetate.

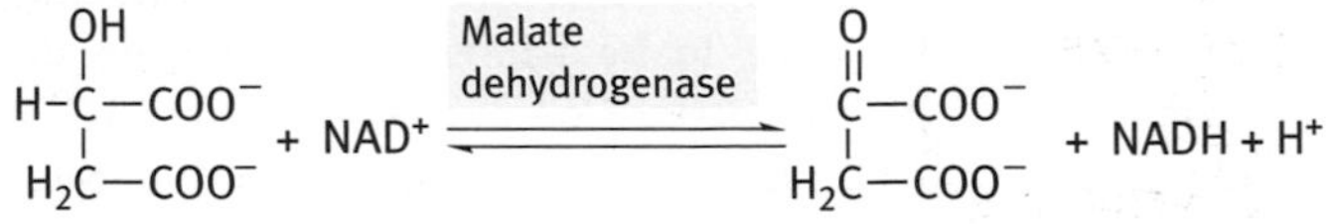

Malate   Oxaloacetate

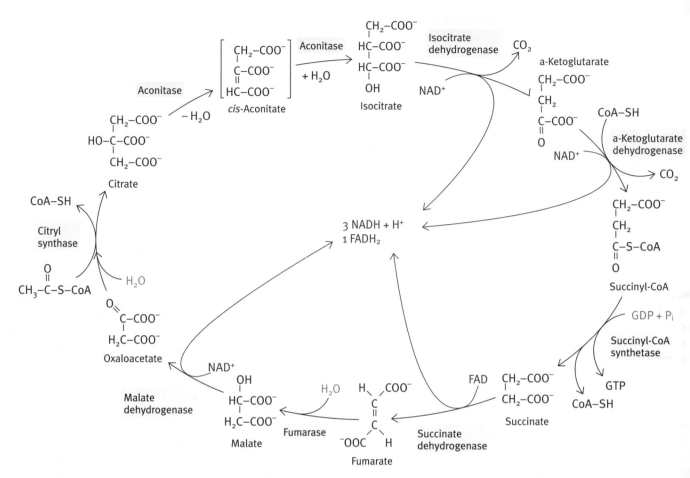

Fig. 9.13 Complete citric acid cycle. Red highlights the production of reducing equivalents from the cycle. Blue highlights the supply of the elements of H_2O to the cycle. (The conversion of citrate to isocitrate involves removed and addition of H_2O but there is no net gain.) Note that the FAD is not free but is attached to the succinate dehydrogenase protein. The involvement of water in the synthesis of citric acid is explained on page 157.

The $\Delta G^{o\prime}$ of this reaction is +29.7 kJ mol^{-1}, which is very unfavourable; the reaction proceeds because the next reaction, the conversion of oxaloacetate to citrate, is strongly exergonic and pulls the reaction over.

The complete cycle is shown in Fig. 9.13.

What determines the direction of the citric acid cycle?

The cycle operates in a unidirectional way as shown in Fig. 9.13. This is due to three of the reactions having a sufficiently large negative ΔG value to be irreversible. These are the synthesis of citrate from acetyl-CoA and oxaloacetate ($\Delta G^{o\prime} = -32.3$ kJ mol^{-1}), the decarboxylation of isocitrate to α-ketoglutarate ($\Delta G^{o\prime} = -20.9$ kJ mol^{-1}), and the α-ketoglutarate dehydro-

genase reaction ($\Delta G^{o\prime} = -33.5$ kJ mol^{-1}). This results in the cycle operating in one direction even though the equilibrium of the malate dehydrogenase reaction is in favour of the reverse direction ($\Delta G^{o\prime} = +29.7$ kJ mol^{-1}). The overall operation of the cycle reactions has a negative ΔG value.

Stoichiometry of the cycle

The overall equation for the process is

$$CH_3CO-S-CoA + 2H_2O + 3NAD^+ + FAD + GDP + P_i \rightarrow$$
$$2CO_2 + 3NADH + 3H^+ + FADH_2 + CoA-SH + GTP;$$
$$\Delta G^{o\prime} = -40 \text{ kJ mol}^{-1}.$$

If you add up the hydrogen/oxygen atoms *as printed*, they don't balance on the two parts of the equation. The reason for the apparent shortfall of H_2O is explained on page 156.

Topping up the citric acid cycle

The cycle start with oxaloacetate condensing with acetyl-CoA and ends with oxaloacetate so that the latter component is not used up. However, the cycle does not exist in isolation from the rest of metabolism and some of its acids are drawn off for other purposes. The cycle acids occupy a special place in metabolism in that they are not necessarily readily available from the diet in large amounts. Carbohydrates in the diet give rise to large quantities of C_3 acid in the form of pyruvate production, and fats similarly provide large amounts of the C_2 (acetyl groups), but cycle acids (C_4, C_5, and C_6) are not similarly available in such quantity. It is true that certain amino acids can provide cycle acids but, by the same token, cycle acids are withdrawn to synthesize some amino acids (described in Chapter 16) and other metabolites. A complex equilibrium situation exists between metabolic systems. A means of topping up cycle acids to keep the energy-generating mitochondria reactions running properly is essential and there is such as provision in cells. An important reaction for this, called an **anaplerotic** or 'filling up' reaction is that of pyruvate plus CO_2 being converted to oxaloacetate, using energy from ATP hydrolysis.

$$ATP + \begin{array}{c} COO^- \\ | \\ C=O \\ | \\ CH_3 \end{array} + HCO_3^- \longrightarrow \begin{array}{c} O \\ \| \\ C-COO^- \\ | \\ H_2C-COO^- \end{array} + ADP + P_i + H^+$$

The enzyme is called **pyruvate carboxylase** (and quite different from pyruvate *de*carboxylase of yeast, please note). This is the crucial point at which C_3 acid can be converted to C_4 acid. Pyruvate carboxylase is an enzyme of central importance. It requires the vitamin, **biotin**, for its function. Wherever 'activated' CO_2 is needed for synthetic reactions catalysed by a group of carboxylase enzymes, biotin is the cofactor. Biotin is a water-soluble B group vitamin. It becomes covalently bound to its enzyme where it accepts a carboxy group from bicarbonate to form carboxybiotin, the reaction being thermodynamically driven by the conversion of ATP to ADP and P_i. Carboxybiotin is a reactive, but stable, form of CO_2 that can be transferred to another molecule that is to be carboxylated. The $\Delta G^{o\prime}$ for the cleavage of CO_2 from carboxybiotin is $-19.7\,kJ\,mol^{-1}$. In this case pyruvate is the substrate, but other carboxylation enzyme systems are also biotin-dependent.

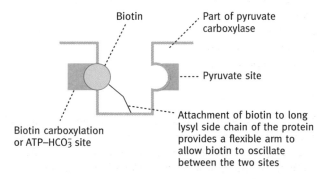

Pyruvate carboxylase has two catalytic sites—one to carboxylate the biotin and the other to transfer the carboxy group from biotin to pyruvate. (In some bacteria, the two activities reside on separate enzymes.) It is believed that the attachment of the biotin to the long lysyl side chain of the protein provides a flexible arm to permit the biotin to oscillate between the two sites (Fig. 9.14).

We will return to pyruvate carboxylase later, for it has metabolic importance other than its anaplerotic role for the citric acid cycle (page 202).

Despite the stated importance of pyruvate carboxylase, *E. coli* cells do not possess it. Since they have the citric acid cycle, how can they manage without the topping up reaction? The probable answer is that *E. coli*, in addition to the normal citric acid cycle, also has a modified form (the glyoxylate cycle) described on page 205.

Fig. 9.14 The role of biotin in the active site of enzymes catalysing carboxylation reactions. The example shown is pyruvate carboxylase.

Stage 3—the electron transport chain that conveys electrons from NADH and FADH₂ to oxygen

Remember that what we are doing is to look at the three major phases involved in glucose (or glycogen) oxidation. Phase 1 was glycolysis, phase 2 was the citric acid cycle, and now we come to the final phase. Energy-wise we haven't achieved much yet—per starting molecule of glucose only a trivial yield of four ATP molecules—two from glycolysis and two from the cycle (via GTP), with one extra if a glucosyl unit of glycogen is the starting compound. Energy-wise, the other products per mole of starting glucose are 10 NADH (two from glycolysis, two from pyruvate dehydrogenase, six from the cycle), and two FADH₂ from the cycle. (Don't forget that one molecule of glucose produces two pyruvate molecules and hence supports two turns of the cycle.)

The oxidation of the NADH and FADH₂ will produce most of the ATP (from ADP and P$_i$) generated by the oxidation of glucose.

The electron transport chain

For reasons that will become apparent, we will now discuss electron transport, pure and simple. Its purpose *is*, definitely, ATP production from ADP and P$_i$, but just for the moment forget all about phosphorylation of ADP to ATP and concentrate on electron movement to oxygen.

Where does it take place?

Electron transport carriers exist in or on the inner mitochondrial membrane. As we have already seen (Fig. 8.4), the inner membrane is folded into cristae, which increases the amount of inner membrane present, the density of cristae in a mitochondrion being related to the energy requirements of the cell.

Nature of the electron carriers in the chain

Heme is the prosthetic group of several electron carriers—called **cytochromes** because of their colour (red). The different cytochromes are called c_1, c, a, and a_3 (in order of their participation in the chain; the role of two b cytochromes is given later). The essentials of the heme structure are shown in Fig. 9.15, and its full structure in Fig. 31.3.

The important thing about the heme molecule is that, as the prosthetic group of the cytochrome electron carriers, the Fe atom oscillates between the Fe^{2+} and Fe^{3+} states as it accepts an

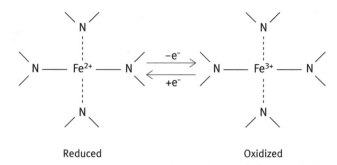

Fig. 9.15 Diagrammatic representation of heme (but take a look at the actual structure in Fig. 31.3).

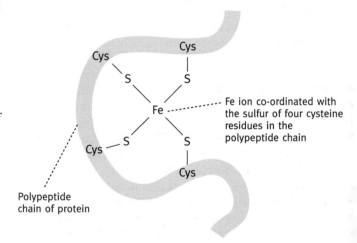

Fig. 9.16 An iron-sulfur centre. There are several types of these, increasing in complexity and numbers of Fe and S atoms. The simplest form is shown here.

electron from the preceding carrier or donates it to the next carrier in the chain. The characteristics of the heme molecule are modified by the specific protein to which it is attached, and variations in the heme side groups occur in different cytochromes and in their attachment to their apoproteins. Thus, it is not a contradiction that different cytochromes have different redox potentials and yet have heme as their prosthetic group.

Another type of electron carrier, based on iron, are the so-called non-heme iron proteins. In these, the iron is bound to the sulfhydryl side group of the amino acid cysteine of the protein and also to inorganic sulfide ions forming **iron-sulfur complexes**, or iron-sulfur centres. The simplest type is shown in Fig. 9.16. As with the cytochromes, the iron atom in these can accept and donate electrons in a cyclical fashion, oscillating between the ferrous and ferric state. Such iron–sulfur centres are associated with flavin enzymes. They accept electrons from FAD—

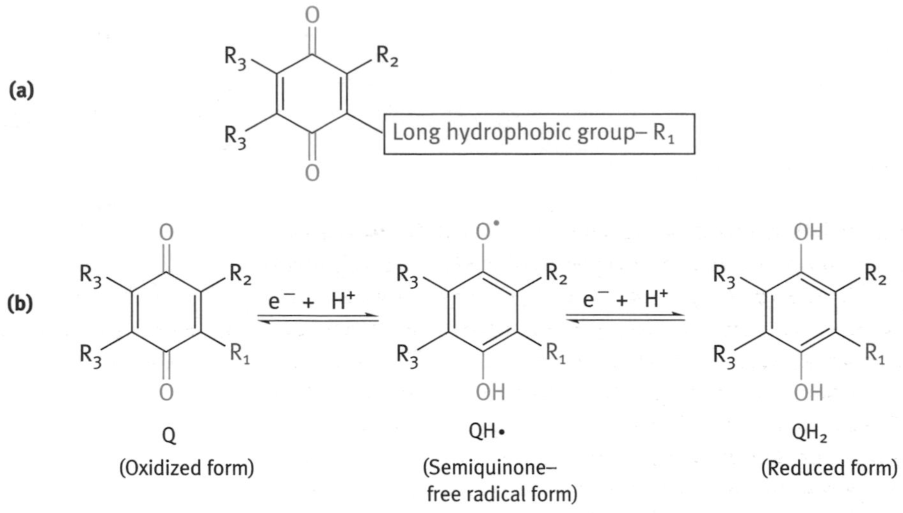

(a)

Long hydrophobic group– R$_1$

(b)

$$Q \quad \xrightleftharpoons{e^- + H^+} \quad QH\cdot \quad \xrightleftharpoons{e^- + H^+} \quad QH_2$$

(Oxidized form)   (Semiquinone–
free radical form)   (Reduced form)

Fig. 9.17 (a) Ubiquinone or coenzyme Q structure (in the oxidized form). **(b)** Oxidized, semiquinone, and reduced forms of ubiquinone (Q). The semiquinone, QH·, can exist as the anion, Q·$^-$. R$_1$, Long hydrophobic group; R$_2$, —CH$_3$; R$_3$, —O— CH$_3$. The full structure of ubiquinone is given on below.

enzymes such as succinate dehydrogenase and the dehydrogenase involved in fat oxidation (described in Chapter 10). Another type of carrier is an FMN-protein. FMN (flavin adenine mononucleotide) consists of the flavin half of FAD (see page 137). It carries electrons from NADH to an iron-sulfur centre. All of the iron-sulfur centres transfer electrons to ubiquinone (see below).

As well as these protein-bound electron carriers, there is one carrier not bound to a protein. This is a molecule, illustrated in Fig. 9.17, called **ubiquinone**, because it can exist as a quinone and is found ubiquitously. It is often referred to as CoQ, UQ, or Q. Q is an electron carrier because it can accept protons as shown in Fig. 9.17; it can exist as the free radical, semiquinone intermediate, thus permitting the molecule to hand over a single electron to the next carrier rather than a pair of electrons. The very long hydrophobic tail on the molecule (as many as 40 carbon atoms long in 10 isoprenoid groups; just have a look at the structure below) make it freely soluble and mobile in the nonpolar interior of the inner mitochondrial membrane.

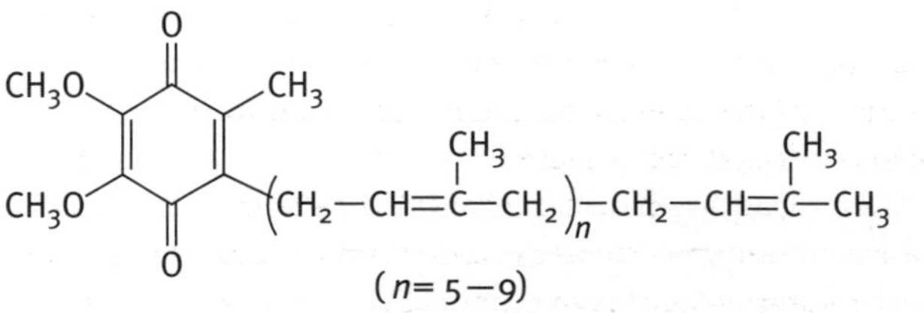

(n = 5 − 9)

Structure of ubiquinone (coenzyme Q)

To summarize, we have in the electron transport chain an FMN-protein, non-heme iron—sulfur proteins, Q not bound

to a protein and freely mobile in the membrane, and heme proteins known as cytochromes. An important point is that one of the latter, cytochrome c, is a small water-soluble protein molecule (molecular weight ~12.5 kDa, just over 100 amino acids) that is loosely attached to the outside face of the inner mitochondrial membrane so that it also is free to move about. All the other proteins of the respiratory complexes are built into the membrane structure as integral proteins in fixed positions.

Arrangement of the electron carriers

In Chapter 8 (page 140) we discussed the redox potentials of electron acceptors and explained that electrons flow from a carrier of higher reducing potential (low redox potential) to one of lower reducing potential (more oxidizing, or higher redox potential). The electron carriers in the chain are of different redox potentials.

The redox potentials are directly related to $\Delta G^{0'}$ values, as already discussed on page 142. The electron carriers are arranged in the electron transport chain such that there is a continuous progression down the free-energy gradient (increasing redox potentials) with the corresponding release of free energy as the electrons move from one carrier to the next (Fig. 9.18). In considering glucose oxidation (the subject of this chapter) the task in this phase is to transfer electrons from NADH and FADH$_2$ to oxygen. The whole scheme involves a somewhat formidable list of steps but the carriers can be grouped into the four complexes shown in Fig. 9.19. These complexes are built into the structure of the inner mitochondrial membrane, interconnected by the mobile electron carriers, ubiquinone and cytochrome c. Ubiquinone takes electrons from complexes I and II and delivers them to complex III. Cytochrome c is the intermediary

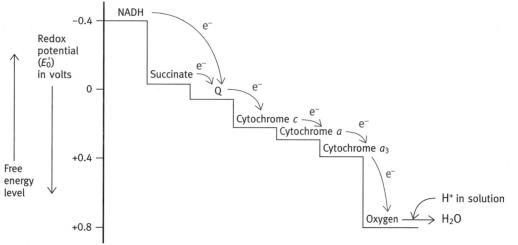

Fig. 9.18 The approximate relative redox potentials of some of the main components of the electron transport system in mitochondria. The arrows indicate electron movements. The role of cytochrome *b* components is shown in Fig. 9.22.

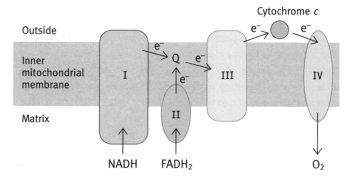

Fig. 9.19 The electron transport chain with electron carriers grouped into four main complexes. Complex I, NADH—Q reductase; complex II, succinate—Q reductase; complex III, QH$_2$—cytochrome *c* reductase; complex IV, cytochrome oxidase. Q, ubiquinone or coenzyme Q. FADH$_2$ is generated in the cycle from succinate by succinate dehydrogenase. Note that the complexes are located in the inner mitochondrial membrane. Coenzyme Q and cytochrome *c* are mobile carriers capable of physically transporting electrons from one site in the membrane to another. Cytochrome *c* is surface-located. Note that FADH$_2$ exists attached to flavoprotein enzymes. The major ones are succinate dehydrogenase and fatty acyl-CoA dehydrogenases involved in fat oxidation; the latter is described in Chapter 10.

between complexes III and IV. Complex I carries electrons from NADH to Q. Complex II carries electrons from succinate via FADH$_2$ to Q; complex III uses QH$_2$ to reduce cytochrome *c*. Complex IV transfers electrons from cytochrome *c* to oxygen. Complexes I, III, and IV are, for convenience, referred to as NADH:Q reductase, QH$_2$:cytochrome *c* reductase, and cytochrome oxidase, respectively. Complex IV, cytochrome oxidase, is a multisubunit structure; electrons are donated to it by

cytochrome *c* on the outer face of the inner mitochondrial membrane. The electrons are transported via two cytochromes, *a* and *a$_3$*, to oxygen. Cytochromes *a* and *a$_3$* have copper sites associated with them that oscillate between the Cu$^+$ and Cu^{2+} states. The final reduction catalysed by cytochrome oxidase is

$$O_2 + 4e^- + 4H^+ \rightarrow 2H_2O.$$

As explained later (Chapter 18), it is vital that the mechanism of cytochrome oxidase ensures that oxygen atoms are fully reduced by four electrons to form water without releasing any intermediate partially reduced states. These would be dangerous reactive free radicals or hydrogen peroxide.

How is the free energy released by electron transport used to form ATP?

If you remember how ATP is generated in glycolysis, you may, in fact should, be puzzled. In glycolysis there is substrate-level phosphorylation (page 151). In this, the phosphorylation (generation of ATP from ADP and P$_i$) is inseparably linked to the reactions in glycolysis. Either the reactions occur, involving ATP synthesis, or they don't occur. You can't have the relevant reactions without ATP generation since this is an intrinsic component of the reactions. The same is true of ATP generation in the citric acid cycle via GTP.

By contrast, we have described the electron transport reactions without even mentioning ATP production. The purpose of electron transport is to generate ATP but electron transport to oxygen proceeds very readily in broken up mitochondria without any ATP generation at all. This situation perplexed biochemists for decades. If electron transport generates ATP in the cell, why can it readily occur without this in broken up mito-

chondria? In short, in cells, or in intact 'healthy' mitochondria, electron transport results in ATP generation but, in broken mitochondria, electron transport occurs very happily without it. It conflicted with the type of substrate-level phosphorylation that was from glycolysis. The solution was discovered by the English biochemist, Peter Mitchell, who, in a private laboratory, in 1961 produced a concept of how electron transport causes ATP synthesis that is so novel that it was at first hardly taken seriously by most, and it took a long time for it to be accepted (and for a Nobel Prize to be awarded to Mitchell in 1978). The concept is based on the simple notion that gradients have the ability to do work. A gradient of water pressure can be used to generate electricity, a gradient of air pressure to drive a windmill, etc. A chemical gradient is no different. Molecules or ions will migrate from a high concentration to a low concentration and, if a suitable energy-harnessing device is interposed, useful work can be done.

Applying this concept, two things are needed for ATP generation coupled to electron transport: firstly, electron transport must create a gradient of some sort; and, secondly, the gradient must be allowed to flow back through a device that uses the energy of the gradient to synthesize ATP from ADP and P_i. Mitchell's concept thirdly required the existence of intact vesicles across whose membrane a gradient could be established. It has to have an inside and an outside, as separate compartments, explaining why broken mitochondria don't make ATP. The membrane must therefore be impermeable to the solute of which the gradient consists.

Mitchell discovered that electron flow caused protons to be ejected from inside the mitochondrion to the outside, thus creating a proton gradient across the membrane; in other words,

the pH of the external solution fell. A membrane (or charge) potential, negative inside and positive outside, is also generated by the proton expulsion and also contributes to the total energy gradient or **proton-motive force** available for ATP synthesis. The inner mitochondrial membrane itself is virtually impermeable to protons. This is a prerequisite for the system, but inserted into the membrane are special proton-conducting channels. Protons flow from the outside through these channels back into the mitochondrial matrix and the energy of this flow is harnessed to the formation of ATP from ADP and P_i.

The proton-conducting channels are knob-like structures that completely cover the inner surface of the cristae and are made up of several proteins (Fig. 9.20). These are, in fact, ATP-synthase complexes that convert ADP and P_i to ATP, the process being energetically driven by the proton flow. The complex is built into the membrane by the F_0 unit (Fig. 9.20), while the actual ATP synthesis occurs in the F_1 unit, both units containing multiple protein molecules. There is evidence that the actual proton-conducting channels consist of strategically placed charged amino acid residues in proteins and the protons hop from one to another.

The overall scheme in mitochondria is shown in Fig. 9.21. This scheme is known as the **chemiosmotic mechanism**. Two questions can be asked,

1   How does the flow of electrons from NADH and $FADH_2$ to O_2 cause protons to be pumped from the matrix side of the inner mitochondrial membrane to outside the membrane?

2   How does the flow of protons into the mitochondrion drive the synthesis of ATP from ADP and P_i?

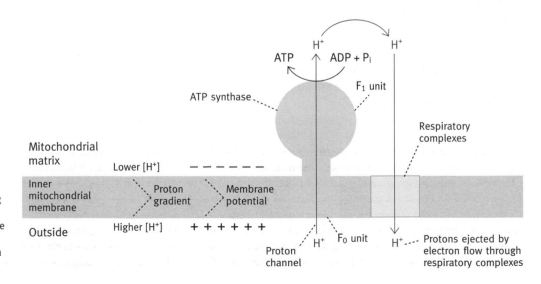

Fig. 9.20 The ATP-generating system of the inner mitochondrial membrane. The F_1 unit consists of nine subunits. The F_0 unit is also a multisubunit structure.

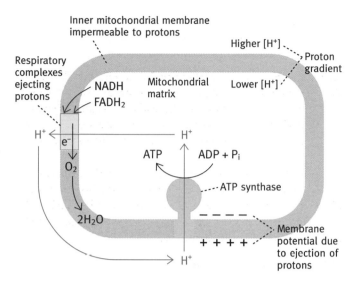

Fig. 9.21 Generation of ATP in mitochondria by the chemiosmotic mechanism. Note that $FADH_2$, produced by the dehydrogenation of fatty acids (described in Chapter 10), enters the same pathway as that utilized by the oxidation of succinate.

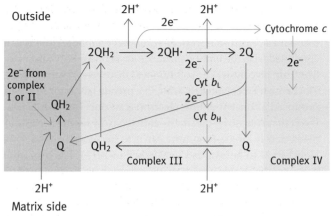

Fig. 9.22 The mechanism of proton translocation by complex III as a result of electron transport in mitochondria. The red arrows represent physical diffusion of components rather than chemical transformations; the latter are indicated by black arrows and electron transport by blue arrows. (However, note that the blue arrow representing electron transport by cytochrome *c* is effected by the physical diffusion of the latter from complex III to complex IV and its return after oxidation.) Cytochrome *b* protein spans the membrane so that electrons are transferred from the outer face to the inner face. The molecules of Q and QH_2 are in equilibrium with a membrane pool of these, but this is not shown in order to simplify the diagram arrangement. The electrons from complex I (as QH_2) arise mainly from NADH generated in glycolysis and the citric acid cycle; those from complex II come from the succinate → fumarate step of the cycle via an FAD-protein. As will be described in the next chapter, electrons from fatty acid oxidation also enter complex III via complexes I and II. Q, Ubiquinone; Q·⁻, Semiquinone anion; Cyt, cytochrome. Cyt b_L and b_H refer to heme sites on cytochrome *b* of low and high redox potentials respectively.

How are protons ejected?

The mechanism by which protons are translocated as a result of electron flow is uncertain for complexes I and IV but is established in the case of complex III. The *principle* of Mitchell's original idea is as simple as it is ingenious. In essence, hydrogen *atoms* are assembled on the matrix side of the inner mitochondrial membrane, using protons from the matrix and electrons from the transport chain. The atoms are assembled on ubiquinone (Q), forming the reduced form, QH_2, which now diffuses to the opposite face of the membrane where the reverse happens—electrons are stripped off the hydrogen atoms and the resultant protons escape to the outside. The essentials of the process are the positioning of the responsible catalytic proteins on opposite sides of the membrane and a mobile carrier to transport the hydrogen atoms across the membrane from one face to another. It is called the **Q cycle**.

That is the principle for proton transport in complex III. The actual mechanism is shown in Fig. 9.22; it looks complicated but is in fact very simple with few reactions involved. Note first of all that the red arrows simply represent physical movement of ubiquinone and its derivatives; the second point is that, in effect, two distinct processes are going on, both of which eject a pair of protons. In the membrane there are pools of Q and QH_2. Let us start with QH_2, arriving from complexes I and II, and whose electrons originate from NADH and $FADH_2$, respectively. In the reduction of Q to QH_2 by complexes I and II, two

protons are taken up from the matrix as shown to the left of the diagram. The QH_2 now migrates to a site, in complex III, *on the external face of the membrane*, where one electron is removed and passed on to reduce cytochrome *c*, which transports it to complex IV. Remember that cytochrome *c* is also mobile. The site on complex IV that accepts electrons from cytochrome *c* is exposed on the external face of the inner membrane and cytochrome *c* is also located on this face of the membrane. A proton is ejected, leaving the half-oxidized quinone anion, Q·⁻. A further electron is now removed from the latter, but in this case the electron is handed on, not to cytochrome *c* but to cytochrome *b*, to which we'll return shortly. That is the end of that half of the story—a molecule of QH_2 from complexes I/II has been oxidized, two electrons passed (one to cytochrome *c* and one to cytochrome *b*), and two protons ejected from the matrix to the outside. The Q so formed now returns to the general pool. There is another part to the story. A *second* molecule

of QH_2 is oxidized in the same way as the first, resulting in the ejection of two more protons. From the two molecules of QH_2 we thus have two electrons passing on to complex IV via cytochrome c and two to cytochrome b. The latter transfers the electrons to a different heme site on the same protein whose redox potential is greater (lower energy) and in this way transports the two electrons to the matrix side of the membrane. They are donated here to a molecule of Q and, with a pair of protons from the matrix, hydrogen atoms on Q are assembled as QH_2. The QH_2 now migrates back to the site of the external face and the merry-go-round starts again. (The diagram shows the same molecule of ubiquinone going round the cycle, for clarity, but molecules of Q and QH_2 will enter and exit the pool in a dynamic equilibrium—it amounts to the same thing.) This somewhat convoluted process actually oxidizes only one molecule of QH_2 but achieves the ejection of four protons. The *net* effect of the reactions *within complex III* can be summarized in the equation

$$QH_2 + 2H^+ \text{ (matrix)} + 2 \text{ cyt } c \text{ (Fe}^{3+}\text{)} \rightarrow$$
$$Q + 4H^+ \text{ (outside)} + 2 \text{ cyt } c \text{ (Fe}^2\text{)}.$$

Figure 9.23 illustrates the fact that the energy for proton pumping is derived from the free energy released as electrons are transported down the electron carrier chain.

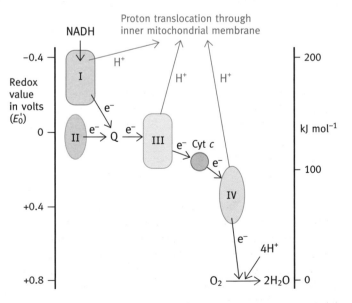

Fig. 9.23 Approximate positions on the redox potential scale of the electron transport complexes. The scale on the right gives the approximate free energy released when a pair of electrons is transferred from components to oxygen. Q, ubiquinone.

How does the proton gradient drive ATP synthesis? —ATP synthase

ATP is synthesized in mitochondria by the enzyme **ATP synthase**. Paul Boyer of UCLA, who won a Nobel prize for his work on the enzyme, referred to it as a 'splendid molecular machine' and added 'All enzymes are beautiful, but ATP synthase is one of the most beautiful as well as one of the most unusual and important'; all of which is true.

ATP synthase is the name of the structures with one part visible as knobs projecting into the matrix on the inside surface of the inner mitochondrial membrane and the other anchored to the membrane itself. They are diagrammatically represented in Fig 9.20. The reaction catalysed is:

$$ADP^{3-} + P_i^{2-} + H^+ \rightarrow ATP^{4-} + H_2O$$

The standard free energy change of this is about 7 kcal (29.3 kJ) and so cannot proceed without a large input of energy which is supplied by the proton gradient established across the membrane by electron transport; protons flow back into the mitochondrial matrix through the ATP synthase. It is a very major metabolic activity; the weight of ATP synthesized in one day exceeds your body weight. Given that the total weight of ATP in the body is only of the order of 5–10 g, this emphasizes the rapid turnover of the molecule, by which we mean its hydrolysis and resynthesis. What we now have to do is to look at the structure of the ATP synthase and examine how it performs its task.

The enzyme complex is found in aerobic organisms wherever the energy derived from electron transport has to be trapped as ATP. It is not confined to mitochondria—chloroplasts in plants use the same device (page 245). So do *E. coli* cells which are roughly the size of a mitochondrion, the cell membrane in this context being equivalent to the inner mitochondrial membrane. Its ATP synthase units project from the cell membrane into the cytoplasm and the electron transport system in the membrane creates a proton gradient from outside to inside by pumping protons from the inside to the outside of the cell.

ATP synthase has been highly conserved during evolution and therefore the enzymes from different sources differ only slightly in detail. Consequently experiments on a variety of bacteria, chloroplasts, and mitochondria have contributed to our current understanding of the structure and function of the ATP synthase. Studies of mutants of *E. coli* in the 1970s allowed the genes concerned and their associated proteins to be characterized and the results extrapolated to the mitochondria.

The ATP synthase complex has basically two parts connected by a short narrow stalk region (Fig. 9.24); the F_0 **unit** is made of several transmembrane proteins located in the lipid bilayer. The

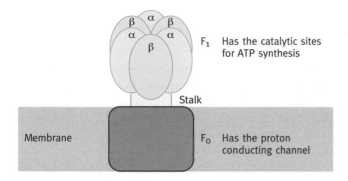

Fig. 9.24 The major components of ATP synthase. The F_0 has multiple subunits and is integral with the lipid bilayer membrane. It is the proton conducting channel. The F_1 is composed of a hexamer ring of alternating α and β subunits enclosing two central subunits γ and ε which project downwards and contact the the F_0 unit.

second part, connected to the F_0 by the stalk, is the knob like structure projecting into the matrix known as the F_1 **unit**; the name originally meant 'coupling factor number 1'. It was the first 'factor' discovered that restored oxidative phosphorylation to mitochondrial membranes stripped of F_1 (explained below)—it coupled electron transport to ATP synthesis.

When mitochondria are disrupted by sonication, pieces of the inner membrane, still carrying the complete ATP synthase complexes, seal up into small closed vesicles in which the F_1 knobs now face outward rather than inward as in an intact mitochondrion (Fig. 9.25). This makes the vesicles a useful experimental system. They are still capable of oxidative phosphorylation when supplied with a suitable oxidizable substrate and pump protons across the membrane, in this case into the interior of the vesicle. Exposure to low ionic strength solutions dis-

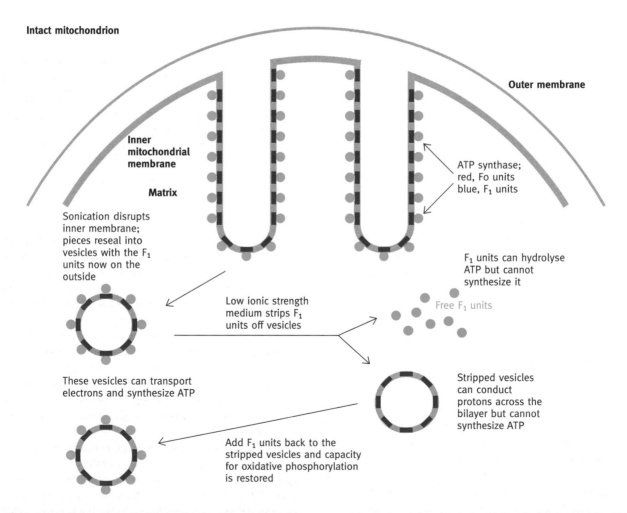

Fig. 9.25 How vesicles are obtained from mitochondria. These vesicles are a useful experimental tool because the F_1 units face outwards and are accessible to ADP and P_i. The free F_1 units and vesicle F_0 units also permit independent study of the two components. The electron transport chain (page 163) is integral with the inner mitochondrial membrane and is therefore present in the vesicles.

lodges the F_1 complexes, leaving the F_0 units in the membrane and the F_1 units in solution. The stripped membrane vesicles can still carry on electron transport coupled to proton pumping but cannot use the proton gradient to synthesize ATP. The solubilized F_1 particles can hydrolyse ATP but not synthesize it. When they are added back to the membranes from which they were stripped, the latter regain their ability to synthesize ATP, provided of course that electron transport is proceeding (see Fig. 9.20). The components of the electron transport chain are all integral with the inner mitochondrial membrane and are therefore present in the vesicles. The F_0 unit is the channel through which protons enter the mitochondrial matrix, and the F_1 units carry the catalytic sites which synthesize ATP, the energy being supplied for this from the passage of protons through the F_0.

Two obvious questions present themselves:

- How is the energy transmitted from the F_0 to the F_1?
- How does the F_1 utilize this energy to drive ATP synthesis?

To discuss this we must now describe the structure of the synthase in more detail.

Components of the F_1 unit and their roles in the conversion of ADP + P_i to ATP

The F_1 unit is essentially a barrel formed by six protein subunits arranged in a ring more or less like the segments of an orange (Fig. 9.24). There are three α **protein subunits**, three β **subunits**, and three '**minor**' subunits (γ, δ, and ε). Each β subunit has a catalytic site which synthesizes ATP from ADP + P_i, so there are three such sites per F_1 unit. The narrow shaftlike cavity of the barrel is occupied by an elongated γ protein subunit and the smaller ε subunit, the shaft projecting on the membrane side beyond the barrel and making contact with the F_0 unit.

Activities of the enzyme catalytic centres on the F_1 subunit

If we consider for the moment an enzymic site on a *single* β subunit, the sequence of events in the synthesis of a molecule of ATP as proposed in Boyer's model is as follows (Fig. 9.26(a)):

- The site is **open** and nothing is bound to it (**the O state**).
- A conformational change in the protein converts the site to a **low affinity** state; ADP and P_i now bind to it loosely but there is no catalytic activity (**the L state**).
- A further conformational change produces a **tight binding state**—the ADP and P_i become tightly bound. This is now catalytically active and ATP is formed (**the T state**).
- A conformational change opens up the site, ATP escapes and the site is back to the original **open** state.

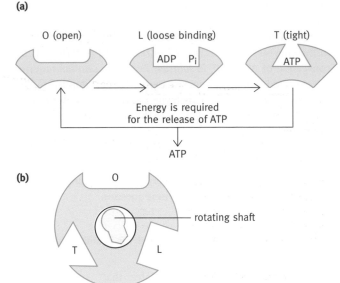

Fig. 9.26 The catalytic sites of ATP synthase as proposed in the Boyer model: (a) the changes that occur in a single site of one β subunit of F_1 during the synthesis of ATP. (b) The three β subunits work in a cooperative manner and the conversions in one site are coordinated with the other two sites. This means that at anyone time an F_1 unit has one subunit in the O state, one in the L state, and one in the T state. The subunits are not depicted. The rotating shaft is shown as a notional assymetric shape to convey that it is believed that it successively interacts with the subunits as it rotates. The actual structure of the shaft within the F_1 barrel is given in Fig. 9.30.

The model postulates that each site progresses sequentially through the three conformations. ATP synthase is a most unusual enzyme in that catalytic activity is dependent on cooperation between the subunits. At any one time one of the three is in the O state, one in the L state and one in the T state (see Fig. 9.26(b)). A site in the T state with a molecule of ATP bound to it will convert to the open O state and release the ATP when the preceding site in the L state is occupied by ADP and P_i which at the same time converts to the T state itself—the ADP and P_i now are tightly bound and ATP synthesis proceeds. It is the conformational changes in the F_1 subunits that constitute the mechanism by which energy is transmitted into ATP production.

You may, perhaps should, be puzzled at this stage for in the third step above, we simply state that ATP is formed from ADP and P_i. The researchers working on this problem found that *when ADP and P_i are firmly bound to the active site of the enzyme* there is little free energy change in the formation of ATP, the equilibrium constant being about 1 as compared with 10^{-5} for components in free solution (see page 6 if you need to be

reminded about equilibrium constants and free energy). This does not conflict with what you have learned about the energetics of ATP, for it applies only to reactants tightly bound to the enzyme; *energy is needed to release the ATP* so that the conversion of ADP + P_i *in solution* to ATP *in solution* requires the expected amount of energy input. This is supplied by the conformational changes which the enzyme undergoes during the catalytic cycle. That brings us to the question of what causes these conformational changes.

The energy for ATP synthesis comes from proton flow through the lipid bilayer via the F_0 complex. The ATP is synthesized on the β subunits of the F_1 knobs, the two being separated by the stalk, which raises the obvious question:

How is the energy of proton flow through the F_0 utilized for ATP synthesis by the F_1?

Boyer's model, as stated, postulates that conformational changes in the F_1 proteins are involved in the energy transmission. Considerable evidence has accumulated suggesting that a rotary mechanism is involved and such a mechanism fits very well with the concept of progressive changes in the catalytic sites of the three β subunits. It is postulated that proton flow through the F_0 creates a rotary motion which drives the shaft formed by the γ and ε subunits to rotate within the space in the centre of the F_1 barrel. It is envisaged that the rotation causes interaction of the shaft with the F_1 subunits in succession. It is believed that these interactions cause the conformational changes and in doing so supply the energy for ATP synthesis, though exactly how the energy transfer occurs is not known. As stated, it is the *release* of the ATP from the catalytic site in the T state that is the energy-requiring step.

Indirect evidence for a rotary mechanism began to accumulate in the 1980s from studies on *E. coli* ATP synthase. It was later observed that there was amino acid sequence homology of the ATP synthase β protein with one involved in bacterial flagellar rotation which also depends on proton flow from outside the cell membrane to the inside.

Yoshida's group in Japan has now directly visualized the rotation of the γ unit (see Fig. 9.27). It is worth looking at this elegant experiment in some detail. As mentioned, the solubilized F_1 units cannot synthesize ATP but they can hydrolyse it to ADP + P_i. In the test tube F_1 is in fact an ATPase (and indeed ATP synthase is sometimes referred to as the F_0/F_1 ATPase). It no doubt strikes you as curious that a structure whose function is to synthesize ATP should readily destroy it in the test tube. The ATPase activity of the isolated F_1 units is due to their catalysis of the reverse reaction. Yoshida's group made use of this to directly visualize the rotation. The F_1 complex was mounted on a glass

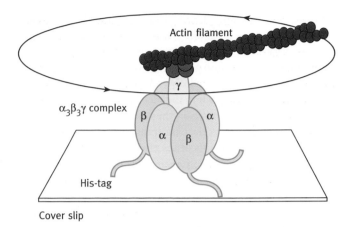

Fig. 9.27 The system used for the observation of the rotation of the γ subunit in the $\alpha_3\beta_3\gamma$ subcomplex of F_1—ATPase. The small purple spheres on the γ-subunit represent streptavidin, a protein used to attach the actin filament. See text for explanation. Note that the actin filament is used only as an experimental tool. Actin has nothing to do with ATP generation.

coverslip with the ends of the β subunits, which normally point away from the membrane, attached to the glass—upside down, as it were (Fig. 9.27). In case the method of doing this intrigues you, histidine residues were added to the N-termini of the β subunit proteins so that the F_1 could be attached to the glass. To do this the glass was coated with a protein to which a nickel salt was bound. The added histidine residues bind to the latter. (The manipulations were done using the technologies of modifying genes and expressing the proteins in *E. coli*. It is not necessary to understand this statement now, but the techniques are described in Chapter 29.)

To make shaft rotation visible, a fluorescent actin filament 'rod', long enough to be visible in the light microscope, was attached to the shaft. (Actin has nothing whatsoever to do with the synthase—it is used simply as a convenient experimental tool.) Time lapse photography showed that in the presence of ATP a significant fraction of the actin rods did indeed rotate invariably in a counterclockwise direction, as viewed from the membrane side. The experiment provided very strong support for the concept that the central shaft of F_1 unit formed by the γ subunit does indeed rotate during ATP synthesis driven by proton flow.

What is the role of the F_0 unit?

The F_0 unit has a more complex structure which is not as fully elucidated as F_1. There is a ring of 9–12 c transmembrane sub-

units together with a large hydrophobic *a* subunit also embedded in the membrane alongside the circle of *c* subunits. (F_0 subunits are denoted by italic letters.) The large *a* subunit is connected to the F_1 head by two elongated *b* subunits and the δ subunit which attaches to the F_1 head. Figure 9.28 shows a diagram of the complete ATP synthase structure and subunit movements.

The way in which rotation of the γ subunit is driven by the flow of electrons, from the outside of the inner mitochondrial membrane through the F_0 and into the mitochondrial matrix, is not completely understood. The γ and ε subunits contact the *c* subunit ring and the evidence suggests that that the ring of *c* units is caused to rotate in a stepwise fashion relative to the *a* subunit. This drives the rotation of the γ and ε subunits. The protons flow through the F_0, via an aspartate carboxyl group present in the centre of a transmembrane α helix of each *c* subunit. The alternate protonation and deprotonation causes a conformational change in each *c* subunit which results in the required movement. The *a* subunit probably plays an important part in the flow of protons and generation of spin. It is estimated that four protons are required to flow through the F_0 for the synthesis of one molecule of ATP from ADP. Figure 9.29 illustrates the mechanism, and Fig. 9.30 gives a ribbon diagram of the structure of two subunits of the F_1 with the γ subunit in the centre. The diagram is a cross-section of the actual three-dimensional structure of F_1 determined by X-ray diffraction. The complete F_1 is one of the largest structures so determined, a remarkable achievement which earned John Walker of Cambridge, England a Nobel prize.

There are some interesting parallels with features of conventional electric motors. The outer casing of an electric motor needs to be secured to prevent it rotating as the shaft turns; the F_1 outer barrel also has to be restrained from turning as the central shaft turns. It is believed that the two β subunits of the F_0 have a role in this (see Fig. 9.29). One end of these subunits is attached to the membrane, and long hydrophilic extensions attach to the outer side of the F_1 barrel via the δ subunit. They act as a stator to strap down the outer casing of the motor. The shaft of an electric motor is usually lubricated to facilitate rotation in its bearings. The central cavity of the F_1 barrel is lined by hydrophobic ('oily') groups which offer minimal frictional resistance to rotation of the shaft.

The picture that has emerged is that ATP synthase is the smallest known rotary motor, driven by protons rather than

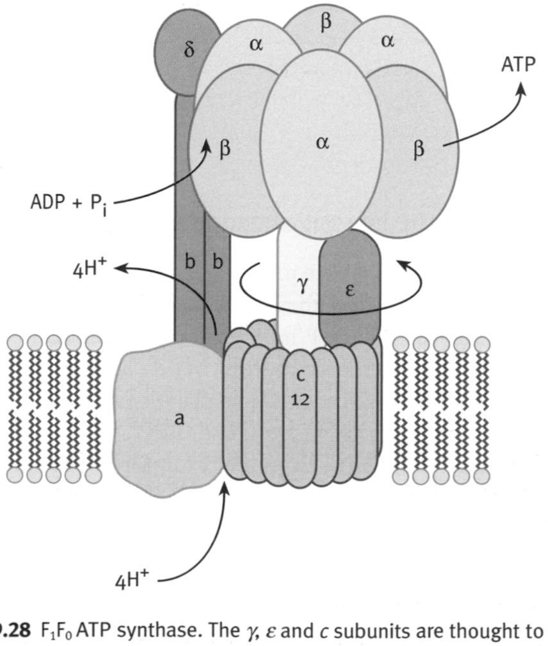

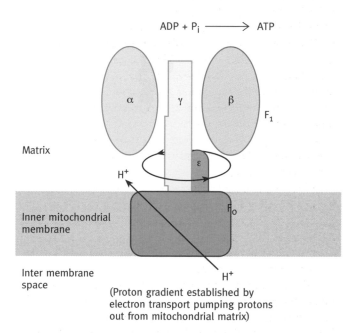

Fig. 9.29 The rotary model of the mechanism of ATP synthesis by ATP synthase. Proton flow through the F_0 unit results in rotation of the γ subunit which somehow causes conformational changes in the F_1 subunits. The diagram represents a cross-sectional view with only two of the six F_1 subunits shown. The representation of the assymetry of the rotating shaft is to represent the interaction of the shaft with successive subunits; a cross-section of the three-dimensional structure of the F_1, with the actual structure of the γ subunit, is shown in the next figure.

Fig. 9.28 F_1F_0 ATP synthase. The γ, ε and *c* subunits are thought to rotate as a unit with respect to the other subunits during proton translocation and ATP synthesis.

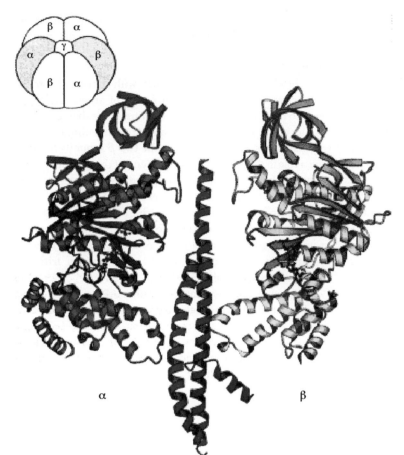

Fig. 9.30 Ribbon diagram of the three-dimensional structure of the F_1 of ATP synthase with the γ subunit in the central cavity. The diagram is a cross-section of the entire head showing only two of the subunits. The shaded parts of the small figure in the top left indicate the subunits of the F_1 depicted.

electrons. The rotating shaft is 1 nm in radius in a stationary outer barrel with a radius of 5 nm. What we do not completely understand at present is how the proton flow through the F_0 induces rotary motion, nor how shaft rotation interacts with the F_1 subunits to induce the sequence of conformational changes believed to occur in the β subunits during the synthesis of ATP. A clear summary and discussion of this fascinating mechanism can be found in the Boyer (1999) reference of the reading list at the end of this chapter.

Despite the complexity of the ATP synthase mechanism in detail, there is a majestic simplicity in the concept that for billions of years all aerobic life on this planet has been driven by the creation of a small pH gradient across a lipid bilayer membrane. It is one of the great concepts in biology.

Transport of ADP into mitochondria and ATP out

Most of the ATP synthesis in most eukaryotic cells occurs in mitochondria while most of the ATP is used outside of the mitochondria. Hence ADP and P_i must enter the mitochondrion and ATP move out. The highly charged molecules cannot diffuse passively across the inner mitochondrial membrane and hence special transport mechanisms exist. ATP–ADP translocase exchanges ATP inside the mitochondrion for an ADP outside the mitochondrion (Fig. 9.31).

Where does the energy for this ATP : ADP exchange come from? As already explained, electron transport generates not only a pH gradient across the inner mitochondrial membrane but also a membrane potential across the inner mitochondrial membrane, positive outside and negative inside, due to the ejec-

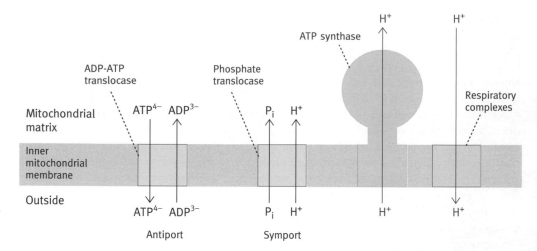

Fig. 9.31 Diagram of transmembrane traffic in mitochondria involved in ATP generation. All of the traffic is via specific transport proteins.

tion of H⁺ ions. ATP carries four negative charges out, while ADP carries only three in. Thus the ATP : ADP exchange tends to neutralize this membrane potential or, in other words, the membrane potential drives the exchange. About 25% of the total energy yield of electron transfer is absorbed in the process. The transport of the P_i needed along with ADP for ATP generation is catalysed by a phosphate translocase in the mitochondrial membrane which is also driven by the proton gradient (Fig. 9.31).

The balance sheet of ATP production by electron transport

It requires three protons to flow through the ATP synthase to generate one molecule of ATP from ADP and P_i, assuming that the latter two are already inside the mitochondrion. The transport of a molecule of ADP into the mitochondrion and that of one of ATP to the cytoplasm requires the energy equivalent of one proton entering the mitochondrion, since maintenance of the charge that drives the exchange requires the extra proton pumping. Hence four protons have to be pumped out of the matrix to drive the production of one molecule of ATP made available in the cytoplasm of the cell. For each pair of electrons transported from NADH to oxygen, the consensus is that 10 protons are pumped out of the mitochondrial matrix. Thus the oxidation of one molecule of NADH (located inside the mitochondrion) will produce 2.5 molecules of ATP. For the oxidation of one molecule of FADH₂ (that is, a pair of electrons from succinate), the value is 1.5. (Earlier estimates were 3 and 2, respectively.) The values are known as **P/O ratios** since a pair of electrons reduce one atom of oxygen.

The molecules of NADH produced in glycolysis, located in the cytoplasm, require separate consideration; these, you will recall, donate their pairs of electrons either to NAD⁺ located inside the mitochondrion or to mitochondrial FAD, depending on which shuttle mechanism (page 153) the cell happens to be using. Thus, a *cytoplasmic molecule* of NADH may give rise to either 2.5 or 1.5 molecules of ATP.

Yield of ATP from the oxidation of a molecule of glucose to CO₂ and H₂O

Starting from free glucose rather than glycogen, the net yield of ATP from the complete oxidation of the molecule is either 30 or 32 depending on which shuttle is used for the cytoplasmic NADH as already described above. To summarize: two from glycolysis at the substrate level (remember that, although four ATP molecules are generated in glycolysis, two were used at the start so the net gain is two). In the citric acid cycle, two molecules of ATP (via GTP in animals) are produced per molecule of glucose at the succinyl-CoA stage (one per turn of the cycle but two acetyl-CoA molecules are produced per glucose). Thus we have four molecules of ATP produced at the substrate level; all the rest come from electron transport.

Glycolysis produces, per molecule of glucose, two molecules of NADH, which are located in the cytoplasm. These will give rise either to a total of five or three molecules of ATP depending on the shuttle used. Per molecule of glucose, pyruvate dehydrogenase produces two molecules of NADH and the citric acid cycle, six. Oxidation of these will produce 20 molecules of ATP. Oxidation of the FADH₂ generated from the succinate → fumarate step produces a further three ATP molecules.

The total is therefore 2 + 5 (or 3) + 2 + 20 + 3 = 32 (with the aspartic-malate shuttle used) or 30 (with the glycerol-3-phosphate shuttle used). Note that these values are estimates—with substrate-level phosphorylation, ATP generation is always a whole number, but there is no such whole-number relationship between proton ejection and ATP generation.

An *E. coli* cell is equivalent in this context to a mitochondrion, the cell membrane equating to the inner mitochondrial membrane and the bacterial cytoplasm to the mitochondrial matrix. In such cells there is no need for shuttle mechanisms to transport NADH electrons to the respiratory pathway. In *E. coli* there is no transport of ATP and ADP needed into the cytoplasm. The ATP yield from the oxidation of a molecule of glucose in *E. coli* is therefore greater.

Is ATP production the only use that is made of the potential energy in the proton-motive force?

The answer is almost, but not quite. In newborn babies, heat production to maintain body temperature is helped by brown fat cells—brown because they are rich in mitochondria that contain the coloured cytochromes. The generation of ATP in mitochondria is dependent on the inner mitochondrial membranes being impermeable to protons, thus forcing the latter to enter the mitochondrial matrix only via the ATP-generating channels. If you made a hole in the membrane the protons would simply flood through it, effectively acting as a short circuit, no ATP would be generated, and the energy would be liberated as heat (which is why broken mitochondria cannot synthesize ATP). In brown fat cell mitochondria, this is essentially what happens, channels permitting non-productive (that is, no ATP synthesis) proton flow being made by a special protein, **thermogenin**. Chemicals that transport protons unproductively through membranes (dinitrophenol is the classical one) also 'uncouple' oxidation from ATP-generation.

Bacteria also harness energy by pumping protons across the membrane to the outside of the cell and generating ATP as described by reversed proton flow (the whole bacterial cell, as mentioned, in this regard, is equivalent to a mitochondrion). However, the proton gradient is also used in uptake of solutes into the cell, the H^+ gradient being used for cotransport (lactose uptake is an example) just as the Na^+ gradient is used in animal cells (page 82). Remarkably also, the cilia of bacteria are rotated by a flow of protons through the protein machinery that rotates the cilium; it runs on 'proticity' rather than the electricity used by an electric motor.

FURTHER READING

Glycolysis

Boiteux, A., and Hess, B. (1981). Design of glycolysis. *Phil. Trans. Roy. Soc. Series B, London.*, **293**, 5–22.

An extensive general review of the pathway and its control.

Electron transport chain

Boyer, P. D., Chance, B., Ernster, L., Mitchell, P., Racker, E., and Slater, E. C. (1977). Oxidative phosphorylation and photophosphorylation. *Ann. Rev. Biochem.*, **46**, 955–1026.

An unusual article in which the leaders in the field wrote separate independent sections. Although dated in detail, it is worth reading for the general principles given, and for its historical interest in this central field of ATP generation.

Slater, E. C. (1983). The Q cycle, an ubiquitous mechanism of electron transport. *Trends Biochem. Sci.*, **8**, 239–42.

Describes the proton pumping system present in both mitochondria and chloroplast.

Babcock, G. T. and Wikstrom, M. (1992). Oxygen activation and the conservation of energy in cell respiration. *Nature*, **356**, 301–9.

In depth review of the cytochrome oxidases and associated proton pumping.

ATP synthase

Abrahams, J. P., Leslie, A. G. W., Lutter, R. and Walker, J. E. (1994). Structure at 2.8 Å resolution of F_1-ATPase from bovine heart mitochondria. *Nature*, **370**, 621–8.

The classical paper giving the three-dimensional structure determination.

Noji, H., Yasuda, R., Yoshida, M., and Kinosita, Jr. K. (1997). Direct observation of the rotation of F_1-ATPase. *Nature*, **386**, 299–302.

A seminal article demonstrating physical rotation of the central subunit of the enzyme.

Boyer, P. D. (1999). What makes ATP synthase spin? *Nature*, **402**, 247–9.

A News and Views article summarizing in a succinct way the rotary mechanism of the enzyme. It relates to the research article below.

Rastogi, V. K. and Girvin M. E. (1999). Structural changes linked to proton translocation by subunit c of the ATP synthase. *Nature*, **402**, 263–8.

Proposes a mechanism based on structural work by which proton flow through the F_0 may produce rotation.

Mitochondrial genetics and ageing

Nagley, P. and Yau-Huei Wei. (1998). Ageing and mammalian mitochondrial genetics. *Trends in Genetics*, **14**, 513–17.

An account of mitochondrial mutations and their possible relationship to ageing.

PROBLEMS FOR CHAPTER 9

1   Explain the chemical rationale for the phosphohexose isomerase reaction.

2   What is meant by substrate-level phosphorylation? Give an example of such a system. How does this differ in basic terms from oxidative phosphorylation?

3   The reaction for which the enzyme pyruvate kinase is named never occurs in the cell. Discuss this.

4   How many ATP molecules are generated in glycolysis, from:
   (a)   glucose;
   (b)   a glycosidic unit of glycogen.

5   In calculating how many molecules of ATP are produced as a result of the oxidation of cytoplasmic NADH, we cannot be sure of the exact answer in eukaryotes. Why is this so?

6   What is the chemical rationale for isocitrate being oxidized before loss of CO_2 occurs?

7   Explain how the citric acid cycle acids can be 'topped up' by an anaplerotic reaction.

8   What is the cofactor involved in carboxylation reactions? Explain how it works.

9   Outline the arrangement of respiratory complexes in the electron transport chain.

10   What characteristic do ubiquinone and cytochrome *c* have in common? What is their physical location in the cell?

11   What is the immediate role of electron transfer in the respiratory chain?

12   It is stated in the text that the yield of ATP from the complete oxidation of a molecule of glucose in eukaryote cells is either 30 or 32 molecules; in the case of *E. coli* it is stated that the yield is greater than this. Why does this difference in statements exist?

13   Explain what is meant by the statement that the citric acid cycle is a water-splitting device.

14   By means of diagrams and brief notes, explain in outline how the proton gradient generated by electron transport is harnessed into ATP production by ATP synthase.

15   The active sites in the F_1 subunits of ATP synthase are said to be cooperatively interdependent, a most unusual situation. Explain what this means.

Chapter summary

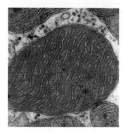

Energy release from fat

In the previous sections you have seen how energy in the form of ATP is produced in the cell by the oxidation of glucose or glycogen. Regulation of the pathways in this chapter are dealt with in Chapter 13 on metabolic control (see note at start of Chapter 7).

Fat is another major source of energy for ATP production. It provides perhaps half of the total energy needs of heart and resting skeletal muscles. By far the largest amount of stored energy occurs as fat for, unlike the situation with glycogen, there appears to be no limit to the amount of neutral fat that can be stored in the body in the adipose cells. It is more highly reduced than carbohydrate and not hydrated as is glycogen, and therefore is a more concentrated energy store.

The subject of fat oxidation and concomitant ATP production is greatly simplified because fat oxidation involves the same citric acid cycle and electron transport system as in glucose oxidation. As already described, the two systems (glucose oxidation and fat oxidation) converge at acetyl-CoA so all we are mainly concerned with in fat oxidation is the relatively simple task of chopping up fatty acids by removing two carbon atoms at a time as acetyl-CoA. This involves NAD^+ and FAD reduction which are oxidized by the same pathways we have already discussed. The remarkable efficiency of using the same machinery for obtaining energy from all classes of foodstuffs should be noted. It is only the preliminary reactions for feeding the material into the cycle that vary between the foods being oxidized.

A few simple points first.

- Before oxidation can occur, free fatty acids must be released by hydrolysis. This also produces a molecule of glycerol per TAG hydrolysed and this is metabolized separately. The glycerol is manipulated to enter metabolism in the glycolysis pathway—it is phosphorylated and oxidized to give dihydroxyacetone phosphate of glycolysis fame. Thus, in starvation, the glycerol moiety *can* give rise in the liver to glucose by the gluconeogenesis pathway (see Chapter 12).

- Free fatty acids for oxidation are obtained by peripheral tissues from that released into the blood by adipose cells when the glucagon level is high. They also enter the cells as a result of lipoprotein lipase attack on chylomicrons or VLDL (page 129).

- Free fatty acids from adipose cells are carried as ionized molecules attached in a freely reversible manner to serum albumin. They readily diffuse into cells so that the amount entering cells increases as their blood level rises as a result of release from adipose cells. As the fatty acids are taken up by cells, more will dissociate from the serum albumin carrier protein.

- Fatty acids are broken down by removing two carbon atoms at a time as acetyl-CoA.

- During conversion to acetyl-CoA the fatty acid is always in the form of an acyl-CoA. The first stage in oxidation of fatty acids is always to convert them to the fatty acyl-CoA compounds, a reaction known as fatty acid activation.

Mechanism of acetyl-CoA formation from fatty acids

'Activation' of fatty acids by formation of fatty acyl-CoA derivatives

The term 'activation' of a carboxylic acid refers to the fact that the thiol ester is a high-energy (or reactive) compound. The activation reaction is

$$R\ COO^- + ATP + CoA{-}SH \rightarrow R\ CO{-}S{-}CoA + AMP + PP_i;$$

$$\Delta G^{0'} = -0.9\,kJ\,mol^{-1}.$$

The free-energy change of this reaction is small (because of the high energy of the thiol ester) but hydrolysis of the PP_i by the ubiquitous enzyme, inorganic pyrophosphatase, makes the overall process strongly exergonic and irreversible ($\Delta G^{0'} = -32.5\,kJ\,mol^{-1}$). (See page 12 if you've forgotten this point.)

There are three different fatty acid activating enzymes specific for short, medium, and long chain acids, respectively—called fatty acyl-CoA synthetases.

Transport of fatty acyl-CoA derivatives into mitochondria

Activation of fatty acids occurs on the outer mitochondrial membrane, which is effectively the same compartment as is the cytoplasm, but their conversion to acetyl-CoA occurs only in the mitochondrial matrix. The acyl group of fatty acyl-CoA is carried in, without the CoA, by a special transport mechanism and then handed over to CoASH *inside* the mitochondrion to become fatty acyl-CoA again. The high-energy nature of the acyl-bond is preserved during the transport—otherwise it could not re-form fatty acyl-CoA inside the mitochondrion without further energy expenditure. To achieve this, outside the mitochondrion, the acyl group is transferred to a rather odd hydroxylated molecule, **carnitine**.

Carnitine

Although a carboxylic ester is usually of the low-energy type, the structure of carnitine is such that the fatty acyl–carnitine bond is of the high-energy type—the acyl group has a high group transfer potential. Presumably this a reason for the evolution of carnitine as the carrier molecule in this transport

system. The fatty acyl-carnitine so formed is transported into the mitochondrial matrix where the reverse reaction occurs—carnitine is exchanged for CoASH and the free carnitine is transported back to the cytoplasm to collect another fatty acyl group.

Fatty acyl-carnitine

The scheme is shown in Fig. 10.1.

The acyl transfer from fatty acyl-CoA to carnitine is catalysed on the cytoplasmic side of the mitochondrion by an enzyme called **carnitine acyltransferase I** and that on the matrix side by **carnitine acyltransferase II**. Genetic defects are known in which there is a carnitine deficiency or deficiency in the carnitine acyltransferase. In some cases this can manifest itself as muscle pain and abnormal fat accumulation in muscles.

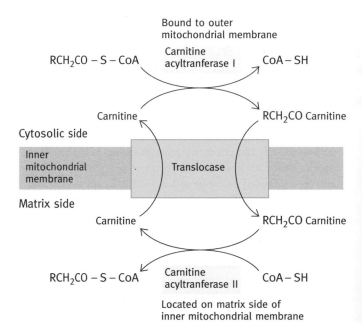

Fig. 10.1 Mechanism of transport of long chain fatty acyl groups into mitochondria where they are oxidized in the mitochondrial matrix. The acyl–carnitine bond is an unusual ester bond in that it has a high group transfer potential—the compound is of the high-energy type so that exchange of carnitine for CoASH inside the mitochondrion occurs without need for energy input. See text for structures.

Conversion of fatty acyl-CoA to acetyl-CoA molecules inside the mitochondrion

There are four separate reactions in this process, three of which are analogous in reaction types to those converting succinate to oxaloacetate in the citric acid cycle—namely, dehydrogenation by an FAD enzyme, hydration, and an NAD^+-dependent dehydrogenation. We suggest that you refresh your memory on these by having a quick look on page 159 at the reaction sequence, succinate $\rightarrow$ fumarate $\rightarrow$ malate $\rightarrow$ oxaloacetate. The corresponding reactions on fatty acyl-CoA derivatives are shown in Fig. 10.2. In fat oxidation, the keto acyl-CoA is cleaved by

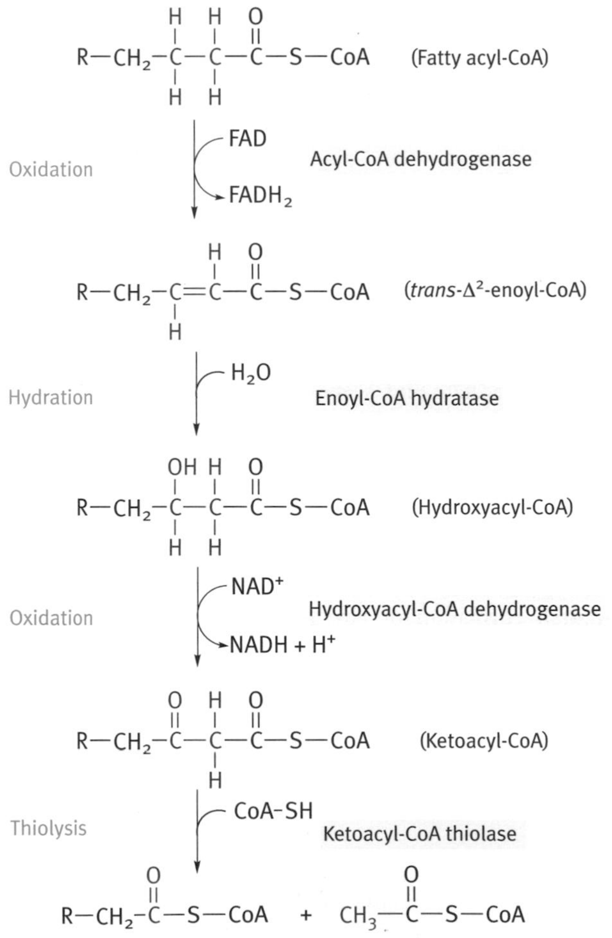

Fig. 10.2 One round of four reactions by which a fatty acyl-CoA is shortened by two carbon atoms with the production of a molecule of acetyl-CoA. Note the similarity of reaction types in the desaturation, hydration, and ketoacyl formation with the succinate $\rightarrow$ fumarate $\rightarrow$ malate $\rightarrow$ oxaloacetate steps of the citric acid cycle.

CoASH, splitting off two carbon atoms as acetyl-CoA but forming a shorter fatty acetyl-CoA derivative. Because the molecule is split by the —SH group of CoASH, the enzyme is called a **thiolase**. The thiolase reaction preserves the free energy as thiol ester of fatty acyl-CoA. When the fatty acyl group has been shortened, by successive rounds of acetyl-CoA production, to the C_4 stage (butyryl-CoA) the next round of reactions produces acetoacetyl-CoA, which is finally split by CoASH into two molecules of acetyl-CoA. A specific thiolase in mitochondria performs this reaction.

$$CH_3COCH_2CO-S-CoA + CoA-SH \rightarrow 2CH_3CO-S-CoA$$

   Acetoacetyl-CoA   Acetyl-CoA

The reaction sequence in each round involves conversion of saturated fatty acyl-CoAs to β-ketoacyl-CoAs. The process is therefore referred to as **β-oxidation of fatty acids**. The NADH and the $FADH_2$ (the latter on the acyl-CoA dehydrogenase) feed electrons into the electron transport chain exactly as already described (see Fig. 9.19 for a summary of this).

Energy yield from fatty acid oxidation

A molecule of palmitic acid (C_{16}) is converted to eight molecules of acetyl-CoA and in this process we generate seven $FADH_2$ molecules and seven NADH molecules. The acetyl-CoA is oxidized by the citric acid cycle and the NADH and $FADH_2$ by the electron transport chain (Chapter 9). NADH oxidation generates 2.5 ATP molecules and $FADH_2$, 1.5 ATP molecules. (Since the NADH is generated in the mitochondrial matrix, it does not have to be carried in by a shuttle mechanism.)

If you count it all up (not forgetting the one GTP per acetyl-CoA from the citric acid cycle) the oxidation of one mole of palmitic acid generates 106 moles of ATP from ADP and P_i (allowing for the two ATP consumed in formation of palmitoyl-CoA). This represents about a 33% efficiency in trapping the free energy in usable form. Put in another way, the oxidation of 256 g of palmitic acid would produce 45 kg of ATP from ADP and P_i. There isn't anywhere near that amount of ATP in the body, of course, for ATP recycles rapidly through ADP and P_i.

Oxidation of unsaturated fat

Olive oil is an oil rather than a solid because of its high content of **monounsaturated fat** (see page 75 for explanation of this). For example, palmitoleic acid has a double bond between carbon atoms 9 and 10. So far as oxidation goes, this is treated by the cell in exactly the same way as palmitic acid for three rounds of β-oxidation. At this point the product is *cis-Δ^3-enoyl-CoA*

$$
\underset{(4)\ (3)\ (2)\ \ \ (1)}{R-\overset{H}{C}=\overset{H}{C}-CH_2-\overset{O}{\overset{\|}{C}}-S\text{-}CoA}\ .
$$

The double bond in this position prevents the acyl-CoA dehydrogenase from forming a double bond between carbon atoms 2 and 3, as is required in β-oxidation of a saturated acyl-CoA.

An extra isomerase enzyme takes care of this by shifting the existing double bond into the required 2–3 position; it generates the *trans*-isomer in doing so.

$$
\underset{(4)\ (3)}{R-\overset{H}{C}=\overset{H}{C}-CH_2-\overset{O}{\overset{\|}{C}}-S-CoA}\qquad \textit{cis-}\Delta^3\text{-enoyl-CoA}
$$

$$\downarrow \text{Isomerase}$$

$$
R-CH_2-\overset{H}{C}=\underset{H}{C}-\overset{O}{\overset{\|}{C}}-S-CoA\qquad \textit{trans-}\Delta^2\text{-enoyl-CoA}
$$

This is now on the main pathway of fat breakdown and so the problem of monounsaturated fat oxidation is solved. (If you look at Fig. 10.2, you will see that, in oxidation of saturated acyl-CoAs, it is the *trans*-isomer that is generated.)

Polyunsaturated fatty acids pose additional problems—for example, linoleic acid has two double bonds (Δ^9 and Δ^{12}). In effect, one (Δ^9) is dealt with as described above while the second (Δ^{12}) is handled by using an additional enzyme at the appropriate stage, again putting the molecule on the normal metabolic path of β-oxidation (the sequence of the steps is not be given here.)

Is the acetyl-CoA derived from fat breakdown always directly fed into the citric acid cycle?

So far we have explained that fatty acids are converted to acetyl-CoA, which then joins the citric acid cycle and is oxidized. This is true for tissues in 'normal' circumstances but it has to be qualified for the liver in a particular circumstance.

As described in Chapter 6, the body can be in a physiological situation where fat metabolism is the main source of energy. This occurs in starvation after exhaustion of glycogen stores; the same can occur in diabetics where inability to metabolize carbohydrate effectively results in an almost analogous glucose 'starvation' irrespective of glucose availability. In this situation, the fat cells are pouring out free fatty acids (page 113) and the liver may produce excessive amounts of acetyl-CoA, 'excessive' is the sense that there is too much produced to be fed into the citric acid cycle. This is exacerbated because lack of carbohydrate metabolism leads to a relative shortage of oxaloacetate, which you will recall is needed to produce citrate from acetyl-CoA. (For an explanation of why oxaloacetate is in short supply in this situation, see page 161.)

As previously mentioned in Chapter 8, the liver cell, *in effect*, joins two acetyl groups together, by a mechanism to be described shortly, to form **acetoacetate** ($CH_3\ COCH_2\ COO^-$) which is partly reduced to β-**hydroxybutyrate** ($CH_3\ CHOH\ CH_2\ COO^-$). The two (**ketone bodies**) are released into the blood.

How is acetoacetate made from acetyl-CoA?

Acetoacetyl-CoA is formed from acetyl-CoA by reversal of the ketoacyl-CoA thiolase reaction

$$2CH_3CO-S-CoA \rightleftharpoons CH_3COCH_2CO-S-CoA + CoA-SH$$

One would have imagined that free acetoacetate would be formed by simple hydrolysis of the acetoacetyl-CoA (thus pulling the equilibrium over). However, it is not so. Instead, a third molecule of acetyl-CoA is used to form **3-hydroxy-3-methylglutaryl-CoA (HMG-CoA)**. This is now decomposed to form acetoacetate and acetyl-CoA. The scheme is shown in Fig. 10.3.

One cannot be sure why the synthesis of acetoacetate proceeds in this way. HMG-CoA is synthesized by many animal cells. It is a precursor of cholesterol, which is an essential constituent for their membranes (page 75). Ketone body formation occurs in mitochondria (where fat conversion to acetyl-CoA occurs). Formation of HMG-CoA for cholesterol synthesis takes place in the cytoplasm where a separate HMG-CoA synthase occurs attached to the endoplasmic reticulum membrane. Acetoacetate can be used by peripheral tissues to generate energy. In mitochondria it is converted to acetoacetyl-CoA by an acyl exchange reaction in which succinyl-CoA is converted to succinate.

Acetoacetate + succinyl-CoA $\underset{\text{transferase}}{\overset{\text{CoA}}{\rightleftharpoons}}$ Acetoacetyl-CoA + succinate

Acetoacetyl-CoA $\xrightarrow[\text{Thiolase}]{\text{CoA-SH}}$ 2-Acetyl-CoA

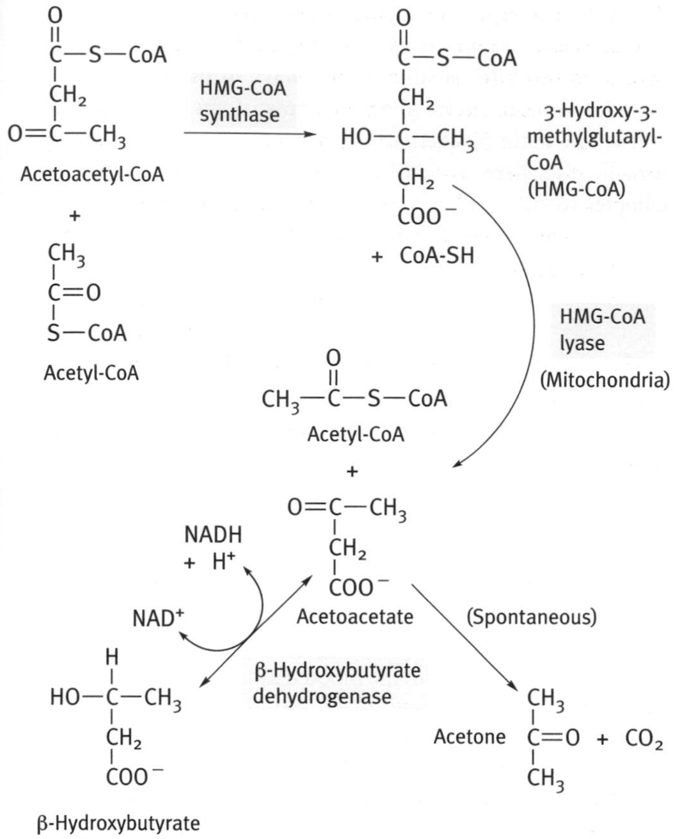

Fig. 10.3 Ketone body production in the liver during excessive oxidation of fat in starvation or diabetes. The process occurs in mitochondria. HMG-CoA is also the precursor of cholesterol but this occurs in the cytosol of rat liver where HMG-CoA synthase also occurs.

Acetoacetyl-CoA is cleaved by a thiolase using CoASH to form two molecules of acetyl-CoA. β-Hydroxybutyrate is also utilized by being dehydrogenated first to acetoacetate.

Oxidation of odd-numbered carbon chain fatty acids

A small proportion of fatty acids in the diet (for example those from plants) have odd-numbered carbon chains, β-oxidation of which produces, as the penultimate product, a five-carbon β-ketoacyl-CoA instead of acetoacetyl-CoA. Cleavage of this by thiolase produces acetyl-CoA and the three-carbon propionyl-CoA

$$CH_3-CH_2-CO-CH_2-CO-S-CoA + CoA-SH \longrightarrow$$

$$CH_3-CH_2-CO-S-CoA + CH_3-CO-S-CoA$$

Propionyl-CoA   Acetyl-CoA

Propionyl-CoA is converted to succinyl-CoA and hence on to the citric acid cycle mainstream by the following reactions. The epimerase catalyses the conversion of D- to L-methylmalonyl-CoA.

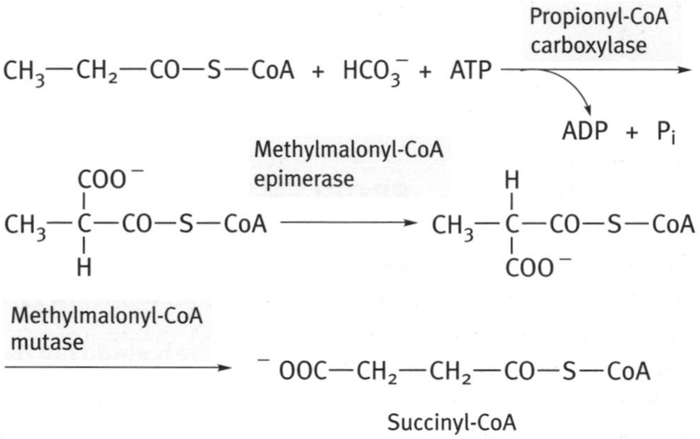

$$CH_3-CH_2-CO-S-CoA + HCO_3^- + ATP \xrightarrow{\text{Propionyl-CoA carboxylase}} ADP + P_i$$

$$CH_3-\underset{H}{\overset{COO^-}{C}}-CO-S-CoA \xrightarrow{\text{Methylmalonyl-CoA epimerase}} CH_3-\underset{COO^-}{\overset{H}{C}}-CO-S-CoA$$

Methylmalonyl-CoA mutase

$$\xrightarrow{} {}^-OOC-CH_2-CH_2-CO-S-CoA$$

Succinyl-CoA

It is the last reaction which is of great interest for it involves the most complex coenzyme of all, deoxyadenosylcobalamin, a derivative of vitamin B_{12}.

Propionate is also formed in the degradation of four amino acids (valine, isoleucine, methionine, and threonine) and from the cholesterol side chain.

Deficiency of either the methylmalonyl-CoA mutase or ability to synthesize the required cofactor from vitamin B_{12} leads to methylmalonic acidosis, usually fatal early in life.

Peroxisomal oxidation of fatty acids

Peroxisomes are small membrane-bound vesicles found in many animal cells. They are discussed in Chapter 17. In the present context, they also oxidize fatty acids, though the bulk of this occurs in mitochondria. β-Oxidation of fatty acids in peroxisomes differs in that the $FADH_2$ produced by oxidation of saturated acyl-CoAs is directly oxidized by O_2, producing H_2O_2 (and heat). There is no electron transport system. The functions of peroxisomes are less well defined than those of other organelles but may be related to handling fatty acids longer than C_{18} which are not oxidized by the mitochondrial system.

We have so far dealt with energy production from glucose and from fat. The remaining third major food component is in the form of the amino acids. It would be logical to deal with energy production from amino acids next but, in fact, it is more convenient at this stage to remain with carbohydrate and fat metabolism and next to deal with glucose and fat synthesis and then with the control of metabolism. The reason for this is that energy production from individual amino acids essentially consists of converting them to compounds on the main glycolytic and citric acid cycle pathways so that there is very little information to be given in terms of energy production *per se*. The main biochemical interests of amino acid metabolism lie elsewhere, as will be seen when we deal with this in Chapter 16.

FURTHER READING

McGarry, J. D. and Foster, D. W. (1980). Regulation of hepatic fatty acid oxidation and ketone body production. *Ann. Rev. Biochem.*, **49**, 395–420.

Despite the regulatory title, the article provides a general review of ketone body formation.

Lardy, H. and Shrago, E. (1990). Biochemical aspects of obesity. *Ann. Rev. Biochem.*, **59**, 689–710.

Very readable account relating obesity, hormones, and metabolism.

PROBLEMS FOR CHAPTER 10

1   Peripheral tissues obtain their free fatty acids for oxidation from the blood. Explain three ways in which free fatty acids become available to cells.

2   Which cells of the body do not use free fatty acids for energy supply?

3   In breaking down fatty acids to acetyl-CoA:
   (a)   What is always the first step?
   (b)   Where does this occur?
   (c)   Where does fatty acid breakdown to acetyl-CoA occur in eukaryotes?
   (d)   How do fatty acid groups reach this site of breakdown?

4   Illustrate similarities in the oxidation of fatty acids to acetyl-CoA with a section of the citric acid cycle.

5   What is the yield of ATP from the complete oxidation of a molecule of palmitic acid? Explain your answer.

6   Explain how a monounsaturated fat (Δ^9) is broken down to acetyl-CoA.

7   Is acetyl-CoA, derived from fatty acid breakdown, always fed into the citric acid cycle? Explain your answer.

8   HMG-CoA is an intermediate in both acetoacetate synthesis and cholesterol synthesis. Where do these two processes occur?

Chapter summary

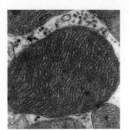

A switch from catabolic to anabolic metabolism—first, the synthesis of fat and related compounds in the body

In the previous chapters on metabolism we have been dealing with catabolic systems that break down fats and carbohydrates to yield energy. We are now starting on the other side of metabolism—the building up of molecules or anabolism. Since fat breakdown will be fresh in your minds from the previous chapter it seems a good idea to start with fat synthesis. We'll deal with carbohydrate synthesis in the next chapter. Note that regulation of the pathways in this reaction can be found in Chapter 13 (see special note at the start of Chapter 7).

Fat synthesis occurs mainly to convert carbohydrate in excess of that needed to replenish glycogen stores to a more suitable storage form, neutral fat. Alcohol and certain amino acids also give rise to fat. As already explained (page 110), if this weren't done, to store the amount of energy contained in our storage fat as glycogen we'd be much larger in size. Fat synthesis occurs in times of dietary plenty.

Mechanism of fat synthesis

General principles of the process

If you refer to the overall diagram of metabolism (Fig. 8.11), you will see that, as well as fatty acids being converted to acetyl-CoA, acetyl-CoA can be converted into fatty acids. The fact that fatty acids are synthesized from acetyl-CoA, two carbon atoms at a time, explains why natural fats almost always have even numbers of carbon atoms. Carbohydrate intake in excess can lead to fat deposition—glucose goes to pyruvate which goes to acetyl-CoA. However, since the latter step catalysed by pyruvate dehydrogenase is irreversible, acetyl-CoA cannot be converted to pyruvate and hence fat cannot be converted into glucose in animals. (Bacteria and plants have, as you will see, a special trick for doing so—see page 205.)

You will by now be familiar with the concept that, in metabolic pathways, at least some reactions are different in the forward and reverse directions. This is true for fat breakdown and synthesis. The two pathways use, for some steps, essentially the same reactions, but there are others that are different in the two directions. These make both pathway directions thermodynamically favourable, irreversible, and separately controllable.

To render synthesis of fatty acids from acetyl-CoA thermodynamically favourable, we must have a reaction that injects energy into the process—a reaction that is irreversible. Refresh your memory on this, if necessary, by reading page 5 again, since it is essential to appreciate this for understanding the very first reaction in fat synthesis. In this, a molecule of CO_2 is added to the acetyl-CoA molecule, using ATP breakdown as energy source, to form malonyl-CoA, but, in the next reaction in fat synthesis, the CO_2 is released again. This seems pointless unless you remember that the process is thus made thermodynamically irreversible. It is to do with energy, rather than chemical change. Malonic acid has the structure HOOC—CH_2—COOH so malonyl-CoA is HOOC—CH_2—CO—S—CoA. The enzymic reaction below is catalysed by acetyl-CoA carboxylase—it inserts a carboxyl group into acetyl-CoA forming malonyl-CoA.

$$CH_3-\overset{\overset{O}{\|}}{C}-S-CoA \ + \ ATP \ + \ HCO_3^-$$

$$\overset{O}{\underset{^-O}{\diagdown}}C-CH_2-\overset{\overset{O}{\|}}{C}-S-CoA \ + \ ADP \ + \ P_i \ + \ H^+$$

The enzyme has a prosthetic group of biotin. All carboxylases using ATP to incorporate CO_2 into molecules, we remind you, have this feature. As an intermediate in the reaction, an activated CO_2-biotin complex is formed (page 161).

To synthesize fatty acids we have to add two carbon units at a time, starting with acetyl-CoA. *Malonyl-CoA is the active donor of two carbon atoms in fatty acid synthesis*, despite the fact that the malonyl group has three carbons. But before this, a few words of explanation about the **acyl carrier protein** or **ACP**.

The acyl carrier protein (ACP) and the β-ketoacyl synthase

As described in Chapter 10, when fatty acids are broken down to acetyl-CoA, all of the reactions occur, not as free fatty acids, but as thiol esters with CoA. When fatty acids are synthesized, all of the reactions also occur on fatty acyl groups bound as thiol esters, but in fatty acid synthesis, instead of CoA being used to bind the reactants, *half* of the CoA molecule is used. We remind you of the structure of CoA.

Phosphate—pantothenate—NHCH$_2$CH$_2$—SH
Phosphate—ribose—adenine
    phosphate

Phosphopantotheine
moiety in box

The half in the box is 4-phosphopantotheine and this is the 'carrier' thiol used in fatty acid synthesis. It is not free, as is CoA, but is bound to a protein called the acyl carrier protein (ACP). You can think of ACP as a protein with built in CoA or, alternatively, as a giant CoA molecule with the AMP part of CoA replaced by a protein.

Another enzyme or domain, **β-ketoacyl synthase**, also has a reactive thiol group in that of the amino acid cysteine that forms part of the structure of the active site. The ACP and β-ketoacyl synthase in mammals are part of a large multifunctional complex.

Mechanism of fatty acyl-CoA synthesis

We will start with the situation shown Fig. 11.1(a) in which the two thiol groups on the fatty acyl-CoA synthase complex are vacant.

$CH_3 CO$—is transferred from acetyl-CoA to the ACP thiol by a specific transferase (Fig. 11.1(b)). This is then further transferred to the β-ketoacyl synthase thiol group (Fig. 11.1(c)) leaving the ACP site vacant. Another transferase transfers the malonyl group of malonyl-CoA to the latter, forming malonyl-ACP (Fig. 11.1(d)). The synthase now transfers the acetyl group to the malonyl group, displacing CO_2 and forming a β-ketoacyl-ACP; in this initial case it is β-ketobutyryl-ACP (Fig. 11.1(e)). The latter is reduced by activities on the same protein complex (described below) to form butyryl-ACP (Fig. 11.1(f)). The resultant situation is precisely analogous to that shown in Fig. 11.1(b) since both have a saturated acyl-ACP complex (acetyl and butyryl, respectively) and a vacant synthase site. A series of reactions now ensues, identical to those in Fig. 11.1 (starting with the reaction b → c), except that acetyl- is now butyryl- and the end-product is hexanoyl-. Five more rounds produces palmityl-. When C_{16} is reached, a hydrolase releases palmitic acid,

$$CH_3(CH_2)_{14}CO-S-ACP + H_2O \rightarrow CH_3(CH_2)_{14}COO^- + ACP-SH.$$

Note that the β-ketoacyl-ACP synthase reaction leading to stage (e) is irreversible because of the decarboxylation involved. This is the purpose of forming malonyl-CoA from acetyl-CoA: to make fatty acid synthesis pathway a one-way process.

If longer chain acids are required to be formed such as stearic acid (C_{18}), a separate system elongates the palmitate.

Organization of the fatty acid synthesis process

The sequence of reactions given above for the conversion of acetyl-CoA to palmitic acid is relatively simple or complex according to your viewpoint but, in animal tissues, the organization of the enzyme activities involved is remarkable from any viewpoint. In *E. coli*, the situation is straightforward, in that there is a series of separate enzymes, as is usual in metabolic pathways, and, after each reaction, the products dissociate and diffuse to the next enzyme. In animals, the situation is different; all of the catalytic activities of the steps following malonyl-CoA formation reside on one giant protein, the activities of which have not been separated. (A pair of such proteins collaborate as the functional dimer complex.) The growing fatty acid chain oscillates from the ACP-thiol group to the β-ketoacyl synthase thiol group but it is never released from the enzyme complex until palmitate synthesis is completed. The elongation step and reductive reactions all occur with the substrate attached to the ACP. The long flexible arm of the 4-phosphopantetheine is

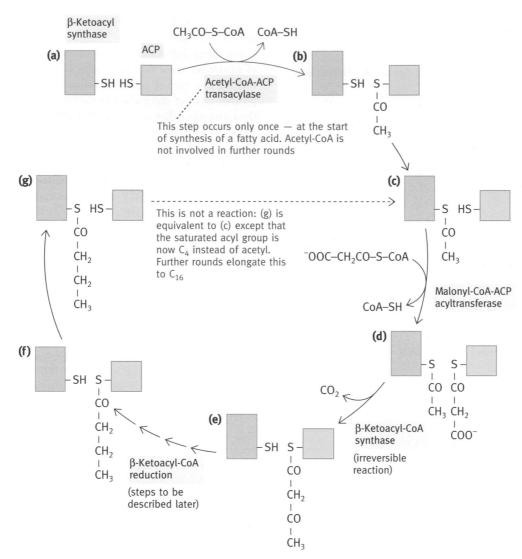

Fig. 11.1 The steps involved in the synthesis of fatty acids. Although the β-ketoacyl synthase of animals is a domain of a single protein molecule, it is shown here as being separate. This is because the functional enzyme unit is believed to be a dimer arranged head to tail and it is also believed that the transfers between the two —SH groups occur between the domains on the two dimer-constituent protein molecules. Note that the thiol of the ACP is on the phosphopantetheine moiety and that of the synthase is on a cysteine residue of the protein. After seven successive rounds of reactions, the resultant palmityl-ACP is hydrolysed to release free palmitate.

presumably needed to permit the different intermediates to interact with the appropriate catalytic centres of the complex. The advantage of a single large complex is that the process of synthesis can be more rapid since each intermediate is positioned to interact with the next catalytic centre rather than having to diffuse away and find the next enzyme.

The reductive steps in fatty acid synthesis

In the above sequence of events involved in saturated fatty acyl-ACP synthesis, the β-ketoacyl group attached to the ACP is re-

duced by a succession of three steps shown in Fig. 11.2. The reductant in these steps is **NADPH** (not NADH), an electron carrier, so far not mentioned in this book, and which will now be described.

What is NADP⁺?

NAD⁺ is
P—ribose—nicotinamide
|
P—ribose—adenine

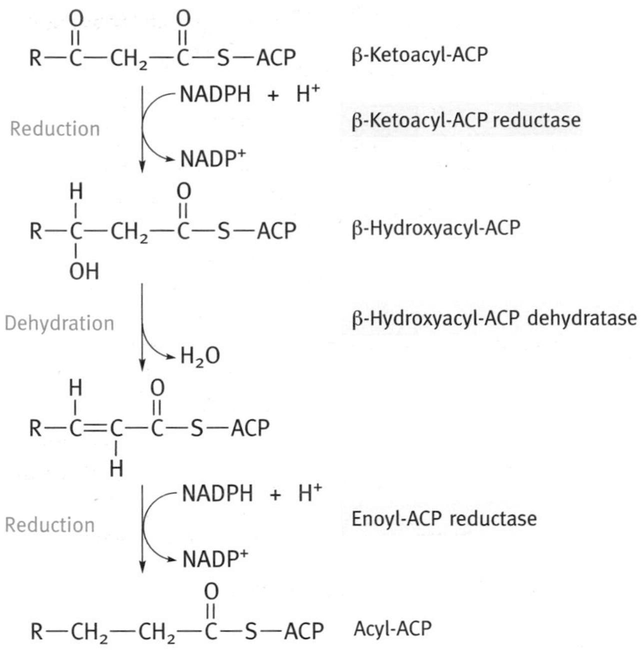

$$R-\overset{\overset{\text{O}}{\|}}{C}-CH_2-\overset{\overset{\text{O}}{\|}}{C}-S-ACP \qquad \text{β-Ketoacyl-ACP}$$

Reduction ⤵ NADPH + H⁺ → NADP⁺   β-Ketoacyl-ACP reductase

$$R-\overset{\overset{\text{H}}{|}}{\underset{\underset{\text{OH}}{|}}{C}}-CH_2-\overset{\overset{\text{O}}{\|}}{C}-S-ACP \qquad \text{β-Hydroxyacyl-ACP}$$

Dehydration ⤵ → H_2O β-Hydroxyacyl-ACP dehydratase

$$R-\overset{\overset{\text{H}}{|}}{C}=\overset{\overset{\text{}}{|}}{\underset{\underset{\text{H}}{|}}{C}}-\overset{\overset{\text{O}}{\|}}{C}-S-ACP$$

Reduction ⤵ NADPH + H⁺ → NADP⁺   Enoyl-ACP reductase

$$R-CH_2-CH_2-\overset{\overset{\text{O}}{\|}}{C}-S-ACP \qquad \text{Acyl-ACP}$$

Fig. 11.2 Reductive steps in fatty acid synthesis.

$NADP^+$ is

P—ribose—nicotinamide
|
P—ribose—adenine
    |
    P

The extra phosphoryl group is attached to the 2′ hydroxyl group of the ribose which is attached to adenine.

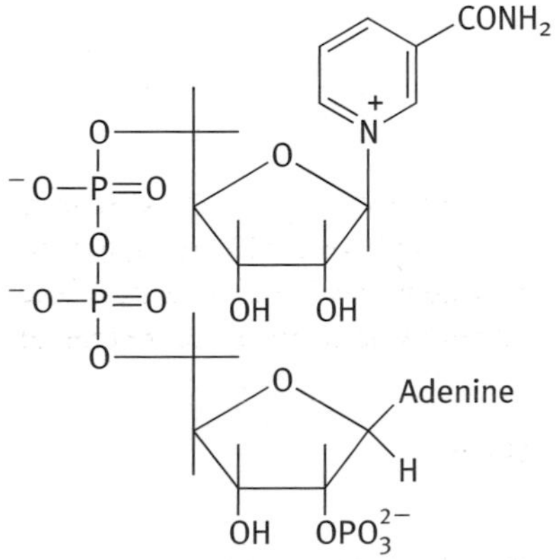

The extra phosphoryl group has no significant effect on the redox characteristics of the molecule—it is, in fact, purely an identification or recognition signal. An NAD^+ enzyme will not react with $NADP^+$ and vice versa (with the possible odd exception of little significance).

For a long time it was thought that this was just an odd quirk of nature, no more than a curiosity. However, eventually it was realized that the use of the two coenzymes constitutes an important form of metabolic compartmentation. To understand this, remember that in energy production the aim is to oxidize reducing equivalents, while in, for example, fat synthesis the aim is to use reducing equivalents for reductive synthesis. The aims are diametrically opposite. The cell keeps these metabolic activities separate by metabolic compartmentation—one, the oxidative part uses NAD^+ as electron carrier; the other, the reductive part, uses $NADP^+$.

Separate mechanisms exist to reduce NAD^+ and $NADP^+$. You already know how NAD^+ is reduced in glycolysis, the pyruvate dehydrogenase reaction, the citric acid cycle, and fat oxidation. How is $NADP^+$ reduced? This is explained in the context of the next question.

Where does fatty acid synthesis take place?

In terms of tissues, the main sites of fatty acid synthesis in the body are the liver and adipose cells (the former predominating) but many other tissues, especially the mammary gland during lactation, also produce fat.

Within a cell, **palmitate synthesis** from acetyl-CoA occurs in the **cytosol**. This is in contrast to **fatty acid oxidation** which occurs in the **mitochondria**. The major source of acetyl-CoA for fat synthesis is the pyruvate dehydrogenase reaction (page 154), which is located in the mitochondrial matrix. Acetyl-CoA cannot cross the mitochondrial membrane to the site of fatty acid synthesis in the cytosol so acetyl residues must be transported from mitochondria to the cytosol. What is done is part of an ingenious scheme. The acetyl-CoA in the mitochondrion is converted to citrate by the citric acid synthase reaction of the citric acid cycle (page 157). The citrate is transported by a mitochondrial membrane system into the cytosol where it is cleaved back to acetyl-CoA and oxaloacetate by a separate enzyme, called ATP-**citrate lyase** or the **citrate cleavage enzyme**. This reaction is coupled to hydrolysis of ATP to ADP and P_i, thus ensuring it goes to completion,

Citrate + ATP + CoA—SH + H_2O → acetyl-CoA + oxaloacetate + ADP + P_i.

That is not the end of the story. The oxaloacetate cannot get back into the mitochondrion, whose membrane is impervious to it and for which no transporter exists. It is therefore reduced in the cytoplasm to malate by NADH (note, *not* NADPH); the malate so formed is oxidatively decarboxylated to pyruvate and

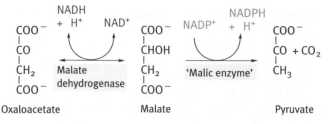

Fig. 11.3 Reduction of $NADP^+$ for fatty acid synthesis. The net effect of the two reactions is to transfer reducing equivalents from NADH to $NADP^+$. The pyruvate so produced in the cytoplasm enters the mitochondrion. The source of the oxaloacetate is shown in Fig. 11.4.

CO_2 by an enzyme (the **'malic' enzyme**) that uses $NADP^+$ (note, *not* NAD^+), thus producing NADPH, which is needed for fat synthesis (Fig. 11.3). The pyruvate is now transported back into the mitochondrion (Fig. 11.4). The pyruvate transported into the mitochondria can be converted back to oxaloacetate by the **pyruvate carboxylase reaction** (previously mentioned on page 161).

$$\text{Pyruvate} + HCO_3^- + ATP$$

$$\downarrow \text{(Pyruvate carboxylase)}$$

$$\text{Oxaloacetate} + ADP + P_i + H^+$$

Citrate leaves the mitochondrion only when it is at a high concentration; this occurs when carbohydrate is plentiful. Citrate does not occur in the cytosol at other times.

Thus, the citrate mechanism not only transports acetyl-groups out of the mitochondrion, it also generates NADPH for fat synthesis. The reduction in the cytoplasm of oxaloacetate to malate by NADH and the oxidation of malate to pyruvate by $NADP^+$ constitutes a neat mechanism for transferring electrons from the NADH metabolic 'compartment' into the NADPH 'compartment' used for reductive synthesis reactions. In addition, citrate *in vitro* activates the first reaction committed to fat synthesis—the acetyl-CoA carboxylase, producing malonyl-CoA, though the physiological significance of this activation has recently been questioned. Control of this enzyme is discussed in Chapter 13.

It will be apparent from the scheme given that, for every acetyl-CoA produced in the cytosol from citrate, we generate *one* NADPH molecule. However, each cycle of the fatty acid synthase process required *two* NADPH molecules (see Fig. 11.2). Yet another mechanism for producing the extra NADPH is required—this will be explained in Chapter 14 (page 235) for

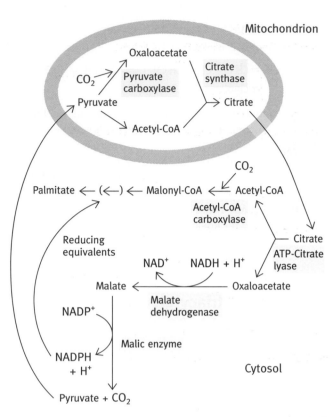

Fig. 11.4 Source of acetyl groups (acetyl-CoA) and reducing equivalents (NADPH + H^+) for fatty acid synthesis. The ATP-citrate lyase reaction involves the breakdown of ATP to ADP and P_i.

it belongs to a different metabolic topic, the pentose phosphate pathway.

Synthesis of unsaturated fatty acids

As explained (page 75), the body requires unsaturated fatty acids for the production of polar lipids for membrane synthesis and other roles. A separate enzyme system in liver can introduce a single double bond into the middle of stearic acid, generating oleic acid.

$$CH_3CH(CH_2)_7 CH = CH(CH_2)_7 COOH$$
$$\text{Oleic acid, } 18{:}1(\Delta^9)$$

However, the system cannot make double bonds between this central double bond and the methyl end of the molecule. Linoleic acid which has two double bonds and linolenic acid with three thus cannot be synthesized by the body.

$$CH_3(CH_2)_4 CH=CH\ CH_2\ CH_2=CH(CH_2)_7COOH$$

Linoleic acid, $18:2(\Delta^{9,12})$

$$CH_3\ CH_2\ CH=CH\ CH_2\ CH=CH\ CH_2\ CH=CH(CH_2)_7\ COOH$$

α-Linolenic acid, $18:3(\Delta^{9,12,15})$

Since they are needed for membrane components and for synthesis of eicosanoid regulatory molecules (see below) these two acids must be obtained from the diet; plants have enzymes that can desaturate in the terminal half of the fatty acids. The liver can elongate linoleic acid and introduce an extra double bond to give the C_{20} arachidonic acid ($20:4$, $\Delta^{5,8,11,14}$, see below) with four double bonds. It can also elongate acids to the C_{22} and C_{24} acids involved in the lipids of nerve tissues. All of these transformations occur as the CoA derivatives.

Synthesis of triacylglycerol and membrane lipids from fatty acids

Neutral fat is synthesized for storage purposes. For attachment of a fatty acid in ester bond to glycerol, the acid must be activated to the form of acyl-CoA, a reaction catalysed by acetyl-CoA synthetase, as already described

$$RCOOH + CoA{-}SH + ATP \rightarrow R\ CO{-}S{-}CoA + AMP + PP_i.$$

A carboxylic acid ester has a lower free energy of hydrolysis than has a thiol ester and hence activation of the acid makes triglyceride synthesis from the fatty acyl-CoA exergonic. Surprisingly perhaps, the acceptor of acyl groups is not glycerol but glycerol-3-phosphate, which arises mainly by reduction of the glycolytic intermediate, dihydroxyacetone phosphate (DHAP), though phosphorylation of glycerol can also occur in the liver but not in adipose cells, the latter point being of significance (page 205). The reaction is

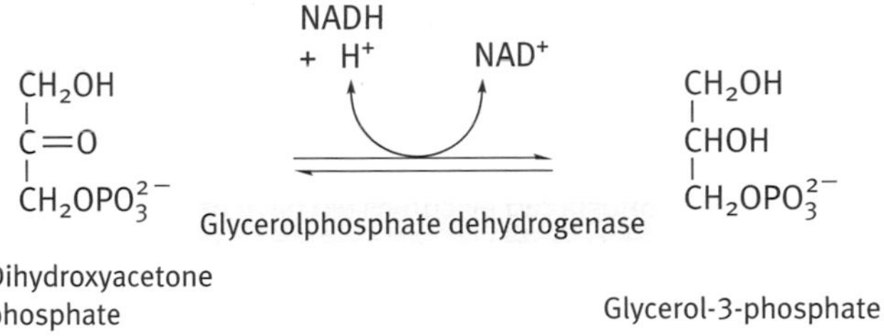

The steps in triacylglycerol synthesis are shown in Fig. 11.5.

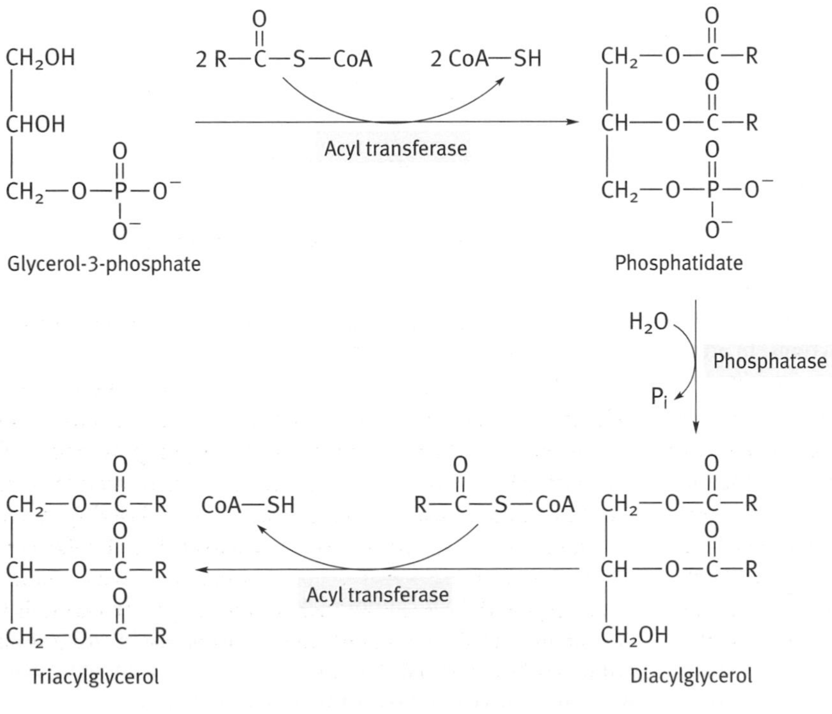

Fig. 11.5 Reactions involved in the synthesis of triacylglycerol (neutral fat) from glycerol-3-phosphate.

Synthesis of glycerophospholipids

If you turn back to page 71 you will recall that membranes contain glycerol-based phospholipids in which a polar alcohol is attached to the phosphoryl group of phosphatidic acid.

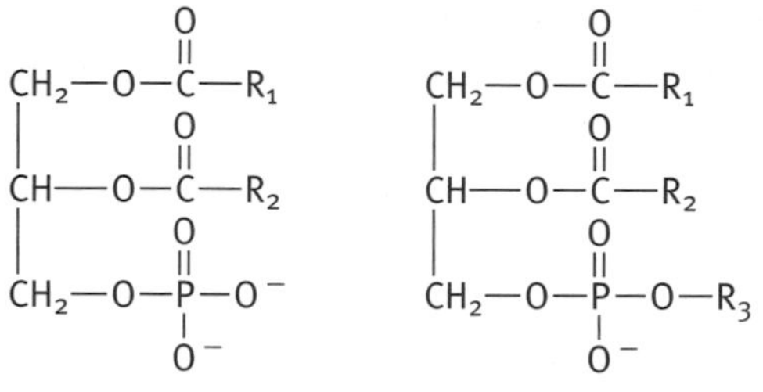

Phosphatidate   Phospholipid

R_3 may be serine, ethanolamine, choline, inositol, or diacylglycerol. The most prevalent membrane lipids in eukaryotes are phosphatidylethanolamine and phosphatidylcholine so we will consider their synthesis first; ethanolamine is $NH_2\,CH_2\,CH_2OH$ and choline is

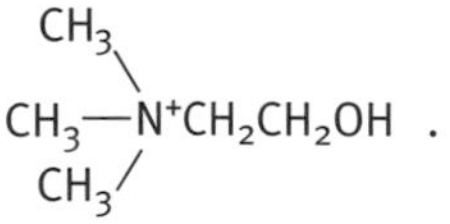

Both can be written as the alcohols 'R—OH'. The alcohols are 'activated' to participate in the synthesis; this occurs in two steps. First, a phosphorylation by ATP,

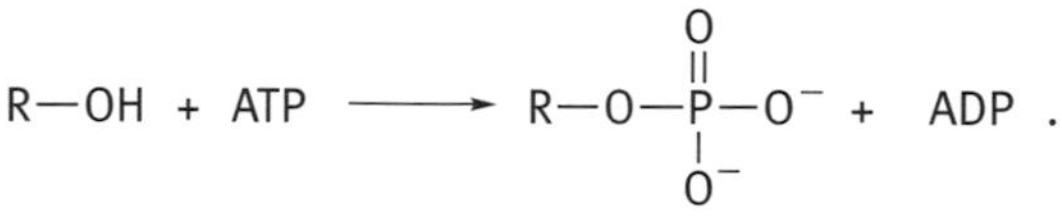

Secondly, a reaction with **cytidine triphosphate** or **CTP**. CTP is the same as ATP except that it has the base cytosine (C) in place of adenine (cytosine-ribose is called cytidine). We will come to its precise structure later in the book. All cells have CTP for it is needed for RNA and DNA synthesis (Chapter 22). Whether CTP involvement rather than ATP is due to an accidental quirk of evolution or whether there is a good reason for it is unknown, but the fact is, that CDP-compounds are always the 'high-energy' donors of groups in this area (just as for equally unknown reasons it is always UDP-glucose for processes such as glycogen synthesis in animals). The reaction is, in fact, very reminiscent of UDP-glucose formation from glucose-1-phosphate and UTP (page 119).

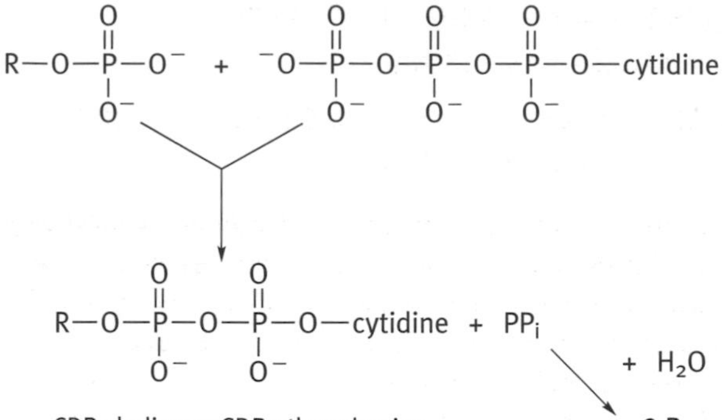

CDP-choline or CDP-ethanolamine   2 P_i

Hydrolysis of PP_i to P_i makes the reaction strongly exergonic.

The final reaction in phosphatidylethanolamine and phosphatidylcholine synthesis is as follows:

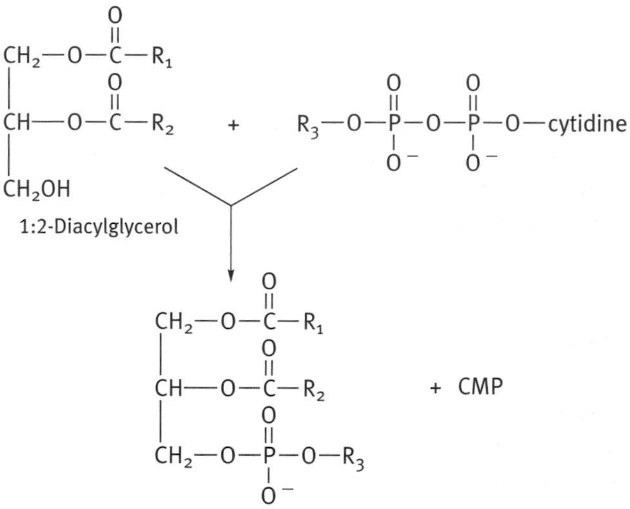

Phosphatidylcholine or phosphotidylethanolamine

The diacylglycerol comes from hydrolysis of phosphatidic acid by a phosphatase (page 192).

In the reactions above we join an alcohol (ethanolamine or choline) to CDP and then transfer phosphoryl alcohol to another alcohol (diacylglycerol); the head group is 'activated'. Energetically, it would be the same to join diacylglycerol to CDP and transfer a diacylglycerol phosphoryl-group to ethanolamine or choline—that is, to activate the diacylglycerol instead of the polar alcohol or head group. This is what happens in phosphatidylinositol and cardiolipin (page 72) synthesis in eukaryotes. In the latter, instead of inositol, the reactant is another molecule of diacylglycerol.

$$CH_2-O-\overset{\overset{O}{\|}}{C}-R_1$$
$$CH-O-\overset{\overset{O}{\|}}{C}-R_2 + CTP \longrightarrow$$
$$CH_2-O-\overset{\overset{O}{\|}}{P}-O^-$$
$$O^-$$

Phosphatidate

$$CH_2-O-\overset{\overset{O}{\|}}{C}-R_1$$
$$CH-O-\overset{\overset{O}{\|}}{C}-R_2$$
$$CH_2-O-\overset{\overset{O}{\|}}{P}-O-\overset{\overset{O}{\|}}{P}-O-cytidine \quad + PP_i$$
$$O^- \qquad O^-$$

↓ + Inositol

$$CH_2-O-\overset{\overset{O}{\|}}{C}-R_1$$
$$CH-O-\overset{\overset{O}{\|}}{C}-R_2 + CMP$$
$$CH_2-O-\overset{\overset{O}{\|}}{P}-O-inositol$$
$$O^-$$

Phosphatidylinositol

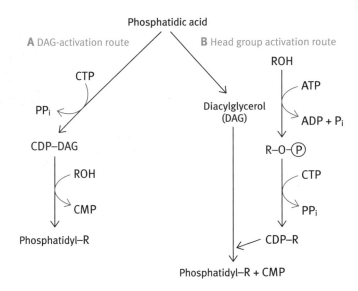

Fig. 11.6 The two pathways (**A** and **B**) for synthesis of glycerophospholipid. (ROH, The polar head group to be attached to the phospholipid.) Different cells may use different routes for synthesizing a given phospholipid. In mammals, phosphatidylcholine (PC), phosphatidylserine (PS), and phosphatidylethanolamine (PE) are synthesized by route **B**, but the three are interconvertible by decarboxylation of PS to PE, methylation of PE to PC, and head-group exchange between PE and PS and free ethanolamine or serine. Synthesis of cardiolipin and phosphatidylinositol in mammals follows route **A**. In *E. coli* phospholipid synthesis occurs by route **A**.

The situation is summarized in the scheme in Fig. 11.6. This scheme illustrates the principle of the two ways in which glycerophospholids can be synthesized. It should be emphasized that the field is a complex one in which extensive interconversion of the phospholipids occurs. For example, serine and ethanolamine can be exchanged, PE can be methylated to PC, PS can be decarboxylated to PE. The particular route of synthesis of a given phospholipid may vary in different organisms.

Site of membrane lipid synthesis

In eukaryotic cells, the main site of phospholipid synthesis is the membrane of the endoplasmic reticulum (ER). Fatty acyl-CoAs and glycerophosphate in the cytoplasm are converted to phosphatidic acid by an enzyme located on the cytoplasmic surface of the ER. The phosphatidate is converted to phosphatidyl derivatives of serine, choline, ethanolamine, and inositol as described earlier, again by enzymes on the cytoplasmic surface of the ER. The newly synthesized phospholipids are inserted into the outer leaflet of the lipid bilayer of the ER membrane (probably at the fatty acyl-CoA stage) and the synthesis completed *in situ*. In this way, new lipid membrane is synthesized—but there is a problem. The newly synthesized phospholipids are in the outer leaflet of the ER lipid bilayer, but there has to be a transfer of phospholipids to the inner leaflet to maintain the balance (Fig. 11.7). Simple flipping of phospholipids from the outer to the inner layer would involve the polar head group traversing the hydrophobic centre of the bilayer, which is energetically not feasible, so there has to be some special mechanism for achiev-

ing this. There is evidence for a so-called 'flippase' which actively does this in erythrocytes, but relatively little is known of how the transfer is achieved in the main site of synthesis, the ER. There is another problem too. The new membrane synthesized in the ER has to be transported to other sites such as the plasma membrane for cell growth and division to occur. There are two possible mechanisms for this; first, spherical membrane vesicles are budded off the ER and these may migrate to their target membrane and fuse with it thus adding new membrane to say the plasma membrane. This is part of a bigger story of how new membrane proteins are delivered to the correct membrane, for different membranes differ in both their phospholipid content and proteins. We deal with this later (page 410), because it would be premature here. Phospholipid transfer proteins are known which pick up phospholipids from one membrane and transport them to another membrane, and this is the second possible way in which newly synthesized phospholipids are delivered to target membranes. This mechanism could transfer newly synthesized material from the ER to the plasma or other membrane. However, it is a very complex field and there is insufficient information to understand just what part each of the

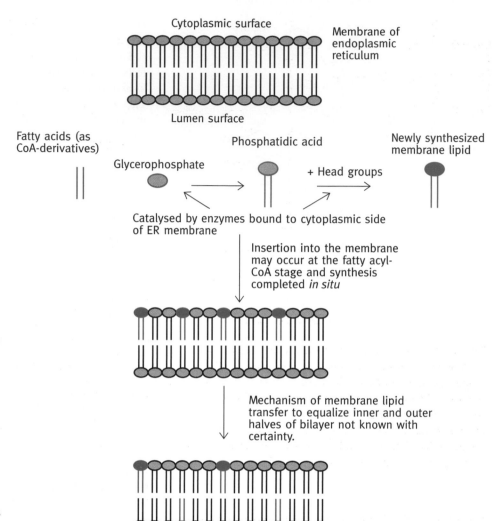

Fig. 11.7 Synthesis of new membrane lipid bilayer. Enzymes which synthesize new membrane lipids are located at the cytoplasmic surface of the ER; the newly synthesized lipids are located in outer half of the bilayer. Transfer to the inner half to maintain bilayer equality may be by a 'flippase' but this is not fully understood. The new membrane is transported from the ER to other membranes in the cell in vesicle form via phospholipid transport proteins, but here again the overall picture of what happens in this respect is not clear. See text.

two routes plays in the synthesis of lipid bilayer in the different membranes of the cell.

Synthesis of prostaglandins and related compounds

The Greek word *eikosi* means 20; this is the basis for the name of a group of compounds called **eicosanoids**, all of which contain 20 carbon atoms and are related to polyunsaturated fatty acids. Although present in the body at very low levels, they have a wide range of physiological functions.

There are three main groups named after the cells in which they were first discovered. These are the **prostaglandins**, the **thromboxanes**, and the **leukotrienes**. Prostaglandins were first discovered in semen and so named because it was thought that they originated from the prostate gland (actually it is from the seminal vesicles). However, it is now known that very many tissues produce prostaglandins. Thromboxanes were discovered in blood platelets or thrombocytes and leukotrienes in leukocytes.

The prostaglandins and thromboxanes

A number of different prostaglandins exist, varying in their detailed structures, the main subclasses being PGA, PGE, and PGF. A subscript number indicates the number of double bonds in the side chains attached to the cyclopentane ring structure. If

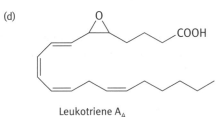

(a) Arachidonic acid

(b) Prostaglandin E$_2$

(c) Thromboxane A$_2$

(d) Leukotriene A$_4$

Fig. 11.8 Structure of prostaglandins and related compounds. (a) Arachidonic acid; (b) prostaglandin E$_2$ (PGE$_2$); (c) thromboxane A$_2$ (TXA$_2$); (d) leukotriene A$_4$ (LTA$_4$). Probably only those specially interested in this area would need to memorize these structures.

you look at Fig. 11.8(b), you will see the structure of PGE$_2$ as an example. In humans the compounds with two double bonds are most important and are synthesized from arachidonic acid, whose structure is shown in Fig. 11.8(a). Other prostaglandins are derived from related unsaturated fatty acids. Arachidonic acid is present in cells, mainly as the fatty acid component of a phospholid. For prostaglandin synthesis it is released by a phospholipase.

The first step in synthesizing prostaglandins is carried out by a **cyclooxygenase**, which forms the ring structure from arachidonic acid (Fig. 11.8(a)). This enzyme is inhibited by **aspirin** (acetylsalicylic acid), which covalently modifies the enzyme by acetylating an essential serine group on the enzyme. Thromboxanes (Fig. 11.8(c)) are formed by further conversion of cer-

tain of the prostaglandins, so that aspirin inhibits this formation also.

The prostaglandins have a wide variety of physiological effects. They are released immediately after synthesis and act as local hormones on adjacent cells by combining with receptors. They cause pain, inflammation, and fever; they cause smooth muscle contraction and are involved in labour; they are involved in blood pressure control; they suppress acid secretion in the stomach. Thromboxanes affect platelet aggregation and thus blood clotting.

Aspirin, by its inhibition of cyclooxygenase, suppresses many of these effects; in particular, a low dose (100 mg per day) inhibits platelet thromboxane formation resulting in decreased platelet aggregation. This therapy has been found to reduce the risk of heart attacks by reducing the danger of blood clot formation blocking the coronary arteries.

Leukotrienes

The structure of one of the leukotrienes is given in Fig. 11.8(d). Leukotrienes arise from arachidonic acid by a different route involving a lipoxygenase reaction. They cause protracted contraction of smooth muscle, are a factor in asthma symptoms by constricting airways, and play a role in white cell function.

Synthesis of cholesterol

Most cells of the body are believed to be capable of cholesterol synthesis; cholesterol is an essential component of cell membranes and the precursor of bile salts and steroid hormones. The liver and, to a lesser extent, the intestine are the most active in its synthesis and from the liver a two-way flux 'equilibrates' the cholesterol of the liver and rest of the body, as described in Chapter 7.

Remarkably perhaps, the sole starting material for synthesis of this large molecule is acetyl-CoA from which, by a long metabolic pathway, the details of which probably few biochemists carry in their heads, cholesterol is produced. The first few stages of the synthesis are of special interest for this is where control of cholesterol occurs and we will confine this account to the production of mevalonic acid, the first metabolite committed solely to cholesterol synthesis. This involves the reactions shown in Fig. 11.9 which occur in the cytosol.

HMG-CoA is also the precursor of acetoacetate, one of the ketone bodies (page 182), but for this purpose is synthesized inside the mitochondrion.

The regulation of cholesterol levels in cells and the 'statin' drugs have already been described (see page 130).

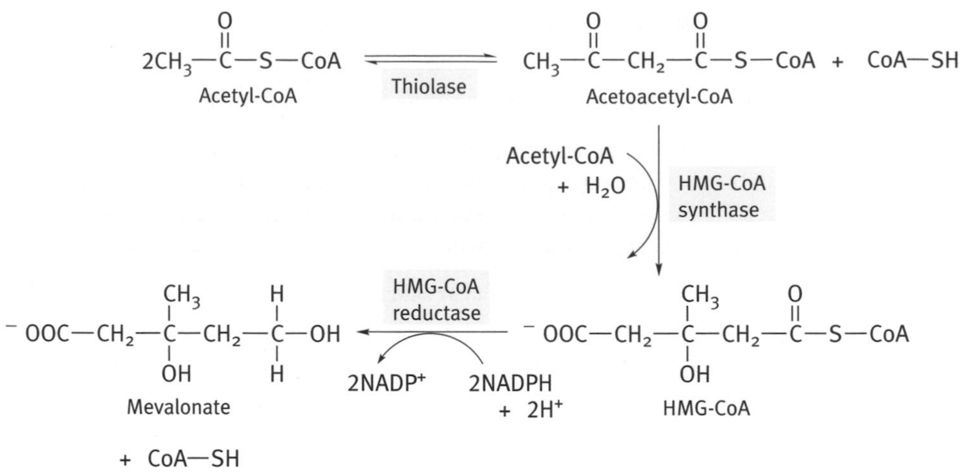

Fig. 11.9 The synthesis of mevalonate from acetyl-CoA. HMG-CoA, 3-hydroxy-3-methylglutaryl-CoA.

FURTHER READING

Fatty acid synthase

Smith, S. (1994). The animal fatty acid synthase: one gene, one polypeptide, seven enzymes. *FASEB J.*, **8**, 1248–59.

Illustrated article on the structure and function of this remarkable protein, with each component activity discussed, and how they are organized in the dimer complex. An interesting article on a classical topic.

Membrane lipid synthesis

Dawidowicz, E. A. (1987). Dynamics of membrane lipid metabolism and turnover. *Ann. Rev. Biochem.*, **56**, 43–61.

Very readable review on membrane lipids, their site of synthesis, transport and assembly into membranes.

Rooney, S. A., Young, S. L., and Mendelson, C. R. (1994). Molecular and cellular processing of lung surfactant. *FASEB J.*, **8**, 957–67.

Reviews a more physiological role of phospholipid—that of lining the alveoli.

Kent, C. (1995). Eukaryotic phospholipid biosynthesis. *Ann. Rev. Biochem.*, **64**, 315–43.

A straightforward, clear account of the various pathways for the synthesis of the various membrane phospholipids.

PROBLEMS FOR CHAPTER 11

1   An early step in fatty acid synthesis is that acetyl-CoA is carboxylated, to be followed immediately by decarboxylation of the product. What is the point of this?

2   By means of a diagram, outline the steps in a cycle of elongation during fatty acid synthesis, omitting the reductive steps.

3   Discuss briefly the physical organization of the fatty acyl synthase complex in eukaryotes. How does this differ from that in *E. coli*? What is the advantage of the eukaryotic situation?

4 Write down the structures of NAD^+ and $NADP^+$. Explain the reason for the existence of both NAD^+ and $NADP^+$.

5 What are the main tissues in which fatty acid synthesis occurs in eukaryotes?

6 Palmitate synthesis from acetyl-CoA occurs in the cytoplasm. However, pyruvate dehydrogenase, the main producer of acetyl-CoA used in fat synthesis, occurs inside the mitochondrion. Acetyl-CoA cannot traverse the mitochondrial membrane. How does the fatty acid synthesizing system in the cytoplasm obtain its supply of acetyl-CoA?

7 The synthesis of fatty acid involves reductive steps requiring NADPH. What are the sources of the latter?

8 Describe the synthesis of TAG from fatty acids.

9 What is the role of CTP in lipid metabolism?

10 (a) What are eicosanoids?
    (b) What are they made from?
    (c) Briefly describe their physiological significance.
    (d) What is the relevance of aspirin in this area of metabolism?

11 Describe a type of drug that is used to lower the rate of cholesterol synthesis.

Chapter summary

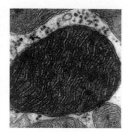

Synthesis of glucose (gluconeogenesis)

The body is able to synthesize glucose from any compound capable of being converted to pyruvate by the process of gluconeogenesis. As already stressed, this excludes fatty acids and acetyl-CoA but includes lactate (which is released into the blood by red blood cells and by muscle during exercise) as well as most of the amino acids. This has the role of permitting the storage of a variety of compounds as glycogen. However, in addition to this 'routine' metabolic role, gluconeogenesis can literally make the difference between life and death, despite the fact the glucose is such a common molecule prevalent in most diets. In case you have forgotten its significance (page 111), in starvation it assumes an overriding importance.

The brain, as stated several times earlier, requires a constant supply of glucose, which means that the blood glucose level must be kept within normal limits or coma and death will ensue. However, the liver's total store of glucose in the form of glycogen is small and, after about 24 hours without food, the entire store is exhausted. Nevertheless, people don't die as a result of a day's starvation. Supply of blood sugar in this situation is dependent on the liver—no other organ, other than the quantitatively unimportant kidney, can fill this need. If fasting is prolonged beyond about 24 hours, the liver must therefore synthesize glucose. Synthesis of about 100 g per day is needed in man.

Fat mobilization results in ketone body production by the liver which has a glucose-sparing effect (page 111) as the blood ketone level rises, but it cannot replace synthesis of glucose. Nor does the total energy requirement of the brain fall with starvation. Although the effects of glucose deprivation on brain are the most dramatic, cells such as those of the retina and kidney medulla and red blood cells are virtually (or completely in the latter, since they have no mitochondria) anaerobic and dependent on glycolysis of glucose for energy supply. The body normally has sufficient fat to supply energy for weeks of starvation but fatty acids do not pass through the blood–brain barrier and cannot be used by the brain.

The starting point for the gluconeogenesis pathway in liver is pyruvate. To re-emphasize a crucial point, whereas pyruvate can give rise to acetyl-CoA (page 144) and therefore lead to fat synthesis, the reverse cannot happen in animals. Acetyl-CoA cannot be converted to pyruvate and therefore fatty acids cannot be converted into glucose in a net sense.

We will first describe the pathway by which pyruvate is converted into glucose and after that discuss the broader biochemical implications.

Mechanism of glucose synthesis from pyruvate

If you examine the glycolytic pathway (Fig. 9.7), you will note that three reactions are irreversible because of thermodynamic considerations. These are: (1) phosphorylation of glucose by hexokinase (or glucokinase) using ATP; (2) phosphorylation of fructose-6-phosphate by phosphofructokinase; and (3) conversion of phosphoenolpyruvate (PEP) to pyruvate. Glucose is synthesized via the intermediate compounds found in glycolysis but a way has to be found to bypass the irreversible reactions.

The first thermodynamic barrier in the process is the conversion of pyruvate to PEP. Because the spontaneous conversion of the enol form of pyruvate to the keto form has a very large negative $\Delta G^{o'}$ value (page 152), the PEP → pyruvate reaction in glycolysis is irreversible in animal cells but, as mentioned earlier, because of the convention of naming kinases after the reaction involving ATP, it is called pyruvate kinase nonetheless. In

animals, a roundabout route involving two reactions is used for the conversion of pyruvate to PEP. Two—Ⓟgroups are used in the process, making the process thermodynamically favourable. The route is as follows:

(1) Pyruvate + ATP + HCO_3^- → Oxaloacetate + ADP + P_i + H^+; (catalysed by pyruvate carboxylase)

(2) Oxaloacetate + GTP → PEP + GDP + CO_2; (catalysed by PEP carboxykinase, or PEP-CK)

Sum. Pyruvate + ATP + GTP + H_2O →
PEP + ADP + GDP + P_i + $2H^+$.

The scheme is shown diagrammatically in Fig. 12.1.

It is not known why GTP is used in the second reaction rather than ATP. Energetically it is just the same. You may recall that reaction 1, catalysed by pyruvate carboxylase, which synthesizes oxaloacetate, has an important role in topping up the citric acid cycle (page ••) quite separate from its role in gluconeogenesis. The second enzyme is usually referred to as **PEP-CK**. Its full name is **phosphoenolpyruvate carboxykinase**—because in the reverse direction it carboxylates phosphoenolpyruvate and transfers a phosphoryl group.

Pyruvate carboxylase occurs only in mitochondria, whereas PEP-CK occurs in both mitochondria and the cytosol. Since oxaloacetate cannot be made in the cytosol there is clearly a problem in converting pyruvate to glucose. There are two solutions, both of which operate. Either PEP is formed in mitochondria and transported out or oxaloacetate is reduced to malate and transported out. The malate is dehydrogenated to oxaloacetate and this has the advantage of supplying NADH for gluconeogenesis.

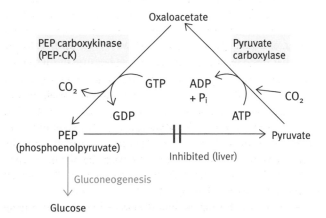

Fig. 12.1 The generation of phosphoenolpyruvate for gluconeogenesis from pyruvate in the liver. Note that this scheme makes sense only if the PEP → pyruvate reaction is prevented. How this is done is described in Chapter 13 on metabolic control (page 226).

Once PEP is formed, the reactions of glycolysis are all reversible until fructose-1:6-bisphosphate is reached. The formation of this in glycolysis from fructose-1-phosphate is irreversible, but the step is bypassed by the simple device of hydrolysing off the phosphoryl group from the 1:6 compound as shown.

Fructose-1:6-bisphosphate

+ H_2O

Fructose-1:6-
bisphosphatase

Fructose-6-phosphate

+ P_i

Similarly, when glucose-6-phosphate is reached, the glucokinase (or hexokinase) reaction of glycolysis is irreversible but, in liver, **glucose-6-phosphatase** produces glucose which is secreted from the cell. The complete gluconeogenesis reactions are shown in Fig. 12.2. Note that whether at any given time glycolysis or gluconeogenesis is taking place depends on control mechanisms described in Chapter 13.

There are thus four enzymes involved in gluconeogenesis that do not participate in glycolysis: pyruvate carboxylase; PEP-CK; fructose-1:6-bisphosphatase; and glucose-6-phosphatase. The levels of these enzymes are about 20–50 times greater in rat liver than in rat skeletal muscle, in keeping with the importance of gluconeogenesis in liver.

What are the sources of pyruvate used by the liver for gluconeogenesis?

In starvation, after the glycogen reserves have been exhausted, the main source of pyruvate originates from the breakdown of muscle proteins. Hydrolysis of the latter produces 20 amino acids. Although the amino acid **alanine** represents only a small proportion of the latter, of the total amino acids leaving the muscle to be transported to the liver, more than 30% is in the

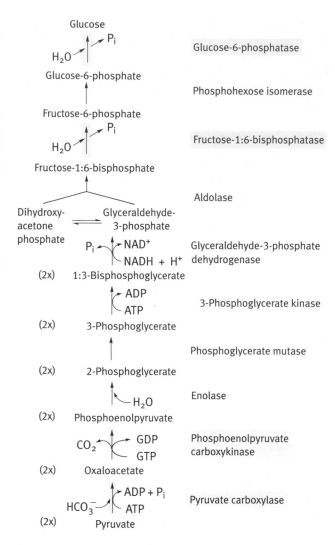

Fig. 12.2 The complete gluconeogenesis pathway from pyruvate to glucose. Reactions in yellow are different from those occurring in glycolysis.

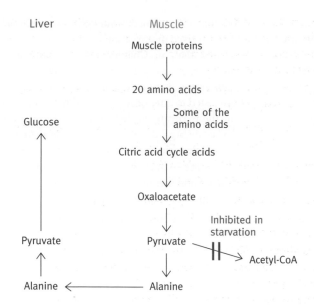

Fig. 12.3 Mechanism by which breakdown of muscle proteins supplies the liver with a source of pyruvate for gluconeogenesis during starvation. Note that the scheme is dependent on the pyruvate dehydrogenase not using the pyruvate (see text). Glutamine is also produced in muscle as a gluconeogenic substrate for the liver in addition to alanine.

form of this amino acid. Alanine is a **glucogenic amino acid**—its carbon skeleton is capable of being converted into glucose by the liver and the same is true of most of the other amino acids. (The metabolism of amino acids is separately discussed in Chapter 14, so do not be concerned with the mechanism of the reactions mentioned in this section.) How is alanine formed from the protein breakdown products in muscle? Several of the amino acids can give rise to extra oxaloacetate by reactions of the citric acid cycle. The oxaloacetate can be converted into pyruvate by the reactions shown in Fig. 12.1 and converted into alanine for release into the blood. The amino group of alanine

comes from the other amino acids. Note, however, that this formation of alanine in muscle depends on the muscle not oxidizing the pyruvate to acetyl-CoA which, in starvation, would defeat the whole object of the exercise. In the situation of starvation the muscle will be preferentially utilizing fatty acids and ketone bodies for energy generation so that acetyl-CoA will be very plentiful. As will be described in Chapter 13, a high acetyl-CoA/CoA ratio leads to the inactivation of pyruvate dehydrogenase. The pyruvate produced from the amino acids is therefore reserved for alanine synthesis. In summary, the net effect is that many of the amino acids derived from muscle protein breakdown are converted to alanine, which migrates in the blood to the liver. (The amino acid, glutamine, is also synthesized in muscle for transport to the liver for glucose synthesis but the principle is exactly the same.) In the liver, the alanine is converted back to pyruvate and thence to glucose. The overall process is summarized in Fig. 12.3.

As the concentration of blood ketone bodies rises during starvation, the brain progressively uses more of these for energy generation and so reduces its utilization of glucose and lessens the demand for gluconeogenesis (which, however, always remains essential). This is important for it requires about 2 grams of muscle protein to be broken down for each gram of glucose

made, a rate of loss that, if continued, would greatly reduce the period that could be survived in starvation.

It should be noted that, in Chapter 16 on amino acid metabolism, a glucose–alanine cycle is presented (Fig. 16.9) In this, alanine is synthesized in muscle and transported to the liver where it is converted to glucose as described here but with the difference that the pyruvate needed in muscle for alanine synthesis comes from the breakdown of glucose. The sequence of events in this cycle is liver glucose → muscle glucose → muscle alanine → liver alanine → liver glucose. This does not result in a net gain of glucose and cannot therefore contribute to the blood glucose supply problem but may be relevant in the transport of amino nitrogen from muscle to liver.

A second source of pyruvate for gluconeogenesis, of less importance in starvation but important in more normal situations, is **lactate**, produced by the anaerobic glycolysis of glucose or glycogen (page 138). As stated, certain cells such as those of kidney medulla, and retina are virtually anaerobic and mature red blood cells, lacking mitochondria, have no oxidative generation of ATP. In normal nutritional situations, a major source of lactate is **muscle glycolysis**—in strenuous muscle activity, the glycolysis rate exceeds the capacity of mitochondria to reoxidize NADH and lactate is produced. This travels via the blood to the liver where it is converted to pyruvate and thence to glucose (or glycogen). This constitutes a physiological cycle, called the Cori cycle after its discoverers (Fig. 12.4). The cycle has two main effects; it 'rescues' lactate for further use and, secondly, it counteracts **lactic acidosis**. The introduction of large amounts of lactic acid into the blood may exceed its buffering power and cause a deleterious fall in pH of the

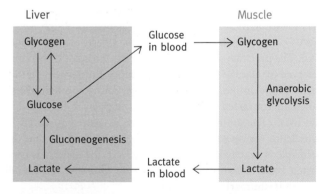

Fig. 12.4 The Cori cycle. Note that this is a physiological cycle involving muscle and liver. The muscle has very low levels of three enzymes essential for gluconeogenesis, which explains why the lactate has to travel from muscle to liver. Excess lactate is produced in muscle during violent action during which glycolysis outstrips the capacity of mitochondria to re-oxidize reduced NAD⁺—that is, during anaerobic glycolysis.

blood. The synthesis of glucose from lactate involves the uptake of two protons (when NADH + H⁺ is used for reduction of 1,3-bisphosphoglycerate).

We will digress briefly to discuss the effect of ethanol metabolism, which has special relevance to gluconeogenesis in the liver.

Effects of ethanol metabolism on gluconeogenesis

Ethanol is oxidized to acetaldehyde which is further oxidized to acetate, part of which enters the blood to be used by other tissues. Acetate is converted to acetyl-CoA by the acetate activating enzyme found in most tissues (page 180) and the acetyl group is then oxidized to carbon dioxide and water by the citric acid cycle or diverted into fat synthesis.

The oxidation of ethanol to acetaldehyde occurs in the liver and is mainly due to alcohol dehydrogenase in the cytosol:

$$CH_3CH_2OH + NAD^+ \rightarrow CH_3CHO + NADH + H^+$$
$$\text{alcohol dehydrogenase}$$

Oxidation of the acetaldehyde occurs in the mitochondrial matrix:

$$CH_3CHO + NAD^+ + H_2O \rightarrow CH_3COO^- + NADH + 2H^+$$
$$\text{aldehyde dehydrogenase}$$

Effect of ethanol metabolism on the NAD/NADH ratio in the liver cell

Because the alcohol dehydrogenase occurs in the cytosol of liver cells, the reduced NAD⁺ has to be reoxidized via the shuttle mechanisms (page 153) which transport reducing equivalents into the mitochondria. All of this sounds quite harmless, and indeed ethanol produced by microorganisms in the large intestine is a normal metabolite; in humans this can amount to a few grams per day. However, much larger quantities of alcohol may be consumed and the rate of shuttle transfer of reducing equivalents into mitochondria may not keep up with the rate of NAD⁺ reduction. The ratio of NADH to NAD⁺ in a liver cell is normally low. With a moderate amount of alcohol the level of NADH is greatly increased. Many of the dehydrogenase reactions in which NAD⁺ participates are close to equilibrium in the cell, which means that the normal ratios of oxidized and reduced substrates are easily affected by the changed NADH/NAD⁺ ratio. In particular, in liver, the lactate dehydrogenase reaction is affected so that the oxidation of lactate arriving at the liver from extrahepatic tissues to form pyruvate is impaired. We remind you of the lactate dehydrogenase reaction:

$$CH_3COCOO^- + NADH + H^+ \rightleftharpoons CH_3CHOHCOO^- + NAD^+$$
(pyruvate) (lactate)

The distortion of the normal reduction pattern of NAD^+ affects the metabolism of liver cells. A serious situation can arise in very heavy drinkers, especially if food intake is restricted during drinking bouts. After 24 hours of food deprivation, the liver glycogen stores are exhausted (page 111) and maintenance of blood glucose levels, and therefore of brain function, depends on gluconeogenesis. The latter depends on an adequate supply of pyruvate which as described is mainly formed from lactate produced by red blood cells and from alanine derived from muscle protein breakdown. However, in the presence of alcohol-induced abnormally high NADH levels in the liver, pyruvate availability is diminished; that formed from alanine may be reduced to lactate, and lactate, arriving from outside, is less efficiently oxidized to pyruvate. Lactate leaks into the blood causing lactic acidosis and the reduced pyruvate availability impairs gluconeogenesis. The situation is exacerbated because gluconeogenesis itself is one of the protections against lactic acidosis, two protons being absorbed per molecule of glucose synthesized.

Synthesis of glucose from glycerol

Another source of carbon compound for gluconeogenesis in the liver is **glycerol** released from **triglyceride hydrolysis** mainly in adipose tissue. This is taken up by the liver and converted to glucose by the route shown in Fig. 12.5. The enzyme glycerol kinase, which phosphorylates glycerol as the first step in the conversion, is very low in amount in adipose tissue and therefore glycerol is not converted to glucose (or TAG) in that tissue,

but the enzyme is present in liver. This system is logical in that, in a situation where blood glucose supply is of vital importance, the glycerol produced in fat cells is transported to the liver and converted to blood glucose. Adipose cells therefore do not use the glycerol themselves; supplying the brain with glucose takes priority over that.

During prolonged starvation, after a small initial drop in blood glucose level, the latter remains constant for several weeks and fatty acids supplied by adipose cells likewise remain at a constant level so that the mechanisms are extremely effective. It is, however, perhaps surprising that the whole business of muscle wastage during starvation could, on the face of it, have been avoided by animals using a simple metabolic trick present in bacteria and plants that permits fat to be converted to glucose. No doubt there are good evolutionary reasons for animals not using it; we will now describe this metabolic trick.

Synthesis of glucose via the glyoxylate cycle

E. coli can survive quite well with acetate as its sole carbon source. A net conversion of acetyl-CoA to C_4 acids of the citric acid cycle and thence to carbohydrate or any other component of the cell can occur, unlike the situation in animals. In plant seeds, energy stored as TAG is converted to glucose on germination. What allows bacteria and plants to do this?

These organisms possess the normal citric acid cycle but, in addition, can bypass some of its reactions by other reactions not present in animals. If you analyse the citric acid cycle, two

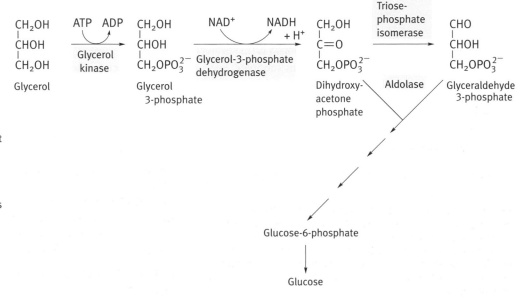

Fig. 12.5 Conversion of glycerol, released by neutral fat hydrolysis, to glucose in the liver. Much of the glycerol is produced in adipose cells but, since the glycerol kinase is not present there, gluconeogenesis from glycerol occurs in the liver. This is a sensible arrangement since it results, during starvation, in blood sugar production from the glycerol.

carbon atoms are added to oxaloacetate from acetyl-CoA, giving citrate (C_6), but then two carbon atoms are lost as two CO_2 molecules in forming succinate (C_4), so that there is no *net* conversion of C_2 to citric acid cycle intermediates.

The **glyoxylate route** bypasses these losses. Isocitrate is directly split into succinate and glyoxylate—a sort of cycle shortcut.

$$
\begin{array}{l}
\text{COO}^- \\
|\\
\text{CHOH} \\
|\\
\text{CHCOO}^- \\
|\\
\text{CH}_2\text{COO}^-
\end{array}
\;\xrightarrow[\text{lyase}]{\text{Isocitrate}}\;
\begin{array}{l}
\text{CH}_2\text{COO}^- \\
|\\
\text{CH}_2\text{COO}^-
\end{array}
\;+\;
\begin{array}{l}
\text{COO}^- \\
|\\
\text{CHO}
\end{array}
$$

Isocitrate   Succinate   Glyoxylate

The C_2 glyoxylate now reacts with acetyl-CoA to produce malate—back on the citric acid cycle.

$$
\begin{array}{l}
\text{CH}-\text{COO}^- \\
\|\\
\text{O}
\end{array}
\;+\;
\begin{array}{l}
\quad\quad\text{O} \\
\quad\quad\|\\
\text{CH}_3-\text{C}-\text{S}-\text{CoA}
\end{array}
$$

Glyoxylate   Acetyl-CoA

$$
\downarrow
\begin{array}{l}
+\ \text{H}_2\text{O} \\[4pt]
\text{Malate} \\
\text{synthase}
\end{array}
$$

$$
\begin{array}{l}
\text{OH} \\
|\\
\text{CH}-\text{COO}^- \\
|\\
\text{CH}_2-\text{COO}^-
\end{array}
\;+\;\text{CoA}-\text{SH}
$$

Malate

The scheme is shown in Fig. 12.6. The net effect is that acetyl-CoA plus oxaloacetate convert to malate plus succinate. Succinate and malate can both be converted to oxaloacetate, one molecule then being available for conversion to glucose. In plants, this process occurs in membrane-bounded organelles, called **glyoxysomes**.

On page 161 we described the enzyme pyruvate carboxylase which is necessary for topping up the citric acid cycle (the anaplerotic reaction). *E. coli* does not possess this enzyme. This is presumably related to the fact that the glyoxylate cycle, pre-

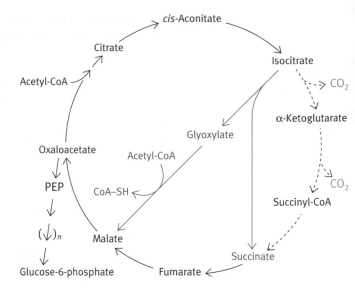

Fig. 12.6 The glyoxylate cycle of plants and bacteria, which permits carbohydrate synthesis from acetyl-CoA. This does not occur in animals. The broken line represents the bypassed section of the citric acid cycle; the red lines are reactions peculiar to the glyoxylate cycle. The two CO_2 molecules are highlighted to emphasize that it is these two decarboxylation reactions that must be bypassed. It will be appreciated that, as a result of the glyoxylate reactions, two molecules of oxaloacetate are produced from citrate, one of which is used to form citrate again and one to produce PEP.

sent in *E. coli*, can generate net increase in cycle acids from acetyl-CoA rendering pyruvate carboxylase unnecessary. The same applies to the need to generate oxaloacetate for carbohydrate synthesis.

It should finally be mentioned that by far the greatest amount of carbohydrate synthesis on Earth occurs during photosynthesis in plants. This involves the fixation of CO_2 into another glycolytic intermediate, bisphosphoglycerate, from which glucose synthesis occurs. The mechanism of this is best left until we deal with photosynthesis in Chapter 15.

We have, in the last few chapters, dealt with energy release from glucose and fat, with fat and glucose synthesis. Now is the time to remind you that fat metabolism and carbohydrate metabolism do not go on independently. They form an integrated metabolic activity, each affecting the other, and the whole lot requiring regulation. This important topic will be dealt with in the next chapter.

FURTHER READING

Felig, P. (1975). Amino acid metabolism in man. *Ann. Rev. Biochem.*, **44**, 933–55.

A most useful review which discusses amino acid metabolism at the organ level, together with the effects of starvation, diabetes, obesity and exercise on it. It also discusses gluconeogenesis at this level.

Snell, K. (1979). Alanine as a gluconeogenic carrier. *Trends Biochem. Sci.*, 4, 124–8.
Reviews the role of muscle in supplying the liver with metabolites for glucose synthesis.

PROBLEMS FOR CHAPTER 12

1   After 24 h starvation, the glycogen reserves of the liver are exhausted but there are still relatively large stores of fat. What is the point of the body going to such trouble to make glucose by gluconeogenesis when virtually unlimited supplies of acetyl-CoA from fatty acids are available to supply energy?

2   Gluconeogenesis requires production of phosphoenolpyruvate (PEP), but pyruvate kinase cannot form this from pyruvate. Why is this and how is the problem overcome?

3   From PEP, gluconeogenesis in the liver produces blood glucose mainly by reversal of glycolytic enzymes. However, two enzymes are involved that do not participate in glycolysis. What are these?

4   Does muscle have glucose-6-phosphatase? Explain your answer.

5   What is the Cori cycle and its physiological role?

6   Adipose cells do not have glycerol kinase, the enzyme which converts glycerol to glycerol-3-phosphate, even though adipose cells produce glycerol from TAG hydrolysis. Liver does have the enzyme. Explain why this appears to be a logical situation.

7   Animals cannot convert acetyl-CoA to carbohydrate. Bacteria and plants can. Explain how they do this.

8   Pyruvate kinase is needed to top up the acids of the citric acid cycle (the anaplerotic reaction). Depletion of the cycle of its acids would impair operation of the cycle. However, *E. coli* does not have this enzyme. Comment on this.

9   In starvation, muscle wasting occurs. What is the relevance of this to gluconeogenesis in the body?

10   Excessive alcohol intake can tend to deprive the brain of glucose if food intake is restricted, as may occur in binge drinking. Explain this.

Chapter summary

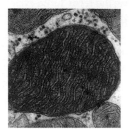

Strategies for metabolic control and their application to carbohydrate and fat metabolism

This chapter should be read in conjunction with the preceding chapters on metabolism. We suggest you keep referring back to the relevant metabolic pathways to refresh your memory. What we will do in this chapter is to put together all the preceding metabolism to see how the various pathways work and interact. In many ways this is the pay-off of all the work you've done so far because it is where the many reactions make collective physiological sense.

We have collected together metabolic control mechanisms into a separate chapter, rather than dealing with the topic when the pathways themselves were discussed, because the pathways themselves are complicated enough without the additional information on control, and because metabolic control makes more sense when the different pathways are considered together.

Although control extends to all facets of metabolism, the control and integration of carbohydrate and fat metabolism is especially important because the flow of chemical change (flux) through these pathways is large and, in the body, the direction of the fluxes is changed as meals are followed by periods of fasting. Control of carbohydrate and fat metabolism illustrate all of the principles of metabolic control. The subject is a fairly complex one, so let's be clear on how this chapter deals with it. Enzymes are responsible for metabolism and therefore metabolic control involves control of enzyme activities. We therefore first deal with the strategies by which enzyme activities are regulated. After this we describe how the pathways are kept in balance by regulatory enzymes which sense the levels of metabolites in other pathways, and automatically adjust their rates of activities to the information.

This is not the end, because the metabolic activities of cells are controlled in a wider sense by circulating hormones and neural signals which operate in the body as a whole according to current needs. We therefore describe how signals external to the cells—extracellular signals produced by other cells of the body—regulate the metabolic pathways, on top of the automatic controls first described.

But before we deal with any of this the first question to consider is:

Why are controls necessary?

It is obvious when you think of the major pathways that we've dealt with—glycogen synthesis, glycogen breakdown, glycolysis, gluconeogenesis, fat breakdown, fat synthesis, the citric acid cycle, electron transport—that they can't all be running at full speed (or even necessarily at all) at the same time. If a metabolic pathway is proceeding in one direction, the reactions in the reverse direction must be switched off, otherwise nothing but heat generation would be achieved. In a single pathway such as glycolysis, its required rate will vary according to the energy needs at the time. The metabolic rate of someone playing squash is about six times greater than that of the person at rest. In short, the metabolism of the main energy yielding materials must be regulated for at least three reasons:

- to avoid the potential problem of futile cycles, or **substrate cycles** as they are now more properly called (explained below)

- to respond to energy production needs as energy expenditure varies

- to respond to physiological needs—metabolic pathways

need to work in different directions after a meal when metabolites are being stored (Chapter 6) as compared with intervals between meals when stored energy reserves are being utilized. The needs are different again in prolonged starvation and in pathological situations such as diabetes where carbohydrate metabolism is abnormal.

The potential danger of futile cycles in metabolism

The process of gluconeogenesis in the preceding chapter provides a good example for illustrating the potential danger in metabolic systems of futile cycles. One wonders how organisms coped with the problem during the evolution of the pathways. Consider the phosphofructokinase (PFK) step in glycolysis and its reversal in gluconeogenesis by fructose-1:6-bisphosphatase (Fig. 13.1). In one direction, fructose-6-phosphate is phosphorylated by PFK to yield the 1:6-bisphosphate, while in the reverse direction the bisphosphatase enzyme hydrolyses the product back to fructose-6-phosphate. This cycle of events would, unchecked, destroy ATP with the generation of heat—more or less equivalent to an electrical short circuit. (Bumblebees deliberately use this trick to warm up their flight muscles on cold mornings before take-off.)

Extending this to the whole of glycolysis and gluconeogenesis pathways, these pathways, uncontrolled, could constitute a giant futile cycle again doing nothing but uselessly destroying ATP. The same applies to glycogen synthesis and breakdown, and fat synthesis and breakdown (Fig. 13.2) or, for that matter, to any synthesis and breakdown pathways.

It is clear from this that the breakdown and synthesis directions of a metabolic pathway must be controlled in a reciprocal manner—activation of one and inhibition of the other. This independent control of the two directions can occur only, as mentioned earlier (page 5), at irreversible metabolic steps, for it is here that there are separate reactions in the two directions catalysed by distinct enzymes which can be separately controlled. A freely reversible reaction is catalysed by the same enzyme in both directions, which cannot be separately controlled. With the realization that controls are operative, the term 'futile cycle' has been replaced by 'substrate cycle' to describe situations such as the fructose phosphate one; they are important control points. Substrate cycling, if allowed to occur at all, may appear to be wasteful. However, a small amount of cycling may be advantageous in making controls on the forward and backward metabolic pathways more sensitive. Suppose you have a pathway in which A is converted to C via B and the reaction A → B occurs by a substrate cycle:

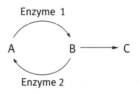

The rate of conversion of A to C can be reduced by inhibiting enzyme 1 or activating enzyme 2. If you do both, the control is more effective; complete shutdown could be achieved without excessively high levels of the inhibitor for enzyme 1. The same considerations apply to increasing the flux of metabolite from C to A. The fructose-6-phosphate → fructose-1:6-bisphosphate cycle has dual controls of the type described.

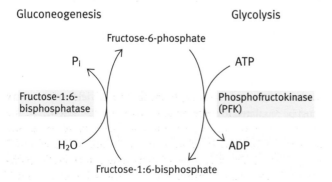

Fig. 13.1 Potential futile cycle at the phosphofructokinase (PFK) step in glycolysis if the reactions were uncontrolled.

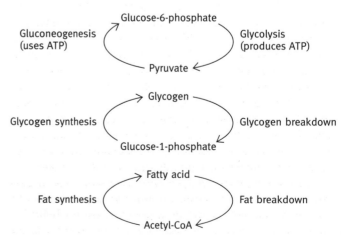

Fig. 13.2 Potential large-scale futile cycles if metabolism were not controlled.

How are enzyme activities controlled?

Metabolic regulation of a pathway means regulation of the rate of one or more reactions in that pathway. There are basically two ways of reversibly modulating the rate of an enzyme activity in a cell. These are:

- to change the amount of the enzyme
- to change the rate of catalysis by a given amount of the enzyme.

There are ways of irreversibly activating enzymes such as the proteolytic conversion of trypsinogen to active trypsin (page 100), but in this chapter we are dealing with control mechanisms which are reversible since this is essential to meaningful metabolic control.

Metabolic control by varying the amounts of enzymes

The level of a protein in a cell can be changed by altering either the rate of its production or the rate of its destruction. Proteins are relatively short lived—the half-life of enzymes in, say, the liver, might range from about an hour to several days.

Metabolic control in animals by changes in enzyme levels is important. The control is a relatively long-term affair, with effects being seen in hours or days rather than seconds. It is at the level of adaptation to physiological needs. There are many instances where enzyme levels are adjusted. For example, the amount of lipoprotein lipase levels in capillaries (page 125) is adjusted to the fat demands of the tissues, its level increasing, for example, in lactating mammary glands. The liver changes its enzyme content within hours in response to dietary changes— whether it has to cope with high fat or high carbohydrate intake. In the well-fed state, hepatic enzymes involved in fat synthesis increase in amount; a few hours of starvation reverses this situation. Intake of foreign chemicals such as drugs results in a rapid increase in hepatic drug-metabolizing enzymes (page 281). A particularly good example is that of enzymes of the urea cycle (page 263). Excess nitrogen is converted into urea for excretion into the urine. All of the enzymes of this pathway change in the rat in proportion to the nitrogen content of the diet.

Control at the level of enzyme synthesis is rapid in bacteria— for example if *E. coli* is exposed to lactose as its sole carbon source, synthesis of the enzyme β-galactosidase, needed to hydrolyse the sugar, starts up immediately; an increase in enzyme activity is detectable in minutes and a thousandfold increase can be measured within hours (page 349). By the same token, when the stimulus for enzyme synthesis is no longer there, reversal of increases in enzyme concentrations occurs relatively slowly since they are returned to lower values only by destruction of the proteins (or dilution by cell growth in bacteria) and this is related to their half-life which may in mammals be as short as 30 minutes to an hour or so. We will return to this subject later when we deal with control of gene expression (Chapter 23). Control of enzyme amount does not come into the instant automatic metabolic control which exists over the activities of pre-existing enzymes in the cell, which we shall now deal with.

Metabolic control by regulation of the activities of enzymes in the cell

For regulation of *activities*, the amounts of given enzymes are not altered, only the rate at which they work. This method of control is instantaneous. It is necessary at this point to deal with some of the properties affecting the rate of an enzyme reaction.

Which enzymes in metabolic pathways are regulated?

In many or most metabolic pathways There are certain enzymes with special regulatory properties. Often, but not always, these are strategically placed—for example in reversible processes at irreversible steps in the metabolic pathways.

The first enzyme of a pathway is often (but by no means always) a strategic place for control. Consider a metabolic pathway such as

A → B → C → D → E → → utilization for cell synthesis

in which E is an end product needed by the cell. An automatic control mechanism is for the first reaction to be controlled by the level of E as shown. This is known as **feedback inhibition**.

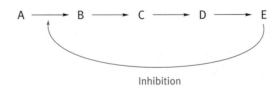

Inhibition

The control occurs through inhibition of enzyme activity (though the slower mechanism of lowering the amount of the enzyme by control of its synthesis may also be used). When the level of metabolite E, is reduced, the inhibition is relieved and E is synthesized. The flux (flow of metabolites through the pathway) is in this sense controlled by the utilization of E, since if more E is produced than can be immediately utilized, its production is automatically diminished.

There are numerous metabolic pathways in bacteria (such as for the synthesis of amino acids) in which such controls exist. Where the pathway branches, the first enzyme of each branch is often subject to regulation so that formation of each end product is separately controlled, and each end product often partially inhibits the first enzyme of the joint pathway.

When it comes to fat and carbohydrate metabolism, the whole system is so complex that the term 'end products' of pathways or 'first' enzymes cannot be defined in quite the same way, so that 'end-product inhibition' is not such a relevant term. Nonetheless, the same principle applies, that key intermediates (products) can control metabolically distant enzymic reactions. The control may be feedback as described, or 'feedforward' in which metabolites activate downstream enzymes which will be needed to cope with the flow of metabolites coming their way.

The nature of regulatory enzymes

There are essentially two main ways in which the catalytic activity of enzymes are modulated (as distinct from change in amount).

- by **allosteric** control; this is instantaneous.
- by **covalent modification** of the protein—mainly by **phosphorylation** and **dephosphorylation**. This is rapid but not instantaneous.

These will now be described.

Allosteric control of enzymes

Allosteric control of enzymes is of central importance in metabolic regulation. The prefix *allo* means 'other'; it refers to the existence on an enzyme of one or more binding sites other than for substrate. The ligands which bind to the allosteric sites are called **allosteric effectors**, or moderators. They need not, and usually don't, have any structural relationship to the substrates of the enzyme. *At a given substrate concentration*, a positive effector increases the activity of the enzyme when it combines at the allosteric site, a negative effector decreases it, so the terms allosteric activator and inhibitor are also used for allosteric ligands.

A few cases are known in which the allosteric effector reduces the catalytic rate of the enzyme so that the V_{max} is reduced, but this is very uncommon as compared with the type in which the allosteric effector alters only the *affinity* of the enzyme for its substrate. Enzymes typically work in the cell at sub-saturating levels of [S], so that an *increase in their affinity will increase their activity and decrease in their affinity has the reverse effect*. At saturating levels of [S], the V_{max} is unchanged by the allosteric effector, even if the affinity is changed, but this situation does not (usually) occur in the cell.

The mechanism of allosteric control of enzymes

We will now discuss the nature of allosteric enzymes in which the substrate binding affinity is moderated by binding of a ligand to an allosteric site, since this is the predominant type.

Such allosteric enzymes have a multi-subunit structure; they are made up of more than one catalytic protein molecule assembled into a single enzyme complex by noncovalent bonds. In some cases the enzyme is a collection of only catalytic subunits but in other there is a complex of catalytic and regularity subunits. The latter have no catalytic activity but are essential for the allosteric control of the enzyme. The individual molecules are variably called **protein subunits**, **protomers**, or **monomers**.

In the case where there are only catalytic subunits, although each is a separate catalytic entity, the subunits interact in the assembled enzyme molecule. A plot of reaction velocity versus substrate concentration of such an enzyme is shown in Fig. 13.3. It differs from the Michaelis–Menten enzyme kinetics described earlier (see page 24) in that the response of the enzyme velocity to changes in substrate concentration is sigmoidal,

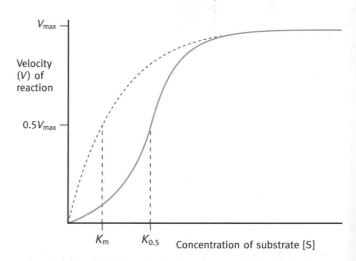

Fig. 13.3 Effect of substrate concentration on the rate of reaction catalysed by a typical allosteric enzyme. The dashed line shows, for comparison, the corresponding curve for a classical Michaelis–Menten enzyme. The term K_m is not strictly applicable to a non-Michaelis–Menten enzyme, and the term $K_{0.5}$ is used instead.

rather than hyperbolic. Why should this be, and how does it happen?

As to the first, when an allosteric activator (positive effector) binds to the enzyme, the sigmoid curve is moved to the left, and with an allosteric inhibitor, to the right, as shown in Fig. 13.4. The former increases the binding affinity of the enzyme for its substrate; the latter reduces it. The sigmoidal shape of the response to substrate concentration means that over a range of substrate concentrations (centred at that giving half maximum velocity), the rate of enzyme catalysis is more sensitive to substrate concentration change than with a Michaelis–Menten type of enzyme. By the same token, it means that increasing or decreasing the substrate affinity of an enzyme displaying sigmoidal kinetics has a greater effect on reaction velocity at a given substrate concentration than with hyperbolic kinetics. Since, as stated, enzymes in the cell usually are working in the sensitive range of substrate concentration, the sigmoidal response to substrate concentration maximises the effect of an allosteric modifier.

What causes the sigmoidal response of reaction velocity to substrate concentration?

Binding of a substrate molecule to *one* of the monomers in an allosteric enzyme has the result that binding of subsequent substrate molecules to the other monomers of the enzyme occurs more readily. This is known as **positive homotropic cooperative binding** of substrate—'positive' because affinity increases, 'homo', because only one type of ligand, the substrate, is involved, cooperative because the different catalytic subunits are interacting. Since only the substrate binding site is involved in producing the sigmoid substrate concentration response, it does not strictly fit the name of allosteric change involving another site, but sometimes any conformational change produced by a ligand is called an allosteric one. It is possible to estimate cooperative interactions in an enzyme molecule by what is known as a **Hill plot** (named after its author). In this, log $V_0/V_{max}-V_0$ is plotted against log [S]. The plot approximates a straight line at its central region whose slope is n, known as the **Hill coefficient**. For an enzyme with only a single catalytic centre, or where there is no cooperativity between sites in an enzyme with more than one centre, the value is 1. Values greater than 1 indicate the degree of cooperativity. The best known example is the binding of oxygen to the four subunits of hemoglobin (page 513) to which the Hill plot applies even though it is not an enzyme. The value is for hemoglobin 2.8; the related single-subunit oxygen-binding protein, myoglobin (page 513), has a value of 1. The allosteric glycolytic enzyme phosphofructokinase, with four subunits, has a value of 3.8.

What is the mechanism of this cooperative effect of substrate binding? There are two theoretical models. In both of these, increase in substrate concentration results in more and more catalytic subunits in a given collection of enzyme molecules being in a high affinity state.

In **the concerted model**, either *all* of the subunits in an enzyme bind substrate with low affinity or *all* with high affinity, the two forms being in spontaneous equilibrium (Fig. 13.5), but with the equilibrium strongly to the low affinity side. Binding of

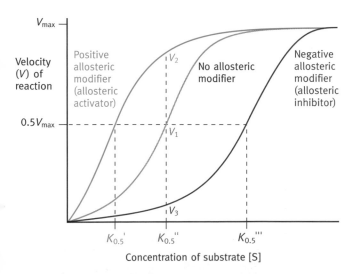

Fig. 13.4 Effect of substrate concentration on a typical allosterically regulated enzyme. v_1, v_2, and v_3 are, respectively, the reaction velocities observed with no allosteric modifier, with an allosteric activator, and with an allosteric inhibitor, at a fixed concentration of S ($K_{0.5}$). The curves are drawn arbitrarily; actual shapes and positions will be a function of the-particular system.

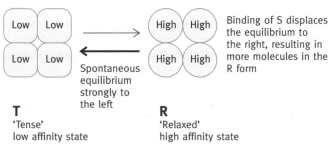

Fig. 13.5 Concerted model of cooperative binding of substrate (S). In this model the enzyme exists in two forms, T and R, the two being in spontaneous equilibrium. Binding of a substrate molecule to the R state swings the equilibrium to the right, thus increasing the affinity of all of the subunits in that molecule. 'High' and 'low' refer to affinities of the enzyme for its substrate.

substrate to a single subunit swings this equilibrium to the high affinity state. Thus, as [S] increases more and more enzyme molecules are in the high affinity state. Note that what we have said so far is concerned with the cooperative binding of substrate; the effect of the allosteric modifier, or effector, which is what *controls* the enzyme, is a separate question. On this question, the model also assumes that allosteric modifiers alter the position of the equilibrium between the low affinity (tense or T) state and the high affinity (relaxed or R) state. If the allosteric modifier binds more tightly to the R state it displaces the T ⇌ R to the right and results in a higher proportion of molecules being in the R state—i.e. it would, at a given substrate concentration, activate the enzyme. As the concentration of the positive allosteric modifier is increased and more of the enzyme converts to the high-affinity state, the substrate–reaction velocity curve tends towards the hyperbolic type. If, by contrast, the allosteric effector is of the negative type, it is assumed that it binds more tightly to the low affinity T state and thus results in more enzyme molecules being in this state and a decreased reaction rate in the collection of enzyme molecules as a whole—it inhibits.

The **sequential model** (Fig. 13.6) differs in that it assumes that, in the absence of substrate, *all* enzyme molecules are in the low affinity state; there is no equilibrium with high affinity state molecules. When one molecule of substrate binds to one of the subunits, it changes its conformation. This has the effect of facilitating a similar change in an adjacent subunit so that a second molecule of substrate binds more easily. Binding of the second molecule of substrate increases still further the ease with which a third adjacent subunit can make the conformational change with a resultant increase of affinity for the substrate of that subunit and so on (Fig. 13.6). Both models explain observed effects; there is nothing to say that the actual mechanism of the conformational change cannot be something intermediate between the two models, and individual cases may differ in this respect.

A different type of allosteric control involves an enzyme whose catalytic subunits are affected by the presence of regulatory subunits in the molecule. The latter respond to an allosteric modifier and cause changes to the activity of the enzyme. This will be discussed later.

The number of catalytic sites in an allosteric enzyme that are involved in cooperativity can be estimated by a **Hill plot** (page 213).

Reversibility of allosteric control

A most important point to note is that allosteric control is instantaneous—in both its application and reversibility. The allosteric ligand attaches to its site by noncovalent bonds and when the concentration of ligand is reduced, it dissociates from combination with the enzyme and everything goes into reverse.

Allosteric control is a tremendously powerful metabolic concept

The vital point about allosteric control is that the allosteric effector need have no structural relationship whatsoever to the substrate of the enzyme regulated (and usually doesn't). The allosteric effector is often from a separate metabolic pathway. This means that any metabolic pathway or area of metabolism can be connected in a regulatory manner to any other metabolic area, the metabolite(s) of one pathway being regulator(s) of another. Moreover, an enzyme can have allosteric sites for more than one effector and receive control signals from several metabolic areas which increases the flexibility of control. The systems controlled are complex; glycogen breakdown feeds glycolysis, which feeds pyruvate into the citric acid cycle and this, in turn, feeds electrons into the electron transport system. This feeds ATP into the energy-utilizing machinery of the cell. Fatty acid breakdown also feeds the citric acid cycle, supplying acetyl-CoA; in reverse, acetyl-CoA formed from pyruvate feeds fatty acid synthesis, and so on.

In such a complex system, each section has to 'know' what others are doing by sensing the levels of key metabolites. Is there enough or too little ATP? Is the citric acid cycle being supplied with too little or too much acetyl-CoA? Is glycolysis too fast or too slow? You can see the chemical chaos that would result

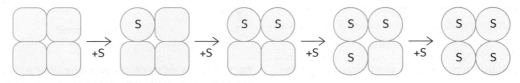

Fig. 13.6 The sequential model for the combination of an allosteric enzyme with its substrate. The binding of a single substrate molecule to a subunit causes a conformational change in the subunit. This facilitates the conformational change of a second subunit when it combines with a substrate molecule and so on for the next subunit. The net effect is to make it 'easier' for successive subunits to undergo the conformational change, which is seen as increasing affinity for the substrate by successive subunits.

unless there was constant second-to-second adjustment of pathways in the light of information reaching each pathway about all other pathways and parts of its own pathway. This is why allosteric control is such an important concept—regulatory enzymes can be designed, in the evolutionary sense, to receive signals from anywhere. Thus glycolysis can be 'informed' on how the citric acid cycle is doing and how the electron transport system is keeping up ATP supply, the information automatically adjusting activities to make the whole harmonious chemical machine. As Jacques Monod, the originator of the allosteric concept, pointed out, you simply could not have anything as complex as a cell without it. He described it as the 'second secret of life' (DNA being the first).

The control network *is* complex. A given enzyme is not just switched on or off by one allosteric signal. It may receive multiple signals from all over the metabolic map, each partially inhibiting or stimulating its rate of reaction. It is a classic case of evolution being a tinkerer; one little adjustment here and another there, by yet another allosteric signal. This gives fine tuning of enzyme activity, and natural selection of the control mechanisms ensures a smooth performance of the whole complicated mass of reactions.

Control of enzyme activity by phosphorylation

The second method of enzyme control is by phosphorylation. To be strictly accurate this section ought to refer to covalent modification of enzymes rather than phosphorylation, because other chemical modifications occur, but phosphorylation is of such overwhelming importance in eukaryotic cells that we will confine ourselves to it. Unlike allosteric control, which is of major importance in both prokaryotes and eukaryotes, phosphorylation is less important in prokaryotes but in eukaryotes it is of absolutely overwhelming importance in many vital areas.

The principle is simple. Enzymes called **protein kinases** transfer phosphoryl groups from ATP to specific proteins. When this happens, the target enzyme undergoes a conformational change such that the enzyme (or an enzyme inhibitor protein) changes its activity (or inhibitory effect). You can imagine that the addition of such a strongly charged group could have an effect on the conformation of a protein molecule. To reverse the process there are **phosphoprotein phosphatases** which hydrolyse the phosphate from the protein (Fig. 13.7).

The phosphorylation occurs on the hydroxyl group of a serine or threonine of the polypeptide chain of the enzyme, and is identified by the protein kinase by the neighbouring sequence of amino acids around the target –OH group. (In Chapter 26 we describe tyrosine group phosphorylation which has great importance in gene control.) We show the serine phosphorylation process below.

$$
\begin{array}{l}
NH \\
| \\
CH-CH_2OH \\
| \\
CO \\
|
\end{array}
+ \; ATP \longrightarrow
\begin{array}{l}
NH \\
| \\
CH-CH_2-O-\overset{\overset{\displaystyle O}{\|}}{\underset{\underset{\displaystyle O^-}{|}}{P}}-O^- \\
| \\
CO \\
|
\end{array}
+ \; ADP
$$

Serine residue
of polypeptide

This control of enzymes by phosphorylation, is a second general mechanism by which enzymes are controlled—allosteric control was the first, phosphorylation and dephosphorylation of proteins the second. We will shortly describe how the two methods of enzyme control apply to specific metabolic systems.

The two classes of controls—intracellular mechanisms and those dependent on extracellular signals

Allosteric control gives an immediate regulatory mechanism, but control by phosphorylation requires that the phosphorylation and dephosphorylation processes themselves are controlled. The balance between phosphorylation and dephosphorylation determines the activity of the target enzyme. What, therefore, controls the protein kinases and phosphatases? The answer lies for the most part *not* in intracellular

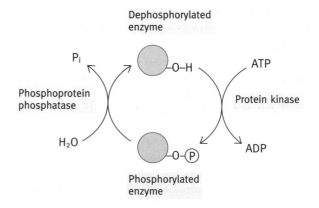

Fig. 13.7 Control of an enzyme by phosphorylation. The phosphorylated enzyme may, in specific cases, be more active or less active than the unphosphorylated enzyme. Note that variations of this may occur in that the activity of an enzyme may be controlled by an inhibitor protein whose inhibitory activity is controlled by phosphorylation.

automatic controls but in controlling agents such as hormones, external to the cell. (There are a few individual exceptions to this in which phosphorylation is part of the intrinsic metabolic controls, but the general point applies.)

To put things in perspective, the internal, allosteric controls coordinate the different pathways and parts of pathways so that metabolic pile-ups and shortages don't occur. To repeat the point, these are the automatic controls in which metabolites 'signal' to other pathways, and other parts of their own pathway. Imagine (again) the chaos there would be if glycolysis produced pyruvate irrespective of how much of it the citric acid cycle could handle, or if fatty acid synthesis took no notice of fatty acid oxidation. A completely isolated cell could carry out this intrinsic control. But, as already implied, in the body there is a second group of controls involving hormones and other signalling molecules (Chapter 26) which instruct cells on what their metabolic direction should be—such as whether to store fuel, or release it. This is essential if cells are to do whatever is beneficial for the body as a whole. Individual cells cannot make these decisions—they must receive signals from other cells. After being so instructed, the internal allosteric controls are still essential to keep the pathways coordinated to avoid bottlenecks and metabolic snarl-ups. It may be likened to the organization of a navy. Each individual ship has its own internal (allosteric) system of discipline which, in all circumstances, maintains it as an organized unit, but what it does as a unit—where it sails and what it does—depends on external signals from higher naval authorities (endocrine glands and the brain).

Phosphorylation controls are largely (but not entirely) responses to extracellular signals. It must be emphasized that phosphorylation controls are not confined to metabolic pathways but, as will be seen from later chapters, are of overwhelming importance in almost all aspects of gene control and cellular reproduction.

Hormonal control of metabolism

In control of carbohydrate and fat metabolism, the hormones of special importance are **glucagon**, **insulin**, and **epinephrine** (also known as **adrenaline**) and these are what we will deal with here; **norepinephrine (noradrenaline)** has similar effects to epinephrine. These are the ones which have immediate and rather dramatic effects and which are invoked on a daily basis as we oscillate between eating and fasting periods between meals (see page 109).

Glucagon, you will recall, is the 'hunger' hormone produced by the pancreas when blood sugar is low, and insulin is pro-

duced when the blood glucose level is high (page 113). Epinephrine, the hormone liberated by the adrenal gland, has a mobilizing effect on food storage reserves. Norepinephrine is liberated from sympathetic nerve endings; the sympathetic system controls involuntary contraction of smooth muscle such as that of viscera but it also innervates adipose cells and provides a means of rapidly liberating norepinephrine in the immediate locality of target organs. Insulin in general does the opposite to whatever epinephrine and glucagon do—it is the signal to store food in times of plenty.

How do glucagon, epinephrine, and insulin, work?

Hormones are a major topic dealt with in Chapter 26, but a brief introduction to them is needed here. They are chemical signals released into the blood so that all tissues are exposed to them. However, only certain cells, called target cells, respond to a given hormone. Glucagon, epinephrine, and insulin do not enter their target cells; they combine with receptor proteins, specific for each hormone. These are transmembrane proteins each with an external receptor part displayed like an aerial on the outside of cells ready to combine with its specific hormone or other signalling molecule. Only target cells have the receptors to specific hormones—the rest have none and are 'blind' to the hormone.

The hormones are quickly eliminated from the blood, usually by the liver; unless the source gland of a given hormone is releasing more, the concentration of the hormone in the blood falls, and that on the receptor dissociates from it (or is internalized and destroyed), terminating the signal. However, while the hormone is bound to the receptor, changes occur inside the cell (Fig. 13.8). This leads us to the **second messenger** concept.

What is a second messenger?

If you regard the hormone as the first messenger, then binding to its external receptor on a cell causes a change in the level of a

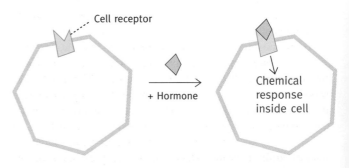

Fig. 13.8 How hormone binding to surface receptor may cause chemical events inside the cell.

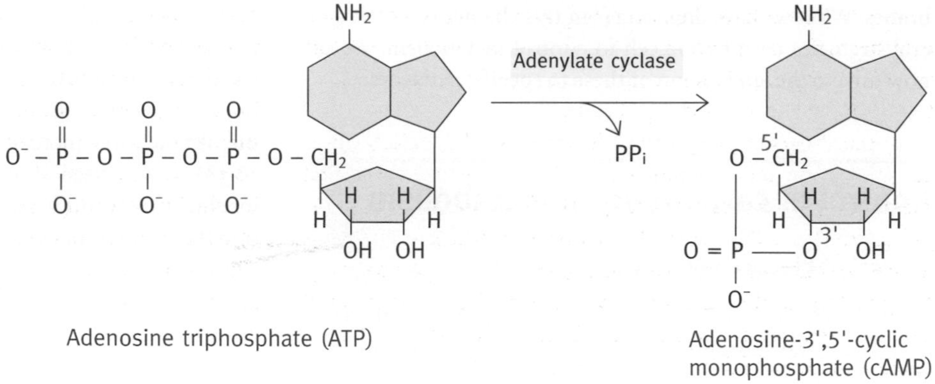

Fig. 13.9 The synthesis of cyclic AMP from ATP by adenylate cyclase.

Adenosine triphosphate (ATP)

Adenosine-3',5'-cyclic monophosphate (cAMP)

second messenger—a small intracellular regulatory molecule which causes cell responses. Not all hormone signalling involves second messengers, but the ones we are dealing with in this chapter do.

What is the second messenger for glucagon and epinephrine?

For these, the second messenger is cyclic-adenosine-3',5'-monophosphate, or **cyclic AMP** (cAMP; not 5'AMP, please note), a molecule you haven't yet met in this book. It is synthesized from ATP by the enzyme **adenylate cyclase** (Fig. 13.9).

Now things start to come together because inside the cell, cAMP is the allosteric activator of a protein kinase (**PKA**) which phosphorylates serine or threonine –OH groups of specific enzymes, thus modulating their activities. The enzymes affected will be given shortly when we deal with specific control systems. Thus the general mechanism is as in Fig. 13.10.

The way in which cAMP activates PKA is shown in Fig. 13.11. In the absence of cAMP, PKA is a tetramer with two catalytic subunits (C) and two regulatory subunits (R). When cAMP binds to the regulatory subunits, the enzymatically active C monomers are released.

An enzyme, **cAMP phosphodiesterase**, hydrolyses cAMP to AMP (Fig. 13.12) inside the cell so that the activation of cellular PKA depends on continual production of cAMP which occurs only as long as the hormone is bound to the receptor and, as already emphasized, the hormones in the blood are rapidly removed and their presence therefore depends on continuous production. Thus everything has the required element of reversibility. If the hormone levels fall, cAMP production ceases, existing cAMP is hydrolysed, and everything goes back to square one, phosphorylated proteins being dephosphorylated

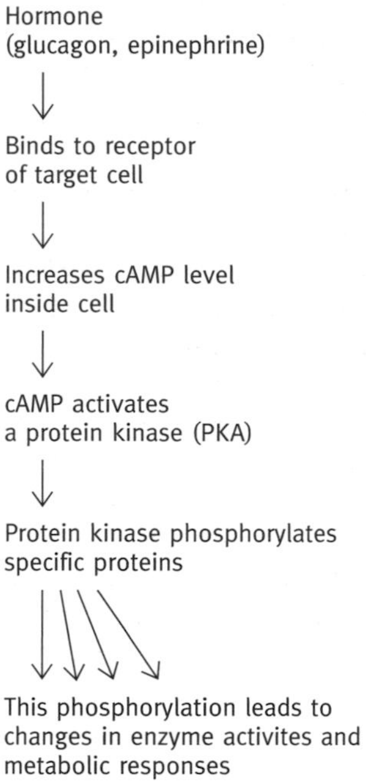

Fig. 13.10 Steps in the hormonal control of metabolism. Not shown are: (1) that cAMP is continually destroyed; (2) that phosphorylation of proteins is reversed by phosphatases. The metabolic response shown occurs only as long as a hormone is bound to the receptor.

by protein phosphatases. The latter are often, or usually, themselves controlled.

What is missing so far in this account is an explanation of *how* the hormone binding switches on cAMP production and how the latter is switched off. We are deferring description of this until Chapter 26 (page 437) because it is part of the more general major topic of signal transduction across cell mem-

branes. What we have done so far in this chapter is to give general strategies used by the cell to control metabolism. We can now turn to the application of these to specific pathways.

Control of carbohydrate metabolism

Control of glucose uptake into cells

Glucose cannot readily penetrate lipid bilayers and therefore membrane transport proteins are needed for its entry into cells. While absorption from the gut is active (page 82), transport into most other cells of the body is passive—the facilitated diffusion type in which a specific transporter protein is provided for glucose molecules to traverse the membrane, driven only by the glucose gradient across the membrane (page 82).

In brain, liver, and red blood cells this uptake of glucose is not

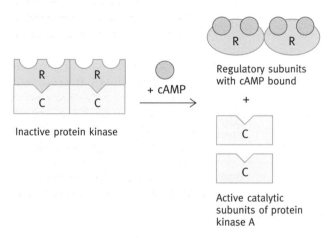

Inactive protein kinase

Regulatory subunits with cAMP bound

Active catalytic subunits of protein kinase A

Fig. 13.11 Activation of cAMP-dependent protein kinase (PKA) by cAMP. R, regulatory subunit of PKA; C, catalytic subunit of PKA.

insulin dependent, but in muscle and adipose cells it is. Inside the latter two insulin-responsive cells, there is a reserve of nonfunctional glucose transport proteins in the form of membrane vesicles. Insulin causes these to fuse with the cytoplasmic membrane, thus increasing the rate of glucose transport (Fig. 13.13). It has been shown in adipose cells that the effect of insulin in recruiting the reserve glucose transporter proteins into the cell membrane is complete in about 7 minutes. If the insulin is removed, the process reverses, the transporters being returned to the nonfunctional intracellular reserve in about 20–30 minutes.

The transport of glucose is by facilitated diffusion not requiring energy. By itself this would simply equilibrate cellular and blood levels, but once in the cell, the glucose is immediately removed by glycogen synthesis or other metabolism and this constitutes a driving force for uptake of the sugar. The first step is the phosphorylation of glucose,

Glucose + ATP → glucose-6-phosphate + ADP

The enzyme doing this is **hexokinase**, but in liver there is also a large amount of an **isoenzyme** known as **glucokinase** which carries out the same reaction as does hexokinase. The name is really a misnomer for it implies that it has a different specificity to sugars than has hexokinase, which is not the case. Isoenzymes are different forms carrying out the identical reaction but with different characteristics, such as affinity for substrate or its susceptibility to product inhibition, tailored to the requirements of the tissue in which they are found. Glucokinase has a much higher K_m for glucose (a lower affinity) than has hexokinase. This is an important regulatory device. Given the glucostatic role of liver, it takes up glucose mainly when its blood level is high. When blood glucose is low, the liver is turning out the sugar, so to take it up again rapidly, would not be logical. The high K_m of liver glucokinase minimizes this; only when the blood sugar is high does glucokinase work

NH$_2$

cAMP phosphodiesterase

NH$_2$

$5'$
O – CH$_2$
H   H
H   H
$3'$
O = P — O   OH
O$^-$
Site of hydrolysis

+ H$_2$O

O
||
O – P – O – CH$_2$
|
O$^-$
H   H
H   H
OH   OH

+ H$^+$

Adenosine-3',5'-cyclic monophosphate (cAMP)

Adenosine monophosphate (AMP)

Fig. 13.12 Reaction catalysed by cAMP phosphodiesterase. The name arises from the fact that cAMP contains a doubly esterified phosphoryl group— that is, a phosphodiester.

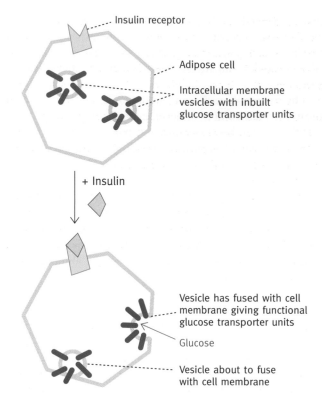

Fig. 13.13 The effect of insulin in mobilizing glucose transporter units in adipose cells and presumably other cells with insulin-dependent glucose entry. The glucose transport is of the passive facilitated diffusion type. Note that liver and brain are not insulin-responsive in this respect.

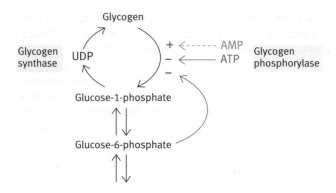

Fig. 13.14 The nonhormonal allosteric controls on glycogen metabolism. See text for explanation. Green lines, allosteric control having positive effects, red, negative ones.

rapidly. When it is low, the brain and other cells dependent on glucose are given priority. There is another reason; hexokinase is inhibited by its product, glucose-6-phosphate, but, at physiological concentrations of the phosphorylated sugar, glucokinase is not. This is important. When the blood glucose is high the liver takes it up and produces concentrations of glucose-6-phosphate for the purpose of glycogen synthesis which would inhibit hexokinase. In keeping with its role, the synthesis of this enzyme is increased by insulin which of course is released in the presence of high glucose levels in the blood. (The role of insulin in regulating gene expression and hence synthesis of specific proteins is very important but it is a major topic dealt with at some length in Chapter 26 on cell signalling.)

Control of glycogen metabolism

Glycogen metabolism plays a central role in the energy economy of the body with liver and muscle being the main organs in-

volved. Glycogen is synthesized by the enzyme **glycogen synthase** after feeding; it is activated by the presence of **insulin** which is secreted by the pancreas when the blood sugar level is elevated. Glycogen breakdown by **glycogen phosphorylase** occurs during the fasting periods between meals when the blood sugar is lowered.

The purposes of glycogen breakdown differ in muscle and liver. In muscle, it provides glucose-6-phosphate (via glucose-1-phospate) which is fed into the glycolytic pathway to generate ATP which is needed for muscle contraction. When a muscle contracts, it hydrolyses ATP to ADP; to regenerate the ATP glycogen is broken down to provide the energy source. The signal for this is AMP which allosterically activates muscle phosphorylase (Fig. 13.14). AMP is the signal rather than ADP because the enzyme adenylate kinase catalyses the equilibrium:

$$2ADP \rightleftharpoons AMP + ATP$$

Because of the position of the equilibrium, AMP is present in greater amounts than ADP and so is a more sensitive indicator that the phosphorylase needs to be activated. In keeping with this, ATP allosterically inhibits phosphorylase. Glucose-6-phosphate does the same (Fig. 13.14). If ATP and glucose-6-phosphate levels are high, breakdown of glycogen is not required. There is, however, another situation in which this 'routine' control of glycogen breakdown is overridden. The activation of phosphorylase by AMP is only partial; it is sufficient to keep things ticking over nicely in normal situations but in an emergency situation—such as being chased by a tiger—what you need is to generate ATP at the maximum possible speed. The first minute or so may make the difference between being eaten and not being eaten, and although glycolysis gives only a low yield of ATP, it occurs very rapidly. Oxidative phosphorylation takes a while to speed up until the blood circulation has

increased to supply extra oxygen. It is therefore logical in an emergency to break down glycogen at the maximum speed to provide glucose phosphate for feeding into glycolysis.

When you first see the tiger, your brain sends a signal to the adrenal glands to liberate epinephrine. This presses the metabolic panic button. It is known as the **flight or fight response**. Binding of this hormone to muscle cell receptors causes the production of the second messenger cAMP (please note, not AMP this time).

We must now discuss how cAMP activates phosphorylase.

Activation of muscle phosphorylase by cAMP

If you reflect on the constant struggle for survival that has dominated evolution, it is not surprising that the activation of phosphorylase in emergency situations has been refined to a high degree and indeed the mechanism is an elaborate one. In muscle, in nonpanic situations, phosphorylase exists in a form known as **phosphorylase b.** This is inactive in the absence of the allosteric activator, AMP, and as explained this partial activation is sufficient for routine needs.

In more demanding situations, phosphorylase b is converted to **phosphorylase a,** *which is maximally active without the presence of AMP.* Phosphorylase b is converted to the 'a' form by a cAMP-activated protein kinase (**phosphorylase kinase**), which phosphorylates serine/threonine –OH groups on the phosphorylase causing a conformational change to the activated 'a' form (Fig. 13.15).

The cAMP activation of the phosphorylase kinase which phosphorylates glycogen phosphorylase is not direct; it activates the protein kinase, **PKA** (A for cAMP), which, in turn, activates phosphorylase kinase (also by phosphorylation) which then activates phosphorylase b, converting it to the 'a' form. The whole scheme from the hormone onwards is shown in Fig. 13.16.

Why so complicated a mechanism? A cell will have only a relatively small number of hormone molecules attached to its receptors. In an emergency, the body requires a huge response to the binding of epinephrine to cells. The attachment of a relatively few molecules of hormones to cell receptors must result in a massive breakdown of glycogen and—if you happen to be being chased by a tiger—as rapidly as possible. What is needed is a rapid, vastly amplified, response, in which the minute signal (hormone binding) cascades into a flood of reaction (glycogen breakdown). The multiple steps in phosphorylase activation constitute an **amplifying or regulatory cascade**. Suppose 1 mol-

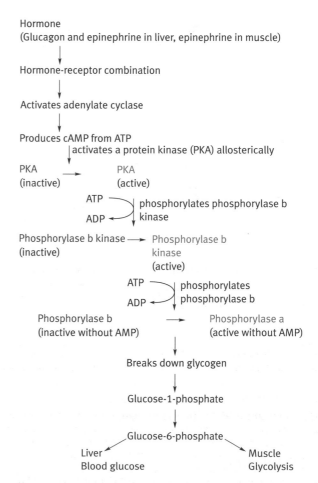

Fig. 13.16 The amplifying cascade mechanism by which hormones activate glycogen phosphorylase.

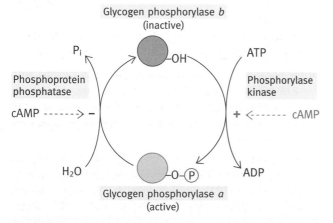

Fig. 13.15 Conversion of glycogen phosphorylase b to glycogen phosphorylase a by phosphorylase kinase and the reverse by protein phosphatase. It is important to note that the effects of cAMP are not exerted *directly* on the reactions shown—see text. The —OH group is that of a serine residue in the protein.

ecule of hormone activates 1 molecule of adenylate cyclase and that the latter produces 100 molecules of cAMP per minute. In this time period, the amplification is a 100-fold. If the cAMP activates a second enzyme which produces an activator at the same rate, the amplification becomes more than 100×100 and so on. In fact, glycogen phosphorylase activation involves four such amplification steps. Regulatory or amplifying cascades are used whenever a massive chemical response is needed from a minute signal. Phosphorylase attacks the end of glycogen chains (page 120) and it wouldn't be much use producing large numbers of active phosphorylase molecules if they couldn't find a glycogen end to work on. This is possibly one reason why glycogen is so highly branched compared with starch.

There is an additional refinement to the muscle control situation. In *normal* muscle contraction (not in the emergency situation involving cAMP), phosphorylase *b* kinase is also *partially* activated by the Ca^{2+} release into the cytoplasm (page 530) which triggers the contraction following a motor nerve signal (Fig. 13.17). This mechanism ensures that in vigorous contraction, even in the absence of epinephrine stimulation, glycogen breakdown keeps pace with energy demand. Since *this* activation of phosphorylase kinase does not involve phosphorylation, it is instantly reversed by the removal of Ca^{2+}, which occurs after cessation of the nerve signal stimulating the contraction. The Ca^{2+} activation is mediated by a protein called **calmodulin**, which is found in almost all eukaryotic cells as part of enzyme complexes (see page 44). Phosphorylase kinase can thus be activated in two ways, one by phosphorylation due to hormonal activation of PKA, and the other by allosteric Ca^{2+} activation not involving phosphorylation.

Reversal of phosphorylase activation

All metabolic controls must be reversible—there has to be a mechanism for switching off as well as on. In the case of phosphorylase, the 'a' form is converted back to the inactive 'b' form by a phosphatase, called **protein phosphatase 1**, which hydrolyses off the phosphoryl group from its serine/threonine residue. The phosphorylase kinase is likewise inactivated by dephosphorylation:

$$ENZ\text{-}O\text{-}P + H_2O \rightarrow ENZ\text{--}OH + P_i + H^+$$

('a' form)   ('b' form)

It is advantageous to switch off the opposing phosphatase action during the conversion of phosphorylase b to the 'a' form since this will result in more rapid and more complete activation of the enzyme, an important consideration in a life-threatening situation. A protein called **phosphatase inhibitor 1**, when phosphorylated by the same protein kinase A (PKA) that is activated by cAMP, inhibits the phosphoprotein phosphatase I. Thus, cAMP simultaneously switches on the activation of phosphorylase and switches off its deactivation (see Fig. 13.18(a)).

A summary of the main facts of this somewhat complicated control system may be useful at this stage (see also Fig. 13.17):

- In the absence of hormone stimulation, phosphorylase is in the unphosphorylated 'b' form, which is inactive unless allosterically (partially) activated by the presence of AMP (not cAMP). This activation does not involve phosphorylation of the protein.

- In normal muscle contraction, Ca^{2+} ions are released into the cell by the motor neuron signal and these *allosterically* partially activate phosphorylase b kinase; this gives a partial activation of phosphorylase. Unlike the cAMP-induced activation of the phosphorylase b kinase, the Ca^{2+} activation of this enzyme does not involve phosphorylation and occurs only as long as the muscle is contracting for the Ca^{2+} is immediately removed on cessation of the neuronal signal.

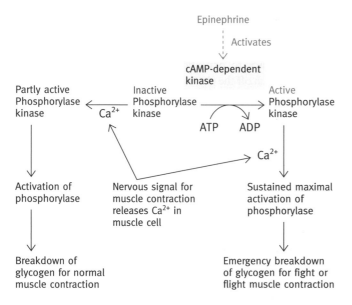

Fig. 13.17 Control of muscle phosphorylase kinase. This has, as well as the mechanism already described for both liver and muscle (to right of diagram), an additional Ca^{2+} activation (cAMP-independent) of phosphorylase kinase. This is to provide energy for normal contraction, which is triggered by Ca^{2+} release. The active (phosphorylated) kinase similarly requires Ca^{2+} for full activity. The kinase contains a protein subunit, calmodulin, to which Ca^{2+} binds, causing the calmodulin to activate the kinase.

(a)

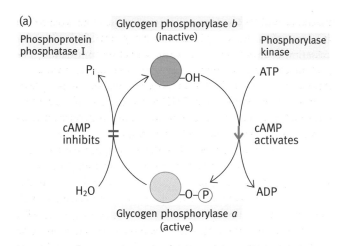

(b)

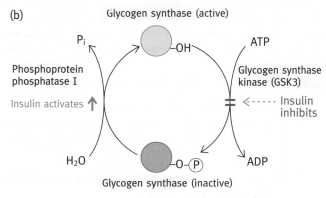

Fig. 13.18 Reciprocal controls on glycogen phosphorylase and glycogen synthase. (**a**) cAMP causes activation of phosphorylase kinase which activates phosphorylase by phosphorylating it. The cAMP-stimulated PKA activates the phosphatase inhibitor protein which prevents inactivation of the phosphorylase by the phosphatase. (**b**) In the presence of insulin, GSK3 is inactivated thus preventing the latter inhibiting glycogen synthase. In addition insulin causes activation of the phosphatase, thus activating the synthase. Red, inactive; green, active.

- Epinephrine in muscle increases the level of cAMP.
- cAMP allosterically activates PKA, which, in turn, activates phosphorylase b kinase. The latter phosphorylates the 'b' form of phosphorylase, producing the active 'a' form, which does not require AMP for activation. The process is an amplifying cascade.
- Phosphoprotein phosphatase I is capable of converting the 'a' form back to the 'b' form but as long as cAMP is present to activate PKA, the latter activates a phosphatase inhibitor protein so that inactivation of phosphorylase occurs only after removal of the hormonal signal.

Control of glycogen breakdown in the liver

The purpose of glycogen breakdown in the liver is not primarily the same as in muscle. The glucose-1-phosphate produced is converted to glucose-6-phosphate. This is not fed into the glycolytic pathway but is hydrolysed to free glucose, which enters the bloodstream to supply the brain and red blood cells. This occurs during fasting periods when the blood glucose level is low. The signal for this to occur in liver is the hormone glucagon, which is secreted by the pancreas in response to low blood glucose. Since insulin is not produced in this situation, the glucagon/insulin ratio is high and glycogen breakdown is the predominant event. Glucagon activates the liver phosphorylase by the same mechanism as that triggered by epinephrine in muscle; the second messenger for glucagon is cAMP as for epinephrine. Muscle does not respond to glucagon because it has no receptors for this hormone.

Liver also participates in the fight or flight reaction. It responds to epinephrine by pouring out glucose into the blood. The rationale is again to give muscles the absolute maximum supply of fuel to generate ATP in the emergency. An important control is that liver phosphorylase is inhibited by free glucose. If there is plenty of this molecule around, why produce more? The glucose also brings about a conformational change in phosphorylase a which makes the latter more susceptible to phosphatase conversion back to phosphorylase b, the rationale being to facilitate switch off the glucose generating mechanism.

We have so far described only the control of glycogen breakdown but obviously control of its synthesis is an essential part of the story for the two must be reciprocally controlled.

Control of glycogen synthase

The most important point is that *glycogen synthesis is activated by insulin*. As stated, this hormone is secreted when the blood glucose level is high, such as occurs after feeding. The presence of insulin in the blood causes activation of glycogen synthase of muscle and liver. *Phosphorylation inhibits glycogen synthase*; it is active only in the unphosphorylated state (the reverse of the phosphorylase situation).

The essence of glycogen metabolism control is that you either want phosphorylase to be active, breaking down glycogen, or you want glycogen synthase to be active in synthesizing glycogen. The hormones glucagon and epinephrine on the one hand, and insulin on the other, determine which one predominates. We have seen above that glucagon and epinephrine raise the cAMP level that activates PKA and this in turn leads to the activation of phosphorylase. At the same time, PKA phosphorylates and helps to inactivate glycogen synthase. This helps to ensure

that when glycogen is breaking down it is not also synthesized; the controls are reciprocal on the two systems.

However, this is not the most important device for inactivating the synthase by phosphorylation. The enzyme has multiple serine phosphorylation sites. One group of sites of special interest to us now, and the most important, control-wise, is a cluster of three serine residues found in the C-terminal end of glycogen synthase. A key observation was that when insulin stimulates glycogen synthase in muscle cells, *it is these three sites that are dephosphorylated* (not the PKA phosphorylated ones). There is a specific protein kinase that phosphorylates the sites, called **glycogen synthase kinase 3 (GSK3)**. *In the presence of insulin, GSK3 is inhibited* (Fig. 13.18(b)).

How does insulin inactivate GSK3?

The mechanism is illustrated in Fig. 13.19. Insulin combines with the insulin receptors on the external surface of target cells. This activates a signalling pathway inside the cell, which results in the activation of yet another protein kinase, **PKB**. (The

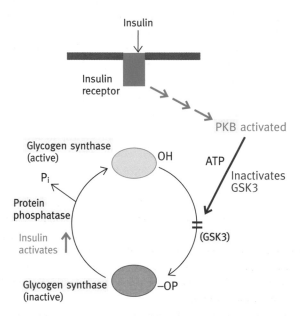

Fig. 13.19 Mechanism of control of glycogen synthase by insulin. The synthase is controlled by phosphorylation which inactivates it and dephosphorylation which activates it. Both PKA and glycogen synthase kinase 3 (GSK3) phosphorylate the enzyme, at different sites, but it is the phosphorylation performed by GSK3 that is reversed by insulin. It does this by activating PKB, a protein kinase which inactivates GSK by phosphorylating it. A protein phosphatase removes the relevant phosphate groups and thus activates the synthase. The mechanism by which insulin activates PKB is dealt with in the chapter on cell signalling.

mechanism of PKB activation by insulin signalling is described in Chapter 26, page 433.) PKB phosphorylates GSK3, which inactivates it. Insulin also causes the activation of the phosphoprotein phosphatase, which is responsible for activating the synthase; this again is due to phosphorylation of the phosphatase. Thus, inhibition of GSK3 and activation of the protein phosphatase cause dephosphorylation and consequent activation of the synthase, both events being insulin dependent.

The control is reversible. When the blood glucose level raized by feeding has been restored to normal by its utilization by tissues, the pancreas ceases to produce insulin and that in circulation is destroyed very quickly. The insulin is internalized with the receptors and the signalling pathway ceases to operate. PKB is inactivated by dephosphorylation; this allows GSK3 also to be dephosphorylated which activates it. The glycogen synthase is now at the mercy of GSK3 which phosphorylates and inactivates it. In the absence of insulin the protein phosphatase needed to activate the synthase is no longer active, and glycogen synthesis ceases.

A summary of glycogen synthase control may help.

- Insulin is needed to activate glycogen synthase.
- In the absence of insulin, the synthase is kept in an inactive state by GSK3 (glycogen synthase kinase) which phosphorylates the enzyme on specific sites of the protein.
- Insulin, via a signalling mechanism described on page 433, activates PKB (protein kinase B).
- PKB phosphorylates and thereby inactivates GSK3.
- Phosphoprotein phosphatase I in the presence of insulin is activated by phosphorylation. This enzyme removes the phosphoryl groups on the 'insulin specific' sites of the synthase thus activating the enzyme.
- Since the PKB has inhibited the GSK, the synthase remains active as long as insulin is present.
- When the insulin disappears, PKB is inactivated, GSK is activated (both of them by dephosphorylation) and the synthase is inactivated by phosphorylation.

Control of glycolysis and gluconeogenesis

Allosteric controls

Figure 13.20 shows the main systems. The rationale of controls is as follows: AMP (not cAMP) indicates an increase in the ratio of ADP to ATP (page 219). As well as activating glycogen phosphorylase, AMP is an allosteric activator of phosphofructokinase (PFK) which is a key controlling enzyme in glycolysis subject to multiple controls. At the same time, it inhibits the fructose-1:6-bisphosphatase. Activation of glycogen

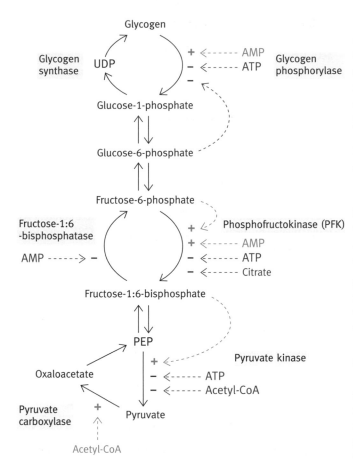

Fig. 13.20 The main intrinsic allosteric controls on glycogen metabolism, glycolysis, and gluconeogenesis. See text for explanation. Green broken lines, allosteric controls having positive effects on activities; red broken lines, allosteric controls having negative effects on activities. UDPG, uridine diphosphoglucose.

breakdown increases the level of fructose-6-phosphate which also activates PFK. Activation of PFK will in turn increase the level of fructose-1:6-bisphosphate which activates pyruvate kinase, an example of feedforward control. Evolution is rarely satisfied by one-hit controls. Thus the activating effects of AMP on PFK are balanced by inhibitory effects of ATP. The cell is thus constantly monitoring the ATP/ADP ratio (via AMP), and adjusting the glycolytic speed accordingly. In addition, when the ATP level is high, the citric acid cycle flux (the passage of metabolites through the pathway) is diminished and citrate accumulates. The latter is transported out of the mitochondria to the cytoplasm where it allosterically inhibits PFK. This again makes sense since if the citric acid cycle is partially shut down, glycolysis, which feeds the cycle, should likewise be inhibited.

The activation of pyruvate carboxylase by acetyl-CoA to produce oxaloacetate needs explanation. Accumulation of acetyl-CoA occurs if the citric acid cycle is deficient in oxaloacetate since the latter is needed to accept the acetyl moiety to form citrate. The acetyl-CoA thus automatically activates the synthesis of oxaloacetate. It is also important in gluconeogenesis (see page 202).

However, as with glycogen metabolism, these internal allosteric controls on glycolysis are over-ridden by extracellular signals from hormones.

Hormonal control of glycolysis and gluconeogenesis

The signal for the liver to produce blood glucose is glucagon, which activates both glycogen breakdown and gluconeogenesis Thus, it is logical for cAMP (whose synthesis is increased by glucagon) *in liver* to switch on glycogen breakdown *and* gluconeogenesis. Both produce glucose, but you don't want to activate glycolysis—it needs to be inhibited because you don't want to activate glucose synthesis and its breakdown at the same time. (Fig. 13.21(a)). The same applies to the effects of epinephrine-stimulated production of cAMP.

In muscle, the situation is quite different. There, epinephrine stimulation of cAMP production (muscle has no receptors for glucagon) is designed to increase generation of ATP; as mentioned earlier, epinephrine is released in the fight or flight reaction in which rapid and vigorous muscular contraction may be called for. It is therefore logical in skeletal muscle, in contrast to liver, for cAMP to switch on glycogen breakdown but, also in contrast to liver, it is *not* logical in muscle to inhibit glycolysis because this would result in producing glucose-6-phosphate from glycogen but preventing any usage of it (see Fig. 13.21(b)), and thwart the primary aim of producing ATP as rapidly as possible. Don't forget that muscle does not produce blood glucose; it has no glucose-6-phosphatase needed for this. The muscle is entirely selfish in this respect; unlike the liver, it looks after itself only. In fact, the reverse (glycolysis speed-up) is required in muscle to prepare for the fight or flight reaction. We thus see that cAMP needs to block glycolysis and activate gluconeogenesis in liver, but activate glycolysis in muscle.

How are these contradictory ends achieved in the two tissues? As mentioned, the most important control point in glycolysis is the phosphofructokinase (PFK) step. In *liver*, cAMP (increased by glucagon or epinephrine) causes inhibition of PFK, thus inhibiting glycolysis; in muscle, cAMP (increased by epinephrine) results in increased glycolysis (see below). These effects of cAMP on PFK are not direct, which leads to the question in the next heading.

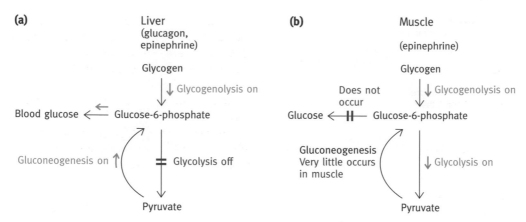

Fig. 13.21 The different control requirements of (a) liver and (b) muscle in response to glucagon and/or epinephrine. Both hormones have a cAMP as second messenger The term glycogenolysis used here is a synonym for glycogen breakdown.

How does cAMP control PFK?

We will consider liver first. When blood sugar is low, glucagon is high and therefore liver cAMP is high. This causes a decrease in the cells of **fructose-2:6-bisphosphate**. This is *not* a misprint for the **fructose-1:6-bisphosphate** that is produced in glycolysis by PFK, but a compound not mentioned previously in this book. It is the most powerful allosteric activator of PFK and thus of glycolysis. It is *solely* a regulator molecule. Raise its level and glycolysis increases; reduce its level and glycolysis decreases. The reverse of the PFK reaction—the hydrolysis of the fructose-1:6-bisphosphate by the bisphosphatase—is by contrast *inhibited* by the 2:6 compound as shown in Fig. 13.22. Recall that gluconeogenesis requires the fructose-1:6-bisphosphatase reaction (page 202) *so that the reduction of the 2:6 compound will activate it.*

You can easily deduce that in liver therefore, cAMP must *reduce* the 2:6 level—thus inactivating glycolysis and activating gluconeogenesis. How is this achieved? The 2:6 compound is produced by a second PFK, called PFK$_2$. cAMP causes *inhibition of* the latter (Fig. 13.23) by activating a protein kinase which phosphorylates the PFK$_2$ protein (Fig. 13.24) This has a double hit effect, because the *phosphorylated* PFK$_2$ not only does not synthesis the 2:6 compound, but actively destroys it by hydrolysis; by contrast the unphosphorylated PFK$_2$ does not destroy the 2:6 compound, but synthesizes it (Fig. 13.24). Thus, the PFK$_2$ protein is a double-headed enzyme in which the synthesis and destruction of the 2:6 compound are reciprocally controlled by a single phosphorylation; it is an astonishing enzyme.

This is fine for liver but not what is required in muscle where PFK must not be inhibited when epinephrine is released, since maximum glycolysis is needed in emergency. The hormonal PFK controls described for liver do not occur in skeletal muscle. Gluconeogenesis does not occur there so there is never the need

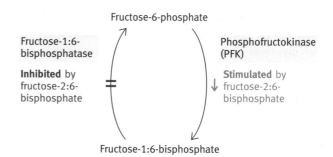

Fig. 13.22 Fructose-2:6-bisphosphate and the control of glycolysis in liver.

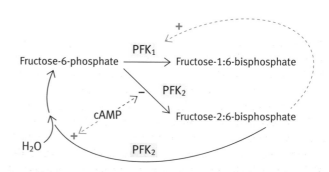

Fig. 13.23 Synthesis of fructose-2:6-bisphosphate by a second form of PFK (PFK$_2$) and its inhibition *in liver* by cAMP. The inhibition is indirect: it activates a protein kinase that phosphorylates PFK$_2$, inactivating synthesis of the 2:6 compound. The phosphorylated PFK$_2$ now destroys the latter. Thus cAMP reduces the fructose-2:6-bisphosphate level in liver.

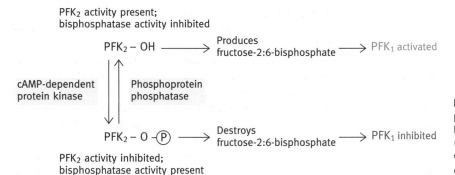

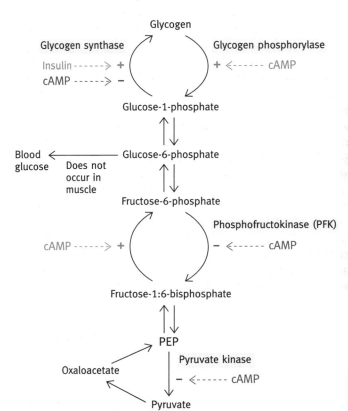

Fig. 13.24 Control of liver PFK_2 by phosphorylation by cAMP-dependent protein kinase. The PFK_2 is a double-headed enzyme: (1) it synthesizes fructose-2:6-bisphosphate when unphosphorylated; (2) it destroys the 2:6 compound when phosphorylated.

to switch from glycogen (or glucose) breakdown to synthesis. Epinephrine increases the rate of muscle glycolysis but this may be due the activation of glycogen breakdown.

Control of gluconeogenesis

We have already seen above that, in liver, glucagon causes inhibition of PFK and activates the fructose-1:6-bisphosphatase so that glycolysis is inhibited and the gluconeogenesis route stimulated.

Pyruvate kinase, whose reaction is near the end of the glycolytic pathway (Fig. 13.25), is another enzyme responding to glucagon. In the liver, but not in muscle, cAMP (and therefore glucagon and epinephrine) causes phosphorylation of the enzyme, resulting in its inactivation. The rationale of this is that if blood glucose is low, the liver needs to produce glucose, by the gluconeogenesis pathway, not to use it. This is a glycolytic switch off, additional to that at the PFK step. Gluconeogenesis is, as stated, a function of the liver mainly in response to glucagon. In Chapter 12 we described how gluconeogenesis from pyruvate requires the following steps, catalysed by pyruvate carboxylase and PEPCK (see page 202) respectively:

1 Pyruvate + ATP + HCO_3^- + H_2O → oxaloacetate + ADP + P_i + H^+

2 Oxaloacetate + GTP → PEP + GDP + CO_2

You can see that there is again a potential futile cycle, as shown in Fig. 13.26, if pyruvate kinase is catalysing the reaction:

PEP + ADP → Pyruvate + ATP

For gluconeogenesis, the pyruvate kinase (PK) in liver needs to be inactivated so that the PEP is sent back up the gluconeogenic path. Inactivation of the liver enzyme by cAMP described above achieves this. In liver, the PFK in the presence of glucagon is inhibited, the fructose-1:6-bisphosphatase active, so gluconeogenesis proceeds to supply blood sugar. Figure 13.25 summarizes this area. Once again muscle has different needs— it does not synthesize glucose, glycolysis must not be switched off, and its pyruvate kinase is not phosphorylated in response to cAMP elevation by epinephrine.

Fig. 13.25 The main external controls in this area of metabolism in liver. Note that 'cAMP' represents the action of glucagon and epinephrine. Its action is not directly on the enzyme controlled (see page 217).

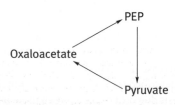

Fig. 13.26 Potential futile cycle.

Control of pyruvate dehydrogenase, the citric acid cycle, and oxidative phosphorylation

Pyruvate dehydrogenase (page 154) occupies a highly strategic position in metabolism, because it is the irreversible reaction producing acetyl-CoA by which pyruvate from glycolysis feeds into the citric acid cycle and into the fat synthesis route. The pyruvate dehydrogenase enzyme complex is regulated in several ways. Acetyl-CoA and NADH, which are products of the enzyme reaction, are end product allosteric inhibitors of the enzyme. CoA-SH and NAD$^+$, which are substrates, allosterically activate it. Thus the acetyl-CoA/CoA–SH ratio and the NADH/NAD$^+$ ratio control pyruvate dehydrogenase. The logic is that if there is a lot of acetyl-CoA and NADH, the production of acetyl-CoA and NADH by pyruvate dehydrogenase should be reduced and vice versa if CoA–SH and NAD$^+$ are high.

Of major importance is the negative control by high ATP levels. This, in effect, is monitoring the 'energy charge'. If ATP is low, feed in more fuel; if high, cut off the fuel supply. The ATP control of pyruvate dehydrogenase is not a direct one. At high ATP/ADP ratios a pyruvate dehydrogenase kinase is activated and this inactivates the dehydrogenase by phosphorylation of the enzyme; this is antagonized by a phosphatase (Fig. 13.27). The kinase is actually part of the pyruvate dehydrogenase complex. The inactivating kinase is additionally allosterically activated by NADH and acetyl-CoA. The rationale of all this is as already stated: if the signals indicate adequate energy production, close off the fuel valve. (Usually protein kinases are activated by extracellular signals, but this is one of the exceptions.)

In the citric acid cycle and electron transport area, the intrinsic controls are somewhat different in that while allosteric controls do exist, the major internal controls are the availability of NAD$^+$ and ADP as substrates. If much of the NAD$^+$ is present in its reduced form (NADH + H$^+$) then the dehydrogenases in the citric acid cycle are restricted in their activity simply by lack of substrate. Since NADH accumulates when electron transport to oxygen cannot cope with available NADH produced by the cycle, this constitutes an important control and the cycle is inhibited. Similarly if the ADP/ATP ratio is low, electron transport is inhibited because oxidation and phosphorylation are tightly coupled, again a desirable control. This tight coupling, called **respiratory control**, is of major importance. If there is no ADP the ATP synthase (page 167) cannot catalyse the uptake of protons into the matrix from the exterior of the mitochondrion (page 165). A build-up of the proton gradient would then be expected to inhibit the electron transport. In short, if there is plenty of ATP the furnaces shut down automatically. In addition to this control by NAD$^+$ and ADP availability, the cycle is allosterically controlled at the citrate synthase step (ATP inhibits), the isocitrate dehydrogenase step (ATP inhibits and ADP activates) and the a-ketoglutarate dehydrogenase step (succinyl-CoA and NADH inhibit). It is all very logical and the cycle metabolites are kept in balance one with another—no pile-ups, no shortages.

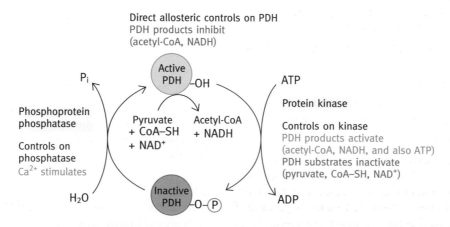

Fig. 13.27 The intrinsic control of pyruvate dehydrogenease (PDH) by direct allosteric control and by phosphorylation and dephosphorylation in the mammalian enzyme complex. The multiplicity of these controls reflects the strategic position of PDH whose activity is the gateway for pyruvate to enter the citric acid cycle and the fat synthesis pathways. Despite their complexity in detail, they largely add up to activation by substrates and inhibition by products. (In this context ATP can be regarded as a product of the result of entry of acetyl-CoA into the citric acid cycle.) It is worth noting that control by phosphorylation of proteins is usually associated with extrinsic controls, rather than with the intrinsic ones here.

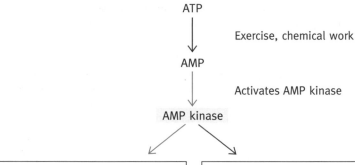

Fig. 13.28 Simplified version of the role of AMP kinase in restricting anabolic reactions when increase in the AMP/ATP ratio indicates that ATP levels are suboptimal.

Regulation of the cellular ATP level by AMP-activated protein kinase

The above section describes how the actual production of ATP by oxidative phosphorylation is controlled. However, ATP level in the cell is of critical importance and for this reason there is a more general control in cells which senses the energy state of the cell. **AMP-activated kinase (AMPK)** is the central effector of this control. The logic is obvious; if AMP is present at increased levels, it indicates a potential deficiency of ATP. In this situation you can't afford to be synthesizing new cell material which is not a vital immediate concern since the risk caused by insufficient ATP could be greater, particularly in a tissue such as heart muscle. Restoration of normal levels of ATP assumes top priority. The AMPK is activated by the increased AMP level, and has widespread effects on metabolism. It essentially shuts down anabolic (synthetic) reactions but increases ATP-generating catabolic processes (Fig. 13.28). This safety mechanism can be triggered for example by vigorous exercise, starvation or oxygen deprivation to a tissue (see Kemp *et al.* ref of the reading list). As already explained (page 210), AMP is a more sensitive indicator of ATP depletion than is ADP. The mechanism by which AMPK controls fat metabolism is dealt with in the next section.

Controls of fatty acid oxidation and synthesis

Nonhormonal controls

These are illustrated in Fig. 13.29. The rationale is that you want *either* fatty acid oxidation *or* fatty acid synthesis but not both at the same time in the same cell, or you have a futile cycle.

The two pathways for fatty acid synthesis and their breakdown suppress each other. Fatty acyl-CoAs (the product of the first step in fatty acid oxidation) allosterically inhibit acetyl-CoA carboxylase, the first committed step in fat synthesis.

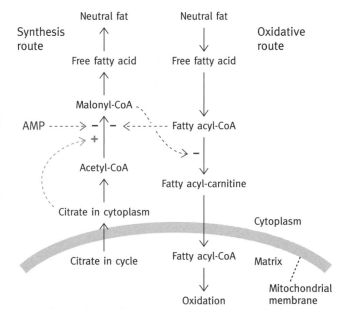

Fig. 13.29 Major intrinsic control points in fat oxidation and synthesis by which the two routes mutually suppress each other. Note that once acetyl-CoA is formed from fat breakdown, its further metabolism is subject to the controls of the citric acid cycle, etc. already described. Dashed lines indicate allosteric effects. Citrate activates acetyl-CoA carboxylase *in vitro*, but whether it is significant in the cell has been questioned.

Malonyl-CoA, the first product committed to fat synthesis, allosterically inhibits transfer of the fatty acyl groups to carnitine which prevents them being transported to the intramitochondrial site of their oxidation. If you are synthesizing fatty acids it is a good idea to stop them being fed into the mitochondrial furnaces. Acetyl-CoA carboxylase is activated *in vitro* by citrate (which also gives rise in the cytoplasm to acetyl-CoA, the sub-

strate of the enzyme, see page 190). Citrate is transported out of the mitochondria only when its level is high, and this happens only in times of food plenty when it is logical to store fat. The *in vitro* activation of acetyl-CoA carboxylase requires concentrations of citrate to do this *in vitro* which are much higher than cytosolic levels so—despite the beautiful logic—there is a question mark over its physiological significance. An important control on acetyl-CoA carboxylase is inactivation by phosphorylation. This is carried out by the AMP kinase described in the section above. As explained there, the rationale of this is that if the AMP level is high then it is not appropriate to indulge in fat synthesis which utilizes ATP.

Hormonal controls on fat metabolism

The most important decisions in this area are whether fat cells should store fat as triacylglycerol (TAG) or release free fatty acids from stored TAG into the blood. Insulin is the signal for the former, glucagon for the latter but in some animals, including humans, epinephrine (adrenaline) and norepinephrine are important in this regard. Fat cells are innervated by norepinephrine-releasing sympathetic nerves and are subject to epinephrine hormone control also. All three agents, glucagon, epinephrine, and norepinephrine use cAMP as second messenger. Fat cells contain a **hormone-sensitive lipase** which carries out the reaction

triacylglycerol + H_2O → diacylglycerol + free fatty acid

The diacylglycerol is further degraded to glycerol and free fatty acids. The hormone sensitive lipase is controlled by phosphorylation, the phosphorylated enzyme being active. A cAMP-activated protein kinase carries this out and a protein phosphatase reverses it (Fig. 13.30).

Insulin antagonizes these effects; it stimulates fat synthesis in liver, restricts release of fatty acids from fat cells, and increases glucose uptake by fat cells. These cells take up free fatty acids produced by the liver and convert them to TAG. For this glycerolphosphate is required (page 205) which is supplied by reduction of dihydroxyacetone phosphate. Since the latter is formed by the glycolytic pathway, uptake of glucose by the adipose cells is needed. As described, the key control enzyme in fat synthesis is acetyl-CoA carboxylase which is inactivated by the AMP-dependent protein kinase (see Fig. 13.28).

Insulin and diabetes

The interrelated nature of metabolic pathways and their control can be illustrated by describing some of the biochemical features seen in diabetes mellitus. In this disease either insulin is deficient in amount or there is a deficiency in the response of cells to the hormone.

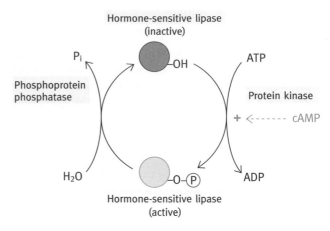

Fig. 13.30 Activation of adipose cell hormone-sensitive lipase by cAMP-dependent phosphorylation. Insulin antagonizes the effects of the catecholamine hormones and glucagon, which increase the cAMP level. Note also that glucagon and epinephrine cause inhibition of fat synthesis in liver and adipose cells, respectively, by preventing dephosphorylation of acetyl-CoA carboxylase. Insulin activates the carboxylase.

Insulin has widespread effects on mammalian metabolism. In general terms it has the opposite effects of those of glucagon. It promotes glycogen synthesis, and increases fatty acid synthesis in the liver and glucose uptake into insulin-responsive cells. When the insulin/glucagon ratio is low, the reverse happens. The mechanisms of these actions have been described earlier in this chapter. Insulin also stimulates the synthesis of specific enzymes involved in response to the hormone. An example is that after feeding, insulin stimulates the synthesis of glucokinase needed for glycogen synthesis from glucose in the liver and also stimulates the synthesis of the lipoprotein lipase which releases fatty acids from chylomicrons for uptake by cells (page 126). Conversely the synthesis of PEPCK, an enzyme required for gluconeogenesis (page 202), is reduced by insulin. However, the especial importance of insulin in biochemistry is because diabetes is one of the most prevalent diseases, affecting perhaps 100 million people.

As mentioned earlier, there are two types of the disease:

- **Type I**, also known as **juvenile onset diabetes** or **insulin-dependent diabetes mellitus**, is caused by autoimmune attack on insulin-producing β-cells of the pancreas so that insulin production is deficient. The insulin/glucagon ratio is therefore low. This type of diabetes usually occurs early in life.

- In **type II** or **maturity onset diabetes (noninsulin-dependent diabetes)**, insulin levels may be normal or even elevated, but cells do not respond normally to the

hormone. The onset usually occurs after the age of about 35, frequently associated with obesity.

In uncontrolled diabetes the blood glucose level may become elevated to the point where it spills over into the urine, carrying with it water and resulting in increased urine volume and thirst. In the type I disease, where there is a high glucagon/insulin ratio, the metabolic events that occur resemble some of those found in the starved state. Some of the biochemical features of the disease are as follows:

- Transport of glucose into muscle and adipose cells is impaired because the glucose transporters are not recruited into their functional sites in the membrane (page 219).

- Utilization of glucose is diminished by these cells, leading to high blood sugar levels; the muscles, being the largest organs of the body, are especially critical in this respect.

- Although glucose entry into liver is not insulin dependent, the high glucagon/insulin ratio reduces glycolysis and favours gluconeogenesis; glycogen breakdown in the liver is favoured, both of these events contributing further to high blood glucose levels;

- The high glucagon/insulin ratio activates lipolysis in adipose cells so that the blood level of free fatty acids is high.

- The liver metabolizes these and produces ketone bodies (page 113) in excess of what can be utilized by peripheral tissues. In uncontrolled type I diabetes the blood levels of acetoacetate and β-hydroxybutyrate are elevated; since these are strong acids, a ketoacidosis may develop. Spontaneous decarboxylation of acetoacetic acid produces acetone which gives a characteristic fruity smell to the breath of people with untreated diabetes. Ketoacidosis does not often occur in type II diabetes.

- Nonenzymic glycation of proteins occurs and this is increased if the blood glucose level is high. The mechanism of this is that the aldehyde group of glucose forms a Schiff base with protein amino groups which undergoes an Amadori rearrangement (well known in chemistry) to form a stable complex. Measurement of the level of glycated hemoglobin is a useful index of how well the blood glucose level is controlled.

A concluding note: metabolic control analysis

It has been said that a metabolic pathway, in the control sense, is a democracy rather than a dictatorship, a statement we hope to explain. We have described the regulatory mechanisms by which the cell and the body adjust the activities of metabolic pathways in a coordinated way that underlies the adaptation

of the body as a whole to changed conditions. We have described the important control enzymes which play the most obvious part in regulation of carbohydrate and fat metabolism. What we have presented are essentially qualitative aspects; these are what most research on metabolic control has dealt with since they are key considerations which together present a beautifully logical control system without which anything as complicated as an animal, plant or even a bacterial cell could not exist.

What more is there to say at the end of this chapter? We will answer that indirectly. Suppose you have a tap which is set at the half-opened position and measure the rate at which water emerges; if now you open the tap fully, the water flow will double (ignoring the small effects of friction). Now consider that a long hosepipe is attached to the tap and you do the same exercise but measure the rate at which water emerges from the end of the hosepipe. The water emerging will increase when you open the tap fully but it will not be doubled because of the frictional resistance of the hosepipe to water flow. If there are one or two kinks in the hosepipe and a few leaks in the pipe, water flow at the end of the pipe will deviate even more from being proportional to the tap opening. The degree of opening of the tap will determine the maximum possible water flux, but will not determine precisely the amount of water emerging from the hosepipe.

Let us now turn to metabolic pathways. In animals these are so complicated and inter-related that it is not always possible to neatly define the beginning and end of a pathway especially as different organs may be involved in a given metabolic system. We could have chosen any pathway, but the purposes of simple illustration we will take the glycolysis pathway from glucose-6-phosphate to lactate as our example. The first key regulatory enzyme in liver and muscle is PFK; the activity of this will determine the rate of the first step, the conversion of fructose-6-phosphate to fructose-1:6-bisphosphate. To continue the analogy, the rest of the glycolytic pathway is a hosepipe with kinks and leaks and other properties that affect the rate of lactate production. Each individual enzyme in the pathway has a say and the pipeline as a whole affects the metabolite flux leading to lactate in this example. In experimental systems it has been found that the rate of lactate production is not directly proportional to the activity of PFK because of the influence of the rest of the metabolic pathway.

In what way do the enzymes which have no direct control mechanisms associated with them affect the metabolite flux? The simplest possibility is that if you speed up a rate-limiting step, some other step may become rate limiting, but more complex considerations apply. A metabolic pathway is a series of enzymes, the product of one being the substrate of the next:

$$S_1 \xrightarrow[v_1]{E_1} S_2 \xrightarrow[v_2]{E_2} S_3 \xrightarrow[v_3]{E_3} S_4 \xrightarrow[v_4]{E_4}$$

Each enzyme in the pathway has its own properties—a K_m, a V_{max}, and susceptibility to inhibition by its own product and by other metabolites not directly involved in its reaction. The V_{max} is a function of the amount of an enzyme and its turnover number (defined on page 24) represents the rate at which the step would be catalysed at saturating levels of substrate(s). Although these are properties of each individual enzyme they do affect the regulation of the whole metabolic pathway.

To return to the example of glycolysis, suppose E_1 is PFK and its activity is increased by the regulatory mechanisms already described, then its product (S_2 in the diagram above because it is the substrate of the next enzyme) is produced more rapidly and may rise in concentration. This may increase product inhibition of E_1 and also increase the activity of E_2 by a simple substrate concentration effect. In turn this may cause S_3 to increase and so on. Another consideration is that metabolites in pathways are drawn off into other pathways. For example the rate of reoxidation of NADH will be an important factor since NAD^+ availability is essential for glycolysis; there are other considerations such as the utilization of glycolytic intermediates for purposes other than lactate production but the example serves to illustrate the complexity of the system. The net result is that the rate of lactate production is the outcome of a very complex set of variables. What all this is saying is that the metabolic pathway *as a whole* has, as it were, a regulatory 'mind of its own'. The flux through the pathway (the rate of carbon metabolites passing through it) responds to control of PFK but the final flux control is shared to some extent by the pathway as a whole. It is as if the system is governed by a democracy (all the enzymes) rather than only by a (PFK) dictatorship. The question now arises as to what in quantitative terms happens to lactic acid production when, say, the V_{max} of the PFK step is changed.

This question can be addressed by **metabolic control analysis**. Quite often information on the properties of individual enzymes are known or, if not, may be readily determined. Metabolic control analysis is a mathematical treatment of such data which permits calculation of how the flux through a pathway will respond to alterations in the intrinsic activities of a single enzyme. In effect it allows simulation of a complex metabolic situation by taking into account the many variables which affect control of the pathway.

Metabolic control analysis can provide a more accurate picture of what will happen in a metabolic situation to the metabolic flux if the activity of a component enzyme is changed or, say, the $[H^+]$ is varied. Apart from its basic interest, this has practical importance. Suppose you are designing a drug to inhibit a metabolic pathway by targeting a specific enzyme. Metabolic control analysis may pinpoint which is the enzyme of choice to attack in that pathway—the one which, if inhibited, would have the greatest effect on the flux of metabolites through the pathway or as it is expressed, the one with the greatest **flux control coefficient**. This is not necessarily the one that might be intuitively selected. Choosing the best target enzyme could result in a drug that is effective at lower concentrations than might otherwise be the case. Alternatively, in biotechnology, to produce a wanted metabolite in maximum amounts, cells may be genetically engineered (page 486) to block a given pathway so as to cause the metabolite to accumulate. Because of the complexity of metabolic interactions, a given blockage may not cause the intuitively expected accumulation. A mathematical simulation of a complex metabolic situation used to guide experimental work may be time and cost efficient, especially in applied biochemistry. It seems likely that metabolic control analysis will assume a more prominent position than has hitherto been the case.

FURTHER READING

Substrate cycles

Newsholme, E. A., Challiss, R. A. J., and Crabtree, B. (1984). Substrate cycles: their role in improving sensitivity in metabolic control. *Trends Biochem. Sci.*, 9, 277–80.

Describes the regulatory roles of these cycles.

Glycogen metabolism — regulation

Hardie, D. G., Carling, D., and Carlson, M. (1998). The AMP-activated/SNF1 protein kinase subfamily: Metabolic sensors of the eukaryote cell? *Ann. Rev. Biochem.* 67, 821–55.

A comprehensive review

Kemp, B. E., Mitchell, K. I., Stapleton, D., Michell, B. J., Zhi-Ping Chen, and Witters, L. A. (1999). Dealing with energy demand: the AMP-activated protein kinase. *Trends Biochem. Sci.*, **24**, 22–5.

A summary of the metabolite sensing mechanism found in all eukaryotes which responds to ATP depletion by shutting down ATP utilizing anabolic pathways and increases catabolic pathways generating ATP.

Cornish-Bowden, A. and Cardenas M. L. (1991). Hexokinase and glucokinase in liver metabolism *Trends Biochem. Sci.*, **16**, 281–2.

Discusses briefly a common misconception about these enzymes.

Ketogenesis — regulation

McGarry, J. D. and Foster, D. W. (1980). Regulation of hepatic fatty acid oxidation and ketone body production. *Ann. Rev. Biochem.*, 49, 395–420.

Despite the regulatory title, the article also provides a general review of ketone body formation.

Quant, P. A. (1994). The role of mitochondrial HMG-CoA synthase in regulation of ketogenesis. *Essays in Biochem.*, **28**, 13–25.

Summarizes the control of ketogenesis and the physiological repercussions of the process.

Insulin signal transduction

Cohen P. (1999). The Croonian Lecture 1998. Identification of a protein kinase cascade of major importance in insulin signal transduction. *Phil. Trans. Series B, Roy. Soc. Lond.*, **354**, 485–95.

A readable review of this work.

PROBLEMS FOR CHAPTER 13

1. What are the two main ways by which the activities of enzymes may be reversibly modulated?

2. Compare the relationship between substrate concentration and rate of enzyme catalysis in a nonallosteric enzyme and a typical allosteric enzyme.

3. What is the advantage of the allosteric enzyme substrate–velocity relationship?

4. If a typical allosterically controlled enzyme is exposed to saturating levels of substrate, what would be the effects of allosteric activators on the reaction velocity?

5. Describe the two models that explain homotropic cooperative binding of substrate to an enzyme.

6. What is the main feature of allosteric control that makes it such a tremendously important concept?

7. What are the salient features of intrinsic regulation and extrinsic regulation by extracellular signals?

8. By means of a diagram, illustrate the main *intrinsic* controls on glycogen metabolism, glycolysis, and gluconeogenesis and explain the rationale.

9. Pyruvate dehydrogenase is a key regulatory enzyme. In general, products of the reaction inhibit the reaction. There are three mechanisms of control involved; what are these?

10. What controls the release of insulin and glucagon from the pancreas?

11. What is a second messenger? Name the second messenger for epinephrine and glucagon and explain how it exerts metabolic effects.

12   How does insulin control the rate of glucose entry into fat cells?

13   Explain how cAMP activates glycogen breakdown.

14   Glucagon activates liver phosphorylase via cAMP as its second messenger. Muscle does the same with epinephrine stimulation. However, cAMP has quite different effects on liver and muscle glycolysis. Explain these.

15   Several hormones that elicit completely different cellular responses nonetheless use cAMP as their second messenger. How can one compound be used for these?

16   Phosphofructokinase is a key control enzyme. What is the major allosteric effector for this enzyme and how is its level controlled in liver?

17   To synthesize glucose in the liver, PEP is needed. However, production of PEP would achieve little in this regard if it were dephosphorylated to pyruvate by pyruvate kinase. How is this futile cycle avoided? Why would this mechanism not be appropriate in muscle?

18   How does glucagon cause fatty acid release by fatty cells?

19   Metabolic pathways often include at least one reaction with a large ΔG value. What advantages accrue from this?

20   The control of glycogen synthase is to a large extent effected by insulin.
(a)   Describe the physiological rationale for this.
(b)   Explain in outline the nature of this control.

21   What part does glucose play in the regulation of glycogen breakdown?

22   Describe the reciprocal controls which arrange that fatty acids are either synthesized or oxidized but not both at the same time.

23   An allosteric enzyme typically exhibits a sigmoid substrate concentration response. Under what conditions would this tend towards a hyperbolic response?

24   Why do you think phosphorylation is such a common way of controlling enzyme activity?

25   Discuss the role of the AMP-activated protein kinase (AMPK).

Chapter summary

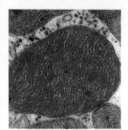

Why should there be an alternative pathway of glucose oxidation—the pentose phosphate pathway?

A completely different pathway of glucose oxidation exists, called the pentose phosphate pathway but sometimes called the 'direct oxidation pathway' or the 'hexose monophosphate shunt'. The need for another pathway will probably puzzle you, since in the preceding chapters we have dealt with a most extensive pathway (glycolysis, the citric acid cycle, and the electron transport pathway) for oxidizing glucose. Hence the question in the chapter title.

Paradoxically, it isn't really a pathway to provide for a different method of glucose oxidation nor is it primarily for energy production, but rather it is to cater for some specialized metabolic needs.

- It supplies ribose-5-phosphate for nucleotide and nucleic acid synthesis (the latter dealt with in later chapters).

- It suplies NADPH for fat synthesis (see page 190).

- It provides a route for excess pentose sugars in the diet to be brought into the mainstream of glucose metabolism.

The oxidative steps

The pentose phosphate pathway has two main parts. First, glucose-6-phosphate is converted to ribose-5-phosphate and CO_2, during which $NADP^+$ is reduced to NADPH. The 6-**phosphogluconate dehydrogenase** generates a β-keto acid which is decarboxylated to a keto-pentose (ribulose); an isomerase converts the latter to the aldose isomer, ribose-5-phosphate. This is the oxidative section (Fig. 14.1). It produces the two components, ribose-5-phosphate and NADPH. Why then does there follow another, nonoxidative series of reactions, which when written down in detail may appear somewhat formidable?

The nonoxidative section and its purpose

Tissue demands for the two products, ribose-5-phosphate and NADPH, vary greatly. For example, fat synthesis requires large supplies of NADPH (page 190). Fat cells and liver cells have large amounts of the pentose phosphate pathway enzymes to supply this, while muscle has very little of these. Production of NADPH for fat synthesis also produces large amounts of ribose-5-phosphate by the reactions given in Fig. 14.1, which may be far more than the cell needs for nucleotide synthesis. Conversely, in a rapidly dividing cell that is not synthesizing fat, large amounts of ribose-5-phosphate are needed to synthesize nucleotides, but there is little requirement for NADPH. The requirements for ribose-5-phosphate and NADPH may, in other cells, be anywhere in between this—equal amounts of the two products or more of one than the other.

The nonoxidative branch is a team act of a pair of enzymes, **transaldolase** and **transketolase**, that, together, interconvert sugars in accordance with the metabolic needs of the cell. These two enzymes detach, from a ketose sugar phosphate, C_3 and C_2 units, respectively, and transfer them to other aldose sugars (Fig. 14.2). It is interesting that transketolase is a thiamin pyrophosphate enzymes as is pyruvate dehydrogenase (page 155). In the latter case the enzyme also acquires a C_2 unit from pyruvate which it transfers to another acceptor, CoA—SH.

Between them, transketolase and transaldolase can work an almost bewildering range of sugar interconversions (which probably few biochemists carry in their heads in detail). If you put together the reactions of the *oxidative* section given above, the following balance sheet emerges.

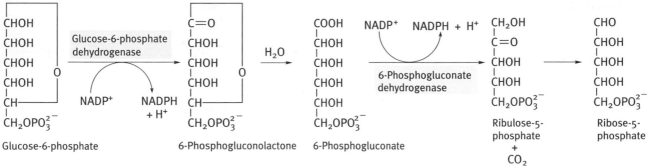

Fig. 14.1 Oxidative reactions of the pentose phosphate pathway.

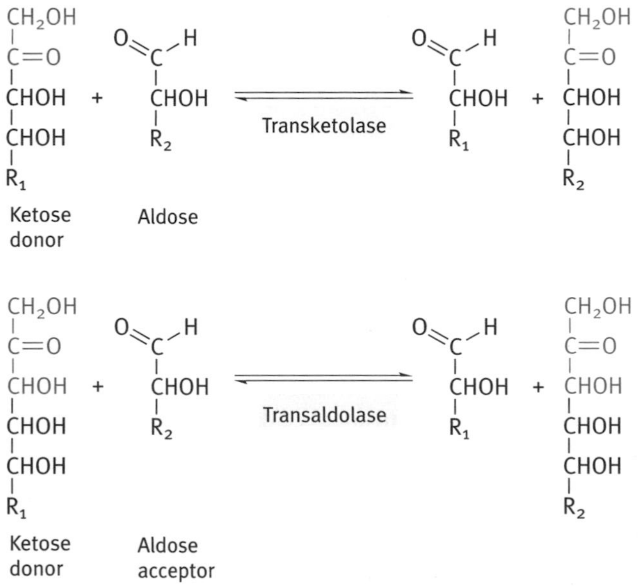

Fig. 14.2 Reactions catalysed by transketolase and transaldolase.

Glucose-6-phosphate + 2NADP$^+$ + H$_2$O →
   Ribose-5-phosphate + 2NADPH + 2H$^+$ + CO$_2$.

In situations where the needs for ribose-5-phosphate and NADPH are *balanced* the nonoxidative section is not required since the oxidative part produces both products in appropriate amounts.

However if, say, a nondividing fat cell requires *more NAPDH than ribose-5-phosphate* the nonoxidative section reconverts (recycles) the latter into glucose-6-phosphate according to the stoichiometry

6 Ribose-5-phosphate → 5 Glucose-6-phosphate + P$_i$.

The mechanism of this process in broad terms is given below, but we suggest you also follow the steps in the detailed reactions in the following section.

First, part of the ribose-5-phosphate (R-5-P), an aldose sugar, is converted into xylulose-5-phosphate (X-5-P), a ketose sugar, since both transaldolase and transketolase use only ketose sugars as group donors. The remaining fraction of the ribose-5-phosphate is the aldose sugar acceptor. The following transformations then occur, reaction 1 being between X-5-P and R-5-P.

(1)   2 C$_5$ ⟷ C$_3$ + C$_7$ Transketolase

(2) C$_7$ + C$_3$ ⟷ C$_4$ + C$_6$ Transaldolase

(3) C$_5$ + C$_4$ ⟷ C$_3$ + C$_6$ Transketolase

The final C$_3$ compound is glyceraldehyde-3-phosphate, two molecules of which are converted to glucose-6-phosphate by the reversal of glycolysis and loss of P$_i$. The net effect of the above three reactions is that three molecules of C$_5$ (indicated in red) are converted into 2.5 molecules of C$_6$ (blue). Note that this set of reactions can also convert dietary ribose to glucose-6-phosphate following its conversion to ribose-5-phosphate by an ATP-requiring kinase.

The pentose phosphate pathway is extremely flexible. Consider another situation where a cell needs to make ribose-5-phosphate for nucleotide synthesis, but has little demand for NADPH. The following overall process caters for this

5 Glucose-6-phosphate + ATP →
   6 Ribose-5-phosphate + ADP + H$^+$.

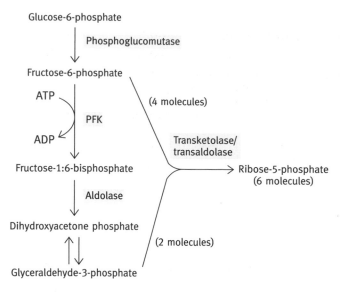

Fig. 14.3 Scheme by which glucose-6-phosphate is converted to ribose-5-phosphate without the production of NADPH. The oxidative part of the pentose phosphate pathway is not involved. See text for the transketolase/transaldolase steps.

The mechanism is as shown in Fig. 14.3. In this, glucose-6-phosphate is converted by the glycolytic pathway partly to fructose-6-phosphate and partly to glyceraldehyde-3-phosphate. These are the C_6 and C_3 products of reaction 3 above. Reversal of the three steps produces xylulose-5-phosphate which is isomerized into ribose-5-phosphate. This scheme does not involve the oxidative part of the pentose phosphate pathway at all.

Reactions involved in the conversion of ribose-5-phosphate to glucose-6-phosphate

$$
\begin{array}{ccc}
\overset{O}{\overset{\|}{C}}\text{—H} & \text{Phosphopentose} & CH_2OH \\
| & \text{isomerase} & | \\
CHOH & & C=O \\
| & \rightleftharpoons & | \\
H—C—OH & & H—C—OH \\
| & & | \\
H—C—OH & & H—C—OH \\
| & & | \\
CH_2—O—PO_3^{2-} & & CH_2—O—PO_3^{2-}
\end{array}
$$

Ribose-5-phosphate (R-5-P)   Ribulose-5-phosphate

Phosphopentose epimerase

$$
\begin{array}{c}
CH_2OH \\
| \\
C=O \\
| \\
HO—C—H \\
| \\
H—C—OH \\
| \\
CH_2—O—PO_3^{2-}
\end{array}
$$

Xylulose-5-phosphate (X-5-P)

Only part of the R-5-P undergoes the above conversion. The remainder is needed for reactions 1–3 shown in the next column and over the page.

$$
\begin{array}{ccc}
CH_2OH & & CHO \\
| & & | \\
C=O & & CHOH \\
| & + & | \\
CHOH & & CHOH \\
| & & | \\
CHOH & & CHOH \\
| & & | \\
CH_2—O—PO_3^{2-} & & CH_2—O—PO_3^{2-}
\end{array}
$$

X-5-P   R-5-P

Reaction 1 ‖ Transketolase

$$
\begin{array}{cc}
 & CH_2OH \\
 & | \\
 & C=O \\
 & | \\
 & CHOH \\
 & | \\
CHO & CHOH \\
| + & | \\
CHOH & CHOH \\
| & | \\
CH_2—O—PO_3^{-} & CH_2—O—PO_3^{2-}
\end{array}
$$

Glyceraldehyde-3-phosphate (G-3-P)   Sedoheptulose-7-phosphate (S-7-P)

Reaction 2 ‖ Transaldolase

$$
\begin{array}{cc}
 & CH_2OH \\
 & | \\
 & C=O \\
 & | \\
CHO & CHOH \\
| & | \\
CHOH + & CHOH \\
| & | \\
CHOH & CHOH \\
| & | \\
CH_2—O—PO_3^{-} & CH_2—O—PO_3^{2-}
\end{array}
$$

Erythrose-4-phosphate (E-4-P)   Fructose-6-phosphate (F-6-P)

$$
\begin{array}{c}
\text{CH}_2\text{OH} \\
| \\
\text{C}=\text{O} \\
| \\
\text{CHOH} \quad + \\
| \\
\text{CHOH} \\
| \\
\text{CH}_2-\text{O}-\text{PO}_3^{2-}
\end{array}
\qquad
\begin{array}{c}
\text{CHO} \\
| \\
\text{CHOH} \\
| \\
\text{CHOH} \\
| \\
\text{CH}_2-\text{O}-\text{PO}_3^{2-}
\end{array}
$$

X-5-P   E-4-P

Reaction 3 ‖ Transketolase

$$
\begin{array}{c}
\text{CH}_2\text{OH} \\
| \\
\text{C}=\text{O} \\
| \\
\text{CHOH} \\
| \\
\text{CHOH} \quad + \\
| \\
\text{CHOH} \\
| \\
\text{CH}_2-\text{O}-\text{PO}_3^{2-}
\end{array}
\qquad
\begin{array}{c}
\text{CHO} \\
| \\
\text{CHOH} \\
| \\
\text{CH}_2-\text{O}-\text{PO}_3^{2-}
\end{array}
$$

F-6-P   G-3-P

Two rounds of reactions 1–3 will produce two molecules of G-3-P.

$$
2 \begin{array}{c}
\text{CHO} \\
| \\
\text{CHOH} \\
| \\
\text{CH}_2-\text{O}-\text{PO}_3^{2-}
\end{array}
\xrightarrow[\substack{\text{(Gluconeogenesis}\\\text{reactions)}}]{} \quad 1 \ \ \text{F-6-P} \ + \ \text{P}_i \ .
$$

G-3-P

Fructose-6-phosphate is converted to glucose-6-phosphate by phosphohexose isomerase.

The scheme can be summarized as

$$2C_5 + 2C_5 \rightleftharpoons 2C_3 + 2C_7$$
$$2C_7 + 2C_3 \rightleftharpoons 2C_4 + 2C_6$$
$$2C_5 + 2C_4 \rightleftharpoons 2C_3 + 2C_6$$
$$2C_3 \longrightarrow 1C_6$$

The net effect of these reactions is

6 Ribose-5-phosphate → 5 Glucose-6-phosphate + P_i.

Where does the complete oxidation of glucose come into all of this?

The oxidative part of the pentose phosphate pathway (see Fig. 14.1) is sometimes referred to as being capable of the direct oxidation of glucose to CO_2, and NADPH + H^+. The stoichiometry of the overall process is

6 Glucose-6-phosphate + 12 NADP$^+$ + 7 H_2O →
  5 Glucose-6-phosphate + 6 CO_2 + 12 NADPH + 12 H^+ + P_i.

In fact, it never occurs, as such, as a metabolic route. A *single* molecule of glucose-6-phosphate is *not* converted to six CO_2 molecules by the pathway. What happens is that six molecules of glucose-6-phosphate can *each* give rise to one molecule of CO_2 and one molecule of ribose-5-phosphate by the reactions given above. If, now, nonoxidative reactions convert the six ribose-5-phosphate molecules back to five glucose-6-phosphate molecules, as described above, then on a balance sheet it looks as if six glucose-6-phosphate molecules have been converted to five molecules of glucose-6-phosphate plus six molecules of CO_2 but, as stated, it hasn't really oxidized one molecule of glucose completely. This set of events generates NADPH without net production of ribose-5-phosphate—a situation required in a cell with rapid fat synthesis but no cell division.

Why do red blood cells have the pentose phosphate pathway?

Erythrocytes do not divide so they have no need for ribose-5-phosphate for nucleic acid synthesis, nor do they synthesize fat. Their energy is derived from anaerobic glycolysis—anaerobic, for they have no mitochondria.

Nonetheless, in patients whose red blood cells lack the first enzyme in the pentose phosphate pathway (glucose-6-phosphate dehydrogenase), a massive hemolytic anaemia may be induced by the antimalarial drug, **pamaquine**. The reason for this relates to the fact that NADPH, generated by the pentose phosphate pathway, is needed for a protective mechanism described on page 283, namely the reduction of glutathione, a thiol compound, which maintains hemoglobin in its reduced (ferrous) state.

An interesting sidelight on glucose-6-phosphate dehydrogenase deficiency is that mutations leading to a defective enzyme confer a selective advantage in areas where a lethal type of malaria is endemic. Possible explanations for this are that the parasite has a requirement for the products of the pentose phosphate pathway and/or that the extra stress caused by the parasite causes the deficient red blood cell host to lyse before the parasite completes its development. It is interesting to compare the selective advantage conferred in this case with that conferred by the sickle-cell trait (page 518) where a potentially lethal disease can give a survival advantage because it gives protection against a more lethal disease, malaria.

FURTHER READING

Beutler, E. (1983). Glucose-6-phosphate dehydrogenase deficiency. In *The metabolic basis of inherited disease* (ed. Stanbury, J. B., Wyngaarten, J. B., Fredrickson, D. S., and Brown, M. S.), pp. 1629–53. 5th edn. McGraw-Hill.

An account of the medical aspects. Comprehensive but very readable.

PROBLEMS FOR CHAPTER 14

1   What are the functions of the pentose phosphate pathway?

2   What is the oxidative part of the pentose phosphate pathway?

3   What enzymes are involved in the nonoxidative part?

4   A nondividing fat cell requires large amounts of NADPH for fat synthesis but very little ribose-5-phosphate. However, the oxidative section produces equal amounts of the two products. Explain how the nonoxidative reactions cope with this problem. (The answer need not give actual reactions.)

5   The pentose phosphate pathway was sometimes referred to as the direct oxidation pathway for glucose. Why was this and why was it a somewhat misleading term?

6   Mature red blood cells have no need for nucleotide or fat synthesis. Why then do they have glucose-6-phosphate dehydrogenase?

Chapter summary

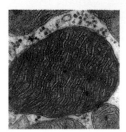

Raising electrons of water back up the energy scale— photosynthesis

From Chapter 9 you will have learned that ATP generation in aerobic cells depends on transporting electrons of **high energy potential** present in food down the energy scale to end up as the electrons present in the hydrogen atoms of water.

Since the amount of food on Earth is limited, if life in general is to continue indefinitely, a way must exist to kick those electrons back up the energy scale. A minor qualification to this statement is that life forms have been discovered in deep oceans around cracks in the Earth's crust from which compounds such as H_2S emerge. H_2S is a reducing agent of low redox potential and its electrons could be transported down the energy gradient releasing energy, provided appropriate biochemical systems are there. Such life could presumably exist as long as H_2S and other such agents are generated in the Earth's crust but, for continuation of the vast bulk of life, electron recycling is necessary. Early life forms must have survived on ready-made 'food', such as H_2S, in the primeval oceans but, given the exponential nature of biological reproduction, such resources would have been exhausted relatively quickly. For that matter, suitable electron 'sinks' might also have been exhausted in the anaerobic atmosphere believed to exist before the onset of photosynthesis.

Overview

Photosynthesis is the biological process that recycles electrons. It has several crucial advantages—water is an inexhaustible source of electrons, sunlight an inexhaustible source of energy, and, in releasing oxygen, an inexhaustible supply of electron sinks is provided to allow the energy to be extracted back from the high-energy electrons of food. The onset of photosynthesis was arguably the most important biological event following the establishment of life. The global energy cycle is shown in Fig. 15.1.

You are probably familiar with the concept of photosynthesis producing carbohydrate (usually starch or sugar) from CO_2 and H_2O, the overall equation (written for glucose production) is

$$6CO_2 + 6H_2O \rightarrow C_6H_{12}O_6 + 6O_2 \cdot \Delta G^{0'} = 2\,820\,kJ\,mol^{-1}$$

To synthesize glucose from CO_2 and H_2O there are two basic essentials looked at from the point of view of energy. First, there must be a reducing agent of sufficiently **low redox potential (high energy)**. If you need to refresh your memory on redox potentials, turn to page 140. In photosynthesis, the reducing agent is NADPH (page 189). (Gluconeogenesis in animals, as described in Chapter 12, uses NADH as the reductant, but, in photosynthesis, note that it is NADPH.) Secondly, there must be ATP to drive the synthesis.

Light energy is directly involved only in transferring electrons from water to $NADP^+$ and in the generation of ATP.

Site of photosynthesis—the chloroplast

Photosynthesis occurs in the **chloroplasts** of green plant cells. They are reminiscent of mitochondria in being membrane-bound organelles in the cytoplasm with an outer permeable membrane and an inner one impermeable to protons. Like mitochondria they have their own DNA coding for part of their proteins. Their protein-synthesizing machinery is prokaryotic in type (page 381) and it is believed that they arose by a symbiotic colonization of eukaryotic cells by primitive prokaryotic photosynthetic unicellular organisms.

Unlike mitochondria, however, chloroplasts contain yet another type of membrane-bound structure—the **thylakoids**—

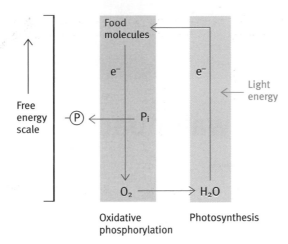

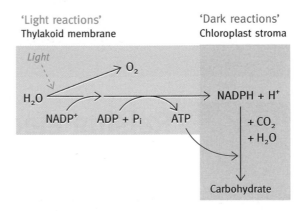

Fig. 15.3 The processes in photosynthesis.

Fig. 15.1 Global 'electron cycling' by oxidative phosphorylation and photosynthesis. Note that the fixation of CO_2 is a process secondary to the raising of electrons from water to a higher energy potential.

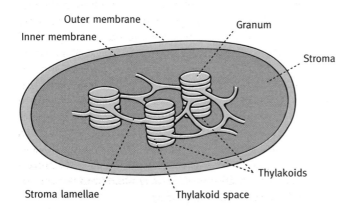

Fig. 15.2 A chloroplast. Grana are stacks of thylakoids.

membrane sacs within the chloroplast. The thylakoids are flattened sacs piled up like stacks of coins into grana, these being connected at intervals by single layer extensions. Inside the thylakoids is the thylakoid lumen; outside is the chloroplast stroma (Fig. 15.2). All of the light-harvesting chlorophyll and the electron transport pathways are in the thylakoid membranes. The conversion of CO_2 and H_2O into carbohydrate molecules is not light-dependent and occurs in the chloroplast stroma. The latter processes are referred to as 'dark reactions', not to imply that they only occur in the dark, but rather that light is not involved in them. In fact, the dark reactions occur mainly in the light, when NADPH and ATP generation is occurring in the thylakoid membranes. This is summarized in Fig. 15.3.

The photosynthetic apparatus and its organization in the thylakoid membrane

It would be useful for you to refresh your memory of the electron transport chain in mitochondria for there are considerable similarities between this and the photosynthetic machinery. In the inner mitochondrial membrane there are four complexes (see Fig. 9.18) that transport electrons. Connecting these complexes by ferrying electrons between them are ubiquinone, the small lipid-soluble molecule, and cytochrome *c*, the small water-soluble protein.

In the thylakoid membrane there are three complexes (Fig. 15.4); connecting the first two is the electron carrier plastoquinone whose structure is very similar to that of ubiquinone. Connecting the second two complexes is plastocyanin, a small water-soluble protein that has a bound copper ion as its electron-accepting moiety; this oscillates between the Cu^+ and Cu^{2+} states as it accepts and donates electrons.

The three complexes are named photosystem II (PSII), the cytochrome *bf* complex, and photosystem I (PSI). PSII comes before PSI in the scheme of things; the numbers refer to the order in which they were discovered rather than to their place in the scheme. The function of the whole array shown in Fig. 15.4 is to carry out the overall reaction

$$2H_2O + 2NADP^+ \rightarrow O_2 + 2NADPH + 2H^+.$$

This involves a very large increase in free energy. What is unique in photosynthesis, so far as biochemical reactions are concerned, is that this energy is supplied by light. For each molecule of NADPH produced, two photons are absorbed. These provide sufficient energy, not only for a molecule of $NADP^+$ to be reduced by water but also there is some to spare and this is used for ATP production. The arrangement is rather beautiful for it supplies the two requirements for carbohydrate synthesis from

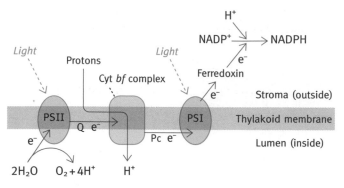

$2H_2O$ $O_2 + 4H^+$ H^+

Fig. 15.4 The diagram gives an overall view of the light-dependent part of the photosynthetic apparatus, without reaction details. The feature to note is that the purpose is to take electrons from water and transfer them to $NADP^+$ and in the process to create a proton gradient across the thylakoid membrane that can drive ATP synthesis by the chemiosmotic mechanism. The gradient is created from two sources— the splitting of water (photosystem II, PSII) and the proton pumping of the Q:cytochrome *bf* complex. Electrons are transported from PSII to the cytochrome *bf* complex by plastoquinone (Q). The cycle is analogous to that in mitochondria (where 'Q' is used for ubiquinone) that results in four protons translocated per QH_2 molecule oxidized, but there is some uncertainty still in the number transported here. Note also that plastocyanin (Pc) is a mobile carrier analogous in this respect to cytochrome *c* in mitochondria and that ferredoxin is located on the opposite side of the membrane.

CO_2 and water—a reducing agent and ATP. We now turn to the light-harvesting machinery present in the photosystems II and I.

What is chlorophyll?

In green plants, chlorophyll is the light receptor. Other receptor pigments exist in bacteria and algae but we shall confine this account to higher plants.

Chlorophyll is a tetrapyrrole much as is heme (page 508) except that it has a magnesium atom at its centre instead of iron and the substituent side groups are different. One of the latter is a very long hydrophobic group that anchors it into the lipid layer. As in heme, there is a conjugated double bond system (alternate double and single bonds right round the molecule) resulting in a strong colour—that is, strongly light-absorbing. In green plants there are two chlorophylls (*a* and *b*) differing in one of the side groups. They both absorb light in the red and blue ranges, leaving the intermediate green light to be reflected. The two chlorophylls have slightly different absorption maxima, which complement each other so that in the red and blue ranges between them they absorb a higher proportion of the incident light. We suggest that you memorize the structure of chlorophyll *a* shown below only in broad outline without bond

or side-chain details. (Chlorophyll *b* is —CH_3 side chain (in red), which is —(

Structure of chlorophyll *a*

When a chlorophyll molecule absorbs light, it is excited so that one of its electrons is raised to a higher energy state; it jumps into a new atomic orbit. In an isolated chlorophyll molecule, after such excitation, the electron drops back to its ground state liberating energy as heat or fluorescence in doing so (and nothing is thereby achieved). But, when chlorophyll molecules are closely arranged together, a process known as **resonance energy transfer** transfers that energy from one molecule to another. In green plants, chlorophyll molecules are packed in functional units called **photosystems**, such that this resonance transfer occurs readily. When a chlorophyll molecule is excited by absorption of a photon, it passes on its excitation to a neighbour and drops back to the ground state itself and so the process continues. The excitation wanders at random from one chlorophyll molecule to another (Fig. 15.5).

This has a function, for amongst large numbers of 'ordinary' chlorophyll molecules there is a special **reaction centre** (probably an arrangement of a pair of chlorophyll molecules in association with proteins). The properties of this reaction centre are such that the excitation of a constituent chlorophyll molecule by resonance transfer results in the excited electron being at a somewhat reduced energy level, as compared with that of other excited chlorophyll molecules, so that resonance transfer *from* this molecule to other chlorophyll molecules does not occur. The excitation energy is, in this sense, trapped in an energetic hole—a shallow hole because the 'trapped' electron is still at a higher energy level than an unexcited electron, sufficient for the excited reaction centre to hand on an electron to an electron acceptor of appropriate redox potential. The latter is the first carrier of an electron transport chain in photosynthesis (to be described shortly).

A photosystem is therefore a complex of light-absorbing

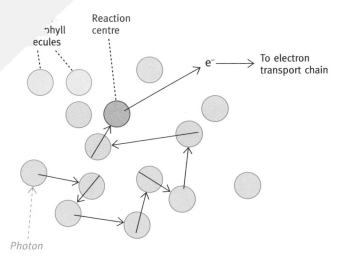

Fig. 15.5 Activation of reaction centre chlorophyll molecule (red circle) by resonance transfer of energy from activated antenna chlorophyll molecules (green circles). The reaction centre of photosystem II is called P680 and of photosystem I, P700.

chlorophylls, reaction centre chlorophyll, and an electron transport chain. In the case of photosystem II (the first one) the reaction centre chlorophyll is called P680 because it absorbs light up to that wavelength and photosystem I is called P700 for an equivalent reason. The chlorophyll molecules feeding excitation energy to the centres are called **antenna chlorophylls** (Fig. 15.5).

Mechanism of light-dependent reduction of NADP$^+$

Figure 15.6 shows the 'Z' arrangement of the two photosystems II and I, with the redox potentials of the components indicated. Why two photosystems? A familiar analogy might help at the outset. An electric torch with a bulb requiring 3 volts to light it up, usually employs two 1.5-volt batteries in series. In photosynthesis, lighting of the bulb is represented by NADP$^+$ reduction, and the batteries by the two photosystems. In fact, as stated, the latter supply more energy than is needed and some of it is sidetracked into ATP generation.

Let us start with a P680 chlorophyll of a reaction centre of photosystem II. In the dark, it is in its ground, unexcited state in which it has no tendency to hand on an electron. When energy of a photon reaches it via antenna chlorophylls, it is excited such that it has a strong tendency to hand on its excited electron. It is, in fact, a reducing agent and it reduces the first component (a chlorophyll-like pigment lacking the Mg^{2+} atom, called pheophytin) of the PSII electron transport chain.

Two molecules of reduced pheophytin then hand on an electron each (one at a time) to reduce **plastoquinone**, the lipid-

soluble electron carrier between PSII and the cytochrome *bf* complex.

$$H_3C \overset{O}{\underset{O}{\bigcirc}} R \xrightarrow[2H^+]{2e^-} H_3C \overset{OH}{\underset{OH}{\bigcirc}} R \qquad R = \text{a long hydrophobic group}$$

Plastoquinone (Q)   Plastoquinol (QH$_2$)

The latter complex contains two cytochromes and an iron sulfur centre (see page 162 if you need to be reminded what this is); the complex transports electrons from QH$_2$ (plastoquinol) to plastocyanin to give the reduced form of the latter. Plastocyanin is a copper-protein complex in which the copper ion alternates between the Cu^{2+} (oxidized) and Cu$^+$ (reduced) forms. At this point we'll leave PSII and move on to PSI but we will return to the reduced plastocyanin very soon.

The chlorophyll at the reaction centre of PSI is P700; when activated by a photon arriving from the antenna chlorophylls it becomes a reducing agent. It passes on its electron to a short chain of electron carriers (we will not give the details of this), which reduces **ferredoxin**, a protein with an iron cluster electron acceptor; ferredoxin is a water-soluble, mobile protein residing in the chloroplast stroma (that is, outside of the thylakoid membrane). It reduces NADP$^+$ by the following reaction, catalysed by the FAD-enzyme called **ferredoxin–NADP reductase**

2 Ferredoxin$_{RED}$ + NADP$^+$ + 2H$^+$ →
    2 Ferredoxin$_{OX}$ + NADPH + H$^+$.

If we summarize what has taken place in PSI, an electron has been excited out of P700 and transported to ferredoxin which, in turn, has reduced NADP$^+$. However, this has left P700 an electron short; it is now P700$^+$, an oxidizing agent. It accepts an electron from plastocyanin (Pc), which, you remember, we left after it had been reduced by PSII. The reaction is

P700$^+$ + Pc$_{Cu^+}$ → P700 + Pc$_{Cu^{2+}}$.

To go back further, remember that we started with light exciting an electron out of P680 (Fig. 15.6), the reaction centre pigment of PSII; this leaves P680$^+$ which must have its electron restored so that it can revert to the ground state, ready for another photon to start a new round of reactions. The electron comes from water.

The water-splitting centre of PSII

P680$^+$ is a very strong oxidizing agent—it has a very strong affinity for an electron (greater than that of oxygen) so that it

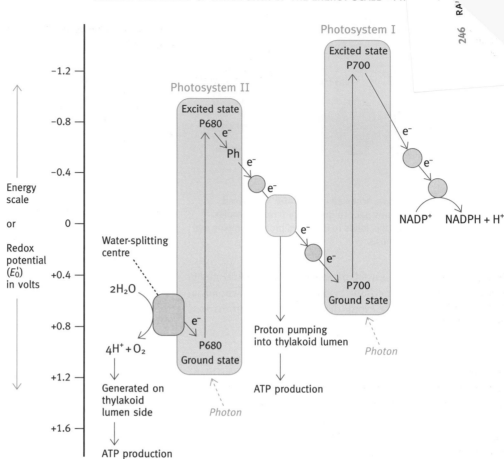

Fig. 15.6 The Z-scheme of photosynthesis. Simplified diagram of the electron flow from H_2O to $NADP^+$. In viewing the diagram, start with a photon of light activating P680 and the transfer of an electron to P700. The electron-depleted P680 then accepts an electron from water and awaits a new round of activation. P700 behaves similarly except that it accepts an electron from the photosystem II transport chain and its donated electron is transferred to $NADP^+$. Ph, pheophytin.

can even extract electrons from water. Four electrons are extracted from two molecules of H_2O with the release of O_2 and four H^+ into the thylakoid lumen. It is necessary to extract all four electrons so as not to release any intermediate oxygen free radicals, which are dangerous to biological systems, just as, in mitochondria, the addition of electrons to oxygen to form H_2O must be complete. In PSII there is a complex of proteins with Mn^{2+}, known as the **water-splitting centre**, which extracts the electrons from water, with the release of oxygen and protons, and passes them on to $P680^+$ molecules, thus restoring them to the ground state (Fig. 15.8(a)). The P680 is now ready for further rounds of reaction.

How is ATP generated?

The **cytochrome *bf* complex** of PSII, which uses QH_2 to reduce plastocyanin, resembles complex III of mitochondria (see Fig. 9.19 for the latter) in that electron transport through the complex causes translocation of protons from the outside of the thy-

lakoid membrane to the inside. By analogy with the Q cycle in mitochondria it would be expected that the cytochrome *bf* complex would translocate four protons per QH_2 oxidized (see page 166). However, experimental data has not yet confirmed this number so there is some degree of uncertainty on this point. In addition, the water-splitting centre generates protons inside the thylakoid lumen, the two effects producing a proton gradient that lowers the pH of the thylakoid lumen to about 4.5. The uptake of a proton in the reduction of $NADP^+$ by ferredoxin in the stroma further contributes to the proton gradient across the membrane.

There is a device that under certain circumstances, leads to an increased proton translocation and therefore a greater potential for ATP synthesis. When virtually all of the $NADP^+$ has been reduced by ferredoxin, the latter donates electrons to the cytochrome *bf* complex (Fig. 15.7) instead. The passage of these through the complex to plastocyanin leads to increased proton pumping by that complex. The extra ATP production is referred to as **cyclic photophosphorylation** driven by **cyclic electron flow**.

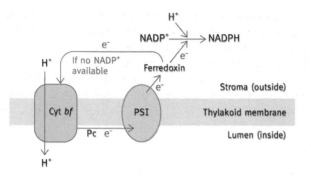

Fig. 15.7 Cyclic electron flow. When all of the NADP⁺ is reduced, ferredoxin transfers electrons back to the cytochrome *bf* complex. Flow of electrons through this leads to increased proton pumping and hence increased ATP synthesis.

The proton gradient is used to generate ATP from ADP and P_i by the chemiosmotic mechanism described for mitochondria. (The whole process of the events described so far is summarized in Fig. 15.8.)

An explanatory note

In mitochondria, the proton gradient is from the outside (high) to the inside (low). The thylakoid membrane is formed from invaginations of the inner chloroplast membrane (cf. the inner mitochondrial membrane) which explains why the proton gradients and the ATP synthases of mitochondria and thylakoid discs look as if they are in opposite orientation.

How is CO₂ converted to carbohydrate?

As already emphasized, the aspect of photosynthesis that is fundamentally different from other biochemical processes is the harnessing of light energy to split water and reduce NADP⁺. From there, although the actual metabolic pathway by which glucose and its derivatives are synthesized is unique to plants, it is nonetheless 'ordinary' enzyme biochemistry and quite secondary to the light-dependent process.

We can dispose of most of the pathway of carbohydrate synthesis very quickly because from the metabolite 3-phosphoglycerate onward it is the same as the process in the liver (Fig. 12.5) except that NADPH is used instead of NADH as the reductant (Fig. 15.9). This leaves the question of how 3-phosphoglycerate is produced in photosynthesis.

Where does the 3-phosphoglycerate come from?

The most plentiful single protein on Earth is the enzyme called 'Rubisco' for short, which stands for ribulose-1:5-bisphosphate carboxylase. It is the enzyme that utilizes CO₂ to produce 3-phosphoglycerate.

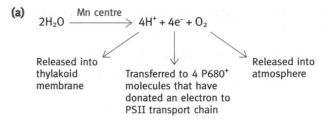

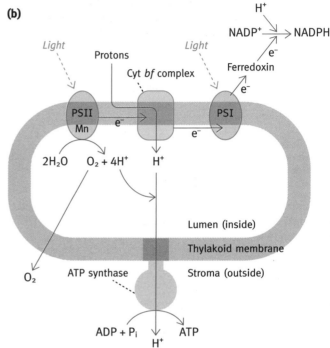

Fig. 15.8 Processes in thylakoid sacs. **(a)** Sum of reactions at the Mn water-splitting centre. **(b)** Routes followed by protons and electrons in thylakoids.

To remind you of terminology, ribose is an aldose sugar and ribulose is its ketose isomer. Ribulose-1:5-bisphosphate is cleaved by Rubisco into two molecules of 3-phosphoglycerate; this fixes one molecule of CO₂.

$$
\begin{array}{c}
\begin{array}{l}
CH_2OPO_3^{2-}\\
|\\
C{=}O\\
|\\
CHOH\\
|\\
CHOH\\
|\\
CH_2OPO_3^{2-}
\end{array}
\qquad
\xrightarrow[\substack{\text{Ribulose-1:5-}\\ \text{bisphosphate}\\ \text{carboxylase}}]{CO_2\ +\ H_2O}
\qquad
\begin{array}{l}
CH_2OPO_3^{2-}\\
|\\
CHOH\\
|\\
COO^-
\end{array}
\;+\;
\begin{array}{l}
COO^-\\
|\\
CHOH\\
|\\
CH_2OPO_3^{2-}
\end{array}
\;+\;2H^+
\end{array}
$$

Ribulose-1:5-bisphosphate   Two molecules of 3-phosphoglycerate

The 3-phosphoglycerate is converted to carbohydrate as already described. This leads to the next question.

Where does the ribulose-1:5-bisphosphate come from?

The answer to this is very simple, in principle. From six molecules of ribulose-bisphosphate (30 carbon atoms in total) and six molecules of CO_2 (six carbon atoms) we get 12 molecules of 3-phosphoglycerate (36 carbon atoms) produced. Two of these latter molecules ($2C_3$) are used to make storage carbohydrate (C_6) by the pathway already outlined in Fig. 15.9.

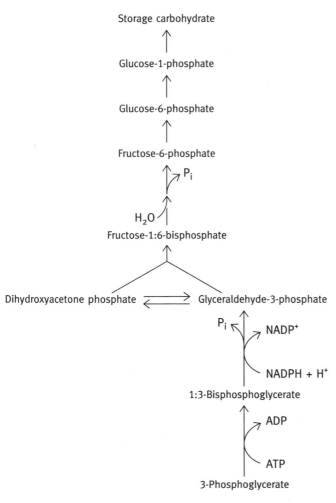

Fig. 15.9 Pathway of starch synthesis in photosynthesis, starting with 3-phosphoglycerate. The process is the same as in gluconeogenesis in liver except that NADPH is the reductant rather than NADH. In starch synthesis, the activated glucose is ADP-glucose rather than the UDP-glucose involved in glycogen synthesis. The special question in photosynthesis is the mechanism by which 3-phosphoglycerate is produced (see text for this).

The remaining 10 phosphoglycerate molecules (30 carbon atoms in total) are manipulated to produce six molecules of ribulose-bisphosphate (30 carbon atoms in total). The manipulations are complex and involve C_3, C_4, C_5, C_6, and C_7 sugars, aldolase reactions, and transfer of C_2 units by transketolase reactions (page 236) reminiscent of the pentose phosphate pathway. The reactions are rather involved and are not presented here; they are summarized below.

$C_3 + C_3 \rightarrow C_6$ Aldolase

$C_6 + C_3 \rightarrow C_4 + C_5$ Transketolase

$C_4 + C_3 \rightarrow C_7$ Aldolase

$C_7 + C_3 \rightarrow C_5 + C_5$ Transketolase

In sum,

$$5C_3 \rightarrow 3C_5.$$

The outcome of it all is that six molecules of ribulose bisphosphate, plus six molecules of CO_2 and six of H_2O, are converted to 12 molecules of 3-phosphoglycerate. From there, the six molecules of ribulose bisphosphate are regenerated, plus the dividend of one molecule of fructose-6-phosphate which is converted to the storage carbohydrate. This whole process known as the Calvin cycle is shown in Fig. 15.10. The stoichiometry of the whole business is given in the following rather lengthy equation which it is suggested need not be memorized.

$$6 \ CO_2 + 18 \ ATP + 12 \ NADPH + 12 \ H^+ + 12 \ H_2O \rightarrow$$
$$C_6H_{12}O_6 + 18 \ ADP + 18 \ P_i + 12 \ NADP^+ + 6 \ H^+.$$

Has evolution slipped up a bit?

The chemical devices in life so constantly astonish one with their elegance and efficiency that it comes almost as a relief to find what (at present, anyway) looks almost like a slip-up in the 'design' of the CO_2-fixing enzyme, 'Rubisco'.

At the dawn of photosynthesis there was no oxygen and much higher CO_2 levels than now, but, as photosynthesis occurred, oxygen accumulated. It so happens that Rubisco, in addition to using CO_2, can also react with oxygen, the two competing with one another, so that strictly the enzyme should be called ribulose-bisphosphate carboxylase/oxygenase. The oxygenation reaction, so far as is known at present, serves no useful purpose and is apparently wasteful in the sense that it destroys ribulose-1:5-bisphosphate and wastes ATP in a reaction pathway known as **photorespiration**, the details of which we are not concerned with here. In high light intensity and temperatures, CO_2, in the immediate neighbourhood of the leaf, is rapidly removed by photosynthesis and the wasteful oxygen reaction is therefore maximized to the detriment of photosyn-

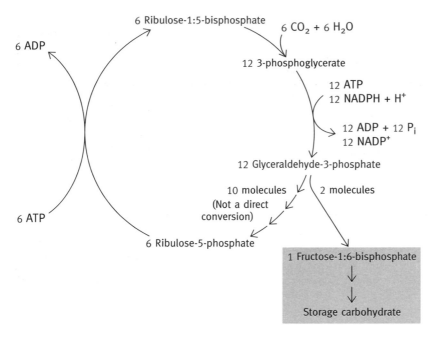

Fig. 15.10 Net effect of the reactions in the Calvin cycle. *Note.* The diagram is simplified in that the conversion of 3-phosphoglyceraldehyde to ribulose-5-phosphate involves part of it being converted to fructose-bisphosphate as an intermediate. The presentation is intended to show the net effect of all the reactions involved. Dihydroxyacetone phosphate, which is in equilibrium with glyceraldehyde-3-phosphate, is also omitted for simplicity.

thetic efficiency. This can reduce CO_2 assimilation by about 30%. Presumably, in earlier evolutionary times, when there was little oxygen and higher CO_2 levels, this would not have been significant, but, today, in the presence of high oxygen levels, the situation is different.

It might have been expected that a new Rubisco that excluded the oxygen reaction would have been evolved but this has not occurred, for reasons that are not apparent. It might be a rare evolutionary slip-up (which seems improbable in a system of such importance), or there may be good reasons for the apparent inefficiency that are not yet appreciated. However, in some plants, a biochemical device has evolved to raise the CO_2 level in cells where Rubisco operates, and thus minimize the oxygenase reaction. This occurs in species of plants that live in high light and high temperature environments where the problem of photorespiration would be maximized—plants such as maize (corn) and sugar cane.

The C_4 pathway

C_3 plants are so called because the first stable labelled product that is experimentally detectable, if they are allowed to photosynthesize in the presence of $^{14}CO_2$, is the C_3 compound, 3-phosphoglycerate, produced by the Rubisco reaction. Some plants, however, *initially* fix CO_2 from the atmosphere into oxaloacetate (Fig. 15.11). The latter is a C_4 compound, the process is referred to as C_4 **photosynthesis**, and the plants as C_4 **plants**.

The anatomical or cellular structure of C_4 plant leaves differs from that of C_3 leaves. In the former, the mesophyll cells at the surface, which are exposed to the atmospheric CO_2, do not contain the Rubisco enzyme but do fix CO_2 very efficiently into oxaloacetate by the carboxylation of phosphoenolpyruvate (PEP) by the enzyme PEP carboxylase (which is not found in animals)

$$CO_2 + H_2O + PEP \rightarrow \text{Oxaloacetate} + P_i.$$

PEP carboxylase has an effectively higher affinity for CO_2, and there is no competition from oxygen. The oxaloacetate is reduced to malate which is transported into neighbouring bundle sheath cells where photosynthesis occurs. Here the CO_2 is released from malate by the 'malic enzyme' (which you have met before in fatty acid synthesis, page 191).

$$\text{Malate} + NADP^+ \rightarrow \text{Pyruvate} + CO_2 + NADPH + H^+.$$

This raises the concentration of CO_2 in the bundle sheath cells 10–60-fold, resulting in a more efficient operation of the Rubisco reaction. The pyruvate returns to the mesophyll cell where it is reconverted to PEP by **pyruvate phosphate dikinase**. This is an unusual reaction in which two —Ⓟ groups of ATP are liberated.

$$CH_3-CO-COO^- \quad + \; ATP \; + \; P_i$$

$$\downarrow$$

$$CH_2{=}C-COO^- \quad + \; AMP \; + \; PP_i \; + \; H^+$$
$$\quad\; |$$
$$\quad OPO_3^{2-}$$

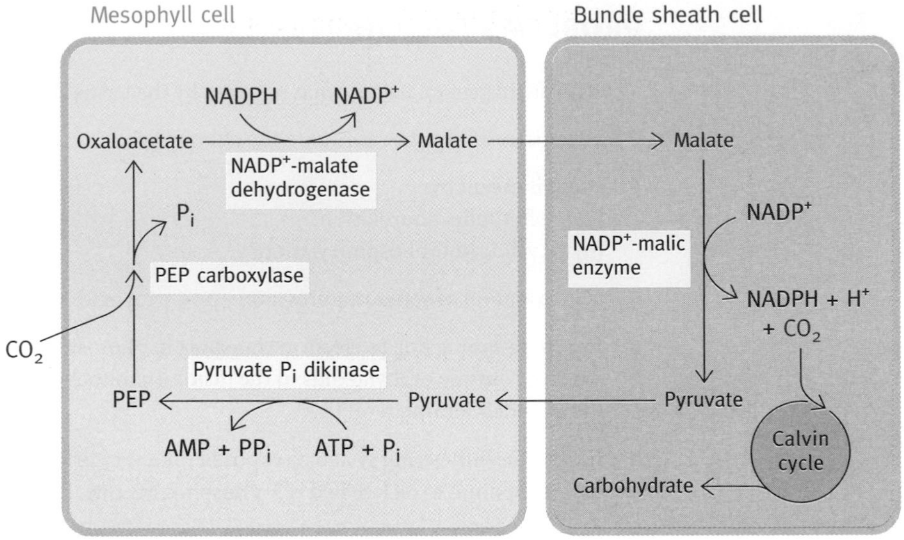

Fig. 15.11 The C_4 pathway for raising the CO_2 concentration for photosynthesis in bundle sheath cells. *Important note.* The pathway of oxaloacetate formation from pyruvate is quite different from that in animals. Direct conversion of pyruvate to phosphoenolpyruvate (PEP) does not occur in animals nor does the carboxylation of PEP. It is important to be clear on this for it has a major effect on metabolic regulation in animals. Note also that the malate dehydrogenase that reduces oxaloacetate uses NADPH, unlike that in the citric acid cycle that is NAD^+-specific.

In animals, PEP can be made from pyruvate only via oxaloacetate by a quite different route (see page 202). Once phosphoglycerate is made in the bundle sheath cells, the Calvin cycle operates exactly as in C_3 plants.

The C_4 route incurs a price for raising the CO_2 concentration in bundle sheath cells, since ATP is consumed in making PEP and transporting acids. However, at higher temperatures, the C_4 route becomes a considerable advantage. C_4 plants such as corn or sugar cane grown in areas of high temperatures and light intensities are prolific producers of carbohydrates.

There is great diversity in the biochemical strategies used by different C_4 plants. For example, the C_4 acid labelled in the presence of $^{14}CO_2$ may be aspartate in some species, and not malate. These plants have high levels of aspartate aminotransferase (Chapter 16) instead of $NADP^+$ malic dehydrogenase. C_4 plants have also evolved three distinct options for decarboxylating C_4 acids in bundle sheath cells, two being located in the mitochondria unlike the $NADP^+$-malic enzyme shown in Fig. 15.11, which is located in the cytoplasm. They are all mechanisms to achieve the same end—increase of CO_2 levels where Rubisco operates.

FURTHER READING

The light-harvesting pathway

Slater, E. C. (1983). The Q cycle, an ubiquitous mechanism of electron transport. *Trends Biochem. Sci.*, **8**, 239–42.

Describes the proton pumping system present in both mitochondria and chloroplast.

The C_1 pathway

Rawsthorne, S. (1992). Towards an understanding of C_3–C_4 photosynthesis. *Essays in Biochem.*, **27**, 135–46.

Discusses the distribution of the biochemical systems among plant species as well as the pathways themselves.

PROBLEMS FOR CHAPTER 15

1   Explain, in general terms, what is meant by the terms 'light' and 'dark' reactions in photosynthesis.

2   What is meant by the term 'antenna chlorophyll'?

3   What is meant by:
   (a)   photophosphorylation?
   (b)   cyclic photophosphorylation?

4   The oxidation of water requires a very powerful oxidizing agent. In photosynthesis what is this agent?

5   Proton pumping due to electron transport in photosystem II (PSII) causes movement of protons from the outside of thylakoids to the inside. In mitochondria protons are pumped from the inside to the outside. Comment on this.

6   If a photosynthesizing system is exposed for a very brief period to radioactive CO_2, in C_3 plants the first compound to be labelled is 3-phosphoglycerate. Explain why this is so.

7   Explain the Calvin cycle in simplified terms.

8   The enzyme Rubisco can react with oxygen as well as CO_2, the oxygen reaction being, so far as we know, an entirely wasteful one. At low CO_2 concentrations such as can occur particularly in intense sunlight, the wasteful oxygenation reaction is maximized. What mechanisms have evolved to ameliorate this problem?

9   In C_4 plants, pyruvate is converted to phosphoenolpyruvate by an ATP-requiring reaction. Have you any comments on this, bearing in mind the corresponding process in animals?

Chapter summary

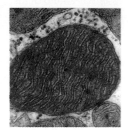

Amino acid metabolism

Digestion of proteins in a normal diet results in relatively large amounts of the 20 different types of amino acids being absorbed from the intestine into the portal blood, which immediately traverses the liver. All cells, except terminally differentiated ones such as erythrocytes, use amino acids for protein synthesis and for the synthesis of a variety of essential molecules such as membrane components, neurotransmitters, heme, and the like. All cells, therefore, take up amino acids by selective transport mechanisms since, in their free, ionized form, they do not readily penetrate the membrane lipid bilayer.

As already stated, in the body there is no dedicated storage of amino acids in the sense that there is no polymeric form of amino acids whose function is simply to be a reserve of these compounds to be called upon when needed. The only 'reserves' are in the form of functional proteins and, since the biggest mass is in muscle proteins, the latter are the main reserves. But, these muscle proteins are part of the contractile machinery and, if broken down, for instance, to provide amino acids for gluconeogenesis in the liver, muscle wasting ensues (see page 202).

This biochemical situation, again as pointed out earlier, can in certain human populations today be an unfortunate one because, during evolution, humans have lost the ability to synthesize 10 of the amino acids. Interestingly enough, the ones that we cannot synthesize are those with many steps in the synthesis and therefore require many enzymes and many genes to code for them—they are, in essence 'expensive' to manufacture.

Why this evolutionary loss occurred is an interesting question that can only be answered in speculative terms. One possibility is that it was simply a measure of economy—that is was more advantageous to 'obtain' 10 of the more complicated amino acids ready-made in the diet and to make only the simpler ones. When the human diet was from the 'wild', if sufficient food could be obtained to sustain life, energy-wise, it would probably contain all the necessary amino acids. In this situation, loss of the ability to synthesize certain amino acids would be pure gain, since you would have been more likely to die from insufficient energy sources than from lack of essential amino acids in the food that was available.

With the development of agriculture, however, vast production of chemical energy in the form of the carbohydrate of cereals permitted large populations to survive but did not necessarily supply adequate amounts of the essential amino acids, for some plant proteins, such as those in corn, lack lysine and tryptophan. What exacerbates the situation is that, when humans have a rich source of protein, containing enough essential amino acids to provide for an extended period, there is no method for storing them. After immediate needs are satisfied, the surplus amino acids are simply oxidized or converted to glycogen or fat. In some circumstances, it is somewhat like burning diamonds to keep warm.

The result of a diet inadequate in even a single amino acid is the condition known as **kwashiorkor**. Since virtually all human proteins contain all 20 amino acids, if a single one of the latter is deficient in amount, the necessary functional proteins cannot be made. This leads to wasting, apathy, inadequate growth, and lowered levels of serum proteins which, by reducing the osmotic pressure of the blood, are probably one cause of edema of the tissues. The condition is a vicious circle since the cells lining the intestine are constantly renewed and, in kwashiorkor, this may be inadequate, as might also apply to the production of digestive enzymes. The result is that what food is available may be inadequately digested and absorbed. The disease affects developing children more than adults because of their greater demand for protein to support growth. This evolutionary loss

of the ability to synthesize certain amino acids and the lack of provision for their storage have turned out to be, apparently, considerable human disadvantages. The conclusion is speculative, however, for it may be that unknown compelling reasons dictated the evolutionary loss of synthetic capability. One suggestion has been made that the intermediates in the synthesis of essential amino acids may have been toxic to higher organisms (brain function impairment has been suggested), but there is little evidence on the matter.

Nitrogen balance of the body

The nutritive aspects of amino acids can be treated on a generalized level through the concept of **nitrogen balance**. If the total intake of nitrogen (mainly as amino acids) equals the total excretion, the individual is in a state of nitrogen balance. During growth or repair, more is taken in than is excreted—this is positive nitrogen balance. Negative nitrogen balance occurs in wasting, where excretion exceeds intake, when, for example, amino acids of muscle proteins are converted to glucose and the nitrogen excreted. The proteins of animals are continually 'turning over'; they are continuously broken down and resynthesized. Although much of the derived amino acids is recycled back into proteins, something of the order of about 0.3% of total body protein nitrogen per day is converted to urea and excreted.

An essential amino acid is one that, when omitted from an otherwise complete diet, results in negative nitrogen balance or fails to support the growth of experimental animals. Some amino acids are essential without qualification such as lysine, phenylalanine, and tryptophan (see Table 16.1 for a list of essential and nonessential amino acids; we suggest that you do not need to memorize this table). Tyrosine is not essential provided

Table 16.1 Classification of dietary amino acids in humans

Nonessential	Essential
Alanine	Arginine‡
Asparagine	Histidine
Aspartic acid	Isoleucine
Cysteine*	Leucine
Glutamic acid	Lysine
Glutamine	Methionine
Glycine	Phenylalanine
Proline	Threonine
Serine	Tryptophan
Tyrosine†	Valine

* Cysteine is produced only from the essential amino acid methionine.
† Tyrosine is produced only from the essential amino acid phenylalanine.
‡ Arginine is required only in the growing stages.

sufficient phenylalanine is available since the latter is convertible to tyrosine; similarly, cysteine synthesis in mammals requires the availability of methionine, another essential amino acid. The nutritive picture, in this respect, is not completely neat and tidy. 'First-class' proteins are rich in essential amino acids. Plant proteins may be poor in certain essential amino acids. Since the amino acid compositions of proteins vary, a mixture of plant proteins is needed to ensure adequate amino acid nutrition in vegetarian diets.

General metabolism of amino acids

The general situation of amino acid metabolism is shown in Fig. 16.1. It is essentially a repeat of that given in Chapter 6 on the broad aspects of metabolism (page 111).

Aspects of amino acid metabolism

In considering how amino acids in excess of the immediate requirements for protein synthesis and other specific needs are broken down for energy production or fuel storage, there are a number of separate questions. We need also to consider how the nonessential amino acids are synthesized and how certain amino acids are metabolized to some other small molecules of

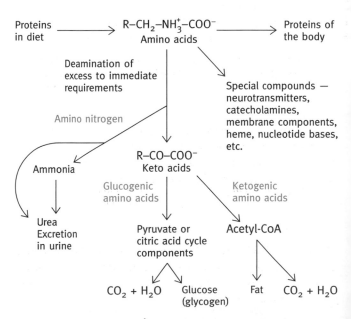

Fig. 16.1 The overall catabolism of amino acids. Note that some amino acids are partly ketogenic and partly glucogenic (see page 257 for explanation of these terms).

physiological importance. The questions we deal with in this chapter are the following:

- How are the amino groups of amino acids removed?—in other words, how does deamination occur? Although there are 20 different amino acids most of them are deaminated by a common mechanism and this is relatively easy to deal with.

- What happens to the keto acids—the carbon–hydrogen 'skeletons'—of the amino acids after deamination? In this case, each amino acid has its own special metabolic route and we will give only a limited treatment of this aspect, mentioning features of special interest or of broader relevance.

- How is the amino group nitrogen, which is removed from amino acids, converted to urea? This is an important area and will be dealt with.

- How are amino acids synthesized? In the animal body this concerns only the nonessential amino acids. If the carbon skeleton (the keto acid) is available, synthesis is often the reverse of deamination. We will deal with only a few examples of these. In bacteria and plants, all amino acids are synthesized by individual pathways deriving their staring products from general metabolic intermediates. Animals depend on this synthetic activity to supply their essential amino acids. The pathways for the synthesis of all amino acids constitute a formidable amount of detailed information which probably few would carry in their heads without a specific need to do so. We will deal only with particular cases of amino acid biosynthesis of special interest.

- Not a question, but there are special aspects of amino acid metabolism. Certain amino acids have metabolic reactions of special interest and more general significance which we collect into this section.

There are also major topics of biochemistry, such as protein synthesis, heme synthesis, and nucleotide synthesis, that have amino acids as their reactants. These are major topics and are dealt with in the relevant later chapters. After this rather long preliminary, let us now start.

Deamination of amino acids

First, a piece of simple chemistry. It concerns **Schiff bases**. A Schiff base results from a spontaneous equilibrium between a molecule with a carbonyl group (aldehyde or ketone) and one with a free amino group.

$$\overset{|}{\underset{|}{C}}{=}O + H_2N{-}R \rightleftharpoons \overset{|}{\underset{|}{C}}{=}N{-}R + H_2O$$

The reaction is freely reversible so a Schiff base can readily hydrolyse. If R$=$H in the above equation, the Schiff base will hydrolyse to give NH_3. The importance of this is that it shows how the removal of a pair of hydrogen atoms from an amino acid can result in **deamination**, and the supply of a pair of hydrogen atoms can result in the synthesis of an amino acid from a keto acid and ammonia (nonionized structures are used below for clarity).

$$\overset{|}{\underset{|}{C}}HNH_2 \xrightarrow{-2H} \overset{|}{\underset{|}{C}}{=}N{-}H \xrightarrow{H_2O} \overset{|}{\underset{|}{C}}{=}O + NH_3$$

Now back to biological deamination. Glutamic acid is an amino acid of central importance in metabolism. It is deaminated by **glutamate dehydrogenase**, unusually working with either NAD^+ or $NADP^+$. (The NH_3 is protonated to NH_4^+ at physiological pH.)

$$\begin{array}{c} COO^- \\ | \\ CH_2 \\ | \\ CH_2 \\ | \\ CHNH_3^+ \\ | \\ COO^- \end{array} + NAD(P)^+ + H_2O \longrightarrow NAD(P)H + H^+ + \begin{array}{c} COO^- \\ | \\ CH_2 \\ | \\ CH_2 \\ | \\ C{=}O \\ | \\ COO^- \end{array} + NH_4^+$$

Glutamate   α-Ketoglutarate

As is described below, glutamate dehydrogenase plays a major role in the deamination, and therefore in the oxidation of many amino acids. In keeping with this, the enzyme is allosterically inhibited by ATP and GTP (indicators of a high energy charge) and activated by ADP and GDP which signal that an increased rate of oxidative phosphorylation is needed.

The α-ketoglutarate can feed into the citric acid cycle, and that disposes of glutamic acid oxidation to CO_2 and H_2O. Since α-ketoglutarate is converted to oxaloacetate in the cycle (Fig. 9.12), glutamic acid can give rise to glucose synthesis in appropriate physiological situations (page 112). It is a glucogenic amino acid. However, there are no corresponding dehydrogenases for the other amino acids. How are they deaminated? A few have special individual mechanisms (see below) but most of them transfer their amino groups to α-ketoglutarate forming glutamate and the corresponding keto acid. The former is then deaminated by the reaction given above. The amino acids are thus deaminated by a two-step process.

The first reaction is called **transamination**.

$$\begin{array}{c} R \\ | \\ CHNH_3^+ \\ | \\ COO^- \end{array} + \begin{array}{c} COO^- \\ | \\ CH_2 \\ | \\ CH_2 \\ | \\ C{=}O \\ | \\ COO^- \end{array} \longrightarrow \begin{array}{c} R \\ | \\ C{=}O \\ | \\ COO^- \end{array} + \begin{array}{c} COO^- \\ | \\ CH_2 \\ | \\ CH_2 \\ | \\ CHNH_3^+ \\ | \\ COO^- \end{array}$$

Let us suppose R=CH$_3$; the amino acid is then alanine. Deamination of alanine proceeds as follows.

1. Alanine + α-ketoglutarate $\rightarrow$ Pyruvate + glutamate
2. Glutamate + NAD$^+$ + H$_2$O $\rightarrow$ α-ketoglutarate
   + NADH + NH$_4^+$

Net reaction: Alanine + NAD$^+$ + H$_2$O

Pyruvate + NADH + NH$_4^+$

The two-step process involving transamination and then deamination of glutamate is called **transdeamination** for obvious reasons. The enzymes involved in reaction 1 above are called **transaminases** or **aminotransferases**. A number of these exist with specific substrate specificities and most amino acids can be deaminated by this route.

The reversibility of transamination means that, provided a keto acid is available, the corresponding amino acid can be synthesized. (The keto acids for essential amino acids are not, however, synthesized in the body.) An example of the importance of this is in the malate–aspartate shuttle (page 154) in which the following reversible reaction occurs.

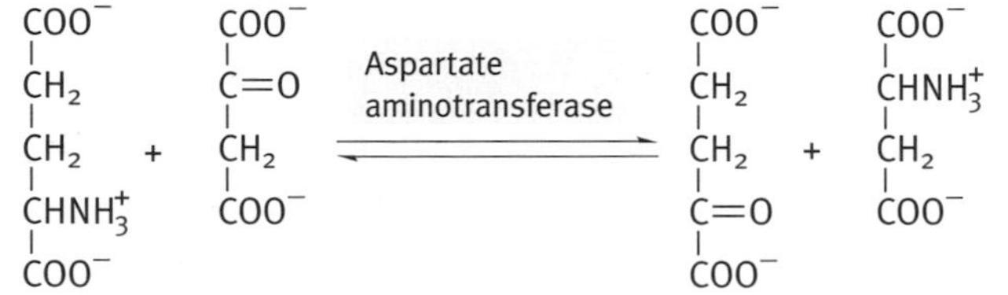

Glutamate Oxaloacetate α-Ketoglutarate Aspartate

Mechanism of transamination reactions

All transaminases have, tightly bound to the active centre of the enzyme, a cofactor, **pyridoxal-5'-phosphate (PLP)** that participates in the transaminase reaction.

Pyridoxal phosphate is a remarkably versatile cofactor, it participates as an electrophilic agent in a wide variety of reactions involving amino acids. In simple terms, PLP acts as an intermediary, accepting the amino group from the donor amino acid and then handing it on to the keto acid acceptor, both phases occurring on the same enzyme. As so often is the case, the cofactor is a B vitamin derivative. Vitamin B$_6$ in the diet consists of three interrelated compounds, pyridoxin, pyridoxal, and pyridoxamine (Fig. 16.2). They can all be converted to pyridoxal phosphate (Fig. 16.2) in the cell. The business end of the molecule is the —CHO group. The general reaction catalysed in transamination is

ENZ—PLP + amino acid 1 $\rightleftharpoons$ ENZ—PLP—NH$_2$ + keto acid 1
ENZ—PLP—NH$_2$ + keto acid 2 $\rightleftharpoons$ ENZ—PLP + amino acid 2

where PLP represents pyridoxal phosphate and PLP—NH$_2$, pyridoxamine phosphate. The mechanism of the reactions is as shown in Fig. 16.3. Both parts of the reactions occur at the active site of the transaminase, the pyridoxamine phosphate remaining attached to the enzyme.

In addition to the general transdeamination reactions, certain amino acids have their own particular way of losing their amino groups.

Special deamination mechanisms

Serine is a hydroxy amino acid; cysteine is the corresponding thiol amino acid. Serine can be deaminated by a **dehydratase reaction** (dehydratases have been described in several places—glycolysis, the citric acid cycle, and fat metabolism). Cysteine can be deaminated by a somewhat analogous reaction in which H$_2$S is removed instead of H$_2$O. In both cases the resultant keto acid is pyruvate (Fig. 16.4).

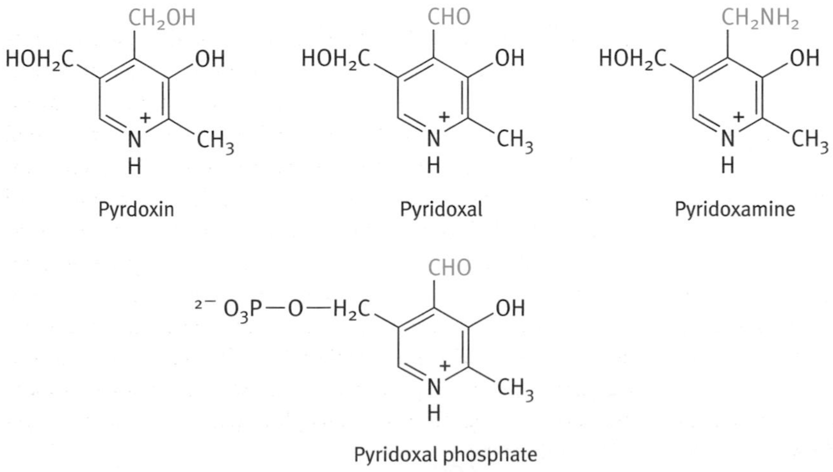

Pyrdoxin   Pyridoxal   Pyridoxamine

Pyridoxal phosphate

Fig. 16.2 Structures of vitamin B$_6$ components and of the transaminase cofactor, pyridoxal phosphate.

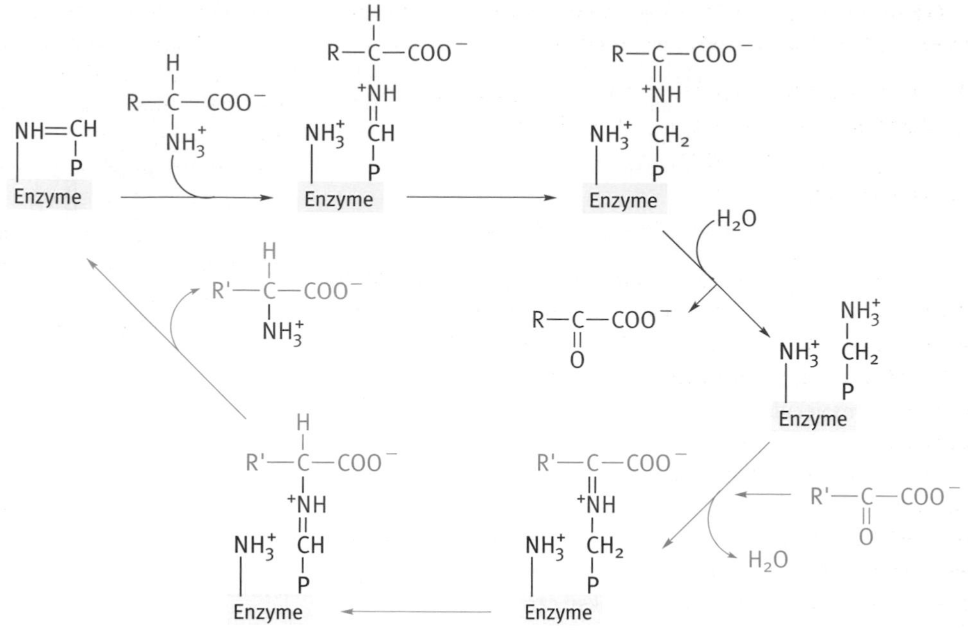

Fig. 16.3 Simplified diagram of the mechanism of transamination. $P-CH=NH-$ in the first figure represents pyridoxal phosphate complexed with a lysine-amino group of the protein. $P-CH_2-NH_3^+$ represents pyridoxamine phosphate. The forward reaction (red arrows) results in the conversion of an amino acid to a keto acid and of pyridoxal phosphate to pyridoxamine phosphate. The reverse reaction (blue arrows) reacting with a different keto acid results in transamination between the (red) amino acid and the (blue) keto acid.

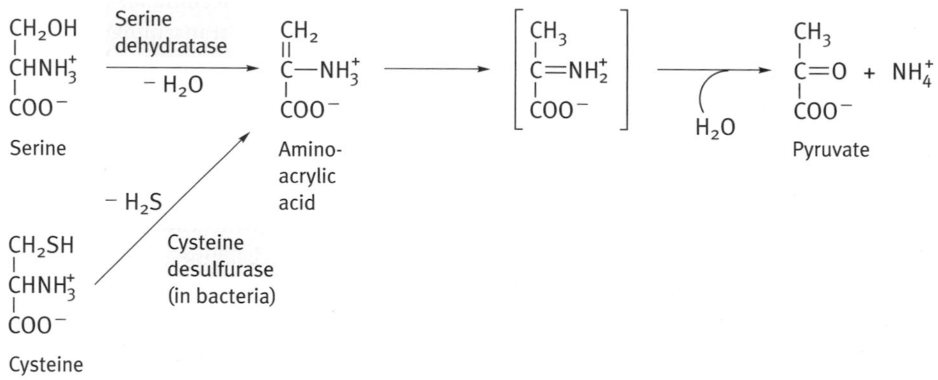

Fig. 16.4 Conversion of serine and cysteine to pyruvate.

Fate of the keto acid or carbon skeletons of deaminated amino acids

As far as general metabolism is concerned, some amino acids are glucogenic and some are ketogenic. The term, 'ketogenic', might seem to imply that an amino acid giving rise to acetyl-CoA results in formation of ketone bodies in the blood, which would conflict with the earlier explanation (page 182) that ketone bodies arise only in conditions of excess fat metabolism. Acetyl-CoA in normal metabolic situations does *not* give rise to ketone bodies.

The universally used term **ketogenic** for certain amino acids is an old one, resulting from the use of starving animals as a test

system for the metabolic fate of amino acids since, in such animals, any increase in acetyl-CoA production causes increases in blood ketone bodies that are easily measured. It does not mean that ketogenic amino acids produce these exclusively in normal animals, but simply that they *can* give rise to acetyl-CoA but not to pyruvate. **Glucogenic** amino acids were detected by increased levels of blood glucose or glucose excretion when administered to diabetic animals, and similar qualifications apply to this term.

Aspartate, like glutamic acid, is converted to a metabolite of the citric acid cycle (oxaloacetate) by the transamination reaction already described and is therefore glucogenic. Alanine and glutamate, producing pyruvate and α-ketoglutarate, respectively, on deamination, are also glucogenic as are serine, cysteine, and others.

Some amino acids are both ketogenic and glucogenic—for example, phenylalanine produces fumarate (an intermediate in the citric acid cycle) and acetyl-CoA. Of the 20 amino acids only two (leucine and lysine) are solely ketogenic. The degradation of four amino acids (isoleucine, methionine, threonine, and valine) has been referred to on page 183.

Phenylalanine metabolism has special interest

Phenylalanine is an aromatic amino acid, an excess of which is normally converted to tyrosine by an enzyme, **phenylalanine hydroxylase** (Fig. 16.5). This enzyme is interesting in that a pair of hydrogen atoms are supplied by an electron donor—a coenzyme molecule called tetrahydrobiopterin (RH_4 in Fig. 16.5).

It may seem odd that a reaction requires *both* oxygen and a reducing agent but it is a mechanism used in other reactions also as you'll see later. The trick is that one *atom* of oxygen is used to form an —OH group on the aromatic ring, but this leaves the other oxygen atom to be taken care of. It is reduced to H_2O by the two hydrogen atoms donated by the tetrahydrobiopterin. A separate enzyme system reduces the dihydrobiopterin formed back to the tetrahydro form, using NADH as reductant, and so the cofactor acts catalytically.

The phenylalanine hydroxlase belongs to a class of enzymes known as **monooxygenases** (because one atom of O appears in the product) or, alternatively, mixed function oxygenases because two things are oxygenated—the amino acid and a pair of hydrogen atoms. (See page 135 for a discussion of the difference between an **oxidation** that removes electrons and an **oxygenation** that adds oxygen.)

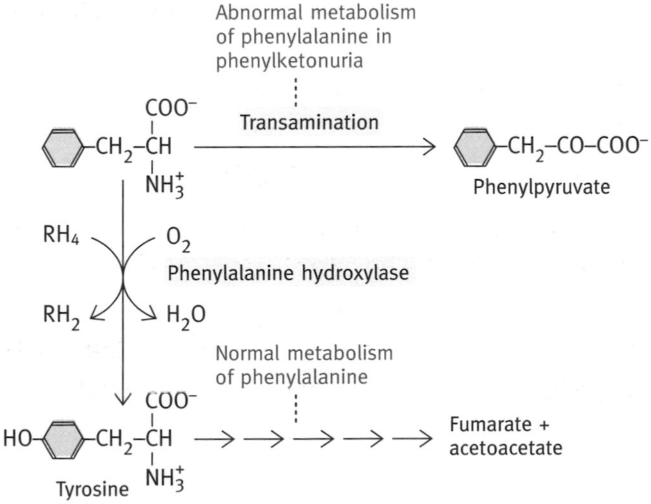

Phenylalanine as such is not normally deaminated, being converted to tyrosine, and only then does deamination occur (Fig. 16.5). However, there is a relatively common genetic abnormality in which the phenylalanine conversion to tyrosine is impaired or blocked due to enzyme deficiency or, rarely, to lack of tetrahydrobiopterin. This causes excess phenylalanine to accumulate and, in this situation, it abnormally participates in transamination producing phenylpyruvate (an abnormal metabolite), which spills out in the urine (Fig. 16.5). The disease is called **phenylketonuria** or **PKU**. The consequences of phenylpyruvate in babies are irreparable mental imparment and early death. If diagnosed at birth (by urine analysis, or preferably, blood analysis, for phenylpyruvate) a child with the disease can be given a diet limited in phenylalanine (but adequate in tyrosine) and development is normal.

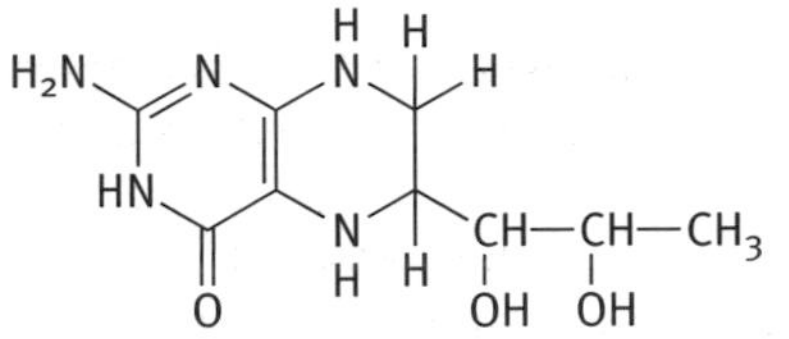

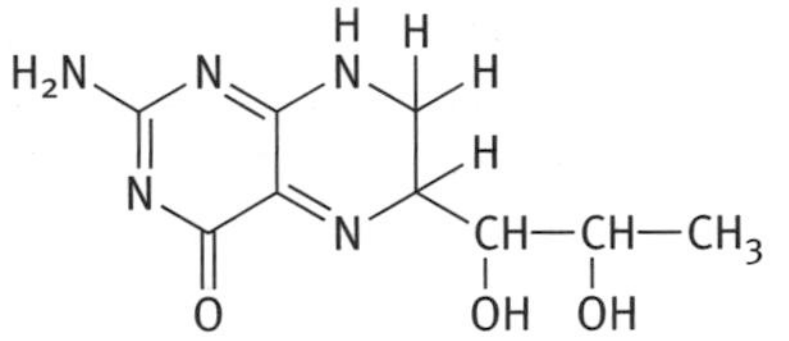

It is not known how phenylpyruvate causes such deleterious brain damage. Curiously enough, another genetic condition, called maple syrup disease, involves accumulation of the keto acids of three aliphatic amino acids, valine, isoleucine, and leucine (whose structures are given on page 36), and this also involves brain impairment. (The name of the disease comes from the keto acids, in the urine, having a characteristic smell.) This disease is much rarer than PKU. Another much-quoted genetic condition is alcaptonuria in which the urine turns black on exposure to air, but this is a benign condition. It is due to a block in the tyrosine degradation pathway in which a diphenol intermediate metabolite, homogentisate, is excreted. The diphenol in air oxidizes to form a dark pigment.

Methionine and transfer of methyl groups

Methionine is one of the essential amino acids. It has the structure

Fig. 16.5 Normal and abnormal metabolism of phenylalanine. RH_4, tetrahydrobiopterin; RH_2, dihydrobiopterin. Structures are given below for reference purposes—have a look at these.

$$CH_3-S-CH_2-CH_2-\underset{\underset{NH_3^+}{|}}{CH}-COO^- \ .$$

The interesting part is the **methyl group**. Methyl groups are very important in the cell—a variety of compounds are methylated, and methionine is the source of methyl groups that are transferred to other compounds. Methionine is a stable molecule—the methyl group has no tendency to leave; however, if the molecule is converted into **S-adenosylmethionine (SAM)**, a sulfonium ion is created and the methyl group is 'activated'—it has a strong group transfer potential making it thermodynamically favourable for it to be transferred (by transmethylase enzymes) to other compounds. ATP supplies the energy for SAM synthesis—in this case three —Ⓟ groups are converted to $PP_i +$ P_i and then the PP_i is cleaved to two P_i (Fig. 16.6).

Transfer of the methyl group of SAM to other compounds generates S-adenosylhomocysteine. The latter is hydrolysed to produce homocysteine—this is methionine with —SH instead of S—CH_3. A reaction sequence exists for transferring the thiol group to serine, producing cysteine (Fig. 16.6). The intermediate compound, cystathionine—a complex between the two amino acids—is mentioned here because a defect in its hydrolysis, to form cysteine, results in the disease cystathionuria.

What are the methyl groups transferred to?

In the body the methyl group of creatine (page 524), phosphatidylcholine (page 71), and epinephrine (page 261) come from S-adenosylmethionine and, in addition so do the methyl groups attached to the bases of nucleic acids. Note, however, that the latter do not include that of thymine which is separately synthesized and is an important topic to be dealt with later (page 297) as also is the subject of nucleic acid base methylation.

Synthesis of amino acids

In the body, as explained, only the nonessential amino acids can be synthesized. We will not include the details of all of these, for the chemistry is quite extensive and much of it relevant only to itself. If needed, pathway details are readily available elsewhere. Our aim here is to deal only with aspects of special interest and to illustrate how amino acids are synthesized from glycolytic and citric acid cycle intermediates. In fact, five of these intermediates (3-phosphoglycerate, phosphoenolpyruvate, pyruvate, oxaloacetate, and α-ketoglutarate) together with two sugars of the pentose phosphate pathway are the precursors of all 20 amino acids in those organisms, such as plants and bacteria, that synthesize all of them.

Synthesis of glutamic acid

You will recall that deamination of this amino acid occurs via glutamate dehydrogenase, an NAD^+ or $NADP^+$ enzyme of central importance. This is reversible, but probably is less important in glutamate formation in animals than is transamination of α-ketoglutarate using other amino acids such as alanine or aspartate as the donor of the amino group. See the reaction for aspartate aminotransferase on page 256. A different energy-requiring route of α-ketoglutarate amination is also used in prokaryotes in situations where the NH_4^+ concentration is very low.

Synthesis of aspartic acid and alanine

These amino acids come from transamination of oxaloacetate

$$\left(R = \begin{array}{c} COO^- \\ | \\ CH_2 \\ | \end{array} \right)$$

and pyruvate ($R=CH_3$) respectively.

$$\underset{\underset{COO^-}{|}}{\overset{\overset{R}{|}}{C}}=O + glutamate \longrightarrow \underset{\underset{COO^-}{|}}{\overset{\overset{R}{|}}{C}}HNH_3^+ + \alpha\text{-ketoglutarate}$$

Synthesis of serine

This is formed from a glycolytic intermediate, 3-phosphoglycerate, which is first converted to a keto acid, 3-phosphohydroxypyruvate.

$$\begin{array}{c} COO^- \\ | \\ CHOH \\ | \\ CH_2OPO_3^{2-} \end{array} \underset{\longrightarrow}{\overset{NAD^+ \quad NADH}{\overset{\curvearrowright}{\qquad}}} \begin{array}{c} COO^- \\ | \\ C=O \\ | \\ CH_2OPO_3^{2-} \end{array}$$

This keto acid is transaminated by glutamic acid to give 3-phosphoserine which is hydrolysed to serine and P_i.

$$\begin{array}{c} COO^- \\ | \\ C=O \\ | \\ CH_2OPO_3^{2-} \end{array} \xrightarrow{Transamination} \begin{array}{c} COO^- \\ | \\ CHNH_3^+ \\ | \\ CH_2OPO_3^{2-} \end{array} \xrightarrow{Hydrolysis} \begin{array}{c} COO^- \\ | \\ CHNH_3^+ \\ | \\ CH_2OH \end{array} + P_i$$

Synthesis of glycine

Glycine is the simplest amino acid of all ($CH_2NH_3^+COO^-$). It is formed by a reaction that is completely new, so far as this book is concerned, involving withdrawal of a hydroxymethyl group

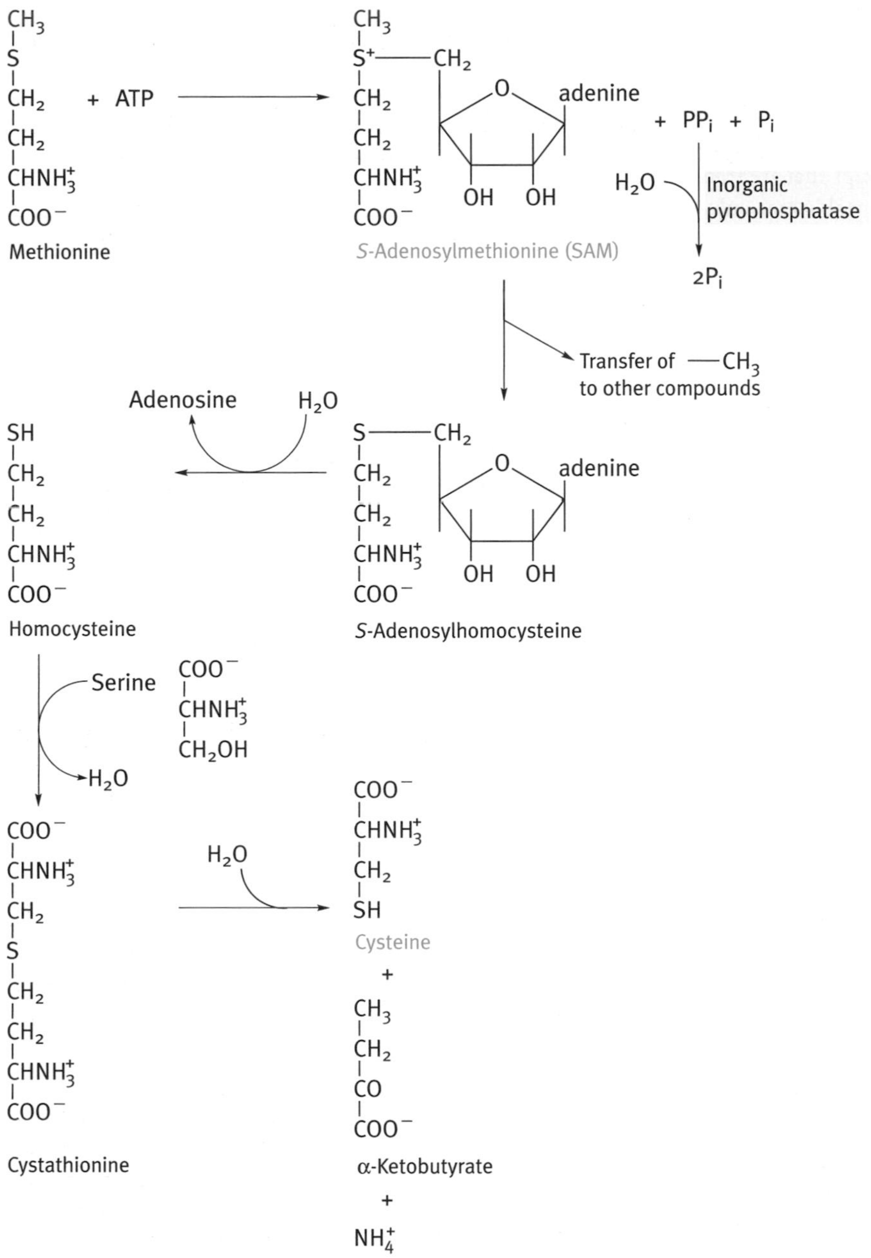

CH_3
|
S
|
CH_2 + ATP ⟶
|
CH_2
|
$CHNH_3^+$
|
COO^-

Methionine

CH_3
|
S^+——CH_2
| O adenine
CH_2
|
CH_2
| OH OH
$CHNH_3^+$
|
COO^-

S-Adenosylmethionine (SAM)

+ PP_i + P_i

H_2O ⟍ Inorganic pyrophosphatase

2P_i

Transfer of ——CH_3 to other compounds

SH   Adenosine H_2O
|
CH_2
|
CH_2
|
$CHNH_3^+$
|
COO^-

Homocysteine

S——CH_2
| O adenine
CH_2
|
CH_2
| OH OH
$CHNH_3^+$
|
COO^-

S-Adenosylhomocysteine

Serine
COO^-
|
$CHNH_3^+$
|
CH_2OH

H_2O

COO^-
|
$CHNH_3^+$
|
CH_2
|
S
|
CH_2
|
CH_2
|
$CHNH_3^+$
|
COO^-

Cystathionine

H_2O ⟶

COO^-
|
$CHNH_3^+$
|
CH_2
|
SH

Cysteine

+

CH_3
|
CH_2
|
CO
|
COO^-

α-Ketobutyrate

+

NH_4^+

Fig. 16.6 The synthesis of S-adenosylmethionine (SAM) from methionine and the synthesis of cysteine from homocysteine.

(—CH_2OH) from serine and adding it to a coenzyme, tetrahydrofolate (again, not yet metioned in this book), whose function is to act as a one-carbon unit carrier. The one-carbon unit transfer area is of importance in nucleotide synthesis. It will be more appropriate to go into this more thoroughly later (page 290).

Synthesis of other molecules from amino acids

There is a whole range of physiologically active small molecules made from amino acids. **Amines** are produced by decarboxylation of amino acids.

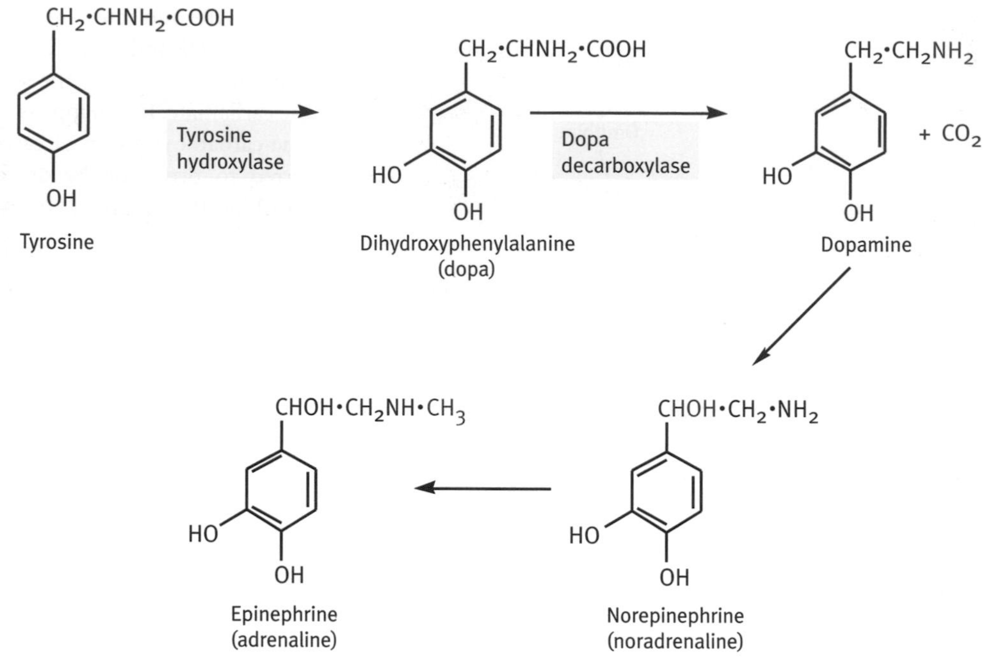

Fig. 16.7 The synthesis and structures of the catecholamines.

$$RCH_2NH_3^+COO^- \rightarrow RCH_2NH_3^+ + CO_2$$

Catecholamines includes the hormones dopamine, epinephrine and norepinephrine, whose synthesis is given in Fig. 16.7. The collective term derives from their structural relationship to catechol.

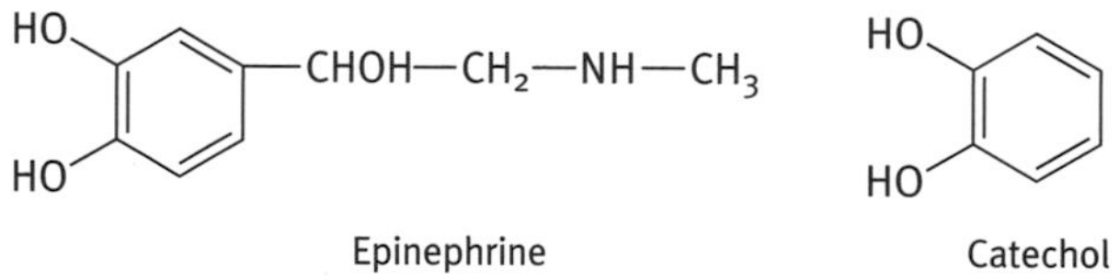

Several neurotransmitters are derived from amino acids. They include γ-aminobutyrate (GABA) and 5-hydroxytryptamine as well as the catecholamines. The hormone, thyroxine (see page 427), is derived from tyrosine. This list is not exhaustive but is illustrative of the general principle that amino acids are the precursors of many compounds in the body.

What happens to the amino groups when they are removed from amino acids?—the urea cycle

The amino groups of catabolized amino acids are excreted in mammals as urea, which is a highly water-soluble, inert, and nontoxic molecule. It is produced in the liver from the guani-

dino group of arginine by the hydrolytic enzyme arginase. The other product is ornithine, an amino acid not found in proteins.

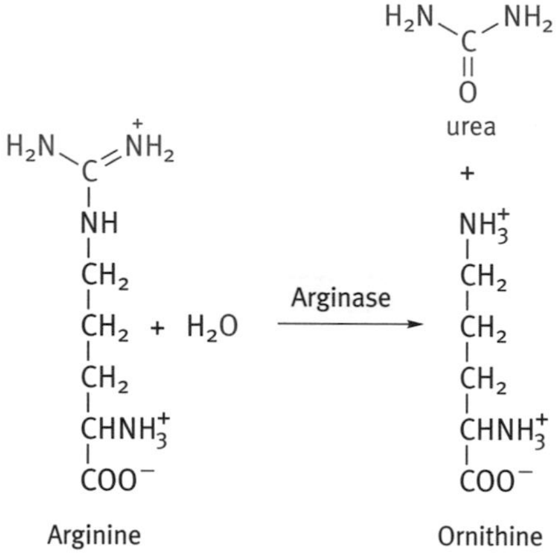

The amino nitrogen of the catabolized 20 amino acids is used to convert ornithine back to arginine. The extra carbon comes from CO_2. This forms a metabolic cycle (Fig. 16.8), the first ever discovered. Krebs (who also discovered the citric acid cycle) together with Henseleit observed that, when arginine was added to liver cells, the increased urea formation caused by this addition far exceeded in amount that of the arginine added—that is, it was acting catalytically and ornithine did the same, suggesting a cyclical process. A cycle in which CO_2 and two nitrogen atoms were added to ornithine to produce arginine required more than one step and it was discovered that citrulline, an amino acid intermediate between ornithine and arginine, is

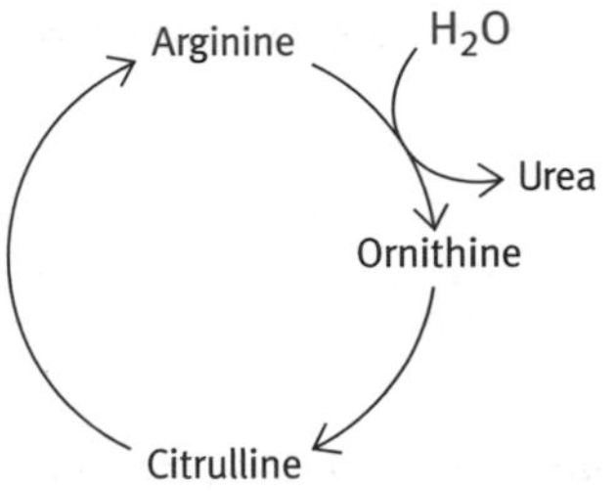

Fig. 16.8 Outline of the arginine—urea cycle. The input of CO_2 and nitrogen into the cycle will be dealt with in the next section.

involved, since it also acts catalytically on urea synthesis when added to liver cells. This discovery led to the famous **urea cycle**. Citrulline is not one of the 20 amino acids used for synthesizing proteins.

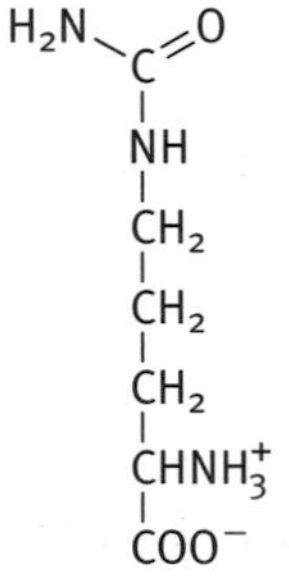

Citrulline

Mechanism of arginine synthesis

We need now to see how ornithine is converted to arginine. Ammonia, CO_2 and ornithine are the reactants for the first step, that of citrulline synthesis which occurs in the mitochondrial matrix. Energy is needed and ATP supplies it. Firstly, ammonia and CO_2 are converted to a reactive intermediate, **carbamoyl phosphate**, which then combines with ornithine to give citrulline. Carbamic acid has the structure NH_2COOH and rbamoyl phosphate is therefore $NH_2-\overset{\overset{O}{\|}}{C}-PO_3^{2-}$. It is a high-energy phosphoryl compound, being an acid anhydride. It is synthesized by an enzyme carbamoyl phosphate synthetase, catalysing the following reaction.

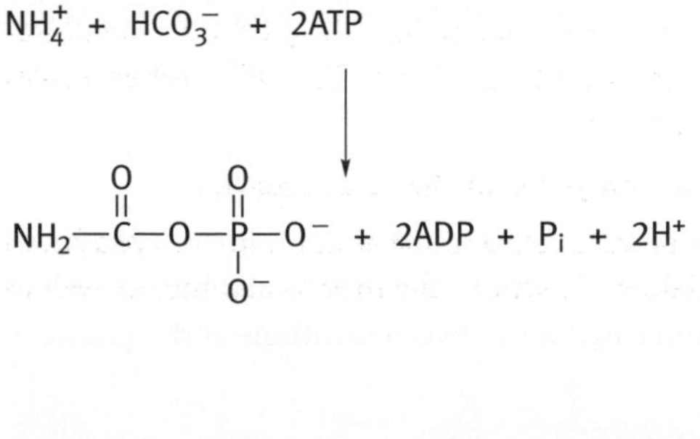

$$NH_4^+ + HCO_3^- + 2ATP$$

$$\downarrow$$

$$NH_2-\overset{\overset{O}{\|}}{C}-O-\underset{\underset{O^-}{|}}{\overset{\overset{O}{\|}}{P}}-O^- + 2ADP + P_i + 2H^+$$

Two ATP molecules are used, the first ATP being broken down to ADP and P_i to drive the production of carbamate from ammonia and CO_2 and the second being used to phosphorylate the carbamate. It all happens on the surface of the one enzyme. The carbamoyl group of carbamoyl phosphate is now transferred to ornithine by ornithine transcarbamoylase giving citrulline. If we represent ornithine as $R-NH_3^+$ The reaction is

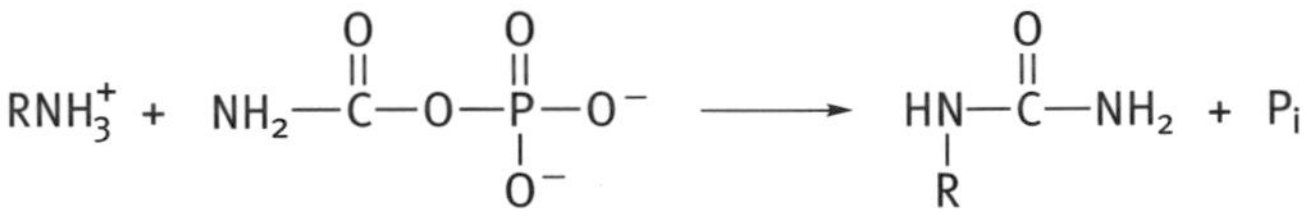

Ornithine   Carbamoyl phosphate   Citrulline

Conversion of citrulline to arginine

The final step in arginine synthesis is to convert the $C=O$ group of citrulline to the $C=NH$ of arginine. This occurs in the cytosol. Ammonia is *not* used here but, instead, the amino group of aspartate is added directly. (You will appreciate that ammonia can be converted to glutamate and that this can generate aspartate by transamination with oxaloacetate. In this way, ammonia and also the amino groups of most amino acids can also be converted to urea via this second stage of the cycle.)

First, aspartate condenses with citrulline as follows:

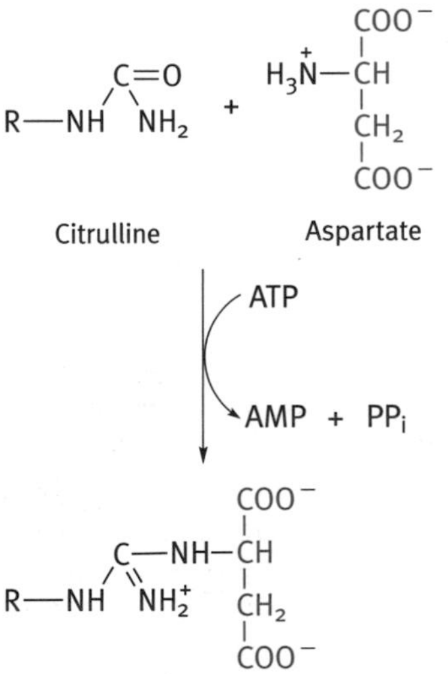

Citrulline   Aspartate

Argininosuccinate

The molecule formed is **argininosuccinate**. The name derives from the fact that the molecule structurally is like an arginine derivative of succinate. It might be somewhat confusing, given this name, that the molecule is now 'pulled apart' by argininosuccinate lyase to yield arginine and fumarate (not arginine and succinate).

$$^-OOC-CHNH_3^+-(CH_2)_3-NH \overset{\overset{\displaystyle COO^-}{|}}{\underset{\underset{\displaystyle COO^-}{|}}{\overset{C-NH-CH}{\underset{NH_2^+}{\|}\quad\underset{CH_2}{|}}}}$$

Argininosuccinate

$$^-OOC-CHNH_3^+-(CH_2)_3-NH-\underset{\underset{\displaystyle NH_2^+}{\|}}{\overset{\overset{\displaystyle NH_2}{|}}{C}} \quad + \quad \overset{\overset{\displaystyle COO^-}{|}}{\underset{\underset{\displaystyle COO^-}{|}}{\overset{CH}{\underset{CH}{\|}}}}$$

Arginine   Fumarate

The whole cycle is therefore as shown in Fig. 16.9. Major control of urea synthesis is exercised by the adjustment of the level of enzymes. These increase with a rich intake of amino acids and in starvation when muscle proteins are degraded. In addition, carbamoyl phosphate synthetase is allosterically

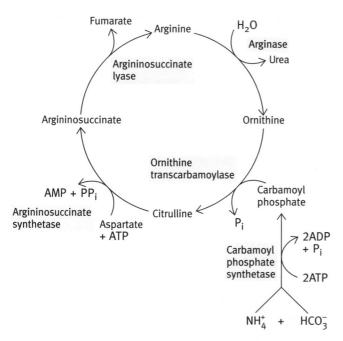

Fig. 16.9 The enzymes of the urea cycle. The levels of urea cycle enzymes are coordinated with the dietary intake of protein. The cycle is allosterically controlled at the carbamoyl phosphate synthetase step. The positive allosteric effector of the enzyme is *N*-acetylglutamate. The conversion of ornithine to citrulline takes place inside mitochondria while the rest of the cycle occurs in the cytoplasm.

activated by *N*-acetyl glutamate whose level reflects that of amino acids.

How is the amino nitrogen transported from extrahepatic tissues to the liver to be converted into urea?

Transport of ammonia in the blood as glutamine

There are two main mechanisms. Ammonia produced from amino acids is toxic, and blood ammonia levels are kept very low since abnormally high levels can impair brain function and cause coma. Free ammonia is not therefore transported, as such, from peripheral tissues to the liver, but is first converted to the nontoxic amide glutamine. This is synthesized from glutamate by the enzyme, glutamine synthetase.

$$\underset{\text{Glutamate}}{\overset{\displaystyle COO^-}{\underset{\displaystyle COO^-}{\overset{|}{\underset{|}{\overset{CH_2}{\underset{CHNH_3^+}{\overset{|}{\underset{|}{CH_2}}}}}}}}} \; + \; NH_4^+ + ATP \longrightarrow \underset{\text{Glutamine}}{\overset{\displaystyle CO-NH_2}{\underset{\displaystyle COO^-}{\overset{|}{\underset{|}{\overset{CH_2}{\underset{CHNH_3^+}{\overset{|}{\underset{|}{CH_2}}}}}}}}} \; + \; ADP \; + \; P_i \; + \; H^+$$

The reaction involves the intermediate formation of an enzymebound γ-glutamyl phosphate. This is a high-energy phosphorylanhydride compound with sufficient free energy to react with ammonia.

The glutamate for this synthesis can be formed from α-ketoglutarate generated in the citric acid cycle followed by transamination with other amino acids. The glutamine is carried in the blood to the liver where it is hydrolysed to release ammonia which is used for urea synthesis.

$$\underset{\text{}}{\overset{\displaystyle CO-NH_2}{\underset{\displaystyle COO^-}{\overset{|}{\underset{|}{\overset{CH_2}{\underset{CHNH_3^+}{\overset{|}{\underset{|}{CH_2}}}}}}}}} \; + \; H_2O \; \xrightarrow{\text{Glutaminase}} \; \underset{\text{}}{\overset{\displaystyle COO^-}{\underset{\displaystyle COO^-}{\overset{|}{\underset{|}{\overset{CH_2}{\underset{CHNH_3^+}{\overset{|}{\underset{|}{CH_2}}}}}}}}} \; + \; NH_4^+$$

Glutaminase in the kidney also liberates ammonia to be excreted along with excess acids from the blood. Glutamine is one of the 20 amino acids found in proteins and is involved in the synthesis of several other metabolites. In the latter, the glutamine amide is used as a nitrogen souce. You will meet examples of this later in the book.

Transport of amino nitrogen in the blood as alanine

From muscle, as much as 30% of the amino nitrogen produced by protein breakdown is sent to the liver as alanine (as well as

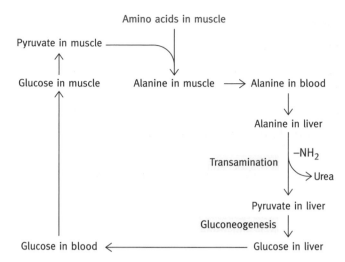

Fig. 16.10 The glucose–alanine cycle for transporting nitrogen to the liver as alanine, and glucose back to the muscles.

glutamine). The amino acids transaminate with pyruvate to yield alanine which is released into the blood. This is taken up by the liver; the amino group is used to form urea (via ammonia and/or aspartic acid. The released pyruvate is converted to blood glucose which can go back to the muscle. The sequence of events is referred to as the **glucose–alanine cycle** (Fig. 16.10).

This alanine transport from muscle to liver has another important physiological role in starvation. As explained already (page 203), after glycogen reserves are exhausted the liver *must* make glucose to supply the brain and other cells with an obligatory requirement for the sugar. The main sources of metabolites for this hepatic gluconeogenesis are the amino acids derived from muscle protein breakdown. Protein breakdown in cells is a controlled process described in Chapter 17. Many of the amino acids can give rise to pyruvate in the muscle which is sent to the liver as alanine. As mentioned on page 203, however, the glucose–alanine cycle itself gives no net increase in glucose and in starvation the amino acids liberated from muscle protein breakdown must be converted to alanine without utilizing glucose as a source of pyruvate.

FURTHER READING

Felig, P. (1975). Amino acid metabolism in man. *Ann. Rev. Biochem.*, **44**, 933–55.

A most useful review which discusses amino acid metabolism at the organ level together with the effects of starvation, diabetes, obesity, and exercise on it. It also discusses gluconeogenesis at this level.

Snell, K. (1979). Alanine as a gluconeogenic carrier. *Trends Biochem. Sci.*, **4**, 124–8.

Reviews the role of muscle in supplying the liver with metabolites for glucose synthesis.

Newsholme, E. A. and Leach, A. R. (1983). *Biochemistry for medical sciences*, pp. 417–29. Wiley.

A very useful description of amino acid metabolism in the different tissues of the body, and its physiological relevance.

PROBLEMS FOR CHAPTER 16

1 Explain how an oxidation can result in the deamination of an amino acid.

2 Which amino acid is deaminated by the mechanism referred to in question 1?

3 How are several of the amino acids deaminated, where the reaction in question 2 is involved? Use alanine as an example.

4 What is the cofactor involved in transamination? Give its structure and explain how transamination occurs.

5 Explain how serine and cysteine are deaminated.

6 What is meant by the terms glucogenic and ketogenic amino acids? Which amino acids are purely ketogenic?

7   Explain the genetic disease phenylketonuria.

8   What is the role of tetrahydrobiopterin in phenylalanine hydroxylation?

9   Methionine is the source of methyl groups in several biochemical processes. Explain how methionine is activated to donate such groups.

10 Outline the reactions of the urea cycle.

11 Why should the level of urea cycle enzymes be increased both in the situation of a high intake of amino acids and in starvation?

12 How are (a) ammonia and (b) amino nitrogen in peripheral tissues transported to the liver for conversion to urea?

Chapter summary

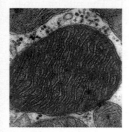

Cellular disposal of unwanted molecules

In all cells there are systems for breaking down unwanted molecules of almost any type. Failure to break down an unwanted component leads to its accumulation within the cell, and lethal genetic diseases are known to be due to such a cause. The destruction of proteins is especially important. The problem is that there has to be selective breakdown—the cell has to define, as it were, which molecules are unwanted and arrange for these to be attacked without putting others at risk. This very demanding requirement transforms what might seem to be a rather mundane topic of hydrolysis of proteins into a highly sophisticated area of mechanisms whose elegance is matched by their importance.

There are a number of reasons why material has to be disposed of in cells. Cells such as phagocytes (a type of white blood cell) engulf bacteria and other objects and these have to be destroyed but, in addition, animal cells in general are constantly taking in particles by endocytosis and these become sequestered into 'digestive' vesicles known as **lysosomes**. Intracellular organelles such as mitochondria are ultimately destroyed to be replaced by new ones, and individual protein molecules also have to be destroyed for a variety of reasons. Some may be abnormal or damaged, but control requirements often depend on destruction of specific perfectly normal proteins.

There are several types of organelles involved in the destructive processes:

- First, lysosomes are vesicles which can destroy any accessible cellular molecule with their collection of about 50 separate hydrolytic enzymes.

- Secondly, **peroxisomes** are vesicles specializing in carrying out oxidative reactions producing hydrogen peroxide.

- Finally, there are **proteasomes**, arguably among the most astonishing organelles of cells, which selectively destroy individual protein molecules within the cell. We will now deal with lysosomes, peroxisomes, and proteasomes in that order.

Lysosomes

These are membrane-bounded organelles in the cytoplasm of all eukaryotic cells. There are hundreds within a liver cell. They are literally bags of enzymes of great potential destructiveness. Their origin is different from that of other vesicles in that they are formed by the fusion of two different types of vesicles, namely **endosomes** and **lysosomal enzyme transport vesicles**. To explain the process we will use **receptor-mediated endocytosis** as the example, though the basic process applies to any endocytosed material. We have already described, in outline, the receptor-mediated endocytosis of chylomicron remnants into liver and of LDL into cells (page 126) but it is a very widespread activity in animal cells. A particle which is to be delivered to a specific cell usually has a protein in it (or is itself a protein) which recognizes and binds to a specific receptor on the target cell. In the case of chylomicron remnants this is apolipoprotein E which binds to a liver cell receptor specific for this protein (page 126). Another example is the disposal of ageing blood proteins, most of which, albumin excepted, are glycoproteins. With time, the terminal sialic acid unit (page 74) is lost, exposing the next unit of the carbohydrate attachment which is galactose. Receptors on liver cells bind to this, causing endocytosis of the protein.

Mechanism of receptor-mediated endocytosis

The receptor–ligand complex is engulfed into the cell where the particle has to be hydrolysed and its component parts made available to the cell, probably by transport out into the cytoplasm. What causes the membrane to endocytose the material, and how are the engulfed remnants subjected to the required digestive process without destroying wanted cellular components?

The membrane receptors are transmembrane components with their cytoplasmic domain exposed on the inside. A protein known as **adaptin** combines with the cytoplasmic domains of a number of the receptors, and clusters them together into a depression of the membrane. The latter has on its cytoplasmic face a basketwork-like coating made up of the protein **clathrin** which attaches to the adaptin molecules; the depressions are known as **clathrin-coated pits** (Fig. 17.1). The pits invaginate more and more and yet another protein called **dynamin** attaches to the neck of the invagination leading to the vesicle being nipped off as a **clathrin-coated vesicle** inside the liver cell. (It is not clear whether the dynamin acts as a tightening noose or is a signal for other proteins to do the nipping off.)

Inside the cell, the coated vesicle is uncoated and the coat

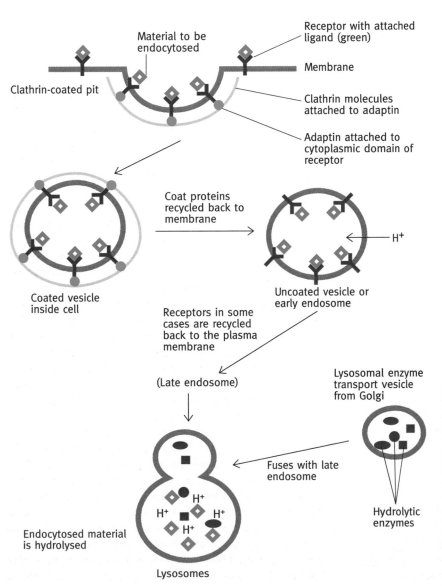

Fig. 17.1 Formation of a lysosome by receptor-mediated endocytosis. Ligand molecules (such as LDL) bind to membrane receptors. Adaptin molecules attach to receptor domains inside the cell and cluster them into a coated pit. Clathrin molecules bind to adaptin and the pit invaginates into a coated vesicle in the cytoplasm. The latter is uncoated and the coat molecules recycled. The uncoated vesicle, now an endosome, is acidified. Acidification releases the receptors (which in many cases are recycled back to the cell membrane) producing a late endosome. A Golgi transport vesicle containing hydrolytic enzymes fuses with the latter, forming a lysosome.

molecules recycled back to the membrane. The uncoated vesicle fuses with other vesicles containing material to be digested forming **early endosomes**. The latter now mature into **late endosomes**, the maturing process involving two things:

- The interior of the endosome is acidified by means of proton pumps in the membrane (Fig. 17.2). This detaches the receptors which may be recycled back to the membrane or in other cases the receptors remain to be digested.

- Finally, transport vesicles budded off from the Golgi membranes containing all of the hydrolytic enzymes fuse with the late endosomes; the result is a **lysosome** in which digestion of the material initially endocytosed is hydrolysed into its component parts.

The lysosomal enzymes include phosphatases, proteinases, esterases, DNAase and RNAase, and enzymes which destroy polysaccharides and mucopolysaccharides. In short, almost all biological molecules can be degraded by them. This of course raises another two big questions: how does the Golgi apparatus package the correct lysosomal enzymes from among many other proteins packaged into different vesicle, and how are the transport vesicles carrying the lysosomal enzymes targeted to the late endosomes? The answers to these questions are largely now understood, but a full discussion of this important topic is left to Chapter 25, after protein synthesis has been dealt with. The cell is protected from destruction by the lysosomal enzymes because they are segregated by the lysosomal membrane. The

enzymes require a pH between 4.5 and 5.0 for activity and this is maintained by the ATP-dependent proton pumps in the lysosomal membrane (Fig. 17.2). If a lysosome were to rupture, the buffering of the cytoplasm would maintain the pH at 7.3 or so, at which lysosomal enzymes are inactive.

We have so far discussed disposal of material endocytosed from outside the cell, but lysosomal digestion of intracellular material also occurs. Mitochondria last in a liver cell for about 10 days and then in an unknown way, they are selected for destruction, when they become enclosed in a membrane-bound vesicle, the membrane originating from the endoplasmic reticulum. The resultant vesicle is called an autophagic vesicle or **autophagosome** which develops into lysosomes as described above (Fig. 17.1) for endosomes generated from endocytotic vesicles.

Lysosomal storage diseases

All types of cell components such as proteins, nucleic acids, and lipid components are destroyed by lysosomal digestion. The importance of this is underlined by the existence of genetic diseases known as **lysosomal storage disorders** which arise because of the lack of one or more specific lysosomal enzymes. Material taken into the lysosomes is not destroyed and the organelles become overloaded with it, sometimes with fatal results.

There is a family of genetic diseases called **sphingolipidoses**. In these, specific lysosomal enzymes are missing, which impair degradation of gangliosides (page 73) of the cell membranes. A classical example is **Tay–Sachs disease** (page 74). Another genetic disease is **I cell disease**. In this almost all of the lysosomal enzymes are missing from the lysosomes so that all manner of molecules accumulate within the vesicles. The absence of so many enzymes could not have been due to mutation of individual genes; it turns out that the mutation is for an enzyme in the Golgi apparatus which is needed to 'label' them for inclusion in the transport vesicles that deliver them to the late endosomes destined to become lysosomes. The disease is often fatal before the age of 10. We refer again to this when we deal with the targeting of proteins to their destinations (page 409).

A rather puzzling role of lysosomes is that they must play some part in the breakdown of glycogen. You will recall (Chapter 7) that glycogen is degraded by glycogen phosphorylase but, in **Pompe's disease**, one of a series of **glycogen storage diseases**, the deficiency is of a lysosomal enzyme which hydrolyses the α-1:4 links of glycogen. In the absence of the enzyme there is a massive accumulation of glycogen in the cells, usually causing death in infancy. It is not known why lysosomes are involved in the breakdown, since they are not implicated in known systems

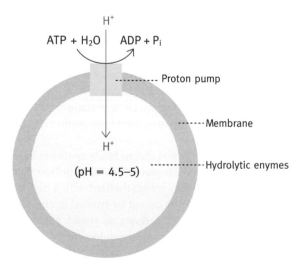

Fig. 17.2 A primary lysosome budded off from the Golgi apparatus. This is essentially a transport vesicle that delivers its load of hydrolytic enzymes to endosomes. The Golgi apparatus and its roles are described in Chapter 25. See Fig. 25.8 for the role of v-SNARES, which are believed to target vesicles to their destinations.

of glycogen metabolism. Possibly degraded glycogen molecules are taken up by lysosomes for final breakdown, but this is speculation.

Peroxisomes

These are vesicles bounded by a single membrane present in mammalian cells; they cannot synthesize proteins and receive their enzymes from the cytoplasm by a special transport mechanism (discussed on page 413). They contain flavoprotein (page 137) oxidase enzymes which attack a number of substrates using molecular oxygen generating not water, but hydrogen peroxide.

$$RH_2 + O_2 \rightarrow R + H_2O_2$$

The substrates include very long chain fatty acids ($>C_{18}$) which are not oxidized by mitochondria, some phenols, and D-amino acids. The fatty acid oxidation is by β-oxidation (page ••) producing acetyl-CoA. The electrons from the first step of the fatty acid oxidation chain, (acyl CoA dehydrogenase) reduce the FAD prosthetic group of the enzyme to $FADH_2$. In mitochondria this is reoxidized by NAD^+ and the NADH is oxidized by the cytochrome chain to produce ATP. In peroxisomes, the $FADH_2$ is re-oxidized by molecular oxygen producing H_2O_2. The NADH produced by the later step in the fatty acid oxidation chain, (hydroxyacyl-CoA dehydrogenase, page 181) is presumably reoxidized by export of the reducing equivalents to the cytoplasm since there is no cytochrome system in peroxisomes. Acetyl-CoA is also believed to be exported. Peroxisomes do not generate ATP by fatty acid oxidation. There is also evidence that the cholesterol side chain oxidation required for formation of bile acids (page 127) occurs in peroxisomes and that synthesis of some complex lipids requires peroxisomes. The products are presumably returned to the cytoplasm. It is intriguing that clofibrate, a drug that reduces TAG, and, less so, cholesterol levels, causes proliferation of peroxisomes. Very little is known of the biogenesis of peroxisomes; in yeast cells there is evidence that vesicular precursors are involved which fuse together and mature into peroxisomes by import of proteins through the membrane (the latter process discussed on page 413). However, why different vesicle intermediates are required, what their origin is, and how the proliferation process is controlled still remain to be elucidated. A number of lethal genetic diseases are known in which peroxisome biogenesis is not normal. An example is the **Zellweger syndrome**, often fatal by the age of 6 months; in this there are abnormally high levels of C_{24} and C_{26} long chain fatty acids and of bile acid precursors.

The metabolic roles of peroxisomes are rather a mixed bag, and they are the least well understood of the organelles, but their essential roles are underlined by the existence of these diseases.

The H_2O_2 generated in peroxisomes is potentially a very dangerous oxidant (see page 283). It is destroyed by the enzyme catalase present in peroxisomes. This is a heme–protein enzyme which catalyses the reaction

$$2H_2O_2 \rightarrow 2H_2O + O_2$$

Proteasomes

The breakdown of proteins by **proteasomes** involves the specific destruction of individual protein molecules of the cell, each one being sequestered in a 'destruction chamber' and degraded. There is no collection of relatively large amounts of material into 'digestive' vesicles as occurs with lysosomal breakdown. Although its purpose does include garbage disposal—the destruction of abnormal protein molecules—there is programmed hydrolysis of normal molecules. Many of these proteins are key players in regulatory systems of the eukaryotic cell, and their precise destruction is an essential part of cellular controls. The importance of the intracellular breakdown of proteins by proteasomes has emerged only relatively recently and has come as a surprise to many biochemists, for not so very long ago the subject tended to be regarded as being fairly uninteresting with little hint of the exquisitely sophisticated molecular mechanisms involved. There has now been a complete turn-around, with the area being regarded as one of the most interesting in the research sense. Its importance is underlined by the fact the proteasome system is an ancient one and most highly conserved in evolutionary terms. Gene mutations which affect the production of correctly structured proteasomes are lethal in yeast cells.

Proteins may be abnormal due to faulty synthesis in which incorrect amino acids are placed in their polypeptide sequences; although proteins are synthesized with a high degree of fidelity (page 390), errors cannot be avoided in any complex chemical process, or, at least, doing so would entail 'uneconomic' measures. Proteins may be misfolded because of an incorrect structure or because they have become irreversibly denatured (which equals unfolding—see page 394 if necessary). They may become damaged by chemical attack from free radicals (page 35) or other agents. Such abnormal proteins cannot be allowed to accumulate.

Important as this disposal of abnormal proteins is, it is far

from being the full story; as already stated, perfectly normal proteins also have to be destroyed in the interests of cellular control. The synthesis of specific proteins is constantly being switched on and off in cells in response to signals, and specific protein breakdown is needed at appropriate times. A striking example occurs in cell cycle control (discussed on page 496) which depends on specific proteins (cyclins) being synthesized when required and destroyed when it is time to move on to the next stage of the cycle. The cyclins activate protein kinases which activate factors required for transition through the phases of the cycle. Failure to destroy the cyclins at the correct time means that inappropriate activation of control factors would occur, resulting in an improperly controlled cell cycle with almost certainly lethal effects. Controlled protein degradation in animals is also necessary for cell differentiation by ensuring the destruction of gene control factors at the appropriate time, and for the ability of the immune system to protect against virus and other infections. Enzymes are also destroyed in a selective manner, with those involved in controlled processes often having half-lives of hours, others lasting for days or weeks. Altogether, you could hardly have a collection of more vital areas of cell biochemistry than those for which proteasomal breakdown of proteins is essential. It might be wondered why control of processes such as mitosis (page 496) should be effected by total destruction of the proteins rather than control by phosphorylation or similar means. Possibly the answer is that such processes are so vital that no risks can be taken and destruction is pretty final. The use of organelles as complicated as proteasomes for the destruction, rather than conventional proteolytic attack, ensures that each molecule is completely degraded with no danger of partially active fragments being left.

What are the problems to be solved by the cell to selectively destroy proteins?

All proteins have basically the same chemical structure, with polypeptide chains potentially susceptible to proteolytic attack, so there have to be mechanisms for selecting those proteins which have to be destroyed quickly and/or at a specific time while others in the same cellular compartment are not to be attacked. There are thus basically two main problems: how are proteins to be destroyed targeted, and how are the selected proteins destroyed without putting at risk proteins which should not be destroyed? The answer to the first is in **ubiquitination**, a labelling process, and to the second, in **proteasomes**, the hydrolytic machines involved. We will now deal with these, the latter first.

The structure of proteasomes

These organelles are large protein structures about 2 million Da in molecular weight; They are present in all eukaryotic cells in both the cytoplasmic and nuclear compartments in large numbers and are readily visible in the electron microscope. A model of the structure of a proteasome is shown in Fig. 17.3. There is a central **20 S core** (S being a size unit—see page 385) consisting of a barrel-shaped cylinder made of four annular rings of protein subunits; the end ones, known as α rings, sandwich the two β rings. The latter contain proteolytic enzymes on the inside of the cylinder; the α rings have no known enzyme activity. At both ends of the 20 S core are **19 S caps**, also known as the regulatory units since they control the selection of proteins to be admitted and unfold the proteins so that they can enter the cavity of the 20 S core where the actual hydrolysis takes place. The dimensions of the proteasome cavity are such that extended polypeptide chains can be accommodated but not folded proteins. Some of the subunits making up the cap structure have ATPase activity which is probably involved in the unfolding of the polypeptide chains of native proteins by the caps.

Proteasomes do not occur in modern bacteria (eubacteria) but are found in archebacteria which are organisms living in hostile environments such as sulfur hot springs at 80° and pH 2, reminiscent of conditions on the primitive earth. These proteasomes are different in that they have only the 20 S core, virtually identical in appearance to the core of those in yeast, but without the end caps and have fewer types of subunits in the rings. It is remarkable that the basic 20 S core structure has been conserved

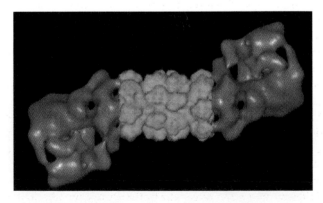

Fig. 17.3 Model of the proteasome. The yellow represents the structure of the 20 S proteasome core and the blue, the 19 S caps.

from archebacteria to humans, spanning billions of years of evolution. It is clearly one of the ancient biochemical systems, comparable in this respect, for example, to glycolysis. The associated ubiquitin system which labels proteins for destruction by proteasomes (described below) is equally ancient.

Proteasomes provide a cavity where proteins destined for destruction can be segregated from the rest of the cell in an unfolded form and degraded to small peptides.

Selection of proteins for destruction by proteasomes— the ubiquitination system

The passport for a protein to enter the interior of a proteasome is **ubiquitination**. **Ubiquitin** is a small protein found universally in eukaryotes but not in prokaryotes, so the name exaggerates somewhat (it is derived from a Latin word meaning 'every-

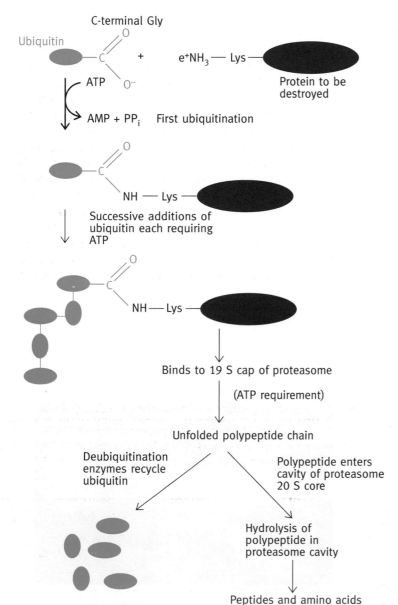

Fig. 17.4 Sequence of events in the targeting and destruction of a protein in proteasomes. Ubiquitination of target proteins is a three-step process but the details of this are omitted for simplicity. The ATP requirement in the proteasome is related to the unfolding of the target protein by the 19 S cap before it enters the hydrolytic cavity of the 20 S core. The attached ubiquitin molecules do not enter the core but are recycled by deubiquitinating enzymes.

where'). Its sequence of 74 amino acids is identical in *all* animal species from insects to humans; even comparing human ubiquitin with that of yeasts and plants there are only three conservative amino acid changes (changes to very similar amino acids). Such conservation means that ubiquitin is a very ancient protein whose function is so important and precise that structural alterations cannot be tolerated. Ubiquitin has a C-terminal glycine residue by which it becomes attached to the ε-NH_2 group of lysine chains of proteins targeted for destruction. Three enzymes are involved in the attachment process, which is ATP dependent. The ubiquitin molecule so attached itself becomes ubiquitinated and so successive ubiquitin molecules are themselves ubiquitinated (Fig. 17.4). The multi-ubiquitinated molecule binds to the proteasome caps and is somehow transferred into the core structure containing the proteases in an unfolded form, this also being ATP dependent. The attached ubiquitin molecules are not fed into the proteasome but are recycled by de-ubiquitinating enzymes which release single molecules of the protein (Fig. 17.4). This leads us to the next obvious question:

What determines which proteins are ubiquitinated?

In other words, what is the selection process? As already explained, different proteins have different lifespans. Those destined for early destruction have built in 'destruction' signals. The simplest is the nature of the N-terminal amino acid. One of the three enzymes involved in linking ubiquitin to a target molecule has two sites for binding substrates, one specific for basic amino acid residues and one for large hydrophobic residues. Proteins with either of these types of N-terminal residue are marked for early destruction. However, the ubiquitination system is more refined than this for in some cases internal amino acid sequences are recognized by the ubiquitinating system. It is not really understood how the selection is made, but a surprising multiplicity of different enzyme systems for attaching ubiquitin to proteins exists in many cells which suggests that a very complex system exists for controlling the selection of proteins to be destroyed. It is likely that there will be much new information on this in the relatively near future.

The role of proteasomes in the immune system

Somatic cells are at risk of virus infection. The body defends itself by destroying infected cells thus aborting virus multiplication. The destruction is brought about by cytotoxic killer T cells (page 459) which recognize infected cells by a remarkable mechanism. The cell continually hydrolyses 'samples' of its cytosolic proteins into peptides eight or nine amino acids long and displays these on its surface for inspection by the killer cells which ignore displayed peptides originating from normal (self) proteins but recognizes those from virus protein synthesized in the cytosol of an infected cell. The cell is then attacked and destroyed. Proteasomes are responsible for producing the peptides of the correct length for display. It is believed that the HIV (the AIDS virus) interferes with this process. A more mechanistic account of this system can be found in Chapter 28 (page 459).

FURTHER READING

Neufeld, E. F. (1991). Lysosomal storage diseases. *Ann. Rev. Biochem.*, **60**, 257–80.

Reviews general principles and specific diseases.

Mallabiabarrena, A. and Malhotra, V. (1995). Vesicle biogenesis: the coat connection. *Cell*, **83**, 667–9.

A minireview that discusses why different coat proteins are used. For the transport from endocytotic vesicles to endosomes, and Golgi to endosomes, a clathrin coat is used. For others, two different coatamers are used. A complex but very interesting story, with more questions than answers.

Clathrin

Marsh, M. and McMahon, H. T. (1999). The structural era of endocytosis. *Science*, **285**, 215–9.

A concise review of clathrin coat assembly.

Proteasomes

Coux, O., Tanaka, K., and Goldberg, A. L. (1996). Structure and function of the 20 S and 26 S proteasomes. *Ann. Rev. Biochem.*, **65**, 801–44.

A comprehensive review of the subject suitable for advanced students taking a special interest in these organelles.

Stuart, D. I. and Jones, E. Y. (1997). Cutting complexity down to size. *Nature*, **386**, 437–8.

A News and Views summary of proteasomes and their structure.

Ciechanover, A. (1998). The ubiquitin–proteasome pathway: on protein death and cell life. *EMBO J.*, **17**, 7151–60.

De Mot, R., Nagy, I., Walz, J., and Baumeister, W. (1999). Proteasomes and other self-compartmentalizing proteases in prokaryotes. *Trends in Microbiology*, **7**, 88–92.

Kirschner, M. (1999). Intracellular proteolysis *Trends in Cell Biology, Trends in Biochem. Sci. and Trends in Genetics*, joint issue. **24**, M42–5.

Part of a special millenium issue of the journals this articles reviews the development of an area of biochemistry from what was a relatively dull subject to one of the exciting ones today.

PROBLEMS FOR CHAPTER 17

1   What are the purposes of lysosomes?

2   What are lysosomal transport vehicles and where are they produced?

3   What evidence is there to indicate that lysosomal digestion is an essential process?

4   Why is Pompe's disease somewhat of a biochemical puzzle?

5   What are peroxisomes and their function?

6   Describe a proteasome. What are its roles in cells? Give some examples of the latter. What is the evidence for their great importance in cells?

7   How are proteins targeted to the proteasomes?

Chapter summary

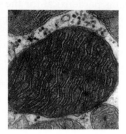

Enzymic protective mechanisms in the body

The immune system (Chapter 27) is the major protective mechanism against attack by disease-causing pathogens. However, the body is threatened by other things. We will, in later chapters, describe DNA repair to cope with damage caused by ionizing radiation, UV light, and mutations (page 330), and the production of heat shock proteins (page 394) to cope with heat and other stresses. These fit into specific chapters but there exists, in addition, a variety of other protective mechanisms involving enzymic reactions and these are the subject of the present chapter. The fact that they are, in a biochemical sense, rather a varied collection should not obscure the fact that all have the same biological role—protection—and are essential to life.

Blood clotting

Coagulation of blood is needed to form blood clots, which plug holes in damaged blood vessels and so prevent bleeding. To be effective, the response has to be very rapid and relatively massive while the initial signal in chemical terms is exceedingly small. A massive amplification of the signal is therefore needed—the response must, in quantitative terms, be vastly greater than the signal.

We have seen earlier (see page 220) that biochemical amplification is achieved by means of a **reaction cascade**. In this, an enzyme is activated that then activates another enzyme and so on. The fact that the enzymes activated are themselves catalysts means there is an amplification at each step. In case this is not clear, if a single enzyme molecule activates 1000 molecules of the next enzyme in 1 minute and each molecule of the second enzyme does the same for the next enzyme and so

on, in a cascade of four steps in a very short time, vast numbers of active molecules of enzyme number four are created. This enzyme at the end of the cascade can rapidly catalyse a massive response.

Blood clotting can conveniently be divided into two parts. There is the cascade resulting in the activation of the enzyme that forms the clot, and there is the mechanism of clot formation itself by that enzyme. The cascade process is based on **proteolysis**—hydrolysis of peptide bonds of inactive precursor proteases activates them (cf. trypsinogen activation to trypsin in digestion, page 100). All of the necessary inactive precursor proteins are present in the blood waiting for the signal of a damaged blood vessel. They all have to be normally inactive.

What are the signals that clot formation is needed?

When a wound occurs, in terms of blood clotting, two things result.

1 The endothelial cell layer lining the blood vessels is damaged, exposing the structures underneath, such as collagen fibres that have a negatively charged or 'abnormal' surface. The blood-clotting response is a strictly localized reaction around the site of damage. Initially, a temporary plug is formed by the aggregation of blood platelets around the hole. For the formation of a clot, a small group of proteins in the blood absorb to the abnormal surface, the net result of which is that two proteases mutually activate each other in an autocatalytic manner. One of these is called factor XII. (The nomenclature is slightly confusing in that some 'factors' are enzymes, some are cofactors for enzymes, and they are not numbered according to the sequence of their appearance in the process.) Factor XII activates a cascade of

three steps resulting in active factor X (another protease). Factor X activates **prothrombin** to the active protease, **thrombin**. Thrombin is what actually causes clotting (or thrombus formation). (We usually expect enzyme names to end in 'ase' but several of the classical proteases end in 'in'— for example, thrombin, pepsin, trypsin, chymotrypsin, and others.)

This particular pathway of blood clotting is called the **intrinsic pathway** because, if blood is put into a glass vessel, the negatively charged glass surface triggers off clotting. Nothing has to be added; therefore the process is intrinsic.

2 Now for the **extrinsic pathway**. This is triggered by the release of a protein complex called tissue factor from damaged cells and tissues. Since something has to be added to blood, it is called the *extrinsic* pathway. This is a very short pathway. A protease is activated and this activates the same factor X as occurs in the intrinsic pathway resulting again in active thrombin formation. The two pathways are set out in Fig. 18.1.

The intrinsic pathway, being longer, is slower to cause clot formation when measured *in vitro*, than the extrinsic pathway, also measured *in vitro*. However, in the disease hemophilia A, in which blood clotting fails to occur, it is the intrinsic pathway that is deficient due to the absence of factor VIII required for factor X proteolytic activation. This seems paradoxical for the rapid extrinsic pathway, not requiring factor VIII, might seem able to do the job. However, it seems that, for normal physiological clotting, both pathways function as one, both being essential. Interactions between the two pathways are possibly involved.

How does thrombin cause thrombus (clot) formation?

In the circulating blood there is a protein called **fibrinogen**. The basic molecular unit consists of short rods made up of three polypeptide chains; two of these rods are joined together by S—S bonds near their *N*-terminal ends, forming the fibrinogen monomer. As shown in Fig. 18.2, at their joining points two of the three chains in each short rod project as negatively charged peptides, called fibrinopeptides. The negative charges mutually repel the monomers and prevent association.

Thrombin cleaves these off, giving a **fibrin monomer**. The fibrin monomer is now able to polymerize spontaneously by noncovalent bond formation. The sites at the end of the fibrin monomer are complementary to sites at the centre of adjacent

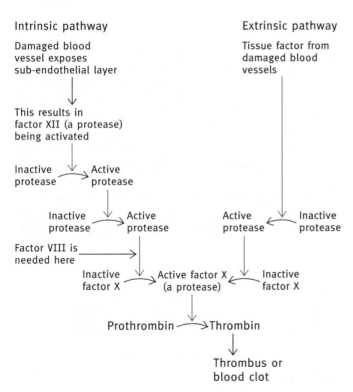

Fig. 18.1 Simplified diagram of intrinsic and extrinsic pathways of blood clotting. The names of the various proteases and other factors involved are omitted for simplicity. (Thirteen factors, numbered I–XIII, are, in fact, known.) The proteases listed are specific for their particular substrate. Factor VIII is the protein missing in patients with hemophilia A.

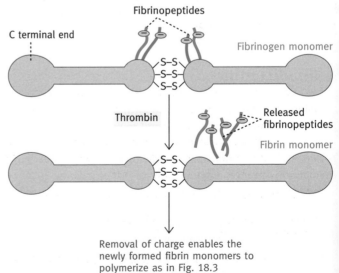

Fig. 18.2 Fibrinogen monomer and its conversion to fibrin monomer. Each half of the fibrinogen monomer is composed of three polypeptide chains, two of which terminate in the negatively charged fibrinopeptides.

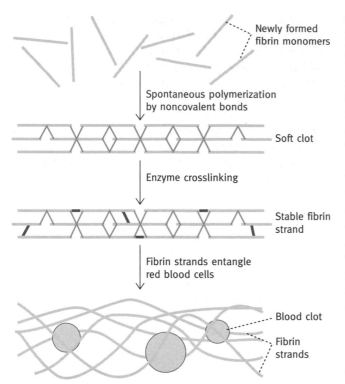

Newly formed fibrin monomers

Spontaneous polymerization by noncovalent bonds

Soft clot

Enzyme crosslinking

Stable fibrin strand

Fibrin strands entangle red blood cells

Blood clot

Fibrin strands

Fig. 18.3 Spontaneous polymerization of fibrin monomers and their enzymic crosslinking to form a stable fibrin strand. The fibrinogen cannot polymerize until thrombin hydrolyses off the fibrinopeptides because (1) the negative charges prevent association and (2) the central sites with which the ends of the fibrin monomers associate are masked by the fibrinopeptides. (Note that the covalent crosslinks in red are arbitrarily represented in position and number.)

molecules so that a staggered arrangement forms from the polymerisation (Fig. 18.3). This gives a so-called 'soft clot'. A more stable 'hard clot' is formed by covalent crosslinking between the side chains of adjacent fibrin molecules.

The covalent crosslinks are curious in that a glutamine side chain on one monomer is joined to a lysine side chain on the next in an enzymic transamidation reaction,

$$—CONH_2 + H_3N^+— \longrightarrow —CO—NH— + NH_4^+$$

(Glutamine   (Lysine   (Crosslink)
side chain) side chain)

The fibrin strands entangle blood cells forming a blood clot.

Keeping clotting in check

Blood clotting is potentially dangerous unless strictly limited to local sites of bleeding. Once started, there is the danger of an autocatalytic process like this getting out of hand, causing inappropriate clotting. An elaborate series of safeguards exists. Proteinase inhibitors (for example, antithrombin) in blood 'dampen down' and prevent the clotting reactions from spreading; heparin, a sulfated polysaccharide (a glucosaminoglycan—see page 55) present on blood vessel walls, increases this inhibitory effect; another protease, plasmin, dissolves blood clots. Plasmin itself is formed from inactive plasminogen; activation of this occurs by a protein, tissue plasminogen activator (TPA) released from damaged tissues. TPA is one of the new therapies for coping with blood clotting. Although present in tissues in minute amounts, its gene and the corresponding cDNA have been isolated and used for massive commercial production of the protein for injection (the technology is described in Chapter 29). This, and other inhibitory mechanisms, limit the reaction to the damaged surface area, the latter being required for initiation of the process. Blood clotting control is complex; inappropriate clotting is responsible for large numbers of deaths.

Rat poison, blood clotting, and vitamin K

The widely used rat poison, warfarin, kills by preventing blood clotting so that the rodents die from unchecked internal bleeding from minor lesions that continually occur. Warfarin is used clinically, for example, after strokes, to minimize the danger of clotting. It is structurally similar to vitamin K (K for the German *Koagulation*) and acts as a competitive inhibitor of vitamin K function, that is it competes with vitamin K for the enzyme site and inactivates the enzyme (have a look at the two structures in Fig. 18.4 but you do not have to memorize these). Vitamin K is needed for prothrombin conversion to thrombin; it acts as a cofactor in an unusual enzyme reaction that adds an extra —COOH group, using CO_2, to a glutamic acid side chain of prothrombin.

$$
\begin{array}{ccc}
| & & | \\
CH_2 & & CH_2 \\
| & + CO_2 \longrightarrow & | \\
CH_2 & & HC—COO^- \\
| & & | \\
COO^- & & COO^-
\end{array}
$$

This group efficiently binds Ca^{2+}

A glutamic acid side chain of prothrombin

Carboxyglutamate

The carboxyglutamate is needed to bind Ca^{2+}, which is essential in the prothrombin → thrombin activation process. The same modification also applies to other factors in the cascade.

(a)

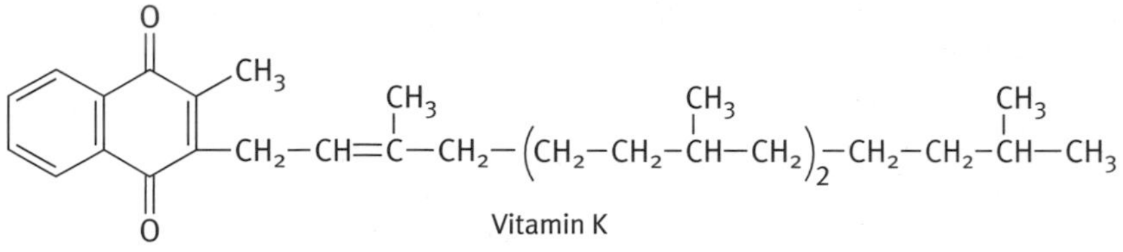

Vitamin K

(b)

Warfarin

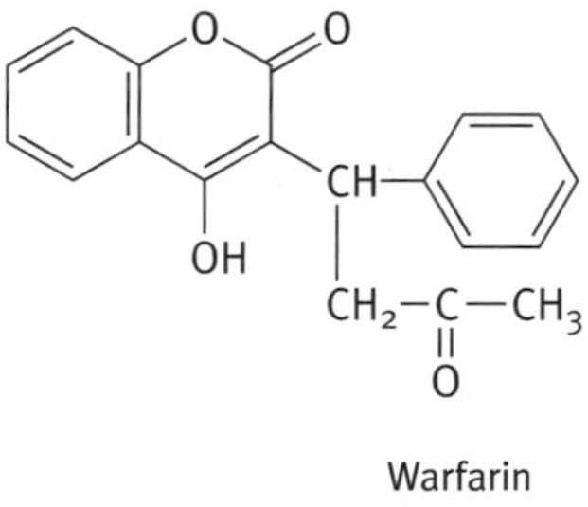

Fig. 18.4 Comparison of the structures of **(a)** vitamin K and **(b)** warfarin. Vitamin K is needed for blood clotting. Warfarin antagonizes vitamin K and prevents blood clotting.

Protection against ingested foreign chemicals

Large foreign molecules are dealt with by the immune system (Chapter 27). Small foreign molecules (which are not indicative of invasion by a living pathogen) are coped with by different systems.

Human beings ingest large numbers of foreign chemicals, collectively referred to as **xenobiotics** (*xeno* meaning foreign). These include pharmaceuticals, pesticides, herbicides, and industrial chemicals as well as complex structures such as the terpenes, alkaloids, and tannins of plants. Many of these are relatively insoluble in water, but soluble in fats, and they therefore tend to partition into the hydrocarbon layer of membranes and the fat globules of fat cells rather than being excreted in the aqueous urine. Unless they are rendered more polar and therefore more water-soluble, they will accumulate in the body with deleterious consequences. To facilitate their excretion, foreign chemicals are metabolized in several ways, but the **cytochrome P450** system (usually referred to simply as P450) in the liver is of central importance. A typical reaction of this P450 system is to add a hydroxyl group to an aliphatic or aromatic grouping. Other, different, enzymes add various highly polar groups, of which glucuronate is the major one, thus facilitating excretion in the aqueous urine.

We now need to look a little more closely at these processes. First the P450 system.

Cytochrome P450

P450 is a heme–protein complex as are the respiratory chain cytochromes (page 162). The name comes from P for pigment and 450 from the absorption maximum of the complex formed with carbon monoxide. (CO is not involved in the reaction—it just happens to give a complex with a spectrum that makes measurement of the amount of P450 easy.) The enzyme is anchored into the smooth endoplasmic reticulum facing the cytoplasm. A foreign compound, AH, is attacked according to the reaction

$$AH + O_2 + NADPH + H^+ \rightarrow A{-}OH + H_2O + NADP^+.$$

It is called a **monooxygenase reaction** because it uses only one atom of oxygen from each O_2 molecule. NADPH is used to reduce the other oxygen atom to water. It is also called a **mixed-function oxygenase** because it both hydroxylates AH and reduces O to H_2O. The electrons from NADPH are transferred to the Fe^{3+} in the heme of P450 by a P450 reductase enzyme also present in the smooth ER membrane. You have already met this type of enzyme—phenylalanine hydroxylase on page 258. (Note the difference between oxidation and oxygenation. Oxidation involves the removal of electrons; oxygenation involves the addition of an oxygen atom from O_2 to the molecule attacked. Oxygenation of hemoglobin is different again, for there a molecule of O_2 is loosely attached—see page 513.)

The amazing thing about the P450 system is the large number of different compounds that it attacks, including many that living organisms could not have significantly encountered before

the advent of the modern chemical industry. One might speculate on the reason for this apparently anticipatory ability of evolution for, without it, it is difficult to see how we could have survived the vast variety of newly produced synthetic chemicals to which we are now exposed. The somewhat weird collection of compounds such as terpenes, alkaloids, etc., found in plants may have been developed as protection for the plants against attacks (for example, grazing) by animals. The latter, therefore, it may be speculated, evolved a detoxifying system that could cope with almost anything present in plants and other food sources. So versatile is this system that it also copes with the chemicals newly devised by humans. The basis of this versatility is that a P450 enzyme has a wide specificity—it attacks a variety of related structures—and different P450 enzymes exist with different but overlapping specificities.

Secondary modification—addition of a polar group to products of the P450 attack

Also present in the smooth ER membrane is a **glucuronidation system**. This transfers a glucuronate group from glucuronyl-UDP to the hydroxyl group generated on a foreign chemical by P450 (Fig. 18.5). UDP-glucuronate is produced by oxidation of UDP-glucose (page 119). This facilitates excretion in the urine by increasing water-solubility of the chemical.

The response of the liver to the ingestion of foreign chemicals

When the liver is exposed to a foreign chemical such as phenobarbital, there is a massive response in which the smooth ER proliferates greatly. At the same time the P450 and glucuronidation systems are induced—the drug switches on the genes for these and the amount in the liver increases many fold. When all the drug is metabolized, the liver reverts to normal. The massive response underlines the threat that the ingestion of such compounds presents. The response has clinical relevance, for the effectiveness of many pharmaceuticals is limited by the rate of their metabolic destruction and hence by the level of the detoxifying enzyme systems, and these, as stated, increase on exposure to the drug. The P450 system is also involved in the disposal of natural compounds such as heme and steroid hormones. Other P450 systems, for example, in adrenal gland mitochondria, are involved in the synthesis of steroid hormones from cholesterol. An ironic twist to the story is that the oxidation of some substances by the P450 increases their carcinogenic effect. Benzpyrene in tobacco smoke is converted by P450 in the liver to a reactive species that can covalently react with DNA and is a carcinogenic mutagen.

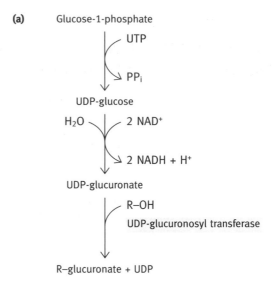

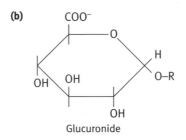

Fig. 18.5 (a) The glucuronidation system. **(b)** Structure of a glucuronide.

Multidrug resistance

Another form of protection of cells against toxic chemicals is the reduction of their accumulation inside cells. Many cells, including those of human tissues, express a glycoprotein in their cell membranes that is an ATP-driven multidrug transporter that transports drugs out of the cell. A remarkable range of chemicals are transported, amongst them several of the anticancer drugs used in chemotherapy. This has caused much interest in the phenomenon but a much wider range of chemicals reacts with the transporter. These include many pharmacological agents and cytotoxic chemicals. Since steroids are also transported and the transporter protein is prevalent in steroid-secreting adrenal cortical cells, the system may have this as a primary biological role. It has recently been found to transport cholesterol out of cells (see page 129). The molecules transported have no chemical similarities but all are amphipathic compounds preferentially soluble in lipids.

Protection of the body against its own proteases

We have seen (Chapter 5) how the digestive system manages to escape the actions of its own proteolytic enzymes. However, proteases exist elsewhere in the body. A particularly important one, in the present context, is the **elastase of neutrophils**. Neutrophils are phagocytic white cells attracted to the site of infection or irritation. When activated at such sites, they secrete elastase which clears away connective tissue from the site. Elastin is the elasticity-conferring connective tissue (page 54) from which elastase gets its name. In the lung, air passages lead to minute pockets, the alveoli, which have a very large surface area needed for the diffusion of gases between blood and air. Neutrophils present in alveoli liberate elastase but this is prevented from destroying the lung structure by α_1-**antitrypsin** (α_1-**antiproteinase**), a protein that is produced by the liver and secreted into the blood. The α_1-antitrypsin inhibits several proteases, including trypsin as the name implies, but is especially effective on elastase. It inhibits by combining tightly with the enzyme and blocking its catalytic site.

α_1-Antitrypsin in adequate levels in the blood is essential for the protection of the lung. The molecule diffuses from the blood into the alveoli. If, due to a genetic defect, the level of α_1-antitrypsin is subnormal, neutrophil elastase can destroy alveoli resulting in much larger pockets in the lung structure and consequent reduction of surface area available for gaseous exchange. The result is emphysema, a symptom of which is extreme shortness of breath.

Smokers are prone to emphysema for two reasons. The irritants in smoke are sufficient to cause neutrophil attraction to the lungs, with a consequent increased release of elastase. Secondly, oxidizing agents in the smoke destroy α_1-antitrypsin; they oxidize the sulfur atom of a crucial methionine side chain to a sulfoxide group ($S \rightarrow S{=}O$) This is sufficient to prevent the α_1-antitrypsin from inactivating elastase and results in proteolysis of lung tissue, resulting in emphysema. Other antiproteases exist but α_1-antitrypsin is the most important one.

Protection against reactive oxygen species

As described in Chapter 8 (page 140), oxygen is an ideal electron sink for the energy-generating electron transport system. Its position on the redox scale means that, in energy terms, electrons from NADH have a long way to 'fall', meaning that the negative free-energy change of the overall oxidation of NADH to produce water is large. O_2 accepts four electrons and four protons to give H_2O the final product of the electron transport chain,

$$O_2 + 4e^- + 4H^+ \rightarrow 2H_2O.$$

The beauty of the system is that O_2 is relatively unreactive and therefore itself does no chemical damage and the product, H_2O, is entirely benign. There is, however, a darker side to the story, for O_2 has the potential to be exceedingly dangerous in the body. During evolution, the switch from energy generation by anaerobic metabolism to the use of oxygen as the electron sink was one of the most important events, but it was in some ways equivalent to jumping on to the back of a tiger for a ride. The danger occurs when a single electron is acquired by the O_2 molecule to give the **superoxide anion**, an extremely reactive corrosive chemical agent,

$$O_2 + e^- \rightarrow O_2^-.$$

The unpaired electron acquires a partner by attacking a covalent bond of another molecule.

Superoxide is formed in the body in several ways. In the electron transport chain, the final enzyme that donates electrons to oxygen, cytochrome oxidase, does not release partially reduced oxygen intermediates in any significant amounts—it ensures that an O_2 molecule receives all four electrons resulting in H_2O formation. However, inevitably, components of the electron transport chain (page 142) may 'leak' a small proportion of electrons to oxygen, one at a time, resulting in O_2^- formation. Moreover, mutations in mitochondrial DNA may block electron transport pathways and deflect electrons into superoxide formation (for example by direct ubiquinone oxidation by oxygen). Mitochondria have few or no DNA repair systems, so that mutations accumulate in them (see page 335).

In addition, there are other oxidation reactions in the body that produce small amounts of dangerous oxygen species such as H_2O_2. These include some oxidases that directly oxidize metabolites using O_2 (see below) quite distinct from the respiratory pathways we have described. Spontaneous oxidation of hemoglobin (Hb) to methemoglobin (Fe^{3+} form) is another source; as a rare event, the oxygen in oxyhemoglobin (HbO_2) instead of leaving as O_2 and leaving behind hemoglobin in the Fe^{2+} form, leaves as O_2^- with the formation of methemoglobin, the Fe^{3+} form. It is also known that ionizing radiation causes superoxide formation.

When phagocytes ingest a bacterial cell there is a rapid increase in oxygen consumption; this is used to oxidize NADPH via a mechanism that deliberately generates superoxide anions.

These are shed into the vacuole and converted to H_2O_2 (see below), which helps to destroy the contained bacterial cell. Excess neutrophils attracted to irritated joints may lead to superoxide release and contribute to arthritic damage.

Although the superoxide anion is present in minute amounts, it can set up chain reactions of chemical destruction in the body. In this, the superoxide anion attacks and destroys a covalent bond of some cell constituent, but in doing so generates a new free radical species with an unpaired election from the attacked molecule that, in turn, attacks yet another molecule of cell constituent producing yet another free radical, and so on. The destruction initiated by the superoxide anion is thus a self-perpetuating chain of reactions.

The biological injuries caused by this process are not finally established but it has been suggested that superoxide damage contributes to ageing, cataract formation, the pathology of heart attacks, and other problems.

Basically, there are three protective strategies—one chemical and two enzymic.

Mopping up oxygen free radicals with vitamins C and E

The chemical strategy is to dampen or quench the chain reactions, initiated by superoxide, by using **antioxidants**. The requirements for such a quenching reagent is that it should itself be attacked by the superoxide anion but generate a radical insufficiently reactive to perpetuate the chain reaction. The main biological quenching agents are **ascorbic acid** (vitamin C) and α-**tocopherol** (vitamin E). The former is water-soluble, the latter lipid-soluble, and so, between them, they protect in both phases of the cell. These two vitamins are not the only antioxidants—some normal metabolites such as uric acid are effective; β-**carotene** is also an antioxidant. **Bilirubin**, produced from heme breakdown by heme oxygenase (page 512), is also an effective antioxidant. It is interesting that heme oxygenase, which catalyses the first step in bilirubin production from heme, is induced by ionizing radiation (which generates oxygen free radicals) and other agents causing oxidative stress. Whether these effects are direct protective mechanisms against oxidative stress by bilirubin generation remains to be proven.

Enzymic destruction of superoxide by superoxide dismutase

Most or all animal tissues contain the enzyme **superoxide dismutase** and, most appropriately, it occurs in mitochondria. It also occurs in lysosomes and peroxisomes (Chapter 17) as well as in extracellular fluids such as lymph, plasma, and synovial fluids. Superoxide dismutase catalyses the reaction.

$$2O_2^- + 2H^+ \rightarrow H_2O_2 + O_2.$$

The hydrogen peroxide is destroyed by **catalase**,

$$2H_2O_2 \rightarrow 2H_2O + O_2.$$

Some oxidases also generate H_2O_2 directly. Flavoprotein oxidases, with FAD as their prosthetic group, generally catalyse reactions of the type

$$AH_2 + O_2 \rightarrow A + H_2O_2.$$

Xanthine oxidase, involved in purine metabolism (Chapter 19), is of this type. H_2O_2 is potentially dangerous because, in the presence of metal ions such as Fe^{2+}, it can generate the highly reactive hydroxyl radical ($OH\cdot$), not to be confused with the hydroxyl anion (OH^-). (H_2O_2 is the result of two electrons being added to O_2, when three electrons are added, $OH\cdot$ and OH^- are formed.)

$$H_2O_2 + Fe^{2+} \rightarrow Fe^{3+} + OH\bullet + OH^-.$$

The hydroxyl radical can attack DNA and other biological molecules. The two protective enzymes, catalase and superoxide dismutase, are therefore important. Another enzyme that destroys H_2O_2 is glutathione peroxidase, described in the next section. Since brain has little catalase, this latter enzyme possibly is the main one for protection against H_2O_2 in that organ.

The glutathione peroxidase–glutathione reductase strategy

Glutathione is a thiol tripeptide (γ-glutamyl-cysteinyl-glycine), which is abbreviated to GSH (Fig. 18.6). It is found in most cells where, because of its free thiol group, it functions as a reducing agent, for example, to keep proteins with essential cysteine groups in the reduced state. This reaction with proteins is nonenzymic, but another protective action of GSH is to

Fig. 18.6 Structure of reduced glutathione (GSH) and of the oxidized form (GSSG). Glu, Cys, and Gly are abbreviations for glutamate, cysteine, and glycine, respectively, using the three-letter system.

inactivate peroxides via the action of **glutathione peroxidase** (GSSG),

$$H_2O_2 + 2GSH \rightarrow GS-SG + 2H_2O.$$

Organic peroxides (R—O—OH) are also destroyed in this way. The GSSG is subsequently reduced by NADPH, the reaction being catalysed by **glutathione reductase**,

$$GSSG + NADPH + H^+ \rightarrow 2GSH + NADP^+.$$

Red blood cells depend for their integrity on GSH, which reduces any ferrihemoglobin (methemoglobin) to the ferrous form, as well as destroying peroxides. This explains why red blood cells have the pentose phosphate pathway (Fig. 14.1) to supply the NADPH needed for the reduction of GSSG. Usually, patients with defective glucose-6-phosphate dehydrogenase, the first enzyme in the pentose phosphate pathway, have enough enzymic activity for normal function. However, when extra stress is placed on the cell, for example, by the accumulation of peroxides due to the action of the antimalarial drug pamaquine, the supply of NADPH can no longer be maintained. The integrity of the cell membrane is impaired and hemolysis results in such patients from taking the drug.

In this chapter we have concentrated on protective mechanisms in animals. It is becoming increasingly evident that plants too have their protective devices—perhaps more varied because they have no immune system and cannot run away or flick off predators. A random example is the possession of an enzyme, in some plants, of chitinase, used to attack chitin in the cuticle of insects.

FURTHER READING

Blood clotting

Scully, M. F. (1992). The biochemistry of blood clotting: the digestion of a liquid to form a solid. *Essays in Biochemistry*, **27**, 17–36.

Very readable review of clotting, limiting of the process and medical aspects.

Cytochrome P450

Coon, M. J., Ding, X., Pernecky, S. J., and Vaz, A. D. N. (1992). Cytochrome P450; progress and predictions. *FASEB J.*, **6**, 669–73.

A crisp overview of different aspects of this important field.

Eastabrook, R. W. (1996). The remarkable P450s: a historical review of these versatile hemoprotein catalysts. *FASEB J.*, **10**, 202–4.

Summarizes the metabolic roles of P450s in broad terms and outlines the current research interests in the field. Points out that there are over 2000 research papers per year on the subject.

Free radicals

Babior, B. M. (1987). The respiratory burst oxidase. *Trends Biochem. Sci.*, **12**, 241–3.

Reviews the deliberate production of superoxide by phagocytes.

Rusting, R. L. (1992). Why do we age? *Sci. Amer.*, **267**(6), 130–41.

Contains a section on the evidence that free radicals may be a significant factor in ageing.

Harris, E. (1992). Regulation of antioxidant enzymes. *FASEB J.*, **6**, 2675–83.

A review of how the level of enzymes destroying oxygen free radicals are adjusted to needs.

Rose, R. C. and Bode, A. M. (1993). Biology of free radical scavengers: an evaluation of ascorbate. *FASEB J.*, **7**, 1135–42.

Clearly discusses the properties required for free radical scavengers and looks at ascorbate in detail.

Fridovich, I. (1994). Superoxide radical and superoxide dismutase. *Ann. Rev. Biochem.*, **64**, 97–112.

Concise review of the field.

Uddin, S. and Ahmad, S. (1995). Dietary antioxidants protection against oxidative stress. *Biochem. Education*, **23**, 2–7.

A useful review. The point is made that free iron and copper ions can catalyse OH• formation and this is presumably why they are always tightly bound to proteins.

PROBLEMS FOR CHAPTER 18

1   Blood clotting involves cascades of enzyme activations. What is the rationale for such a long process?

2   Explain how thrombin triggers the formation of a blood clot.

3   The spontaneous polymerization of fibrin monomers forms a 'soft clot'. How is this converted into a more stable structure?

4   Why is vitamin K needed for blood clotting?

5   What is the function of cytochrome P450?

6   Why should NADPH be involved in an oxygenation reaction?

7   What is the function of glucuronyl-UDP in disposing of water-insoluble compounds?

8   What is multidrug resistance?

9   Why does smoking cause emphysema?

10   What is superoxide?

11   What mechanisms exist for guarding against the deleterious effects of superoxide?

Chapter summary

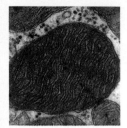

Nucleotide metabolism

Several of the preceding chapters have been mainly concerned with energy production from food. In this, ATP occupies the central position, but GTP, CTP, and UTP are also involved in aspects of food metabolism. However, the involvement of these nucleotides described so far has all been concerned with the phosphoryl groups of the molecules. The nature of the bases, whether A, G, C, or U, has been important only for recognition by the appropriate enzymes but otherwise has not been directly relevant to the metabolic processes. For this reason we have not previously given information about the bases themselves.

We are now about to start, in the next chapter, a new main area of biochemistry in which information transfer is a main purpose—we refer to nucleic acids and protein synthesis and for this the structures of the bases of nucleotides become important. In this chapter, the structures, synthesis, and metabolism of nucleotides are dealt with. This is an essential prerequisite for the subsequent chapters.

Structure and nomenclature of nucleotides

The term 'nucleotide' originates from the name of nucleic acids, originally found in nuclei, and which are polymers of nucleotides. A **nucleotide** has the general structure.

phosphate—sugar—base.

A **nucleoside** has the structure

sugar—base.

Thus, AMP and the corresponding nucleoside, adenosine, have the structures shown.

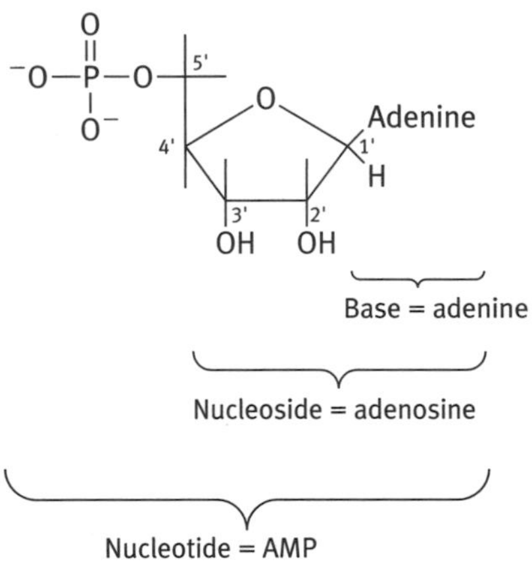

Strictly speaking, the AMP shown here should be written as 5′ AMP. The prime (′) indicates that the number refers to the position on the ribose sugar ring, to which the phosphate is attached, rather than to the numbering of atoms in the adenine ring. However, it is a common practice to assume that the phosphate is 5′ unless specified, since this is the most common position. Thus 5′ AMP is often called AMP, whereas if the phosphate is on the carbon atom 3 of the ribose, this is always specified as 3′ AMP (a compound you'll meet later).

The sugar component of nucleotides

The sugar component of a nucleotide is always a pentose, most often ribose, or 2′-deoxyribose, which are always in the D-configuration, never the L-form.

D-Ribose

D-2'-deoxyribose

In **RNA** the sugar is always ribose (hence the name, **ribonucleic acid**) and in **DNA**, deoxyribose (hence, **deoxyribonucleic acid**). A nucleotide containing ribose is a ribonucleotide but this is not usually specified; unless otherwise stated, a named nucleotide such as AMP is taken to be a ribonucleotide. A deoxyribonucleotide *is* always specified; for example, deoxyadenosine monophosphate or dAMP, etc. (with the one occasional exception mentioned below).

The base component of nucleotides

Nomenclature

We are primarily concerned with five different bases—adenine, guanine, cytosine, uracil, and thymine, all often abbreviated to their initial letter.

A, G, C, and U are found in RNA;
A, G, C, and T are found in DNA.

The ribonucleotides are AMP, GMP, CMP, and UMP, but older and still used terms are adenylic, guanylic, cytidylic, and uridylic acids, respectively (or adenylate, guanylate, cytidylate, and uridylate for the ionized forms at physiological pH). The deoxyribonucleotides are dAMP, dGMP, dCMP and dTMP. The latter often is called TMP, or thymidylate, without the d-prefix because T is found only in deoxynucleotides (with rare exceptions of no significance here).

When the intention is to indicate a nucleotide without specifying the base, the abbreviations NMP or 5'-NMP are often used, or dNMP or 5'-dNMP for deoxynucleotides.

Deoxy UMP exists only as an intermediate in the formation of dTMP; it does not occur in DNA (except as a result of chemical damage to the DNA, when it is promptly removed—see page 332).

The corresponding ribonucleosides (base-sugar) are, respectively, adenosine, guanosine, cytidine, uridine, and d-adenosine, etc. for the deoxyribose compounds. The names for the nucleosides, cytidine and uridine, sound like those of free purine bases (cf. adenine and guanine), while the name of the free base, cytosine, sounds like that of a nucleoside (cf.

adenosine), so be careful here. Other so-called 'minor' bases exist and are found in transfer RNA (page 382). Hypoxanthine is one of these. It is also the first base produced by the pathway of purine biosynthesis as hypoxanthine ribotide or inosine monophosphate (see below); hypoxanthine riboside is called inosine.

Structure of the bases

The first point is that:

A and G are **purines**;
C, U, and T are **pyrimidines**.

These names originate from their being derivatives of the parent compounds, purine and pyrimidine, respectively (neither of which occur in nature).

Purine

Pyrimidine

These rings in future structures will be represented by the simplified forms below.

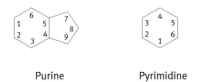

Purine

Pyrimidine

The structures of the nucleotide bases are represented in Fig. 19.1.

Of especial importance, note that *T is simply a methylated U*. It will be useful to fix in your mind that T is essentially the same as U except that it is 'tagged' by a methyl group. T is found only in DNA; U only in RNA. The significance of this will be apparent later (page 322).

Attachment of the bases in nucleotides

The bases are attached to the sugar moieties of nucleotides at the N-9 position of purines and the N-1 position of pyrimidines. The glycosidic bond is in the β-configuration, that is, it is above the plane of the sugar ring. The structures of AMP and CMP are given in Fig. 19.2.

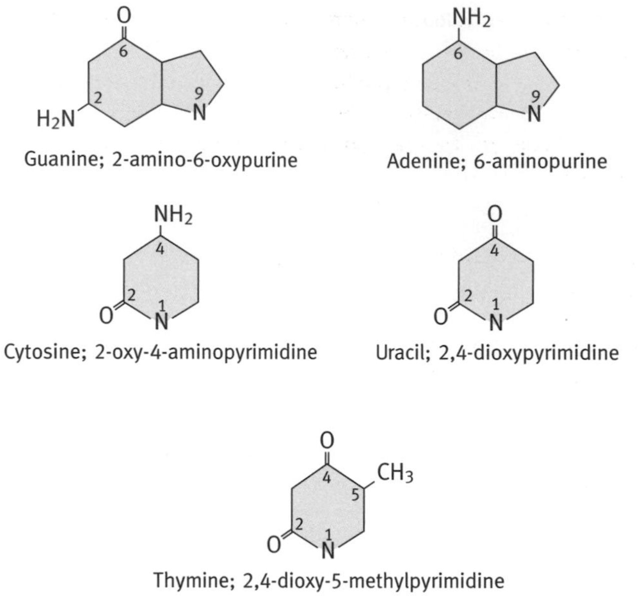

Guanine; 2-amino-6-oxypurine

Adenine; 6-aminopurine

Cytosine; 2-oxy-4-aminopyrimidine

Uracil; 2,4-dioxypyrimidine

Thymine; 2,4-dioxy-5-methylpyrimidine

Fig. 19.1 Diagrammatic representation of structures of purine and pyrimidine bases found in nucleic acids (Full structures in Chapter 20). Other minor bases are found in tRNA, described in Chapter 24.

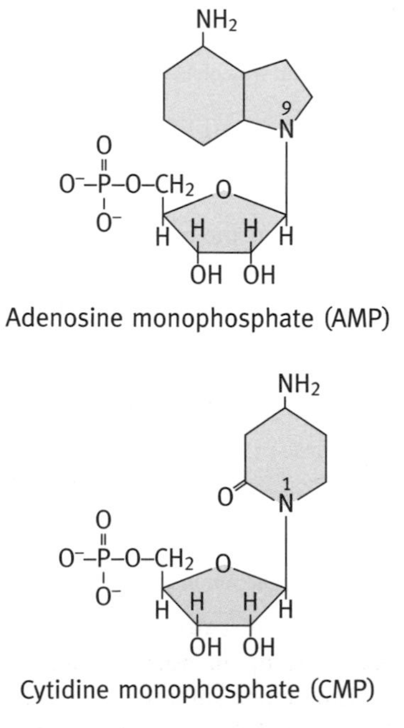

Adenosine monophosphate (AMP)

Cytidine monophosphate (CMP)

Fig. 19.2 Structures of a purine and a pyrimidine nucleotide.

Synthesis of purine and pyrimidine nucleotides

Purine nucleotides

Most cells can synthesize purine bases *do novo* from smaller precursor molecules. In the *de novo* synthesis of purine nucleotides, bases are *not* synthesized in the free form but rather the purine ring is assembled piece-by-piece with all the intermediates attached to ribose-5-phosphate so that, by the time a purine ring is assembled, it is already a nucleotide. This refers only to the *de novo* synthesis of purines because free purine bases released by degradation of nucleotides *are* utilized for nucleotide synthesis by the separate **salvage pathway** to be described later. The mechanism of **ribotidation** (the addition of ribose-5-phosphate) is the same in both pathways, as well as in pyrimidine nucleotide synthesis (but the compound that is ribotidated differs in each case, as described below). Which brings us to PRPP, the metabolite that is involved in all ribotidation.

PRPP—the ribotidation agent

PRPP is 5-**phosphoribosyl-1-pyrophosphate**. It is formed from ribose-5-phosphate (produced by the pentose phosphate pathway, page 236) by the transfer of a pyrophosphate group from ATP by the enzyme **PRPP synthetase**.

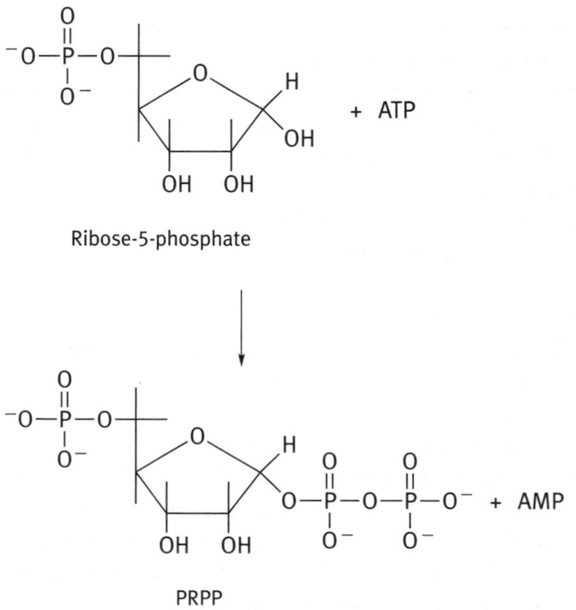

Ribose-5-phosphate

PRPP

The PRPP is an 'activated' form of ribose-5-phosphate; appropriate enzymes can donate the latter to a base forming a nucleotide, and splitting out —℗—℗. Hydrolysis of the latter to $2P_i$ drives the reaction thermodymically.

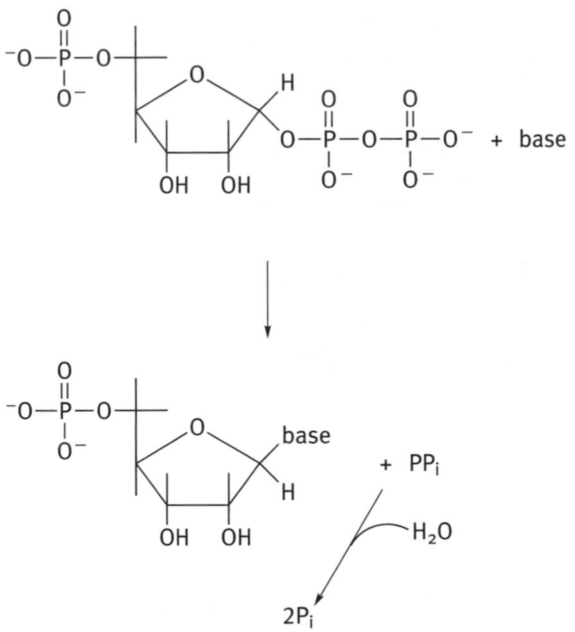

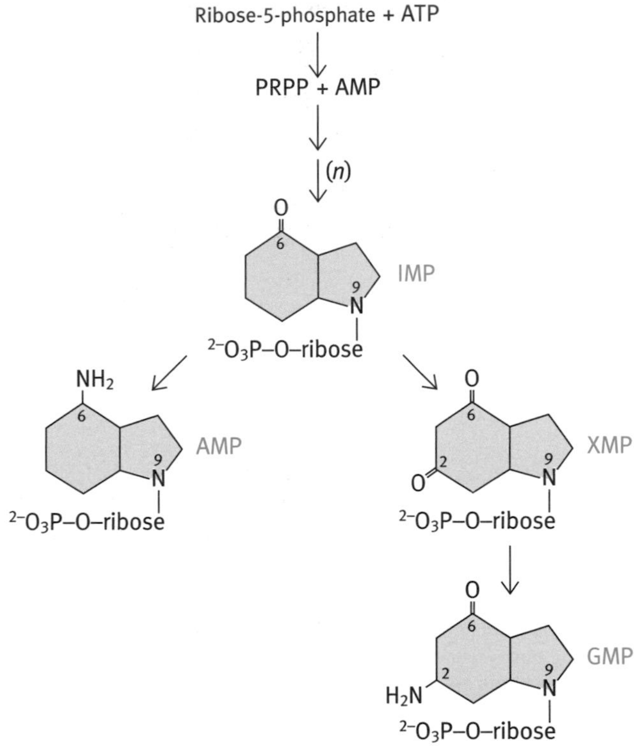

Fig. 19.3 Diagram of the purine *de novo* pathway of GMP, AMP, and XMP synthesis. The base in IMP or inosine monophosphate is hypoxanthine. The base in XMP or xanthosine monophosphate is xanthine. The complete pathway of AMP and GMP synthesis can be seen in Figs 19.4 and 19.5.

In this reaction the configuration at carbon atom 1 is inverted so that the base is in the required β position. Note that in the *de novo* pathway (see reaction 1 of this pathway in Fig. 19.4) the base is simply $-NH_2$, derived from glutamine and the product, 5-phosphoribosylamine; in the purine *salvage* pathway it is a purine (see page 294). We usually associate the term nucleotide with a purine or pyrimidine base but it can be applied to any base attached in the appropriate manner to the sugar phosphate.

To return from this general point to the purine *de novo* pathway in particular, after the formation of 5-phosphoribosylamine, there follows a series of nine reactions resulting in the assembly of the first purine nucleotide in which hypoxanthine is the base (see Fig. 19.3). This nucleotide is IMP or inosinic acid.

IMP is a branch-point since its hypoxanthine base is converted either to adenine or guanine yielding AMP and GMP, respectively. The pathway is summarized in Fig. 19.3, but you should also look at all of the reactions of the pathway as set out in Figs 19.4 and 19.5, which it is suggested do not need to be learned in detail. We will refer to some reactions of specific interest. You will see that six molecules of ATP are consumed

in the synthesis of one purine nucleotide molecule. The ATP utilization refers only to —℗ groups; there is no loss of the adenine nucleotide of ATP so that the pathway results in a net synthesis of AMP. We do not want to go into all of the reactions shown in Fig. 19.4, but the reactions numbered 3 and 9 of this pathway have a general importance and we need to divert to deal with these in some detail. This concerns a type of reaction not dealt with in this book before—**one-carbon transfer**.

The one-carbon transfer reaction in purine nucleotide synthesis

Reactions 3 and 9 of the pathway involve the addition of a formyl (HCO—) group to intermediates in the pathway (see Fig. 19.4). The donor molecule in both cases in N^{10}-formyltetrahydrofolate, a molecule not mentioned before in this book. **Tetrahydrofolate** (FH_4) is the carrier in the cell of formyl groups. It is a coenzyme derived from the vitamin folic acid (F) or pteroylglutamic acid (we suggest that you need to

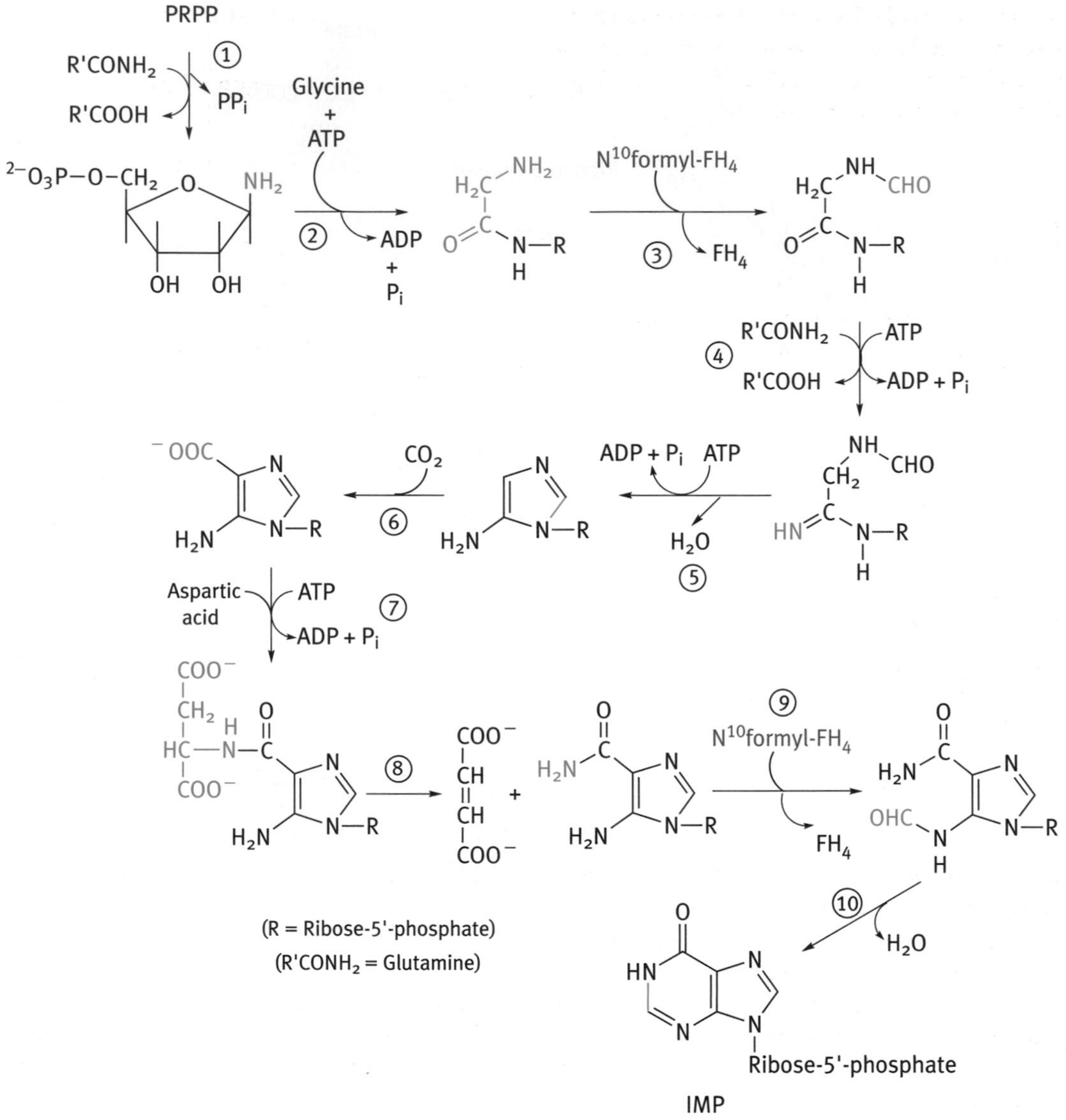

Fig. 19.4 Pathway for the *de novo* synthesis of the purine ring from PRPP to inosinic acid, given for reference purposes. (The circled reaction numbers are referred to in the text.) Blue indicates the structural change resulting from the latest reaction.

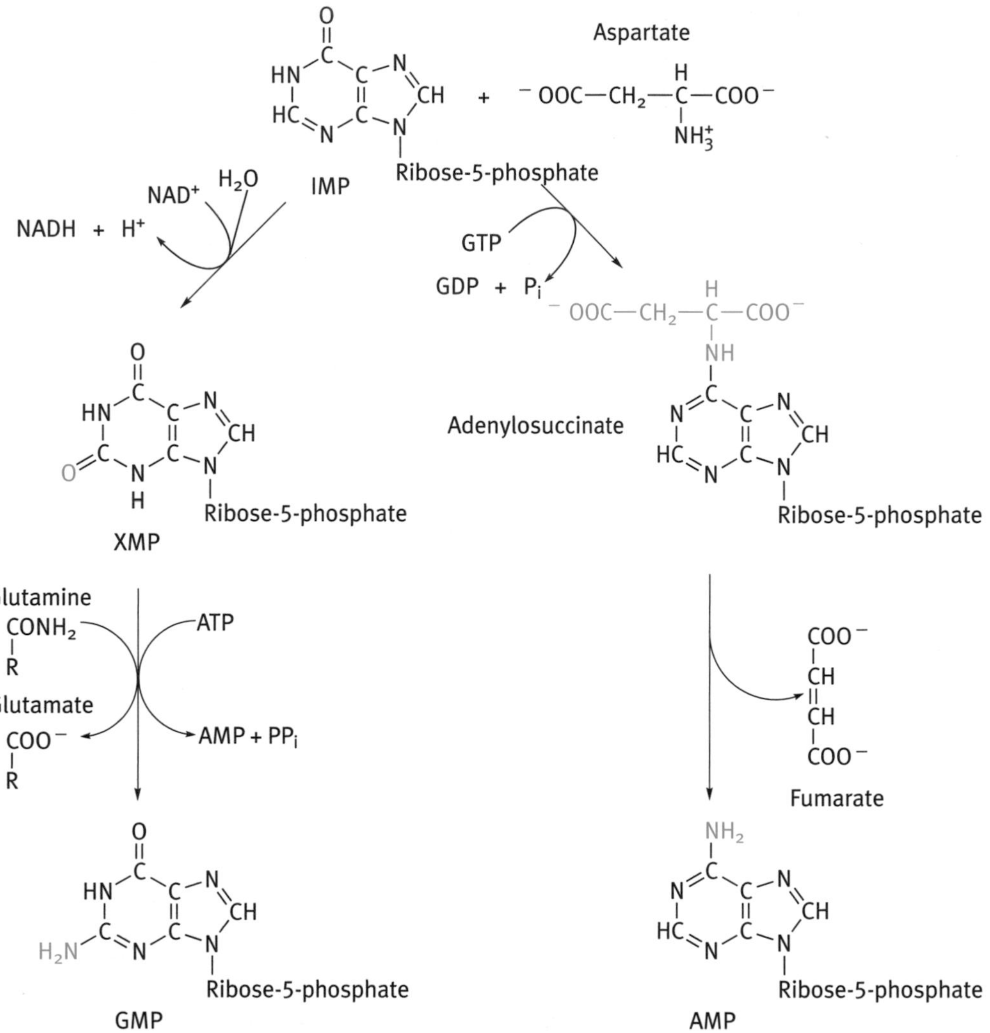

Fig. 19.5 Pathways for the synthesis of GMP and AMP from inosinic acid (IMP). The colour shows the change resulting from each reaction.

memorize only the relevant part of this and related structures below, not the whole molecules).

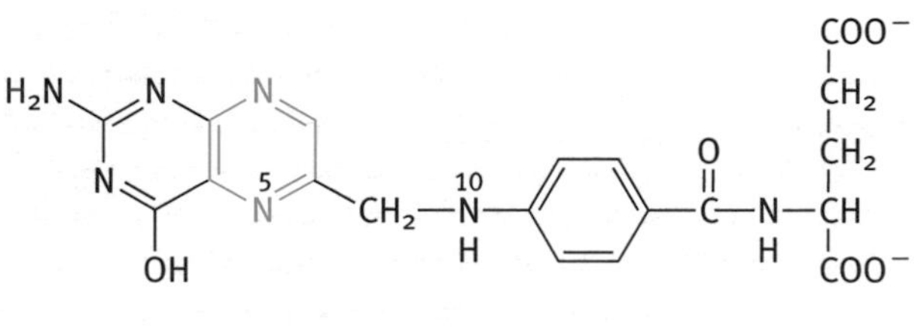

Folic acid (pteroylglutamic acid)

The vitamin (F) is reduced to FH_4 by NADPH in two stages.

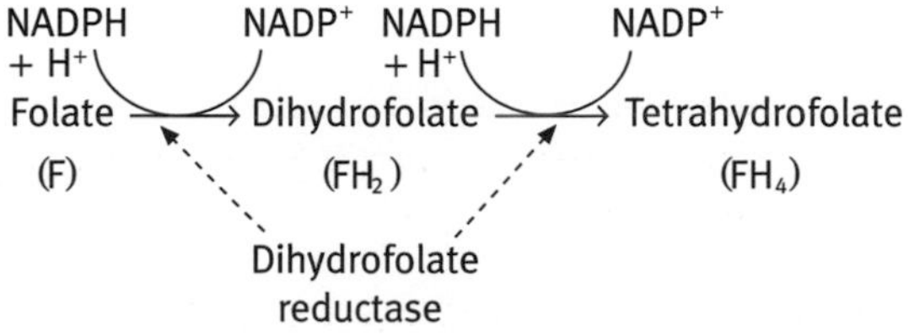

The structures of FH_2 and FH_4 are given below.

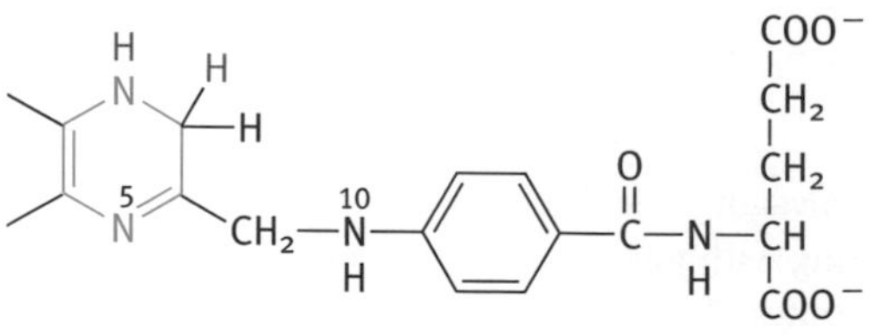

Dihydrofolate (FH$_2$)

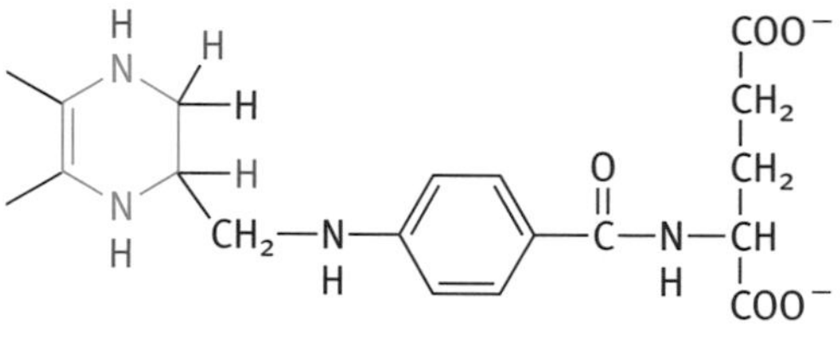

Tetrahydrofolate (FH$_4$)

Have a look at the structure of FH$_4$ and notice that the N-5 and N-10 atoms are placed such that a single carbon atom can neatly bridge the gap between them. For our present purposes we can therefore represent FH$_4$ as

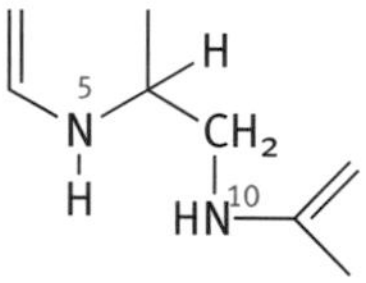

and N^{10}-formyl FH$_4$ as

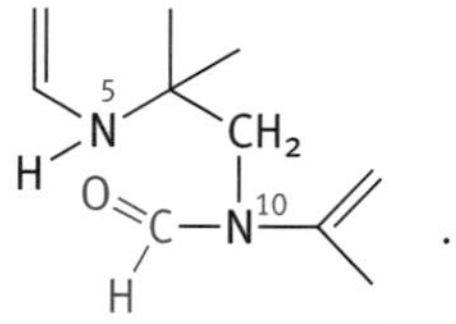

N^{10}-formyltetrahydrofolate
(formyl FH$_4$)

This is the donor of the formyl group in reactions 3 and 9 of the purine biosynthesis pathway; specific formyl transferase enzymes catalyse the reactions.

Where does the formyl group in formyl FH$_4$ come from?

The answer is the amino acid serine (which is readily synthesized from the glycolytic intermediate 3-phosphoglycerate).

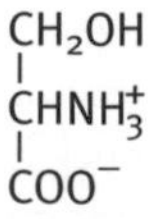

Serine

An enzyme, **serine hydroxymethylase**, transfers the hydroxymethyl group (–CH$_2$OH) to FH$_4$ leaving glycine and forming N^5, N^{10}-methylene FH$_4$.

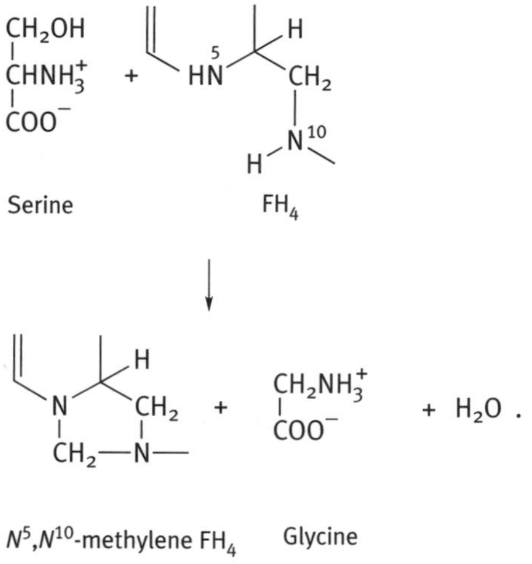

Serine   FH$_4$

N^5,N^{10}-methylene FH$_4$ Glycine

The product methylene FH$_4$ is not quite what we want for formylation because the –CH$_2$– group is more reduced than a formyl group. It is therefore oxidized by an NADP$^+$-requiring enzyme forming the methenyl derivative, which is hydrolysed to N^{10}-formyl FH$_4$, the formyl group donor.

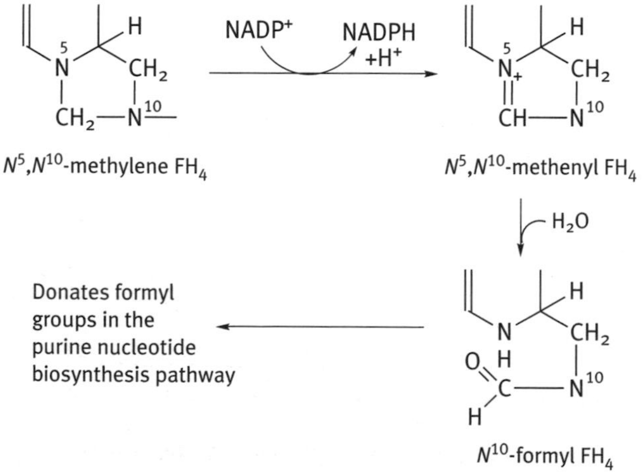

N^5,N^{10}-methylene FH$_4$ N^5,N^{10}-methenyl FH$_4$

Donates formyl groups in the purine nucleotide biosynthesis pathway

N^{10}-formyl FH$_4$

How are ATP and GTP produced from AMP and GMP?

Most of the synthetic reactions of the cell involve nucleoside triphosphates. As you will see later, these are needed for nucleic acid synthesis. It is a simple but especially important concept that enzymes (kinases) exist in the cell to transfer —Ⓟ groups between nucleotides *at the high-energy level*. There is little free-energy change involved so that —Ⓟ groups can be shuffled around from nucleotide to nucleotide with ease. The main source of —Ⓟ is, of course, ATP, for remember that the energy generating metabolism constantly regenerates ATP from ADP and P_i. Newly formed AMP and GMP are phosphorylated by kinase enzymes as shown.

AMP + ATP → 2ADP   Adenylate kinase

GMP + ATP → GDP + ADP   Guanylate kinase

GDP + ATP → GTP + ADP   Nucleoside diphosphate kinase
or   or
ADP   ATP

The purine salvage pathway

We have emphasized that the *de novo* synthesis of purines does not involve free purine bases—purine nucleotides are produced. However, as already indicated, there is a separate route of purine nucleotide synthesis in which **free bases** are converted to nucleotides by reaction with PRPP. The free bases originate from degradation of nucleotides—they are salvaged and hence the name of the pathway. Two enzymes are involved—these are phosphoribosyltransferases, one of which forms nucleotides from adenine and the other from hypoxanthine or guanine. The latter enzyme, known as **HGPRT** (for **hypoxanthine–guanine phosphoribosyltransferase**), catalyses the reaction

$$\text{Guanine or Hypoxanthine} + \text{PRPP} \rightarrow \begin{matrix} \text{GMP} \\ \text{or} \\ \text{IMP} \end{matrix} + \text{PP}_i.$$

The enzyme salvaging adenine may be of lesser importance than that dealing with guanine and hypoxanthine in humans, for the main routes of nucleotide breakdown produce the free bases hypoxanthine (from AMP) and guanine (from GMP) as shown in Fig. 19.6.

What is the physiological role of the purine salvage pathway?

Since purines are energetically 'expensive' to make, a mechanism for re-utilizing free purine bases is economical since it can reduce the amount of *de novo* synthesis a cell has to carry out. Moreover, certain cells such as erythrocytes have no *de novo* purine synthesis pathway and must rely on the salvage pathway.

The physiological importance of purine salvage is underlined by the genetic disease of infants called the Lesch-Nyhan syndrome in which the enzyme HGPRT is missing. This results in neurological problems including mental retardation and self-mutilation. Brain possesses the *de novo* pathway only at low levels, so purine nucleotide synthesis is very sensitive to the salvage defect. Lack of the salvage reaction leads to a hepatic *overproduction* of purine nucleotides by the *de novo* pathway in these patients because the level of PRPP rises (due to lack of utilization by the salvage reaction) and stimulates the *de novo* pathway. This explains why in these patients excessive uric acid production occurs as in gout (see below), which may result in kidney failure caused by the urate crystals. The connection between the biochemical defect and the neurological symptoms is not clear in the Lesch–Nyhan patients. While uric acid over-

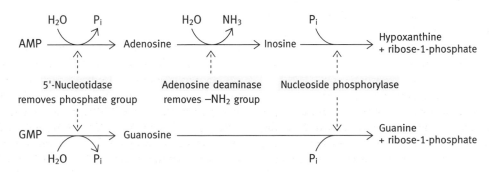

Fig. 19.6 Production of free purine bases hypoxanthine and guanine by nucleotide breakdown. Patients lacking adenosine deaminase in lymphocytes have an immune deficiency that formerly could be treated only by keeping the affected child in a sterile plastic bubble. The disease was the first to be successfully treated by gene therapy in which the normal gene for adenosine deaminase was inserted *in vitro* into bone marrow stem cells and returned to the patient. (See page 287.)

production is treatable with allopurinol (see below), this does not relieve the neurological problems. Nor do patients with gout develop the neurological symptoms.

The sources of free purine bases for salvage are probably several. Although the diet contains purines in the form of nucleic acids, there is evidence that most of them are destroyed by epithelial cells of the intestine and not absorbed. By contrast, purines injected into the bloodstream are utilized by cells. The liver is a major site of purine synthesis and some evidence exists that it releases the bases into the blood for use by other cells, such as reticulocytes, that do not have the complete *do novo* pathway. It is also probable that the salvage pathway recycles, within cells, purine bases released from breakdown of nucleic acids. Lysosomal destruction of cellular components containing nucleic acids (see page 267) would presumably release free bases. Although, as discussed further below, there is no doubt as to the importance of the purine salvage pathway, information on the traffic of free bases in the body is not complete.

The recycling of preformed purine bases has the obvious advantage of energy-saving provided, of course, that the *de novo* pathway synthesis is correspondingly reduced. This is achieved in two ways: (1) salvage reduces the level of PRPP and hence of the pathway; and (2) the AMP and GMP produced by salvage exert feedback inhibition on the pathway (see below).

Formation of uric acid from purines

Nucleotide degradation leads to the production of free hypoxanthine and guanine. Part of this is salvaged back to nucleotides but part is oxidized to produce **uric acid** (Fig. 19.7). The enzyme, xanthine oxidase, that produces uric acid is present mainly in the liver and intestinal mucosa. Gout is due to a raised level of urate in the blood, leading to the deposition of crystals in tissues. Although gout is traditionally associated with rich living, the main source of uric acid is probably excess *de novo* production of purine nucleotides due, in some patients, to a high level of PRPP synthetase activity. Also, as described above, deficiency of the HGPRT enzyme leads to the overproduction of purine nucleotides. The drug **allopurinol** used in the treatment of gout, mimics the structure of hypoxanthine. It is converted to a potent xanthine oxidase inhibitor, alloxanthine, by the xanthine oxidase itself, a process known as **suicide inhibition**. These structures are shown below. This inhibition results in xanthine and hypoxanthine formation rather than that of uric acid (see Fig. 19.7). These products are more water-soluble than uric acid and more readily excreted, thus preventing the deposition of insoluble uric acid crystals in tissues that results in the clinical symptoms of gout.

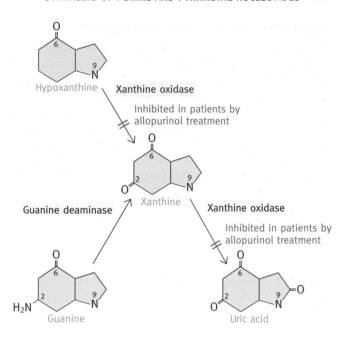

Fig. 19.7 Conversion of hypoxanthine and guanine to uric acid. XO, xanthine oxidase. The drug allopurinol is closely related in structure to hypoxanthine. It is converted to alloxanthine, which now inhibits xanthine oxidase. The conversion is carried out by xanthine oxidase itself.

Allopurinol   Alloxanthine   Enol form
                                       of hypoxanthine

Control of purine nucleotide synthesis

As with all metabolic pathways, there must be regulation or chemical anarchy would prevail. The *de novo* pathway is a classical example of **allosteric feedback control** (see page 211). The first step of a pathway is the logical place for control. In the *de novo* pathway this is the PRPP synthetase. This enzyme is negatively controlled by AMP, ADP, GMP, and GDP. The next enzyme, which catalyses the first *committed* step to synthesis of purine nucleotides (reaction 2, Fig. 19.4), is inhibited also, as shown in Fig. 19.8. However, this isn't quite the end of the story, because the *de novo* pathway produces IMP, and then the IMP goes in two directions—to AMP and GMP. These latter feedback-control their own production, as shown in Fig. 19.8.

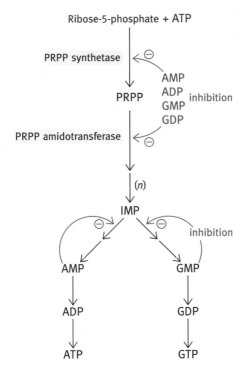

Fig. 19.8 Simplified scheme of the control of the purine nucleotide biosynthesis pathway.

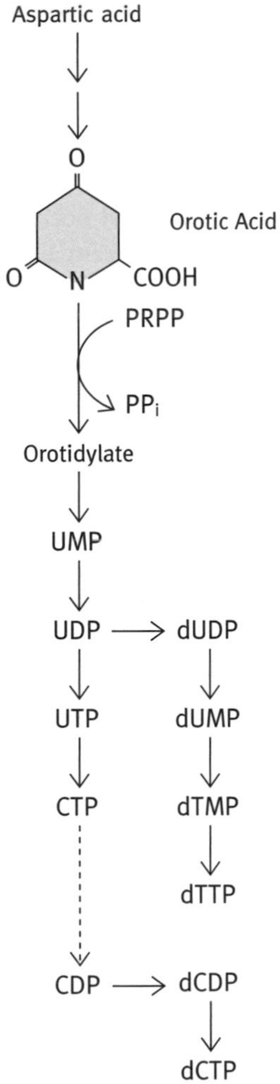

Both negative and positive regulatory loops serve to ensure a balanced production of ATP and GTP since both are required for nucleic acid synthesis (Chapter 20).

Synthesis of pyrimidine nucleotides

Most cells of the body synthesize pyrimidine nucleotides *de novo* but, unlike bacteria, mammals do not appear to have significant pyrimidine salvage pathways for free bases, analogous to that for purines. The nucleoside, thymidine, however, is readily phosphorylated to TMP by thymidine kinase and, in that sense, salvage of this nucleoside does occur.

The **pyrimidine pathway** is summarized in Fig. 19.9 and given in full in Fig. 19.10 for reference purposes. It starts with aspartic acid and produces a ring structure compound, orotic acid. Orotic acid is converted to the corresponding nucleotide by the PRPP reaction and this is converted to UMP. UTP is produced by kinase enzymes much as in the purine pathway. CTP is produced by amination of UTP.

In *E. coli*, control of pyrimidine nucleotide synthesis is mainly at the aspartate transcarbamoylase step. In mammals the pathway is controlled at the carbamoyl phosphate synthase

Fig. 19.9 Summary of pyrimidine nucleotide synthesis. (It is assumed here that CTP is converted to CDP.) The formation of deoxynucleotides is described later in this chapter. The complete pathway is given in Fig. 19.10.

step; it is inhibited by pyrimidine nucleotides and activated by purine nucleotides. The latter control serves to keep the supply of all the nucleotides required for nucleic acid synthesis in balance.

How are deoxyribonucleotides formed?

For DNA synthesis dATP, dGTP, dCTP, and dTTP are required (Chapter 20). The reduction of ribonucleotides to deoxy compounds occurs at the diphosphate level with NADPH as the reductant (the electrons being transported to the reductase by a complex pathway, not described here).

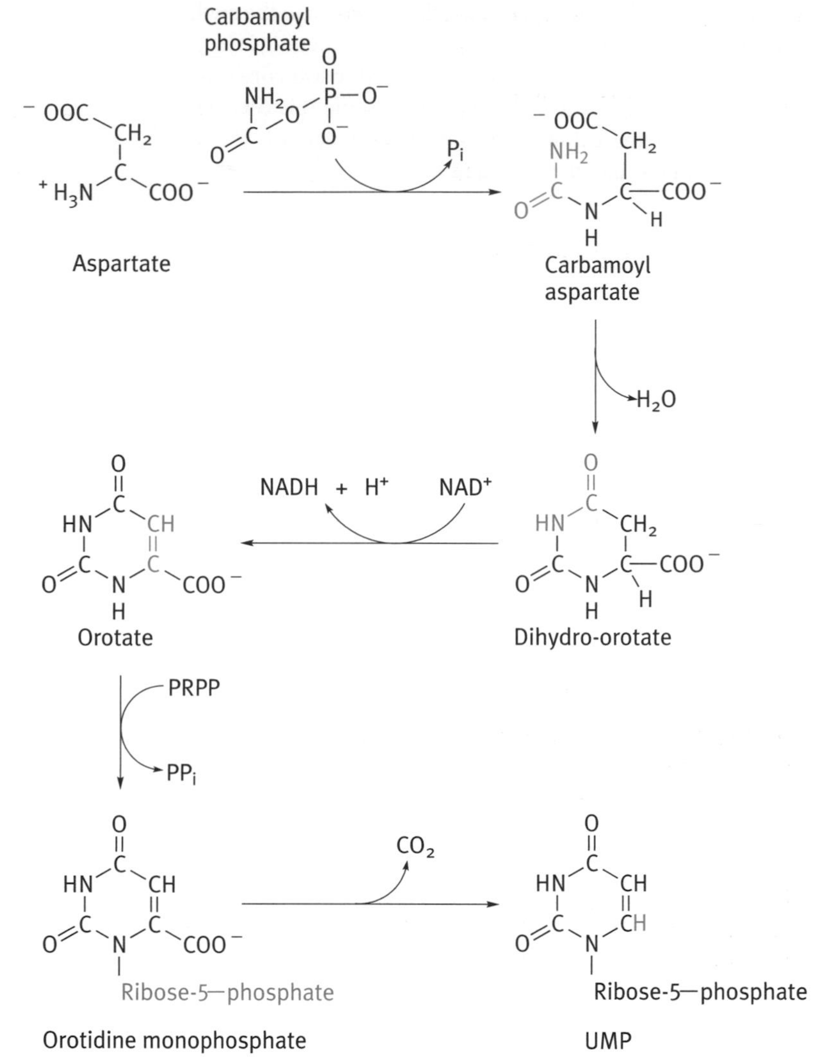

Fig. 19.10 Pathway for the *de novo* synthesis of pyrimidine nucleotides. Blue indicates the structural change resulting from the latest reaction.

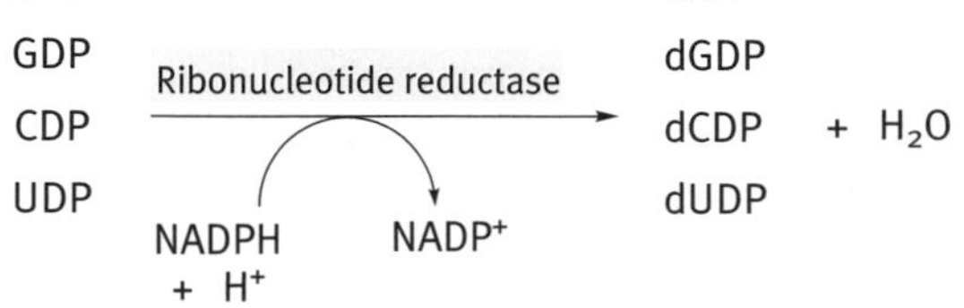

The resultant dADP, dGDP, dCDP, and dUDP are converted to the triphosphates by phosphoryl transfer from ATP. However, dUTP is not used for DNA synthesis since DNA, you will recall, has thymine (the methylated uracil) as one of its four bases, but never U. The dUTP is converted to dTTP. This is done in three steps: first, dUTP is hydrolysed to dUMP,

$$dUTP + H_2O \rightarrow dUMP + PP_i.$$

The dUMP is converted to dTMP and then to dTTP by phosphoryl transfer from ATP. An appropriate system of allosteric feedback controls exists to keep the production of the four deoxynucleotide triphosphates in balance.

Conversion of dUMP to dTMP

The methylation of dUMP is of especial interest. The enzyme involved is called **thymidylate synthase** and it utilizes the coenzyme N^5, N^{10}-methylene FH$_4$. Recall (page 293) that, in purine synthesis, the methylene group of the latter is oxidized to produce a formyl group. In thymidylate synthesis, the methylene

group is transferred and, at the same time, *reduced* to the methyl group of thymine; the reaction is catalysed by thymidylate synthase. The reducing equivalents for the reduction come from FH_4 itself, leaving it as FH_2 (note how versatile this coenzyme is). In the scheme below, only the relevant part of N^5, N^{10}-methylene FH_4 is shown (see page 293).

dUMP N^5,N^{10}-methylene FH_4

dTMP   Dihydrofolate (FH$_2$)

The FH_2 produced in this reaction is reconverted to FH_4 by dihydrofolate reductase. The FH_4 can be reconverted to methylene FH_4 by reaction with serine (page 293).

For thymidylate synthesis to continue, FH_2 must be reduced to FH_4 by dihydrofolate reductase. The antileukemic drugs **methotrexate** (amethopterin) and **aminopterin** (two of the so-called 'antifolates') inhibit FH_2 reductase by mimicking the folate structure.

The structure of methotrexate (amethopterin) is shown here for interest (you do not need to memorize it); aminopterin is similar but lacks the N^{10}-methyl group.

Methotrexate (Amethopterin)

Cancer cells, like those of leukemia, require rapid dTMP production to synthesize DNA and are thus selectively inhibited. The scheme is outlined in Fig. 19.11.

Tetrahydrofolate, vitamin B$_{12}$, and pernicious anemia

N^5, N^{10}-Methylene FH_4 in addition to the above reaction, can be reduced to N^5methyl FH_4 which supplies the methyl group to convert homocysteine to methionine (see page 240), a reaction catalysed by **methionine synthase**. The latter requires coenzyme B$_{12}$ as cofactor; it is believed that in the absence of this, as in the disease pernicious anemia, the available supplies of FH_4 become trapped as methyl FH_4 and are unavailable for reactions involved in purine nucleotide biosynthesis and other reactions. The **methyl trap hypothesis** would account for the fact that vitamin B$_{12}$ deficiency produces a megaloblastic anemia identical to that seen in folate deficiency. The neurological symptoms of pernicious anemia may be related to methylmalonic acidosis, referred to earlier (page 183).

Although several vitamin B$_{12}$-related enzyme reactions occur in bacteria only the two, methylmalonyl-CoA mutase (page 183) and methionine synthetase, are known in humans. Lack of vitamin B$_{12}$ in pernicious anaemia is due to the absence of a gastric glycoprotein, the intrinsic factor, required for absorption of the vitamin in the intestine, rather than to a dietary deficiency.

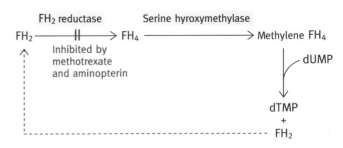

Fig. 19.11 Site of action of the anticancer agent, methotrexate. The relationship of the methotrexate structure to folic acid can be seen by comparing the structure on page 292.

FURTHER READING

Antifolate drugs

Huennekens, F. M. (1994). The methotrexate story: a paradigm for development of cancer therapeutic agents. *Adv. Enzyme Regulation*, 34, 392–419.

Gives history of the antifolate drug, the structure of dihydrofolate reductase and development of methotrexate resistance in cells.

PROBLEMS FOR CHAPTER 19

1   Describe the method of ribotidation.

2   What is the cofactor involved in the two formylation reactions involved in purine nucleotide synthesis? Give the essential structure of its formyl derivative.

3   What is the origin of the formyl group? Explain how this is generated.

4   What is the function of hypoxanthine–guanine phosphoribosyltransferase (HGPRT)?

5   The Lesch–Nyhan syndrome is a severe genetic disease in children. Discuss its biochemistry.

6   How does the drug allopurinol reduce uric acid formation?

7   Draw a diagram illustrating the main allosteric controls on purine nucleotide synthesis.

8   Several compounds in the body are methylated using methionine as methyl group source. Is this true of thymidine monophosphate synthesis? Explain your answer.

9   How does the antileukemic drug methotrexate inhibit cancer cell reproduction?

10   The symptoms of pernicious anemia resemble, in some respects, those of folate deficiency. What is a possible reason for this?

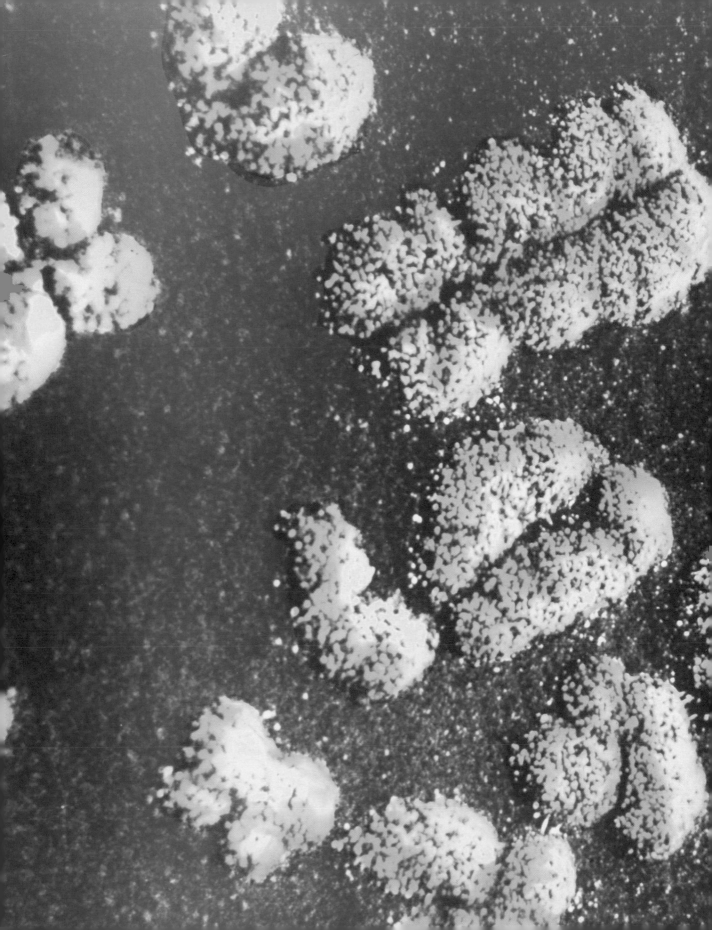

Part 4

Information storage and utilization

Colorized scanning electron micrograph of a number of human mitotic chromosomes.

Chapter summary

DNA—its structure and arrangement in cells

In this chapter we will deal with the structure of DNA and what a gene is. We will then, in the following chapter, describe DNA synthesis, followed in Chapter 22 with the way in which DNA directs protein synthesis. The central point is that the function of DNA is to carry information needed for the synthesis of proteins whose amino acid sequences have been devised by millions of years of evolution to carry out the myriad processes on which life depends.

being the remainder, whereas most of the RNA is found in the cytoplasm although, with the exception of that made in mitochondria and chloroplasts, it originates in the nucleus. Although we have already discussed nucleotides in Chapter 19 which dealt with their synthesis, we will repeat some of the material here both for convenience and because of its importance. With that introduction let us turn to DNA.

What are nucleic acids?

DNA was first isolated from cell nuclei; it is an acid because of its phosphate groups—hence the term nucleic acid. It contains a sugar, 2-deoxy-D-ribose and therefore is called deoxyribonucleic acid, or DNA for short. There is another acid of similar structure found in cells in which the sugar is D-ribose. It is therefore called ribonucleic acid or RNA for short (this is dealt with in Chapter 22). The two sugars are shown below. 2-Deoxy-D-ribose lacks the oxygen on the carbon 2 position; it is usually simply referred to as deoxyribose.

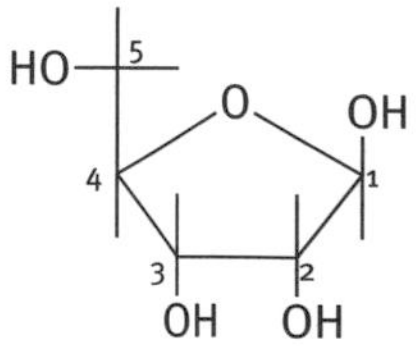

D-Ribose 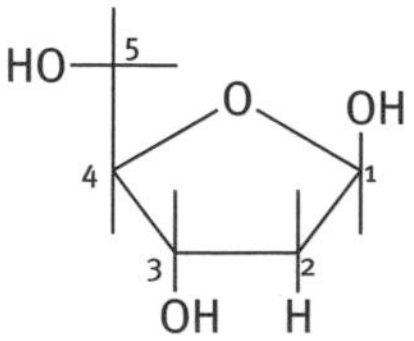   2-Deoxy-D-ribose

In eukaryote cells, the bulk of the DNA is confined inside the nuclear membrane, that in mitochondria and chloroplasts

The primary structure of DNA

DNA is a polynucleotide. A nucleotide has the structure

phosphate-sugar-base.

The structure of a deoxyribonucleotide is shown in the nonionized form for simplicity.

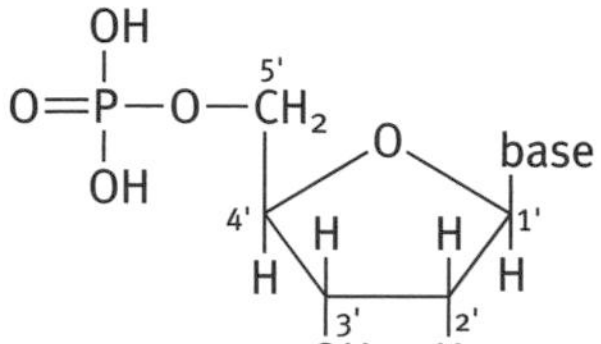

The sugar, as stated above, is 2-deoxy-D-ribose (deoxyribose). To specify a position in the deoxyribose moiety, a prime (′) is added to distinguish it from the numbering of the base ring atoms. Thus the sugar carbon atoms are 1′, 2′, 3′, 4′, and 5′ (pronounced 'five prime', etc.) and indicated outside the ring. The sugar is in the furanose, five-membered ring form. The nomenclature of nucleotides is described on page 287.

What are the bases in DNA?

Unlike the situation in proteins where you have 20 different amino acids, in DNA there are only four different bases, but their structures are just a bit more difficult to remember than those of amino acids. The bases are adenine, guanine, cytosine, and thymine—abbreviated to A, G, C, and T. A and G are purines; C and T pyrimidines. The numbering of atoms in the bases is given inside the ring structures.

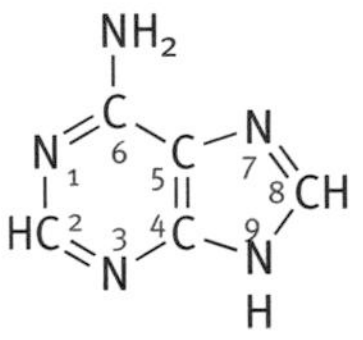

Adenine (A)

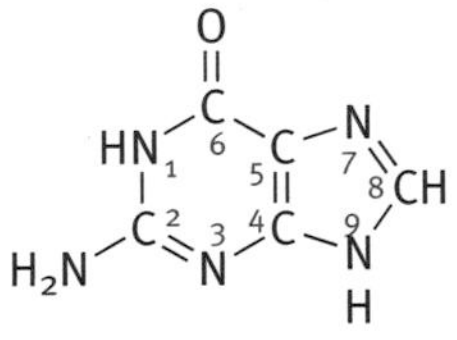

Guanine (G)

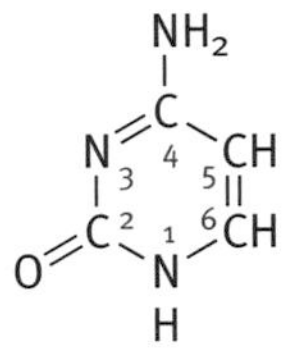

Cytosine (C)

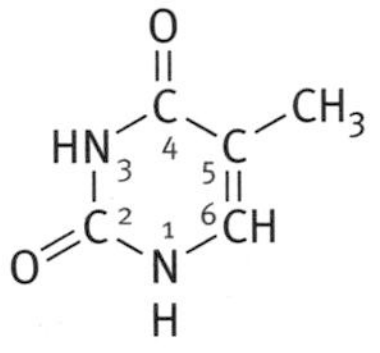

Thymine (T)

Note the methyl group of thymine. In subsequent structures we will represent these diagrammatically.

Tautomeric forms of the bases can exist (keto–enol and amino–imino) but in DNA the bases are essentially always in the form shown with —C=O and —NH₂ groups the

predominant form at neutral pH rather than C—OH and

C=NH groups.

Attachment of the bases to deoxyribose

The glycosidic link is between carbon atom 1 of the sugar and nitrogen atoms at positions 9 and 1, respectively, of the purine and pyrimidine rings. The linkage is β (that is, above the plane of the ring).

The structure, base–sugar, is called a **nucleoside;** if the sugar is deoxyribose it is a **deoxyribonucleoside.** All the deoxyribonucleosides have specific names. where the base is adenine it is deoxyadenosine; the guanine derivative is deoxyguanosine. Deoxycytidine and deoxythymidine are the deoxynucleosides of cytosine and thymine respectively. The structures are shown

below in diagrammatic form. The heterocyclic ring structures are not shown in detail but the characteristic side groups are given.

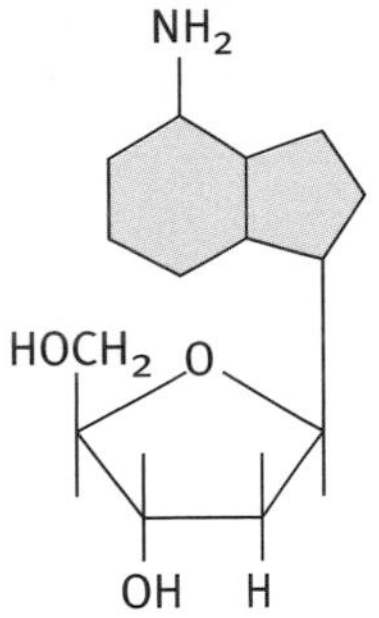

Deoxyadenosine

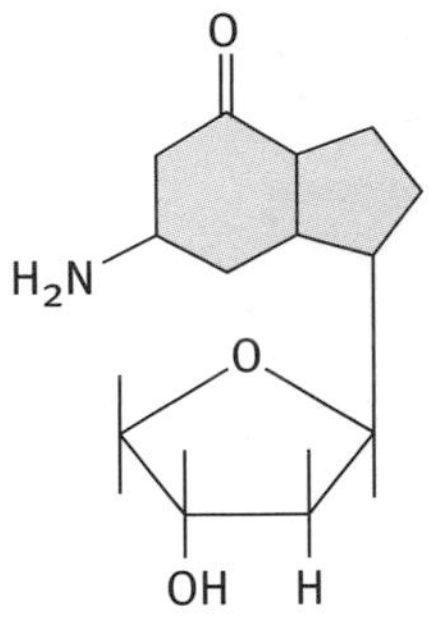

Deoxyguanosine

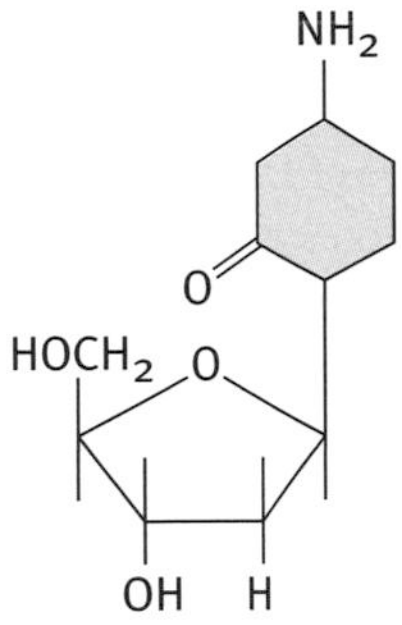

Deoxycytidine

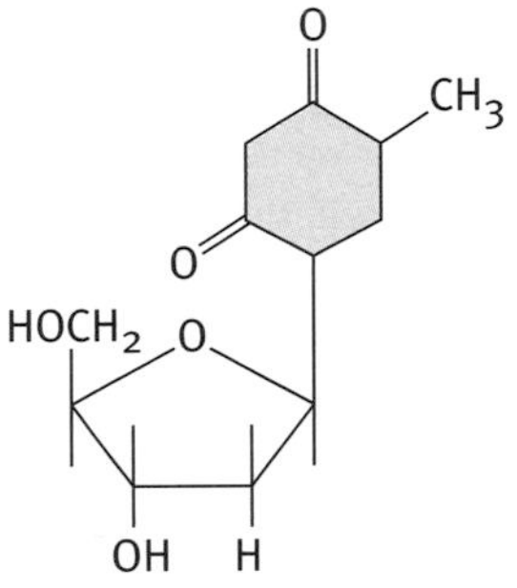

Deoxythmidine
(or, simply, thymidine)

What are the physical properties of the polynucleotide components?

The nucleotides in DNA are the 5′ phosphate compounds, dAMP, dGMP, dCMP, and dTMP. The latter may be referred to as TMP since the ribose analogue is rarely encountered.

The phosphoric acid —OH groups are both ionized at physiological pH since one of the —OH groups has a pK_a of around 2 and a second of around 7. This means they are highly hydrophilic. DNA is strongly negatively charged. The sugar with its hydroxyl group is also strongly hydrophilic. The bases are different—and this is important—in that they are almost water-insoluble. Their flat faces are essentially hydrophobic but at the edge of each there is hydrogen-bonding potential.

Structure of the polynucleotide of DNA

Suppose we have two such nucleotides and notionally eliminate a water molecule as shown; the resultant structure is a **dinucleotide** (the nonionized forms are shown here for clarity).

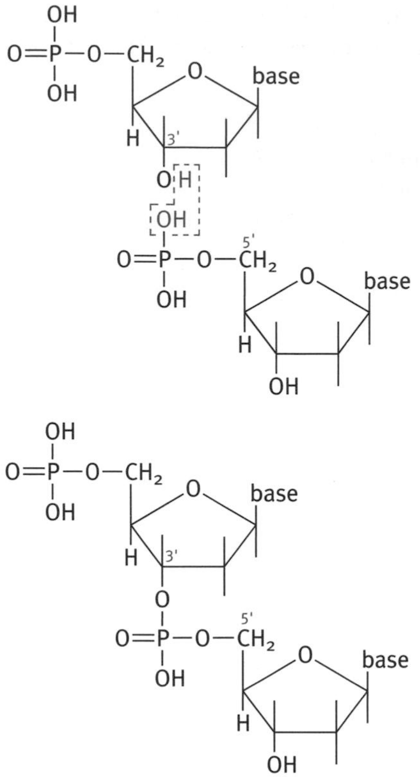

A dinucleotide

sugar–phosphate–sugar groups with a base attached to each sugar residue on the 1′ position; coded information is carried in the sequence of the bases. DNA therefore has the primary structure.

Backbone section	Informational or coding section of the structure
2′-deoxyribose	—base
phosphate	
2′-deoxyribose	—base
phosphate	
2′-deoxyribose	—base

or, in structural terms,

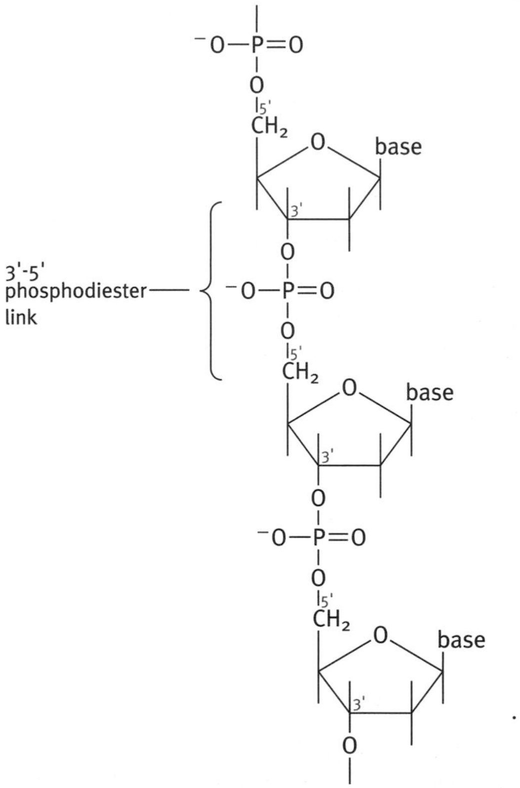

The formation of a dinucleotide involves a large positive $\Delta G^{0\prime}$ value, so be clear that synthesis cannot occur by the direct condensation of two nucleotides. In a mononucleotide, the phosphate group is a **primary phosphate ester** by which we mean that there is only a single ester bond. In the dinucleotide, a **phosphodiester link** is formed—it is **diesterified**, the phosphate being linked to two groups.

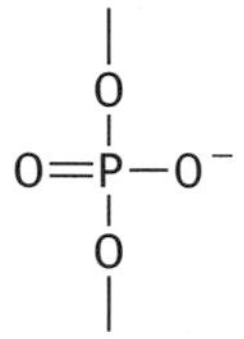

In the dinucleotide shown above it is a 3′,5′ - phosphodiester—the phosphate bridging between the 3′-OH of one nucleotide to the 5′-OH of the next. Nucleotides can be added in the same way indefinitely, giving a polynucleotide. DNA is in its primary structure a polynucleotide of immense length. In dealing with proteins you will recall that there is a **polypeptide backbone** (page 38) with amino acid residues attached. A polynucleotide has a backbone of alternating

Why deoxyribose? Why not ribose?

The cell has several nucleotides (for example, AMP) with ribose as the sugar component, and the nucleic acid, RNA (described in Chapter 22), has ribose as its sugar. In evolution it is probable that ribonucleotides predated deoxyribonucleotides and RNA predated DNA. The cell nevertheless goes to considerable

energetic expense to convert ribonucleotides to the deoxyribonucleotides required for DNA synthesis. One reason for this at least seems logical. DNA is the repository of genetic information gathered over countless millions of years and it is stored in chemical form in DNA molecules. This raises what might seem at first sight to be an insuperable obstacle to life—namely, that chemical molecules always have some degree of instability—they spontaneously break down, while DNA molecules have to remain largely unchanged, and certainly intact, for untold numbers of generations. The presence of the 2'-OH group of ribose makes a ribopolynucleotide less stable than the corresponding deoxyribose molecule—that is, DNA is more stable than RNA. This is because the 2'-OH group is suitably placed for a nucleophilic attack in the presence of OH$^-$ ions on the phosphorus atom, thus causing breakage of the phosphodiester link by forming a 2', 3' cyclic phosphate as shown; in DNA, lacking the 2'-OH group, this does not happen.

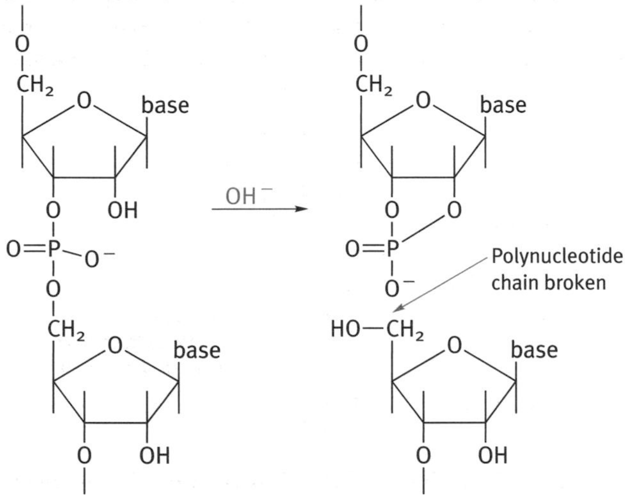

The OH$^-$ ion facilitates the reaction because it can generate a 2'-O$^-$, from the 2'-OH group, which attacks the phosphorus atom and converts the phosphodiester group into a 2', 3' -cyclic nucleotide, thus breaking the polynucleotide chain. Hydrolysis of the cyclic nucleotide produces a mixture of 2' and 3' nucleotides at the breakpoint.

The difference in stability is illustrated by the fact that dilute NaOH will completely destroy RNA at room temperature while DNA is unaffected. This is not the only chemical stability problem to be coped with—the subject of a later separate section on the repair of DNA (page 330). DNA is therefore a more stable repository of genetic information than is RNA. The fact that some viruses can get away with RNA for this role does not contradict this concept, as is explained later.

The DNA double helix

There will be few readers who have not heard of the double helix. DNA almost always exists as a double strand—only in a few viruses is it not double-stranded. In other words, you have two polynucleotide molecules paired together. What holds them together? The answer is **complementary base pairing**.

'Complementary' refers to base complementarity. It means that A and T in DNA chains are complementary in shape so that, when they are opposite one another in the two chains, they automatically form two hydrogen bonds between them. (We remind you that hydrogen bonds are short-range ones so that precise positioning of the pairing atoms is essential.) G and C are also complementary in shape and they can form three hydrogen bonds between them. Other combinations are *not* complementary so that G will not pair in the same way with A or T, etc. Only A–T and G–C pairing takes place in DNA, this being known as **Watson–Crick base pairing**. It is stressed that this is a completely automatic process requiring no catalysis. Because hydrogen bonds are weak, they are easily broken, for example, by thermal energy (page 14). Mild heat will cause unpairing; cool the substance and re-association occurs.

The geometry of base pairing is shown in Fig. 20.1. Note that the base pairs always include one purine (larger molecule) and one pyrimidine (smaller) so that the base pairs are essentially the same size. The reality of base pairing as a spontaneous process is shown by the phenomenon of **hybridization**. If a molecule of DNA is cut up into thousands of short double-stranded pieces, each say, about 20 to a few hundred nucleotides long and then the mixture is heated to an appropriate temperature (about 95°C), the two strands of each piece of DNA will

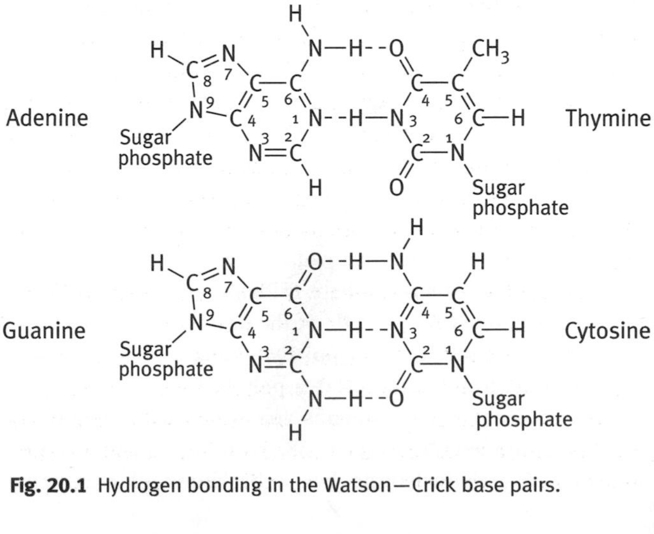

Fig. 20.1 Hydrogen bonding in the Watson—Crick base pairs.

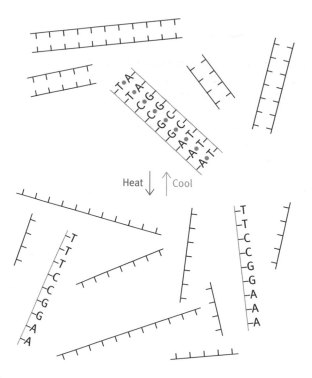

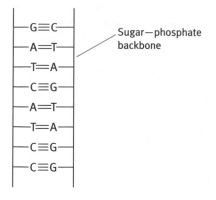

Heat ↓ ↑ Cool

Fig. 20.2 Spontaneous hybridization of pieces of complementary DNA. Base sequences are shown on a single piece of DNA to illustrate the fact that hybridization depends on them.

separate—referred to as **DNA melting**. This is because heat disrupts the hydrogen bonding between them, resulting in thousands of single-stranded DNA pieces each of different base sequence. However, if the solution is cooled, the pieces will slowly re-associate with their original partners (Fig. 20.2). This hybridization technique, sometimes called **annealing**, lies at the heart of gene molecular biology (see Chapter 29). There is a thermodynamic driving force for hybridization of the pieces, since the formation of hydrogen bonds and associated weak forces releases energy. The most stable state in which free energy is minimized is that in which the bases are paired since this gives the maximum number of hydrogen bonds. This experimental illustration of hybridization shows that the G–C, A–T base pairing occurs spontaneously and is a real observable phenomenon.

Because of base complementarity, if you analyse the different base contents of different DNAs, the amount of G equals that of C and that of A equals T. Different DNAs vary in their percentages of [A + T], and of [G + C] because their base compositions are different, reflecting their different genetic information. For the first approximation then a stretch of DNA might be represented as shown below, the long solid line representing the

sugar–phosphate backbones and the attached bases interacting by hydrogen bonding.

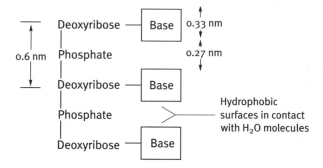

Note that, in a piece of DNA rich in G + C, the two strands will be more strongly held together than in a piece rich in A + T.

However, weak forces have a profound influence on the conformation of large molecules and this is no exception—the above structure, with ladder-like straight strands—is not 'allowed' under normal solution conditions. One important reason is shown in the diagram below.

The length of the phosphodiester link is 0.6 nm (1 nanometer (nm) = 10^{-9} m) while the bases are about 0.33 nm thick, so that in the straight ladder-like structure there would be a gap between them. The faces of bases are, as mentioned, hydrophobic, and in the 'straight' structure above, they would be exposed to H_2O molecules, an unstable situation. What can be done to bring the hydrophobic bases together to exclude H_2O from their faces? It can be achieved by collapsing them together by sloping the phosphodiester link; or, perhaps more accurately, hydrophobic forces cause the bases to collapse together, as shown in Fig. 20.3(a).

The base pairs still lie flat, stacked on top of each other—a phenomenon known as **base stacking**. The hydrogen-bonding face at the edge is still exposed so that it can bond to its partner strand. However, the 'skewed ladder' structure shown is not the form in which DNA occurs, for such a structure has

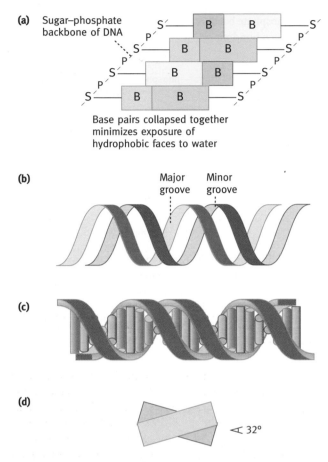

(a) Sugar–phosphate backbone of DNA

Base pairs collapsed together minimizes exposure of hydrophobic faces to water

(b) Major groove   Minor groove

(c)

(d) 32°

Fig. 20.3 **(a)** How a skewed arrangement of the ladder collapses the bases together. **(b)** Outline of the backbone arrangements in the DNA double helix. **(c)** As (b). but showing the base pairs in the centre of the helix. (Note that each coloured band is a base *pair*.) **(d)** Diagram of two successive base pairs in a double helix showing the twist imposed by the double helix. A more realistic model corresponding to (c) is shown in Fig. 20.5.

stereochemically unacceptable features. Instead, each of the two DNA chains forms a helix, with the bases inside and the hydrophilic sugar and phosphate groups outside (Fig. 20.3(b). (c)).

The arrangement still permits base-pair stacking and the exclusion of water from between them, but the stacking cannot be exactly vertical. Instead, successive base pairs must rotate slightly relative to one another, as illustrated in Fig. 20.3(d), such that approximately 10 base pairs are required to rotate through one complete turn.

The helices are right-handed—as you move along a strand or a groove you continually turn clockwise; alternatively, imagine you are driving in a screw, holding the screwdriver in your right hand. The turning motion gives the direction of twist. The structure of the double helix is such that there is a major and a

minor groove (see Fig. 20.3(b)). Although the grooves are shown in the illustrations, it is not necessarily easy to picture the three-dimensional arrangement from these. If you have access to a physical model of DNA it would be helpful to walk around it and follow the two grooves, for they are important. Any particular base pair can be viewed from both the major and the minor grooves but only their edges are visible. The major grooves provides easier access for proteins to 'recognize' (by which we mean attach to) the base pair edges. A given base pair 'looks' quite different when viewed from the two grooves (see Fig. 20.4), the significance of which will become apparent in Chapter 22 where gene regulation is discussed.

The DNA conformation described above is known as the **B form** (Fig. 20.5) and is the normal form that exists in cells. However, DNA *can* adopt different configurations in special circumstances. When dehydrated, the double helix is more squat in shape and the bases are tilted; this is known as the A form. It may exist in spores. Another form is known as Z (because the polynucleotide backbone zigzags); in this the double helix is left-handed (cf. right-handed in B DNA). It has been observed to occur in short synthetic DNA molecules with alternating purine and pyrimidine bases provided the solution is of high ionic strength. Whether either of the A or Z forms has biological significance is not known, but the possibility of DNA adopting modified configurations in localized sections of the chromosome is not excluded.

An important property of the double helix structure is that it can bend. A molecule of DNA may be a million times longer than the widest dimension of a cell or nucleus; to pack it in, flexibility is clearly needed.

It needs to be mentioned that, while DNA is almost always in the double helix form, single-stranded DNA does occur, for example, in certain bacterial viruses. In such situations the molecule takes up a complex internally folded structure to satisfy thermodynamic considerations. The main thing to note is that in such cases the life cycle involves a double helix form of DNA (Chapter 28) so that the basic principles of genetic information with complementary base pairing are the same in all cases. The existence of single-stranded forms is in this sense a specialized idiosyncrasy—not a fundamental difference.

DNA chains are antiparallel; what does this mean?

By antiparallel we mean that the two chains of a double helix have opposite polarity—they run in opposite directions. It may not be immediately clear what is meant by the polarity or direction of a DNA strand. It is worth spending a little time on this so that you are totally comfortable with the concept, because a lot of biochemistry requires an understanding of it. Two antiparallel strands of DNA are illustrated in Fig. 20.6.

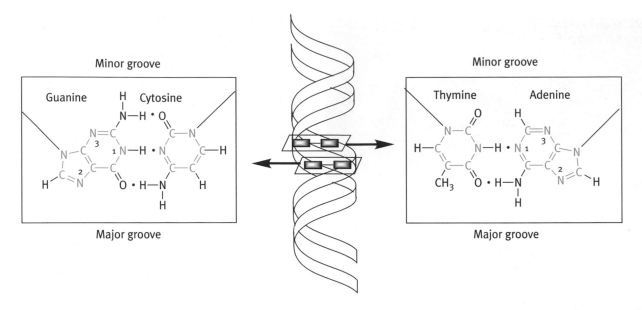

Fig. 20.4 The edges of a given base pair in DNA look different when viewed from the major and minor grooves. DNA binding proteins designed to recognize specific sequences of base pairs in DNA can identify (bind to) the characteristic chemical groupings of different base pairs without unwinding the DNA. The importance of this is discussed in Chapter 22 which deals with gene control.

The first point is that any single linear strand of DNA has (obviously) two ends. One end has a 5′-OH group on the sugar nucleotide that is *not* connected to another nucleotide. (It may have a phosphate on it.) This is the 5′ end. The other end has a 3′-OH group that it *not* connected to another nucleotide (though it also may have a phosphate group esterified to it). This is the 3′ end. At one end of a piece of double helix there is always one 5′ end and one 3′ end. You can never have two 5′ or two 3′ ends together. This is what antiparallel means. If a piece of DNA is circular, there are no free ends but there is still inherent polarity in the individual strands, as you can see from the sugar moieties.

You can read off the direction by looking at the 5′ and 3′ positions. Thus the structure on the left runs 5′ → 3′ down the page and that on the right runs 5′ → 3′ up the page (by convention, the 5′ → 3′ direction is always specified).

Incidentally, in case it seems contradictory, as you move down a DNA strand in the 5′ → 3′ direction the actual phosphodiester bonds that you traverse are 3′ → 5′ (see above diagram). *Remember that a 5′ → 3′ direction in a DNA strand means that you are processing (in a linear molecule) from a terminal 5′-OH group towards a terminal 3′-OH group* or, if you prefer it put it another way, you are travelling in the direction from the 5′ group of the deoxyribose towards the 3′ group of the same sugar as illustrated in the diagram above. This will all become familiar to you, as will its importance, as you progress through the next two chapters. The antiparallel arrangement fits the stereochemical requirements of the double helix structure.

There are conventions in writing down the base sequence of DNA and it is important that you understand these. It is usual to represent a polynucleotide structure simply by a string of letters representing the bases of component nucleotides, but sometimes the phosphodiester link is indicated by the letter p inserted between the bases (for example, CpApTpGp, etc.). Suppose we have a piece of double-stranded DNA whose base sequence is

5′ CATGTA 3′
3′ GTACAT 5′.

Sometimes it is useful to write both strand sequences, but usually it is not necessary to write both sequences since, given one, the complementary sequence is automatically specified. So you will find that the structure of a gene is often given as a single

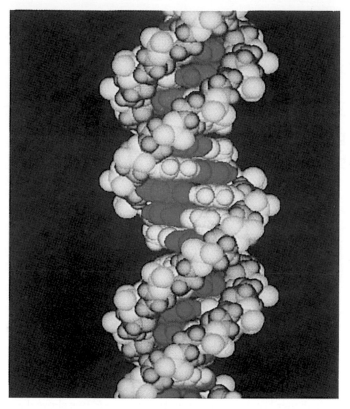

Fig. 20.5 A model of B DNA. Space-filling atomic model of a DNA segment with two major grooves and one minor groove.

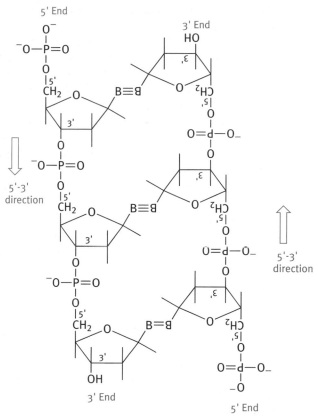

Fig. 20.6 Two antiparallel strands of DNA. B, Base.

base sequence, despite there being two strands. There is a convention that a single base sequence is written with the 5′ end to the left and there is no need therefore to specify the 5′ and 3′ ends of a sequence. Thus, if the structure illustrated above is part of a gene, it would be written as CATGTA.

How large are DNA molecules?

The variation can be enormous but the main thing is that even the shortest is a very long molecule in terms of the number of base pairs and the longest is gigantic. The DNA of a virus may be as small as a few thousand base pairs in length; *E. coli* DNA has about four million base pairs (it is circular in form). The human genome (all the chromosomes taken together) totals 1–2 metres of DNA in length and contains 6 billion base pairs. There are about 10^{13} cells in a human being and, if you add up the total length of DNA in a person, it comes out to an astronomical length, of the order of a diameter of the solar system.

Clearly, the more information that has to be recorded the greater the amount of DNA needed for this. You would have imagined, therefore, that the more complex the organism, the greater would be the DNA content of its cells. Surprisingly, it doesn't' always work out like that. A broad bean cell has more DNA than a human cell—among vertebrates, amphibia have the most. In some comparisons, the DNA contents of organisms do correspond to perceived complexity. The apparent anomalies in DNA content are related to the fact that a good deal of DNA is not in the form of functional genes, as described later.

How is the DNA packed into a nucleus?

There are many cases in biochemistry where a problem facing life is so staggering that the solution boggles the imagination to a point where it would be reasonable to wonder if it could pos-

sibly work—except for the fact that it obviously does. The packing of DNA is one of these.

In a human cell, there are 46 chromosomes giving, as already mentioned, a total length of DNA per cell of 1–2 metres, packed into a nucleus millions of times smaller in diameter. A very elaborate packing procedure indeed is needed. DNA in eukaryotic cells exists as **chromatin**—a DNA–protein complex. The main proteins are **histones**; these are small basic proteins rich in arginine and lysine, giving them positive charges that can form ionic bonds with the negative charges on the phosphate groups on the outside of the double helix of DNA. The amino acid sequences of eukaryotic histones are highly conserved throughout evolution. Indeed, one of the histones differs only in two amino acids in the entire molecules found in peas and cows and the changes are very conservative (valine for isoleucine, lysine for arginine). This extreme conservation presumably means that this protein must precisely combine with a structure, or structures, equally invariant. Possibly the histone has to participate in a complex structure with multiple essential contacts and any mutation to be tolerated must not abolish any of them.

We will start with the four histones called H2A, H2B, H3, and H4. These form an octamer protein complex called a **nucleosome core** around each of which the DNA wraps two turns (146 base pairs) with an intervening stretch linking successive nucleosomes (Fig. 20.7(a), (b)). If, in the test tube, chromatin is treated with a microbial enzyme capable of hydrolysing DNA into nucleotides, the intervening DNA links between nucleosomes are hydrolysed but the 146 base pairs wrapped around the nucleosome core are protected from the enzyme—the latter cannot readily get at the DNA to hydrolyse it. If the remaining DNA is then extracted, it is found to contain these protected segments. This was the clue that led to the discovery of nucleosomes. This arrangement condenses (packages) the DNA somewhat, but nowhere near enough. The nucleosomes are about 10 nm in diameter and form the 10 nm fibre shown in Fig. 20.7(b). A fifth histone protein known as H1 is involved; it is not part of the nucleosome itself. This 10 nm is now condensed further to form a fibre, known as the 30 nm fibre, illustrated in Fig. 20.7(c). The nucleosome packing in this fibre is often depicted as a solenoid arrangement of the nucleosomes, but this is but one of several models; another model has a zigzag arrangement. Since the exact structure has not yet been determined, the diagram of the 30 nm fibre has been left without detail of packing. An electron micrograph of such a fibre is shown in Fig. 20.8. The condensation of the DNA achieved so far is about 100-fold. The 30 nm fibres now form long loops that are attached to a central chromosomal protein scaffolding (Fig. 20.7(d)). This looped structure forms yet more densely packed structures, not yet fully understood, involving folding and/or

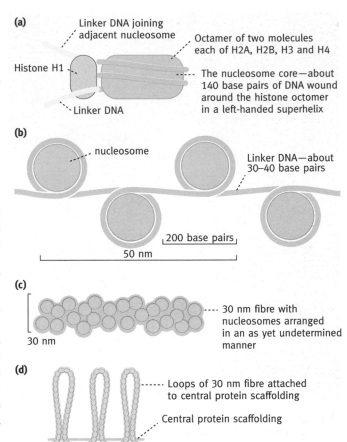

Fig. 20.7 Order of chromatin packing in eukaryotes. **(a)** Diagram of a nucleosome. **(b)** Beads on a string form. **(c)** A 30 nm fibre of chromatin (see Fig. 20.8 for an electron micrograph of a 30 nm fibre). **(d)** Loops of the 30 nm fibre are attached to a central protein scaffold in a 360° array. It is believed that these loops are yet further condensed, perhaps by supercoiling and ultimately into the extremely compact metaphase chromosome. The latter condensation stage is not illustrated.

coiling and achieving a 10 000-fold packing of the original DNA.

We thus have several levels of organization: (1) winding around nucleosomes; (2) nucleosomes packed into a 30 nm fibre about 3–5 nucleosomes thick; (3) the fibres form loops, thousands of nucleosomes long, attached to a central scaffolding; and (4) the loops form yet other coils and/or folds (see Fig. 20.7(a)–(d)).

As will be described in later chapters, to be functional all this packed DNA has to be accessible to enzymes that read its informational content and to enzymes that replicate the entire DNA.

In an *E. coli* cell, the DNA is a circle of double helix. Its length

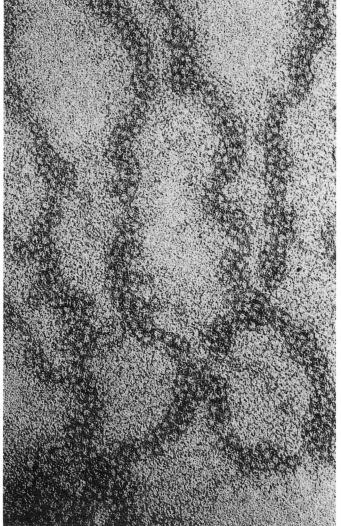

Fig. 20.8 Electron micrograph of a 30 nm fibre of chromatin. After Fig. 28.19 in Lewin B. (1994), *Genes V*, Oxford University Press, Oxford. (The electron micrograph was provided to B. Lewin by Barbara Hamkalo.)

is about 1000 times that of the cell. Prokaryotes have no nuclear membrane and there is no well-established structure corresponding to the eukaryote nucleosome organization. Nonetheless, proteins reminiscent of histones are found in *E. coli* that are postulated to play a somewhat analogous role. However, the proteins bind much less firmly to the DNA than do eukaryote histones so that nucleosome-like complexes have not been demonstrated. Looped DNA structures held in position by proteins are believed to occur in *E. coli*, but details in this area are uncertain at present.

How does the described structure of DNA correlate with the compact eukaryote chromosomes visible in the light microscope?

Chromosomes in stained dividing eurkaryote cells appear in the light microscope as compact solid structures such as those illustrated in Fig. 20.9.

The compact structures in dividing cells are **metaphase chromosomes** in which the DNA is in its most condensed state for the purpose of enabling the replicated chromosomes to be translocated into each daughter cell, a process described later (page 541). It is 'packaged DNA' for delivery to new cells. The **centromere** consists of a highly condensed section of DNA comprising repeated sequences. This section holds together the pair of chromatids and, on its two outward facing sides, the **kinetochore** assembly of proteins is attached. The microtubules of the mitotic spindle attach at the kinetochore to cause the daughter chromosomes to separate. (Microtubules are described in Chapter 33.) Once cell division is achieved, the chromosome becomes unpacked into the **interphase chromosome**. The DNA of a metaphase chromosome is functionally inert—it is the unpacked form that is functional (but still much more packed than a simple DNA thread). The degree of 'unpacking' of DNA in an eukaryotic cell is important. To use information in DNA the molecule must be accessible to enzymes and other proteins and not concealed in tightly condensed structures. Certain parts of eukaryotic chromosomes contain DNA believed not to be involved in gene structure—that is, DNA that does not carry coded information for the synthesis of proteins (see below). The DNA in these regions even in interphase (the phase between cell division events) chromosomes remain condensed (called **heterochromatin**) while the functional regions are less condensed (called **euchromatin**). The function of heterochromatin is not understood (with the exception of the centromere).

Proteins other than histones exist in chromatin, these 'nonhistone' proteins appear to be involved in the formation of the long loops of DNA attached to a central protein scaffolding. The essential message in all of this is that chromatin structure is complex; it varies in its degree of packing and the degree of packing of the DNA in chromatin is relevant to gene expression. The term **gene expression** refers to protein production directed by a gene, but there are exceptions to this statement in that there are genes that code for specific RNA molecules, as described later.

What is a gene in molecular terms?

A **gene** is the unit of heredity. In molecular terms it has no independent existence—it is simply a stretch of DNA, part of a huge

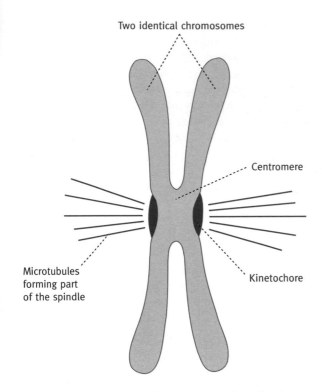

Two identical chromosomes

Centromere

Microtubules
forming part
of the spindle

Kinetochore

Fig. 20.9 Chromosome at metaphase consisting of two daughter chrornatids. See Fig. 33.7.

molecule, and carries coded information for the sequence of amino acids of one polypeptide chain (but see also page 346 for a more detailed description of a gene). For a protein with a single polypeptide chain, a gene therefore codes for that protein. The information resides in the base sequence of the DNA—there is no physical or molecular discontinuity between one gene and the next—only coded information. For a more familiar analogy a chromosome can be compared with an entire magnetic tape, a gene with a piece of music recorded on it, and the base sequence with the magnetic signals. The above describes what might be called the 'standard' typical protein-coding gene, which is the overwhelmingly predominant type.

Some variations on the 'standard' gene

Sometimes there are multiple copies of the same gene—this is true of some histone proteins. This situation can arise from gene duplication during evolution. Multiple copies of genes can provide for more rapid production of a given protein; there are multiple copies of histone genes that occur in clusters in some organisms—hundreds of repeating copies in the case of the sea urchin. In evolutionary terms, this also gives the evolutionary

potential of modification of one of the copies to code for an alternative protein. **Pseudogenes** also exist—essentially copies of functional genes but ones that are not expressed or that make a nonfunctional protein or protein fragment.

A group of genes exist that do not code for proteins. These are the ribosomal RNA genes and transfer RNA genes. They code for RNA molecules that are essential for the synthesis of proteins (see Chapter 24).

The usual 'standard' gene is a stretch of DNA fixed in position in the chromosome. DNA sequences are not necessarily fixed in position; chromosomes sometimes have elements that are mobile—they move from one place in the DNA to another. They are called **transposons** or jumping genes (page 335). Chromosomes can also acquire new resident genes from retroviruses (page 470).

Gene composition may not always be completely fixed; under certain conditions a DNA section containing a given gene may be amplified so that there is a vast number of copies of the gene and a vast, corresponding increase in the production of the protein coded for by that gene—as much as 1000-fold. This has been observed experimentally in cultured mouse cancer cells treated with the drug methotrexate. Methotrexate is an 'antifolate' used in treatment of leukemia and is described on page 298.

Repetitive DNA

In *E. coli* DNA has, speaking in general terms, one copy of a given section of DNA per genome, as indeed you might expect if all the DNA were in the form of genes. This is not true of eukaryotic DNA where only about 60% of the DNA is in the form of genes. The rest is known as **repetitive DNA**. This takes two forms; one class known as **satellite DNA** consists of short sequences as small as seven nucleotides tandemly repeated possibly millions of times per genome. If they are rich in A and T the lower density causes them to separate from the bulk of DNA in density gradient experiments giving rise to the name of satellite DNA; it is not a satellite in the cell. This form constitutes perhaps 10–20% of the total DNA. The other form consists of DNA sequences, not tandemly repeated (not joined together in the DNA) but scattered as single units throughout the genome. SINES are the shorter versions (short interspersed elements), about 300 base pairs; about a million of these are present. The best known are the **Alu sequences**, making up 10% of the total DNA, so named because they have a site for the Alu I restriction enzyme. LINES are long interspersed elements, a few thousand bases in length repeated many thousands of times. What is

the function of repetitive DNA? It is one of the mysteries of biochemistry.

Where are we now?

To avoid losing sight of the main thread amongst all the necessary detail, we have earlier dealt with protein structure and started the longish trail to understanding how proteins are synthesized. We next must deal with DNA synthesis for two reasons—it comes logically after DNA structure and will help in understanding the next steps towards protein synthesis. So, in the next chapter we move on to DNA synthesis. After that we will move on to the way information in the genes is used to direct the synthesis of proteins.

FURTHER READING

Ptashne, M. (1987). *A genetic switch: gene control and phage λ.* Cell Press and Blackwell Scientific Publications.

A classic book full of insights into DNA structure and function, as well as phage molecular biology.

Callandine, c. R. and Drew, H. R. (1992). *Understanding DNA: the molecule and how it works.* Academic Press.

A very clear discussion of DNA structure, supercoiling and DNA organization in the cell. Written in an interesting style.

Rennie, J. (1993). DNAs new twists. *Sci. Amer.,* **266** (3), 88–96.

A most interesting account of the 'unorthodox' genes—jumping genes, fragile chromosomes, expanded genes, edited mRNA, non-standard genetic code in mitochondria and chloroplasts.

Chambers, D. A., Reid, K. B. M., and Cohen, R. L. (1994). DNA: the double helix and the biomedical revolution at 40 years. *FASEB J.,* **8,** 1219–26.

Reviews a meeting to mark the 40th anniversary of the double helix. Biochemical nostalgia, but which summarizes the landmarks in the whole area and looks to the future. Strongly recommended for all students (and staff too!).

Nucleosomes

Hewish, D. R. and Burgoyne, L. A. (1973). Chromatin substructure. The digestion of chromatin DNA at regularly spaced sites by an nuclear deoxyribonuclease. *Biochem. Biophys. Res. Comms.,* **52,** 504–10.

This describes the experiment referred to in the text in which chromatin was found to be digested in a nonrandom manner. It was subsequently explained by R. D. Kornberg's discovery of nucleosomes.

Kornberg, R. D. and Lorsch, Y. (1999). Twenty-five years of the nucleosome, fundamental particle of the eukaryote chromosome. *Cell,* **98,** 185–94.

Reviews the nucleosome story.

PROBLEMS FOR CHAPTER 20

1   Write down the structure of a dinucleotide.

2   Ribonucleic acid (RNA) almost certainly evolved before deoxyribonucleic acid. Why do you think DNA evolved?

3   The flat faces of the bases of DNA are hydrophobic. Explain the structural repercussions of this fact on the structure of double-stranded DNA.

4 What is the main form of double-stranded helical DNA called? Is it a right- or left-handed helix? Approximately how many base pairs are there in a stretch of DNA that completes one rotation of the helix?

5 Explain what is meant by DNA chains in a double helix being antiparallel.

6 Explain in everyday language what is meant by a $5' \rightarrow 3'$ direction in a linear DNA molecule.

7 If you see a DNA structure simply written as CATAGCCG, what exactly does this means in terms of a double-stranded structure and the polarity of the two chains? Explain your answer.

8 (a) What is a nucleosome?
   (b) Describe an experiment which was an important clue to their existence.

9 What are Alu sequences?

Chapter summary

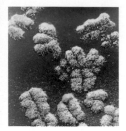

DNA synthesis and repair

In this chapter we will deal mainly with the *E. coli* cell, though the eukaryotic situation is described where information is available. DNA synthesis is a very complex process, and much more is known of the mechanism in *E. coli* than in eukaryotic cells. Sufficient is known of the latter to be sure that the processes are basically the same in both, even if not in absolute detail.

Every time a cell divides, its entire content of DNA must be duplicated or, as is more usually stated, the chromosome(s) must be replicated, so that a complete complement of DNA can be given to each daughter cell. A human cell has about 6 billion base pairs in its total DNA. The magnitude of the task of faithfully replicating these needs no emphasis. Even a single incorrect base in a gene may cause a protein with impaired function to be produced.

Overall principle of DNA replication

We will go into the question of *how* DNA is synthesized in due course, but for the moment let us look at it at a general level.

A chromosome is double-stranded DNA. Its replication is described as **semiconservative** in that the two original strands, called parental strands, are separated and each acts as a template for synthesizing a new strand; each new double helix has one old and one new strand. This was established in the classic experiment shown in Fig. 21.1.

The basis of the replication is that of complementarity in that a G will base pair with C, and A with T, so that a base on the parental strand automatically specifies which base is to be in-

corporated into the new strand as its partner. How this is specified to the synthesis machinery is described later in this chapter. Since this copying process depends on Watson–Crick hydrogen bonding of base pairs, it follows that strand separation is essential to unpair the bases and make them available for base pairing with incoming nucleotides. The *E. coli* chromosome is circular. It contains about 4 million base pairs. The strands are initially separated at one particular point called the **origin of replication** and *two* **replication forks**, moving in opposite directions, synthesize DNA at the rate of about 1000 base pair copies per second, with separation of parental DNA and synthesis of new DNA occurring at the same time (Fig. 21.2). The two forks meet at the opposite side of the circle. As a result of the replication, at the end there are two newly completed interlinked rings, a seemingly impossible situation. The prediction that this would occur was previously regarded as militating against the model of DNA replication. A remarkable insight of Crick was that evolution would have produced a mechanism to separate the circles. The topoisomerase type II described later (pate 320) does precisely this.

Eukaryotic chromosomes are linear. In eukaryotes, a replication fork synthesizes DNA at about an average 50 base pairs copied per second—too slow by a huge margin for a single replicon (explained below) to synthesize the vast lengths of DNA in a chromosome in the time allowed for it in cell division. To cope with this, there are hundreds of origins of replication along the chromosome from which replicative forks work in both directions (Fig. 21.3); thus, the process is analogous to the *E. coli* situation.

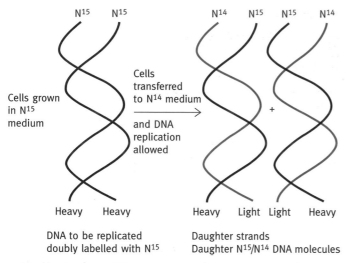

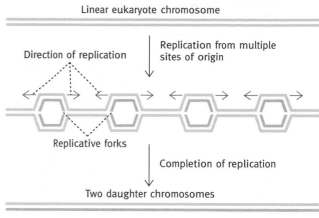

Fig. 21.1 Demonstration of semiconservative DNA replication by Meselson and Stahl. The DNA of cells was labelled by growing them in a medium in which the nitrogen source was N^{15}, so that both strands of DNA were 'heavy'. They were then transferred to N^{14} medium so that all subsequent DNA chains synthesized would be 'light'. The density gradient analysis indicated that, one generation after the transfer, each DNA molecule contained one 'heavy' and one 'light' strand. This is known as semiconservative replication. Continuation of the experiment for further generations confirmed the result. The red strands are newly synthesized.

Fig. 21.3 Diagram of multiple bidirectional replicative forks in a eukaryotic chromosome.

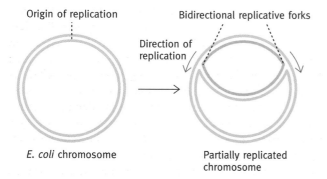

Fig. 21.2 Bidirectional replication of the *E. coli* chromosome. Parental strands are blue; newly synthesized strands are red.

Control of initiation of DNA replication in *E. coli*

Before cell division occurs in *E. coli* there must be a complete duplication of the chromosome; cell division follows about 20 minutes later. Exactly how cell division and DNA replication are coordinated is not understood. Protein synthesis and a critical

enlargement of the cell are required. As already seen (Fig. 21.2), in *E. coli* there is a single point of origin of DNA synthesis called *ori*C at which replication commences bidirectionally. The entire chromosome is called a single **replicon**, the latter term referring to a stretch of DNA whose replication is under the control of a single origin of replication.

The origin of replication has a specific base sequence, very rich in A—T pairs, presumably to facilitate strand separation. (We remind you that A—T pairs have two hydrogen bonds and G—C pairs three and, therefore, the former are less tightly bound together.) At the time of initiation, a protein referred to as DnaA binds in multiple copies to this region and causes strand separation. This permits the main unwinding enzyme (helicase), which works at each replicative fork, to attach and begin progressive unwinding of the strands in both directions. The **helicase** (or DnaB, the protein coded for by the gene *dnaB*) is referred to later when we describe the mechanism of DNA synthesis.

Initiation and regulation of DNA replication in eukaryotes

Eukaryotic cells have a more complex cell cycle (Fig. 21.4) than that of *E. coli* cells, in that DNA synthesis is confined to a definite period of time called the S (for synthesis) phase. Different cells vary a great deal but, in cultured animal cells, the **S phase** take about 8 hours out of the total cycle of 24 hours. Be-

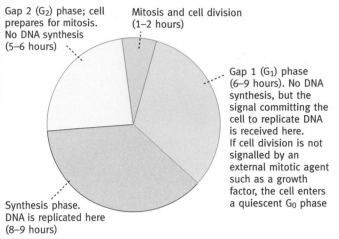

Gap 2 (G₂) phase; cell prepares for mitosis. No DNA synthesis (5–6 hours)

Mitosis and cell division (1–2 hours)

Gap 1 (G₁) phase (6–9 hours). No DNA synthesis, but the signal committing the cell to replicate DNA is received here. If cell division is not signalled by an external mitotic agent such as a growth factor, the cell enters a quiescent G₀ phase

Synthesis phase. DNA is replicated here (8–9 hours)

Fig. 21.4 The eukaryotic cell cycle. The duration of the cell cycle varies greatly between different cell types. The times given here are for a rapidly dividing mammalian cell in culture (24 hours to complete the cycle). See Chapter 30 for a more detailed account of the cell cycle.

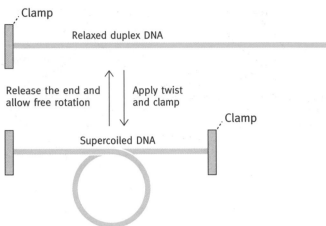

Clamp

Relaxed duplex DNA

Release the end and allow free rotation | Apply twist and clamp

Clamp

Supercoiled DNA

Fig. 21.5 The twisting of a piece of DNA that is not free to rotate induces supercoiling to accommodate the twisting strain. Cellular DNA is effectively clamped and is not free to rotate. If, somehow, free rotation is allowed, the supercoil will relax. If the applied twist is in the direction of unwinding, the supercoil will be negative; it will be positive if the applied twist is in the opposite direction.

fore the S phase is the **G₁ phase**, (G for gap in DNA synthesis). To proceed to cell division, a mammalian cell requires a mitogenic signal from outside itself. This takes the form of protein signalling molecules or **growth factors** that attach to surface receptors on cells and transmit the signal to grow to the interior of the cell. The latter process (that of cell signalling) is of immense importance and is the subject of Chapter 26. As explained, an animal chromosome may have hundreds of replicative origins from which bidirectional forks duplicate a replicon or stretch of DNA. What controls initiation of replication is not understood; the multiple replicons do not necessarily fire off at the same time. It is vital that each replicon fires off only once for each cell division.

Unwinding the DNA double helix and supercoiling

DNA strand separation by helicase presents topological (physical) problems, and to emplain these we must deal with the subject of DNA supercoiling.

Since duplex DNA has two strands in the form of right-handed helices, it has an inherent degree of twist, there being one turn of the helix per approximately 10 base pairs. An isolated short piece of linear DNA that is completely free to rotate on its own long axis automatically adopts this strain-free configuration, known as the **relaxed state**. Suppose instead that

you clamped one end of the duplex so that it was not free to rotate and you gave an extra twist to the other end, so that the coil of the double helix is tightened somewhat—the number of turns of a given stretch of DNA is increased; that is, the number of base pairs per turn is decreased. It is now **positively supercoiled** or **overwound**. If you twisted in the opposite direction, the coil would be opened up—the number of turns per unit stretch would be reduced, or the number of bases per turn increased. The DNA would be **negatively supercoiled** or **underwound**. Both the underwound and overwound states are under tension and one way of accommodating the strain is for the DNA double helix to coil upon itself forming a **coiled coil** or **supercoil** (Fig. 21.5). You can illustrate supercoiling very easily with a piece of double-stranded rope. Have someone hold one end or clamp it somehow, so that the rope cannot rotate freely, and twist the rope on its axis. Coils will form to take up the twisting strain. If you release the end of the supercoiled rope it will immediately spin back to the relaxed state. To determine whether a coil is positive or negative, look along the coil from either end. If the uppermost strand is turning to the left it is positive, if to the right it is negative (see Fig. 21.7).

What has this to do with DNA replication? DNA in the cell is not free to rotate on its own long axis; in *E. coli* the closed circle chromosome effectively 'clamps' the DNA. In eukaryotes, the DNA is of such vast length, arranged in fixed loops (see Fig. 20.7) and attached to protein structures, that once again free ro-

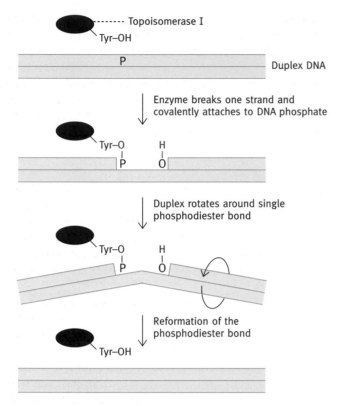

Fig. 21.6 The reaction catalysed by type I topoisomerases. The Tyr-OH represents a tyrosine residue of the enzyme.

tation is impossible. But, separation of DNA strands demands that the duplex rotates. This causes overwinding—it generates positive supercoils ahead of the replicative fork and, as the helix tightens, further separation is resisted. If unrelieved, the tension would bring strand separation and DNA replication to a halt.

A very simple experiment will convince you of this. If you take a short piece of double-stranded rope and pull the ends apart, the rope will spin rapidly, thus preventing the accumulation of positive supercoils. The rope strands will easily separate completely. Now take a long piece of the same rope coiled on the floor, or have someone hold one end of a reasonably long piece, so that it cannot freely rotate and try to pull the strands apart. Positive supercoils will rapidly snarl up the separating fork and prevent any further separation. This would be the situation in DNA replication in the cell if something weren't done about it.

It follows that, for DNA synthesis to proceed, the positive supercoils ahead of the replicative fork must be relieved and this, of necessity, involves the transient breakage of the polynucleotide chain.

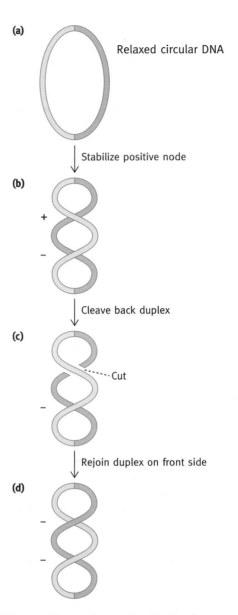

Fig. 21.7 Diagram of the reaction catalysed by *E. coli* gyrase to insert negative supercoils into circular DNA. This action can convert positively supercoiled DNA into negatively supercoiled DNA. Steps **(a)–(d)** are referred to in the text.

How are positive supercoils removed ahead of the replicative fork?

A group of enzymes, known as **topoisomerases,** catalyse the process. They act on the DNA and isomerize or change its topology. There are two classes of topoisomerases, called types I and II. We will deal with the principles of their mechanisms first, and then explain their roles in DNA replication.

In type I, the enzyme breaks one strand of a supercoiled double helix, which permits the whole duplex to rotate on the single phosphodiester bond of the partner strand, effectively introducing a swivel into the DNA. After rotation has occurred, the enzyme reseals the duplex (see Fig. 21.6). It is important to note that the enzyme *does not hydrolyse the phosphodiester bond it attacks*—it simply transfers the bond from the deoxyribose-3'OH to the —OH of one of its own tyrosine side chains. Since no energy loss is involved in the cutting of the chain, the process is freely reversible. Note that this enzyme does not use ATP. A supercoiled DNA molecule is in a state of tension—it is at a higher energy level than the relaxed state whether the supercoiling is positive or negative. A topoisomerase type I can only relax supercoiled DNA—it cannot insert supercoiling. As it happens, the particular topoisomerase I of E. coli can relax only negatively supercoiled DNA so it cannot solve the unwinding problem discussed so far. But, as will become evident shortly, it has an important role.

The relaxation of the positive supercoiling ahead of the replicative fork in E. coli is done in a different way—by the active insertion of negative supercoils, which 'neutralize' as it were the positive supercoils. This is done by topoisomerase type II, called gyrase, described below.

A type II topoisomerase breaks two strands of the DNA double helix transferring the bonds to itself, making the breakage of the polynucleotide chains a freely reversible process. The enzyme physically transfers the DNA duplex of a coil through the double gap, ATP being required for this piece of work, possibly through its hydrolysis effecting a conformational change in the protein. In E. coli, the topoisomerase II is called **gyrase**; it introduces negative supercoils in the DNA. The mechanism of this is illustrated in Fig. 21.7. In this figure, the insertion of negative supercoiling into a simple relaxed circle of DNA (Fig. 21.7(a)) is given for simplicity, but the same principle applies to the DNA ahead of the replicative fork in E. coli where positive supercoiling has occurred as a result of the strand separation. As stated, insertion of negative supercoiling in this region is equivalent to relaxation of the positive supercoiling. The gyrase binds to the DNA such that the latter is wrapped around the protein. In doing so it creates overwinding (positive supercoiling) in the local region associated with the protein (Fig. 21.7(b)). However, since no covalent bonds have yet been broken, there can have been no net change to the structure as a whole and, therefore, the local positive supercoiling caused by attachment to the protein must be compensated for by a negative supercoiling elsewhere in the molecule of DNA, thus keeping the net change in the supercoiling to zero. The enzyme now breaks both strands of the DNA (Fig. 21.7(c)) and, by passing the double strand from underneath to the upper side, followed by resealing of both breaks, converts the positive node to a negative one (Fig.

21.7(d)). In short, negative supercoiling has been inserted. The physical movement of the DNA through the gap is driven by ATP hydrolysis. The negative supercoiled state is also under tension and, therefore, the *insertion* of negative supercoils is an energy-requiring reaction. As already state the bonds broken are not hydrolysed but simply transferred temporarily to the enzyme itself without loss of free energy, the process of bond breaking is therefore freely reversible.

What are the biological implications of topoisomerases?

It is clear that is has been essential during evolution for living cells to cope with DNA topological problems and, as described, types I and II topoisomerases are the two mechanisms for doing so. However, once again, evolutionary tinkering has occurred and in various organisms the precise properties of the enzymes differ somewhat; at least 11 variants have been identified in different organisms.

In E. coli, the topoisomerase I relaxes only negative supercoils, not positive ones, while gyrase (type II) actively inserts negative supercoils and, since this will relax positive supercoils, permits DNA synthesis to proceed.

In eukaryotes, the topoisomerase I can relax both positive and negative supercoils; the type II topoisomerase can relax positive supercoils but cannot introduce negative supercoils (see Table 21.1). Thus, in prokaryotic and eukaryotic DNA replication, the potential snarl-up of strand separation through the accumulation of positive supercoils is averted.

However, *DNA supercoiling is important in more than DNA replication.* When DNA is carefully isolated from cells it is found to be negatively supercoiled. In relaxed DNA, the double helix

Table 21.1 Summary of E. coli and eukaryote topoisomerases, types and action

Type*	Action on DNA	Effect on supercoiling
Topoisomerase I		
E. coli	Cuts one DNA strand	Relaxes negative supercoils
Eukaryotes	Cuts one DNA strand	Relaxes positive and negative supercoils
Topoisomerase II (gyrase)		
E. coli	Cuts two DNA strands; ATP-dependent	Relaxes positive supercoils; inserts negative supercoils; separates interlinked circles
Eukaryotes	Cuts two DNA strands; ATP-dependent	Relaxes positive supercoils but cannot insert negative supercoils

*Note that variants of these occur in other organisms.

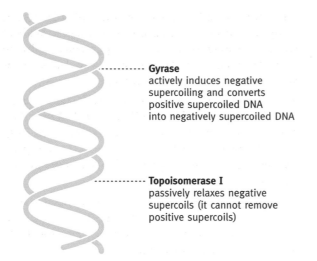

Gyrase
actively induces negative supercoiling and converts positive supercoiled DNA into negatively supercoiled DNA

Topoisomerase I
passively relaxes negative supercoils (it cannot remove positive supercoils)

Fig. 21.8 Diagram of the balancing act of gyrase and topoisomerase I in *E. coli* to achieve the correct degree of negative supercoiling required for DNA replication and transcription. Gyrase is ATP-driven. The amounts of the two enzymes synthesized (and hence the balance between them) is thought to be controlled by the degree of supercoiling of the gene promoters.

has one turn per 10.5 base pairs; in cellular DNA it has about one turn per 12 base pairs—it is underwound. The degree of supercoiling is roughly comparable in the DNA of different cells, which suggests that it is of importance and that its generation is controlled. It is known that negative supercoiling is essential in *E. coli*, both for replication and transcription (Chapter 22). The reason for this is that, in such a state, DNA strand separation occurs more readily than in the relaxed or positively supercoiled state. It has also been suggested that protein binding to negatively supercoiled sections of DNA could relieve the strained conformation and, in doing so, increase the release of energy due to the binding. This would facilitate the attachment of such proteins. As will become evident in the next chapter, DNA binding proteins are of supreme importance in gene function.

To summarize the *E. coli* situation, the required negative supercoiling of DNA is introduced by gyrase (topoisomerase II) and the *E. coli* topoisomerase type I relaxes negative, but not positive coils. The balance between the two opposing activities (gyrase and topoisomerase I) maintains the correct degree of supercoiling (Fig. 21.8); the correct balance is essential to the life of the cell and the synthesis of the two enzymes is controlled to achieve this, the degree and sign of supercoiling in the DNA of the respective genes being a controlling factor. The essential nature of gyrase to bacteria is underlined by the fact that the antibiotic **nalidixic acid** that inhibits their multiplication does so

by inactivating gyrase. Since eukaryotes do not possess this enzyme (see below) nalidixic acid can be used clinically to treat certain infections in humans.

It is particularly interesting that, in a thermophilic bacterium living at extremely high temperatures, a 'reverse gyrase' has been found. A possible reason is that positively supercoiled DNA is less likely to undergo the strand separation that occurs at high temperatures.

Eukaryotic DNA, like that of prokaryotes, is underwound or negatively supercoiled in the cell. However, unlike the situation in prokaryotes, no eukaryotic topoisomerase is known that can actively insert negative supercoils into DNA. How then is this achieved?

When chromatin is assembled, the DNA winds around nucleosomes in such a manner that, in the local region in contact with the protein, it is in an underwound state. (This is achieved by a left-handed coil; it can be difficult to grasp and retain such topological details in three dimensions and, since it is not essential in the present context, we omit description of the handedness of such coils.) Since this nucleosome winding does not involve any bond breakage and since the chromosomal DNA cannot freely rotate, it follows that there cannot have been any *net* change in the supercoiling of the DNA. Therefore, the local negative supercoiling at the nucleosome must be compensated for by positive supercoiling elsewhere so that the net change in the structure is zero (Fig. 21.9(b)). The eukaryotic topoisomerases I or II now relax the positively supercoiled section, thus achieving the insertion of a negative supercoil (Fig. 21.9(c)). A prokaryotic type of gyrase is thus not needed; the fact that prokaryotes do not have the nucleosome structures correlates with the need for their own type of gyrase.

Thus far, we have helicase unwinding the strands and topoisomerases allowing the separation to proceed right along the chromosome by relaxing positive supercoiling induced by the turning force applied to the double helix during unwinding. This is not the end of the unwinding mechanism, for **SSB** is involved.

What is SSB?

The answer is DNA single-strand *b*inding protein, which has a high affinity for single-stranded DNA but with no base sequence specificity—it binds anywhere on the separated strands. The energy release in binding helps to drive strand separation to completion; the single-stranded form is prevented from reannealing. When the strands function as templates for new DNA synthesis, it has to be swept out of the way by the replicating machinery.

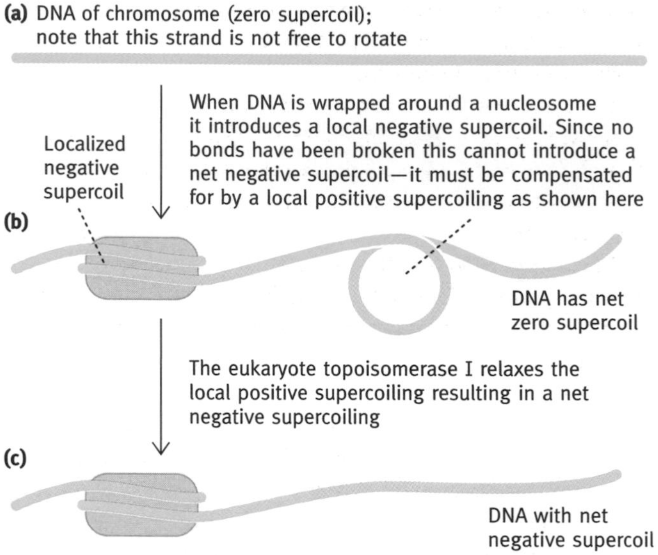

(a) DNA of chromosome (zero supercoil);
note that this strand is not free to rotate

When DNA is wrapped around a nucleosome
it introduces a local negative supercoil. Since no
bonds have been broken this cannot introduce a
net negative supercoil—it must be compensated
for by a local positive supercoiling as shown here

Localized
negative
supercoil

(b)

DNA has net
zero supercoil

The eukaryote topoisomerase I relaxes the
local positive supercoiling resulting in a net
negative supercoiling

(c)

DNA with net
negative supercoil

Fig. 21.9 A mechanism by which eukaryotic DNA becomes negatively
supercoiled despite the absence of any enzyme capable of actively
inserting negative supercoils such as the prokaryotic gyrase. Steps
(a)–(c) are referred to in the text.

The situation so far

So far we have dealt with the rather broad aspects of DNA replication—its semiconservative nature based on Watson–Crick
base pairing, with the biological aspects of the cell cycle, with
the initiation of replication, and with the mechanism of unwinding. It says a lot for the complexity of DNA replication that,
after all this, we have only got as far as separating the DNA
strands and nothing so far has been said about how new DNA is
actually synthesized in the replication fork. We now want to
turn to this; the simplest part by far is the chemistry by which
nucleotides are linked together to for the new DNA chain. The
enzyme(s) that catalyse this are called **DNA polymerases** for
they polymerize nucleotides into DNA.

The basic enzymic reaction catalysed by DNA polymerases

First let us consider the basic chemical reaction. A series of facts
first.

- There are three DNA polymerases in *E. coli* called Pol I, II,
  and III—named in order of their discovery.

- The DNA synthesis occurring in the replicative fork is
  catalysed by Pol III or its eukaryotic equivalents, but Pol I
  also plays an essential role in DNA replication as well as in
  repair. Less is known of Pol II, but it is believed to be associated with certain types of DNA repair.

- The substrates for DNA polymerases are the four deoxyribonucleoside triphosphates dATP, dCTP, dGTP, and dTTP.
  These are synthesized in the cell as described earlier (page
  294). The regulatory mechanisms in their synthetic pathways ensure that they are produced in adequate amounts
  and in coordinated concentrations.

- The polymerase must have a DNA template strand to copy.
  'Copy' is used in the complementary sense—a G on the
  template strand is 'copied' into a C in the new strand, and
  likewise A into T, C into G, T into A.

- A most important fact to fix in your mind: *a DNA polymerase can only elongate (add to) a pre-existing strand called
  a primer.* This **primer** may only be 2–10 nucleotides long
  but without it nothing happens. **DNA polymerases cannot
  start a chain—they cannot join together two free
  nucleotides.**

- As illustrated in Fig. 21.10, the polymerase attaches a nucleotide to the 3′ free OH group of the end of the primer
  strand, liberating inorganic pyrophosphate. Hydrolysis of
  the latter increases the negative $\Delta G^{0'}$ value for the synthesis
  thus helping to drive the reaction (page 12). Incorporation
  of a nucleotide into the new strand of DNA involves the formation of hydrogen bonds with its template partner with
  the liberation of energy (page 14), thus adding to the thermodynamic drive of the process.

- Which of the four deoxyribonucleoside triphosphates is accepted by the DNA polymerase is determined by the base
  on the parental strand being copied.

- *DNA synthesis always proceeds in the 5′ → 3′ direction
  with respect to the growing strand.* Be sure that you know
  what this means—that the growing DNA chain is being
  elongated in the 5′ → 3′ direction—a nucleotide is added
  to the free 3′-OH of the preceding terminal nucleotide.
  At the risk of overemphasizing the point, for it is important, when we talk of synthesis being in the 5′ → 3′ direction
  we always refer to the direction of elongation—the polarity
  of the *new* strand. We are *not* referring to the template
  strand, which has the opposite polarity. The polarity of
  DNA strands has been explained in the previous chapter
  (page 308).

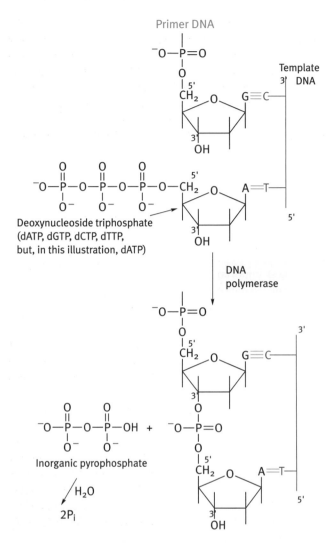

Fig. 21.10 The reaction catalysed by DNA polymerase. The diagram shows the addition of an adenine deoxynucleotide from dATP to the 3′ end of the primer DNA strand, the base selected for addition being determined by the base on the template strand. Note that the synthesis is in the 5′ → 3′ direction—the chain is being lengthened in the 5′ → 3′ direction.

Problems in DNA synthesis

How does a new strand get started?

As is now fixed in your mind, DNA polymerases cannot initiate new chains and yet, at each origin of replication, new chains must be initiated.

The solution to the question in the above heading is rather surprising in that DNA chains are initiated by RNA, which is

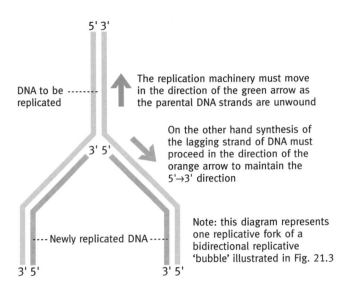

Fig. 21.11 The polarity problem in DNA replication.

slightly unfortunate because we don't want to deal with RNA and its synthesis until the next chapter. However, for the moment, RNA has the same structure as single-stranded DNA except that the sugar is ribose and the base thymine (T) is replaced by uracil (U). RNA is synthesized by RNA polymerases by essentially the same basic chemical mechanism as outlined above for DNA except that ATP, CTP, GTP, and UTP, are used. But, in the present context, the vital difference is the RNA polymerases *can initiate new chains*. They can take two nucleotides and link them (a template being required here also); DNA polymerase cannot do this.

When a small piece of RNA primer (perhaps five nucleotides) has been synthesized by a special RNA polymerase called **primase**, DNA polymerase takes over and extends the chain.

We now come to yet another problem due to the antiparallel nature of DNA.

The polarity problem in DNA replication

It might be useful to turn back again to page 308 to make sure you understand DNA polarity. Consider these facts. DNA is replicated at each replicative fork which steadily progresses along the chromosome. In *E. coli*, there are two DNA polymerase III molecules involved in each fork, one for each strand, the two enzymes being linked together into a single asymmetric holoenzyme dimer. As shown in Fig. 21.11, the polymerase molecules that are replicating the two strands must physically move in the same direction (that is, up the page as it were). However, one parent strand runs 5′ → 3′ in the reverse direction

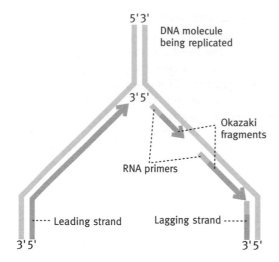

Fig. 21.12 Diagram of a replicative fork. The leading strand is synthesized continuously, while the lagging strand is synthesized as a series of short (Okazaki) fragments.

of fork movement and the other runs $5' \rightarrow 3'$, the opposite way, in the direction of fork movement. Since DNA polymerase can synthesize only in the $5' \rightarrow 3'$ direction, the template strand must run $5' \rightarrow 3'$ in the *opposite* direction to that of synthesis. This is fine for the synthesis of one new strand (the left-hand one in Fig. 21.11), but what about the other? It runs $5' \rightarrow 3'$ in the direction of fork movement and hence DNA synthesis must run the opposite way. A seemingly impossible problem is illustrated in Fig. 21.11 in which the polymerase on the right-hand strand *must* physically move up the page, as it were, but synthesize DNA in the direction of down the page. The left-hand strand, with no problems, is called the **leading strand** and the other the **lagging strand**. (As a brief aside, relevant to Fig. 21.11, be careful not to fall into the trap of imagining that a linear piece of DNA is replicated from the end. It does not happen in that way but starts from the centre of a replicative 'bubble' as shown in Fig. 21.3. This misconception may easily arise because, to avoid having to show both replicative forks of a 'bubble', diagrams of the replicative fork usually show an inverted Y-shaped structure. This point is also emphasized in Fig. 21.3).

First, how can the lagging strand initiate? The leading strand has no problem. Primase lays down a single primer at the origin of initiation and the DNA polymerase proceeds from there, but the same will not suffice for the lagging strand. The solution is that, as the DNA unwinds, there is repeated initiation by primase (an RNA polymerase), followed by many short stretches of DNA synthesis, each about 1000–2000 bases long. The net result is illustrated in Fig 21.12 but do not be worried about how this is achieved, for it still looks to be an impossible situation.

The short stretches of DNA attached to RNA primers on the lagging stand are called **Okazaki fragments** after their discoverer. This still leaves the original problem of how a DNA polymerase can synthesize DNA backwards while moving forwards—a physical or topological problem. It also leaves the lagging strand not as a single piece of DNA but a series of disconnected short pieces attached to RNA primers that must be made into uninterrupted DNA. Let us deal with the physical problem first.

Mechanism of Okazaki fragment synthesis

The basic principle in solving the physical or topological problem of how the lagging strand is synthesized in the $5' \rightarrow 3'$ direction, without the replicative machinery moving backwards, is extremely simple. The lagging template strand is looped, so that for a short distance it is oriented with the same polarity as the leading strand template. The replicative machinery can therefore proceed in the direction of the fork and synthesize both new strands.

Although the principle is simple, the mechanical problem of how the loop system can move along the entire length of the parental strand and permit the synthesis of Okazaki fragments is not simple because it requires that the loop is reformed and enlarged at regular intervals, and a new RNA primer laid down at each stage.

This model, put forward by Arthur Kornberg, is shown in Fig. 21.13. As the loop enlarges and the replicative machinery moves forward, the polymerase will meet the 5'-RNA end of the previous Okazaki fragment. At this point the polymerase must detach and the loop fall away and a new one started. To understand this model we must first discuss the replicative machinery at the replication fork.

Enzyme complex at the replicative fork in *E. coli*

The functional complex of proteins and protein subunits at the replication fork is illustrated in Fig. 21.14. The key enzymes are the **helicase** to unwind the double helix, attached to which is the **primase**, which synthesizes RNA primers at intervals on the lagging strand, and the extremely complex **Pol III**. The helicase unwinding activity is ATP driven and moves along a DNA strand (probably by allosteric conformational changes) and, in doing so, separates the two strands of the double helix. Finally, **SSB** attachment stabilizes the single strands. As indicated, in *E. coli* there are two connected molecules of Pol III in the replicative fork, one synthesizing the leading strand and the other the lagging strand. They have the same core enzyme, but the holoenzyme dimer is asymmetric with extra subunits present

on the laggng strand side. Pol III has high processivity—once it locks on to a DNA strand it does not fall off but can go on replicating its template strand indefinitely without the risk of dissociating. A special mechanism has been devised to guard against premature dissociation. We will now describe this.

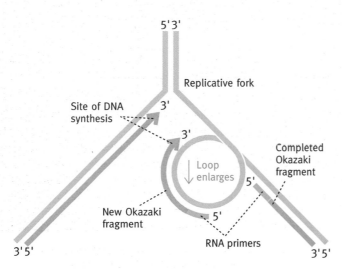

Fig. 21.13 The 'loop' model for Okazaki fragment synthesis. (Red arrowheads indicate DNA synthesis.) This model requires the loop to fall away when the new Okazaki fragment meets the old one. A new loop has then to be made. Both strands can be synthesized in this model in the required 5′ → 3′ direction as the replication machinery moves in the direction of the fork. It is not possible to specify the precise mechanical details of the looping mechanism.

The DNA sliding clamp and the clamp loading mechanism

The sliding clamp is a ring shaped protein structure surrounding the DNA. The annulus has a hole big enough for double-stranded DNA to slide through it easily but it cannot fall off the DNA (Fig. 21.15). This clamp structure is found both in *E. coli* where it is known as the **β protein** and in eukaryotes as **proliferating cell nuclear antigen** (**PCNA**), or in other words a protein in the nucleus necessary for proliferation. Although the structures of the two clamps look at first sight to be the same (Fig. 21.15), the *E. coli* one is a dimer whereas the PCNA has a three-subunit structure. The convergent evolution of different structures for the same function emphasizes its vital role. An additional protein complex is needed to load the clamps onto the DNA. These have been found in *E. coli* and in eukaryotes. In *E. coli* it is known as the **γ complex**. As explained, at the initiation of replication, a special RNA polymerase, the primase, lays down a short stretch of RNA against the template DNA strand. The γ complex, with a bound molecule of ATP, recognizes the short stretch of RNA/DNA hybrid and attaches a sliding clamp around it. It does this by seizing a circular clamp from solution which it opens and places it around the RNA primer/DNA hybrid. The ring then snaps shut, this step being associated with ATP hydrolysis and release of the loading protein. The face of the clamp has a site for binding the Pol III DNA polymerase which is recruited from solution. The Pol III is now firmly attached to the DNA so that it cannot fall off but is free to move along the DNA and replicate the template strand (Fig. 21.16). The clamp loading complex places a clamp wherever there is a primer RNA laid down, a fact that is

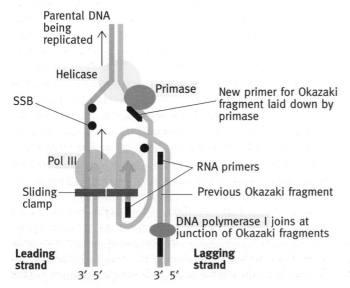

Fig. 21.14 The loop model. This is to solve the problem of how the polymerase can move forward (upwards on the page) but still synthesize DNA in the required 5′ → 3′ direction when the template strand polarity demands that synthesis is in the opposite direction (downwards on the page). The action of DNA polymerase I is described in Fig. 21.16 and in the text below.

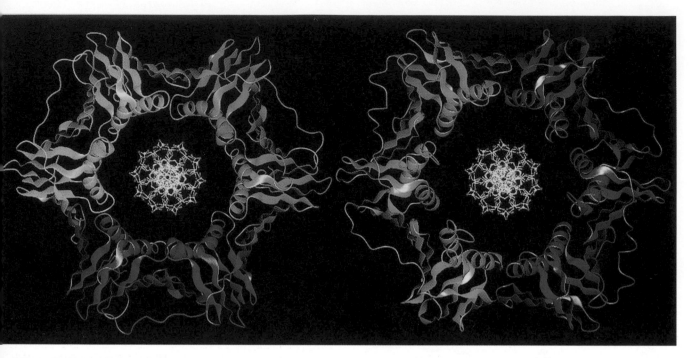

Fig. 21.15 Ribbon representations of the yeast and *E. coli* sliding 'clamps'. **(a)** The yeast clamp that confers processivity on DNA polymerase δ is a trimer (see page 264). **(b)** The *E. coli* clamp that attaches to DNA polymerase III is a dimer. The individual subunits within each ring are distinguished by different colours. Strands of β sheet are shown as flat ribbons and α helices as spirals. A model of B DNA is placed in the centre of each structure to show that the rings can encircle duplex DNA.

very relevant to the synthesis of the lagging strand as you will see shortly.

The coordination of leading and lagging strand synthesis is obviously a very important and complex problem. As stated, when the Pol III on the lagging strand meets the primer for the next Okazaki fragment, it must detach and re-initiate at the primer laid down by the primase. The sliding clamp is probably removed and reattached at the next Okazaki fragment primer. Now to the other problem—how the Okazaki fragments are tidied up into continuous DNA.

What happens to the Okazaki fragments?

In *E. coli*, when the Pol III reaches the RNA primer of the preceding Okazaki fragment, it disengages from the DNA, leaving a nick at the DNA/RNA junction. This is where DNA polymerase I (Pol I) comes in. Pol I is an astonishing enzyme with three separate catalytic activities on the same molecule.

If we look at the problem, as illustrated in Fig. 21.12, the separate pieces of DNA, the Okazaki fragments, must be converted into a continuous DNA molecule. Each piece starts with RNA which must be removed, replaced with DNA, and the

separate DNA pieces joined up. The Pol I attaches to the nicks, or breaks, between successive Okazaki fragments, and adds nucleotides to the 3′-OH of the preceding fragment, moving in the 5′ → 3′ direction; as with Pol III (or any DNA synthesis), nucleotide additions are always in the 5′ → 3′ direction. Since, as Pol I moves, it encounters the RNA of the next Okazaki fragment, the nucleotides of this are chopped out. Thus, as it were, the front end of Pol I chops nucleotides out, and a site further back adds nucleotides to fill the gap with DNA. The 'front' activity is a 5′ → 3′ **exonuclease** activity—'exo' because it works on the end of the molecule, 'nuclease' because it hydrolyses nucleic acids, and 5′ → 3′ because it nibbles away at the 5′ end of the RNA and moves in the direction of the 3′ end of the molecule. Note that Pol III does *not* have a 5′ → 3′ exonuclease activity like Pol I. Thus, Pol III cannot chop out the RNA primer when it meets the preceding Okazaki fragment. It disengages from the DNA at this point and hands over the job to Pol I.

That is still not the end of the matter. At the 'rear' end of the Pol I molecule there is a 3′ → 5′ exonuclease as well as the polymerase activity. This exonuclease hydrolyses off the nucleotide at the 3′ end of the DNA chain that Pol I has just synthesized and it chops in the direction of the 5′ end—exactly the reverse of the

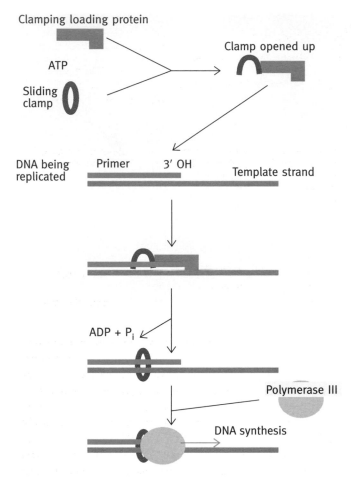

Fig. 21.16 The steps in loading a sliding clamp for DNA synthesis in *E. coli*. The clamps exist in solution as complete rings. The clamp loading protein, in the presence of ATP opens up the ring, binds to the DNA wherever a primer laid down by primase awaits elongation and snaps the clamp around the DNA to which a Pol III molecule attaches. Based on Fig. 1 of Kelman, Z. and O'Donnell, M. (1995), *Ann. Rev. Biochem.*, **64**, 171, with permission of Annual Reviews.

'front-end' $5' \rightarrow 3'$ exonuclease. Hence, it is a $3' \rightarrow 5'$ exonuclease. Put in another way, it is chopping backwards.

This is almost beyond reasonability—an enzyme whose 'rear end' both chops out DNA and synthesizes DNA. But there are other vital facts. *The $3' \rightarrow 5'$ exonuclease hydrolyses the terminal nucleotide of the newly formed DNA strand only if it is unpaired with its corresponding base on the template strand and the DNA polymerase moiety will add nucleotides only if the preceding one* (Fig. 21.16) *is properly base paired.* Thus, the enzyme checks the preceding base for correctness and cuts it out if unpaired, and replaces it. This is called **proofreading**—it checks the fidelity of what it has synthesized.

The DNA polymerase I has (unlike Pol III) low processivity—it does not hold on to the DNA template strand firmly and detaches relatively soon after the RNA has been replaced. It does not have the annular clamp to hold it on to the DNA. This is essential for otherwise it would go on replacing long stretches of the newly synthesized DNA. When it detaches, a nick is left in the chain, which is healed by a separate enzyme, called **DNA ligase**.

This enzyme synthesizes a phosphodiester bond between the 3'-OH of one DNA fragment and the 5'-phosphate of the next, a process requiring energy. In some prokaryotes and eukaryotes, ATP supplies this. The mechanism is that the enzyme (E) accepts the AMP-group of ATP, liberating pyrophosphate, and transfers AMP to the 5'-phosphate of the DNA. Finally the DNA-AMP reacts with the DNA-3'-OH, releasing AMP and sealing the brcak.

$$E + ATP \rightarrow E\text{-}AMP + PP_i;$$

$$E\text{-}AMP + \text{\textcircled{P}} - 5'DNA \rightarrow E + AMP - \text{\textcircled{P}} - 5'DNA;$$

$$DNA3' - OH + AMP - \text{\textcircled{P}} - 5'DNA \rightarrow$$

$$DNA3' - O - \text{\textcircled{P}} - 5'DNA + AMP.$$

The AMP is linked to the enzyme and to the DNA via its 5'-phosphate group.

In *E. coli*, instead of ATP, NAD^+ is used. This is most unusual role for NAD^+ which you have met only as an electron carrier. However, NAD^+ can donate an AMP group just like ATP and, for some reason, *E. coli* uses this route. What happens then, in summary, is the following.

The DNA Pol I binds at the attachment site shown in Fig. 21.17—at the nick. The polymerase adds DNA nucleotides to the 3' end of the left-hand fragment in the diagram, checking each addition, and moves in the $5' \rightarrow 3'$ direction. The RNA, and some DNA, is nibbled away and the gap replaced by DNA. The Pol I detaches and a ligase joins the two fragments of DNA together. Thus, a series of Okazaki fragments becomes a continuous new DNA strand.

Proofreading by polymerase III

The base sequence of the newly formed DNA must be reproduced correctly, for even a single incorrect base could cause a lethal mutation. But, with such vast numbers of bases, mistakes are unavoidable.

When Pol III encounters, say, a G in the template strand, it is highly selective in that in the vast majority of cases only dCTP is accepted into the active site of the enzyme and added to the nascent DNA chain and so on with the other bases. The mechanism by which Pol III selects the correct nucleotide for the

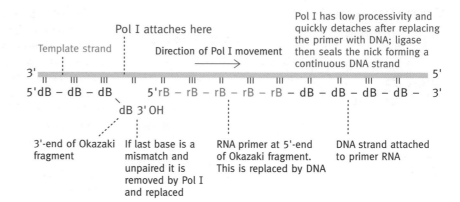

Fig. 21.17 Polymerase I actions in processing Okazaki fragments. dB, Deoxynucleotide; rB, ribonucleotide.

template base is most likely to be due to the fact that all Watson–Crick base pairs have the same geometrical shape, which is different from that of any other base pairs and which is precisely recognized by the enzyme. (Have a quick look at the illustration of this in Fig. 21.18.) Hydrogen bonding of correct pairs and base stacking also help the selection process since bond formation involves energy release.

There is a limit to the accuracy of such mechanisms; although Watson–Crick pairing is the predominant situation, in any collection of, for example, dATP molecules, a few will be in the imino form (=NH instead of $-NH_2$) and this *can* base pair with C. The net result is that Pol III *can* make an error of DNA copying of about one base in $10^5–10^6$. This would give much too high a rate of mutation. In *E. coli* the actual final error rate in DNA replication is much lower, about 10^{-10}.

To achieve a satisfactorily small error rate, Pol III also has a proofreading activity. It has a $3' \rightarrow 5'$ ('rear end') exonuclease activity that removes the last incorporated base if it is improperly paired, much as does Pol I. The major events involved in *E. coli* replication (as well as eukaryote replication, to be described shortly) are illustrated in Fig. 21.19.

Methyl-directed mismatch repair

Despite the proofreading mechanisms described above to achieve fidelity of DNA replication, in a system involving such vast numbers of nucleotide additions mistakes inevitably occur. Another backstop mechanism operates in *E. coli* to correct the mistakes that do occur in the base sequence of newly synthesized DNA. If a mismatch has escaped the polymerase proofreading correction, the error will cause some distortion in the duplex chain illustrated diagrammatically below.

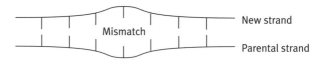

The error will be in the new strand and the repair system has to recognize this and nick it as the first step in error correction. The important point is that, in such a mismatch, the base on the parental (template strand), by definition, is correct and it is the complementary base on the new strand that is incorrect. The repair system must discriminate between the two strands, for if it replaced the template strand base it would confirm the mutation. It must remove the base from the new strand and correct it. How is this strand discrimination made? In *E. coli*, wherever there is a GATC sequence in the DNA, the adenine of this sequence is methylated by an enzyme in the cytoplasm known as the Dam methylase. This does not affect base pairing or DNA structure. It takes some time after its synthesis for the new strand to be methylated and so, for this brief period, it is unmethylated. Thus the parental strand is methylated but the just-synthesized new strand is not. The repair system is little less complex than that for DNA synthesis; first the protein Mut S recognizes the mismatch distortion in the helix; Mut H, linked to Mut S by Mut L, finds an *unmethylated* GATC on one of the duplex strands that identifies it as being newly synthesized and nicks it at this site (Fig. 21.20). This may be thousands of bases from the error. Then helicase, and SSB, and an exonuclease cooperate to remove the entire section from the nick to beyond the error and polymerase III replaces it with DNA. DNA ligase completes the repair by sealing the nick. To correct one base, thousands of nucleotides may be replaced (Fig 21.20). This error correction system increases fidelity of replication as much as a thousandfold. There is evidence that mismatch repair occurs in eukaryotes. In humans, proteins corresponding to Mut S

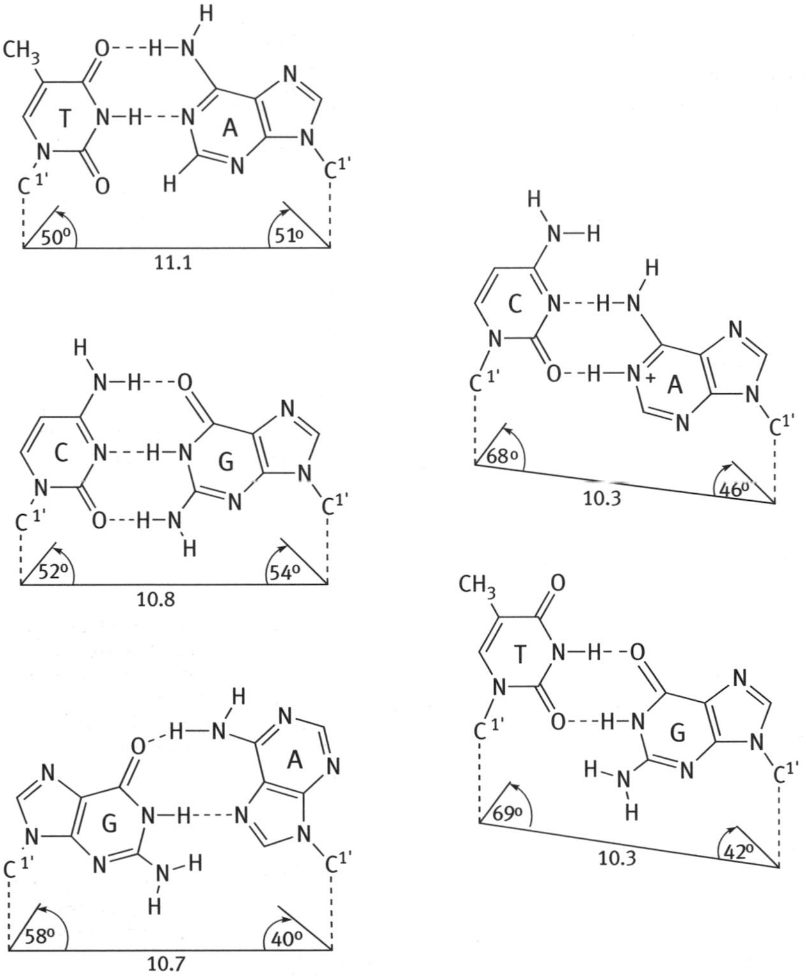

Fig. 21.18 Geometric characteristics of Watson–Crick and mismatched base pairs. The figure is based on X-ray crystallography of duplex B DNA oligonucleotides. The striking geometrical identity of the Watson–Crick pairs is not matched by the A–C and G–T wobble pairs or by the G (*anti*)–A (*syn*) pair. The term 'wobble pair' is explained on page 383. Distances (in nm) are those between the deoxyribose C–1 atoms of each pair, and angles are those of the glycosidic bonds.

and Mut L proteins of *E. coli* are known. Mutations in the genes for these are associated with increased risk of cancer.

Double strand break repair poses special problems; it is discussed on page 338.

Repair of DNA damage in *E. coli*

The mechanisms described above ensure that DNA is replicated with the almost incredible degree of accuracy needed to ensure continuity of life. However, there is still a major problem. The base sequence of DNA must be essentially immortal, for much of it carries information that must remain unchanged for vast time spans. But DNA is an ordinary molecule in chemical stability terms, and chemical changes are occurring all the time and at a rate that would result in large numbers of mutations

per day, in each cell, if there weren't constant repair. Some of the damaging changes are spontaneous. The glycosidic link that binds the purine and, to a lesser extent, pyrimidine bases to the deoxyribose moieties is fairly unstable and depurination and depyrimidation occur—large numbers of purines (and a lesser number of pyrimidines) break off the DNA every day in a human cell. In addition, cytosine and adenine chemically deaminate to become uracil and hypoxanthine, respectively (see page 289 and Fig. 19.7 for structures). Also, DNA is subject to 'insults' by a variety of agents. Ionizing radiation, oxygen free radicals (page 282), carcinogens, UV light, all can alter the DNA and thereby cause mutations.

An important general principle is that it will be relatively infrequent for damage to happen to *both* strands of a duplex DNA molecule at exactly the same place in both chains (though it does happen). When only one strand is affected at a given place, there is always the other strand to act as template and 'direct' the

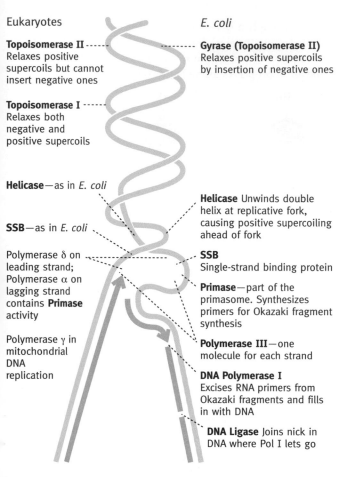

Eukaryotes

Topoisomerase II
Relaxes positive
supercoils but cannot
insert negative ones

Topoisomerase I
Relaxes both
negative and
positive supercoils

Helicase—as in *E. coli*

SSB—as in *E. coli*

Polymerase δ on
leading strand;
Polymerase α on
lagging strand
contains **Primase**
activity

Polymerase γ in
mitochondrial
DNA
replication

E. coli

Gyrase (Topoisomerase II)
Relaxes positive supercoils
by insertion of negative ones

Helicase Unwinds double
helix at replicative fork,
causing positive supercoiling
ahead of fork

SSB
Single-strand binding protein

Primase—part of the
primasome. Synthesizes
primers for Okazaki fragment
synthesis

Polymerase III—one
molecule for each strand

DNA Polymerase I
Excises RNA primers from
Okazaki fragments and fills
in with DNA

DNA Ligase Joins nick in
DNA where Pol I lets go

Fig. 21.19 Events involved in DNA replication in *E. coli* and in
eukaryotes. (Initiation of replication and assemblage of the replicative
complex at the fork involves other proteins.)

repair of the damaged part. A variety of repair systems exist in
cells to repair DNA damage for, of all things, this is the area
where maintaining the integrity of a molecule is of paramount
importance. When other molecules such as proteins are damaged they are destroyed, but DNA must be repaired at all costs.
There are different types of repair systems.

- **Direct repair**. Exposure of DNA to ultraviolet light can result in the covalent linking of two adjacent thymine bases
  (on the same strand), forming a T dimer.

T dimer

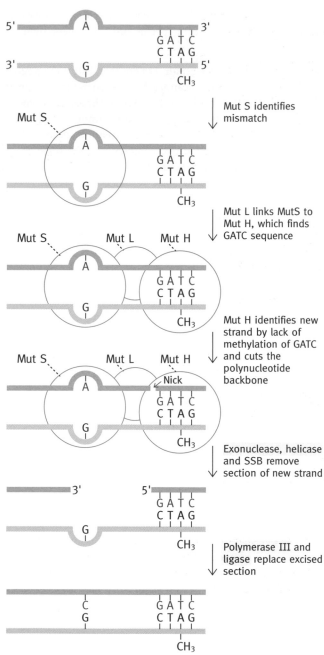

Fig. 21.20 Methyl-directed pathway for mismatch repair. If the GATC
sequence is distant from the error, bending of the DNA could bring the
two into proximity.

In *E. coli*, the abnormal bonds are cleaved by a light-activated
mechanism that restores the two thymine moieties to their
original form. The system is widespread and probably of importance in plants. Alkyl groups on bases (which are formed by
some mutagenic agents) may be directly removed by a 'suicide

enzyme' that accepts the alkyl group and in so doing destroys its own action. It is more of a specific protein reagent than an enzyme.

- **Nucleotide excision repair.** Lesions that distort the double helix such as a T dimer can also be repaired by the excision of a short stretch of nucleotides, including the lesion, followed by its correct replacement, the opposite strand serving as the template for this. In *E. coli*, an unusual endonuclease (called *excinuclease*, or the *uvr*ABC complex after the three genes coding for the enzyme) cuts the DNA on both sides of the lesion and removes a singe-stranded section of 12–13 nucleotides (Fig. 21.21). DNA polymerase I attaches to the nick and adds nucleotides to the 3' end of the nicked chain; ligase heals the nick. The system depends on it being possible to recognize which strand of the DNA is faulty.

- **Base excision repair and AP site repair.** Deamination converts cytosine to uracil and adenine to hypoxanthine. DNA glycosylases recognize the abnormal bases and hydrolyse them off, leaving AP (apurine or apyrimidine) sites in which the deoxyribose has no base attached to it (Fig. 21.22). AP sites can also be formed spontaneously, since the purine–deoxyribose link especially is somewhat unstable. Repair of AP sites involves nicking of the polynucleotide chain adjacent to the lesion followed by replacement of the section containing the latter by DNA polymerase I and sealing by ligase.

The need to remove uracil formed from cytosine probably explains why DNA has T, instead of U. Remember that T is, in essence, simply a U tagged for identification purposes with a methyl group (page ••). If DNA *normally* contained U, it would be impossible to distinguish between a U that should be there and an 'improper' U, formed by deamination of C.

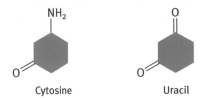

Using T in DNA, instead of U, solves the problem. (As described in the next chapter, U can be used in RNA, because RNA has a

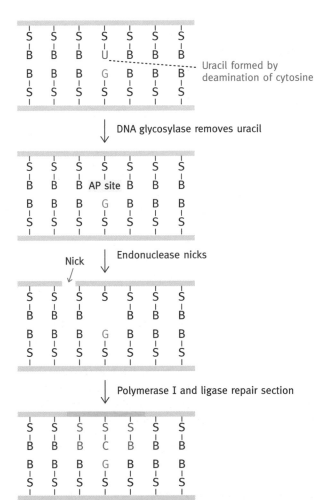

Fig. 21.22 AP site formation and repair. In the example given, the site is created by removal of a uracil by a glycosylase, but sites are also formed by spontaneous hydrolysis of purine bases (and, to a lesser extent, of pyrimidine bases) from the nucleotide. S, sugar; B, base.

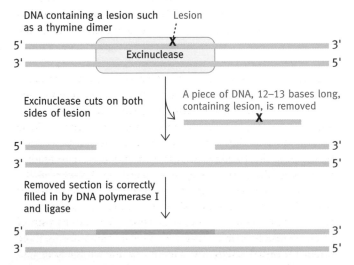

Fig. 21.21 The pathway of nucleotide excision repair in *E. coli*.

relatively short lifetime, is much smaller, and errors do not have the same long-term consequences. Hence RNA is not repaired.)

There is rather a nice correlation; in *E. coli* cells there can be lots of nicks in DNA requiring Pol I to attach to and carry out repair. However, there are only a small number of replicative forks requiring Pol III. In an *E. coli* cell there are only about 10–20 molecules of Pol III but many more molecules of Pol I.

The machinery in the eukaryotic replicative fork

The general principles of DNA replication in *E. coli* apply to eukaryotes, but there are differences in detail; different polymerases synthesize the leading and lagging strands, unlike the Pol III dimer in *E. coli*.

In place of the *E. coli* polymerases I, II, and III, there are five DNA polymerases in eukaryotes, which are given Greek letter designations. Polymerases α and δ synthesize nuclear DNA (cf. Pol III in *E. coli*). Pol α has primase associated with it, but no demonstrated $3' \rightarrow 5'$ 'rear end' exonuclease. How fidelity is achieved is an open question. Polymerase δ has a $3' \rightarrow 5'$ exonuclease activity. It seems that polymerases α and δ form the replicative assembly in eukaryotes—δ for the leading strand. The δ polymerase of yeast has associated with it an annular sliding clamp, known as the PCNA (for proliferating cell nuclear antigen), as does the *E. coli* enzyme, conferring processivity on it. In Fig. 21.15 the structures of *E. coli* and yeast 'clamps' are shown in ribbon diagrams. Despite their almost identical appearance, the two have very little amino acid sequence similarity—the *E. coli* clamp is a dimer while the yeast one is a trimer, suggesting they arose by independent evolutionary designs.

Two of the other polymerases (ε and β) in eukaryotes have repair function. The last, polymerase (γ) is for mitochondrial DNA replication.

The nucleosomes (page 311) must be displaced or otherwise coped with at the replicative fork, but immediately behind the fork the nucleosomes reform so that the replicated DNA is immediately reassembled into these structures.

The problem of replicating the ends of eukaryotic chromosomes

Eukaryotic chromosomes are linear and this poses a problem not encountered in the replication of circular chromosomes.

Consider the replication of the chromosome shown in Fig. 21.23 as a very short one for diagrammatic convenience. Let us assume, again for simplicity, that it is replicated (in a bidirec-

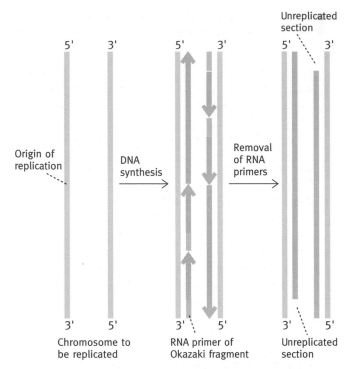

Fig. 21.23 The shortening of linear chromosomes by replication. For diagrammatic convenience the bidirectional replication of a very short piece of DNA is represented. It should be noted that primer removal from Okazaki fragments is a continuous process—it is represented here, for clarity, as occurring as a separate event. A typical chromosome will have multiple origins of replication. The red lines represent new DNA synthesis; the green lines the RNA primers of Okazaki fragments.

tional manner, don't forget) from a single initiation site in its centre (but remembering that a eukaryotic chromosome has many such sites in actual fact). The 5' end of each strand is fully replicated by leading strand synthesis. This is not true of the 3' ends, which are not fully replicated by lagging strand synthesis, because the synthesis of the end Okazaki fragments requires RNA primers to be laid down as shown on the 3' ends of the template strands. When the primers are removed it leaves these ends unreplicated and no mechanism exists by which it could be replicated by the DNA synthesizing machinery that we have described so far. To fill in the missing parts would require RNA primers but there are no templates against which they could be laid down. This means that, with cell division, chromosomes would become progressively shorter; a more potentially disastrous situation could hardly be imagined. The mechanism of DNA synthesis means that incomplete replication of the double-stranded DNA cannot be avoided. The solution adopted is that, at the ends of eukaryote chromosomes stretches

of special DNA are attached in which the DNA has no informational content but is sacrificial DNA, called **telomeric DNA** (the ends of the chromosomes containing it are called **telomeres**). The ends will still not be replicated by the DNA synthesis machinery so far described, but it no longer matters for the 'real' chromosome is fully replicated with the primer at the 3′ end of the template strand being laid down against telomeric DNA. In rapidly dividing cells, moreover, the telomere is added to so that the chromosome is never at risk. The telomeres are also believed to prosect the chromosome ends from exonuclease attack.

How is telomeric DNA synthesized?

A telomere consists of repeating short stretches of bases—it varies between species. In humans there are hundreds of repeats of the TTAGGG sequence. The enzyme **telomerase** adds these sequences one after the other to the 3′ end of pre-existing telomeric DNA and so extends the latter (all chromosomes have telomeres to start with). Telomerase is a type of enzyme you have not met in this book before. It has two remarkable features: (1) it uses RNA as the template for DNA synthesis—it is a **reverse transcriptase** (see page 470 for a fuller explanation); and (2) it carries its own template in its structure. This RNA carries the sequence complementary to the TTAGGG sequence (in humans). The RNA template hybridizes to the end of a repeating unit (Fig. 21.24), which positions the template bases for the repeating unit correctly, and a unit is then added, one base at a time. When this is done the enzyme moves along and hybridizes to the end of the new repeating unit and thus the telomere is constructed in a discontinuous manner. The complementary strand to the telomeric extension, added as described, is synthesized so that the telomere is double-stranded, but there is some uncertainty as to how this occurs; a DNA polymerase is presumably involved.

The necessity for telomeres has been demonstrated by the use of **YACs** or **yeast artificial chromosomes**. These contain the three types of DNA essential for chromosome replication—centromeres, sites of origin, and telomeres. It was shown that YACs are correctly maintained for generations when inserted into yeast cells but that, when they lack telomeric ends, they disappear in time from the cells.

There is not the same problem in prokaryotes, for in a circular chromosome there is always a DNA template available for priming. The existence of telomeres in eukaryotes has profound potential significance. Telomerase has been found in germ cells, which are continuously rapidly dividing, and it might be expected that continual lengthening of the telomeres would be necessary to compensate for the continuing shortening. However, in somatic cells where cell division is slower, additions to

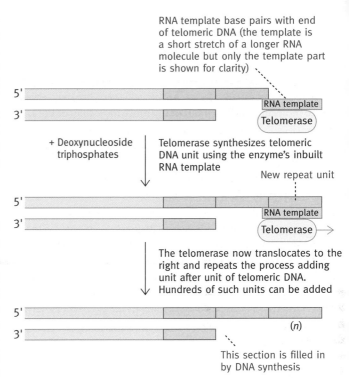

Fig. 21.24 Mechanism by which telomerase synthesizes telomeric DNA. The blue lines indicate the chromosomal or informational DNA (the 'real' chromosome) and the red lines the pre-existing telomeric DNA at one end of a chromosome. The telomerase has an inbuilt short RNA molecule that contains the sequence complementary to the repeating unit characteristic of the species. The enzyme becomes positioned with the RNA pairing with the terminal bases of the pre-existing telomere and adds one repeating unit of TTAGGG (in the case of humans) one base at a time. Synthesis is, as always, in the 5′ → 3′ direction. The enzyme moves so that the RNA template is now paired with the end bases of the new repeating unit and a further unit is added, and so on. The newly synthesized telomeric DNA acts as the template for filling in the opposite strand, probably involving a DNA polymerase, so that the telomere is double-stranded.

the telomeres would be less appropriate and, indeed, telomerase is believed not to be present in such cells. If this is so, the implication could be that somatic cells receive their initial 'ration' of telomeric DNA to suffice for the lifetime of that cell and its progeny. This leads to the speculation that the telomeres of most cells become shorter with age and this could be an important factor in determining the possible lifespan of species. Somatic cells can be transformed to the rapidly dividing state if they become cancerous; such cells in effect become immortal—there is no limit to their ability to divide in culture. It is interesting that telomerase has been reported to be present in cancer cells. This is discussed more fully in Chapter 30.

DNA damage repair in eukaryotes

The dealkylating 'suicide enzymes' described for *E. coli* occur in human cells. Excision repair also occurs in eukaryotes. As an example, in humans with the genetic disease xeroderma pigmentosum, normal excision repair of pyrimidine dimers does not occur. A thymine dimer is formed by covalent joining of two adjacent thymine bases (on the same strand) as a result of UV light on skin. In patients with this disease, exposure to sunlight causes cancerous skin lesions. Excision repair is known to be of general importance in humans. Genes for proteins corresponding to Mut S and Mut L of *E. coli* (page 329) have been found; mutations in these genes have been found to be associated with the development of cancers.

Eukaryotic repair systems play a major role in maintaining integrity of the genome. Whether ageing in humans is associated with decreased repair and a consequent increase in mutations is an interesting question that has attracted research effort. See also the section on mitochondrial mutations below.

Is the mechanism described above the only way in which DNA is synthesized?

As is so common, evolution has exploited many avenues to achieve the same ends. Some viruses, for example, use a protein instead of RNA for DNA chain synthesis priming. However, as already indicated, retroviruses have quite a fundamentally different mechanism in that the RNA genome is copied into DNA. This will be described in Chapter 28 on viruses. It is particularly interesting that mitochondria have their own DNA, which must be replicated for mitochondrial multiplication. There are few or no DNA repair systems in mitochondria and this presumably reflects the facts that the chromosome is very much smaller with fewer chances for errors and that, since there can be large numbers of mitochondria per cell, a small proportion of faulty ones has less significance. The apparent absence of repair, however, means that mutations in mitochondrial DNA may accumulate with time and possibly be related to the process of ageing (see page 282). A large number of genetic diseases are now known to be caused by mitochondrial mutations.

Transposons or jumping genes

There is a phenomenon that is different from anything described so far in this chapter, in which a piece of DNA jumps from one place in the chromosome to another place in the same or different chromosome. The existence of **jumping genes** was first detected by Barbara McClintock in genetic studies on maize; she concluded that gene control elements moved from one place to another in the genome and influenced the expression of genes, giving rise to phenotypic variations. This work was largely ignored for about 30 years until studies in *E. coli* confirmed that genes do, in fact, move around, and a Nobel Prize was awarded to McClintock.

The simplest type of transposon in *E. coli* is called an **insertion sequence (IS)**. It consists of a stretch of DNA with a gene for transposase, an enzyme that catalyses the transposition event. From one point of view, an IS might be regarded as the ultimate in futility—a gene that does nothing but code for its own movement from place to place; it could be regarded as a parasite. An IS can jump to any place in a chromosome—there are no specific insertion sites for this class of mobile gene. If a transposon is inserted into another gene the latter is inactivated because the gene sequence is thereby disrupted, but the frequency of transposition is, however, low. In addition to the gene for the transposase, the insertion sequence has at the ends **inverted terminal repeats**—these consist of short stretches of bases oriented as shown in Fig. 21.25. The insertion process results in a short sequence of the recipient DNA being replicated to give non-inverted repeats.

There are many variations on the simple case given above. One variation is that, instead of the transposon itself transferring to a new location, it is replicated and the copy inserted elsewhere, leaving the original one where it was. The replication involves a reverse transcriptase—an enzyme that copies RNA into DNA (described in more detail on page 470). The creation of multiple identical sequences in the genome in this way could provide the opportunity for homologous recombination. It is possible that chromosomal rearrangement caused by mobile genes is of significance in the evolutionary process.

There are viruses that integrate into the chromosome of the host cell—HIV is one and lambda bacteriophage another. The former integrates anywhere; the latter only at a specific site in the *E. coli* chromosome. We will leave a discussion of these to Chapter 28 which deals with viruses.

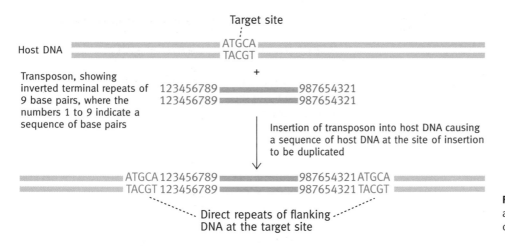

Fig. 21.25 Diagram of the insertion of a transposon (IS type) into the host chromosome.

Homologous recombination

Genetic recombination involves the *in vivo* rearrangement of the DNA of chromosomes within cells. Don't confuse this with recombinant DNA technology described in Chapter 29. The latter does involve the creation of recombined DNA molecules, and indeed recombination of genes may be involved, but the recombinant molecules are created artificially in the test tube. Genetic recombination occurs naturally in living cells. The main type is **general or homologous recombination** between separate chromosomes (DNA duplexes) at a point where there are homologous sections of DNA—where the base sequences of the two are largely the same. There is another quite different type of recombination known as **site specific recombination** involved in chromosomal rearrangements for specific purposes (see page 454).

Homologous recombination has the important consequence of generating genetic diversity in organisms. We can illustrate the importance of this using *E. coli* as an example. Suppose two *E. coli* cells have genes *A* and *B* on their chromosomes and that in one cell gene *A* mutates and in the other gene *B* mutates, both mutations conferring a survival advantage. Both strains of cells containing the respective genes will compete, and one may be eliminated as a result. Thus one of the improved genes will survive but one may be lost in the evolutionary sense. This would be the situation if genes were rigidly fixed in one chromosome with no genetic recombination possible. However, as a result of sexual transfer of one *E. coli* chromosome (or part of a chromosome, as is more common) to another cell the two may come into contact. By homologous recombination the two improved genes *A* and *B* can be put onto one chromosome. The resulting daughter cells containing this will have the benefit of both improved genes; in short, a new strain better equipped for survival will have evolved. The opposite consideration also applies. If a chromosome develops an unfavourable mutation in one of its genes, without recombination the rest of the genes on that chromosome are stuck with it and the chromosome could be lost to evolution because of the burden of the unfavourable gene. With recombination, the continual re-assortment of genes means that 'bad' genes can be eliminated without sacrificing the 'good' ones, in an evolutionary sense. In *E. coli* there are about 4000 genes on the chromosome; with recombination available, a single base mutation in a cell might put a gene at risk; without recombination it might put 4000 genes at risk. Homologous recombination to produce genetic diversity also occurs in eukaryotic meiosis (described later).

The result of homologous recombination is illustrated in Fig. 21.26; in this, you see two chromosomes (both are DNA duplexes) with a short stretch of DNA homologous in base sequence in the two. They come together at this point and exchange sections of duplex DNA producing new combinations of genes on the two chromosomes as shown in the diagram. The recombinant process produces a small patch of heteroduplex DNA at the cross-over point in which the two strands in that section come from the two parent duplexes. This hybrid DNA patch could indeed cause limited genetic recombination, but the main event is that the two arms on either side of the cross-over point are exchanged, producing extensive swapping of genes between the two parent chromosomes.

Mechanism of homologous recombination in *E. coli*

We can conveniently divide the process of homologous recombination into two parts. First there is the linking together of two

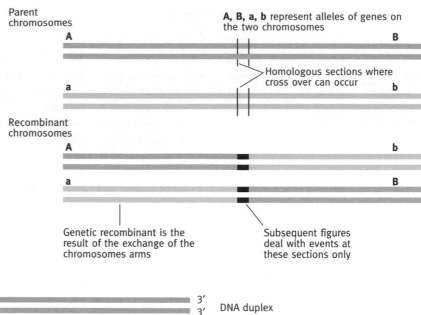

Parent chromosomes

A, B, a, b represent alleles of genes on the two chromosomes

Homologous sections where cross over can occur

Recombinant chromosomes

Genetic recombinant is the result of the exchange of the chromosomes arms

Subsequent figures deal with events at these sections only

Fig. 21.26 Homologous recombination which occurs via cross-over junctions as described in the text. Genetic recombination is the result exchange of chromosome arms or sections. This is caused by the formation of hybrid DNA sections at the site of cross-over junctions (described later). It is important not to confuse the exchange of DNA strands at the cross-over junction with the exchange of chromosome arms. The possibility of this confusion is heightened because, for space reasons, subsequent diagrams are confined to the cross-over mechanism with the long arms not shown. A, a, B, and b represent allelic genes on the two chromosomes. Gene alleles represent the same gene but differ in base sequence such that their phenotypic expression is different.

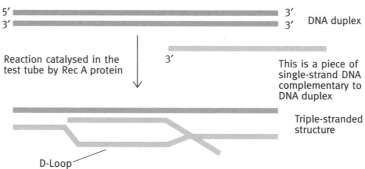

DNA duplex

Reaction catalysed in the test tube by Rec A protein

This is a piece of single-strand DNA complementary to DNA duplex

Triple-stranded structure

D-Loop

Fig. 21.27 Single-strand invasion. This is to illustrate a somewhat remarkable reaction that can happen *in vitro* between a free piece of single-stranded DNA and a piece of duplex DNA, one strand of which is homologous with the single strand. The single-stranded DNA base pairs with one strand of the duplex, displacing it, and forming the D-shaped loop. The process requires that the single-stranded invading molecule is homologous to the duplex DNA, otherwise base pairing could not occur. This reaction requires the presence of *E. coli* rec A protein. ATP is hydrolysed in the process of invasion. Note that this reaction is a test tube model of what happens in recombination; free single strands of DNA are not involved in the cellular process.

chromosomes by DNA strand exchange and secondly there is the separation of the pair in such a way as to produce genetic recombinants.

The molecular mechanism of homologous recombination is more fully established in *E. coli* than in eukaryotic cells, but sufficient is known to make it likely that the two classes of cells have much in common in this respect. We will first describe what is a basic reaction in homologous recombination—**single-strand invasion**. This has been established in *in vitro* experiments using proteins isolated from *E. coli* and pieces of DNA. It involves the invasion of a DNA duplex molecule by a single strand of DNA complementary in base sequence to one of the duplex strands. The 3′ end of the single strand inserts itself into

the duplex and, by hybridizing to the complementary strand, displaces the original duplex strand forming a triple-stranded D loop structure, so called because of its resemblance to the shape of the letter (Fig. 21.27).

Single-strand invasion is catalysed in *E. coli* by an enzyme, RecA, though other proteins are involved. RecA⁻ mutants, lacking the gene for this protein, are deficient in recombination. Multiple molecules of RecA, each with an attached ATP molecule, bind to the single invading DNA strand. This now invades the duplex and searches for a complementary sequence, with which it base pairs forming the D loop structure. ATP hydrolysis by the RecA causes loss of affinity of the enzyme for duplex DNA from which it dissociates. We now have to see how this

single-strand invasion reaction participates in recombination within cells.

Formation of cross-over junctions by single-strand invasion

A model to account for homologous recombination was put forward by Holliday. The first stage is the pairing of chromosomes alongside each other, with homologous sections of DNA in contact. Since single-strand invasion requires a strand of DNA with a free 3′ end, one strand of each DNA duplex was postulated to be nicked by an endonuclease and the nicked strands mutually invade the homologous duplexes leading with their 3′ ends. This results in the formation of limited stretches of hybrid DNA where the invading strands have displaced the original partners. The nicks are sealed resulting in the formation of the cross-over junction shown in Fig. 21.28, known as a **Holliday junction**. As explained in the figure, the cross-over junction randomly moves along the chromosomes just as long as there is homology. This increases the length of the hybridized sections of the invading strands (see lower part of Fig. 21.28 and the legend) and has the result of moving the cross-overs along. This movement is known as **branch migration**.

Separation of the duplexes

The DNA duplexes bound together by the Holliday junction must now be separated. If you examine Fig. 21.28 it may seem

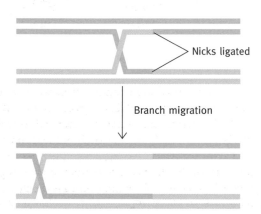

Fig. 21.28 A cross-over junction after mutual strand invasion and ligation of nicks. Note that once a limited amount of exchange has occurred, as shown in the top figure, you can see that the extent of the cross-hybridization can extend, so long as the sections exhanged are homologous so that hybridization can occur. In doing so, the cross-over junctions are moved; this is known as branch migration. The migration can occur to the limits of the homologous sections. It should be emphasized that the diagram represents only the small area which will form the hybrid DNA section shown in Fig. 21.26, not the long arms of the chromosome.

that the obvious thing to do is to cut the crossed-over strands and ligate the breaks; this will separate the duplexes and produce a heteroduplex patch at the junction site, but the genes on the arms will be exactly as before on the two chromosomes. Genetic recombination is not achieved. (If the base sequences in the homologous patches were not identical then there will be some gene conversion or alteration but this is confined only to the local heteroduplex area.) To obtain genetic recombination, the duplexes are separated by cutting the *noninvading* strands which has the result of swapping the chromosome arms (see Fig. 21.26). To achieve this the cell rearranges the Holliday junction in three dimensions, which involves turning over half of the molecule to uncross the strands, a process known as **isomerization**. This does not involve any chemical change and is somewhat akin to solving a string puzzle. The result is a structure in the form of a cross, the essential point being that the noninvading strands can now be cut to separate the duplexes, followed by appropriate healing of nicks; this results in exchange of chromosome arms and achieves genetic recombination. In the case of a circular chromosome two such cross-over junctions are needed to exchange sections.

The Holliday model described above provides the basic concepts. One of the unsolved difficulties in the model presented so far is that it requires a nick to be made in single strands of the two chromosomes at exactly the same position. There is no known mechanism by which this could be achieved, but it might be solved by an alternative model. There is evidence that homologous recombination is initiated by double-strand breaks in the chromosome followed by a modified Holliday mechanism. A model for this has been proposed in which single-strand sections of DNA are created by partial removal of stretches of DNA at the double-strand nicks by exonucleases; the single strands invade the intact partner homologous chromosome. Limited repair synthesis restores the removed sections. The model would additionally explain how a double-strand break could be repaired. This is important because double-strand breaks do occur accidentally, due to ionizing radiation for example. In all of the DNA repair mechanisms we have described so far, the lesion is always on a single strand so there is always the other to direct the repair. This of course does not apply to double-strand breaks, which if not repaired would be lethal. The homologous recombination method uses a homologous chromosome to direct the repair of such breaks. The details of the proposed model (which are somewhat difficult to remember) can be found in the Feather stone and Jackson reference of the reading list. It perhaps should be emphasized that at this stage it is a model with much to be elucidated in detail.

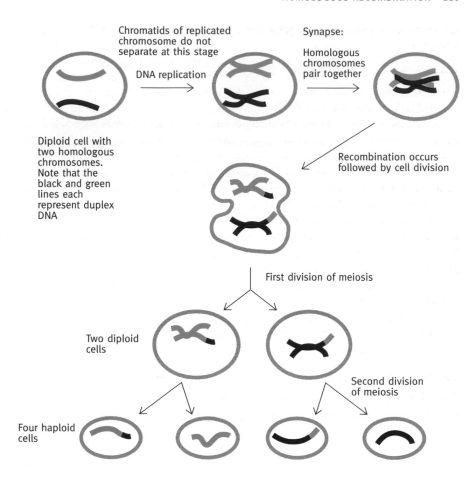

Fig. 21.29 Meiosis.

Recombination in eukaryotes

In **meiosis**, in which haploid sperm and eggs are produced in animals, and corresponding gametes in plants, chromosomes linked by **chiasmata** (strand cross-overs) are seen. We remind you that higher eukaryotic cells are, apart from specialized situations, diploid—they have two copies of each chromosome. (Yeast, used a lot in recombinant research is haploid but diploid after mating which precedes sporulation.) At meiosis the gametes receive only one copy of each—they are haploid. On fertilization the resultant cell becomes diploid once more with one copy from sperm and egg respectively and thereafter **mitosis** at cell division (page 540) maintains this state. As shown in Fig. 21.29, a cell about to undergo meiosis replicates its DNA so that each chromosome now has the two **chromatids** united at the centromere giving the familiar 'X' shape. (Each of the chromatids on separation becomes a normal chromosome.) Instead of these chromatids separating at this stage, as in mitosis, the homologous double chromosomes pair together and recombination occurs by two of the four chromatids of the chromosome pairs exchanging sections. The recombined double chromosomes are separated by the first cell division of meiosis giving two diploid cells. A second cell division separates the chromatids producing four haploid cells. The mechanism of recombination at meiosis is not established at the molecular level. However, a eukaryote homologue of *E. coli* RecA, known as RAD51 protein, has been identified in humans. It is similar to RecA structurally and in its biochemical properties. It is believed to be involved both in homologous recombination and in double strand break repair. From this, and the finding of RecA homologues in yeast, it would seem that homologous recombination and chromosome break repair have been strongly conserved throughout evolution, which underlines their importance.

FURTHER READING

Lewin, B. (1999). *Genes VII*. Oxford University Press.

Replication

Li, J. J. (1995). Once and once only. *Current Biology*, 5, 472–5.

Discusses the essential link between the eukaryotic cell cycle and initiation of sites of origin of DNA replication.

Wyman, C. and Botcham, M. (1995). A familiar ring to DNA polymerase processivity. *Current Biology*, 5, 334–7.

Reviews the sliding clamps in DNA replication. Beautifully illustrated.

S. Waga and B. Stillman (1998). The DNA replication fork in eukaryotic cells *Ann. Rev. Biochem.* **67**, 721.

A comprehensive review.

Baker, T. A. and Bell, S. P. (1998). Polymerases and the replisome: machines within machines. *Cell*, **92**, 295–305.

O'Donnell, M. (1999). Processivity factors. *Current Biology*, **9**, R545.

A one-page quick guide to essential facts on the circular clamps that ensure that DNA polymerases can progress for long distances.

Telomeres

Greider, C. W. and Blackburn, E. H. (1996). Telomeres, telomerase and cancer. *Sci. Amer.*, **274**(2), 80–5.

Discusses the problem of DNA end-replication, DNA shortening and how telomerase protects chromosomal end segments. Possible relevance to ageing and cancer discussed.

Barinagam M. (1997). The telomerase picture fills in. *Science*, **276**, 528–9.

A brief summary of the identification of the catalytic subunit of telomerase.

Johnson, F. B., Marcniak, R. A., and Guarente, L. (1998). Telomeres, the nucleolus and ageing. *Current Opinion in Cell Biology*, **10**, 332–8.

Discusses the relationship between telomere lengthening and replicative senescence. Includes discussion of the relationship to human ageing.

Nagley, P. and Yau-Huei Wei. (1998). Ageing and mammalian mitochondrial genetics. *Trends in Genetics*, **14**, 513–17.

An account of mitochondrial mutations and their possible relationship to ageing. Presents an alternative to the telomere theory of ageing.

Repair

Echols, H. and Goodman, M. F. (1991). Fidelity mechanisms in DNA replication. *Ann. Rev. Biochem.*, **60**, 477–511.

The article is an interesting advanced review on base insertion selectivity by DNA polymerases, proofreading by DNA polymerase, and mutations generated by damage to DNA.

Demple, B. and Karran, P. (1983). Death of an enzyme: suicide repair of DNA. *Trends Biochem. Sci.*, **8**, 137–9.

Reviews dealkylation DNA repair enzymes.

Prolla, T. A. (1998). DNA mismatch repair and cancer. *Current Opinion in Cell Biology*, **10**, 311.

Mutations of mismatch repair proteins are associated with the development of tumours in mice and in humans. Gives a concise summary of the field.

Featherstone, C. and Jackson, S. P. (1999). Double strand break repair. *Current Biology*, **9**, R759–61.

A 'primer' concise summary.

Mitochondrial DNA abnormalities

Hammans, S. R. (1994). Mitochondrial DNA and disease. *Essays in Biochemistry*, **28**, 99–112.

A discussion of diseases arising from abnormalities in mitochondrial DNA.

PROBLEMS FOR CHAPTER 21

1   What is a replicon?

2   In separating the strands of parental DNA during replication, what topological problem occurs?

3   How is the problem referred to in the preceding question solved, both in *E. coli* and eukaryotes?

4   By means of diagrams, explain the actions of topoisomerases I and II.

5   Eukaryotes have no topoisomerase capable of inserting negative supercoiling into DNA and yet eukaryotic DNA is negatively supercoiled. Explain how this is brought about.

6   What are the substrates for DNA synthesis?

7   Why is TTP used in DNA synthesis—why not UTP as in RNA?

8   (a)   Can a DNA chain be synthesized entirely from the four triphosphate substrates? Explain your answer.
    (b)   In which direction does DNA synthesis proceed? Explain your answer so as to be totally unambiguous.

9   What are the thermodynamic forces driving DNA synthesis?

10   *E. coli* polymerase I is a complex enzyme. Describe its different activities and explain their roles in DNA synthesis.

11   Discuss the mechanism by which DNA polymerase III of *E. coli* achieves a high standard of fidelity in DNA synthesis.

12   The proofreading activity of *E. coli* polymerase III is important, but insufficient to give a sufficiently high fidelity rate. If an improperly paired nucleotide is incorporated giving a mismatch, it has to be replaced. This demands that the repair system recognizes which of the two bases in the mismatch is wrong. How is this done and how is the problem fixed? Does this mechanism exist in humans?

13   Explain what a thymine dimer is, how it is formed, and how it is repaired.

14   Explain how eukaryotic chromosomes become shortened at each round of replication.

15   Explain how the DNA shortening problem in replication is coped with.

16   What is meant by processivity? Which DNA polymerase does not have it and why?

Chapter summary

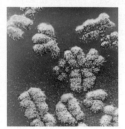

Gene transcription—the first step in the mechanism by which genes direct protein synthesis

The information in DNA, encoded in the sequence of the four bases, is used to direct the assemblage of the 20 amino acids in the correct sequence so as to produce the protein for which a given gene is responsible. In case you wonder how four bases can code for 20 amino acids, and to anticipate the next chapter, the system is that a sequence of three bases codes for an amino acid. Since with four bases $4 \times 4 \times 4$ triplets are possible, there is no shortage of coding ability. A gene does not participate directly in protein synthesis; in eukaryotes the DNA is enclosed inside the nuclear membrane while the protein-synthesizing machinery is outside in the cytoplasm and the two never meet. How then does such a gene direct protein synthesis? It does so by sending out copies of its coded information to the cytoplasm. (In *E. coli* the copy is immediately in contact with the cytoplasm.) Since the information is in a sequence of bases it follows that the copy must also be a nucleic acid, but in this case it is RNA rather than DNA. There are several types of RNA and this one, for obvious reasons, is called **messenger RNA** or **mRNA** for short, or it is often referred to as 'messenger'.

We had better, therefore, now describe mRNA.

Messenger RNA

The structure of RNA

RNA stands for ribonucleic acid. It is a polynucleotide essentially the same as DNA but with these differences:

- The sugar is ribose, not the deoxyribose of DNA. It has an —OH in the 2′ position.

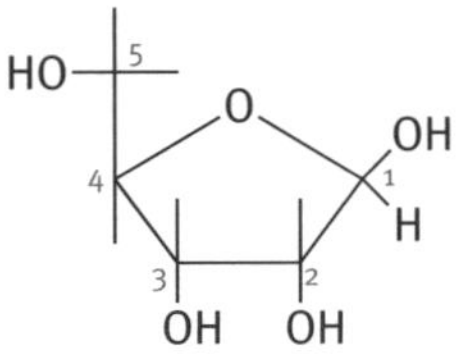

D-Ribose

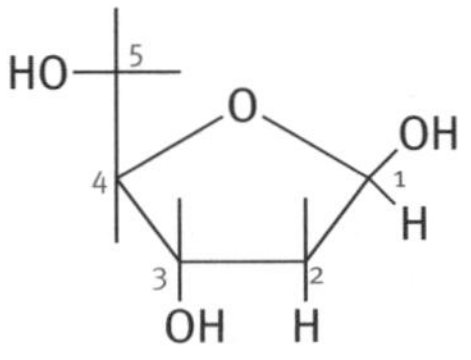

2′-Deoxy-d-ribose

- mRNA is single-stranded, not a duplex of two molecules as is DNA. From this, you will infer that mRNA is a copy of only one of the two strands of the DNA of a gene, a point elaborated upon later.

- Its four bases are A, C, G, and U. There is no T. As we described earlier (page 288), U and T are identical in base-pairing properties—they both pair with A. You should regard T as U that has been tagged with a methyl group without changing its properties (see page 332 for the probable reason for the methyl group).

Were it not for these differences the structure of a single strand of DNA, shown on page 305, could be that of single-stranded RNA. There are the same $3' \rightarrow 5'$ phosphodiester bonds between successive nucleotides. There is a 5′ end lacking a nucleotide substitution on the 5′-OH and the 3′ end lacks a nucleotide attachment on the 3′-OH. The 2′-OH group makes the molecule more unstable than DNA (page 306). In dilute alkaline solution at room temperature, RNA is destroyed while DNA is unaffected.

How is mRNA synthesized?

The building block reactants for RNA synthesis are ATP, CTP, GTP, and UTP, which are produced in all cells (Chapter 19).

mRNA is synthesized from these by a single enzyme in *E. coli*, DNA-dependent RNA polymerase, commonly referred to simply as **RNA polymerase**. In eukaryotes three such polymerases exist. The synthesis requires that the duplex DNA strands are separated so as to provide a single-stranded template for directing the sequence of nucleotides to be assembled into mRNA. The two strands are transitorily separated at the site of mRNA synthesis, and then come together again after the polymerase has passed. In effect, a separation 'bubble' moves along the DNA. The basic process of synthesis is much the same as in DNA synthesis in that the base of the incoming ribonu-cleotide is complementary to the base on the DNA template (Fig. 22.1).

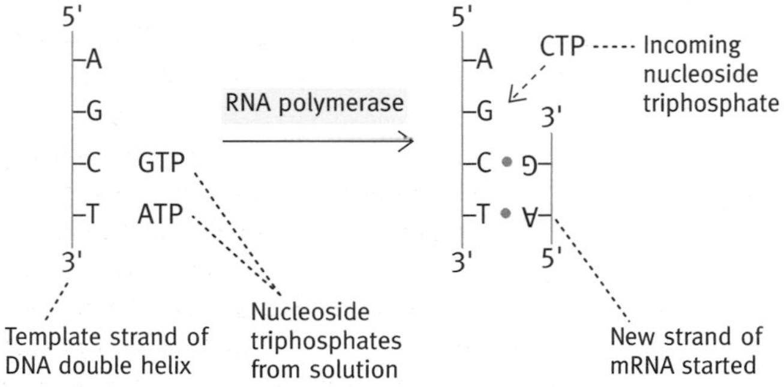

The RNA polymerase works its way along the template, joining together the nucleotides in the correct order as determined by the DNA template. RNA synthesis is always in the $5' \rightarrow 3'$ direction (as is the case also in DNA synthesis). That is, new nucleotides are added to the 3'-OH and so the chain elongates in the $5' \rightarrow 3'$ direction. The template is antiparallel, running in the opposite ($3' \rightarrow 5'$) direction. The chemical reaction catalysed by the polymerase involves the transfer of the α-phosphoryl group (that is, the first one attached to the ribose) of the nucleotide triphosphates to the 3'-OH of the preceding nucleotide, splitting off inorganic pyrophosphate. The latter is hydrolysed to two P_i molecules, thus making the reaction, shown in Fig. 22.2, strongly exergonic.

An important point is that RNA polymerase *can* initiate new chains—it does not need a primer; it can synthesize the entire mRNA molecule from the four nucleoside triphosphates, provided a DNA template is there. This is quite different, we remind you, from the situation in DNA synthesis where primer is always required for DNA polymerase activity (page 324).

Some general properties of mRNA

In a typical chromosome there are thousands of different genes. An mRNA molecule is coded for by a single gene (or, in prokaryotes, often by a small group of genes) and, therefore,

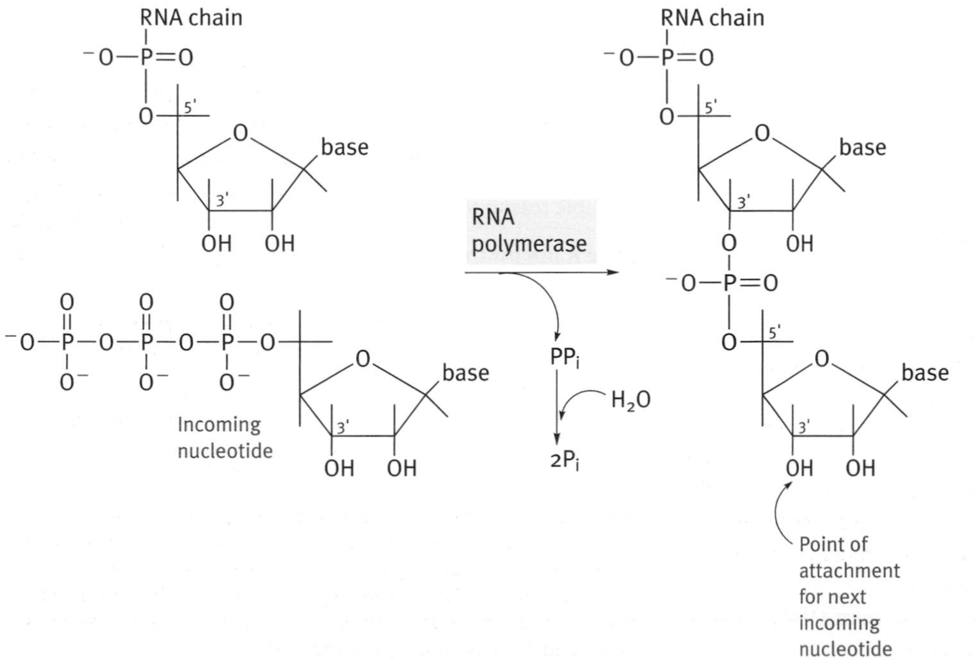

Fig. 22.1 Copying mRNA from a DNA template strand. The non-template strand is not shown. Note that the separation of the two strands is transitory. A bubble of DNA strand-separation moves along the DNA as the polymerase progresses along it.

Fig. 22.2 The reaction catalysed by RNA polymerase.

large numbers of different mRNA molecules are formed in the cell. It will be clear that, while the DNA molecule is vast in length, an mRNA molecule is minute in comparison. In the cytoplasm the mRNAs direct the synthesis of proteins for which their respective genes code.

DNA is immortal in cellular terms, but mRNA is ephemeral with a half-life of perhaps 20 minutes to several hours in eukaryotes and about 2 minutes in bacteria. Thus, for expression of a gene (the term 'expression' means that the protein coded for is actually being synthesized), a continuous stream of mRNA molecules must be produced from that gene. The gene, as it were, 'stamps out' copy after copy, RNA polymerase being the stamping machinery. This might seem wasteful but it gives the important benefit of permitting control of the expression of individual genes. Destruction of mRNA is the main 'off switch' in protein synthesis once a gene ceases to produce mRNA (how the gene is controlled is a major topic discussed later in this chapter).

The genes of an *E. coli* cell are in contact with the cytoplasm since there is no nuclear membrane. In eukaryotes, mRNA always corresponds to single genes (except in certain viruses) but in prokaryotes it may carry the coded instructions for the synthesis of several proteins, all joined together in a single RNA molecule. This is called **polycistronic mRNA** after the genetic term **cistron**, which effectively refers to the coding for one polypeptide. Such clustered genes giving rise to a single mRNA results in the coordinated expression of several genes whose proteins function as a unit, such as the enzymes of a metabolic pathway. There are other very significant differences between mRNA production in prokaryotes and eukaryotes that will be described later.

Some essential terminology

The flow of **information** in gene expression is

(transcription)   (translation)

DNA — — —→ mRNA — — —→ protein.

(Please be careful to note that the broken arrows represent *information* flow, not chemical conversions. DNA cannot be converted into RNA nor RNA into protein.)

The 'language' in DNA and RNA is the same—it consists of the base sequences. In copying DNA into RNA there is transcription of the information. Hence mRNA production is called gene transcription, or more often, simply **transcription**, and the DNA is said to be transcribed. The RNA molecules produced are called transcripts. The 'language' of the protein is different—it consists of the amino acid sequence whose structures and chemistry are quite different from those of nucleic acid.

The synthesis of protein, directed by mRNA, is therefore called **translation**. If you copy this page in English you are transcribing it. If you copy it into Greek you are translating it.

We have so far talked of mRNA synthesis as 'copying' the DNA. The template DNA strand is 'copied' only in the complementary sense as you have seen from its method of synthesis, which is dependent on Watson–Crick base pairing of incoming ribonucleotides to the template bases. A in the template becomes U in the copy and so on. The terminology in this area can be confusing, especially if you consult different texts. The DNA strand that acts as the template for mRNA synthesis is, not surprisingly, called the **template strand**, the other one the **nontemplate strand**. There's nothing ambiguous about that, but other terms are commonly used—'coding' and 'noncoding' strands and 'sense' and 'nonsense' strands, and the terms have quite different and contradictory usages in different texts. There is no 'authorized' version, but it is important to know what is being implied by the terms used. In this book we will base nomenclature on the mRNA. This has the information for the sequence of amino acids in a protein; it therefore carries the sense or message of the gene to the translational machinery. The base sequence of the mRNA is the same (apart from the T → U switch) as that of the nontemplate DNA strand and, *therefore, we will call the nontemplate strand, the coding or sense strand. The template strand therefore is defined here as the nonsense or noncoding strand* (Fig. 22.3). *In viruses the template strand is often called the minus strand* (−) *and the nontemplate the plus strand* (+).

A note on where we go from here

So far we have dealt with gene expression only in general terms in that we have referred to the gene only as supplying template DNA. However, there are many important and interesting questions such as how the genes that are to be expressed are selected out of the huge collection available in a cell. In transcribing a gene, which is part of a huge chromosome, how does the RNA polymerase 'know' where to start copying the DNA and where

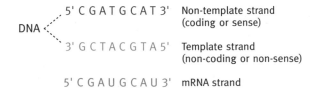

Fig. 22.3 Relationship of transcribed mRNA to template and nontemplate strands of DNA terminologies. There is no consensus in texts about the terminologies, but the system used here refers everything to the mRNA as carrying the sense of the information (see text). In viruses a frequently used terminology is: template, minus (−) strand; nontemplate, plus (+) strand.

to stop? How is the rate of gene expression controlled? A protein coded for by one gene may be produced in large amounts and another in tiny amounts or not at all; and some genes may be expressed at one time and not at other times. How are these situations achieved? To answer these questions we now must look at the structures of the actual genes, by which, of course, we mean the base sequences of genes.

In this book we have, so far, largely concentrated on the biochemistry of animal cells for simplicity and also because the differences between eukaryotes and prokaryotes tend to be more at the level of detail in many areas of biochemistry rather than involving different fundamental principles. However, when it comes to gene expression, the differences are sufficiently great to warrant separate treatments. We will deal with *E. coli* first.

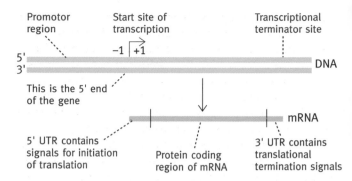

Fig. 22.4 Geography of a prokaryotic gene and its mRNA. Note that the 5' end of a gene is referring to the nontemplate strand or sense strand. In an adjacent gene, the other strand could be the nontemplate strand with transcription occurring from right to left.

Gene transcription in *E. coli*

What is specified by the term 'gene' in prokaryotes?

In Chapter 20 (page 313) we described what a gene is in molecular terms—namely, a section of a large DNA molecule. We now have to refine this description for reasons that will become clear.

A gene functions, as already mentioned, only by acting as a template for the transcription of RNA molecules whose base sequence corresponds to the nontemplate DNA strand. A gene is a specific section of DNA that is transcribed into RNA. There is a handful of genes whose function is to code for RNA molecules that are not messengers—these are the ribosomal and transfer RNAs whose role in protein synthesis is described in the next chapter. The rest of the vast array of genes produce mRNA molecules whose base sequences code for the amino acid sequences of specific proteins. How mRNA functions is the story of the next chapter. However, an mRNA molecule has sections at each end that are not translation into protein. The 5' **untranslated region (UTR)** contains encoded signals necessary for initiation of translation and the 3' untranslated region signals for its termination (again described in the next chapter), so that a gene includes sections of DNA that are transcribed into these regions as well as that for the protein itself.

This does not complete the list of DNA regions associated with the gene, for, in addition, there is a region of DNA adjacent to the 5' end of the gene, called the **promoter**, that is essential for transcription of the gene, but is not itself transcribed into RNA. At the opposite (3') end is a **terminator region** necessary, at the name implies, for termination of transcription and that also is not itself transcribed into RNA. A typical gene is illustrated in Fig. 22.4. There is a 'start' site where transcription begins; it pro-

vides the template for the first nucleotide of the mRNA. The first template nucleotide is given the number +1 and the nucleotide 5' to this −1 and so on. The start site is illustrated by an arrow (→) that indicates the direction of transcription. Nucleotides 5' to this are referred to as 'upstream' and 3' to this as 'downstream'. You will notice that Fig. 22.4 includes one end of the gene being labelled 5'. This brings us to the next question.

What do we mean by the 5' end of a gene?

A typical gene has two strands of DNA of opposite polarity, and **duplex DNA** has therefore no intrinsic polarity. How then can we refer to one end of a gene being 5'? The 5' end of a gene is the end containing the promoter. Since RNA synthesis always proceeds in the 5' → 3' direction, the 5' end of a gene corresponds with the 3' end of the template strand (but note that there are no physical ends—the DNA strand continues to the next gene).

A point to note is that, when we illustrate a *single* gene on a chromosome in a diagram, the 5' end is usually placed to the left. However, the template strand on one gene may not be the same DNA strand as the template strand of the next gene; strand usage can switch from one gene to the next. If the opposite strand is used as template on a second adjacent gene, then transcription would occur in the opposite direction, since synthesis occurs in the 5' → 3' direction.

Phases of gene transcription

There are three phases—initiation, elongation, and termination. Initiation is by far the most complex.

Initiation of transcription in *E. coli*

In the promoters there are short stretches of bases that are called 'boxes' or 'elements'. These are accepted terms but the stretches

of bases are not in any sense boxes nor do they have anything to do with atomic elements. Probably the term 'box' derives from the practice of drawing a rectangle around a sequence of bases to specify it. In a typical *E. coli* promoter there are two boxes—the **Pribnow box** (named after its discoverer) centred at nucleotide −10 and the other box centred at −35. (The numbering is explained above.) The consensus sequences of the boxes are shown in Fig. 22.5. A **consensus sequence** for a box is obtained by determining the sequence of, for example, the Pribnow box in a number of different genes, for they are often not exactly the same. You then look at the first nucleotide position in the box and count up which base is most often used by the different genes and so on for all the other positions. In fact, the consensus sequence itself might never actually occur in any gene but the variation from it will be small. Single sequences are always given in the 5′ → 3′ direction but, of course, in a gene there is the second DNA strand. Thus, although we say the Pribnow box has the sequence TATAAT, it really is

5′ − − TATAAT − − 3′

3′ − − ATATTA − − 5′

It is the *double strand* that is recognized by the proteins that control transcription (see below).

Correct initiation of transcription is obviously important. Synthesis of an mRNA must commence at the correct nucleotide on the template. The question is how the RNA polymerase is positioned in the correct place to start transcribing a gene. The −35 and Pribnow boxes are the signals for this. DNA-dependent RNA polymerase of *E. coli* is a large protein complex. In the 'core' enzyme there are four subunits, and it has an affinity for any stretch of DNA to which it attaches at random, but it cannot recognize the correct initiation site. When it is joined by another protein from the cytoplasm, the sigma protein (σ) or **sigma factor**, the resultant **holoenzyme** loses much of its affinity for random DNA but binds tightly to the −35 and Pribnow boxes and initiation of transcription can start. (We remind you that, although the DNA bases are Watson–Crick paired in the centre of the duplex, their edges are still 'visible' in the DNA grooves and can be recognized by proteins designed to do so; see page 308 and Fig. 20.4) This aligns the enzyme in the correct starting position, and the correct orientation (that is,

pointing in the right direction). It can now separate the DNA strands and initiate. A 'bubble' of separated DNA strands is formed (about one and a half turns of the helix in length), thus making the template strand bases available for pairing with incoming bases of NTPs. The enzyme now synthesizes the first few phosphodiester bonds from nucleoside triphosphates and initiation is thus achieved. At this point the protein flies off (to be used again) and the polymerase, now released, moves down the gene synthesizing mRNA at the rate of about 40 nucleotides per second until it reaches the terminator. It is not known what causes the protein to fly off.

Untwisting the DNA

Gene transcription requires strand separation of the DNA double helix, but only in a transitory way, the strand re-annealing once the section has been transcribed. It is presumably no coincidence that the Pribnow box with its high A and T content and therefore weak hydrogen bonding is where initial strand separation occurs. DNA unwinds ahead of the polymerase and rewinds behind it. The negative supercoiling (under-winding) of the DNA facilitates the process (see page 322). Thus a temporary unwound 'bubble' passes along the gene with the polymerase. The newly synthesized mRNA is paired with the template strand for a short distance but then detaches. Figure 22.6 indicates how the RNA polymerase moves along the DNA double helix.

Termination of transcription

At the end of some transcribed prokaryote genes are sequences that result in the transcribed RNA having a **stem loop structure**. This needs to be explained. Although mRNA is single-stranded, it still has the thermodynamic obligation to maximize

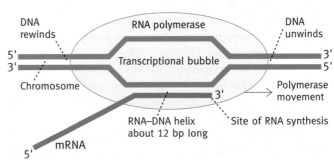

Fig. 22.6 DNA transcription by *E. coli* RNA polymerase. The polymerase unwinds a stretch of DNA bout 17 base pairs in length forming a transcriptional bubble that progresses along the DNA. The DNA has to unwind ahead of the polymerase and rewind behind it. The newly formed RNA forms a RNA–DNA double helix about 12 base pairs long.

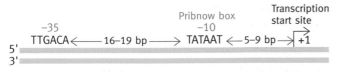

Fig. 22.5 Consensus sequences of *E. coli* promoter elements.

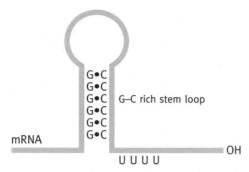

mRNA detachment from DNA template
facilitated by weak A–U pairing to the latter

Fig. 22.7 The stem loop structure of an RNA transcript involved in the Rho-independent termination of gene transcription (see text).

base pairing within itself. The newly synthesized mRNA is attached to the DNA template strand by base pairing. Near the end of the gene, the sequence of bases in the mRNA produced is such that the stem loop structure shown in Fig. 22.7 forms because the base sequence here permits formation of G–C pairs, a stable structure because of the triple bonding between G and C. Immediately following the stem loop structure in the mRNA transcript is a string of U residues, giving weak bonding of the RNA to DNA. This probably is to facilitate detachment of the mRNA and termination of transcription. The stem loop structure would prevent binding of the mRNA to the template at this point, since, if its bases are preferentially internally paired, they cannot pair with the template DNA. The string of Us would facilitate final detachment of the mRNA because of the weak A–U hydrogen bonding. Whatever the precise mechanism, this hairpin structure followed by several U residues gives precise termination. However, it should be added that, while the requirements for termination are known, it is still not completely certain how termination of this type occurs.

There is a second method of termination of transcription in many prokaryote genes. This requires the assistance of a protein called Rho for the detachment of mRNA from the DNA–RNA hybrid. The **Rho factor** attaches to the newly transcribed mRNA and moves along it behind the RNA polymerase. At the termination site, the polymerase pauses, possibly because of a difficult to separate G–C rich section of the DNA, and allows the Rho factor to catch up with the polymerase. The Rho factor has an unwinding (helicase) activity specific for unwinding the RNA–DNA duplex formed by transcription and so, having caught up with the polymerase, it detaches the RNA transcript and terminates transcription.

The mRNA directs protein synthesis; in fact, in *E. coli* it starts to do this before the full mRNA molecule is completed because the transcription occurs in contact with the cytoplasm.

However, we are not finished with *E. coli* gene transcription for there is still the question of gene control.

The rate of gene transcription initiation in prokaryotes

Genes that are **constitutively expressed** are those that are 'switched on' all the time and whose rates of expression are not selectively modulated. Such genes code for enzymes and other proteins that are needed at all times and in amounts that do not vary from time to time. However, amongst these constitutive proteins some will be required in much larger amounts than others. The major influence on the rate of gene expression in bacteria is the rate of mRNA production and this is largely determined by the frequency of initiation of transcription of a given gene. This varies because genes have promoters of different 'strengths'. A 'strong' promoter will initiate transcription frequently and thus cause many mRNA transcripts of the gene to be made and hence a lot of the specific protein. A 'weak' promoter has the reverse effect. The strength of a promoter is a function of the precise base sequence of the Pribnow and −35 boxes, the distance between them, and the nature of the bases in the +1 to +10 region. The greater the affinity of these regions for the polymerase, the stronger the promotion though this may not be the sole determinant.

Control of transcription by different sigma factors

A particularly neat method of controlling whole blocks of genes in prokaryotes is by using different sigma factors. Under certain conditions, the usual sigma protein is replaced by a different one which causes the RNA polymerase to initiate at a different set of genes. Such conditions include: (1) sporulating bacilli, in which a new sigma protein is produced after the signal of adverse conditions in the environment is received; this causes expression of a set of genes leading to sporulation; (2) a special σ factor that is produced in nitrogen starvation; (3) after a heat shock (a sudden rise in temperature), *E. coli* transitorily increases the synthesis, stability, and activity of a different protein (σ^{32}) that normally is present at a nonfunctional level. This factor directs the transcription of genes for a set of 'heat shock proteins' that protect the cell against the consequences of the heat shock. We will refer to what some of these proteins do in the next chapter (page 394).

How are individual *E. coli* genes regulated in a variable fashion? The lac operon

Bacteria live in continually changing environments to which they must adapt for survival. It is relevant to the topic in hand to remember how fierce is the competition for survival amongst bacterial populations. A typical *E. coli* cell under optimal condi-

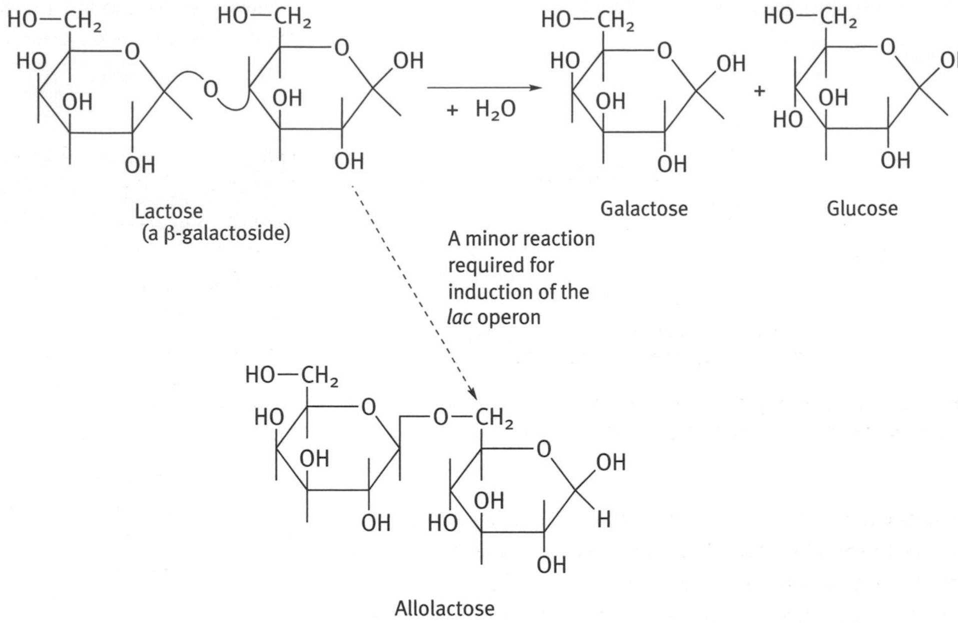

Fig. 22.8 The reactions catalysed by β-galactosidase. The term lactase is often applied to the enzyme catalysing this reaction in digestion. See text for explanation of allolactose formation.

tions will divide in about 20 minutes. If a strain lops 1 minute off that time, it will outgrow (wipe out) other strains very quickly. Wastage and inefficiency in biochemical processes is not tolerated in the face of such competition.

Production of enzymes consumes resources and energy, and it would not do if the *E. coli* cell produced enzymes that were unnecessary at the time. As mentioned already, some enzymes are constitutive—they are produced in all circumstances without any 'signal' being needed. Glucose-metabolizing enzymes come into this category for glucose is the commonest sugar and other sugars are shunted on to the glucose pathways. The cell assumes (as it were) that these enzymes will always be needed so the promoters of the genes coding for them have no 'on' and 'off' switches—they are always 'on'.

However, the *E. coli* cell may encounter other sugars—the disaccharide lactose present in milk is an example. If it is the sole source of carbon available, ability to utilize it would make survival possible. The enzyme needed to utilize this sugar is β-galactosidase, so called because lactose is a β-galactoside, and it must be hydrolysed to free galactose and glucose before it can be metabolized (Fig. 22.8). An additional transport protein, β-galactoside permease, is needed to transport the lactose into the cell. A third protein, **galactoside transacetylase**, is believed to be involved in protection of the cell against nonmetabolizable, potentially toxic β-galactosides that may be imported, though less is known of this. The three proteins are normally made in minute amounts (basal levels) because they are not required unless lactose is encountered. When lactose is encountered as the sole energy source, there is an almost instant burst

of synthesis of the three proteins. The cell can then use the lactose as a carbon and energy source. However, if, in addition to lactose being present, there is also glucose, then production of the three enzymes would be wasteful since this merely leads to production of more glucose inside the cell when there is plenty of it available anyway. The cell therefore 'ignores' the lactose signal and does not produce the enzymes. The regulation, which we will now describe, is at the gene transcription initiation level.

Structure of the *E. coli* lac operon

First a few terms: production of a protein in response to a chemical signal is called **induction** of that protein and the responsible chemical, the inducer. Prevention of the production of a protein is called **repression**. As mentioned earlier, eukaryotic genes are single units but many prokaryotic genes are grouped together, the individual groups being under transcriptional control of a single promoter. The RNA polymerase transcribes through the entire group thus creating a polycistronic mRNA molecule with the coding instructions for several proteins.

Such a group of genes, with its single promoter control, is called an **operon** (the promoter being part of the operon). β-Galactosidase, lactose permease, and transacetylase genes belong to such an operon. The three genes are often referred to as *z*, *y*, and *a*, respectively. There is also an *i* gene (i for inducibility) that codes for a protein called the *lac* **repressor** and there is a stretch of DNA called the operator region to which the *lac* repressor protein can bind. Finally, there is a stretch of DNA to which a cyclic AMP receptor protein (**CAP**) can bind. The latter

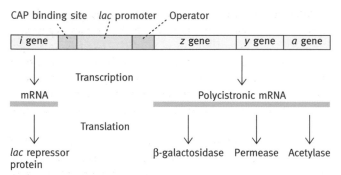

Fig. 22.9 Diagram of the *lac* operon. Note that the *i* gene is an independent gene that codes for the *lac* repressor protein. Similarly, there is a completely independent gene producing the catabolite gene activator protein (CAP) to which cAMP can bind.

stands for **catabolite gene activator protein**. It is given this general name because it is involved in the induction of other enzymes involved in catabolism of substrates. The approximate arrangement of these elements in the operon is shown Fig. 22.9. We can now systematically go through the control mechanism noting the following points. The *lac* promoter *on its own* is a weak one so that the RNA polymerase does not readily bind to it and initiate transcription. Without extra help the *lac* operon is not transcribed except at a low basal level. The extra help in polymerase binding is given by the attachment of the protein, CAP, to the adjacent site. Binding of CAP causes the double helix to bend at the site of attachment. When CAP is attached to the DNA, the promoter is a strong one. However, CAP does not attach unless cAMP is bound to it—it is an allosteric protein. *cAMP is produced by the cell only when glucose levels are low.* When glucose is at a high level, cAMP is scarce in the *E. coli* cell, CAP does not bind, the RNA polymerase therefore does not bind effectively, and *lac* operon transcription is minimal. This situation is illustrated in Fig. 22.10(a).

Does this mean that the *lac* operon is always transcribed when glucose is scarce? The answer is no. As explained, there is no point in doing so unless lactose is present. In the absence of lactose the *lac* repressor protein is attached to the operator and blocks the RNA polymerase from transcribing the genes (Fig. 22.10(b)). The *lac* repressor is also an allosteric protein. In the absence of lactose in the environment, it has strong affinity for the operator, but, if lactose is present, a small amount is able to enter the cell via the basal level of permease, and it is hydrolysed, during the course of which a small amount is converted to the lactose isomer, allolactose (see Fig. 22.8). This binds to the repressor protein, causing an allosteric change in the protein, which then dissociates from the operator and unblocks the operon.

Situation (a). High glucose; no cAMP; no lactose; no transcription of *lac* operon

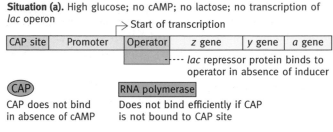

Situation (b). Low glucose; high cAMP; CAP–cAMP complex binds CAP site; RNA polymerase can now bind to promoter; no lactose; repressor protein blocks operator; no transcription

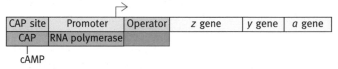

Situation (c). Low glucose; high cAMP; lactose present; repressor protein–allolactose complex detaches from operator; transcription of operon proceeds

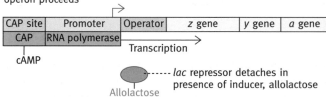

Fig. 22.10 Expression of the *lac* operon. **(a)** In the presence of high glucose there is no cAMP to cause CAP to bind and this binding is necessary for the attachment of RNA polymerase to the promoter. **(b)** With low glucose but no lactose, although CAP binds and assists the RNA polymerase to bind, transcription still does not occur because the *lac* repressor is bound to the operator blocking polymerase movement. In **(c)** the inducing allolactose binds to the repressor, causing its release from the operator, and transcription can proceed. (In the presence of lactose, a small amount of a lactose isomer, allolactose, is produced, which is the actual inducer—see text.) CAP, Catabolite gene activator protein.

Why is allolactose the inducer rather than lactose? A possible (and speculative) answer is that allolactose is produced as a minor side-product of the *β*-galactosidase reaction and is formed in sufficient amount to induce only when significant quantities of lactose are present. This would prevent traces of lactose firing off the synthetic process when the amount of sugar is insufficient to warrant this. Having lactose as the inducer but with a low affinity for the repressor protein would not achieve the same thing since, presumably, induced *β*-galactosidase would keep intracellular levels of lactose very low irrespective of the external concentration. To generate the inducer, allolactose, by the very process that destroys lactose would be an attractive mechanism. Additionally, when lactose was ex-

hausted, allolactose would be destroyed by β-galactosidase, for which it is a substrate, and thus a mechanism for terminating the induction would be provided. The system would be constantly testing the rate of inflow of lactose to decide whether the induction should continue.

With the unblocking of the operator, the RNA polymerase is now free to move down the operon, producing the polycistronic mRNA (Fig. 22.10(c)); this is translated into production of the three enzymes.

The *lac* operon was the first understood example of prokaryote operon control, but it applies to many pathways. In *E. coli* the **tryptophan operon** (*trp* operon), which contains the five structural genes needed to produce three enzymes involved in tryptophan synthesis, is similarly controlled by a *trp* repressor protein. In this case, tryptophan bound to the latter causes it to block the operator site. This prevents most of the mRNA transcription but a second mechanism in the presence of high levels of tryptophan stops transcription by those RNA polymerase molecules that do manage to get through the repressor block.

Control of transcription by attenuation

This second mechanism, called **attenuation**, depends on the fact that, in prokaryotes, translation of mRNA starts as soon as a small stretch of a messenger molecule has been synthesized, and therefore the polymerase is closely followed by the ribosome doing the translation. (The latter is dealt with in the next chapter so, for the moment, the **ribosome** is a large complex that moves along the messenger adding amino acid residues.) The operon first codes for a peptide only 14 amino acids in length; it is called the **leader peptide** but note that this is quite different from the leader peptide attached to certain proteins as described in the next chapter. It is destroyed immediately after its production. The attenuation leader peptide contains two tryptophan residues next to one another. If there is plenty of tryptophan, the ribosome adds these and moves along the mRNA; if tryptophan is scarce, the ribosome halts in the attenuation region. The crucial point is that, in the absence of a ribosome in this position (that is, when adequate tryptophan levels enable the ribosome to translate right through), the partially completed mRNA molecule internally base pairs to form a termination hairpin structure—the transcription is aborted; if a ribosome is in this region (because it is stalled in the absence of tryptophan), it interferes with the base pairing such that the polymerase is allowed to complete its transcription of the operon. Thus the level of transcription is adjusted according to the level of tryptophan in the cell; the lower this is, the higher the proportion of polymerase molecules that are allowed to produce a complete mRNA molecule.

Attenuation is known to control six operons concerned with specific amino acid biosynthesis. (Only the *trp* operon amongst these has the repressor control as well.) All have the same ingenious method for assessing the level of the particular amino acid in the cell, and this is made sensitive by the occurrence of multiple residues of the particular amino acid in the leader peptide; in the case of the histidine operon there are seven histidine residues.

One of the features of the *lac* operon control (and that of other operons) in bacteria is that, once the facts are known, the mechanism is readily understandable. It all works with military precision—CAP binds and enables the polymerase to bind, a repressor protein binds and blocks transcription, and the inducer causes it to become unblocked. It is the sort of 'mechanical' control that one might have thought of oneself Gene transcription control in eukaryotes is more complex (see next chapter) and is not remotely like anything one could have planned oneself.

The structure of DNA binding proteins

From what has been said in this chapter, it is clear that proteins that bind to specific sites on DNA play a central role in life. There are numerous repressors and transcriptional factors involved in differentiation, embryonic development, and gene control in general and all depend on their ability to recognize, and bind to, DNA.

Because of the vital importance of this, a large research effort has gone into elucidating exactly how such proteins bind to their appropriate sites on the DNA. It has emerged that most DNA binding proteins can be grouped into a small number of families on the basis of their structural characteristics and we will describe these shortly but, before this, a general overview may be useful. The main families of such proteins are the helix–turn–helix proteins, the homeodomain proteins, the leucine zipper proteins, and the zinc finger proteins.

The structures or motifs that give rise to these names are only small sections of the proteins that are characteristic of each family. When you see a diagram of a helix–turn–helix protein, for example, it usually refers only to this small part of the total protein.

The DNA binding proteins bind to double-stranded DNA; site-specific binding involves weak bond formation between the amino acid side chains of the protein and the DNA bases, though additional stabilizing bonding to the sugar–phosphate–sugar backbone may occur. The contacts of the protein recognition motifs with the bases occur predominantly in the major groove of the double helix where the edges of the bases are exposed. In many cases, the contact is made by a recognition α helix that readily fits into the major groove.

(a)

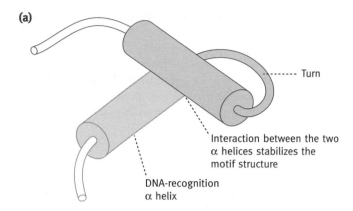

Turn

Interaction between the two
α helices stabilizes the
motif structure

DNA-recognition
α helix

(b)

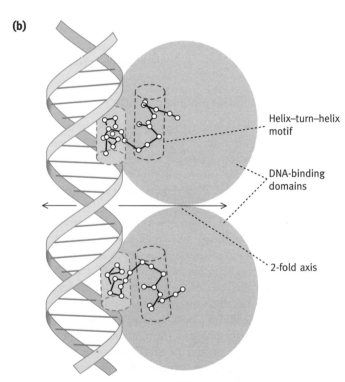

Helix–turn–helix
motif

DNA-binding
domains

2-fold axis

Fig. 22.11 (a) The helix–turn–helix motif of a DNA-binding protein monomer. **(b)** Dimer helix–turn–helix protein binding to DNA in major grooves.

A point of particular importance is that often there are two adjacent sites on the DNA necessary for specific attachment of a protein. These may be identical, formed by a palindromic arrangement of the bases (see below), or they may be dissimilar. These two situations are catered for by the protein being a dimer, made of two identical subunits (a **homodimer**) or different subunits (a **heterodimer**), respectively. In the case of zinc finger proteins, there may be multiple DNA binding motifs on the same protein.

The essence of variable gene control is that the DNA binding proteins, such as transcriptional factors and repressors, often bind only when they are 'instructed' to do so and are detached at other times (or vice versa). We have already met situations in which transcription factor activation is achieved by the regulatory protein being allosterically modified by the binding of a ligand (see the *lac* operon control above). In eukaryotes, the 'instruction' usually comes from a chemical signal external to the cell, such as that provided in the form of hormones and growth factors. The mechanisms by which such signals are transmitted to activate transcriptional factors to bind to the DNA is a major topic, which will be dealt with in Chapter 26.

With that preamble, we will now deal with the different families of DNA binding proteins.

Helix–turn–helix proteins

This was the first type to be identified; it occurs commonly in prokaryotes and many examples exist. We will use the lambda repressor as an example; lambda is an *E. coli* bacteriophage (described on page 471). We will not go into the role of the repressor protein here for it is a very complex story, but it binds to a region of the lambda DNA and is involved in the decision of whether an infecting virus enters the lytic or lysogenic cycles (explained on page 471). For the present, its importance is that its mechanism of binding to DNA has been fully elucidated.

The helix–turn–helix (HTH) motif is, as explained, the small section of the protein that makes the binding contact to the operator site on the DNA. (It is not a protein domain for it could not exist as such in isolation; motif means a recognizable structural feature.) The motif has two α helices linked by a β turn; one of the two (the recognition helix) sits in the major groove of DNA (Fig. 22.11(a)). The operator, or **DNA recognition site**, is an imperfect palindrome. A palindrome is a phrase or word that reads the same forwards or backwards. An example is 'Madam I'm Adam'—it is symmetrical about the letter 'I'. Or, as is usually stated, palindromic DNA binding sites have a **dyad symmetry** (dyad means two units treated as one, or a group of two). The structure of the operator (not to be remembered) is

5' TATCACCGCCAGTGGTA3'
3' ATAGTGGCGGTCACCAT5'   .

The two halves of the palindrome represent identical binding sites, although the coloured nucleotides show that the palindrome is not perfect. The repressor protein is a dimer, the recognition helices of the two HTH motifs fitting into them (Fig. 22.11(b)).

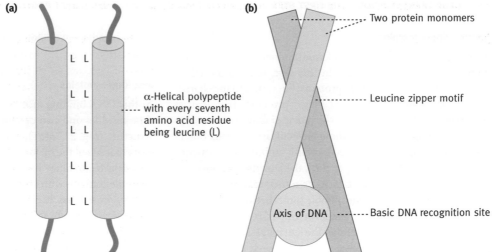

Fig. 22.12 **(a)** The leucine zipper motif. The hydrophobic leucine residues are opposed, not intercalated as in a zipper. **(b)** The leucine zipper protein attached to DNA, looking down the axis of DNA. The two arms of the protein lie in adjacent sections of the wide groove. The attachment sites are rich in basic amino acids.

(a)

α-Helical polypeptide with every seventh amino acid residue being leucine (L)

(b)

Two protein monomers

Leucine zipper motif

Axis of DNA

Basic DNA recognition site

(a)

Cys

His

Zn

Cys

His

(b)

Zn

(c)

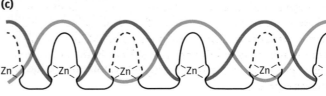

Zn   Zn   Zn   Zn   Zn

Fig. 22.13 **(a)** The zinc finger structure. **(b)** Model of the interaction between the target region of DNA and a zinc finger protein. One side of the finger structure is in the form of an α helix that attaches to the major groove of DNA. **(c)** Multiple zinc finger motifs attached to DNA.

Leucine zipper proteins

These are found in many eukaryotic transcription factors. Note that the name does *not*, in this case, refer to the DNA-recognition motif as in HTH proteins, but rather to the characteristic structure found in these proteins that causes dimerization of two subunits (which may or may not have identical recognition sites). The whole molecule is an α helix but, in the 'leucine zipper' motif, every seventh amino acid is leucine. Since there are 3.6 residues per turn, the leucines all appear on the same side of the α helix forming a hydrophobic face. Two such subunits attach by hydrophobic forces between the leucine side chains (Fig. 22.13(a)). The term 'zipper' is a misnomer resulting from the initial belief that the leucine residues interdigitated. The actual DNA-binding region of each monomer is a region rich in positively charged arginine and lysine residues. The dimer attaches to the DNA in a 'scissor' grip, the two arms being in adjacent major groove sections of the duplex (Fig. 22.13(b)).

Zinc finger proteins

In this DNA binding motif, a zinc atom attached to two histidine residues and two cysteine residues stabilizes a finger-like structure (Fig. 22.13(a)), one part of which is a recognition α helix that sits in the major groove of DNA (Fig. 22.13(b)). Transcription factors may have more than one such finger, making multiple contacts with DNA (Fig. 22.13(c)). One such gene regulatory protein has 30 zinc fingers. Not all zinc fingers necessarily bind specifically to DNA; some are nonspecific and help stabilize the specific binding. Zinc finger proteins are common in many different eukaryote gene regulatory proteins. Intracellular receptors to which steroid hormones bind (page 428) have variants on the zinc finger theme.

FURTHER READING

Lewin, B. (1999). *Genes VII.* Oxford University Press.

DNA-binding proteins

Brennan, R. G. and Matthews, B. W. (1989). Structural basis of DNA-protein recognition. *Trends in Biochem. Sci.*, **14**, 287–90.

Reviews the structures of proteins involved in the control of gene transcription and the way they interact with DNA.

DNA-binding proteins—zinc fingers

Klevit, R. E. (1991). Recognition of DNA by Cys_2, His_2 zinc fingers. *Science*, **253**, 1367 and 1393.

Crisp, concise summary.

Rhodes, D. and Klug, A. (1993). Zinc fingers. *Sci. Amer.*, **263**(2), 56–65.

Discusses structures, functions, and distribution of these transcription factors.

Klug, A. and Schwabe, J. W. R. (1995). Zinc fingers. *FASEB J.*, **9**, 597–604.

A complete account of the structure of this protein motif and its role in DNA binding.

DNA-binding proteins—leucine zippers

McKnight, S. L. (1991). Molecular zippers. *Sci. Amer.*, **264**(4), 32–9.

A very clear account of these transcription factors.

Ellenberger, T. E., Brandl, C. J., Struhl, K., and Harrison, S. C. (1992). The GCN4 basic region of leucine zipper binds DNA as a dimer of uninterrupted α helices: crystal structure of the protein-DNA complex. *Cell*, **71**, 1223–37.

A research paper, but readable and worth looking at for the molecular illustrations.

PROBLEMS FOR CHAPTER 22

1   In what ways does RNA synthesis differ from DNA synthesis?

2   By means of notes and diagrams, describe the components of an *E. coli* single gene and its associated flanking regions.

3   Describe the process of initiation of transcription of a gene in *E. coli*.

4   Describe two methods by which, in *E. coli*, gene transcription is terminated.

5   What factors determine, in a constitutive gene of *E. coli*, the strength of a promoter?

6   Describe how the *lac* operon is controlled.

7   In terms of structural motifs, there are several families of transcription factors. What are these?

Chapter summary

Eukaryotic gene transcription and control

The basic processes involved in eukaryotic mRNA production

The basic enzymic reaction by which RNA is synthesized in eukaryotes is the same as in prokaryotes, so Fig. 22.1, showing the assembly of RNA in prokaryotes from the four ribonucleoside triphosphates under the direction of a DNA template, also applies here. The DNA-dependent RNA polymerase which transcribes genes coding for proteins in eukaryotes is called **RNA polymerase II**, a very large multi-subunit enzyme. (We will leave **RNA polymerases I and III** to later in the chapter where we describe genes from which RNA molecules that are not messengers are transcribed.) In prokaryotes, genes are transcribed and the RNA molecules produced in this way (the primary transcripts) are fully functional mRNAs which are immediately translated. The situation in eukaryotic cells is very different, because the primary transcript is greatly modified before it is a functional mRNA. Not only that, but the processes of initiation of the transcription of a gene and its control are quite different in eukaryotes. In fact the processes of initiation and control are inseparable from one another and quite complex. For this reason we will first deal with the actual production of mRNA molecules in eukaryotes and then return to the major topic of control and initiation.

Capping the RNA transcribed by RNA polymerase II

The RNA of the eukaryotic gene primary transcript immediately undergoes a modification at its 5′ end, called **capping**, as soon as RNA synthesis is initiated. At the 5′ end of the RNA there is a triphosphate group, since the first nucleotide triphos-

phate incorporated into the RNA simply accepts a nucleotide on its 3′–OH group, leaving the initial triphosphate unchanged. The terminal phosphate of this is removed and a GMP residue is added from GTP (Fig. 23.1). The 5′–5′ triphosphate linkage (see Fig. 23.1) is very unusual in nature. The G is then methylated in the N-7 position and also at the 2′—OH of the second, and sometimes of the third, nucleotide. The cap is believed to protect the end of the mRNA from exonuclease attack and it is involved in initiation of translation as described in the next chapter.

These are not the only differences between mRNA production in prokaryotes and eukaryotes, because most eukaryotic genes are split genes.

What are split genes?

A mRNA molecule in both prokaryotes and eukaryotes is proportionate in length to the size of the protein it codes for (or, in polycistronic mRNAs, to the several proteins coded for), apart that is from the 5′ and 3′ untranslated regions which are involved in initiation and termination of translation of the messenger into protein (Chapter 24) and are short anyway. This correspondence between the length of the *primary* RNA transcript (the RNA molecule transcribed from the gene without any modification) and the polypeptide it codes for is true for prokaryotes but not for eukaryotes. In the case of almost all eukaryotic genes the primary transcript is very much longer—perhaps even 10 times longer—than would be expected on the basis of the size of the protein it codes for. This is because the RNA coding for protein in the primary RNA transcript is split up into several parts, linked together by intervening stretches of RNA which do not code for amino acid sequences. In the DNA

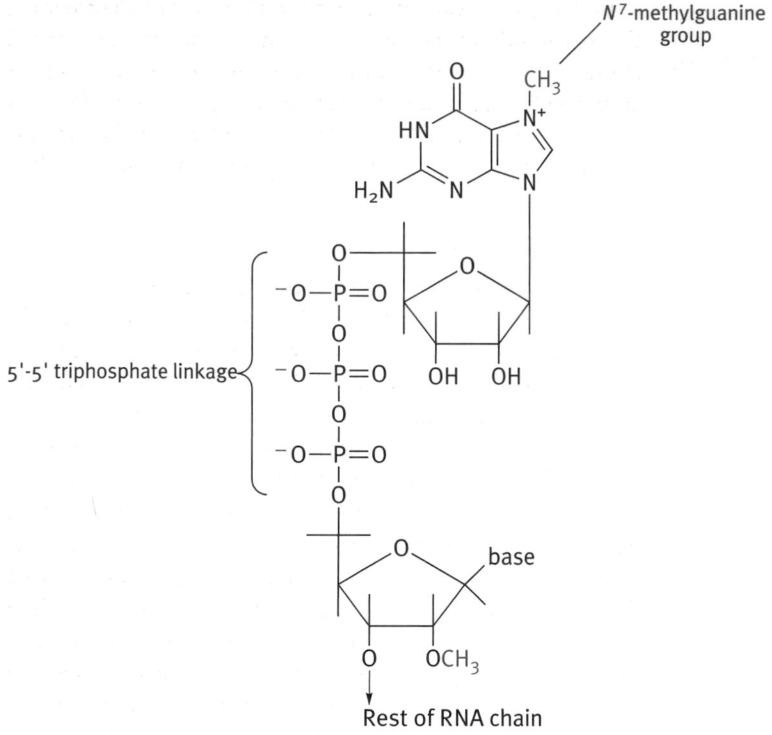

Fig. 23.1 Structure of the 5'-cap in eukaryotic mRNA. The terminal nucleoside triphosphate of the primary RNA transcript is converted to a diphosphate followed by a reaction with GTP in which pyrophosphate is eliminated. This is followed by methylation reactions. (A third methyl group may be added to the 2'—OH of the next nucleotide of the primary transcript.) The capped primary transcript is then processed to mRNA—see text for details.

from which such RNA molecules are transcribed, the sections coding for the intervening sequences, with no protein-coding content, are called **introns** and the coding stretches are called **exons**. There can be from 2 to 500 or so introns in a gene (Fig. 23.2(a)) and introns can vary in length from about 50 to 20 000 base pairs. Exons are usually less than 1000 base pairs in length. The primary transcript is processed to eliminate the introns and link together the exons into one mRNA molecule. This is known as **RNA splicing** (Fig. 23.2(b)).

Mechanism of splicing

Removing the unwanted RNA introns of a primary transcript and joining up the exons into mRNA looks a formidable task but the key to it is the **transphospho-esterification reaction**. In this, a phosphodiester bond is transferred to a different —OH group. There is no hydrolysis and no significant energy change.

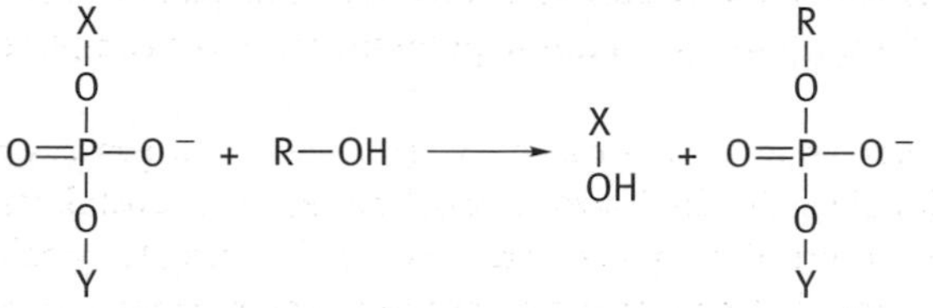

$$X-O-P(=O)(O^-)-O-Y + R-OH \longrightarrow X-OH + R-O-P(=O)(O^-)-O-Y$$

If X–Y is an RNA chain, it would be broken.

Let's now turn to RNA processing or splicing. The exon–intron junctions are 'labelled' by consensus sequences; all introns begin with GU and end with AG (though the full consensus sequences are longer than this). The ROH of the above diagram is actually the 2'—OH of an adenine nucleotide in a short sequence (seven bases long in yeast) of the intron chain (Fig. 23.3) known as the **branch site**. The reason for this name will be seen from the lariat structure in Fig. 23.3. The 2'—OH group attacks the 5' phosphate of the G nucleotide at the splice site, forming a lariat structure. This breaks the chain at the 3' end of exon 1, thus producing a free 3'OH and the 3'—OH attacks the 5' end of exon 2, joining the two exons.

In most eukaryotes, the splicing reaction in the nucleus is catalysed by very complex protein–RNA structures called **spliceosomes**. They contain RNA molecules, 100–300 bases long in higher eukaryotes, called **small nuclear RNAs (snRNAs)**. They are associated with proteins in structures known as **small ribonucleoprotein particles (snRNPs)**, each containing multiple protein subunits. There are five snRNPs in a spliceosome, known as **U1, U2, U4, U5, and U6**, and altogether there are about 40 proteins in this group of snRNPs. Additional proteins known as **splicing factors** are also needed. The U1 and U2 snRNPs bind by base pairing to the 5' splice site and the branch site respectively and then associate with each other. A trimer of

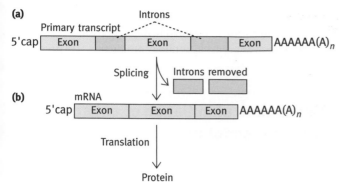

Fig. 23.2 Primary polymerase II transcript of a eukaryotic gene: (a) introns after capping and addition of polyA tail; (b) excision of introns to form the mature mRNA is called splicing.

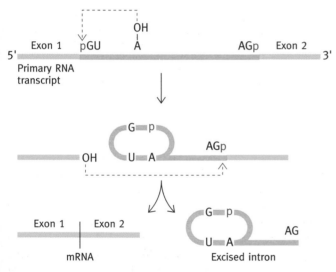

Fig. 23.3 Mechanism of RNA splicing. Note that, for clarity, the process is shown in two stages.

U4, U5, and U6 then associates with the complex to form the spliceosome. U6 catalyses the transphospho-esterification reactions but this has to be triggered by the release of U4 which masks the catalytic centre. Release of U4 requires ATP hydrolysis. The various interactions between the snRNPs themselves and with the RNA to be spliced are believed to cause the reaction sites to be appropriately aligned and may also serve to create the catalytic site which brings about the reactions. It is not established what has the catalytic site(s), but it is probably an RNA structure. Since you are used to proteins catalysing everything this statement may seem surprising, but we will go into this aspect more fully in the final section of this chapter (page 372).

Faulty splicing can lead to genetic diseases. In β thalassemia, the β subunit of hemoglobin is not produced in normal amounts, because the G at the 5′ splicing sequence of an intron is mutated to an A and primary transcripts are therefore not properly processed to mRNA. The disease **lupus erythematosus** is caused by an autoimmune attack (page 423) on the snRNPs.

What is the biological status of introns?

An important biological aspect of the existence of introns is that they may have facilitated evolution. At this point you might re-read page 45 on protein domains. As research on protein structure identified domains, and gene research identified exons, it was found that the latter often code for discrete protein domains. As stated earlier (page 46) new proteins are believed to have evolved by 'domain shuffling'—the concept is that the same domain is used repeatedly, for a partial function of an enzyme for example, combined with a variety of other domains, so that a new family of proteins is assembled from pre-existing domains. The separation of the parts of a gene into exons coding for discrete protein domains would facilitate domain shuffling. This would provide a more rapid means of producing novel proteins by recombination events than would point mutations in DNA leading to single amino acid changes.

The role of introns in exon shuffling may need explanation. Exon shuffling will be the result of chromosome rearrangements. The DNA coding for a protein domain must be intact and the existence of introns on either side could facilitate successful transposition of the exon. If a cross-over rearrangement occurred in the middle of a domain coding region, it would probably destroy the protein domain structure coded for and be a nonviable change since, as explained, a protein domain has a structure which could be expected to exist on its own as a discrete entity. Since introns have no protein coding sense, gene cross-overs could occur at introns placed on either side of an exon and therefore lead to a potentially successful exon shuffle since the domain coding region would remain intact.

What is the origin of split genes?

There are two views on this question. One is that the prokaryotic type, noninterrupted genes, are primitive and that introns were inserted later in evolution. This gives no explanation of the origin of introns which, in this hypothesis, remains a complete mystery.

The alternative view is called the exon theory of genes, or more descriptively, the 'introns early' model. This postulates that ancient primitive genes were very small, coding for small proteins, and that on either side of their coding regions were untranscribed flanking regions. Modern genes are postulated to

have formed by the fusion of several of these mini-genes. Introns would then represent the fused untranscribed flanking regions of adjacent mini-genes. The fact that RNA molecules can have the property of self-splicing (see page 372) removes the requirement for a complex splicing mechanism early in evolution. On this hypothesis, prokaryotic uninterrupted genes are regarded as a later development.

As stated, the 'introns early' view has developed as a result of the correspondence sometimes observed between exons and protein domains (page 46), which, in turn, led to the concept of exon shuffling in which new proteins are assembled from new combinations of domains.

However, it has been pointed out that if introns had been inserted randomly into ancient noninterrupted genes, some parts of genes coding for protein domains might, by chance, have introns nicely placed on either side of them so that successful exon shuffling could occur. Assuming that such shuffling conferred an evolutionary advantage, then more modern genes arrived at by exon shuffling would automatically exhibit the correspondence between protein domains and exons. On this view, it is the shuffling which has produced the exon/domain correspondence, not the other way about, and the correspondence of exons with domains says nothing about the origin of introns.

To examine this, several ancient genes (ones coding for proteins conserved over long evolutionary times) were examined. It was found that there was no correspondence between the intron–exon structure and structural features of the protein. The workers concluded that this is incompatible with the 'introns early' view and indicates that uninterrupted genes are primitive. (Note that the occurrence of exon shuffling is not questioned by either hypothesis; this is generally accepted and is a different issue from that of the origins of introns which made shuffling possible.) It has been pointed out that interrupted genes in prokaryotes would pose a difficult splicing problem in that mRNA is immediately translated by ribosomes which initiate at the 5′ end of the messenger before synthesis of the 3′ end has been completed. The jury is still out.

Alternative splicing or two (or more) proteins for the price of one gene

There is one other well-established advantage that split genes confer—alternative splicing. In typical splicing, all the exons of the primary RNA transcript are linked together to form the mature mRNA leading to the formation of a single specific protein. However, there are many known cases where the splicing can occur in different patterns so that a particular group of exons form one mRNA, whereas a different group from the same gene transcript forms another mRNA, leading to different proteins.

The mechanism may be employed to produce variant forms (isoforms) of a protein required in different tissues or at different times such as occurs in antibody production (see page 457).

Mechanism of initiation of eukaryotic gene transcription and its control

Although initiation of eukaryotic gene transcription obviously precedes transcription and processing, we have left discussion of the subject until now because it is inseparable from control of gene activity.

Unpacking of the DNA for transcription

We have already mentioned the very great degree of packing of eukaryotic DNA necessary to fit its huge length into the nucleus (page 311). Before (or during) transcription, the DNA structure is selectively loosened up at those parts of the DNA that are to be transcribed. We know that selective loosening up actually occurs from two lines of evidence.

- First, in insect salivary gland polytene chromosomes, the loosening can be directly seen in the light microscope. These chromosomes are unique and classically important. Salivary gland cells of the larval form of the fruit fly, *Drosophila,* are very large. Whether for this or other reasons, a most unusual chromosomal structure occurs. The chromosomes are replicated about 1000 times without cell division occurring and the elongated chromosomes, still with much condensation packing, lie side by side precisely aligned lengthways. The packing arrangement along the DNA gives rise to bands visible in the light microscope; in a single chromosome these would not be visible but multiplied a thousand times they become so. During larval development successive banks of genes are expressed and it is seen that during transcription, the DNA of specific bands becomes loosened—forming what are called **chromosome puffs** (Fig. 23.4).

- The second means of knowing that transcription of eukaryotic genes is associated with loosening of the DNA packing is that transcriptionally active regions of chromatin become more susceptible in *in vitro* experiments to attack by added DNase—an enzyme which hydrolyses DNA. It is known that globin genes, for example, become hypersensitive to DNase attack only in chromatin from those cells which synthesize globin and only at the developmental time at which the genes are actively transcribed.

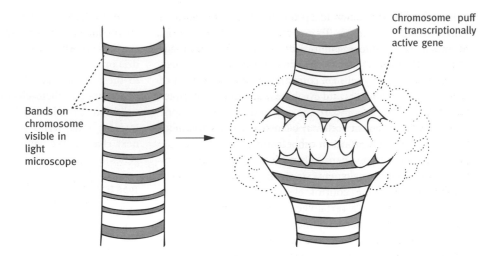

Fig. 23.4 Chromosome puffs at transcriptionally active chromatin on insect salivary gland polytene chromosomes.

A general overview of the differences in the initiation and control of gene transcription in prokaryotes and eukaryotes

In prokaryotes the RNA polymerase, attached to a sigma factor, binds to the DNA at a gene promoter (page 347). The elements of the different prokaryotic gene promoters are essentially the same, differences in their affinity for the polymerase being due to subtle variations in the base sequence of elements and their spacings. In the case of the lac operon described earlier, an additional protein (CAP) is required which binds to the DNA at a site adjacent to the polymerase-binding site and is essential for efficient initiation and this applies to many operons. The rate of gene transcription is a function of the affinity of the polymerase for the promoters. This determines the number of transcripts initiated per unit time; a 'strong' promoter has a high affinity for the enzyme.

The eukaryotic situation looks less orderly, indeed almost chaotic. The gene promoter has control elements (boxes) but gene control regions vary greatly in the elements they contain. The polymerase does not bind by recognizing sites on the DNA, as happens in prokaryotes. Instead, a large number of other proteins assemble on the promoter and the polymerase is recruited from the surrounding medium. In doing so it becomes correctly positioned on the DNA but, it is emphasized, it does not itself initially join to the DNA as in prokaryotes; as stated, it has to be recruited from the medium by a protein complex assembled on the promoter.

The control requirements in eukaryotic cells are vastly different from those of prokaryotes. An *E. coli* cell has about 4000

genes. During the lifetime of a single cell it may be necessary to transcribe all of those genes. For the most part what is required is an 'on–off' switch for controlled genes or, for constitutive genes, a permanent 'on' condition with the actual rates of initiation being determined in both cases by the affinity of the polymerase for the promoter.

In a eukaryote such as a mammal the situation is much more complex; a human cell has about 100 000 genes. There are many different types of cells—liver, muscle, brain, epithelial, blood, bone, and so on. Many proteins are common to all—those for glycolysis are an example, coded for by what are known as **housekeeping genes**—but each type of cell has its own cohort of proteins needed for specific cell functions. Liver cells have liver-specific proteins not present in brain and muscle cells, and *vice versa*. However, the DNA of all cells in the animal is the same (ignoring special cases such as the gene rearrangements in B and T cells of the immune system described on page 454). In a liver cell, liver-specific genes must be activated while those coding for proteins specific for brain must be ignored by the transcription apparatus. All cells of the body arise from a single fertilized egg cell and differentiation into specific cell types involves gene control.

Even in mature cells when the differentiation process to produce specific cell types has been achieved, a quite different set of control problems exist which have no counterpart in prokaryotic cells. As already explained in Chapter 13 (page 215) an animal cell cannot make all decisions itself. Its activities must be such that they correspond to the needs of the animal as a whole. An obvious instance of this is that cell division should not proceed independently (as happens in cancer)—the cell must receive a signal from elsewhere before it proceeds to division. Also,

the rate of synthesis of individual proteins varies over short time periods according to needs. To give an example, after feeding, enzymes devoted to storage of foodstuffs rise in amounts (page 211) and this involves hormonal activation of specific genes. The activities of many cells are controlled by a whole battery of hormones, growth factors and cytokines (described in Chapter 26), which constantly bombard the cells and control appropriate genes. A given gene in a cell may be instructed by a multiplicity of signals from hormones or other factors.

The key to eukaryotic gene control is that there can be any number of controlling elements on the DNA associated with a particular gene. Some control elements determine tissue-specific expression—why certain genes are activated only in the appropriate tissues; others respond to the different signals arriving at the cell from other cells of the body. Many different control elements exist in a cell. The question therefore is how these multiple control elements result in efficient control of single genes. To put this in a more general way, the cell has the difficult job of making sense out of multiple signals as to what the rate of transcription of a given gene should be. There is nothing analogous to this in prokaryotes.

To anticipate what we shall shortly be describing more fully, the important concept you need to become familiar with is that of **transcription factors** (TFs). A plethora of different TFs in eukaryotic cells control a multitude of different genes. TFs are proteins which attach to control elements of eukaryotic genes and are responsible for controlling transcription at the level of initiation. Tissue-specific expression of genes depends on the presence of the cognate TFs in the cells. (The term 'cognate' implies a relationship—a cognate TF is one related to, or having affinity with, the element in question—the particular TF designed to be capable of binding to that DNA sequence.) Some TFs are activated only when the cell receives the appropriate signal such as from a hormone; a wide variety of external signals work in this way, by activating specific TFs. The system gives the very great flexibility to gene control in eukaryotes which is needed to respond to the many different instructions that may be showered on individual genes.

After that preview, the first thing to discuss is the anatomy of a eukaryotic gene promoter.

Types of eukaryotic genes and their controlling regions

As mentioned earlier, in prokaryotes there is only one type of RNA polymerase involved in transcription (page 344). In eukaryotes there are three different RNA polymerases, designated I, II, and III (nothing to do with order of importance). They are specific for transcribing types of genes also known as I, II, and III respectively. The important ones in our present context are the type II genes which include all nuclear genes coding for proteins and are therefore numerically dominant. (Nuclear genes are those residing in the cell nucleus rather than in mitochondria or chloroplasts.) The rest of this chapter is devoted largely to these. Type I and type III genes code for RNA molecules which are not themselves translated into proteins but are involved in protein synthesis. They are ribosomal RNAs and transfer RNA, terms you may not fully understand until the next chapter. We will however refer to transcription of their genes later in this chapter.

Type II eukaryotic gene promoters

Figure 23.5 shows the DNA components of a type II promoter. There are two sections.

- The **basal elements** (sometimes known as core elements—terminology in this field tends to be slightly variable) refer to the **initiator** (Inr), a short sequence at the start site rich in pyrimidine bases and the **TATA box**, at −25 base pairs, with a consensus sequence TATAAAA reminiscent of the Pribnow box of prokaryotes. It is the only element with a fixed position relative to the start site. Most, but not all, type II genes have the TATA box. (We remind you that the terms 'box' and 'element' are interchangeable, with the latter now being used more frequently but the former still used commonly for some sequences. The term box probably arose from the practice of delineating a DNA sequence in a gene by drawing a rectangle around it.)

- The **upstream control elements** are found at variable positions within the range of −50 to −200 or so base pairs. The most commonly found ones are the CAAT box (pronounced CAT) with a consensus sequence GGCCATCT and the GC box (GGGCGG). It seems that all genes require at least one of such common control elements but different gene promoters are quite different in which are present. A given gene promoter may have any one or a mixture of them and there may be several copies of one or more of them; there is no standard 'recipe'. Figure 23.6 shows examples of the variety of control elements found in promoters. In addition to these there may be any number (as appropriate to the particular gene) of control elements concerned with tissue specific expression, hormonal control and control by many other factors.

Enhancers and their mechanism of gene control: alternative models

The eukaryotic promoter is not strictly defined but is usually taken to be the region roughly 200 base pairs upstream from the start site which includes the upstream control elements.

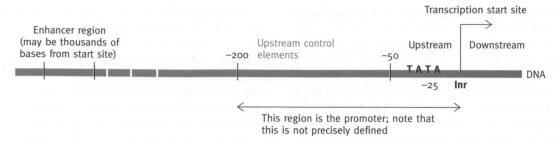

Fig. 23.5 Eukaryotic type II gene control elements. Examples of common transcription elements are the CAA T box and the GC box. Different promoters can have any mixture of the general transcription elements including multiple copies of individual ones and no one element is present in every gene (see Fig. 23.6 for an illustration of this). Inr, initiator (a pyrimidine-rich stretch).

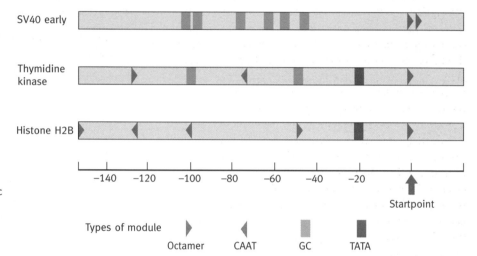

Fig. 23.6 The elements of three eukaryotic gene promoters Promoters contain different combinations of TATA boxes, CAAT boxes, GC boxes, and other elements.

Another control region exists for many genes called the **enhancer region** which contains a number of control elements close together—usually they are the same elements that exist in the promoter. Enhancers are not included in the definition of promoter because in the latter, the individual elements are in a relatively fixed distance from the start site (as stated, within about 200 bases) whereas the enhancer can be thousands of bases distant and moreover remains effective even if placed in the reverse orientation. The exclusion of enhancers from the definition of the promoter is an arbitrary one. The enhancer is generally believed to interact with the distant promoter by looping around. This concept is presented in Fig. 23.13 (page 368) and will be found in most accounts of the subject. However, it is important to note that the mechanism of enhancer action is not understood and this model is currently being challenged by an alternative one. The popular model is described as 'rheostatic' because it envisages that the enhancer increases the rate of transcription of a given gene. It is not

challenged that the enhancer does indeed increase the amount of mRNA from a studied gene, but this refers to the total amount of mRNA produced in experiments in which large numbers of cells are studied. Studies in which *single cells* are examined have suggested that an enhancer may not increase the *rate* of transcription of the gene but that it increases the *probability* of a gene being activated by the factors at the promoter already described. This is known as the **stochastic model**. It will be appreciated that that if, in a population of cells, the enhancer increases the number of cells expressing the gene it will have the same effect on total mRNA production as increasing the rate of transcription by the rheostatic model, so in terms of quantitative gene control the two models are not in conflict. At present it is not known which model is operative and indeed it is possible that both occur. A discussion of this interesting topic is given in the reference by S. Fiering *et al.* in the reading list, which also points out that the stochastic model could have repercussions on cell differentiation programs.

It may be useful at this stage to summarize the information on control elements so far.

- The promoter comprises the first 200 bases or so upstream from the start site.

- It has **basal elements** in the region to −50 consisting of the Inr at the start site and in most genes a TATA box at −25 bases upstream.

- At about −50 to −200 are located the upstream control elements. These include a variable collection of the common control elements such as the CAAT box and the GC box and any number of other control elements involved in tissue specific expression and control by hormones and other signals.

- Often at great distances away from the promoter there may be **enhancers** which are a group of elements usually similar to those in the promoter itself but which are arbitrarily excluded from the definition of the promoter.

We now come to the protein factors that bind to the various DNA sequences of the promoter and enhancer regions.

Transcription factors

We now have to turn to what the role of the specific DNA control elements is. In all cases they are sites of attachment for specific proteins. There are two broad categories of these.

- First, the **general TFs** present in all cells. These are components of the **basal transcription machinery** (described below) which form a complex attached to the basal elements.

- Second there are the sequence specific TFs which bind to the upstream control elements and the enhancer elements. They activate genes, so that the term **activator** is frequently used for them rather than TF.

Activators which bind to the common elements such as the CAAT box and the GC box are found in all cells. As already mentioned, all genes require at least one of these common factors (CAT and GC boxes are the most common elements in this regard.) They do not require any activation but are always capable of binding to their cognate elements. In addition there are TFs or activators for a whole variety of control elements found as appropriate only in specific cells to bring about patterns of gene activation according to the nature of those cells. To give an example, there is a factor that binds to the GATA element. This is found in developing red blood cells and is required for activation of genes coding for proteins specific to red blood cells, such as hemoglobin. A tissue such as muscle does not have it and therefore hemoglobin is not produced there.

Many TFs exist in an inactive form in the cell; they cannot stimulate transcription until they are activated (as mentioned earlier, common TFs binding to CAT or GC boxes do not require activation). Activation may be phosphorylation (or dephosphorylation) or other change causing a conformational change in the protein which now can bind to the DNA sequence in question. The activation is usually the result of signals arriving at the cell from other cells. In Fig. 23.7(a) a steroid hormone is shown to enter the cell directly (due to its lipid solubility) and on binding to a soluble receptor protein causes a conformational change in the latter so that it is now an active TF which activates certain genes. Figure 23.7(b) shows that cAMP which is elevated as a result of the action of certain hormones (page 436) activates protein kinase A which phosphorylates an otherwise inactive transcription factor. Figure 23.7(c) shows how many hormones bind to membrane receptors and induce a signal cascade inside the cells which results in activation of specific transcription factors often by phosphorylation of the protein. (Cell communication by hormones and other agents is an important subject which is dealt with much more fully in Chapter 26, starting on page 423.)

An active transcription factor has two domains (Fig. 23.8)— the DNA binding site and the activation domain which is a binding site for other proteins (the latter yet to be described).

How do activators promote transcriptional initiation?

Before we can answer that we must first deal with another important aspect:

The role of chromatin in eukaryotic gene control

Eukaryotic genes *in vivo* are in the form of the protein–DNA complex known as **chromatin**, not as naked DNA (see page 311). In this, two turns of DNA are wrapped around nucleosomes made of octamers of histone proteins, 146 base pairs being intimately in contact with the octamers. Individual nucleosomes are separated by linker DNA which varies somewhat in length in different species but averages about 50 base pairs, so the whole length of DNA per nucleosome is about 200 base pairs (see Fig. 20.7).

Until recently, chromatin was regarded as an inert structure whose sole function was to condense the DNA to fit into the nucleus. However, as already mentioned, it has long been known from *in vitro* experiments that when a gene is activated, the DNA in that area becomes hypersensitive to hydrolysis by added DNAse indicating that in some way the chromatin opens up making the DNA itself more accessible to the enzyme. From this and other work it has been established that the 'default' state of chromatin (i.e. the state in the absence of any action to coun-

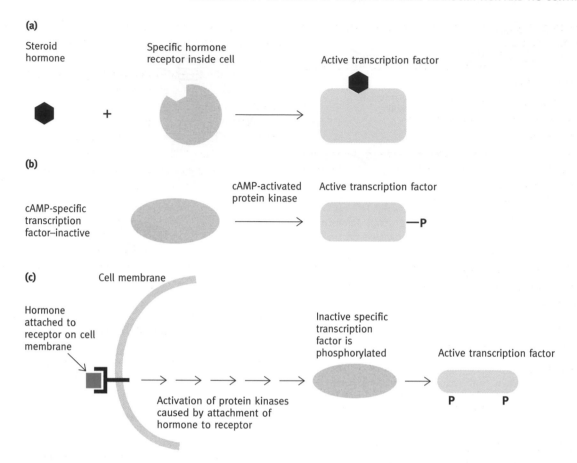

Fig. 23.7 Examples of transcription factor activation: (a) A steroid hormone enters the cell; it attaches to a receptor specific for that hormone and causes a conformational change in the receptor protein which is now an active transcription factor. This activates the gene(s) which are controlled by the particular hormone. A whole family of steroid hormone receptors exists. (b) cAMP is produced as a result of epinephrine binding to cell receptors (page 436). The cAMP activates a protein kinase which phosphorylates the inactive transcription factor which is activated. (c) Protein hormones such as insulin do not enter the cell but bind to receptors on the cell surface. This results in a sequence of events which ends in the phosphorylation of the appropriate inactive transcription factors and activates them. Note that there are many different hormones which bind to different specific receptors and activate different transcription factors. These bind to specific response elements of different genes. Thus each hormone can exert control over appropriate genes (activation of transcription factors inside cells by steroid binding and receptor-mediated signal transduction is dealt with in Chapter 26 on cell signalling).

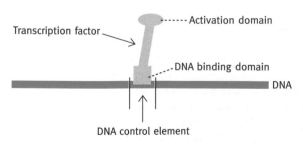

Fig. 23.8 Transcription factor domains.

teract it) is a 'shut down' condition—the genes are inactive. The reason for this is that gene promoters are blocked by nucleosomes which prevent assembly of the basal initiation machinery on the promoters. Gene control in eukaryotes involves 'opening up' or unblocking of the promoters. It requires modification of the chromatin structure known as **chromatin remodelling** (Fig. 23.9). This term means the *effective* removal of nucleosomes from the promoter site of the gene to be activated. It is not known whether a nucleosome physically leaves the DNA or just changes its attachment so as to permit the transcription

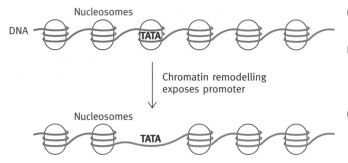

Fig. 23.9 Chromatin remodelling. The principle is that the promoter in chromatin is blocked by nucleosomes. Gene activation requires exposure of the promoter; this may require the physical removal of one or more nucleosomes, or it could be some change in the relationship of the nucleosome(s) to the DNA which effectively gives accessibility to the promoter. Use of the term 'chromatin remodelling' reflects the current uncertainty about exactly what happens at the molecular level.

complex to assemble on the promoter The term 'remodelling' avoids implication of what exactly is happening in molecular terms. It is not known how many nucleosomes need to be remodelled—a nucleosome and its linker is about 200 bases in length which is about the size of a promoter so perhaps one would suffice, but in one particular case it is known to be three or four.

How do transcription factors open up gene promoters?

First, one or more transcription factors attach to cognate elements on the promoter and/or the enhancer (in Fig. 23.10(a) only a single one is illustrated). The activation domain of each of the bound factor(s) is free to bind to other proteins.

One of the latter so far not mentioned now enters the picture—the **coactivator.** This is a protein complex which binds to the activation domain of the bound TF or, to put it in another way, the bound TF recruits the coactivator from the nucleoplasm. The coactivator is not a TF and does not bind to DNA but is essential for gene transcription. Its attachment to the DNA-bound TF positions it close to the promoter and its blocking nucleosomes (Fig. 23.10(b)). Several coactivators have been identified named p300, the related CPB, and PCAF. Factors are casually named as discovered so that the names are often somewhat cryptic; just to mention one, CBP stands for CREB binding protein and CREB in turn stands for cAMP response element binding protein, i.e. CREB is a transcription factor activated by a cAMP-induced phosphorylation (Fig. 23.7(b)). It is particularly interesting that it has been shown that when CREB is activated it readily binds to the coactivator CBP.

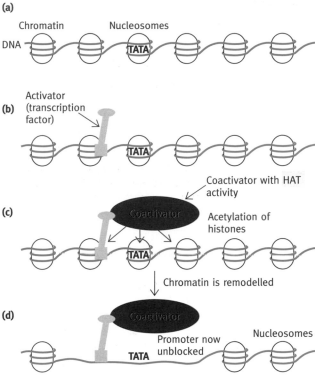

Fig. 23.10 Remodelling of chromatin—the first step in gene transcription: (a) Chromatin with a promoter blocked by nucleosomes. (b) An activator (a transcription factor) attaches to its site. (c) A coactivator with histone acetyltransferase (HAT) activity attaches to the bound activator and acetylates the histones of the blocking nucleosome(s). (d) The blocking nucleosomes are removed or remodelled. Several different activators are involved in control of a typical eukaryotic gene (see Fig. 23.5 for illustration of this) and several or all may be involved in recruiting the coactivator. In caseof the activation of the thyroid hormone gene, 3–4 nucleosomes are removed from the DNA.

An exciting discovery was that some coactivators have an enzymic activity known as **histone acetyltransferase (HAT)** (Fig. 23.11). It catalyses the transfer of the acetyl group of acetyl-CoA to the ε-NH$_2$ group of lysine residues in the N-terminal domains of the histone octamer subunits which form the nucleosomes (Fig. 23.10(c)). These domains are exposed on the surface of the nucleosomes almost like short tails so that they are readily accessible to HAT activity. Acetylation eliminates the positive charge on the amino groups and is believed to loosen the attachment of the negatively charged DNA to the nucleosome. It appears to be a factor in chromatin remodelling (Fig. 23.10(d)) and therefore for opening up the gene prior to initiation.

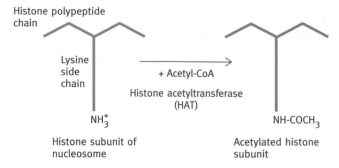

Fig. 23.11 Acetylation of the lysine residues of the N-terminal tails of subunits of the histone octamers of nucleosomes. The acetylation reduces the positive charge on the protein and is believed to result in lessening the attachment of DNA to the nucleosome leading the chromatin remodelling described in the text.

Oddly enough it is not the only mechanism known for re-modelling chromatin; for separate ATP-dependent remodelling, machines have been discovered first in yeast and later in humans. They may work independently, associated with TFs, or they may cooperate with the HAT mechanism; little is known about this. This opening up of the promoter by remodelling leads us to the next question:

How is transcription initiated on the opened promoter?

The next step is to assemble the **basal initiation machinery** on the promoter. Let us look at the components of this first. In all cells there are general transcription factors required for initiation of all genes. One of these is a large complex of proteins called TFII D (transcription factor for type II genes, the D indicating which of several it is). It binds to the TATA box of the promoter (Fig. 23.12). RNA polymerase II is a large multi-subunit enzyme which exists in the nucleoplasm. It joins up to the TFIID complex on the TATA box; several other general initiation factors are involved such as TFIIB which plays a role in linking up the polymerase to the TFIID. Since the TATA box is in a fixed position relative to the start site the position of TFIID is fixed and therefore the polymerase in joining up to it is automatically correctly positioned at the start site. As already mentioned, rather disconcertingly some genes don't have a TATA box; in these cases it is the Inr that positions the basal transcription machinery.

The basal machinery is not assembled on the promoter in isolation (Fig. 23.12 is only to describe the components of the machinery) but rather the entire cohort of transcription factors and coactivators that we have mentioned so far are involved.

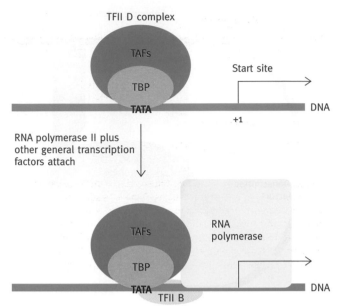

Fig. 23.12 Diagrammatic representation of the components of the basal initiation complex. TF II D is a complex of the TATA box binding protein (TBP) and a number of TAFs (TBP associated factors). RNA polymerase attaches to the preinitiation complex and forms the basal transcription complex. This figure is to illustrate the components of the basal complex but this does not exist in isolation; it is part of a very large assembly and interacts with transcription and other factors (see Fig. 23.13).

The actual process of assembling the final initiation complex illustrated in Fig. 23.13 is believed to be as follows:

- TFs bind to their cognate sites and recruit the coactivator from solution.

- The HAT activity of the coactivator acetylates the histones of the nucleosomes blocking the promoter site including the enhancer site.

- The acetylation causes the removal or remodelling of the nucleosome blocking the promoter so as to permit assembly of the transcriptional machinery on the promoter.

- The TFII D binds to the TATA box and the RNA polymerase is recruited by the TFII D to complete the initiation complex.

Transcription now proceeds. It should be noted that interactions between the transcription factors and the coactivator and between the both of these and the basal machinery components, including the polymerase are probably involved in the assembly of the final complex. The initiation complex shown in Fig. 23.13 is fairly well established in its main features, though new details are likely to be discovered. There are many interac-

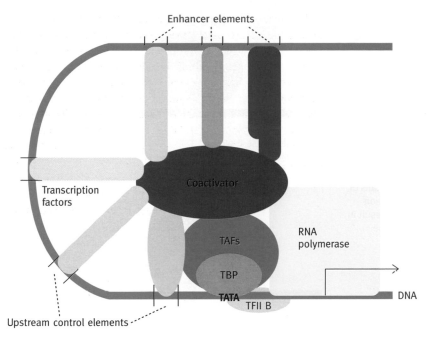

Enhancer elements

Transcription factors

Coactivator

TAFs

RNA polymerase

TBP

TATA   TFII B

DNA

Upstream control elements

Fig. 23.13 The completed initiation complex of a eukaryotic gene to illustrate multiple controls on gene transcription initiation. The sizes, positions, and interactions of proteins are arbitrarily drawn to illustrate that transcription factors may have multiple contacts with coactivators and the basal transcriptional machinery. The unlabelled shapes in pink represent various transcription factors (activators) bound to their cognate control elements on the DNA. Enhancer elements may be thousands of bases distant. See text for a discussion of the mechanism of enhancer action.

tions between components of the initiation complex and this leads to a major question as follows:

What is the mechanism by which transcription rates are controlled by the multiple transcription factors?

In Fig. 23.13 it is shown that the many TFs attached to their control elements interact with coactivators which interact in turn with the basal transcription machinery and other interactions occur. It is a very complex cluster of proteins with many, but as yet incompletely defined, interactions.

How do these interactions control the rate of initiation? There is no certain answer to this, but interactions between the protein components help to stabilize the basal transcription complex. (We remind you that bond formation even of the noncovalent type releases free energy; see page 14.) This would increase the likelihood of the latter assembling on the promoter and result in the observed synergism between factors on the initiation of transcription. The latter would be expected if protein–protein interactions stabilize the complex, because increased stabilization would increase its chances of persisting so that statistically there would be an increase in the number of transcriptions occurring per unit time. As discussed below, there has to be some mechanism(s) for inactivating gene transcription once the need for it has gone and there may be a constant battle between activation of a gene and its inactivation, activation involving assembly of the complex and inactivation

its disassembly. Increasing amounts of noncovalent bonding between components would, by lowering the free energy of the complex, presumably favour activation. Synergism between the effects of the interactions would make the system sensitive to increasing numbers of TFs involved in the initiation. The reasoning is speculative and other reasons for the cooperative effect of TFs on transcription may emerge.

A separate possibility is that the interactions may cause conformational changes in components of the complex (including the polymerase itself) which in some unspecified way increase the rate of transcription initiation. Indeed, as described below the polymerase undergoes conformational changes in detaching from the initiation complex and enters into the phase of elongating the RNA transcript.

How are activated genes switched off?

Controls have to be reversible; activated genes must be switched off when the need for transcription has gone. In the case of induction by ligands such as hormones, removal of the signal ligand will result in the inactivation of the TF; for example a TF activated by phosphorylation will be inactivated by dephosphorylation once the hormone signal is removed. This would favour the disassembly of the initiation complex. It has been discovered that deacetylases exist in cells which remove the acetyl groups on histones by hydrolysis

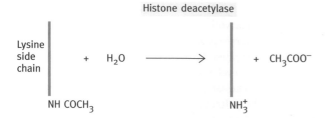

Fig. 23.14 Reaction catalysed by histone deacetylase.

(Fig. 23.14); these could well be involved in reversing gene activation.

A concluding note on initiation: the process of initiating transcriptional eukaryotic genes may be more complex than described above. Phosphorylation and methylation of components may also be involved, but there is little understanding of these matters.

Synthesis of the messenger RNA

The progression from initiation to actually synthesizing mRNA is surprisingly complicated and incompletely understood. It involves the polymerase complex detaching from the initiation complex, the first synthesis of phosphodiester bonds, and establishment of a very firm complex between the polymerase and the DNA. Firm attachment is necessary because if the polymerase becomes detached during the transcription of a gene it cannot re-attach in midstream, and some genes are of enormous length. (The human dystrophin gene has 200 000 kb to be transcribed.)

As the polymerase moves along reading the template strand it faces a problem not found in prokaryotes, namely how to negotiate the nucleosomes. The polymerase is roughly the same size as a nucleosome and would have difficulty in negotiating the curve of the latter. There are various possibilities. The histone octamer may be displaced entirely as the polymerase passes and then reinsert itself after it has passed, or the nucleosome core particle may never actually lose contact with the DNA but detach just enough to allow transcription, which would facilitate its reinsertion. Or part of the histone octamer may be temporarily displaced to allow the polymerase to pass. It is all very speculative at the moment. There is still a lot to be discovered about gene transcription and its control.

Termination of transcription in eukaryotes

It is not understood how termination by RNA polymerase II occurs in eukaryotic genes; nothing is known corresponding to the prokaryotic termination signals. In fact it looks (erroneously, no doubt) as if during evolution the design of the process was somewhat fouled up. The RNA polymerase II when it reaches the 3′ end of the gene, synthesizes a sequence AAUAAA (coded for by the template strand) but then goes on to transcribe well beyond this and then somehow terminates. A separate enzyme cuts the RNA transcript a short distance beyond the above sequence and then another enzyme, not dependent on a template, adds adenine nucleotides (using ATP as their source)—as many as 200 to form a polyA tail. Histone mRNA does not have this tail, so it can't be essential for translation. Evidence has been obtained that the polyA tail is involved in stability of the mRNA molecule in the cell, as described below.

mRNA stability and the control of gene expression

Although regulation of gene transcription is of overriding importance in the control of gene expression, the stability of individual mRNAs is also of considerable significance. In most situations, the rate of synthesis of a protein is a reflection of the mRNA level for that protein except where translation of a messenger is controlled, such as occurs in the erythrocyte aminolevulinate synthase and globin (of hemoglobin) as described on page 511. The level of an mRNA in a cell is a function of its synthesis and breakdown rates. At a given rate of mRNA production, a messenger with a longer half-life will be present at a higher steady state level within the cell than a less stable one, resulting in a higher rate of synthesis of the cognate protein. Mechanisms that determine the half-life of an mRNA can thus provide a way of regulating gene expression.

Prokaryotic mRNAs in general have an ephemeral existence with half-lives of 2–3 minutes. Rapid messenger turnover permits rapid responses to changing circumstances. In mammals, the half-life of individual mRNAs ranges from about 10 minutes to 2 days in extreme cases. The mRNA for globin, which is regarded as a stable one, has a half-life of about 10 hours. Regulatory proteins tend to be coded for by short lived messengers so that changes in the rate of transcription of their genes have rapid effects on their synthesis; the relevant messengers usually have half-lives of less than 30 minutes. A given cell thus contains mRNAs with widely different rates of degradation and this may change in individual mRNAs according to the prevailing conditions in the cell.

Determinants of mRNA stability and their role in gene expression control

Role of the polyA tail

Almost all eukaryotic mRNAs have a polyA tail added to their 3′ ends before they emerge from the nucleus (page 369 and Fig. 23.15 (a)), histone mRNAs being exceptions. The polyA tail is believed to protect the messengers from rapid destruction. A polyA binding protein binds to the tail and it protects the mRNA from destruction by a $3′ \rightarrow 5′$ exonuclease. Oddly enough it also protects the other end of the mRNA. The polyA tail presumably loops around and brings the polyA binding protein into contact with the 5′-methylguanosine cap and shields the latter from attack by a decapping enzyme. Removal of this cap exposes the 5′ end of the mRNA from attack by a $5′ \rightarrow 3′$ exonuclease. mRNA destruction is preceded by deadenylation. When the tail is reduced to less than 15 adenylate residues the polyA binding protein cannot attach thus triggering destruction from both ends of the mRNA molecule. It is also believed to have effects on the transport and translation of the messengers and these may in turn affect the rate of their destruction.

Structural stability determinants of mRNAs

Histone mRNAs lack a polyA tail and their stability is determined by a stem loop at the extreme 3′ end of the molecules

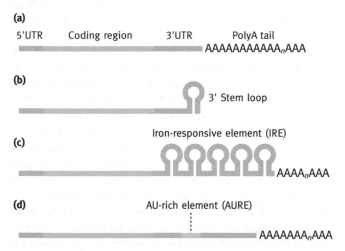

(a)

5′UTR   Coding region   3′UTR   PolyA tail

AAAAAAAAAAA$_n$AAA

(b)

3′ Stem loop

(c)

Iron-responsive element (IRE)

AAAA$_n$AAA

(d)

AU-rich element (AURE)

AAAAAAA$_n$AAA

Fig. 23.15 Some structures present in the 3′ untranslated regions (UTR) of mammalian mRNAs that influence the half-lives of the molecules in the cell: (a) the polyA tail found in the majority of eukaryotic mRNAs; (b) the 3′ stem loop found in histone mRNAs; (c) the iron-responsive element (IRE) of transferrin mRNA; (d) the AU-rich element (AURE) found in a large number of unstable eukaryotic mRNAs.

(Fig. 23.15 (b)). Synthesis of histones is required only in the S phase of the eukaryotic cell cycle (page 319) when DNA synthesis and nucleosome (page 311) assembly are occurring. Histone genes are transcribed during S phase but this ceases in the G_2 phase when the level of histone mRNAs rapidly falls. The latter is partly due to cessation of transcription but, in addition, the half-life of the mRNAs decreases from 40 to 10 minutes, a change dependent on the presence of the stem loop referred to above. The experimental transfer of this feature to globin mRNA produces a hybrid mRNA molecule, which after insertion into a cultured cell, is destabilized at the end of the S phase as if it were a histone mRNA. The destabilization of the histone messengers requires the presence of free histone (protein) monomers, which accumulate as soon as DNA synthesis (and therefore nucleosome assembly) ceases at the end of the S phase. Prompt switch-off of histone synthesis is necessary because the histone monomers are toxic to cells. The fourfold reduction in the histone messenger half-life means that, after cessation of histone gene transcription, mRNA is exhausted but without destabilization, it would take almost 9 hours for this to occur (Fig. 23.16). The reduction in half-life is thus essential, for otherwise histone synthesis could proceed during the G_2 phase of the cell cycle.

The synthesis of β-tubulin is also regulated by such a feedback mechanism modulating the mRNA stability. (β-tubulin is involved in microtubule formation described in Chapter 33 but all that matters at this stage is that it is an important protein.) The presence of excess free tubulin monomers destabilizes the mRNA for this protein; the mechanism by which this occurs is not understood, but the regulatory rationale of the arrangement is self-evident.

The synthesis of the transferrin receptor protein is another case where mRNA stability regulates synthesis of the protein. The receptor is responsible for the transport of iron into cells (discussed on page 509). In the 3′ untranslated region of the mRNA is a group of five stem loops called an **iron-responsive element (IRE)**. In the absence of iron an **IRE binding protein** attaches to the IRE and stabilizes the mRNA, thus increasing receptor synthesis with a consequent increase in the import of iron into the cell. In iron abundance, the iron complexes with the protein, which then no longer binds to the IRE. It is believed that this exposes the region to attack by an endoribonuclease, resulting in reduction of the import of iron into the cell.

Some short-lived mRNAs contain 'instability' sequences in their nucleotide sequence that mark the molecules for rapid destruction. A structural feature of many unstable mRNAs is the **adenine/uracil-rich element (AURE)** found in the 3′ untranslated regions of the molecules (Fig. 23.15 (d)). These elements

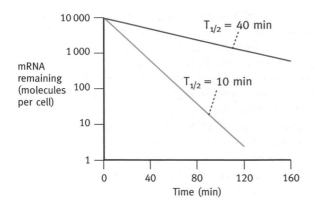

Fig. 23.16 The effect of a fourfold reduction in the half-life of an mRNA on the amount of the latter remaining in a cell at increasing times after the cessation of synthesis of that mRNA.

vary from one messenger to another but all contain AU sequences at least nine bases in length. They cause messenger destabilization, possibly by virtue of AURE-binding proteins. It is not known how the latter function and the effect of AUREs may be indirect. A particularly interesting example is the mRNA derived from the c-*fos* gene, which codes for a leucine zipper class of transcriptional factor and contains an AURE in its 3′ untranslated region (the terminology of c-*fos*, v-*fos* and of oncogenes is dealt with on pages 471), the corresponding v-*fos* gene is strongly oncogenic (cancer producing); the mRNA derived from it lacks the destabilizing AURE found in the nononcogenic c-*fos* mRNA and has a longer half-life in the cell. This results in an inappropriately prolonged gene-activating signal to the cell. As explained in Chapter 26 this can be a factor in the development of cancer. The determinants described so far confer instability on mRNAs but there is evidence that the mRNA for hemoglobin, which is a relatively stable messenger, has a stabilizing element.

It is also known that several hormones modulate the stability of specific messengers in target cells. The mechanisms of mRNA stability control seem to be varied and complex and are likely to be recognized as of increasing interest in the gene control area.

Gene transcription in mitochondria

In the earlier chapters on energy generation from food oxidation, the central role of mitochondria was described. Mitochondria are replicating organelles of eukaryotic cells, with their own DNA and protein synthesizing machinery (the same is true of plant chloroplasts). They divide to maintain the number appropriate to the cell—about 1000 in rat liver and 10^7 in a frog's oocyte. The DNA of mitochondria is usually a circular duplex—less than 20000 nucleotides in mammals, but there can be 5 or 10 copies in each mitochondrion.

The mitochondrion is far from self-contained, for its DNA codes for only a small fraction of mitochondrial proteins. The rest are coded for by the nuclear DNA of the cell; the proteins are synthesized in the cytoplasm and transported into the mitochondria, a topic dealt with in Chapter 25. In mitochondria *both* strands of the DNA are completely transcribed into single long transcripts. (The polymerase molecules doing this move in opposite directions since synthesis is always in a 5′ → 3′ direction.) The single primary transcripts are then processed, in a not totally understood way, into mRNAs, the tRNAs and rRNAs (see below for explanation of the latter two). To produce more of the rRNA, many shorter transcripts encompassing the rRNA genes are made—again explained below.

As will be described in the next chapter, mitochondrial translation (protein synthesis) has prokaryotic characteristics as compared with the eukaryotic features of cytoplasmic translation. This has given rise to the belief that mitochondria evolved from engulfed prokaryotic cells, which became symbiotic (a similar theory holds for the chloroplasts in plant cells). The mitochondrial transcriptional system really does appear to be confused as to whether it is prokaryotic or eukaryotic. In mammalian mitochondria, the mRNAs are polyadenylated (eukaryotic) but not capped (prokaryotic), and there are no introns in the genes (prokaryotic).

In trypanosome mitochondria, mRNAs are produced which then have to be 'edited'. By a mechanism too complex to describe here, RNA transcripts have extra U residues inserted at specific places to produce the correct coding sequence for proteins. These Us are *not* coded for in the DNA. However, after this discovery—and more shocking still—it was found that mRNA editing may occur even in nuclear transcripts. Thus a gene for as respectable a protein as mammalian apolipoprotein B (page 105), coded for by a nuclear gene, has two mRNA transcripts. One of these has to have a DNA-coded C converted to a U to form a translational stop codon (see next chapter). A simian virus (which is eukaryote in its gene characteristics) has a mRNA edited by insertion of two G residues not specified by the DNA. Whether there are fundamental reasons requiring these unsettling (to us) variations from the normal, or whether it is another example of evolutionary tinkering and the policy of accepting anything that works, is unclear.

Genes that do not code for proteins

Exceptions to the rule that genes code for proteins are genes that code for special RNA molecules which are not messengers. In the next chapter you will learn that proteins are synthesized by small bodies called ribosomes and the amino acids to be incorporated into proteins are carried on small RNA molecules called **transfer RNA** (tRNA), many distinct species of which exist. Ribosomes contain RNA molecules collectively called **ribosomal RNA** (rRNA). Prokaryotic ribosomes have three different rRNA molecules in them and eukaryotes have four. Relatively vast amounts of rRNA are made, accounting for perhaps half of total transcription. Since mRNA is unstable whereas tRNA and rRNA are long lived, the majority of RNA in a cell is in these forms. In *E. coli,* the three rRNAs plus a few tRNAs are transcribed together and the large precursor molecule is cut up into the appropriate pieces. The other tRNA molecules are also transcribed in larger precursor molecules which are processed into the tRNAs. In eukaryotes three of the four rRNAs are produced from the primary transcript of one gene and processed into the final rRNA molecules. Unlike prokaryotes there are three RNA polymerases:

- RNA polymerase I transcribes most of the rRNA.
- RNA polymerase II, which transcribes protein coding genes, has already been described.
- RNA polymerase III transcribes the tRNA and smaller rRNA.

Since so much rRNA and tRNA is needed, multiple copies of their genes exist. In eukaryotes there may be hundreds or thousands of rRNA genes tandemly arranged head to tail—they exist in the nucleolar regions of the nucleus. The genes transcribed by polymerase III are unusual in that regulatory elements or boxes occur, not only in the DNA 5′ to the start site but also in the region which is transcribed, namely, downstream from the start site.

Ribozymes and self-splicing of RNA

This topic is concerned with a different method of RNA splicing. We have left it to the end of this chapter rather than including it with the earlier description of splicing because it is of importance beyond that of splicing alone. Also, to have included it earlier would have been disruptive to the story of eukaryotic transcription and mRNAs are not involved.

In the earlier description of splicing, we described the spliceosome, a complex with scores of protein components and several RNA components. You can imagine the shock it gave to the biochemical world when it was discovered that some RNA gene transcripts were accurately self-spliced without any help from proteins. It was the first case known of a specific biochemical reaction occurring as the result of catalytic activity brought about by a molecule other than a protein. It is a topic of major interest both in its own right and also because of its potential for providing therapeutic tools; it also is of considerable relevance to ideas on the development of life.

In the protozoan *Tetrahymena,* one of the rRNAs (described on page 385 but you do not need at this stage to know what it is) is made as a precursor transcript containing an intron which has to be spliced out to produce the mature rRNA. The two exons on either side of the intron become joined together by the splicing to form the mature molecule.

During a study of the splicing of the isolated precursor RNA in the laboratory of Tom Cech, it was found that the intron was spliced out with the two exons properly joined without any protein being needed. A divalent metal ion (Mg^{2+} is used *in vitro*) and guanosine (or a 5′ guanine nucleotide) are essential. This caused great excitement for as stated, it was the first known biochemically specific reaction occurring without a protein being involved. It should be noted, however, that this self-splicing is not a true catalytic reaction for because the molecule itself is changed. This is described as the molecule being *cis*-acting. A true catalyst such as an enzyme brings about reactions in successive molecules of substrate without itself being changed; it is *trans*-acting. However, a discovery by the laboratory of Sydney Altman showed that RNA can act as a true *trans*-acting catalyst. An enzyme called **ribonuclease P,** found in *E. coli,* and elsewhere, processes tRNA (page 382) precursors by a specific hydrolytic reaction. This enzyme has an RNA component attached to a protein, but the RNA by itself is capable of catalysing the hydrolysis in a truly catalytic manner. Because of its similarity to an enzyme, the term **ribozyme** was coined. No corresponding molecules involving DNA exist. Cech and Altman shared a Nobel prize for these discoveries.

The mechanism of the self-splicing in *Tetrahymena* is shown in Fig. 23.17. The guanosine 3′—OH (G—OH) attacks the phosphodiester bond thus releasing the left-hand exon with a 3′ —OH group. The latter now makes an attack on the second phosphodiester bond releasing the intron and splicing the two exons. The chemistry of the two reactions involved is as illustrated in the diagram of transphospho-esterification on page 359.

The intron is fairly large (414 nucleotides) and has a complex three-dimensional structure, the integrity of which is essential

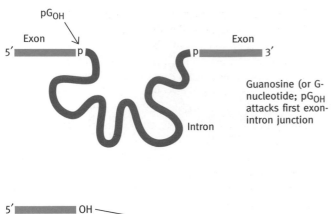

pG$_{OH}$

Exon   Exon

5′   p   p   3′

Guanosine (or G-nucleotide; pG$_{OH}$ attacks first exon-intron junction

Intron

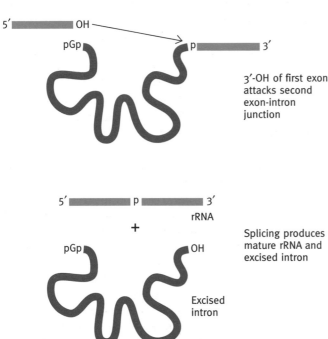

5′   OH

pGp   p   3′

3′-OH of first exon attacks second exon-intron junction

5′   p   3′

rRNA

+

pGp   OH

Splicing produces mature rRNA and excised intron

Excised intron

Fig. 23.17 Mechanism of the self-splicing reaction of the *Tetrahymena* rRNA precursor. G$_{OH}$ represents guanosine, GMP, GDP, or GTP. A metal ion is needed; Mg^{2+} is usually used for *in vitro* experiments.

mitochondria have ribozyme capability. Most spectacularly, a reaction of central importance in protein synthesis is catalysed by an RNA component of the ribosomes (see page 389) of all species so far as is known.

Why should most splicing occur by the elaborate spliceosome apparatus and others self-process? The answer to this is not known, but there is a group of pathogens in which it is possible to see a biological reason for the use of self-cleavage of RNA molecules rather than enzyme-catalysed cleavage, as will now be described.

Self-cleaving in small circular RNA pathogens

A group of pathogens which infect plants and one (the hepatitis delta virus, HDV) which infects humans, have small single-strand RNA circular genomes and RNA self-cleavage is involved in their replication. Two are **viroids** (page 472) which are small naked RNA circles. These are the avocado sunblotch viroid (ASBV) and the peach latent mosaic viroid (PLMV). Some are known as satellite RNAs, which need a helper virus to replicate, and some are viruses. The latter includes HDV. This is a deficient virus in that it can replicate its RNA after infecting a liver cell but its protein envelope cannot be formed unless there is a co-infection by the hepatitis B virus. Without the latter helper virus, the HDV cannot form infectious virions. HDV infection on its own does not spread to many cells because it cannot form infectious progeny but it may exacerbate an existing infection by the hepatitis B virus and is of medical significance.

Why is self-cleavage capability so consistently prevalent in this group? It is related to the fact that they reproduce their RNA genomes by the **rolling circle method** as illustrated in Fig. 23.18. In this, the template RNA circle is transcribed round and round continuously by RNA polymerase so that the emerging RNA transcript is a continuous linear molecule (**concatamer**) in which successive copies of the genome are covalently linked together but automatically self-cleave into monomers as they are transcribed. The linear monomers so produced circularize to produce the infective form. It is necessary for the monomers to be separated by cleavage of the concatamer at exactly the correct points.

A probable reason for self-cleavage being used here is that no ribonuclease is known which has the degree of specificity required to separate the monomers (one specific phosphodiester bond amongst about 250 or more has to be cut) so that the presence of a host enzyme capable of doing this in the infected cell is unlikely. The normal strategy of viruses which require particular enzymes, not present in host cells, for their multiplication, is to code for the enzyme themselves so that the host cell synthesizes it for them. However, viroids are not known to

for the self-cleaving. It is believed that the structure provides a binding site for the cofactor (G–OH). This type of catalytic activity was not anticipated to occur in RNA since it does not have the dissociable groups of proteins which are involved in enzyme catalysis (page 47) nor was it expected to be able to form binding sites. Both of these expectations have proved to be wrong.

A number of ribozymes are known but they are not common and seem to occur for the most part unpredictably. Enzyme reactions still account for the overwhelming majority of all biochemical reactions. RNA transcripts in the newt and certain

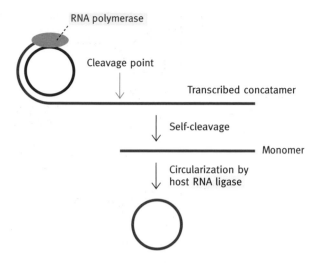

Fig. 23.18 The rolling circle method of RNA replication. The circle in black is the template strand. This mechanism is used by all of the known small single-stranded circular RNA pathogens. (Although the infective pathogen is a plus strand, its replication proceeds via a minus strand form. Plus and minus strands in viruses are discussed on page 467, but this concept is not needed in the present context.)

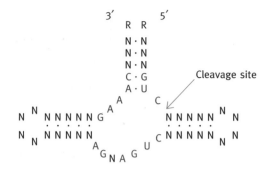

Fig. 23.19 A consensus sequence of hammerhead structure (given for reference purposes). The bases in red are conserved among various hammerheads and are necessary for activity. The R groups are to indicate that this structure is a small section of a larger RNA molecule which has folded up into the hammerhead structure at the cleavage site. RNA molecules containing hammerheads range in size from 247 to 457 nucleotides. The C next to the cleavage site is occasionally A but never U or G.

code for any proteins and so could not direct synthesis of a required enzyme.

The RNAs in the different pathogens have different types of structures leading to self-cleavage; three are known but the hammerhead mechanism is one of the best known and we will confine our description to this.

The hammerhead ribozyme mechanism of self-cleavage

Following the discovery of the *Tetrahymena* self cleavage referred to above, it was observed in the laboratory of Bob Symons in Adelaide, that whenever the concatamer transcript of the ASBV was thawed from deep freeze storage, the molecule had progressively broken down into clearly defined segments, not in a random fashion. It was found that it accurately self-cleaved in the presence of Mg^{2+} to form monomers. The base sequence around the cleavage site was found to be such that it could be predicted that, by base pairing, it could be folded into a structure called a **hammerhead** (because of its shape). A typical hammerhead is shown in Fig. 23.19. Transphospho-esterification is involved, but the hammerhead mechanism is different from that in the *Tetrahymena* ribozyme mechanism already described. No guanosine cofactor is needed; the cleavage is achieved by forming a 2′,3′-cyclic phosphodiester, as shown in Fig. 23.20. This simply involves an attack of the 2′–OH on the vicinal 3′-phosphodiester group.

The concept of gene shears

One of the reasons for the great interest in ribozymes, apart from their considerable basic interest, is that they offer a possible new tool in biology and medicine. The ability to prevent the expression of selected genes in organisms could have important applications in knocking out the expression of genes which are harmful. Ribozymes cannot attack DNA but they could in principle prevent the expression of DNA genes by cutting the mRNA coded for by specific genes; a single cut is all that is required. The gene selected could be, for example, one that is oncogenic or it could be a gene essential for an infective virus to reproduce or, if it is an RNA virus, by attacking the viral genome itself.

In the concatamer cleaving reactions mentioned so far, the RNA molecule alters itself but does not attack other molecules—it works in *cis* but not in *trans*. It is not a catalyst as an enzyme is; as stated, the latter is unchanged by its reaction and so is capable of attacking successive molecules of substrate. It is, however, possible to create artificial hammerhead structures which act in a truly catalytic way; such a structure is shown in Fig. 23.21. The key to this is that two appropriately designed pieces of RNA will spontaneously hydrogen bond together to form the self-cleaving hammerhead structure and since the cleavage products are bound only by noncovalent bonds they can spontaneously separate so that another substrate molecule can attach to the artificial ribozyme and be cleaved. The ribozyme is then *trans*-acting in a truly catalytic manner much like an enzymic reaction. In Fig. 23.21 the molecule to be cleaved (highlighted in blue) is the substrate of the ribozyme.

How does this allow the engineering of gene shears to cut specific mRNA molecules? If you examine Fig. 23.21 again, you

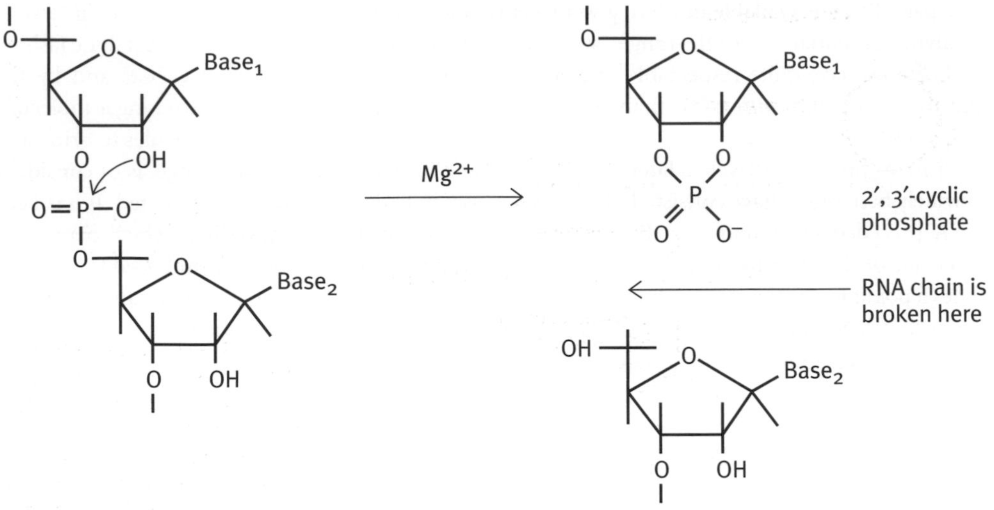

Fig. 23.20 Self-cleavage reaction by a hammerhead ribozyme.

2', 3'-cyclic phosphate

RNA chain is broken here

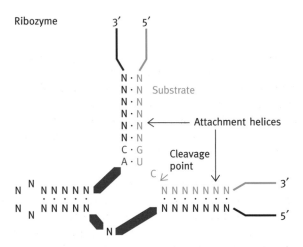

Ribozyme

Fig. 23.21 Simplified diagram of an artificial ribozyme based on the hammerhead structure. The substrate is in blue. After cleavage, the two halves of the substrate can separate, enabling the ribozyme to attack another substrate molecule; the ribozyme acts in *trans* in a truly catalytic manner. The substrate could be part of an mRNA or a viroid/virus RNA genome which would be cleaved and functionally destroyed. All that is required for this is that the GUC sequence of the substrate is present and that the ribozyme's base sequence is designed to be complementary to the bases of the substrate which form the attachment helices. These artificial ribozymes are the basis of 'gene shears' described in the text. N–N' represents any complementary base pairs. The bars in red represent the essential bases given in Fig. 23.19.

see that the two base-paired stems needed to attach the 'substrate' to the ribozyme do not have any specific base sequence requirements and provided that there is the GUC sequence to provide the cleavage site, reaction will occur. (The C is not shown as one of the essential conserved bases because occasionally it is replaced by an A—but never a G or U.) Suppose you wish to make a ribozyme which will cleave a particular mRNA (whose sequence must of course be known), you pick out a sequence in the mRNA around a GUC and design your ribozyme sequence so that it will base pair to form the attachment stems needed to bind the mRNA at the selected sequence which will then be cleaved. RNA synthesizing machines can be used to make partial sequences, which facilitates the making of the synthetic ribozymes. There are obviously problems of delivery of the specific ribozymes to cells but potential routes for this exist including that of arranging for the appropriate RNA sequence to be transcribed from DNA introduced into the target cells. The concept is the subject of a lot of current research.

Ribozymes and hypotheses about the origin of life

Life in the most basic terms consists of informational molecules of RNA or DNA which must be replicated, and of catalysts to support the replication. In modern cells, the catalysts are predominantly protein enzymes. The dissociable side groups of proteins and especially those involved in general acid—base catalysis such as histidine, serine, aspartic, and glutamic acids

(see page 47), not available in RNA, give proteins much greater catalytic potential in both range and rate enhancement. (Ribozymes have quite respectable rate enhancements of 10^6, but this is still 1000 times less than achievable in comparable reactions by enzymes)

Proteins presumably were a later development, and early life is widely believed to have consisted of an 'RNA world' in which RNA preceded DNA and was the replicative 'informational' molecule of the primitive cell. The existence of ribozymes provides a possible solution to the chicken and egg problem; which came first, RNA or catalytic agents? The discovery that RNA itself can bring about reactions suggests that it could have had both roles. There is a known case in which an RNA molecule can take short pieces of RNA and polymerize them into longer ones. The large intron split out of the *Tetrahymena* rRNA described above has further catalytic ability of its own though it is not known that this has any role within the cell. It self-cleaves, resulting in a loss of 19 nucleotides, the product being known as the 'linear minus 19 intervening sequence' (L–19 IVS). This, acting in a true catalytic fashion, can both hydrolyse RNA of the structure CCCCC and by transphospho-esterification 'polymerize' it into longer molecules. The discovery of the potential of RNA molecules to bring about chemical reactions, even if of a limited range, is of considerable relevance to the question of how life originated. It has been speculated that early in evolution, perhaps small peptides binding nonspecifically to RNA may have enhanced the catalytic activities of the latter by stabilizing the structure into a particular conformation. Initially the peptide may have had no catalytic role themselves but the evolution of functional groups in the peptide such as participate in modern enzyme catalysis may have taken over the catalytic role from the RNA because of their greater efficiency and versatility. As the peptide increased in size to the point where it could assume a catalytic conformation on its own the RNA component may have been dispensed with, leaving a protein enzyme.

FURTHER READING

mRNA editing

Hodges, R. and Scott, J. (1992). Apolipoprotein B mRNA editing; a new tier for the control of gene expression. *Trends Biochem. Sci.*, **17**, 77–81.

An interesting account of how mRNA editing leads to the production of two forms of apolipoprotein B from one mRNA transcript.

mRNA stability

Russell, J. E., Morales, J., and Liebhaber, S. A. (1997). The role of mRNA stability in the control of globin gene expression. *Progress in Nucleic acid Research*, **57**, 249–87.

A comprehensive review dealing with mRNA stability in general as well as its relevance to globin synthesis.

Introns and exons

Gilbert, W. (1987). The exon theory of genes, *Gold Spring Harbor Symp. Quant. Biol.*, **LII**, 901–5.

Describes the 'introns early' theory of assemblage of genes from minigenes.

Go, M. and Nosaka, M. (1987). Protein architecture and the origin of introns. *Cold Spring Harbor Symp. Quant. Biol.*, **LII**, 915–24.

Considers the two theories of the origins of introns.

Trotman, C. N. A. (1998). Introns early: slipping lately? *Trends in Genetics*, **14**, 132–3.

A short 'Comment' on the origin of the introns controversy.

Protein modules or domains

Doolittle, R. F. (1995). The multiplicity of domains in proteins. *Ann. Rev. Biochem.*, **64**, 287–314.

A fascinating discussion of domains, domain shuffling, exons, and introns.

Splicing

Breitbart, R. E., Andreadis, A., and Nadal-Ginard, B. (1987). Alternative splicing: a ubiquitous mechanism for the generation of multiple protein isoforms from single genes. *Ann. Rev. Biochem.*, **56**, 467–95.

Fairly detailed but gives a good overview of the biological role.

Orgel, L. E. (1994). The origin of life on earth. *Sci. Amer.*, **271**(4), 52–61.

Growing evidence supports the idea that the emergence of catalytic RNA was a crucial early step.

RNA self-cleavage

Symons, R. H. (1989). Cell cleavage of RNA in the replication of small pathogens of plants and animals. *Trends Biochem. Sci.*, **14**, 445–50.

Covers viroids and the classical 'hammerhead' RNA self-cleaving structure.

Symons, R. H. (1994). Ribozymes. *Current Opinion in Structural Biology*, **4**, 322–30.

A general review.

Fedor, M. J. (1998). Ribozymes. *Current Biology*, **8**, R441–3.

Concise summary.

Chromatin remodelling

Pollard, K. J. and Peterson C. L. (1998). Chromatin remodelling: a marriage between two families. *BioEssays*, **20**, 771–80.

Extensive review of two families of remodelling enzymes.

Gregory, P. D. and Horz, W. (1998). Life with nucleosomes: chromatin remodelling in gene regulation. *Current Opinion in Cell Biology*, **10**, 339–42.

Kornberg, R. D. (1999). Eukaryote transcriptional control. *Trends in Cell Biology, Trends in Biochem. Sci. and Trends in Genetics* (joint issue), **24**, M46–8.

Millenium review including chromatin remodelling. A short overview of the subject including an account of what still needs to be done.

Control of eukaryotic transcription

Ogbourne, S. and Antalis, T. M. (1998). Transcriptional control and the role of silencers in transcriptional regulation in eukaryotes. *Biochem. J.*, **331**, 1–14.

A review article.

Fiering, S., Whitelaw, E., and Martin, D. I. K. (2000). To be or not to be active: the stochastic nature of enhancer action. *BioEssays*, **22**, 381–7.

Brown, C. E., Lechner, T., Howe, L., and Workman, J. L. (2000) The many HATs of transcription coactivators. *Trends Biochem. Sci.*, **25**, 15–8.

A discussion of the multiple histone acetyltransferases.

Transcription in mitochondria

Clayton, D. A. (1984). Transcription of the mammalian mitochondrial genome. *Ann. Rev. Biochem.*, **53**, 573–94.

If differs from nuclear transcription. General review of the topic.

PROBLEMS FOR CHAPTER 23

1   Describe, in broad terms, the main ways in which the formation of mRNA in eukaryotes differs from that in prokaryotes.

2   By means of a diagram, explain the mechanism of splicing. What possible biological significance does the existence of introns have?

3   In what ways does eukaryotic initiation of transcription differ from that in prokaryotes?

4   What part does acetyl-CoA play in the initiation of eukaryotic gene transcription?

5   What is the experimental evidence that chromatin loosening is involved in activation of eukaryotic genes?

6   What is one probable way by which an activated eukaryotic gene is inactivated?

7   The control of gene initiation by the acetylase and deacetylase implies that these enzymes are somehow targeted to specific gene promoters. How is this done?

8   Discuss the two viewpoints on the nature of introns.

Chapter summary

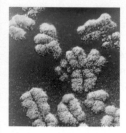

Protein synthesis

In the previous chapter we dealt with production of mRNA. In this chapter we deal with the way in which mRNA directs the assembly of amino acids into the finished protein coded for by the gene from which the mRNA was transcribed. The basic steps in protein synthesis are the same in prokaryotes and eukaryotes, but there are sufficient significant differences that we have to treating them separately. We will deal with the process in *E. coli* first and then look at the ways in which protein synthesis differs in eukaryotes. First of all we outline the concepts that apply to both.

Essential basis of the process of protein synthesis

mRNA is a long molecule with 4 different bases in its component nucleotides. Proteins are synthesized from 20 different species of amino acids and the sequence of bases in the messenger specifies the sequence of amino acids in the protein. A 1-base code (in which a single base represents an amino acid) could code for only 4 amino acids; a 2-base code could code for 16, still not enough. A 3-base code is the minimum requirement and this is the situation. A triplet of bases called a **codon** represents each amino acid on the messenger RNA. With 4 bases 64 different triplets or codons ($4 \times 4 \times 4$) are possible.

The genetic code

The identification of which codon triplets correspond to which amino acids is complete and the list constitutes the **genetic code**. This is almost universal, but not quite. In mitochondria,

derived from symbiotic early prokaryotes, there are one or two minor variations with corresponding variations in their translation apparatus; the same is true of some protozoans, but apart from these the code is universal. Although there may originally have been a simpler code it is now fixed, because a single alteration in the code would mean that many, or most, proteins of a modern cell could not be made. Crick referred to the genetic code as 'a frozen accident'. If you know the base sequence of a mRNA you can work out the amino acid sequence of the protein it codes for. The reverse is not completely true (see below). Thus, to give a single example, the codon UUU means phenylalanine. The process by which protein is synthesized is called *translation* because the language of nucleic acid bases is translated into the different language of protein amino acids.

With 64 different triplet codons available for the 20 amino acids, you might think that evolution would have picked out 20 codons to be used and ignored the rest. However, this would mean that chance mutations would frequently result in unusable triplets corresponding to no amino acid, which would thus render genes inoperative since the synthesis of the protein from its messenger would halt when an unusable triplet in the mRNA was encountered.

An alternative policy has been adopted. Three codons have been reserved as 'stop' signals, which indicate to the protein synthesizing machinery that the protein is complete. These three codons (UAA, for example, is one) have no amino acids assigned to them in the genetic code. If, by chance, a mutation produces a stop signal in a mRNA coding region (known as a nonsense mutation), the protein will not be produced from that gene. However, with only 3 stop triplets, the chance of this is much less than if there were 44 of them. Of the

remaining 61 codons, all code for amino acids, which means that an amino acid is likely to have several different codons, giving what is known as a **degenerate code**. The assignment of codons to amino acids, the genetic code itself, is shown in Table 24.1.

Table 24.1 The genetic code

5′ base	Middle base				3′ base
	U	C	A	G	
U	UUU Phe	UCU Ser	UAU Tyr	UGU Cys	U
	UUC Phe	UCC Ser	UAC Tyr	UGC Cys	C
	UUA Leu	UCA Ser	UAA Stop*	UGA Stop*	A
	UUG Leu	UCG Ser	UAG Stop*	UGG Trp	G
C	CUU Leu	CCU Pro	CAU His	CGU Arg	U
	CUC Leu	CCC Pro	CAC His	CGC Arg	C
	CUA Leu	CCA Pro	CAA Gln	CGA Arg	A
	CUG Leu	CCG Pro	CAG Gln	CGG Arg	G
A	AUU Ile	ACU Thr	AAU Asn	AGU Ser	U
	AUC Ile	ACC Thr	AAC Asn	AGC Ser	C
	AUA Ile	ACA Thr	AAA Lys	AGA Arg	A
	AUG Met†	ACG Thr	AAG Lys	AGG Arg	G
G	GUU Val	GCU Ala	GAU Asp	GGU Gly	U
	GUC Val	GCC Ala	GAC Asp	GGC Gly	C
	GUA Val	GCA Ala	GAA Glu	GGA Gly	A
	GUG Val	GCG Ala	GAG Glu	GGG Gly	G

*Stop codons have no amino acids assigned to them.
†The AUG codon is the initiation codon as well as that for other methionine residues.

Only two amino acids, methionine and tryptophan, have single codons (AUG for methionine). The rest have more than one—leucine has six. Codon assignment is not random. Where several codons exist for one amino acid, they tend to be closely related and varying mainly in the third base. For example, those for isoleucine are AUU, AUC, and AUA. Not only that, but codons for similar amino acids tend to be similar. For example isoleucine and leucine are very similar aliphatic hydrophobic amino acids (page 36); their codons include CUU (leucine) and AUU (isoleucine). This has important genetic consequences, since it means that many mutations involving single base changes have no effect on the protein synthesized (changing AUU to AUC still represents isoleucine) or else substitute a very similar amino acid (changing CUU to AUU substitutes isoleucine for leucine). Isoleucine and leucine are so similar in size and hydrophobic properties that the substitution may not impair the function of the protein. Thus the arrangement of the genetic code provides a 'genetic buffering' action whereby the effects of many single base change mutations on the proteins synthesized are minimized.

How are the codons translated?

It is important at this stage to be clear that protein synthesis occurs only on particles called **ribosomes**—we'll see shortly what these are and how they function. But for now, and remembering firmly that *it all happens on ribosomes*, we can give the concept of how codons on mRNA are translated into the amino acid residues of proteins.

There is no physical or chemical resemblance or relationship between an amino acid and its codon that could lead to their direct association. It was therefore predicted that there must be **adaptor molecules** to associate amino acids with particular codons and, since the hydrogen bonding potential of codons was most likely to be of importance, the adaptor molecules were postulated to be small RNA molecules. Almost at the same time these were discovered as the **transfer RNA** (**tRNA**) molecules.

Transfer RNA

These are small RNA molecules, less than 100 nucleotides in length. Diagrammatically they have a cloverleaf structure (Fig. 24.1(a)). Internal base pairing forms the stem loops. The really important parts (from our present viewpoint) are the three unpaired bases, which form the **anticodon** (explained below), and the 3′-CCA terminal trinucleotide flexible arm to which an amino acid can be attached. Two of these tRNA molecules at a time have to be positioned side by side on the ribosome with their anticodons attached to adjacent codons on the mRNA. In real life the tRNA molecules are folded up into quite a narrow shape diagrammatically represented in Fig. 24.1(b).

The tRNA molecules have an anticodon and an amino acid accepting site. The anticodon is a triplet of bases complementary to a codon (Fig. 24.2). It is located at a hairpin bend so that the three bases are unpaired and therefore available for hydrogen bonding. Thus if a codon on the mRNA is UUU (coding for phenylalanine), the anticodon corresponding to this on a tRNA molecule will be AAA. This particular tRNA molecule must have put on to it *only* phenylalanine and not any other amino acid; we will come to how this is done shortly. There are 61 codons, each of which represents an amino acid. To translate these, it might be expected that there are 61 different tRNA molecules each with its own anticodon complementary to 1 codon and each accepting the 1 amino acid represented by its codon. In fact, there are fewer than 61 tRNA species—obviously there must be at least 1 for each of the 20 amino acids, but some tRNA molecules can recognize several codons. In these cases each of the codons recognized by a single tRNA must, of course,

(a)

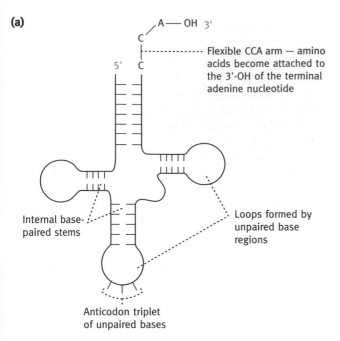

Flexible CCA arm — amino acids become attached to the 3'-OH of the terminal adenine nucleotide

Internal base-paired stems

Loops formed by unpaired base regions

Anticodon triplet of unpaired bases

(b)

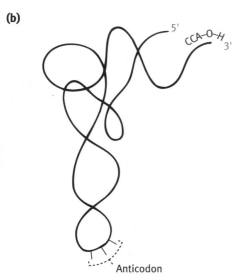

Anticodon

Fig. 24.1 (a) The cloverleaf structure of transfer RNA. (b) The folded structure of tRNA molecules.

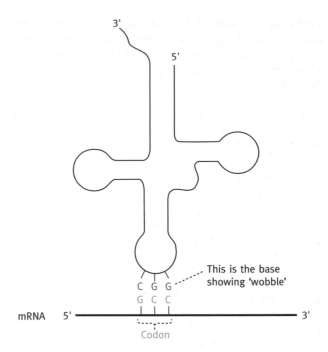

This is the base showing 'wobble'

mRNA   5'   3'

Codon

Fig. 24.2 Base pairing of an anticodon of a tRNA molecule to an mRNA codon. To achieve antiparallel pairing the tRNA molecule is flipped over. This is why a tRNA structure is presented in one way in Fig. 24.1 (with 5' by convention to the left) but in this paired form in the reverse way; the mRNA is written with the 5' end to the left. (In the example shown, the tRNA can base pair with codons GCC and GCU, both of which code for alanine.)

represent the same amino acid. The arrangement means that the cell needs to make fewer tRNA molecules. How is this achieved? The answer is **wobble pairing**.

The wobble mechanism

In view of the importance of complementarity in DNA replication and transcription where Watson–Crick base pairing is ab-

solutely sacrosanct, it may be somewhat disconcerting that, in codon–anticodon base pairing, the rules are bent a little. This applies *only* to the first base of the anticodon. The 'improper' pairing of this particular base is the result of flexibility in the adjacent structure in the tRNA (see Fig. 24.2), such that a U in this position will pair with A or G on the codon, and a G with C or U. This is known as wobble pairing.

The term 'first base of the anticodon' needs explanation. When an anticodon sequence of a tRNA molecule is given *on its own*, we follow the convention is of writing it in the 5' → 3' direction as is the case for any nucleotide sequence. In the anticodon GGC, G is the first (5') base. However, codon–anticodon interaction is antiparallel; (see page 308 if you want an explanation of this term). If therefore we want to show an anticodon interacting with its codon, the latter by convention is written in the 5' → 3'direction so that the anticodon has to be the opposite direction with the 'first' base to the right. This is why, in diagrams, a tRNA molecule *on its own* as in Fig. 24.1 is shown with the 5' end to the left, but when it is base-paired on a codon, it is flipped over as in Fig. 24.2. Thus, the anticodon

CGG will base pair as shown, the first (5′) base of the latter being printed in red:

3′ CGG 5′ anticodon

5′ GCC 3′ codon

The wobble mechanism permits the same anticodon to pair 'improperly', as follows:

3′ CGG 5′ anticodon

5′ GCU 3′ codon

Since GCC and GCU both code for the amino acid alanine, wobble pairing does not alter the amino acid sequence of the protein synthesized, but it enables a single tRNA to translate both. In short, as stated, it permits the cell to manage with fewer species of tRNA molecules.

Evolutionary tinkering has produced other ways of enabling more flexible codon–anticodon interactions without jeopardizing translational accuracy. One is to use the nonstandard base hypoxanthine (Fig. 19.7) in the anticodon, because this will pair with C, U, or A in codons.

How are amino acids attached to tRNA molecules?

It will be clear to you that the system depends on a tRNA molecule having attached to it the particular amino acid specified by the codon which is complementary to the anticodon on that tRNA molecule. Thus a tRNA molecule with the anticodon AAA must be 'charged' only with phenylalanine, since UUU is the complementary codon for that amino acid. If any other amino acid were to be attached to that tRNA, a mistake would be made in the synthesis of protein molecules since phenylalanine would be replaced by the other amino acid. tRNA specific for phenylalanine is depicted as tRNA^{Phe} and so on for each of the 20 amino acids, using the first 3 letters of amino acid names as abbreviations. (Note that tRNA^{Phe} specifies only the tRNA; it does not mean that it has Phe attached. For the latter situation the term Phe-tRNA^{Phe} is used.) Enzymes that attach amino acids to tRNAs are called **aminoacyl-tRNA ligases**. Each cell must have at least 20 different species of these enzymes which each can attach a specific amino acid to an appropriate tRNA. This means that each enzyme recognizes both a specific tRNA and the appropriate amino acid, and joins them together. There is no general pattern for the way in which a synthetase recognizes its tRNA—the anticodon is recognized in some cases, but in others a specific base or several bases elsewhere in the molecule are recognized.

The overall reaction, which involves ATP breakdown to supply energy, is as follows:

Amino acid + tRNA + ATP → aminoacyl-tRNA + PP$_i$ + AMP

The process is sometimes referred to as **amino acid activation**.

The inorganic pyrophosphate (PP$_i$) is hydrolysed to 2P$_i$ and this drives the reaction to the right (see page 12 if necessary). However, there is more to this reaction. The accuracy of the translation of mRNA into protein depends on the accurate selection by these enzymes of the correct amino acid. After an amino acid is loaded on to a tRNA, the aminoacyl-tRNA enters into the protein synthesizing machinery but the latter has no known means of checking that a particular mRNA is carrying the correct amino acid. It 'assumes', as it were, that it is correct, so it is very important that the enzyme attaching the amino acid does not have a high error rate. The active site of an enzyme is usually highly specific for its substrate, but there are limits to the accuracy. It is relatively easy for an enzyme to distinguish between amino acids with markedly different characteristics, but much harder to do this between amino acids such as valine and isoleucine that have very similar structures.

Valine   Isoleucine

The difference in binding energies due to a single –CH$_2$ group is not enough to give a sufficiently high degree of selectivity between isoleucine and valine. The aminoacyl tRNA synthetase specific for leucine would attach valine to tRNA^{Leu} at a rate that would result in an unacceptable rate of errors in mRNA translation unless there was a corrective mechanism. A 'proofreading' mechanism has been evolved, based on the fact that the overall reaction above occurs in two stages as follows:

1. Amino acid + ATP → aminoacyl-AMP + PP$_i$

2. Aminoacyl-AMP + tRNA → aminoacyl-tRNA + AMP

The aminoacyl-AMP does not leave the enzyme. In the case of the isoleucine-specific enzyme, when tRNA^{Ile} attaches, a conformational change exposes an additional catalytic site that hydrolyses valyl-AMP, while isoleucyl-AMP is not so attacked, probably because it is too large to enter the site. By this means, the error rate in isoleucine attachment is reduced. Other aminoacyl synthetases also do their proofreading by hydrolysing incorrect aminoacyl tRNAs. Another example is the valine-specific synthetase; threonine is the same size as valine and may be incorrectly activated to the aminoacyl-AMP intermediate, but the proofreading (hydrolytic) site of the enzyme

preferentially binds the hydrophilic –OH group of threonine rather than the hydrophobic –CH$_3$ of valine. The double selectivity—preferential loading of the correct amino acid onto the tRNA and preferential hydrolysis of aminoacyl tRNAs carrying the incorrect amino acid—reduces the error rate to one in several thousand. Not all aminoacyl tRNA synthetases have proofreading mechanisms; they are needed only where structurally similar amino acids occur. An occasional mistake in protein synthesis is not serious, since huge numbers of protein molecules are made and all are ultimately destroyed (the faulty ones probably more quickly). Such mistakes do not have the same potential long-term consequence as do errors in DNA synthesis, but all the same, a high degree of accuracy is necessary. Protein synthesis is energetically expensive, with the expenditure of four phosphoric anhydride groups being required for the formation of a single peptide bond of a protein.

The tRNA molecule has a terminal 3′ trinucleotide sequence of CCA, the terminal adenosine having a free 3′-OH group on the ribose moiety. It is to the latter that the amino acid is attached by an ester bond (see Fig. 24.3). This CCA trinucleotide forms a flexible arm, which can position the aminoacyl group on the appropriate reactive site on the ribosome (see below). The ester bond formed by the aminoacyl group has essentially the same energy level as that of a peptide bond so there is no thermodynamic problem in transferring the aminoacyl-ester group to the -NH$_2$ of another aminoacyl group to form a peptide bond. In other words, the energy required for the formation of a peptide bond is inserted into the process by the aminoacyl-tRNA synthetase using ATP as the energy source.

Ribosomes

Ribosomes are multi-subunit particles present in cells in large numbers. Their name comes from their content of RNA or ribonucleic acid, which accounts for about 60% of their dry weight. A ribosome consists of two subunits—in *E. coli*, a large one containing two molecules of RNA and a small subunit with one RNA molecule. Eukaryotic ribosomes, which are somewhat larger, are described later. You have already met mRNA and tRNA; **ribosomal RNAs (rRNAs)** are different again. Because of internal base pairing, they assume highly folded compact structures. Figure 24.4 illustrates the complex shape of an RNA molecule of an *E. coli* ribosome. The rRNAs are associated with many proteins forming solid particles—34 proteins in the case of the large subunit and 21 for the small.

With very large structures such as a ribosome, their sizes are measured in terms of the rate at which they sediment in an ul-

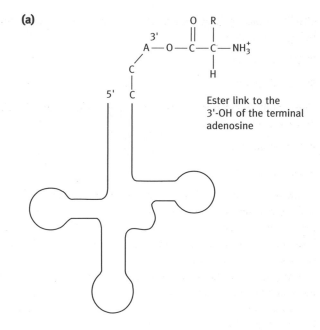

(a)

Ester link to the 3′-OH of the terminal adenosine

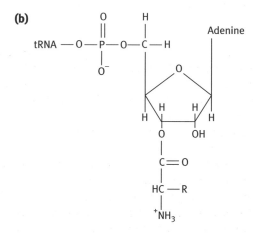

(b)

Fig. 24.3 (a) Diagram of tRNA molecule showing the CCA base sequence at the 3′ end, where the amino acid is attached by ester link. (b) Structure of the terminal nucleotide with the attached amino acid.

tracentrifuge—expressed as Svedberg units or S values. (An ultracentrifuge spins so fast that large molecules in solution move towards the bottom of the tube.) An *E. coli* ribosome is 70 S, the subunits being 50 S and 30 S (the values are not simply additive, since S depends both on size and shape).

The overall *principle* of protein synthesis can be given very simply. The ribosome becomes attached to the mRNA near the 5′ end and then moves down the mRNA towards the 3′ end, assembling the aminoacyl groups of charged tRNA molecules

Central domain

3' major
domain

5' end

3' end

3' minor domain

5' domain

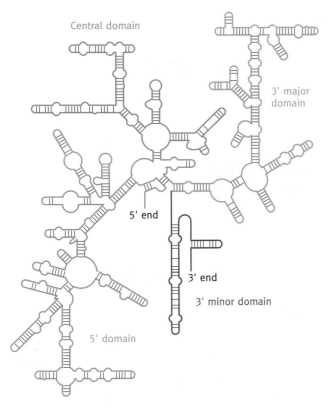

Fig. 24.4 16 S RNA.

into a polypeptide chain according to the sequence of codons in the mRNA. The N-terminal amino acid of the polypeptide is the first to emerge and the C-terminal one is added finally. At the end of the mRNA, the ribosome meets a stop codon, the protein is released, the ribosome detaches from the mRNA, and dissociates into its subunits. Note that there are no 'special' ribosomes. In a given cell, any ribosome can use any mRNA just as a tape player can play any recorded tape; a slight qualification exists in that mitochondrial and chloroplast ribosomes are different from those in the cytosol.

That is the principle—to understand the process we need to give more detail. Synthesis of a protein molecule can be divided into three phases—**initiation, elongation, and termination.**

Initiation of translation

It is of vital importance that a ribosome begins its translation of a mRNA at exactly the correct point—in other words, that it initiates translation correctly. mRNAs have 5′ and 3′ untranslated

regions and the coding region lies in between. The ribosome therefore must recognize the *first* codon of the coding sequence and start translating there. Absolutely precise initiation is essential, for the correct translation of a mRNA depends on the ribosome being in the correct **reading frame**. This may need explanation. Suppose the *coding region* of a mRNA starts with the sequence:

5′ AUGUUUAAACCCCUG ----------- 3′

The first five amino acids are specified by the codons AUG, UUU, AAA, etc. There is nothing to indicate what constitutes a codon, other than that the first three bases of the coding part of the mRNA encountered constitute codon 1, the next three are codon 2, and so on. There are no commas or full stops between them. The message therefore depends on starting to read *exactly* at AUG. Suppose an error of one base is made and we start instead at the second base. The codons then read would be:

(A)   UGU, UUA, AAC, CCC, UG -----

In other words, the codons translated would be totally different and the amino acids inserted would correspondingly be totally different. This error would be a **reading frameshift**; mutations deleting or adding one or two bases from a mRNA cause reading frameshifts and result in the amino acid sequence of the polypeptide synthesized after the frameshift being incorrect. The resultant polypeptide is useless garbage instead of the one specified by the gene.

Initiation of translation in *E. coli*

As mentioned, mRNA is not translatable in its entirety; there is a stretch of RNA 5′ to the translational start site that is the ribosome positioning section. Initiation is *not* at the extreme 5′ end of the mRNA and therefore the first codon must be identified in some other way.

The start site is the codon **AUG** (less frequently **GUG**), but oddly enough, an AUG codon can occur *anywhere* in the mRNA since it codes for the amino acid methionine. For a long time, biochemists wondered how the ribosome initiated only at the correct AUG and not at one of the others further along the mRNA. Why wasn't there a special codon always used for initiation? It was discovered that there are two *different* tRNAs both specific for methionine—they have the same anticodon, but *one tRNA is used exclusively for initiation* and the other exclusively for adding methionine in the elongation process. The tRNA that *initiates* translation can base-pair with either of the codons AUG or GUG, owing to wobble (see earlier). Normally, wobble involves the 5′ base of the anticodon, but in this one case, in the

tRNA used for initiation, it is the 3′ base of the anticodon that is the wobble base (on the left hand as written when it is shown as bonded to its codon). What then determines the different roles of the two methionine-specific tRNA species? The answer is that the *initiating* aminoacyl tRNA species has structural features that are recognized by an **initiating protein** factor (**IF2**) in the cytosol and delivered by this to the initiation complex. The aminoacyl -tRNAs involved in elongation following initiation are recognized by a different cytoplasmic factor (described below) which delivers them to the ribosome. This latter factor does not bind to the initiating tRNA. There is another quirky difference in *E. coli*—the methionine that becomes attached to the initiating tRNA is formylated on its $-NH_2$ group by a trans-formylase using N^{10}-formyltetrahydrofolate (page 293) as formyl donor. Prokaryotic proteins, for reasons that are not clear, are synthesized with *N*-formylmethionine as the first amino acid residue. The formyl group, and frequently the methionine also, are removed before completion of the synthesis. The initiating tRNA is usually called tRNA$_f$^{Met} (f for formyl) and the charged version $_f$Met-tRNA$_f$^{Met}, often shortened to $_f$Met-tRNA$_f$, while the tRNA for methionine is called tRNA$_m$^{Met} (or tRNA$_m$). This then clears up the mystery of how the correct methionine-specific tRNA is used for initiation and elongation—specific protein factors involved in initiation and elongation recognize only $_f$Met-tRNA$_f$^{Met} and Met-tRNA$_m$^{Met} respectively.

In the cytoplasm there is a pool of 30 S and 50 S ribosomal subunits in equilibrium with 70 S ribosomes. There are three cytosolic initiating factors that are involved in initiation—they bind, participate in the process, and then are released for further use. Initiating factors **IF1** and **IF3** in the cytoplasm bind to the 30 S subunit. IF3 prevents premature reassociation with the 50 S subunit at this stage (Fig. 24.5(a)). The function of IF1 is not known. These factors are found loosely bound to the 30 S subunit and are easily removed by washing. **IF2**, as explained above, is necessary to bind $_f$Met-tRNA$_f$ in the cytoplasm and deliver the latter to the initiation complex.

In the presence of mRNA, $_f$met-tRNA$_f$, and GTP, a complex of these with a 30 S subunit (with its attached IFs 1 and 3) is formed. As shown in Fig. 24.5(b), there is a sequence of bases on the mRNA known as the **Shine-Dalgarno sequence**, which is complementary to a section of the 16 S rRNA. Binding of the two by base pairing correctly positions the mRNA on the small ribosomal subunit. The $_f$met-tRNA$_f$ is delivered to the partial P (for peptidyl) site on the subunit with its anticodon paired with the AUG codon by IF2 (Fig. 24.6). This complex now associates with a 50 S subunit. The event is accompanied by the hydrolysis of GTP and the release of GDP, P_i, IF1, IF2, and IF3 (Fig. 24.6). We now have a complete 70 S ribosome positioned on the mRNA with the $_f$Met-tRNA$_f$ in the P site with its anticodon base

(a)

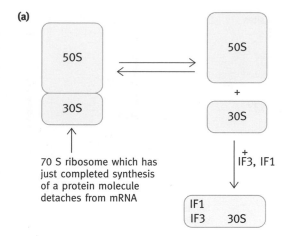

(b)

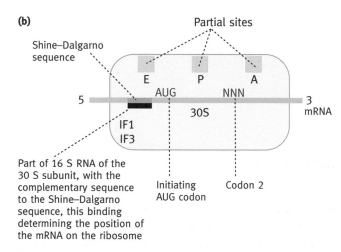

Fig. 24.5 (a) Dissociation of 70 S ribosome into 50 S and 30 S subunits. The initiating factor IF3 prevents 50 S subunit association. In initiation of translation, the IF3 must be released before the 50 S subunit can join. (b) The role of the Shine–Dalgarno sequence in positioning the 30 S ribosome of *E. coli* on the mRNA at initiation. E, exit site; P, peptidyl site; A, aminoacyl site. The E, P, and A sites are only partial sites; when the 50 S subunit joins (see text) the sites are completed.

paired with the initiating AUG codon. The A site is vacant, awaiting delivery of the second amino acid on its tRNA. Initiation is complete.

Many bacterial mRNA molecules are polycistronic; the *lac* mRNA is one such example (page 350). In this, there are three regions in the one mRNA molecule coding for three different proteins. In this situation, each cistron has a Shine–Dalgarno sequence adjacent to it so that each can be initiated independently (Fig. 24.7).

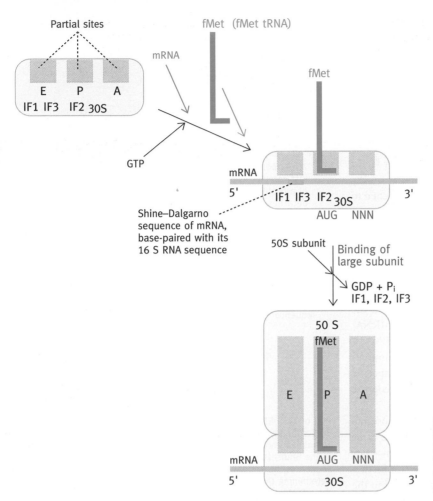

Fig. 24.6 Initiation of translation in *E. coli*. The initiating tRNA, tRNA$_f^{Met}$, is represented by the blue line, the anticodon being the horizontal short line. The binding of fMet-tRNA to the 30S subunit requires IF2. NNN represents any codon (N for any nucleotide).

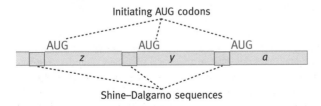

Fig. 24.7 The structure of a polycistronic prokaryote mRNA. z, y, and a refer to coding regions of the *lac* mRNA (see Fig. 22.9).

Once initiation is achieved, elongation is the next step

Cytoplasmic elongation factors

There are two soluble protein elongation factors in the cytoplasm. A preliminary understanding of their characteristics will be helpful. Both bind a molecule of GTP, and only in this state do they bind to ribosomes. Both are latent GTPases. When attached to the ribosome they hydrolyse the bound GTP molecule to GDP and P$_i$ and release of the latter causes a conformational change in the proteins. The GTPase activity is low, so this occurs only after a short pause. In their GDP-bound state, the two elongation factors detach from the ribosome. In the cytoplasm their GDP is exchanged for GTP so that the factors are ready to participate in a new round of elongation.

The two factors are **EF-Tu** (elongation factor, temperature unstable) and **EF-G**, also known as **translocase**. EF-Tu has the task of delivering the incoming aminoacyl-tRNA to the ribosome, EF-G of moving the ribosome along the mRNA in the 5′ → 3′ direction to the next codon, once an aminoacyl group has been added to the growing peptide.

It is useful to keep in mind this central concept of a pair of factors *alternately* hopping onto the ribosome in their GTP form, performing their tasks and detaching in their GDP form

to be recycled for subsequent rounds of elongation. They cannot both be present on the ribosome at the same time. With that introduction we can move to the details of peptide bond synthesis.

Mechanism of elongation

We suggest that you follow the steps in Fig. 24.8 carefully as you read the next part. Starting with the initiation complex (state a), we have an fMet-tRNA$_f$ in the P site and the A site is vacant. Note that *only* in the initiation process does the P site accept a tRNA charged with an amino acid—in this case N-formylmethionine; *all* subsequent aminoacyl-tRNAs enter the A site. Aminoacyl-tRNAs (other than the initiating species) are complexed in the cytosol with the elongation factor EF-Tu, carrying a molecule of GTP bound to it. This factor attaches to the ribosome only if both GTP and an aminoacyl-tRNA molecule are attached. The EF-Tu–GTP–aminoacyl-tRNA complex binds to the ribosome such that the aminoacyl-tRNA occupies the A site with its anticodon positioned at the mRNA codon (Fig. 24.8, state b). EF-Tu exists in high concentration in the cytosol, sufficient to bind all of the aminoacyl-tRNA. As stated, on the ribosome EF-Tu is a slow GT-Pase and *after a short pause* its bound molecule of GTP is hydrolysed to GDP and P$_i$, the latter being released. This is the signal for the EF-Tu to detach from the aminoacyl-tRNA, presumably because of conformational changes occurring, and the EF-Tu-GDP leaves the ribosome (to be regenerated by a GDP-GTP exchange process in the cytosol), giving state (c).

The aminoacyl groups on the two tRNA molecules on the P and A sites are in the vicinity of the catalytic site of **peptidyl transferase**, which transfers the $_f$Met group from the tRNA in the P site to the free amino group of the aminoacyl-tRNA in the A site, producing a dipeptide attached to the tRNA (state d). Despite its name, 'peptidyl transferase' is an RNA molecule with catalytic properties, not a protein. The peptidyl transferase reaction is shown below, using the first peptide bond synthesis as an example.

tRNA—O—CO—fMet + NH$_2$—CH—CO—O—tRNA $\longrightarrow$
(P site) |
                             R'   (A site)

tRNA—OH + fMet—CO—NH—CH—CO—O—tRNA .
(P site) |
                               R'   (A site)

It is rather important to be clear on what is happening in this reaction. The aminoacyl group on the tRNA in the P site is transferred to the free amino group of the aminoacyl-tRNA in the A site. We emphasis this because as the synthesis of the polypeptide proceeds, the acyl group in the P site is actually the partially completed polypeptide chain and this is transferred to the incoming amino acid. This is why the process is called the peptidyl

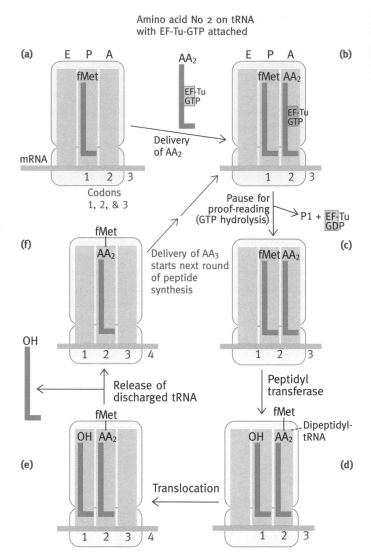

Fig. 24.8 The elongation process in protein synthesis following translational initiation. tRNAs are shown as blue lines, the anticodon being represented by the short horizontal section; AA2 and AA3 represent amino acids. The positioning of EF-Tu–GTP on the tRNA is purely diagramatic. The reason for the naming of the enzyme peptidyl transferase is not evident from the diagram, but if you do the next round of synthesis you will see that in all subsequent rounds of synthesis it is a peptide that is transferred to the incoming aminoacyl-tRNA as is also explained in the text.

transferase reaction. One would instinctively imagine that the new amino acid residue would be added to the partially completed polypeptide, but the reverse happens. If the protein being synthesized is 200 amino acids long, the final round involves transfer of the 199-long peptide to the final aminoacyl-tRNA giving a protein–tRNA complex. The dog is added to the tail as it were, not the other way round.

After this first peptide bond synthesis, we have the A site filled with a peptidyl-tRNA and the P site containing an uncharged tRNA (d). The ribosome is now repositioned one codon further along the mRNA, a process called **translocation**. As a result of this, the discharged tRNA is moved to a third **exit site (E)** and the peptidyl-tRNA in the A site is moved to the P site giving state (e) in which the A site is vacant. The movement of the ribosome relative to the mRNA requires EF-G (translocase) and GTP hydrolysis. Peptide synthesis and translocation alternate. From state (e), the discharged tRNA is released to be used again giving state (f). This is equivalent to state (a) with the P site occupied by the dipeptidyl-tRNA and the A site, now positioned on codon 3, vacant. It is therefore the start of the next round of peptide synthesis with the delivery of the third aminoacyl-tRNA to the ribosome, as indicated by the red arrows. Protein synthesis involves round after round of this until the ribosome reaches the stop codon at the end of the mRNA coding sequence.

In Fig. 24.8 the mechanism by which peptidyl transfer and translocation occurs (stages c–e) is not specified. More recent work has indicated that during the process, each tRNA is bound to two sites on the ribosome (see Fig. 24.9). The discharged tRNA straddles the P and E (exit) sites—the anticodon end of the molecule is still in the P site but the other end is in the E site. Similarly, the tRNA in the A site (now carrying the peptide) straddles the A and the P sites.

Two alternative models have been proposed to account for this attachment of tRNAs to two sites at once.

- Model I (Fig. 24.9) envisages that one end of the aminoacyl-tRNA swings over as shown, peptide transfer occurs and the discharged tRNA swings one end to the E site. Translocation straightens things up, and we are back to the situation shown in Fig. 24.8 (state e).

- In the alternative model II, the large subunit moves but the small one remains stationary. This creates hybrid P/E and A/P binding sites as shown in Fig 24.9. After peptidyl transfer, the hybrid sites are occupied as shown. Translocation of the small subunit then produces the same situation as in model I, ready for the next aminoacyl-tRNA addition.

Both models have the important feature that the *nascent peptide remains in a fixed position relative to the large subunit*, as shown. This would eliminate the problem of how a tRNA physically moves with a relatively huge polypeptide attached to it. Model II would additionally explain why ribosomes always have two subunits. The models also remove the problem of how the tRNAs can move on the ribosome during translocation without the danger of them diffusing away, since in both models at least one end of the tRNAs is always attached to a site.

How is fidelity of translation achieved?

The fidelity of translation depends on the correct aminoacyl-tRNA being bound in that site in each round of elongation. The basis of this must be due to codon–anticodon interaction

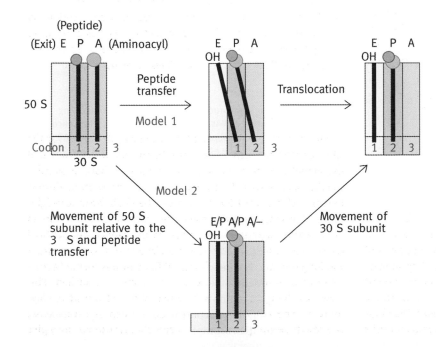

Fig. 24.9 Alternative models for peptide synthesis on the ribosome and translocation. Both reflect the observation that the tRNAs straddle the P/E and A/P sites on the ribosome. In model I, the AA–tRNA swings to make the straddle; in model II, hybrid sites are formed on the ribosome by movement of the 50 S subunit relative to the 30 S. Peptide synthesis is presumed to occur such that the peptide always remains in the same position relative to the large subunit as shown. In this illustration, synthesis of the first peptide bond has been chosen as the example so that the red circle represents fMet attached to tRNA on the P site which is aligned with mRNA codon number 1; at the end of the round it is aligned with the second codon. The coloured rectangles represent binding sites: E, exit site; P, peptidyl site; A, acceptor site. E/P and A/P are hybrid sites formed by large subunit movement relative to the small subunit.

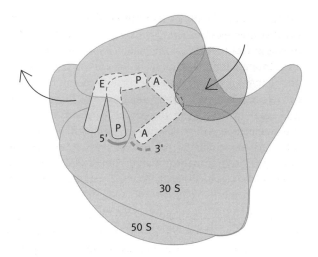

Fig. 24.10 Diagram of a ribosome with plausible locations for ribosome-bound tRNAs in the A/A, P/P, and E states (see Fig. 24.9 for the terminology). The shaded area at the right shows the approximate site of interaction of EF-Tu. The polarity of a fragment of mRNA containing the A- and P-site codons is shown. The arrows indicate the likely path of a tRNA as it transits the ribosome.

selecting the correct aminoacyl-tRNA, but exactly how this selection is effected is not completely known. The binding energy difference between a correct and an incorrect pairing alone is insufficient to account for the translational fidelity achieved. The EF-Tu does not 'know' which aminoacyl tRNA has to react next, and it would seem that it must deliver them to the A site randomly. It is believed that an incorrect aminoacyl-tRNA diffuses away before it reacts since it will not be held in the A site as strongly by hydrogen bonding to the codon as a correct one would be. Peptide synthesis cannot occur until EF-Tu–GDP is released and this cannot happen until the GTP is hydrolysed. The delay in GTP hydrolysis, which occurs, is postulated to provide sufficient time for this 'proofreading' process to be effective.

Termination of protein synthesis

At the end of the mRNA there is at least one of the three stop codons: UAG, UAA, or UGA. When the ribosome reaches one of these, a specific **cytoplasmic protein release factor** attaches and causes release of the finished protein from the tRNA by altering the peptidyl transferase such that it hydrolyses the ester bond between the —COOH of the protein and the —OH of the 3' terminal nucleotide; it transfers the peptide to water. The ribosome detaches from the mRNA and dissociates into subunits

ready for the next round of initiation. Two release factors are known, recognizing different stop codons.

The whole process of polypeptide synthesis *following initiation of translation* may be summarized as follows, starting with the ₁met-tRNA occupying the P site:

- An aminoacyl-tRNA corresponding to the second codon and complexed to EF-Tu–GTP is delivered to the A site from the cytosol.
- If the anticodon corresponds to the codon it will base pair; if it is not the correct aminoacyl-tRNA this will not happen and it dissociates from the ribosome so that an alternative one can enter the site. This is the proofreading.
- GTP is hydrolysed after a pause (during which time proofreading has occurred), P_i is released and the EF-Tu–GDP looses its attachment to the aminoacyl-tRNA which remains in the A site. The EF-Tu–GDP returns to the cytosol where its GDP is exchanged for GTP to be used again.
- The peptidyl group on the peptidyl-tRNA is now transferred to the amino group of the aminoacyl tRNA in the A site, forming a dipeptidyl-tRNA.
- Translocation occurs, the ribosome moving along by one codon so that the peptidyl-tRNA is in the P site while the discharged tRNA is in the E site from which it is released. The A site is now vacant.
- Aminoacyl-tRNA corresponding to the third codon is delivered to the A site by EF-Tu–GTP and the whole process is repeated until the polypeptide is fully synthesized.
- At this point the ribosome encounter a stop codon on the mRNA; a cytosolic **release factor** attaches and causes hydrolysis of the protein from its tRNA. The ribosome–mRNA–tRNA complex dissociates, to be used again.

Physical structure of the ribosome

In the preceding diagrams we have used simple shapes for the ribosome for the purpose of clarity, but a good deal is known of the actual structure. Prokaryotic ribosomes have molecular weights of 2.5×10^3 daltons, eukaryotic ribosomes about double that; in each case the RNA being more than half the total weight. It has been speculated that they originated when life was in an 'RNA world' (see page 375). As already mentioned, the peptidyl transferase activity is catalysed by RNA and not by a protein. It was long thought that the ribosomal RNA simply formed a scaffold to which the many ribosomal proteins are attached. This view has changed, and it is now believed that the RNA plays more active roles, though apart from the peptidyl transferase activity these are speculative. It seems that ribosomes may give

us a glimpse of the 'RNA world' that is believed to have existed early in the establishment of life.

The ribosome has a distinctive shape, as shown in Fig. 24.10. The tRNAs bind to the face of the large subunit, which faces the small subunit, but at the contact faces there is a hollow big enough to accommodate the tRNAs and form a tunnel. The aminoacyl-tRNAs enter the tunnel on the right of the diagram and the discharged tRNAs emerge on the left. The peptidyl transferase catalytic region is located on the face of the large subunit within the cavity, and the mRNA binds to the face of the small subunit facing the cavity. Ribosome are astonishing molecular machines. Despite the fact that they have been studied for several decades now, it seems that more remains to be learned about them.

What is a polysome?

It takes about 20 seconds for a ribosome to synthesize an average protein in *E. coli*, which adds about 15 amino acid residues per second. If only a single ribosome at a time moved along the mRNA molecule, the latter could direct the synthesis of the protein at the rate of 1 molecule per 20 seconds. However, as soon as an initiated ribosome has got under way and has moved along about 30 codons, another initiation can occur. One ribosome after another hops onto the mRNA. They follow one another down the mRNA, each independently synthesizing a protein molecule—a typical case would be about 5 ribosomes per mRNA molecule but this varies with the length of the mRNA. The term **polysome** is thus a shortened version of **polyribosome**.

How does protein synthesis differ in eukaryotes?

The process is basically the same as in eukaryotes as in prokaryotes, but there are important differences. Eukaryotic ribosomes are larger (80 S, with 60 S and 40 S subunits) and have extra rRNA molecules and proteins. Methionine is always the first amino acid in protein synthesis, but the methionyl-tRNA used for initiation is not formylated. The purpose of formylation in prokaryotes is not understood. However, as in prokaryotes, there is a special methionyl-tRNA (Met-tRNA$_i$) for initiation, distinct from that used in elongation. This does not mean that all eukaryote proteins start with methionine, because frequently this amino acid is removed from the polypeptide before it is released. However, assembly of the initiation complex is quite different from that in *E. coli*. We remind you that eukaryotic mRNAs are 'capped' with a methylated guanine nucleotide at the 5′ end. A group of protein factors and Met-tRNA$_i$ attach to the cap, which is joined by a 40 S ribosomal subunit (Fig. 24.11). We thus have a 40 S preinitiation complex at the very 5′ end of the mRNA. As in *E. coli*, there are three initiation factors involved, referred to as eIF I, 2 and 3, the e standing for eukaryotic. GTP is again hydrolysed in the process. There is no Shine–Dalgarno sequence in the eukaryotic ribosomal RNA, so how is the ribosome positioned at the correct starting point on the mRNA? The ribosome in the initiation complex is located at the cap site, which is some distance from the start AUG codon. By an ATP-driven mechanism, the 40 S subunit complex moves along the mRNA until it encounters the initiating AUG triplet and then the 60 S subunit joins up to complete the initiation. GTP is hydrolysed in the process. This has one interesting repercussion. There can, with this scanning mechanism for selecting the AUG initiation codon, be only one initiating site per mRNA molecule. Eukaryotic mRNAs are monocistronic (code for a single polypeptide), unlike prokaryotic mRNAs which are often polycistronic with Shine–Dalgarno sequences provided for the initiation of translation of each cistron.

After completion of initiation, polypeptide synthesis proceeds as already described for *E. coli* involving two cytosolic elongation factors. The counterpart of EF-Tu in eukaryotes is called EF1α and that of EFG (translocase) is EF2.

Protein synthesis in mitochondria

Mitochondria contain DNA of their own and have their own protein synthesizing machinery. The ribosomes of mitochondria are prokaryote-like and use $_f$Met-tRNA$_f$ for initiation. They have other features of interest, such as a slightly different genetic code, and codon–anticodon interactions are simplified so that mammalian mitochondria can manage with only 22 tRNA species. The possibility of such simplification is related to the fact that mitochondria synthesize only a handful of different proteins. The mitochondrion is not autonomous, most of their proteins being synthesized in the cytoplasm and then transported in (page 412). The same is true of chloroplasts which are likewise believed to have originated from incorporated (in this case, photosynthetic) prokaryotes.

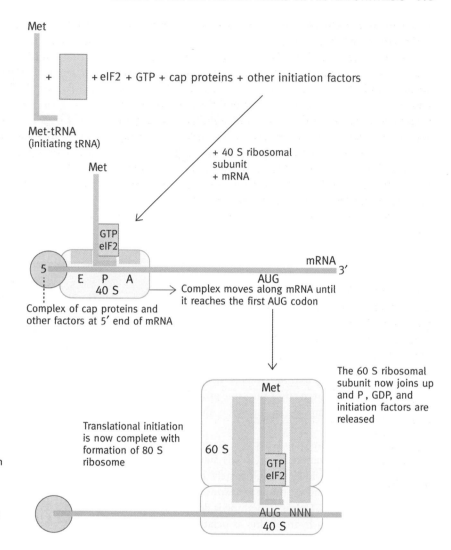

Fig. 24.11 Simplified diagram of intiation in eukaryotes. Note that several eukaryotic initiation factors besides eIF2 are involved. tRNA, is the initiating tRNA, eIF2 is the eukaryotic initiation factor corresponding to IF2 in prokaryotes. NNN represents the second codon, N representing any nucleotide.

Effects of antibiotics and toxins on protein synthesis

Antibiotics are the chemical missiles microorganisms throw at each other in the competition for survival. They attack highly important points in cellular processes, and translation offers many such targets.

In prokaryotes:

- **streptomycin** affects initiation

- **kirromycin** and **fusidic acid** prevent EF-Tu release

- **erythromycin** binds to the 50S subunit and inhibits translocation

- **chloramphenicol** inhibits peptidyl transferase

- **tetracyclines** inhibit the binding of aminoacyl-tRNAs to the ribosome.

In addition to these examples of antibiotic action, **diptheria toxin** inhibits EF2, the eukaryotic translocase (equivalent to EFG in bacteria). **Ricin**, the toxin of the castor bean, is an *N*-glycosidase, which removes a single adenine base from one of the eukaryotic rRNAs and inactivates the large subunit. One molecule of ricin can destroy a cell containing tens of thousands of ribosomes.

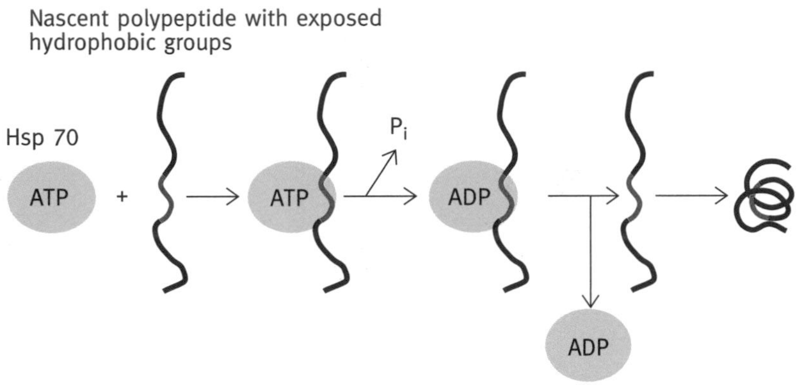

Nascent polypeptide with exposed
hydrophobic groups

Fig. 24.12 Simplified diagram of *E. coli* chaperone Hsp 70 action in assisting folding of a polypeptide. The participation of other cochaperone proteins in the process has been omitted for simplicity. The essence of the process is that the chaperone is a slow ATPase and alternates beween an ATP-bound form and an ADP-bound form, the former having a high affinity for the polypeptide and the latter a low affinity. The red line represents a hydrophobic patch on the polypeptide. Hsp 70 can attach to nascent proteins emerging from the ribosome.

How does the polypeptide chain synthesized on the ribosome fold up?

Chaperones (heat shock proteins)

In Chapter 3 on protein structure, we dealt in a general way with the problem of protein folding, because it was relevant there (page 46). We now want to cover the same ground giving more information on mechanisms, so inevitably there will be some repetition.

The newly synthesized polypeptide chain emerges from the ribosome in a denatured (unfolded) state; as it does so, hydrophobic groups which will ultimately be buried in the interior of the folded native protein are exposed to the aqueous medium of the cytosol. Unless something is done to prevent it, these groups will form hydrophobic associations to whatever other hydrophobic groups are available, either on the same nascent polypeptide or on adjacent polypeptides. Such random, 'improper' associations could prevent the polypeptide from folding up correctly.

To guard against this there is a family of proteins collectively referred to as **chaperones or molecular chaperones**. They are highly conserved in evolution and are present normally in all cells. They were discovered when it was observed that in cells subjected to temperatures higher than normal, certain proteins increased in amounts and were called **heat shock proteins (Hsps)**. Chaperones are Hsps. You might be puzzled as to what heat shock has to do with protein synthesis. Heat denatures (unfolds) native proteins, and newly synthesized polypeptides also are unfolded as they emerge from the ribosome. The heat-denatured proteins have improperly exposed hydrophobic groups, and if they are to be salvaged must be refolded. So the problem is similar to that of newly synthesized proteins.

Mechanism of action of molecular chaperones

The Hsps are classified on a size basis into three groups. The best two known are the **Hsp 70** and the **Hsp 60** groups, and these are what we will describe.

Hsp 70 attaches to the hydrophobic groups of nascent polypeptides. It has a molecule of attached ATP, in which form it has a high affinity for the unfolded chain. Hsp 70 is a slow ATPase. After attaching to a polypeptide, the ATP is hydrolysed after a short period; in the ADP form it has a low affinity for the polypeptide and so releases it, conformational change of the protein being responsible for this. The ATPase is thus a timing mechanism to determine how long the chaperone remains attached to the unfolded polypeptide. That is all it does; it attaches and then detaches (Fig. 24.12). By attaching, the Hsp prevents incorrect hydrophobic associations occurring during which polypeptide synthesis is likely to be completed. When it detaches the correct folding is given the opportunity to occur. If this fails to happen hydrophobic groups will be still exposed and the chaperone reattaches and gives another chance. You will have noticed that this does not explain *how* the folding occurs, and this remains one of the big problems in biology. The chaperone does not instruct the polypeptide on how to fold; this, as emphasized already, is solely determined by the amino acid sequence.

A polypeptide in the cell folds up after synthesis in a couple of minutes or so; the seemingly obvious mechanism is that the polypeptide randomly tries every possible configuration until the minimum free energy is achieved. However, there so many possibilities that it would take millions of years for a protein to fold in this way. It is believed that certain structures rapidly assume their secondary structure, forming what are termed **'molten globules'**, and these somehow facilitate complete folding.

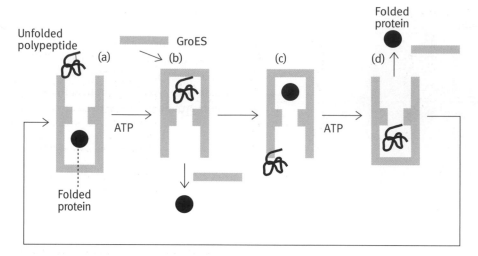

Fig. 24.13 Simplified schematic diagram to illustrate the principle of GroEL action. The 'lid' structure is known as GroES. The chaperonin has two folding chambers which work in alternation. In (a) an unfolded polypeptide has attached via its hydrophobic groups to GroEL. The lower cavity has a folded polypeptide represented as a solid circle waiting for release (this corresponds to the situation in the upper chamber in (c)). In (b) the unfolded polypeptide has entered the hydrophilic cavity and a 'lid' seals it in. Meanwhile the lower cavity has released its folded protein. In (c) the situation is just the same as (a) but upside down. Note that the diagram does not show the ring structure of the chaperonin nor the conformational changes that occur. The mechanism involves the hydrolysis of seven molecules of ATP at the steps indicated. The next figure shows a more realistic structural representtion.

The Hsp 70 type of chaperone assistance is all that is required for most proteins; even large ones probably have several essentially structurally independent domains which fold sequentially as the polypeptide emerges. (See page 45 for a description of what a domain is if necessary.) A number of proteins require a different form of assistance for completion by a molecular chaperone of the Hsp 60 class (sometimes known as **chaperonins** to distinguish them from the Hsp 70 class, but nomenclature is not completely consistent in this area.) The best known of these is a protein complex known as **GroEL**, together with a 'lid' structure known as **GroES**. We mentioned earlier that in Anfinsen's experiment, in which ribonuclease was refolded *in vitro*, the denatured protein was at very low concentration which favoured the refolding because there is less chance of aggregation with other molecules. GroEL provides what is sometimes known as an 'Anfinsen cage' because it literally encloses a protein to be folded in a hydrophilic box secluded from all the other proteins in the cell. It gives the protein a private box with an environment ideal for folding and then ejects it. If it has failed to fold it can try again because its hydrophobic groups are still exposed and will reattach to the GroEL. Hsp 60 is needed typically for proteins imported, in an extended form, into the mitochondrial matrix (page 412) where the concentration of proteins is extremely high, the water content being similar to that found in a protein crystal. As illus-

trated in Fig. 24.13 the unfolded, or partially unfolded, protein attaches to the Hsp 60 entry point by its hydrophobic groups and is allowed into the folding chamber and sealed in with the GroES lid. Since the box is lined by hydrophilic groups it provides the greatest encouragement for correct folding. Once again it is important to note that the GroEL provides no folding guidance to the protein but just optimal conditions for the protein to fold itself; it does not explain how the folding is achieved. To complete the remarkable business the GroEL is double sided, the two folding chambers accepting and ejecting their folding candidates alternately. The mechanism is driven by ATP hydrolysis. GroEL is a multi-subunit structure with two back-to-back rings of seven subunits. In Fig. 24.13 all such details of subunit structure are omitted for simplicity. A space-filling model of the chaperonin, together with a vertical cross-section, is given in Fig. 24.14.

Enzymes involved in protein folding

The chaperone assistance in folding is not entirely the end of the folding story; there are two additional possible problems which necessitate covalent changes to the newly synthesized polypeptide, requiring enzymic intervention. It is again important to note that the enzymes do not give instructions on folding; they

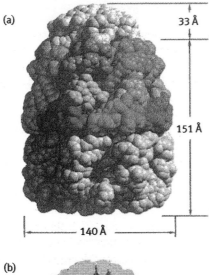

(a)

33 Å

151 Å

140 Å

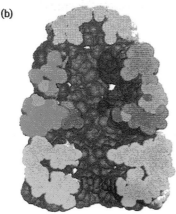

(b)

Fig. 24.14 (a) Space-filling model of GroEL with a GroES cap. This state corrseponds with state (b) in the previous figure. (b) a cross-section of the same.

act randomly to free the polypeptide from restraints which might prevent folding, but the folding process is entirely up to the protein itself.

The first enzymic intervention is by is **protein disulfide isomerase (PDI)**. This enzyme 'shuffles' S—S bonds in polypeptide chains. If an incorrect S—S bond were to be formed, being covalent it would not break spontaneously and would fix the polypeptide in an incorrect configuration. PDI, by breaking and reforming S—S bonds between different cysteine residues (page 42), permits the folding to correct itself. It simply transfers an S— bond from one disulfide bridge to another, using itself as the intermediary attachment site. High concentrations of PDI are found inside the endoplasmic reticulum involved in folding proteins destined for secretion, many of which have disulfide bridges.

Another enzyme is **peptidyl proline isomerase (PPI)**. Wherever a proline group occurs in peptide linkage the configuration can be *cis* or *trans*. The PPI plays the role of 'shuffling' proline residues between the configurations in order to permit the whole protein to assume the correct configuration.

Prion diseases and protein folding

A group of fatal neurological degenerative diseases exist known as prion diseases which affect humans and animals. These include **Creutzfeldt–Jakob disease** and **kuru** in humans. The latter, known as the **laughing disease** because of the facial grimaces it causes, used to be transmitted in certain New Guinea tribes by cannibalism. In sheep, the disease is known as **scrapie** because the animals scrape off their wool by rubbing against fence posts; in cattle there is **bovine spongiform encephalopathy (BSE)**, commonly known as **mad cow disease**. The diseases can be transmitted by consumption of infected tissue or, rarely, can be an inherited trait. The diseases were first believed to be caused by 'slow viruses' because they are infectious and can take years to develop. However, all attempts to find nucleic acid in infectious material purified from brain failed and the evidence strongly militates against the possibility of any infectious agent such as a virus being involved. Despite this, the infectious agents appear to replicate. Infection is associated with a form of a normal protein found in brain. The disease producing unit is called a **prion** (for **proteinaceous infectious particle**) and the protein itself as **PrP^{sc}** (for prion protein, scrapie) while its *normal* counterpart is called **PrP^c** (for the normal constitutive prion protein). The function of the latter in the brain is not known. The two have the same polypeptide *and are coded for by the same gene*, but their folded conformations are different. PrP^c has a high α-helix content, is soluble and protease sensitive. PrP^{sc} by contrast has a high β-sheet content, is resistant to proteases, and readily forms aggregates, which result in the **amyloid plaques** found in prion diseases. It is believed that the different conformation of the protein is responsible for the disease. The question arises then as to how an improperly folded protein can be infectious and reproduce itself, since you will appreciate that none of the biochemical mechanisms for protein production that we have discussed allows a protein molecule to direct its own replication. There is, however, strong evidence that PrP^{sc} somehow causes PrP^c (the normal protein) to convert to the abnormal form. This has been demonstrated *in vitro* by incubating the two together when the conversion was demonstrated.

Models for the mechanism of this conversion include the intervention of a chaperone and an energy source for an unfolding of PrP^c and a refolding under the influence of a PrP^{sc} molecule. An alternative 'nucleation' model is that molecules of PrP^c are trapped by an aggregate of PrP^{sc} and conformational rearrangement of PrP^c takes place.

The conversion of PrP^c to PrP^{sc} is a rare event in the absence of infection by PrP^{sc}, so spontaneous occurrence of the disease is rare. It is believed that mutations in the gene for the normal PrP may increase the probability of this, which would explain the hereditary origin of some cases of the disease. Once some PrP^{sc} is formed, it would then trigger the autocatalytic formation of more. As stated earlier, it is still not known what the normal function of the prion protein is in the brain, nor is it understood how the abnormal form results in the diseases.

A note on where we are in the book

In the preceding chapters we have discussed the structure of DNA, how it is synthesized, and its role in coding for proteins. We have then discussed how the information in the gene is converted into mRNA, and in this chapter how proteins are synthesized using this information. As explained, apart from the synthesis of a few proteins in mitochondria (and chloroplasts) all proteins are synthesized in the cytosol. However, in eukaryotic cells, different proteins function and are located in different parts of the cell, in the cytosol, in the nucleus, in mitochondria, and other membrane-bound organelles, in different membranes, or are released from cells. Thus there is a very large and important problem of how they are transported from the ribosome on which they are synthesized to their many destinations. It is a fascinating business and the answers are known to a remarkable degree; it is the subject of the next chapter.

FURTHER READING

Transfer RNA

Schimmel, P. and de Pouplana, L. R. (1995). Transfer RNA: from minihelix to genetic code. *Cell*, **81**, 983–6.

This review addresses the question of why particular nucleotide triplets correspond to specific amino acids. tRNA can be thought of as comprising of two informational domains: the acceptor-TψC minihelix encoding the operational RNA code for amino acids, and the anticodon-containing domain with the trinucleotides of the genetic code.

Ribosomal RNA

Noller, H. F. (1991). Ribosomal RNA and translation. *Ann. Rev. Biochem.*, **60**, 191–227.

A review covering the structure and function of rRNA with a useful section on interactions of antibiotics with rRNA. Removes any idea that rRNA is only an inert scaffolding for the ribosome. Very readable.

Elongation factors and translational fidelity

Thompson, R. C. (1989). EF-Tu provides an internal kinetic standard for translational accuracy. *Trends Biochem. Sci.*, **13**, 91–3.

Describes the 'pause' mechanism for translational fidelity.

Powers, T. and Noller, H. F. (1994). The 530 loop of 16S rRNA—a signal to EF–Tu? *Trends in Genetics*, **10**, 27–31.

A penetrating discussion of how fidelity in protein synthesis is achieved.

Wilson, K. S. and Noller, H. F. (1998). Molecular movement inside the translational engine. *Cell*, **92**, 337–49.

An authoritative review of how the ribosome works. It gives more information than most students would require but is listed for anyone taking a special interest in the area.

Translational control

Vassalli, J. D. and Stutz, A. (1995). Awakening dormant mRNAs. *Current Biology*, **5**, 476–9.

Discusses the role of polyA tail length in activating oocyte mRNAs for translation.

Chen, J.-J. and London, I. M. (1995). Regulation of protein synthesis by heme-regulated eIF-2α kinase. *Trends Biochem. Sci.*, **20**, 105–8.

A comprehensive review of the control of hemoglobin synthesis by this mechanism. (Hemoglobin synthesis and its control is dealt with in Chapter 31 of this book.)

Molecular chaperones and protein folding

Hortl, F.-U., Hlodon, R., and Langer, T. (1994). Molecular chaperones in protein folding: the art of avoiding sticky situations. *Trends Biochem. Sci.*, **19**, 20–5.

An excellent general review of protein folding.

Horwich, A. L. (1995). Resurrection or destruction? *Current Biology*, **5**, 455–8.

Discusses how chaperones are involved both in rescuing proteins and directing them towards proteolytic destruction.

Netzer, W. J. and Hartl, F. U. (1998). Protein folding in the cytosol: chaperonin-dependent and independent mechanisms. *Trends Biochem. Sci.*, **23**, 68–73.

A comprehensive but easily readable account of protein folding and the mechanism of chaperone action.

Bukau, B. and Horwich, A. L. (1998). The Hsp 70 and Hsp 60 chaperone machines. *Cell*, **92**, 351–66.

An authoritative clear review, with structural models; would be suitable for students particularly interested in this area.

Pfanner, N. (1999). Protein folding: who chaperones nascent chains in bacteria? *Current Biology*, **9**, R722.

Summarizes the actions of Hsp 60 and Hsp 70.

Dobson, C. M. (1999). Protein misfolding, evolution and disease. *Trends Biochem. Sci.*, **24**, 329–32.

Reviews diseases in which protein misfolding leads to amyloid or fibrillar aggregates.

Gottesman, M. E. and Hendrickson, W. A. (2000). Protein folding and unfolding by *E. coli* chaperones and chaperonins. *Current Opinion in Microbiology*, **3**, 197–202.

Summarizes the structures and functions of these molecular machines.

Prion diseases

Weissmann, C. (1995). Yielding under the strain. *Nature*, **375**, 628–9.

A concise summary of the molecular basis of prion diseases.

Prusiner, S. B. (1995). The prion diseases. *Sci. Amer*, **272**(1), 30–7.

Excellent general review of the field.

Thomas, P. J., Qu, B.-H., and Pedersen, P. L. (1995). Defective protein folding as a basis of human disease. *Trends Biochem. Sci.*, **20**, 456–9.

Discusses the possibility that a large number of diseases, in addition to prion diseases, may be due to protein folding abnormalities.

Taubes, G. (1996). Misfolding the way to disease. *Science*, **272**, 1493–5.

A research news item presents the general hypothesis that protein misfolding may cause amyloid diseases such as Alzheimers as well as the prion diseases. Introduces the concept that aggregation of proteins into insoluble complexes may be more prevalent and more important than hitherto suspected.

Hunter, N. (1999). Prion diseases and the central dogma of molecular biology. *Trends in Microbiology*, **7**, 265–6.

A 'Comment' article summarizing the nature of prions.

Manson, J. C. (1999). Understanding transmission of the prion diseases. *Trends in Microbiology*, **7**, 465–7.

A 'Comment' article summarizing this topic.

Hope, J. (1999). Prions. *Current Biology*, **9**, R763–4.

A quick guide to the essential facts.

PROBLEMS FOR CHAPTER 24

1   There are 64 codons available for 20 amino acids. Why do you think 61 codons are actually used to specify the 20 amino acids?

2   Despite the facts stated in question 1, there are fewer than 61 tRNA molecules. Explain why this is so.

3   In diagrams, when a tRNA molecule is shown base paired to a codon, the molecule is shown flipped over as compared with the same tRNA shown on its own. Why is this so?

4   At which points in protein synthesis do fidelity mechanisms operate?

5   Describe the participation and, where known, the role of GTP in protein synthesis.

6   Protection studies have indicated that, in *E. coli*, tRNA molecules (with their aminoacyl or peptidyl attachments) straddle A, P, and E sites on the ribosome. Explain why this occurs and the possible mechanisms.

7   The mechanism of initiation of translation in eukaryotes is not compatible with polycistronic mRNA. Explain why.

8   Explain the role of chaperones in protein synthesis.

9   What diseases are associated with improper protein folding?

10   (a)   If you are given the base sequence of the coding region of a mRNA can you deduce the amino acid sequence of the protein it codes for?

(b)   If you are given the amino acid sequence of a protein can you deduce the sequence of the coding region of the mRNA which directed its synthesis?

Explain your answers.

11   Protein synthesis is energetically expensive. Work out how many joules of energy are expended in the synthesis of a single peptide bond of a protein during the elongation process assuming that the free energy change in the hydrolysis of a single phosphoryl group is 29.3 kJ. Briefly explain the role of the various energy-requiring steps and the nucleotide triphosphates involved but without detail of mechanisms.

12   Consider an mRNA which codes for a protein 200 amino acid residues in length. What would the resultant polypeptide be from translation of this messenger if codon number 100 was mutated so that its first base was deleted or if the first and second bases were deleted? What if all three bases were deleted? Explain your conclusions.

13   In a ribosome, is the RNA there simply as an inert scaffold on which to hang proteins? Discuss two pieces of evidence which bear on this problem.

14 Chaperones are also known as heat shock proteins. What is the connection between the two names?

15 Explain the different between transcription and translation.

16 Explain what a codon is; how many codons are used to specify amino acids in proteins? Why doesn't the cell use one codon for each amino acid and forget about the rest?

17 In what ways can a single base mutation in an mRNA affect the polypeptide resulting from the translation of that messenger?

Chapter summary

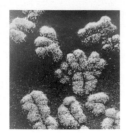

How are newly synthesized proteins delivered to their correct destinations?—protein targeting

With the exception of the ribosomes in mitochondria and chloroplasts, which are prokaryote-like, there is only a single type of ribosome in a eukaryotic cell and they are all located in the cytoplasmic compartment. Which protein a given ribosome synthesizes at any one time is solely a function of the mRNA that it happens to be translating, but, once synthesized, proteins have a number of different destinations. Eukaryotic cells have different compartments separated by membranes and since the latter contain integral proteins, membranes themselves are separate compartments. The term **cytoplasm** refers to the protoplasm inside the external (plasma) membrane but excluding the nucleus, mitochondria and, in plants, chloroplasts. The **cytosol** refers to cytoplasm excluding ribosomes, proteasomes, and membrane-bound organelles.

So far as protein delivery goes, cytoplasmic proteins present no problems—they are synthesized in the cytoplasm, released from the ribosome, and stay there. However, proteins destined for other compartments present intriguing problems for, to reiterate the point, synthesis of all proteins other than the handful synthesized in mitochondria and chloroplasts occurs in the cytoplasm. How do the integral proteins (page 77) of the plasma membrane and other membranes get to be there? Blood serum proteins are released by liver cells whose plasma membranes are obviously designed not to leak proteins. How are they selectively released? The same applies to release of any of the many extracellular proteins such as the digestive enzymes or insulin from the pancreas and the connective tissue proteins from fibroblasts (page 52) to name only a few. Most mitochondrial proteins are synthesized in the cytoplasm coded for by nuclear genes. How are these selectively transported into the mitochondria? Lysosomes and peroxisomes (Chapter 17) are membrane-bound vesicles full of enzymes, but they cannot synthesize proteins. How are the different enzymes transported into the correct vesicle? The nucleus has its own cohort of proteins such as the enzymes responsible for synthesizing and transcribing the DNA, quite different from that of the cytoplasm. There is also traffic of proteins out of the nucleus. As you will learn from Chapter 26 on cell signalling, some gene control proteins exist in the cytosol but on receipt of extracellular signals, these migrate into the nucleus to regulate transcription of genes. It is not just a question of how proteins are able to cross membranes but also how specific proteins are selected from the whole mixture of proteins in the cell to be delivered to, and transported across, the correct membrane. The mechanisms of protein targeting have been substantially elucidated. The aim of this chapter is to describe those mechanisms.

A preliminary overview of the field

An overview without any details will be useful.

- **Cytoplasmic proteins** are released from the ribosome on completion of their synthesis and they stay there (Fig. 25.1).

- Proteins destined to go into **mitochondria, peroxisomes, or the nucleus** are released from the free cytoplasmic ribosomes which have synthesized them and are then transported into the appropriate organelle but by a different mechanism in each case; this is known as **posttranslational transport** (Fig. 25.1).

- The synthesis of **extracellular (secreted) proteins, lysosomal proteins, proteins inside the lumen of the endoplasmic reticulum (ER)**, and all **integral membrane proteins**

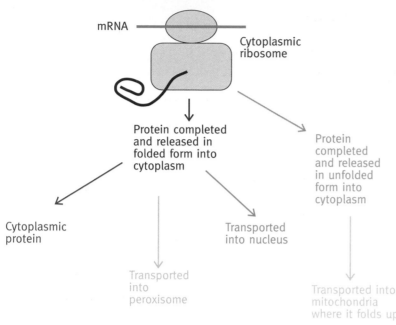

mRNA

Cytoplasmic ribosome

Protein completed and released in folded form into cytoplasm

Protein completed and released in unfolded form into cytoplasm

Cytoplasmic protein

Transported into nucleus

Transported into peroxisome

Transported into mitochondria where it folds up

(The method of transport into the three organelles is different in each case.)

Fig. 25.1 Preliminary overview summary of events in posttranslational targeting of proteins to the cytoplasm, peroxisomes, nucleus, and mitochondria. The essence of the process is that free ribosomes synthesize complete polypeptide chains, release them in the cytoplasm, and then these proteins are transported into their targeted destinations but by different mechanisms in each case as will be described. This contrasts with cotranslational transport depicted in Fig. 25.6 in which proteins are transported across the ER lipid bilayer as they are synthesized.

commences on free ribosomes but these become attached to the ER membrane and the proteins are transported into the lumen (or into the ER membrane in the case of membrane proteins) *as they are synthesized.* This is known as **cotranslational transport.**

Proteins in this category make their way to the Golgi apparatus from which they are packaged into transport vesicles that bud off from the Golgi membranes and deliver their cargo. In each case the vesicle migrates to its appropriate target membrane and fuses with it. Secretory vesicles eject their contents from the cell by exocytosis (page 76), transport vesicles for lysosomal enzymes deliver their contents to endosomes by fusing with the latter to form secondary lysosomes (page 268), but vesicles carrying integral proteins fuse with their target membrane and so new target membrane is produced (Fig. 25.2).

The rest of this chapter describes the mechanisms by which the targeting of proteins is achieved but some information on the ER and the Golgi apparatus will be useful since these remarkable organelles play such a central role.

Structure and function of the ER and Golgi apparatus

The ER is a membranous structure that pervades to varying degrees all eukaryotic cells. It consists of a complex of flattened sacs linked by tubular connections so that its lumen is one con-

tinuous cavity separated from the cytoplasm by the ER membrane. Its size varies enormously in different cells, depending on the metabolic functions of the cell. Part of the ER when seen in the electron microscope is studded with attached ribosomes and so is referred to as **rough ER**; the rest has no attached ribosomes and is called **smooth ER**, but the two types merge into each other on the same membrane (Fig. 25.3). They are not physically discrete membrane structures but have separate functions and different integral proteins. The ER membrane is continuous with the membrane surrounding the nucleus.

The ribosomes are not a permanent fixture on the rough ER but are those which happen at the time to be translating an mRNA coding for a protein destined for secretion, inclusion in a lysosome, or incorporation as an integral membrane protein. The proteins made by the bound ribosomes enter the lumen of the ER (if they are to be secreted or enter lysosomes) or become fixed into the ER membrane as they are made (if their fate is to be integral membrane proteins); they are never liberated into the cytosol. When the ribosome has completed the synthesis of a protein molecule, the ribosomal subunits detach and re-enter the general cytosolic pool to be replaced on the ER membrane by other ribosomes. There is nothing special about ER-bound ribosomes except that, as stated, they just happen to be translating mRNAs for the proteins specified above. Why they attach is the subject of the next section. Inside the lumen of the ER are

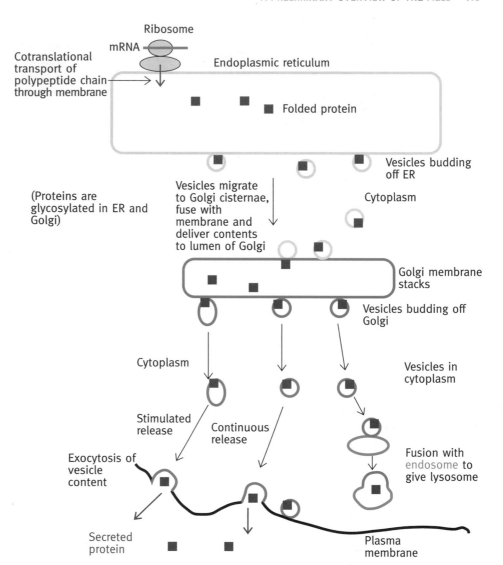

Fig. 25.2 Preliminary overview of how proteins are secreted from cells and how enzymes are delivered to lysosomes. The essence of the process is that the initial transport of proteins through the lipid bilayer into the ER lumen is cotranslational —the polypeptide traverses the membrane as it is synthesized. This is quite different from the transport into peroxisomes, mitochondria, and nuclei which occurs posttranslationally (see Fig. 25.1.). The targeting of new membrane proteins to their appropriate sites is basically by a similar mechanism; the proteins are inserted into the ER membrane as they are synthesized and then sections of membrane, complete with the new proteins, are packaged as vesicles. These migrate to and fuse with the target membrane, thus delivering both new lipid bilayer and membrane proteins. Note that, as described later, proteins are glycosylated as they pass through the ER and Golgi. The many questions as to the mechanism of all this are dealt with below.

enzymes which **glycosylate** the proteins (page 44) as they move through ER into the smooth section where they are enclosed in small vesicles which bud off the ER and transport them to the **Golgi apparatus** (see Fig. 25.4).

The ER and the Golgi are completely closed structures, there being no physically evident entry or exit sites. The cytoplasmic face of the smooth ER is the major site for synthesis of new lipid bilayer (see page 195), a function which correlates with it budding off transport vesicles. The Golgi apparatus, named after its discoverer, is also referred to as the Golgi complex, membranes, stacks, or just the Golgi.

The Golgi apparatus consist of 4–6 (more in plant cells) membranous flattened structures enclosing spaces known as

cisternae. They resemble a stack of large platelike vesicles placed near the nucleus. (What you see in electron micrographs is a cross-section of the plates.) The side facing the ER is the transport vesicle reception area known as the *cis* cisternae; transport vesicles carrying newly synthesized proteins from the smooth ER arrive there, fuse with the *cis* membranes, and deliver their contents into the Golgi cisternae. Proteins move through the Golgi stacks via transport vesicles; there are no direct connections between the plates. Proteins are modified in successive cisternae, progressing towards the final one on the *trans* side where they are packaged into vesicles and despatched to their destinations having been sorted out by the Golgi. The latter thus takes newly synthesized proteins arriving from the ER, identifies

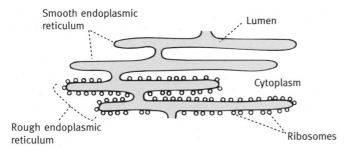

Fig. 25.3 Diagrammatic representation of the endoplasmic reticulum.

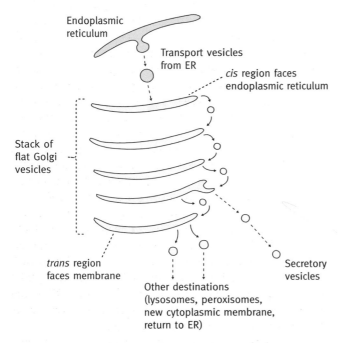

Fig. 25.4 (a) The central role of the Golgi apparatus in posttranslational sorting and targeting of proteins. In addition, newly synthesized membrane lip!ds are transported to the appropriate destination by transport vesicles. (b) Electron micrograph of part of a cell in which prominent Golgi apparatus is evident. Photograph kindly provided by Professor W.G. Breed, Department of Anatomy, University of Adelaide.

them, modifies them, packages them into membrane vesicles addressed to their proper destinations, and finally despatches them. The Golgi apparatus is to secretory, lysosomal, and membrane proteins what a mail sorting office is to posted mail. There is even a 'return to sender' service; as proteins move through the ER, proteins that are needed in the ER are swept along with them and need to be sent back in appropriately addressed transport vesicles.

The vesicles are released into the cytoplasm where they migrate to their destinations. The molecular processes by which it all happens are fairly well understood, if not in complete detail.

That is the end of the general overview, and we can now get down to the molecular processes which are fascinating if a little complicated in parts. We start with the ER-mediated processes, including the synthesis and delivery of integral membrane proteins. This will be followed by the transport of proteins into mitochondria and peroxisomes (in which the ER and Golgi are not involved) and finally with transport of proteins across the nuclear membrane which is a quite different story from all the rest. However, there is a pervading concept in some of the molecular processes which is worth spending a little time on first.

The important role of the GTP/GDP switch mechanism in protein targeting

As we go through the protein targeting mechanisms, GTP hydrolysis to GDP + P_i will be encountered several times. GTP attaches to specific proteins called **GTPase proteins**, and then is hydrolysed; subsequent to this the GDP is exchanged for GTP again. GTPase proteins are involved in several important areas of biochemistry (you will find a description of different types of these on page 436, but this is not needed for now).

You are used to ATP breaking down to perform chemical or other work, but GTP is broken down on GTPase proteins without any such work apparently being performed. This is not a waste of energy, for the GTP/GDP conversion is a control switch mechanism. The work done is to change the conformation of the protein; there is one form in the GTP state and another in the GDP state. The conformational change allows the next step in a process to occur—it is a molecular switch. The hydrolysis is always slightly delayed because the GTPase activity is low or has to be activated by a **GTPase activating protein** (**GAP**) so that the switch mechanism is also a timing device—it allows something essential to happen before the switch operates and the next step in a process initiated. The concept will be more readily appreciated when we come to processes to which it applies.

How are proteins secreted through the ER membrane?

For his work on this, Gunther Blobel at the Rockefeller Institute, New York was awarded a Nobel Prize in 1999. When a protein is to be secreted or has to end up in the external plasma

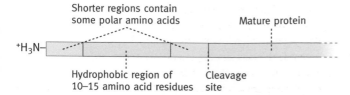

Shorter regions contain
some polar amino acids

Mature protein

$^+H_3N-$

Hydrophobic region of
10–15 amino acid residues

Cleavage
site

Fig. 25.5 A typical leader sequence attached to the N-terminal end of a protein destined to be transported through the ER membrane. Such leader sequences show the same general pattern of amino acids but no specific amino acid sequence. Acidic amino acids are not present.

membrane, inside lysosomes, or in the lumen of the ER itself, it has on its N-terminal end a **signal peptide sequence** of about 25 ± 11 amino acids. If the coding information for such a signal sequence is artificially added to the mRNA for a protein that normally stays in the cytoplasm, the protein will be transported into the ER as it is synthesized. The signal peptide amino acid sequences of different proteins have a pattern, rather than a fixed sequence, as shown diagrammatically in Fig. 25.5. There is a short, positively charged N-terminal section, a central hydrophobic region 10–15 amino acid residues in length and a short more polar region in the form of an α helix. Substitution of a single charged residue into the hydrophobic central region is enough to destroy the signal sequence. When the polypeptide traverses the ER membrane, the signal sequence is cleaved off by a **signal peptidase** on the inner face of the membrane.

Mechanism of cotranslational transport through the ER membrane

It will help if you follow each numbered step in Fig. 25.6 as you read the following section. A free ribosome in the cytoplasm translating an mRNA for an ER targeted protein synthesizes the signal sequence first (since it is at the N-terminal end of the polypeptide). This is immediately recognized by a **signal recognition particle** (SRP), an RNA–protein complex in the cytoplasm which binds to the hydrophobic section of the nascent signal peptide and arrests further elongation of the polypeptide chain, probably by blocking the ribosome (step 1 in Fig. 25.6). The ribosome migrates to the ER membrane (step 2) which has on it **SRP receptors** or SRP **docking proteins** *to which GDP is bound*. These receptors are found only on the ER membrane. The SRP of the cytosolic ribosome–SRP complex attaches to this (step 3 of Fig. 25.6) which positions the ribosome on the membrane. In the ER membrane are protein assemblies known as **translocons**. These are channels formed by several subunits of a protein complex which span the membrane, but in the absence of a ribosome these are in a closed condition. This is essential, for otherwise there would be leakage of small

molecules from the cytosol into the ER. It may be that the ribosome in attaching to the membrane causes the assembly of the translocon subunits, but there is some uncertainty as to whether a channel has to be newly assembled for each transfer or pre-existing assemblies are used; both may occur. The essential point is that a ribosome attaching to the SRP docking protein on the membrane becomes associated with a closed translocon channel. This attachment triggers step 4, the exchange of GDP on the SRP receptor for GTP catalysed by an exchange enzyme present in the cytoplasm. In this form, the receptor (the docking protein) binds more tightly to the SRP which in turn causes the SRP to release the signal peptide, all believed to be the result of the conformational change brought about by the binding of GTP in place of GDP. The signal peptide is now free to insert into the translocon (step 5) which resembles a gated channel that opens when the signal peptide attaches (gated channels are dealt with on page 83). The ribosome is tightly bound to the channel so that leakage of small molecules across the ER membrane is avoided; the peptide emerges from the interior space of the ribosome (see Fig. 22.12) so that there is a completely enclosed channel from ribosome to the ER lumen. The SRP is still tightly bound to its receptor but the latter is a slow GTPase; it hydrolyses GTP to GDP but, because of the low enzymic activity, there is a time delay before the hydrolysis occurs. Why does the docking protein have a time delay before hydrolysing the GTP? A probable reason is that it gives time for the signal peptide to be inserted into the translocon. If the hydrolysis were instantaneous, the SRP–GDP (which you will remember has a high affinity for the signal peptide) would recapture it before it was inserted. In the GDP bound state the receptor has a reduced affinity for the SRP which is released into the cytosol for further use (step 6).

The signal peptide inserts into the translocon channel in a hairpin-looped fashion and positions itself as shown in Fig. 25.6. It is probably held there by its hydrophobic central region, though it is not clear what it binds to in the hydrophilic channel of the translocon. One possibility is that membrane lipid is accessible to it between the interfaces of the subunits which form the channel, though this is speculative.

The release of the SRP enables the ribosome to recommence synthesis of the polypeptide which traverses the membrane via the translocon channel as it is synthesized (also shown in step 6). The signal peptide is positioned in the channel such that, as the polypeptide emerges from the membrane to the inside of the ER lumen, the cleavage site is exposed at the internal face of the membrane as illustrated in step 6. The signal peptidase, which is responsible for the cleavage (step 7), has a hydrophobic patch which attaches it on the membrane so that as the signal peptide emerges from the protection of the membrane it

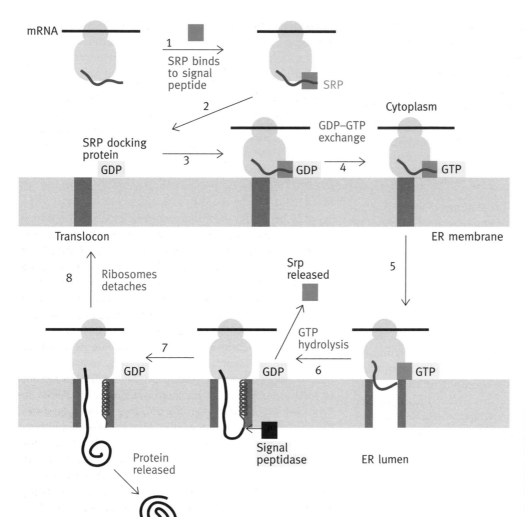

Fig. 25.6 Sequence of events by which proteins are cotranslationally transported into the lumen of the ER. The numbered steps are referred to in the text. The signal peptide is shown in red and the polypeptide which constitutes the mature protein is shown in black.

encounters the peptidase. The area around the cleavage site has an extended structure so that it is readily accessed by the active centre of the peptidase lying in wait.

On completion of the polypeptide synthesis, the protein is released into the lumen of the ER (step 8), and the ribosome dissociates into its subunits to rejoin the cytoplasmic pool for further use. The signal peptide is destroyed by oligopeptidases and the translocon 'gated channel' no longer has its effector (the signal peptide) and so closes.

Folding of the polypeptides inside the ER

The lumen of the ER contains **chaperones** (page 394) which attach to unfolded and partially folded polypeptides. Their function is to hold the chain in a conformation which prevents the polypeptides from going down an unproductive folding route leading to aggregation. A key component in this system is the ER isoform of **Hsp 70** which interacts with the polypeptide chain as it emerges from the translocon channel so that the incoming polypeptide folds cotranslationally. To assist with this process the lumen contains **disulfide isomerase** and **prolyl peptide isomerase** whose role in assisting correct folding has already been dealt with (page 395).

Glycosylation of proteins in the ER lumen and Golgi apparatus

In Chapter 3 (page 78) we described proteins, particularly membrane and secreted proteins, which have complex oligosaccharides added to them. The attachment points are either the

amide $-NH_2$ of asparagine side groups (*N*-glycosylation) or the $-OH$ of serine and threonine residues (*O*-glycosylation). The additions are made in a series of steps. Inside the ER, *N*-glycosylation is carried out. The first step of this is particularly interesting in that a 'core' oligosaccharide of 14 sugar units is assembled in the cytoplasm and transported through the membrane attached to a long hydrophobic chain (which can exceed 100 carbon atoms in length) called **dolichol phosphate**. A transferase enzyme on the inside of the ER membrane transfers the group to the nascent polypeptide chains as they enter the ER lumen. *O*-Glycosylation occurs in the Golgi cisternae.

How are proteins sorted, packaged, and despatched by the Golgi apparatus?

Proteins to be returned to the ER

In most cases the sorting mechanism must include membrane receptors which recognize (attach to) certain structural features or specific sequences of the proteins to be sorted. These are not yet elucidated fully but in one category it is known. In those proteins (such as protein disulfide isomerase) which have to be returned to the ER, the 'label' is a peptide sequence of four amino acids in the protein, Lys–Asp–Glu–Leu (KDEL in the single-letter abbreviation system for amino acids, see Table 3.1).

Proteins destined for lysosomes

In the case of enzymes destined for inclusion in lysosomes, the signal is on the carbohydrate part of the glycoprotein. In the Golgi the carbohydrate attachment is enzymically modified so that it terminates in a mannose-6-phosphate residue which attaches to membrane receptors and leads to their inclusion in lysosomal enzyme delivery vesicles. In **I cell disease** (discussed on page 269) there is a deficiency of the enzyme leading to tagging of lysosomal enzymes with mannose-6-phosphate so that they are not packed and transported to the lysosomes, resulting in the failure of lysosomes to degrade all types of material.

Proteins to be secreted from the cell

The Golgi packages proteins for secretion into vesicles which migrate towards the plasma membrane (see Fig. 25.4). There are two types of secretion—Some proteins are released continuously as produced. Serum proteins are an example; these are continuously released from the liver without any signal being required. In this process, the vesicles fuse with the cell plasma membrane as they arrive there and release their contents by exocytosis.

- In the case of digestive enzymes, from the pancreas for example, enzyme release is required only when food enters the gut. In this case the vesicles from the Golgi containing these enzymes are larger secretory vesicles also known as **secretory granules**. These store the enzymes until wanted, at which point a neuronal, or hormonal, stimulus causes a release by exocytosis.

How are transport vesicles budded off from the Golgi membranes?

The Golgi membrane has internal receptors to which the mannose-6-phosphate of glycosylated proteins destined for lysosomes attach and the region buds off vesicles containing the lysosomal enzymes. These vesicles fuse with other 'sorting vesicles' with a low internal pH (created by proton pumps in their membranes). The low pH dissociates the mannose phosphate receptors from the glycoproteins and they are returned (again by vesicle budding) back to the Golgi while the enzymes are delivered to a lysosome.

Lysosomal vesicles are formed by the clathrin-coated pit process already described for plasma membrane endocytosis (page 268), but the general mechanism used by the Golgi stacks to form transport vesicles for other proteins is different. At the membrane site where this type of vesicle is to be formed, the area becomes coated with **coatamer proteins** and a protein known as **ARF** as illustrated in Fig. 25.7. It is believed that these cause the curvature of the membrane required for the budding off of a vesicle. A coated vesicle is formed as shown in Fig. 25.7. We need to digress here for a moment to explain the name ARF. It derives from the discovery of a protein involved in the action of cholera toxin (discussed on page 438). The toxin causes addition of ADP-ribose to a protein using NAD^+ as the donor (an unusual role for the coenzyme). ARF is the **ADP ribosylation factor** needed for this reaction. ADP ribosylation is not involved in vesicle formation, but ARF is involved in this also. It is a protein with GTPase activity; in its GTP form, ARF binds to membranes. There are several examples like this of evolution using a protein for more than one function.

To return to the main theme, when a coated vesicle reaches its target membrane (with which it has to fuse) the vesicle must be uncoated. This is caused by the activation of the GTPase of the ARF protein which hydrolyses GTP to GDP resulting in the coatamer and ARF proteins dissociating from the vesicle. Presumably the ARF undergoes a conformational change on hydrolysis of the attached GTP molecule as is believed to occur with all such GTPase proteins. The uncoating of the vesicle

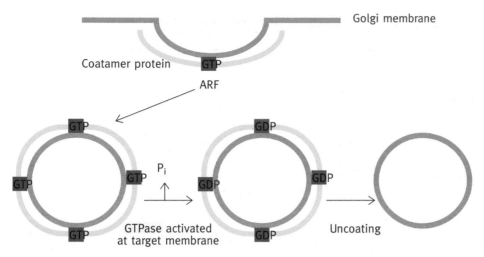

Fig. 25.7 Formation of transport vesicles from the Golgi membrane. The membrane at the site of vesicle formation becomes coated with subunits of a coatamer protein and of a GTP–ARF protein. When the vesicle arrives at its target membrane the ARF–GTPase is activated to form ARF–GDP. This triggers the uncoating of the vesicle followed by fusion with its target membrane. In the case of a secretory vesicle the contents are ejected by exocytosis.

leads to the fusion of the vesicle with its target membrane. If the latter is the plasma membrane then its contents are secreted by exocytosis.

How does the vesicle find the correct target membrane?

The vesicle has a receptor type molecule called a v-SNARE (v for vesicle) on it which binds to a complementary t-SNARE (t for target) on the target membrane (Fig. 25.8). These SNAREs are long helical proteins, which can associate and bring the two membranes together, ready for fusion. Following uncoating, other proteins then join and bring about the fusion of the vesicle and target membrane. This method of targeting vesicles to its destination is a general one, not just confined to protein secretion. If you refresh your memory of nerve impulse transmission on page 86, when a signal arrives at a synapse, vesicles containing neurotransmitter substances fuse with the synovial membrane and eject their contents into the synapse. This occurs also by courtesy of v-SNAREs attaching to membrane t-SNAREs. It is fascinating that the neurotoxic proteins of the tetanus and clostridium microorganisms snip off these snares and interfere with nerve impulse propagation.

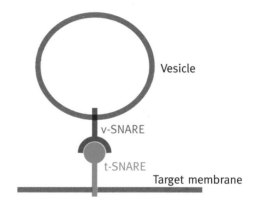

Fig. 25.8 The targeting of vesicles to target membranes is by complementary 'SNAREs' on the two. Binding leads to uncoating of the vesicles followed by fusion of the two membranes. This is a general mechanism for vesicle targeting. The release of neurotransmitters from presynaptic membranes of nerve cells depends on such SNAREs. The botulinus and clostridium neurotoxins hydrolyse the SNAREs from the membranes thus preventing release of neurotransmitters on which transmission of nerve impulses depends.

How are membrane integral proteins inserted?

Integral membrane proteins are also synthesized on the rough ER, in fact placed in the ER membrane as they are synthesized, and then transported to the plasma membrane in the form of vesicles. The new proteins remain *in situ* in the vesicle membrane as they will appear in the plasma or other target membrane and in the same orientation as they were synthesized in the ER. The internal face of the latter in this respect corresponds to the exterior face of the plasma membrane when the vesicle is incorporated into the latter. New membrane lipid synthesis also occurs in the smooth ER (page 194) and hence new membrane originates in this organelle.

An interesting question is why, in the case of a secretory protein, the polypeptide goes right through the rough ER membrane while an integral protein becomes anchored into the membrane. In the latter case the polypeptide chain contains a **stop transfer signal** also known as an **anchor signal** which, as the name suggests, anchors the polypeptide in the membrane.

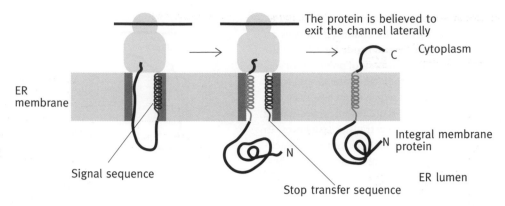

Fig. 25.9 How integral membrane proteins are inserted into the ER membrane. The initial steps are exactly as depicted up to step 7 in Fig. 24.6 but the ribosome synthezises another sequence, shown in blue, which is a stop transfer or anchor sequence. This arrests the further movement of the polypeptide so that when the protein is fully synthesized and exits the channel, the protein is left as an integral membrane protein. As described in the text, models have been proposed for a mechanism by which the integral protein may be oriented across the membrane in the opposite orientation, with the N-terminus in the cytoplasm.

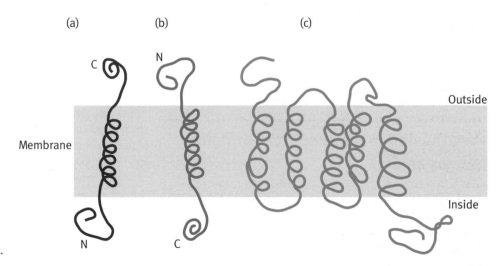

Fig. 25.10 Different orientations of integral membrane proteins. See text.

The simplest situation is a variant of the secretion process already described (Fig. 25.6). If the ribosome translates a stop transfer signal as the polypeptide is transported through the ER membrane, the hydrophobic signal arrests the transport of the polypeptide. After completion of the protein it is believed that it exits laterally from the channel into the lipid bilayer. This exposes the hydrophobic signal to the membrane lipids so that the transmembrane section of the integral membrane protein is now anchored into the lipid bilayer (Fig. 25.9). The details of how the stop transfer signal arrests the polypeptide transfer in the hydrophilic channel are not fully understood, nor how the protein becomes free from the channel. This sequence of events produces a transmembrane protein with its N-terminus inside

the ER (Fig. 25.10(a)). Sometimes the stop transfer sequence is at the C-terminus of the protein so that the transfer is halted only at the very end, the protein then having almost no N-terminal part projecting into the cytoplasm.

Other orientations occur, which leads to the question:

How is the membrane protein given the correct orientation?

It is postulated that an integral membrane protein in the reverse orientation could be produced by **noncleavable signal peptide** positioned within the polypeptide that also acts as a stop transfer signal and is therefore called a **signal anchor sequence**. It is envisioned that the signal sequence inserts into the channel in a

hairpin-looped fashion, resulting in the situation shown in Fig. 25.11. Release of the protein from the channel produces an integral protein with the C-terminal end inside the ER and the N-terminal end in the cytosol. The synthesis of polytopic or serpentine proteins which criss-cross the membrane several times is a complex task, the mechanism of which is not established. A model has been proposed involving a succession of alternating noncleavable signal sequences and stop transfer sequences. In this it is envisaged that successive detachments and reattachments of the ribosome thread the polypeptide across the membrane in the required way as it is synthesized, but much remains to be established about the way in which integral proteins are correctly orientated in the membrane, especially in the case of these polytopic proteins.

The newly synthesized membrane proteins are, as described already, transported as vesicles to the Golgi and thence to the plasma membrane or other destinations. The *trans* face of the ER membrane (the lumenal side) corresponds to the outside of the plasma membrane when the vesicle carrying the proteins to their destinations fuse to produce new plasma membrane complete with its proteins. The orientation of the latter in the bilayer is preserved throughout the transport process via the Golgi.

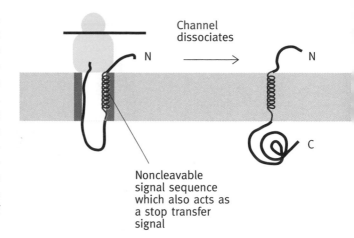

Fig. 25.11 How a transmembrane protein may be synthesized with the orientation shown. See text for explanation. In this the signal sequence is also a stop transfer signal. It inserts into the channel in a looped fashion.

Posttranslational transport of proteins into organelles

To remind you, all of the transport so far described has been co-translational. Transport into mitochondria, (and chloroplasts in plants), peroxisomes and the nucleus is posttranslational. The polypeptides are synthesized, released from the ribosome and then transported, all three by different mechanisms. Secretion from prokaryotic cells is mainly posttranslational.

Transport of proteins into mitochondria

A mitochondrion contains many hundreds of proteins, only a handful of which (13 in humans) are synthesized within the mitochondrion, coded for by mitochondrial genes. All of these are subunits of larger complexes of the oxidative phosphorylation mechanism of the inner mitochondrial membrane, such as cytochrome oxidase and ATP synthase. The rest are coded for by genes in the nucleus, the mRNAs for which are fully translated on free cytoplasmic ribosomes and released into the cytoplasm. There they have to be selected from other cytoplasmic proteins,

delivered to receptors on the mitochondrial membrane and transported to their destinations which may be the outer mitochondrial membrane, the inner membrane, the intermembrane space, or the mitochondrial matrix (Fig. 25.12).

Proteins destined for mitochondrial matrix targeting are synthesized as **preproteins** containing targeting sequences. One of the best characterized is an N-terminal **targeting sequence** of 15–35 amino acids which is not present in the mature mitochondrial proteins. The targeting sequences consist not of defined amino acid residues but of mixtures of hydrophobic, hydroxylated, and basic amino acids which form an amphipathic α helix with hydrophobic side chains on one side and basic, positively charged groups on the other side. An important feature is that as the polypeptide is synthesized on the ribosome, it does not fold up because molecular chaperones such as **Hsp 70** (see page 394) attach to the extended chain (Fig. 25.12) and hold it in this form. The polypeptide–chaperone complex migrates to a mitochondrial receptor which is part of the translocase of the outer mitochondrial membrane (**TOM**). This is a protein complex consisting of a protein (TOM 40) forming an aqueous channel through which the preprotein will traverse the membrane. A number of other TOM proteins are the receptors to which the preproteins attach; the receptors transitorily associate with the channel protein and deliver the preproteins to it for transport. In the case of proteins destined for the mitochondrial matrix, the preprotein is transported in an extended form (without the chaperone) through the outer membrane, via a general import pore consisting of a number of TOM components and then across the intermembrane space and deliv-

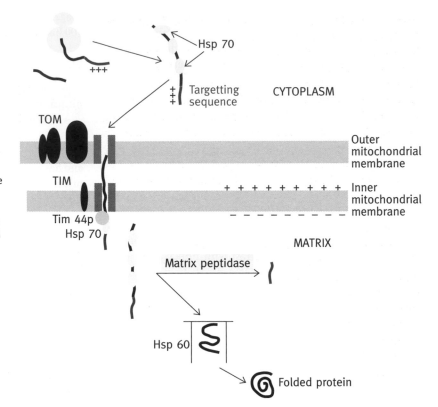

Fig. 25.12 Targeting of proteins to the mitochondrial matrix. Proteins are synthesized as preproteins with an amphipathic targeting sequence at the N-terminal end (red). They are released into the cytoplasm but held in the unfolded state by chaperones (yellow). The preprotein attaches to the TOM complex and is transported through to the TIM complex and into the matrix where a peptidase cleaves off the targeting sequence and the protein folds up assisted by the chaperone Hsp 70 and the chaperonin Hsp 60. The Hsp 60 provides an isolated cage amongst the densely packed matrix proteins inside which the transported polypeptide can fold up (see page 394 for the mechanism of action of Hsp 60 and Hsp 70). The TOM and TIM complexes contain a variety of protein subunits. Import of the polypeptide into the matrix is assisted by TIM 44p, Hsp 70, and other subunits which form the translocation motor.

ered to the translocase of the inner mitochondrial membrane (**TIM**). For transport by TIM the inner membrane must have a charge potential (negative inside) As the preprotein emerges into the matrix an ATP-driven translocation motor, composed of the mitochondrial isoform of Hsp 70, the subunit TIM 44, and other protein subunits, attaches. This drives the import process to completion. A matrix peptidase hydrolyses off the targeting peptide and the protein becomes fully folded, a process involving the Hsp 70 machine and/or the Hsp 60 chaperone machine. The latter forms a secluded chamber amongst the densely packed proteins of the matrix inside which the protein can fold properly (A description of chaperones and their mechanisms can be found on page 394.) Hydrophobic proteins destined for the inner mitochondrial membrane require the attachment of chaperone-like proteins for transport across the intermembrane space. Lack of one of these proteins in humans can lead to a disease causing blindness and deafness. (See ref. of reading list.)

Plant chloroplasts also import about 90% of their total proteins from the cytoplasm. The mechanism is similar to that for mitochondria with the targeted proteins synthesized on cytoplasmic ribosomes as preproteins but the targeting peptide, known as the **transit peptide**, is 30–100 amino acids long and is not positively charged as is the mitochondrial targeting sequence. Possibly correlating with this there is no requirement for a negative membrane potential on the inner membrane as there is for mitochondria.

Targeting peroxisomal proteins

Peroxisomes and their functions have been described in Chapter 17 (page 270). They are small organelles with a single membrane; they are not self-dividing and their origin from vesicles is not elucidated in detail. They contain about 50 different enzymes which are synthesized on free cytoplasmic ribosomes. There are two **peroxisome-targeting signals** (**PTS1 and PTS2**). PTS1, the most common, is a C-terminal tripeptide Ser–Lys–Leu (SKL in the single-letter abbreviations). After transport into the peroxisome matrix, it is not cleaved, and so is part of the mature protein. PTS 2 is an N-terminal nonapeptide, which is cleaved. A variety of cytoplasmic proteins called **peroxins** are needed for transport of the targeted proteins. It has been

demonstrated that a folded protein can be transported into the peroxisome which might imply a very large membrane pore, but how the transport occurs is not known nor whether there is a single mechanism only. There are many uncertainties about the mechanism of peroxisomal protein targeting; the interest in the field is heightened by the existence of genetic diseases that are due to defects in the import process.

Nuclear–cytoplasmic traffic

The eukaryotic nucleus is surrounded by inner and outer membranes which are continuous with the ER membrane. Its function is to separate the DNA and genetic apparatus from the cytoplasm. This is in sharp contrast with prokaryotes which have no nuclear membrane, the DNA simply being in contact with the cytoplasm. The difference represents a major evolutionary divide. The existence of a separate nuclear compartment means that there is intensive traffic across the membrane between the nuclear and cytoplasmic compartments.

The transport of proteins through the nuclear membrane is drastically different from anything in the sections already dealt with. We are dealing not only with delivery of proteins to the nucleus but also traffic in the reverse direction and also with RNA (complexed with proteins) as well as proteins. The nuclear membrane is studded with large numbers of pores of huge size and elaborate construction. Although, as described earlier, the channel complexes (translocons) in the ER membrane can be seen in the electron microscope, there is nothing on the same scale as the nuclear pores.

The traffic is complex and large in volume. Proteins such as those synthesizing and transcribing DNA function in the nucleus and have to be delivered to it. The nucleus manufactures all the RNA of the cell (apart from that in mitochondria and chloroplasts) most of which is required in the cytoplasm so that it has to be transported out. The biggest proportion of this is rRNA which remarkably, is not transported as such; the ribosomal proteins are transported in, the ribosomal subunits are assembled and transported out. By contrast, **snRNP**, a ribonucleoprotein, is a nuclear component involved in mRNA splicing (page 358). It has small nuclear RNAs (snRNAs) bound to protein, but the snRNA component is transported out to be equipped with its protein in the cytoplasm and then the complete snRNP is transported back into the nucleus, the reverse of what happens with ribosomal subunit assembly.

Particles of large size are moved across the membrane in large numbers. When a cell is synthesizing DNA, histones must be supplied from the cytoplasm to allow nucleosome assembly (page 311). It has been calculated that each pore must transport 100 histone molecules per minute.

Why should there be a nuclear membrane?

This is a slight, but relevant, diversion from the mechanism of nuclear cytoplasmic traffic.

An interesting question is why, in eukaryotes, evolution has chosen to go to the trouble of having its DNA inside a nuclear membrane. An *E. coli* cell manages perfectly well without one and has none of the associated transport problems. The eukaryotic nucleus separates the act of transcription of DNA in both time and space from translation of the mRNA in the cytoplasm. This provides opportunities for the mRNA to be modified before translation; the most important of these may have been the splicing mechanism which, in allowing the existence of split genes, made possible exon shuffling (page 359) a factor likely to have been important in fostering evolution. In addition, differential splicing allows the production of different proteins from a single gene (page 360). In *E. coli* ribosomes attach to the mRNA even before the synthesis of the latter has been completed and detached from the gene, presumably making splicing more difficult or impossible.

The nuclear membrane also provides the opportunity to regulate transcription by regulating entry into the nucleus of proteins activating or participating in transcription. Gene regulation in eukaryotes and especially in multicellular organisms is much more complex than in prokaryotes (explained on page 361) and the cell cycle is more strictly controlled than in prokaryotes (page 497). In the latter, DNA replication is not strictly coordinated with cell division, but in eukaryotes it is essential that genes are replicated once and once only before each mitosis. Much of eukaryotic gene control is the result of signals (hormones, for example) synthesized by other cells arriving at the cell and exerting gene control. The control of genes and the mechanisms of cell–cell signalling are major topics dealt with in Chapters 22 and 26 respectively. For the moment, such gene control almost always involves specific proteins being transported from the cytoplasm into the nucleus on arrival of a signal (Fig. 25.13). After that diversion we will turn to the mechanism of nuclear–cytoplasmic traffic.

The nuclear pore complex

Each nucleus has several thousand pores forming aqueous channels between the cytoplasm and nucleoplasm. They are huge structures, almost organelles, with a total size of 10^8 daltons and built up of over 30 protein subunits. The pores have been isolated free of the membrane by detergent treatment and their structure studied. The main body of the pore consists of an

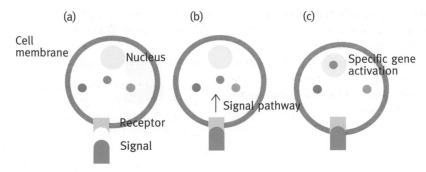

Fig. 25.13 Outline of the role of nuclear import in the control of genes by cell signalling: (a) A hormone or other chemical signal arrives at a cell surface and combines with its specific receptor. (b) The receptor triggers a series of events which selects a specific cytoplasmic protein (blue) to migrate into the nucleus (this is a major subject dealt with in Chapter 26). (c) The protein inside the nucleus causes specific gene activation.(also a major subject dealt with in Chapter 23). It should be noted that this is an exceedingly simplified scheme to give the bare outlines of gene control by hormones, etc. as related to nuclear transport. Although it applies to most situations, lipid-soluble hormones such as steroids, thyroxine, and vitamin D diffuse through the membrane and bind to cytoplasmic receptors or to receptors already in the nucleus. Nonetheless the majority of signalling agents do not enter the cell. (All of these matters are dealt with in Chapter 26.)

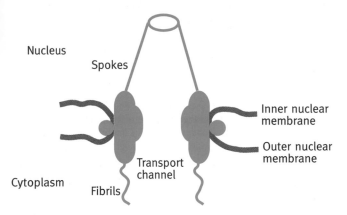

Fig. 25.14 Simplified diagram of the main body of a nuclear pore. See text for the mechanism of transfer through the pore.

annular ring of protein subunits at each end of the pore and these rings are connected by 8 spokes with gaps between them. The inner subunits form a transporting cylinder and the outer 'knobs' anchor the pore into the nuclear membrane, the inner and outer nuclear membranes being continuous at the site of the pore (see Fig. 25.14). In addition to the main pore structure, there is a basket-like projection into the nucleoplasm and fibrils into the cytoplasm, giving it a rather fantastic appearance. The fibrils are the site of receptors to which bind the complexes due for transport. Possibly they bend and feed the latter into the pore.

Molecules up to about 40 000 daltons are able to enter the nucleus by diffusion. The free diffusion of molecules may occur through the main channel, which in that case must in some unknown way expand to allow transport of much larger molecules. Alternatively, there may be an 'leakage' pathway through the spokes which permits this. The actual process of transportation through the pore is not understood but it is temperature sensitive and dependent on ATP.

Nuclear localization signals

Protein molecules larger than the free diffusion cut-off point cannot enter the nucleus unless they have a **nuclear localization signal** (**NLS**) which results in them being actively transported into the nucleus. A nuclear localization signal is a sequence of amino acids in the protein that may be located anywhere in the polypeptide chain. A number of these have been identified ranging from 6–18 amino acid residues in length, rich in arginine and lysine, and sometimes with a central spacer of largely noncritical residues. The 'prototype' of a class of NLS, known as **conventional** because it signals only the inwards transport into the nucleus, is that found in the T antigen, a protein of the SV40 virus; it is transported into the nucleus as part of the infective process. The NLS is PKKKRKV (see page 34 for the single-letter abbreviations of amino acids). Another class of NLS is known as **bipartite**; that found on nucleoplasmin, a chromatin assembly protein, is KR-10aa spacer-PAAIKKAGQAKKKK (given only to illustrate the type of thing, not for memorising). Mutation of an NLS results in a normally nuclear-located protein remaining in the cytoplasm, whereas addition of such a signal to a cytoplasmic protein results in it being transported into the nucleus. Transcription factors are examples of proteins with an NLS. As stated, a conventional NLS signals for import of the protein into

the nucleus but not for its export. Separate **nuclear export signals (NESs)** have been identified for this. However, exceptions occur. There are some proteins which rapidly shuttle in and out of the nucleus, perhaps the best known being the protein which binds mRNA in the nucleus and transports it via the nuclear pore into the cytoplasm as a **heteroribonucleoprotein complex (hnRNP)**. The mRNA is destroyed in the cytoplasm so the mRNA binding protein shuttles back into the nucleus to pick up more. It has a signal of 38 amino acid residues which is both an import signal and an export signal.

Mechanism of nuclear–cytoplasmic transport and the role of guanine nucleotide binding proteins

Apart from the free diffusion of smaller molecules through the pores, all nuclear–cytoplasmic transport depends on combination with proteins known as importins and exportins. There are different proteins known as **importins,** each specific for certain cargo molecules to which they bind via the nuclear localization signals on the latter (Fig. 25.15, step 1). They dock them to the nuclear pore complexes (step 2) and transport them into the nucleus (step 3). It is not known how the importin with its cargo of protein moves through the pore. What causes the importin to release its cargo once in the nucleus? The importin, which is a hetrodimer of α/β subunits, undergoes conformational change when a small protein, **Ran,** a **guanine nucleotide binding protein** (page 436) combines with it. **Ran** *exists in the nucleus combined with GTP*; when this attaches to the importin–cargo complex (step 4) it causes a conformational change in the importin which now releases its cargo into the nucleoplasm (steps 5 and 6) The **Ran–GTP,** coupled to the importin, is presumed to be recycled back to the cytoplasm via the nuclear pore, though how this is achieved is not clear (step 7). Ran has GTPase *activity but this requires activation by a* **GTPase activating protein (GAP***) which occurs only in the cytoplasm, not in the nucleus.* Thus, *in the cytoplasm Ran–GTP is hydrolysed to Ran–GDP* (step 8), resulting in release of the importin into a conformation able to pick up a new cargo to transport into the nucleus. The Ran–GDP is believed to be recycled back into the nucleus by a nuclear transport factor (NTF2) where the GDP–GTP exchange enzyme converts it to Ran–GTP. *The exchange enzyme does not occur in the cytoplasm. The system depends on the Ran protein being in the GTP form in the nucleus and in the GDP form in the cytoplasm.* To re-emphasize the crucial point, this is achieved because of the asymmetric compartmentalization of the exchange enzyme and the GTPase activating protein.

The reverse transport of proteins carrying a nuclear export signal (NES) out of the nucleus into the cytoplasm occurs by a

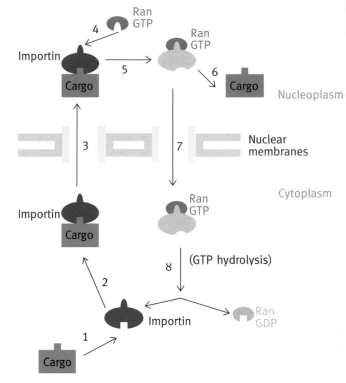

Fig. 25.15 Mechanism of protein import into the nucleus. See text for explanation. The deep red colour represents importin in its form capable of attaching to a protein to be transported; the pale red form cannot do so.

cycle (Fig. 25.16) which is the mirror image of the import cycle. In the nucleus there is a group of **exportin** proteins, each specific for proteins carrying the appropriate NLS. Attachment of the exportin to the NLS of its target cargo occurs only when Ran–GTP is attached to the exportin (steps 2 and 3). Note that this is the reverse of the situation with importin which releases its cargo when Ran–GTP attaches. The Ran–GTP–exportin–cargo moves through the pore (step 4), again by an unknown ATP-dependent mechanism, into the cytoplasm where the GAP activates the Ran–GTPase (step 5) causing hydrolysis of the GTP and dissociation of the complex into Ran-GDP and exportin with the release of the cargo (steps 5 and 6). The Ran–GDP is recycled back to the nucleus (step 7); the nuclear transport factor (NTF2), as mentioned above, is believed to be necessary for this. The exportin must also be recycled back to the nucleus.

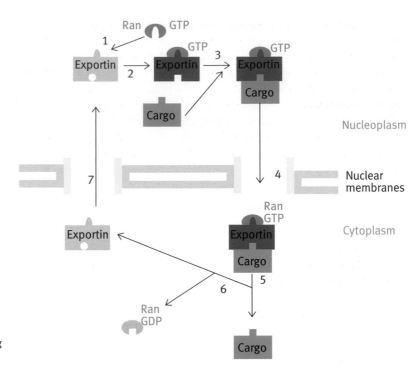

Fig. 25.16 The mechanism of export of proteins from the nucleus. See text for explanation. The deep red colour represents exportin in a form capable of carrying a cargo; the pale red form cannot do so.

Regulation of nuclear transport by cell signals and its role in gene control

In the next chapter we deal with cell signalling, which has a vital role in virtually all aspects of the life of eucaryotic cells including control of cell division which means control of genes. The central feature of this is that signals arriving at cells cause changes within the recipient cells which lead to regulation of the activities of specific genes. A strategy of central importance, already mentioned, is for the extracellular signal to activate a signal transduction pathway inside the cell, one of whose main effects is to cause proteins involved in specific gene control, resident in the cytoplasm, to migrate into the nucleus where they cause changes in the activity of genes and regulate the most fundamental cellular processes. Thus, nuclear–cytoplasmic traffic is a key part in this tremendously important area of cell signalling. Figure 25.13 gives a very simple presentation of this concept.

As already explained, a protein to be transported into the nucleus must have the appropriate NLS so the question arises of why they aren't they carried into the nucleus in the absence of any signal. The answer is that such proteins have their NLS rendered ineffective until an extracellular signal arriving at the cell causes changes in the protein which now make it available for transport.

There are several strategies employed to bring this about. One is to mask the NLS on a given protein by the binding of another protein molecule which masks the NLS so that the importin has no access to it (Fig 25.17(a)). An example of this is described in the chapter on cell signalling (page 427), but for the moment it is enough to say that certain steroid hormones combine with the protein to be transported, causing a conformational change that results in detachment of the masking protein. The NLS is thereby made available for combination with the importin, resulting in import of the protein into the nucleus where it is a transcription factor for those genes that are activated by the steroid hormone.

An alternative strategy is to phosphorylate proteins. Protein phosphorylation has a dominant role in signal transduction pathways (Chapters 12 and 26). The addition of a phosphoryl group has large effects on proteins. It adds a strong negative charge and may cause a conformational change. In some cases the ability of importin to bind to the NLS is prevented by the phosphorylation—in other words, the NLS is masked by phosphorylation (Fig. 25.17(b)). Masking is not the only way in which phosphorylation is known to control nuclear import. As stated, many signal transduction pathways cause activation of protein kinases so this presents an effective control mechanism for linking cell signalling to nuclear targeting as a means of gene control. Some NLSs have only a low

(a) Masking of nuclear localization signals by another protein

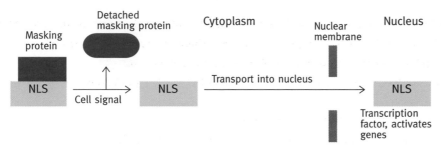

(b) Masking of NLS by phosphorylation

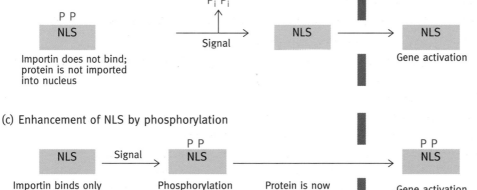

(c) Enhancement of NLS by phosphorylation

Fig. 25.17 Mechanisms of signal-mediated regulation of import of proteins into the nucleus. Extracellular signals to cells are of central Importance in gene control, and nuclear import regulation plays a vital role in this. The essence of it is that proteins required for specific gene control reside in the cytoplasm until a signal causes them to be imported into the nucleus. This requires that the proteins in question are not transported into the nucleus (or at a slow rate) until a signal causes them to be transported. There are several mechanisms for this, three of which are illustrated.

binding affinity for importins, so that the proteins carrying them are transported at a slow rate; phosphorylation of the protein may increase the binding affinity of the two proteins and increase the rate of transport into the nucleus (Fig. 25.17(c)). Dephosphorylation by protein phosphatases (page 215) provides a way of reversing the process when the cell signal is terminated.

The realization that the nuclear–cytoplasmic transport traffic plays a major role in cell signalling and gene control enormously extends its importance and interest. As is almost always the case in biochemical systems, much remains to be learned, but the essentials of nuclear cytoplasmic transport have been well established by the quite spectacular research progress made in the past decade or so.

FURTHER READING

Protein targeting

Von Heijne, G. (1995). Membrane protein assembly: rules of the game. *BioEssays*, **17**, 25–30.

Interesting discussion of how proteins can be 'stitched' into membranes with multiple transmembrane helices.

McNew, J. A. and Goodman, J. M. (1995). The targeting and assembly of peroxisomal proteins: some old rules do not apply. *Trends Biochem. Sci.*, **21**, 54–8.

Reviews evidence that folded proteins may be imported into peroxisomes as such.

Schmid, S. L. and Damke, H. (1995). Coated vesicles: a diversity of form and function. *FASEB J.*, **9**, 1445–53.

Reviews the different ways in which coated vesicles are budded off from the Golgi and other membranes.

Schekman, R. and Orci, L. (1996). Coat proteins and vesicle budding. *Science*, **271**, 1526–33.

An excellent review of vesicle transport and their role in sorting of proteins.

Pfanner, N. and Meijer, M. (1997). Mitochondrial biogenesis: the TOM and TIM machine. *Current Biology*, **7**, R100–3.

Reviews the field with good illustration of the whole system.

von Heijne, G. (1998). Life and death of a signal peptide. *Nature*, **396**, 111–12.

A News and Views article with a fascinating description of the precision with which the signal peptide cleavage is carried out as the transported polypeptide emerges into the lumen of the ER. Also valuable as a succinct summary on ER transport.

Edwardson, J. M. (1998). Membrane fusion: all done with SNARE pins. *Current Biology*, **8**, R390–3.

All about SNARE proteins and transport vesicle targeting.

Mattaj, I. W. and Englmeir, L. (1998). Nucleocytoplasmic transport: the soluble phase. *Ann Rev. Biochem*, **67**, 265–306.

A comprehensive review of the subject.

Koehler, C. M., Merchant, S., and Schatz, G. (1998). How membrane proteins travel across the mitochondrial intermembrane space. *Trends Biochem. Sci.*, **24**, 428–32.

Deals with the special proteins needed to conduct hydrophobic proteins across the aqueous intermembrane space. Deficiencies in these lead to blindness and deafness.

Matlack, K. E. S., Mothes, W., and Rapaport, T. A. (1998). Protein translocation: tunnel vision. *Cell*, **92**, 381–90.

A major review, probably too comprehensive for most students but readable and a valuable reference for anyone particularly interested in the field of translocation across the ER membrane.

Mattaj, I. W. and Conti, E. (1999). Snail mail to the nucleus. *Nature*, **399**, 208–10.

A very concise News and Views article describing the mechanism of nuclear-cytoplasmic transport.

Wieland, F. and Harte, C. (1999). Mechanisms of vesicle formation: insights from the COP system. *Current Opinion in Cell Biology*, **11**, 440–6.

Discuss formation of transport vesicles.

Blobel, G. and Wozniak, R. W. (2000). Proteomics for the pore. *Nature*, **403**, 835–6.

The 'pore' referred to is the massive nuclear pore complex. This News and Views article describes work by Rout and colleagues which gives a complete structural picture of the pore. Equally important is the emphasis given to the modern technologies used by the researchers including the use of protein databases.

Protein glycosylation

Abeijon, C. and Hirschberg, C. B. (1992). Topography of glycosylation reactions in the endoplasmic reticulum. *Trends Biochem. Sci.*, **17**, 32–6.

Reviews the different protein glycosylation reactions occurring in the endoplasmic reticulum.

PROBLEMS FOR CHAPTER 25

1 What is meant by cotranslational and posttranslational transport of a protein across a membrane? Give examples.

2 In protein targeting, GTP hydrolysis is often involved although this is not associated with any chemical synthesis or performance of other obvious work. What is the function of this?

3 The transport of proteins through nuclear pores is critically dependent on the asymmetric distribution of two specific protein-catalysed activities between the nucleoplasm and cytoplasm. Explain this.

Chapter summary

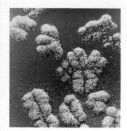

Cell signalling

Cell signalling brings together several major biochemical concepts which have been separately dealt with in earlier chapters in the book. Important in this category are conformational changes in proteins (page 23), protein kinases (pages 215 and 248), second messengers (page 216), gene control in eukaryotes (page 261), transcription factors (page 262), and the cell cycle (page 497). In this chapter there are brief recapitulations of essentials where these seem useful and there are page references to fuller treatments of individual topics.

Eukaryotic cells respond to signals released by other cells. In unicellular yeast cells this is limited to mating signals, but in complex animals such as mammals the situation is different. The activities of individual cells are controlled so that they are integrated with, and appropriate to, the needs of the animal as a whole. An end result of cellular activities uncoordinated with those of the whole animal is cancer. In recent years it has emerged that the extracellular controls operating on mammalian cells are far more complex than was ever imagined. A large number of different signals are used to coordinate their activities. The purpose of this chapter is to describe the molecular mechanisms by which the various signalling molecules are detected by cells and how these result in appropriate responses.

To put the area into perspective, we briefly remind you of allosteric control of metabolism (page 212). This, found in all cells, prokaryotic and eukaryotic, is the type of modulation of enzyme activities in which metabolites of biochemical pathways feed back on to enzymes of the same or different pathways and coordinate them. These controls are entirely intracellular and could operate in a completely isolated animal cell. What they can't do is to make the decision of what the direction of the metabolism should be; a liver cell can either be storing glycogen or breaking it down, to give one example. Which should it be doing? The liver cell itself cannot decide this; it is determined by chemical signals produced by other cells, in the example used, by insulin and glucagon produced by the pancreas (page 113).

Metabolic control is only the tip of the cell signalling iceberg. The most fundamental cellular processes are determined by extracellular signalling. Unless appropriate signals are received, normal (noncancerous) cells will not grow, divide, and differentiate. Apart from its intrinsic interest, the field is of great medical importance. The fact that extracellular signals control so many vital activities of so many different cells in animals provides the potentiality for therapeutic intervention in some of the most intractable human diseases. A large proportion of existing pharmacological drugs and of the research into production of new drugs is targeted to cell signalling systems. Of special interest, cancer is a disease in which the signalling pathways operate in an abnormal fashion so that the cell divides uncontrollably. The elucidation of signalling pathways in cells has revealed many of the abnormal proteins underlying cancer, and the study of cancer in turn has made major contributions to the elucidation of the signalling pathways by supplying ready-made signalling mutants.

Communication between cells is dependent on the release from cells of signalling molecules, which migrate to other cells, and deliver stimuli to those equipped to receive the signals, called **target cells.** A target cell for a given signal has specific protein receptors to which the signal molecule binds, resulting in biochemical responses within the cell. These responses usually include modulation of the expression of specific genes as well as more direct effects on metabolism, not involving gene control (see Figs. 26.1(a) and (b)).

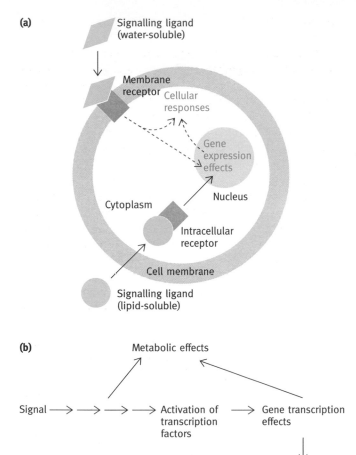

(a)

Signalling ligand
(water-soluble)

Membrane
receptor

Cellular
responses

Gene
expression
effects

Nucleus

Cytoplasm

Intracellular
receptor

Cell membrane

Signalling ligand
(lipid-soluble)

(b)

Metabolic effects

Signal ⟶ ⟶ ⟶ ⟶ Activation of ⟶ Gene transcription
transcription   effects
factors

Cell growth
Cell division
Cell differentiation
Apoptosis

Fig. 26.1 (a) Outline of receptor-mediated signalling. Water-soluble signalling molecules cannot pass through the lipid bilayer. They bind to external domains of receptors. The lipid-soluble signalling molecules such as steroids and thyroxine enter the cell directly and bind to intracellular receptors. Note that some intracellular receptors reside in the nucleus as described later; cytoplasmic steroid receptors move into the nucleus upon ligand binding. (b) A more detailed account of cellular responses to signals.

Questions to consider are therefore how such signals outside of the cell membrane result in something happening inside the cell membrane (**membrane signal transduction**) and how the signal generated at the cytoplasmic side of the membrane is transmitted to the site of control. In the case of gene control this means transmission of the signal from membrane receptor to nucleus. A small number of lipid-soluble signalling molecules directly traverse the lipid bilayer and enter the cell where they encounter intracellular receptors involved in gene

regulation (Fig. 26.1(a)), but the difference from membrane receptor signalling is secondary rather than a difference in concept.

Variations of this general picture exist in situations in which cell–cell contact delivers the signal. An example is activation of the immune system helper T cells by antigen presenting cells (page 458) but again, there is no difference of principle—it's just that the signalling molecule is attached to a cell rather than being free.

What are the signalling molecules?

From the viewpoint of cell signalling mechanisms, the chemical structures of the molecules are irrelevant—they are simply molecules of the right shape for binding by noncovalent bonds to their receptors with great specificity. Nothing happens to them in the course of signalling though they are subsequently destroyed to terminate the signal. Briefly, to illustrate their variety, in chemical terms they include **proteins, large peptides, small peptides, steroids, eicosanoids** (page 195), **thyroid hormone, epinephrines, nitric oxide,** and **derivatives of vitamins A and D.** Examples of the first five groups respectively are **insulin, glucagon, vasopressin, sex hormones,** and **prostaglandin.**

A biological classification is as follows (nitric oxide being in a class of its own):

- neurotransmitters
- endocrine hormones
- growth factors and cytokines
- vitamins A and D derivatives.

This classification is important in terms of nomenclature and physiology, but they are all signalling molecules which bind to cellular receptors and elicit responses; so that they have basic roles in common and conceptually are all the same. With that qualification we will now describe the basis of that classification.

Neurotransmitters

A variety of these are involved in nerve function but we will mention only a few relevant to our immediate topic. The sympathetic (involuntary) nervous system, which innervates fat cell depots for example, secretes epinephrine (adrenaline) and nor-epinephrine on receipt of a nerve impulse. Thus, a fat cell may be regulated by epinephrine from the adrenal glands via the blood (Chapter 13) or from nerve endings. The difference is that the latter delivery route is faster and the signal is released precisely at the target cell site. The motor neurons innervating

voluntary striated muscle trigger contraction by the release of acetylcholine from the nerve endings (page 85). All of these work by binding to external cell receptors.

The transmitters are rapidly destroyed to prepare the neurons for the next impulse.

Endocrine hormones

These are the 'classical' signalling molecules, most of which have been known for a long time. They are important both in metabolic control and in control of the expression of specific genes. Hormones are produced by endocrine glands which secrete the hormones into the circulation. They reach their target cells by flooding the whole body and so reach cells distant from the secreting gland. The system works because only the **target cells** have **receptors** capable of picking up the signal. A large number of hormones are known and their biological effects well docu-

mented. Table 26.1 lists the principal hormones, for reference purposes.

Much of the endocrine system is under a hierarchical control system with the **hypothalamus** being at the top of the chain of command. The hypothalmus is a small part of the brain that produces hormones which stimulate release of anterior pituitary hormones, known as tropic hormones because they cause target endocrine glands (thyroid, adrenal cortex, and the gonads) to release *their* hormones, the 'final' ones in the chain which elicit responses from cells. Feedback loops in which end products (the final cell-targeted hormones) inhibit the first step (release of hypothalamic hormones) maintain appropriate levels of circulating hormones. There are exceptions to this control system, a notable one being release of the pancreatic hormones, insulin and glucagon, which is controlled by the level of blood glucose (page 216).

Table 26.1 The principal hormones

Secreting organ	Hormone	Target tissue	Function
Hypothalamus	Hormone-releasing factors	Anterior pituitary	Stimulation of circulating hormone secretion
	Somatostatin (also from pancreas)	Anterior pituitary	Inhibits release of somatotropin
Anterior pituitary	Thyroid-stimulating hormone (TSH)	Thyroid	Stimulates T_4 and T_3 release
	Adrenocorticotropic hormone (ACTH)	Adrenal cortex	Stimulates release of adrenocorticosteroids
	Gonadotropins (luteinizing hormone, LH and follicle-stimulating hormone, FSH)	Testis and ovary	Stimulates release of sex hormones and cell development
	Somatotropin (growth hormone)	Liver	Stimulates synthesis of insulin-like growth factors, IGFI and IGFII
	Prolactin	Mammary gland	Required for lactation
Posterior pituitary	Antidiuretic hormone (ADH) or vasopressin	Kidney tubule	Promotes water resorption
	Oxytocin	Smooth muscle	Stimulates uterine contractions
Thyroid	Thyroxine (T_4), triiodothyronine (T_3)	Liver, muscles	Metabolic stimulation
Parathyroid	Parathyroid hormone	Bone, kidney, intestine	Maintains blood Ca^{2+} level, stimulates resorption and dietary Ca^{2+} uptake
	Calcitonin (also from thyroid)	Bone, kidney	Inhibits resorption of Ca^{2+}
Adrenal cortex	Glucocorticosteroids (cortisol)	Many tissues	Promotes gluconeogenesis
	Mineralocorticosteroids (aldosterone)	Kidney, blood	Maintains salt and water balance
Adrenal medulla	Catecholamines (epinephrine, norepinephrine)	Liver, muscles, heart	Mobilize fatty acids and glucose into bloodstream
Gonads	Sex hormones (testosterone from testes estradiol, progesterone from ovaries)	Reproductive organs, secondary sex organs	Promote maturation and function in sex organs
Liver	Somatomedins (insulin-like growth factors: IGFI, IGFII)	Liver, bone	Stimulate growth
Pancreas	Insulin	Liver, muscles	Stimulates gluconeogenesis, lipogenesis, protein synthesis
	Glucagon		Stimulates glycogen breakdown, lipolysis

Cytokines and growth factors

It has been suggested that these should be regarded as developmental regulatory factors, because they are more understandable if regarded as signals which may induce cell growth and division or inhibit it, and may affect differentiation or instruct the cell to undergo programmed cell death (**apoptosis**) rather than as simply growth factors. The cytokines and growth factors control such fundamental processes because they control specific gene transcription. Many such factors exist; this is an intensively researched area and of great medical interest. There are no 'authorized' definitions that distinguish between cytokines and growth factors, and the terms are sometimes used interchangeably. The many factors controlling blood cell development, including those involved in immune responses (page 452) together with the interferons (page 435), are referred to as cytokines. Cytokines and growth factors are regulatory proteins or polypeptides secreted by many cell types which, unlike the cells producing endocrine hormones, are not specialized for this function, but are typical of whatever tissue they belong to, such as hepatocytes and lymphocytes. Most cytokines/growth factors are **paracrine** in their action—they diffuse only short distances to act only on local cells, while some are **autocrine** in action (Fig. 26.2)—these, such as **interleukin 2**, which stimulates T cell proliferation (page 458), act on the cells secreting them. The cytokine **erythropoietin** produced by the kidney medulla controls the proliferation of erythrocytes (page 452). It is released when the oxygen tension in the blood is low and is different in that it travels to the bone marrow in the blood.

The names given to cytokines and growth factors often depend on the way they were discovered. The first known growth factor was **platelet-derived growth factor (PDGF)**. Blood platelets lyse at the sites of damage in blood vessels to initiate clotting and the released PDGF stimulates cell division and repair. However, other cells also produce the factor so this is not its only role. **Epidermal growth factor (EGF)** stimulates the growth of skin cells. **Interleukins** are cytokines produced by leucocytes (white blood cells) to affect other leucocytes. **Colony stimulating factors (CSFs)** are so named because they were discovered in experiments in which they stimulated the growth of colonies of white cells on culture plates. Some are used clinically to control white cell production. For example, people with leukaemia given bone marrow transplants first have their bone marrow cells ablated by radiation and chemotherapy. After the transplant there is a period during which there are insufficient neutrophils, with the risk of serious infections as a result since these white cells combat bacterial infections. Treatment with G-CSF (granulocyte-colony stimulating factor) stimulates neu-

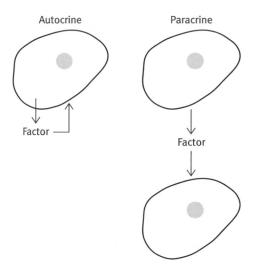

Fig. 26.2 Autocrine signals affect the cell producing them; paracrine signals diffuse only a short distance to affect nearby cells.

trophil (granulocyte) production and thus reduces the risk period.

Growth factors and the cell cycle

It will be useful if we describe the eukaryotic cell cycle here, though it is covered more extensively in the chapter on cancer molecular biology (page 496). Eukaryotic cell division involves a progression of phases from one division to the next. DNA synthesis is confined to a definite period of time called the S (for synthesis) phase (see Fig. 30.2(a)) typically lasting about 7 hours out of the total cycle of 24. Before and after the S phase are the G_1 and G_2 phases respectively (G denoting gap in DNA synthesis). The crucial checkpoint known as the **restriction point** controls the transition from the G_1 to the S phase. Once past this checkpoint the cell is committed to duplicate its chromosome and proceed to mitosis. If a mitogenic (growth) signal is not received in the G_1 phase, the cell is shunted into a quiescent (G_0) phase until a signal arrives. The mitogenic signal is delivered by a growth factor or cytokine which thus assume critical importance in cell division control.

Vitamin D_3 and retinoic acid

We usually associate vitamins as being enzyme cofactors, or components of these, but vitamins A and D are different. Retinoic acid, derived from vitamin A, is important as a signalling molecule in embryonic development and normal cell growth, and vitamin D_3 (calcitriol) in control of genes involved in calcium absorption from the intestine.

We will now move on to the question of how cells detect signals.

Intracellular receptor–mediated responses

As already mentioned, a handful of hormones are lipid soluble and easily traverse the cell membrane. These are steroid hormones, thyroid hormone, vitamin D_3, and retinoic acid. All regulate expression of specific genes in target cells at the level of initiation of gene transcription (page 360). They **include glucocorticoids, estrogen,** and **progesterone** (see list of hormones on page 425.) A number of the lipid-soluble signalling molecules are shown in Fig. 26.3 for reference only; as stated earlier, their chemical structures have no relevance to the mechanism of their action other than that they can combine with their specific receptors which exist *inside* the cell rather than in the membrane. A superfamily of related steroid/thyroxine receptors exists.

We will use the glucocorticoid receptor as an example. Glucocorticoid hormones have diverse effects on metabolism, including increasing gluconeogenesis. The receptor exists in the cytoplasm, attached to a complex of heat shock proteins (Hsps, described on page 394) which mask its nuclear localization signal (NLS), a peptide sequence on the protein—see page 415. When a glucocorticoid hormone binds to a specific site on the receptor, the latter undergoes a conformational change and the Hsps dissociate from it, revealing the NLS so that the receptor is transported into the nucleus. There, it combines as a dimer at the specific glucocorticoid response element (Fig. 26.4) on the DNA causing activation of appropriate genes. In the case of those members of the steroid family of receptors which reside in the nucleus, the nuclear localization signal is not obscured by the Hsp complex so that the whole structure is transported as a unit into the nucleus as soon as it is synthesized but, until the Hsp is released by hormone attachment, it does not bind to DNA and activate genes. (Nitric oxide is a lipid-soluble signal with an intracellular receptor but for convenience it will be dealt with later.)

We now turn to the predominant type of signalling molecules which are water soluble and cannot traverse the lipid bilayer.

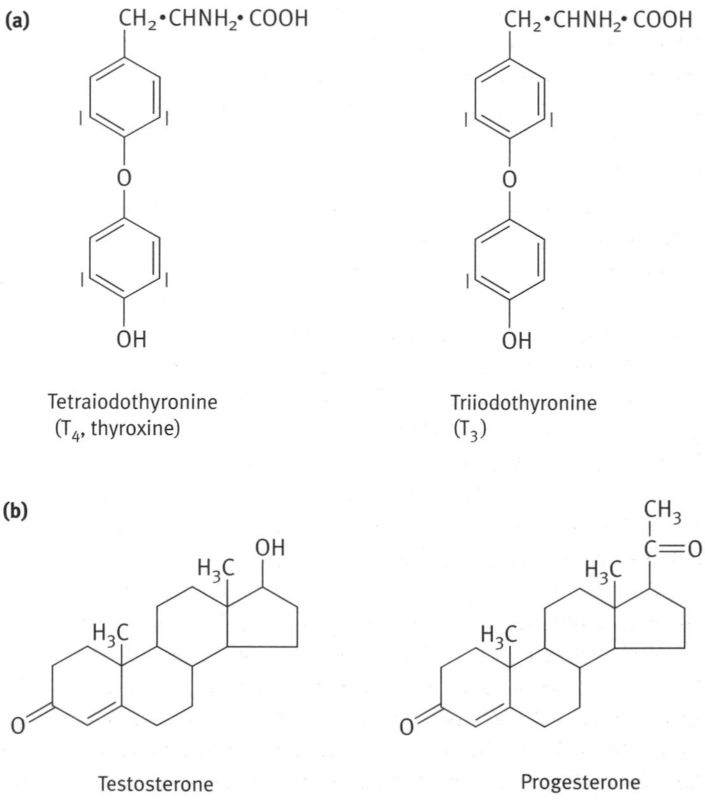

Fig. 26.3 (a) The structures of the thyroid hormones; (b) the structures of two steroid hormones.

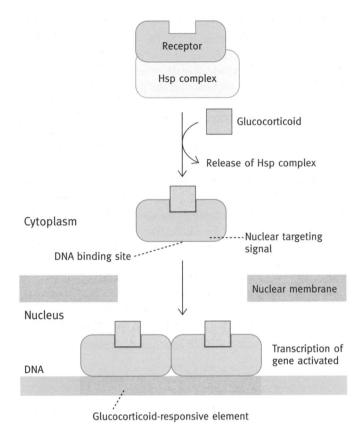

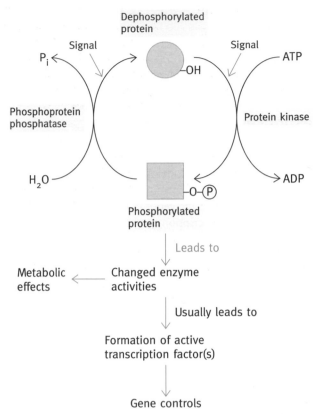

Fig. 26.4 Gene activation by steroid hormones. The example shown is the glucocorticoid receptor, one of a superfamily of steroid/thyroxine receptors, all of which involve complexes with heat shock proteins (Hsps) and binding to DNA sites by zinc fingers. In the case of those members of the superfamily that reside in the nucleus, the Hsp complex does not obscure the nuclear targeting signal. The locations of the latter and the various binding sites are drawn arbitrarily in the diagram.

Fig. 26.5 Central principle of control by many extracellular signals. The diagram illustrates protein kinase and protein phosphatase action. Phosphorylation is generally on serine or threonine residues of proteins. The phosphorylation brings about a conformational change in the protein which changes its activity in some way, resulting in a cellular response. The process is reversed by removal of the phosphoryl group by a protein phosphatase action. In the case of some membrane receptors, described later in the text, phosphorylation of their cytoplasmic domains on tyrosine —OH groups of protein side chains is involved in their activation. Note that an arrow may represent several steps.

Membrane receptor-mediated signalling systems

An overview

Most of the signalling in the body occurs via membrane receptors. All have an external domain to which the signal binds, a transmembrane domain, and a cytoplasmic domain. When its cognate signal molecule (cognate means 'having affinity to') is bound to a receptor, the cytoplasmic domain undergoes a conformational change which activates a signalling pathway. This in turn activates specific genes and/or controls metabolic systems. *Key processes in cell signalling are protein phosphorylation*

by protein kinases and reversal by protein phosphatases (Fig. 26.5). Addition of a phosphoryl group to a protein causes conformational changes which result in the protein participating in the relevant signalling pathway. Protein kinases and phosphatases have already been described in Chapter 13 since they play a major role in metabolic control also. There are two types of protein phosphorylation involved in signal transduction as described below. One involves phosphorylation of serine/threonine residues on proteins, the other on tyrosine residues. The latter is illustrated in Fig. 26.6.

Broadly, there are two classes of membrane receptors. In one class, the tyrosine kinase receptors, binding of the signal mol-

ecule causes dimerization of the receptors which are laterally mobile in the membrane (Fig. 26.7(a)) followed by autophosphorylation of tyrosine groups of their cytoplasmic domains, the latter themselves being tyrosine kinases. (The insulin receptor is an exception in that it is effectively permanently dimerized—page 433.) In the cytoplasm are different proteins involved in control processes which bind to phosphorylated receptors and act as adaptor molecules; they form the link between the activated receptors and the signalling pathways that convey the signals, usually to the nucleus. Around the phosphorylation sites on a receptor are amino acid sequences to which these proteins specifically attach, so that the correct adaptor proteins bind to the correct receptor and thus the correct signalling pathway(s) are activated. A given receptor may be connected to a number of signalling pathways and thus have multiple control effects in the cell. It turns out that these adaptor proteins exist as a number of structural families. Although individual proteins have their own specific binding sites, the binding site domains in a family have related structures. One of

the best-known groups of such proteins is those with receptor binding sites known as **SH2 domains** because they are homologous (similar in general structure but not identical) and this is associated with the ability to bind to phosphorylated tyrosines. SH2 refers to the prototype Src (pronounced sark) homology domain of the Rous sarcoma virus. As stated the domains are adaptor regions that allow this family of proteins involved in signal transduction to bind to receptors. The proteins are adaptor molecules in that they form the link between the receptor(s) and one or more signalling pathways.

The second broad class of receptors (Fig. 26.7(b)) do not dimerize and do not become phosphorylated. Binding of a signal molecule causes a conformational change in the cytoplasmic domain of the receptor. These also have adaptor molecules which undergo a conformational change and link the receptor to signalling pathways.

What these signalling pathways are and how they are activated is the subject of most of the rest of this chapter. To help you keep track of the text as we deal with the individual signalling systems they are listed below, but don't be concerned if you do not understand some of the terms at this stage.

Tyrosine kinase-associated receptors

– Ras signal transduction pathway

– phosphatidylinositide 3-kinase (PI 3) pathway (used by insulin)

– JAK/STAT protein-associated receptors

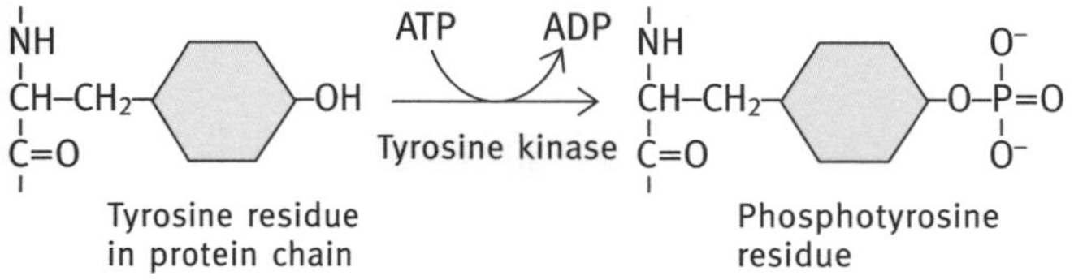

NH
|
CH–CH$_2$ ⬡ –OH $\xrightarrow[\text{Tyrosine kinase}]{\text{ATP} \quad \text{ADP}}$ NH
|
C=O
|

CH–CH$_2$ ⬡ –O–P=O
|
C=O
|

O$^-$

O$^-$

Tyrosine residue in protein chain

Phosphotyrosine residue

Fig. 26.6 Tyrosine phosphorylation by tyrosine kinase.

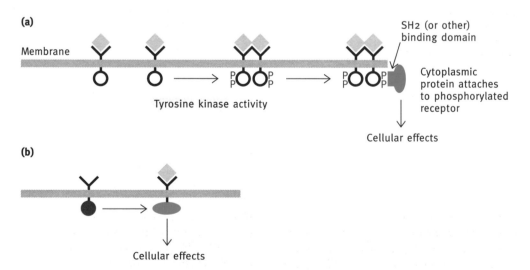

Fig. 26.7 (a) Diagram illustrating the tyrosine kinase type of receptors; the phosphorylation of the receptors causes attachment of the appropriate cytoplasmic protein by its SH2 domain; the attached protein then activates a signalling pathway. (SH2 domains are common, but other binding domains exist—see text). (b) In this type of signalling, the receptor is not phosphorylated but attachment of the signal molecule causes a conformational change in the cytoplasmic domain of the receptor. This leads to the activation of one or more signalling pathways.

G-protein-associated receptors

- cAMP pathway
- phosphatidylinositol cascade pathway
- vision—the light transduction pathway

Signalling pathways mediated by cGMP as second messengers

- membrane receptor mediated pathways
- nitric oxide signalling (This strictly should be in the earlier section on lipid-soluble signals using intracellular receptors, but it is more convenient to include it here.)

Tyrosine kinase-associated receptors

The Ras pathway

This widespread signalling pathway from membrane to genes in eukaryotes is used by a wide variety of growth factors. A protein known as **Ras** exists in all eukaryotic cells. It is part of an ancient major signalling pathway from growth factor specific tyrosine kinase-associated receptors, to activation of gene transcription factors. *There are no small molecular weight second messengers in this pathway—all of the components are proteins.*

The protein Ras, and other components of the associated intracellular signalling pathway, were known to be important in cell regulation, because mutations resulting in abnormal forms of some of the proteins are **oncogenic**—they are associated with transformation to the cancerous state and uncontrolled cell division. Ras was discovered as the oncogenic protein (v-ras) coded for by the **rat** sarcoma virus that causes muscle tumours (sarcomas) in rats. Its normal counterpart is found in all eukaryotic cells and an abnormal form of this (not of Src virus origin) in many human cancers

The signalling pathway involves the receptor stimulating a series of three cytoplasmic protein kinases, the final one of which, when activated by phosphorylation, migrates into the nucleus and by phosphorylating specific transcription factors activates specific genes. Figure 26.8 gives a simplified outline of the pathway to provide a preliminary orientation which will help in understanding the more detailed account which follows.

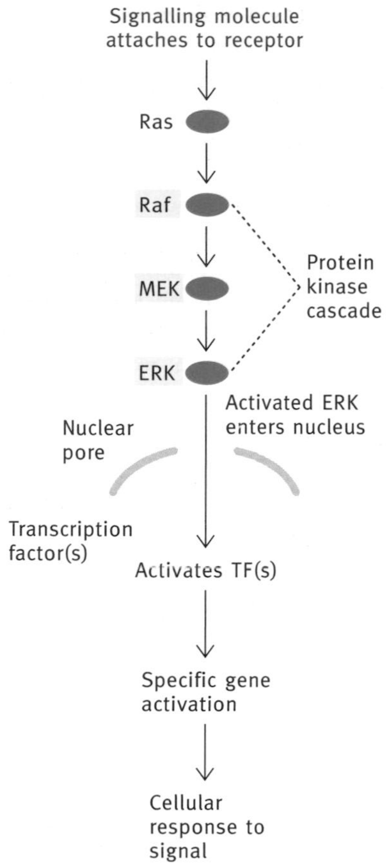

Mechanism of the Ras signalling pathway

We will use the receptor for the **epidermal growth factor (EGF)** to illustrate this (see Fig. 26.9). The receptor for EGF is a monomer with an external binding domain, a single transmembrane domain and a cytoplasmic domain, which is a tyrosine kinase. When EGF binds, the receptors dimerize in the membrane

Fig. 26.8 Simplified overview of the Ras signal tranduction pathway. Raf, MEK, and ERK are protein kinases, each of which phosphorylates the next component of the pathway. The nomenclature and further details of the pathway are given in the text and subsequent figures.

brane thus bringing their kinase domains together and they phosphorylate each other as already shown in Fig. 26.7(a).

In the cytoplasm is a **growth receptor-binding protein (GRB)** associated with a protein called **SOS**, the latter name being derived from a *Drosophila* (fruit fly) genetic mutant known as 'son of sevenless'. These are adaptor proteins which undergo conformational change when they bind to the phosphorylated receptor. After SOS was identified in *Drosophila* as being involved in the signalling pathway from the receptor it was discovered that proteins homologous to SOS are found in all eukaryotic cells. GRB/SOS binds to the phosphorylated receptor *but not to the unphosphorylated form* by means of the SH2 domain (described above) of the GRB. The activated GRB/SOS complex in turn activates Ras, the mechanism of which we will now describe.

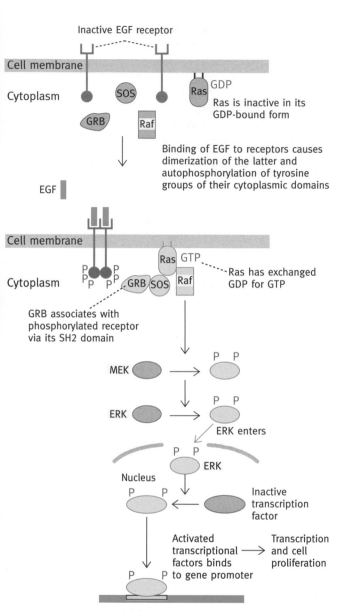

Fig. 26.9 The Ras pathway of signal transduction. See text for explanation. In this diagram signalling by EGF (epidermal growth factor) is used as an example. Activation of the various proteins involves changes in conformation but these are not indicated to keep the presentation reasonably simple. The nomenclature is explained in the text.

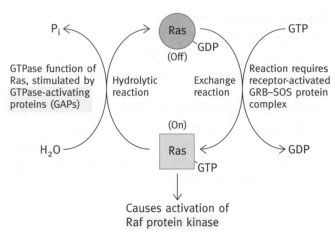

Fig. 26.10 The control of the Ras protein.

The Ras GTP/GDP switch mechanism

Ras has a switch mechanism widely used in control pathways; it belongs to a family known as **small monomeric GTPase proteins**. They have a molecule of GDP *or* GTP bound to them. *Ras is active in the GTP form and inactive in the GDP form* (Fig.

26.10), the nucleotides causing conformational changes in the protein. To activate Ras (for Ras to activate the next component) the GDP must be exchanged for GTP, but this exchange occurs only when Ras is in contact with GRB–SOS bound to the receptor (which, we remind you, occurs only when, in this example, EGF is attached to the receptor). Signals have to be switched off when the activating signal disappears or a stimulus would last indefinitely; when the EGF is no longer present (due to destruction of the growth factor in the medium and consequent lowering of its concentration) the signal has to be terminated. The Ras protein switches itself off by hydrolysing its attached activating GTP molecule to GDP. If the receptor is still activated, Ras will be reactivated by the exchange process; if not, Ras will be inactivated until more EGF arrives. The GTPase activity of Ras is very low, so that the switch-off is not immediate. One gets a glimpse here of the complexity of signalling controls, because other proteins called **GTPase activating proteins** (**GAPS**) stimulate the rate of GTP hydrolysis by the Ras-type proteins and thus speed up the switch-off. Why have this complication? Because the GTPase activity of Ras is so low, GAPS provide yet another means of control. This accounts for the fact that GAP mutations can be oncogenic. If the GTPase activity of Ras is not activated by a functional GAP, the Ras does not switch itself off and an inappropriate activation of the signalling pathway occurs. Ras in its GTP form activates a cascade of three protein kinases. It activates Raf presumably by combining with it and inducing a conformational change. GAPs may themselves be subject to control.

The protein kinase cascade of the Ras pathway

The cascade is as shown in Fig. 26.9, the first kinase (Raf) activating the second by phosphorylation and the second phosphory-

lating the third which migrates into the nucleus and activates specific transcription factors by phosphorylating them. Why so many components? The answer is probably that it amplifies the signal as one component after another is activated. The use of amplifying cascades has been discussed in glycogen breakdown and blood clotting (pages 220 and 277 respectively).

As already mentioned, the first protein kinase in the Ras pathway (after the receptor tyrosine kinase) is **Raf** whose name derives from a viral oncogene; it is activated by Ras-GTP and phosphorylates **MEK** (nomenclature explained below), the second protein kinase, which phosphorylates **ERK**, the last of the three protein kinases. Activated ERK now *enters the nucleus* and phosphorylates target transcription factors, which results in the transcription of specific genes, the synthesis of their cognate proteins, and the desired cellular response to the EGF signal such as cell proliferation. Raf and ERK are serine/threonine type specific kinases (page 215). MEK uniquely, has dual specificity. To summarize, the activation sequence, also illustrated in Fig. 26.9, is as follows:

- EGF attaches to its receptor; the receptor dimerizes and is autophosphorylated on its tyrosine —OH groups.

- GRB/SOS attaches to the phosphorylated receptor by the SH2 domain of GRB.

- Receptor-bound GRB/SOS activates Ras; Ras activates Raf, a protein kinase.

- Raf phosphorylates and thus activates MEK, also a protein kinase.

- MEK phosphorylates ERK, the final protein kinase.

- ERK migrates in to the nucleus and phosphorylates specific target transcription factor(s) appropriate to the EGF signal; the latter activates gene transcription; specific proteins are synthesized; these promote cell proliferation.

Nomenclature of the protein kinases of the Ras pathway

A digression on nomenclature is needed. Raf, MEK and ERK are collectively known as **m**itogen-**a**ctivated **p**rotein **k**inases (**MAP kinases**), a mitogen being something that stimulates growth. Since there are several different MAP kinase pathways (see below), *individual* MAP kinases in a pathway are given identifying names. In the case of the Ras pathway these are Raf, MEK, and ERK. MEK stands for **M**ap kinase/**E**RK; that is a mitogen-activated protein kinase whose substrate is ERK. **ERK** stands for **e**xtracellular signal-**r**egulated protein **k**inase. As mentioned different MAP kinase pathways have kinases analogous to Raf, MEK, and ERK but with different specificities so these have individual names also. The next short section on nomenclature can be skipped if you prefer.

It is useful to be able to collectively refer to all kinases of the Raf type, and similarly to all protein kinases of the MEK type and also to all of the ERK type found in the different signalling pathways. In the literature the terms **MAPKKK, MAPKK,** and **MAPK** are used; to understand them it is best to start with the last and work backwards. Thus ERK is a MAP kinase or MAPK. MEK is a kinase that phosphorylates ERK (a MAP kinase), and therefore is a MAP kinase-kinase or MAPKK. Raf is a kinase that phosphorylates MEK (a MAP kinase-kinase), and therefore is a MAP kinase-kinase-kinase or MAPKKK.

Inactivation and modulation of the Ras pathway

The removal of EGF or other factors from receptors terminates the activation of the receptor but the whole pathway needs to be switched off also. We have already described how Ras is inactivated by its intrinsic GTPase activity. How are the activated MEK and ERK kinases inactivated? A number of **protein phosphatases** are known which inactivate specific protein kinases by dephosphorylating them. It has been discovered that, somewhat surprisingly, the protein kinases and the cognate phosphatases are physically bound together (Fig. 26.11) implying that no sooner has a phosphorylation taken place than it is reversed. It is difficult to visualize how sufficient signal is allowed to get through the pathway, but possibly a competing push–pull situation is less likely to get out of control—perhaps analogous to continually touching the brakes of a car going down a steep hill. A signalling pathway out of control is characteristic of cancer cells (page 500).

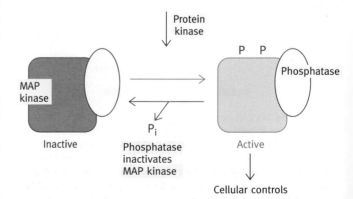

Fig. 26.11 The association of a MAP kinase with a phosphatase to provide a rapid molecular switch mechanism. The MAP kinase is activated by phophorylation and the associated phosphatase reverses this by removing the phosphoryl groups. The implication is that MAP kinase pathway signalling involves the rapid oscillation of the kinase components between the active and inactive states. It may be to give stability of control.

Multiplicity of MAP kinase pathways

As already mentioned, multiple pathways analogous to the Ras type exist, each with a protein kinase cascade terminating in a MAP kinase which enters the nucleus and activates its target transcription factors. There are thus multiple Ras type pathways operating in parallel with specific protein kinases corresponding to Raf, MEK, and ERK in other pathways and targeted at different transcription factors (Fig. 26.12).

These multiple pathways can be envisaged as conveying signals from different receptors to the nucleus so that activation of specific genes appropriate to the particular signal occurs, and this is what appears to happen. However, there are aspects not understood for it is known, for example, that quite different receptors can activate Ras. Insulin and EGF both do this, but the cell responses to the two signals are quite different. Somehow the cell is able to send signals from receptors to the correct end points despite this; there may be subtle factors such as the precise relative locations of components in the cell, but this is speculative and the answers are not known at present. It is known that groups of kinases in different pathways are linked together by **scaffold proteins** which thus would channel a given signal down a given pathway, but this does not give a complete answer. The problem is not helped by the fact that the multiple pathways appear to have some cross-talk between them; on the other hand this cross-talk between pathways is probably essential for the coordination of the many signals which the cell somehow achieves very efficiently. At present it seems to be a very formidable problem.

The phosphatidylinositide 3 (PI 3)-kinase pathway and insulin signalling

This, we remind you, is another example of the class of signalling pathways involving tyrosine kinase-associated receptors. The PI 3-kinase pathway has a variety of control roles involved in cell proliferation, differentiation, and other cellular activities including metabolic control; it is activated by receptors responding to a very wide range of hormones, growth factors, and neurotransmitters. There is great medical interest in the pathway which is heightened by the discovery of an oncogenic (cancer producing) form of PI3-kinase and by the fact that **insulin** is one of its customers. We will use the latter to illustrate the pathway. Insulin's effects have been discussed earlier (page 229) but briefly, it controls glucose uptake into muscle and adipose cells, increases protein synthesis, and controls the synthesis of specific enzymes. Of special metabolic interest, it controls **glycogen synthase** (page 223).

The **insulin receptor** is unusual in structure in that it resembles two receptors of the EGF type covalently dimerized (see Fig. 26.13). On binding of insulin, the cytoplasmic domain of the receptor, which is a tyrosine kinase, becomes phosphorylated on tyrosine —OH groups. The adaptor protein that binds to this phosphorylated receptor is called the **insulin receptor substrate (IRS)** and this is phosphorylated by the receptor tyrosine kinase. An enzyme, **phosphatidylinositol 3-kinase (PI3-kinase)** binds to the phosphorylated IRS and is thereby brought in an activated form close to the plasma membrane. The substrate of PI 3-kinase is a membrane component, **phosphatidylinositol-4,5-bisphosphate (PI(4,5)P2)**. As a result of the PI3-kinase action, a **second messenger, PI(3,4,5)P3 (PIP3)**, is formed (Fig. 26.14). A second messenger, we remind you (and see page 216), is a small molecule produced in response to an activated receptor which mediates a signalling

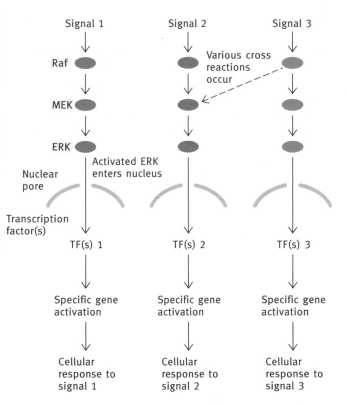

Fig. 26.12 Multiple signal pathways of the Ras type. The red and blue shapes represent protein kinases corresponding in function to Raf and MEK and ERK but are protein kinases of other signalling pathways. It should be noted that this is a simplified scheme for it appears that the different pathways interconnect (indicated by the broken arrow) rather than control being a series of isolated linear pathways. It is not clear how signals from different receptors are channelled to the correct MAP kinase(s) (see text). Scaffold proteins (not shown) have been shown to group together some of the kinases of individual pathways which may be relevant to this.

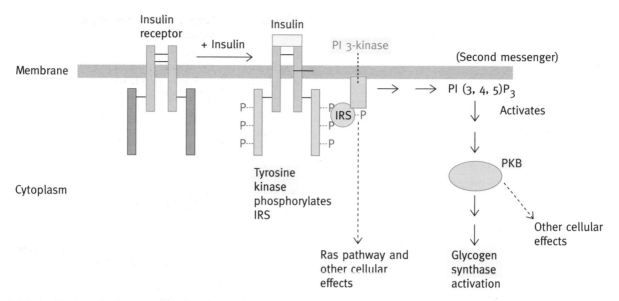

Fig. 26.13 Simplified insulin signalling pathway. The attachment of insulin to its receptor activates tyrosine kinase of the receptor's cytoplasmic domain which phosphorylates itself. IRS attaches to the phosphorylated receptor and itself becomes phosphorylated. PI 3-kinase binds to the phosphorylated form of IRS, thus coming into contact with the membrane where it leads to the conversion of phosphatidylinositol (PI) to a second messenger PI(3,4,5)P_3 by the reactions shown in Fig. 26.14. This activates protein kinase B whose action specifically leads to glycogen synthase activation. The broken arrows indicate that other signalling pathways are activated leading to gene controls in addition to glycogen synthase activation.

pathway by activating protein kinases or other enzymes. (The signalling molecule is the first messenger.) The second messenger causes activation of **protein kinase B (PKB)**, also known as **Akt**. The mechanism of this is that a membrane enriched with the second messenger (PIP3) attracts the inactive cytoplasmic PKB to attach to the membrane. There it encounters a membrane-located kinase called **PDK1** which phosphorylates the PKB, thus activating it. Among its other actions, PKB activation results in the activation of glycogen synthase and therefore of glycogen synthesis by a mechanism described on page 223. A summary may be useful here:

- Insulin attaches to its receptor.
- The receptor is autophosphorylated on its tyrosine —OH groups.
- The IRS attaches by its SH2 domain to the phosphorylated receptor and is itself phosphorylated.
- PI3-kinase binds to the phosphorylated IRS and is activated. Its action on PI(4,5)P2, a membrane phospholipid, leads to the production of a second messenger (PI(3,4,5) P3).
- Cytoplasmic inactive PKB is attracted to bind to the membrane enriched with the second messenger.

- The PKB is phosphorylated by the kinase PDK1 located in the membrane and is thus activated.
- The activation of glycogen synthase is described more fully on page 223.

Insulin has many effects in addition to those on glycogen metabolism control; IRS in the presence of insulin can activate Ras and thus affect gene control by the pathway described in the previous section and PKB probably has gene controlling roles also. The IRS thus acts as an adaptor to connect the receptor for different signalling pathways. As stated earlier, PI 3-kinase signal transduction pathways, activated by a large number of different tyrosine kinase receptors have roles far beyond insulin signalling including control of cell division, cell shape determination, and differentiation. Its important role in these fundamental processes is underlined by the discovery of an oncogenic form of the enzyme.

The JAK/STAT pathways; direct intracellular signalling pathways from receptor to nucleus

This is the third type of tyrosine kinase—associated signalling receptors.

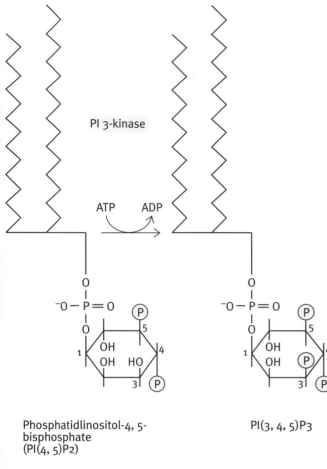

Fig. 26.14 Production of the second messenger PI(3,4,5)P₃ from the membrane component PI(4,5)P₃ by PI 3-kinase, the latter being insulin activated.

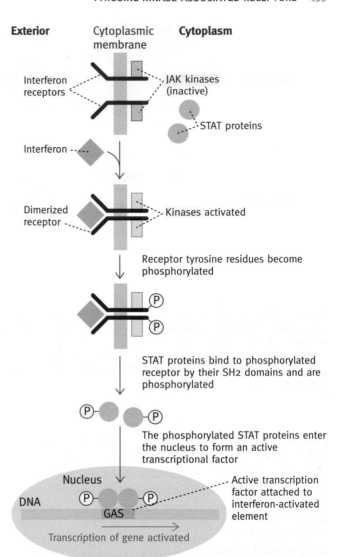

Fig. 26.15 The signalling pathway by which γ-interferon activates specific gene transcription. STAT, signal transducer and activator of transcription; GAS, γ-interferon-activated sequence element. Note that, unlike the EGF receptor (Fig. 26.9), which is a tyrosine kinase, in this pathway the dimerized receptor activates cytoplasmic tyrosine kinases; these are known as Janus kinases (JAK kinases) after the Roman god with two faces. (JAK kinases have two kinase sites.) α-Interferon and several cytokines have analogous signalling pathways but may differ somewhat in detail.

Interferons, the protective proteins released by cells infected by viruses were the first to be associated with this type of pathway, but many cytokines are known to use it. Ligand binding causes dimerization of receptors and association of a **JAK kinase** with the cytoplasmic domain of each receptor in the dimer (see Fig. 26.15). The activated kinases phosphorylate tyrosine residues of the receptor dimer. This, in turn, causes two **STAT proteins** in the cytosol to bind to the phosphorylated receptor. (STAT stands for **signal transducer and activator of transcription**.) The proteins have **SH2 domains** which bind to the phosphorylated receptors. The protein kinase, still bound to the receptor, phosphorylates the two STAT proteins. This is a second type of phosphorylation; the first was phosphorylation of the receptor which attracts STAT proteins and the second phosphorylation of the STAT proteins. To do this the JAK kinase has two catalytic sites; JAK stands for **Janus kinase** after Janus

the Roman god with two faces. The phosphorylated STAT proteins move into the nucleus, where together with a third protein component, they assemble into an active transcription factor and promote transcription of specific genes. Figure 26.15 illustrates the pathway using one of the interferons as an example.

The interesting feature of this type of pathway is its directness; in essence inactive cytoplasmic factors are activated and migrate into the nucleus where they switch on genes. The Ras pathway is much more complicated. Different receptors specific for a variety of cytokines and activating a variety of specific STAT proteins could account for a wide variety of signalling controls affecting gene activities.

The G-protein-coupled receptors

Overview

We now turn to a different type of signal transduction which does not involve tyrosine kinase. The G-protein-coupled receptors constitute a very large superfamily of receptors and examples of the receptor type are found universally in organisms from yeast and insects to humans. Many different receptors are known because as well as responding to many hormones and growth factors they are used to detect odours, light, and other signals. They are of great medical interest; perhaps half of current pharmaceutical drugs target G-protein-coupled receptors.

The general principle can be stated simply. Each receptor has associated with it on the cytoplasmic side a specific **heterotrimeric G protein, a** protein made up of three different subunits, α, β, and γ. There is a very great diversity of G proteins which can couple to a diversity of control pathways, so that the system is of enormous flexibility. The G proteins all have a common pattern even though individual G proteins control quite separate control pathways and each is specific for a given receptor. The α subunit of G proteins has bound to it a molecule of GTP or GDP (this is what gives rise to the name G protein) whose actions resemble that of the GTP/GDP switch mechanism already described for the small monomeric GTPase proteins (page 431) typified by Ras. Despite this similarity, the term G protein is used only for the heterotrimeric G proteins to be described in this section and the term 'small monomeric GTPase proteins' for proteins of the Ras type. The *G protein receptors have no enzymic activity of their own*; the conformational change to their cytoplasmic domains that occurs on binding of the signal molecule to the external domain of their receptor causes a conformational change in the associated G proteins. This in turn causes activation of a signal pathway involving the formation of second messengers which bring about the cellular responses as outlined earlier (see Fig. 26.7(b)). After that overview, we will now give more details of this class of signalling system.

Structure of G protein receptors

All are proteins whose polypeptide chain cris-crosses the membrane seven times (Fig. 26.16); this type of protein is called ser-

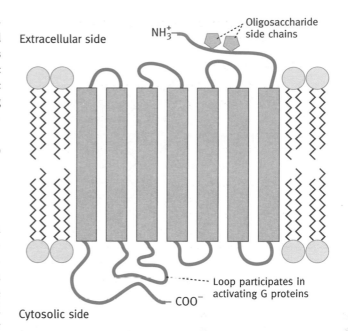

Fig. 26.16 The structure of the β_2-adrenergic receptor.

pentine or, less descriptively, polytopic. In the case of epinephrine, the hormone binds to a cleft formed by the transmembrane helices while the heterotrimeric G proteins associate with a polypeptide loop on the cytosolic side of the membrane.

Epinephrine signalling—a G protein pathway mediated by cAMP as second messenger

Earlier (page 220) we dealt with control of metabolism by cAMP activation of protein kinases; in this section we deal only with the way in which the level of cAMP inside a cell itself is modulated and how cAMP regulates specific gene expression. A wide array of hormones (*first messengers*) use cAMP as *second messenger*, including adrenocorticotropic hormone (ACTH), antidiuretic hormone (ADH), gonadotropins, thyroid stimulating hormone (TSH), parathyroid hormone, glucagon, the catecholamines epinephrine and norepinephrine, and somatostatin (page 425). This list, which is not exhaustive, is given to illustrate that cAMP has different effects in different cells. The system works because the response to cAMP in a given cell is appropriate to the signal which that cell recognizes via its specific receptors. Thus, in cell A, signal X elevates cAMP levels which produce cellular responses appropriate to signal X. Cell B does not have receptors for X but does for signal Y which also elevates cAMP but, in cell B, this causes responses appropriate to signal Y. To give examples, cAMP mobilizes glycogen breakdown in liver cells and triglyceride breakdown in fat cells. The two aspects of these effects, which we will now discuss, are how signal

binding to receptors controls cAMP levels in the cell and how cAMP affects gene transcription.

Control of cAMP levels in cells

cAMP is produced from ATP by **adenylate cyclase** (Fig. 26.17(a)), an enzyme which is an integral cell membrane protein (Fig. 26.18(a)). Let us consider, as a typical example, the way in which adenylate cyclase is activated by epinephrine to produce cAMP in liver and skeletal muscle, the metabolic effects of which have been covered earlier (page 220). The receptor in this example is called the β_2-**adrenergic receptor**, after the alternative name for epinephrine, adrenaline. Associated with the cytoplasmic face of the receptor protein is a G protein which has the three subunits, α, β, and γ. As mentioned, the α subunit has a site on it which can have bound to it, *either* GTP or GDP. When the site on the α subunit is occupied by GDP, nothing happens. This is the situation in the absence of hormone, illustrated in Fig. 26.18(a).

When a molecule of epinephrine binds to the receptor, the latter undergoes a conformational change. This, in turn, affects the G protein bound to it, also by conformational change. The change causes the G protein (called G_s in this system—'s' for stimulatory) to exchange its GDP for GTP. *The G protein cannot make this exchange unless it is attached to a receptor to which the hormone is bound.* The a subunit–GTP complex detaches, migrates to, and activates by binding to it, an adenylate cyclase enzyme molecule which now produces cAMP (Fig. 26.18(b) and (c)) by the reaction shown in Fig, 26.17(a).

The hormone thus switches on cAMP production using the G protein as an intermediary. Activation of cAMP production by the GTP α subunit must be limited in time—it must be switched off, for otherwise a single hormone stimulus would last indefinitely, long after the response would be appropriate. When the receptor is no longer bound by epinephrine, cAMP production must cease; destruction of cAMP by phosphodiesterase in the cell completes the reversal (see Fig. 26.17(b)).

The α subunit of the G protein, in its GTP-bound form (which activates the adenylate cyclase), has enzyme activity—it is a GTPase. It hydrolyses its attached GTP molecule to GDP (Fig. 26.18(d)). The activity is low so that the hydrolysis occurs only after a delay. As soon as GDP is formed, the α subunit reverts to its original state; it detaches from the adenylate cyclase which is now inactivated. The α subunit joins up to its old partners, the β and δ subunits, and the G protein–GDP complex is reassembled, in contact with the receptor (Fig. 26.18(e)). If the

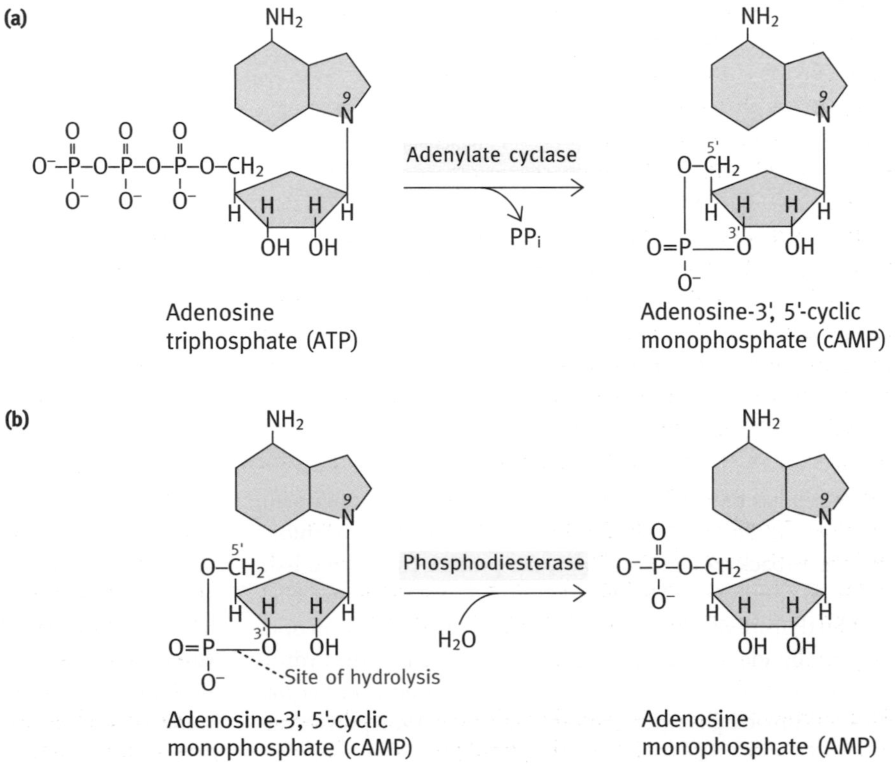

Fig. 26.17 (a) Formation of 3′, 5′-cyclic AMP by adenylate cyclase; (b) hydrolysis of 3′, 5′-cyclic AMP by phosphodiesterase.

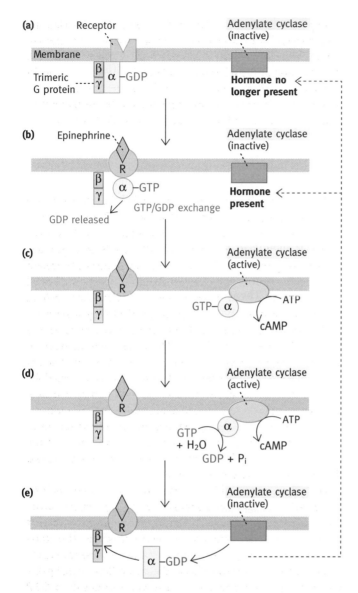

(a)
Receptor
Adenylate cyclase (inactive)
Membrane
Trimeric G protein
β γ α —GDP
Hormone no longer present

(b)
Epinephrine
R
β γ α —GTP
GDP released
GTP/GDP exchange
Adenylate cyclase (inactive)
Hormone present

(c)
R
β γ
GTP— α
ATP
cAMP
Adenylate cyclase (active)

(d)
R
β γ
GTP + H₂O
α
GDP + Pᵢ
ATP
cAMP
Adenylate cyclase (active)

(e)
R
β γ
α —GDP
Adenylate cyclase (inactive)

Fig. 26.18 The control of adenylate cyclase activity by a hormone such as epinephrine. The geographical locations of the subunits shown are hypothetical, the essentials being that the α-GTP complex activates adenylate cyclase. The steps (a)–(e) are referred to in the text.

ton, the light goes on, and the button slowly comes back out and switches off the light after a minute or two. You have to keep on pressing the button at intervals to keep the light on. The G protein is a timing device to limit the period of activation of adenylate cyclase. The system has an amplifying effect, for one hormone-bound receptor can activate one G protein molecule after another and therefore one molecule of hormone binding to a receptor results in numbers of adenylate cyclase molecules being activated and large numbers of cAMP molecules being produced.

The importance of the GTP hydrolysis switch-off is dramatically seen in cholera. Gut mucosal cells secrete Na⁺ into the intestinal lumen and this activity is activated by cAMP. The cholera toxin causes inactivation of the GTPase activity of the G protein α subunit. Thus, once a hormone stimulus activates the adenylate cyclase, it cannot be switched off; it is frozen into the α-GTP state. The prolonged cAMP production results in massive loss of Na⁺, accompanied by water molecules, causing massive diarrhoea and possible death from fluid and electrolyte loss.

So far we have described only how the GTP α subunit activates adenyl cyclase, but most recent evidence indicates that the $\beta\gamma$-dimer of the heterotrimeric G protein has separate controlling effects, possibly by activating MAP kinase and other pathways so that one signal may cause multiple responses. When the α subunit in its GDP form detaches from the adenyl cyclase it recombines with the $\beta\gamma$ and the heterotrimeric G-protein recombines with the receptor.

Different types of epinephrine receptors

In the case given in Fig. 26.18 the GTP α subunit stimulates adenylate cyclase activity. Another type of receptor for epinephrine (known as the α_2 receptor) operates similarly, except that the GTP α subunit (G_i—for inhibitory) inhibits adenylate cyclase. Thus, one hormone (epinephrine) can exert different effects on different tissues, according to the type of receptor present on the cells of the tissues. It illustrates the way in which the use of different G proteins associated with particular receptors can control many different control pathways.

How does cAMP control gene activities?

In Chapter 13, we described how cAMP activates **protein kinase A (PKA)**. PKA, as well as being involved in metabolic control (page 220), is part of an important pathway of gene control, as shown in Fig. 26.19. Activated PKA migrates into the nucleus. The promoters of several cAMP-inducible genes contain **cAMP responsive elements (CREs)**. A CRE binding protein (CREB), when phosphorylated by PKA, dimerizes and becomes an active transcription factor of the leucine zipper type (page 353).

latter still has bound hormone, the whole cycle can start again (back to Fig. 26.18(b)). If the receptor no longer has hormone attached (back to Fig. 26.18(a)), the process comes to a halt. Thus, to continue cAMP production, the a subunit runs back and forth from receptor to adenylate cyclase, the duration of its stay on the adenylate cyclase being that required for the hydrolysis of its attached GTP molecule. The situation is rather like that of a timed light switch on a staircase where you press a but-

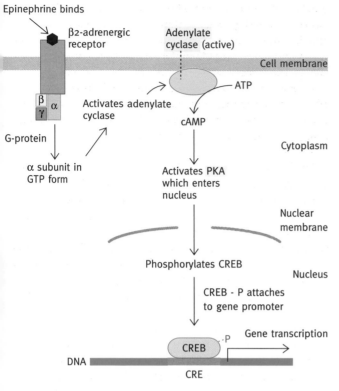

Fig. 26.19 The β_2-adrenergic receptor function. Binding of epinephrine to the receptor causes the heterotrimeric G protein to convert to the GTP form, detachment of the α subunit–GTP which activates adenylate cyclase. The cAMP produced activates protein kinase A (PKA) which is transported into the nucleus. There, the PKA phosphorylates cAMP-response element binding protein (CREB) which in this form is a transcription factor for specific genes. This activates the genes, resulting in the appropriate responses to the epinephrine. See the text for the mechanism by which the G protein activates adenylate cyclase. CRE, cAMP response element of gene promoter.

Mutations which permanently activate the α subunit lead to excessive PKA activation and can be oncogenic.

Desensitization of the G protein receptors

Most, or all, cellular responses to extracellular signal diminish with prolonged exposure to the signal molecule. In some cases synthesis of the receptor diminishes, or the receptors are endocytosed and destroyed in lysosomes (page 267) as happens for example with the insulin receptor. The reduction in receptor numbers is known as **downregulation**. Many of the G protein receptors are desensitized to prolonged stimulus by a family of enzymes called **G protein receptor kinases (GRKs)** which phosphorylate the receptors and cause inactivation. This is not to be confused with the tyrosine kinase receptors in which activation

is coupled with phosphorylation. The GRKs phosphorylate *activated* receptors to which their cognate signal molecules are bound—they inactivate receptors which are in the activated state. Because of the widespread role of G protein receptors in so many important processes, the GRKs are of considerable importance. There is evidence that the GRKs are themselves subject to controls, and this has potential medical interest.

G-protein-coupled receptors which work via a different second messenger; the phosphatidylinositol cascade

So far in this section on G-protein-coupled receptors we have dealt with epinephrine receptors that use cAMP as second messenger. However, other G-protein-coupled receptors exist which use quite different signalling pathways. Each G-protein-coupled receptor, we remind you, has its own specific G protein (so that a large number of different G proteins exist) but they all have the same general heterotrimeric structure that we have described. The α subunit in all cases is activated by attachment of the cognate signalling molecule to the receptor causing a GDP/GTP exchange and the α-GTP detaches and migrates to activate its target enzyme. The GTPase switch-off mechanism operates as already described. All this so far is as described for the adrenergic receptor, but in the signalling system we will now describe, what is different is that a quite different enzyme is activated by the α-GTP, leading to the formation of a second messenger, different from cAMP. Examples of hormones that use the mechanism include the thyrotropin releasing hormone and the gonadotropin-releasing hormone.

The start of this story is the phospholipid **phosphatidylinositol-4,5-bisphosphate (PIP$_2$)**, whose structure is given in Fig. 26.20. This signalling pathway is different from the PI3-kinase one already described which also starts with PIP$_2$ (Fig. 26.14). Once again, as indicated, a specific G protein links the receptor stimulus to the intracellular signal pathway. Binding of the hormone to a receptor causes the a-GTP subunit to exchange its GDP for GTP; it then migrates to, and activates, a membrane-bound enzyme, **phospholipase C (PLC)** which splits PIP$_2$ into **IP$_3$ (inositol trisphosphate)** and **DAG (diacylglycerol)** (Fig. 26.20). The G protein hydrolyses GTP to GDP and P$_i$, thus inactivating itself after an interval, much as occurred with the adrenergic system described earlier.

The IP$_3$ causes release of Ca^{2+} from the lumen of the endoplasmic reticulum (ER) where the ion is stored in high concentration relative to that in the cytoplasm. It opens ligand-gated Ca^{2+} channels in the membrane (page 83), allowing the ion to escape back into the cytoplasm. Ca^{2+} is continually returned, by a Ca^{2+}/ATPase pump, to the ER lumen and the IP$_3$ is enzymically removed. Thus, the combination of a hormone, or other signal,

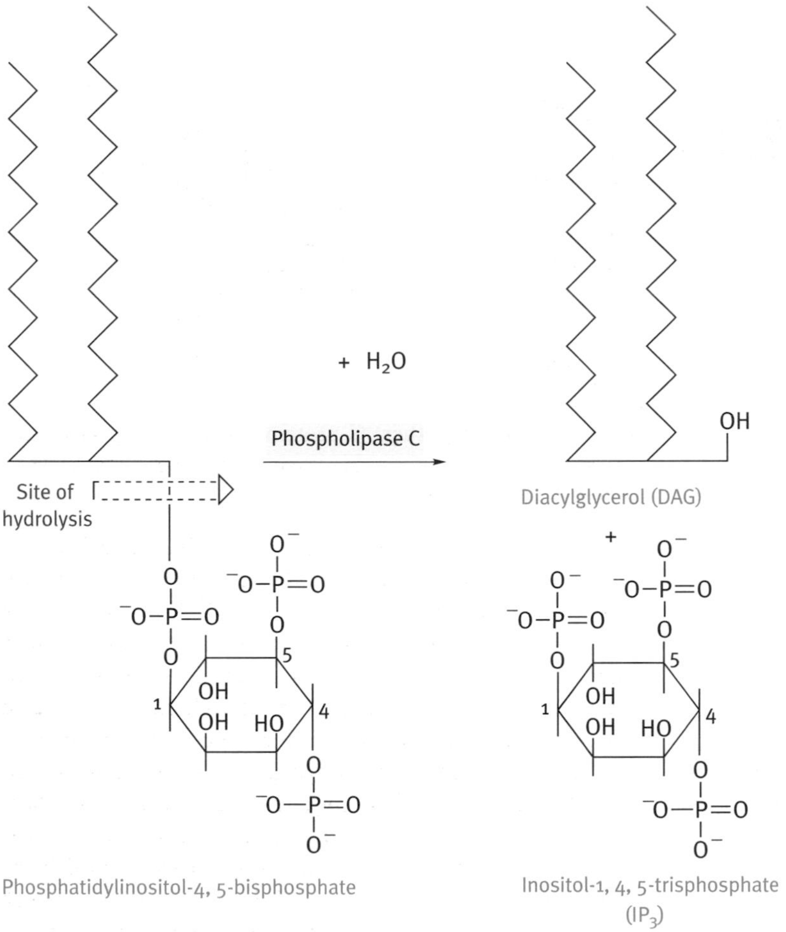

Site of hydrolysis

Phospholipase C

+ H$_2$O

Diacylglycerol (DAG)

Phosphatidylinositol-4, 5-bisphosphate

Inositol-1, 4, 5-trisphosphate (IP$_3$)

Fig. 26.20 Hydrolysis of phosphatidylinositol-4,5-bisphosphate(PIP$_2$) to diacylglycerol (DAG) and inositol trisphosphate(IP$_3$).

with a receptor associated with the phosphatidylinositol cascade results in increases of intracellular DAG and Ca^{2+} (Fig. 26.21) and reversal of the signal when hormone is no longer attached to the receptor.

DAG is the physiological activator of **protein kinase C (PKC)** which is also activated by Ca^{2+}. It is a protein kinase of central importance which has multiple effects by phosphorylating target proteins. As one example, it is involved in phosphorylating transcription factors in the nucleus but also has controlling roles in cell division regulation as is illustrated by the tumour-promoting effect of **phorbol esters**. These are analogues of DAG (have a quick look at Fig. 26.22) capable of activating PKC, leading to cell division. It may seem incongruous that DAG, a naturally occurring molecule, has the same effect in activating PKC as does an agent which is a promoter of tumour formation. The difference is that DAG is rapidly destroyed and activates PKC only when this is required, whereas phorbol esters are longer-lived and deliver an inappropriately prolonged signal.

DAG and Ca^{2+} are both needed for maximal activation of PKC but quite apart from this, Ca^{2+} is an important second messenger on its own.

Other control roles of calcium

Ca^{2+} ions are extremely important in control of a wide variety of cellular processes. Given that there are large amounts of calcium in the body in the bones and the blood, it seems remarkable at first sight that this should be so. The general strategy is to keep the cytoplasm of cells very low in Ca^{2+} concentration by Ca^{2+}/ATPases which pump it either to the outside, the mitochondria, the ER lumen, or in the case of skeletal muscle into a special sac called the **sarcoplasmic reticulum** (page 530). Appropriate signals open ligand-gated Ca^{2+} channels (page 83)

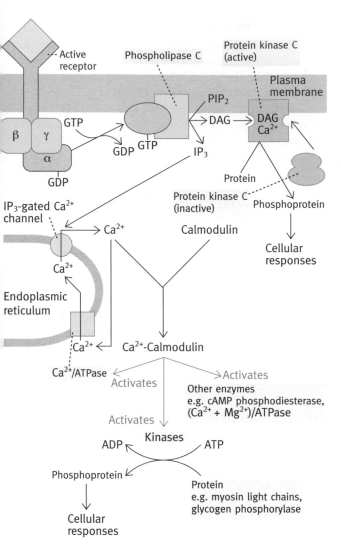

Fig. 26.21 The phosphatidylinositol cascade: interactions of DAG, IP_3, and Ca^{2+} as second messengers. Note that the G_α protein binds to phospholipase C (PLC) only in the GTP-complexed form. On hydrolysis to GDP the process reverses. This illustrates the versatility of G-protein-associated receptors. Different receptors are associated with different G proteins whose α subunits, when in the GTP form, control the activities of different enzymes—in this case G_α activates phospholipase C. (Compare with Fig. 26.18, where the G protein α subunit activates adenylate cyclase.)

which release the ion back into the cytoplasm where it acts as a regulator. The system is very efficient, because the steep concentration gradient across the relevant membranes means that there is an instant delivery of Ca^{2+} to the cytoplasm.

A general second messenger role of the ion involves combination with a widely distributed protein called **calmodulin**. This has four sites for binding Ca^{2+} with high affinity, causing a

Fig. 26.22 Phorbol esters are analogues of diacylglycerol, the natural activator of PKC (The complete structure of the phorbol ester is given only to illustrate this point.)

conformational change in the protein. Calmodulin is sometimes found in association with the enzymes it controls, or it may be free and attach to enzymes in its Ca^{2+} bound form. The Ca^{2+} causes a conformational change in the calmodulin which alters the activity of the enzyme it is associated with. The important point is that a number of calmodulin-Ca^{2+} activated protein kinases exist and Ca^{2+} can therefore, via this route, exert multiple cellular effects. The target proteins of the calmodulin-activated kinases include glycogen phosphorylase, myosin light chains (page 525); the full list is much longer than this. The uptake of calcium by the intestine is controlled by calcitriol, a molecule related to vitamin D. It binds to receptors which activate genes involved in the absorption of calcium from the intestine.

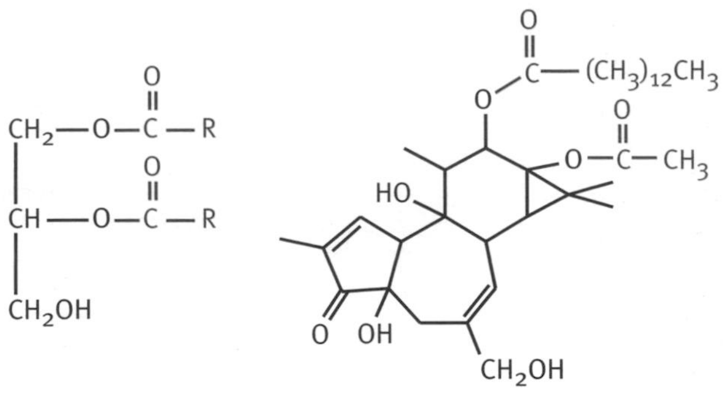

Vision: a process dependent on a G-protein-coupled receptor

The versatility of G protein signalling pathways using different specific G proteins is shown by this example, in which light is the signal. The basic problem of vision is to convert the stimulus of light photons into chemical changes which result in impulses in the optic nerve carrying signals to the brain.

The retina of vertebrates has two types of cells for light detection: **rods** for black and white and dim light vision, and **cones** for colour vision. The rod cell (which is the one that we'll discuss) has three sections. The middle part has the mitochondria, nucleus, etc. One end of the cell makes a synapse with a bipolar cell which connects with the optic nerve. At the other end is a cylindrical rod-shaped section in which there is a stack of membranous discs (as many as 2000) embedded in the cytoplasm (Fig. 26.23). These contain the light-detection machinery.

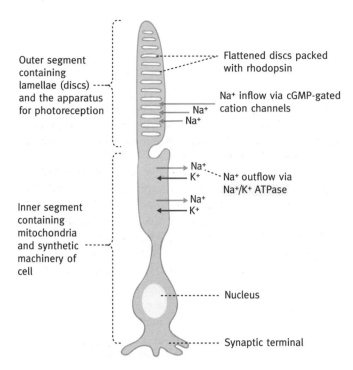

Fig. 26.23 Structure of a rod cell.

Transduction of the light signal

The process of light detection is complex in detail; here we deal only with the essential principles. In the dark, the cone cells have a relatively high level of cGMP synthesized by a guanylate cyclase (see Fig. 26.27). (Note that, in this instance, cGMP is not a second messenger, since it is not produced as a result of receptor activation.) In the cell membrane there are ligand-gated Na^+ channels which are kept open by cGMP (see page 83 for ligand-gated channels).

The constant inflow of Na^+ in the dark means that the cell membrane has a lower potential across it—the equilibrium established between the inflow and pumping out of Na^+ results in a potential of $-30\,mV$ (quite different from the situation in a neuron where, at rest, Na^+ channels are closed and a membrane potential of $-70\,mV$ exists; see page 86 if you want to refresh your memory on membrane potentials).

The light receptor in the discs is rhodopsin, a complex of the protein opsin and 11-*cis*-retinal. As with other G-protein-associated receptors it has seven transmembrane helices. Light causes activation of rhodopsin due to a conformational change in the visual pigment to become all-*trans*-retinal. (Have a quick look at the structures in Fig. 26.24 and the relationship of the retinal molecule to vitamin A and β-carotene.)

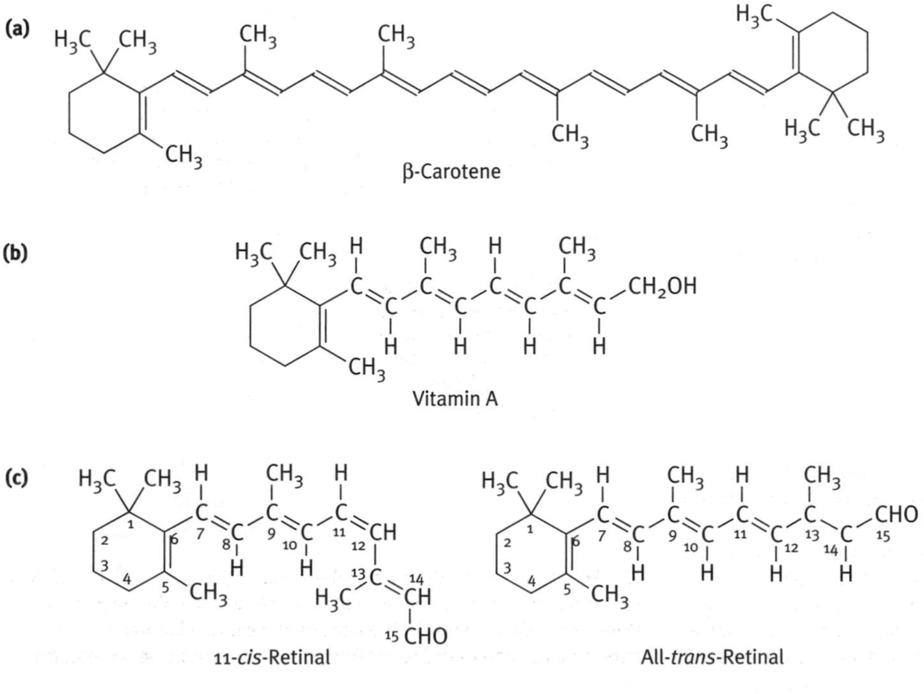

(a)

β-Carotene

(b)

Vitamin A

(c)

11-*cis*-Retinal   All-*trans*-Retinal

Fig. 26.24 (a) Structure of β-carotene, which is the principal accessory photosynthetic pigment in plants and a precursor of vitamin A in animals; (b) structure of vitamin A; (c) structures of 11-cis-retinal and all-*trans*-retinal. These structures are given for reference purposes only.

When this happens there is a conformational change in the cytoplasmic domain of the receptor protein, rhodopsin. This causes a heterotrimeric G protein called **transducin** to exchange its bound GDP for GTP. By now you are familiar with what will happen—the α-GTP subunit detaches and activates a membrane bound enzyme, in this case **cGMP-phosphodiesterase** which destroys cGMP by hydrolysis (Fig. 26.25). This enzyme has a similar action to the cAMP phosphodiesterase shown earlier in Fig. 26.17(b). Lowering the cGMP level results in closure of the cation channels, the inflow of Na^+ ceases and the membrane potential increases to $-70\,mV$. This causes a nerve impulse in the optic nerve to flow to the visual centre in the brain.

The recovery of the cell after illumination is complex. A primary event is the inactivation of the α GTP subunit by its GTPase activity which results in the reassembly of the heterotrimeric G protein, transducin. Recovery of the cell involves inactivation of the activated rhodopsin, lowering of the Ca^{2+} level in the cell which stimulates cGMP synthesis, and recycling of the rhodopsin by a complex pathway.

To summarize the whole process (Fig. 26.26), in light, cGMP levels fall, cation channels close, hyperpolarization of cell membrane leads to visual signal in the optic nerve; after illumination, cGMP levels are restored, Na^+ channels open, and the system is restored to be ready for the next photon.

Signal transduction pathways using cGMP as second messenger

We now leave G-protein-coupled receptors and turn to two different signalling systems, which have in common only that they have cyclic GMP (cGMP) as a second messenger. It is convenient to deal with them together.

The visual process as described above involves cGMP. As stated, it is not there acting as a second messenger. cGMP is analogous to cAMP and can be formed from GTP by two different guanylate cyclase enzymes in the reaction shown in Fig. 26.27. In one of these, the hormone combines with a membrane receptor whose *inner domain is a guanylate cyclase* and this is activated by a conformational change resulting from the signal

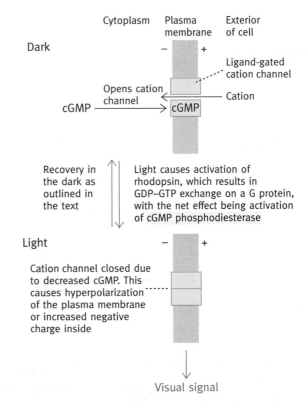

Fig. 26.26 Simplified diagram of the visual process. The action of light on rhodopsin results in the activation of cGMP phosphodiesterase. Decreased cGMP results in the closure of a cation channel, hyperpolarization of the cell membrane, and an optic nerve impulse.

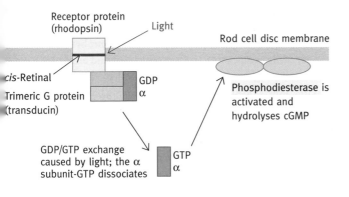

Fig. 26.25 The G-protein-coupled receptor involved in vision. The receptor consists of seven transmembrane helices with 11-*cis*-retinal as the chromophore. The cytoplasmic domain on light stimulation causes the trimeric G protein transducin to exchange its GDP for GTP. The α-GTP subunit migrates to the membrane and activates a phosphodiesterase which hydrolyses cGMP. This results in closure of cation channels and cessation of Na^+ inrush. The consequent increase in membrane polarization is converted into a signal in the optic nerve.

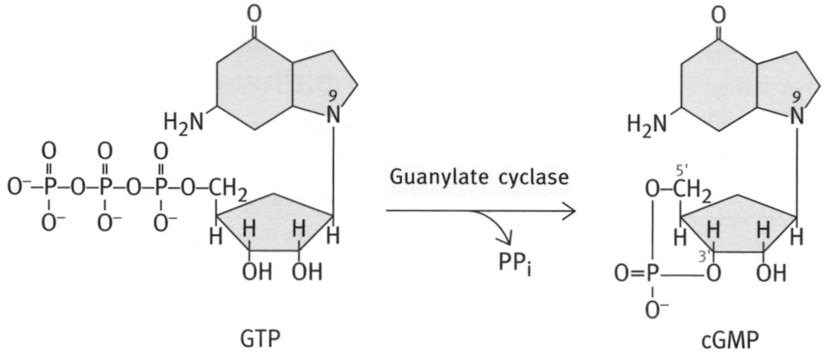

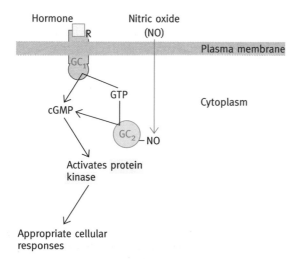

Fig. 26.27 Formation of 3′, 5′-cyclic GMP by guanylate cyclase.

GTP

cGMP

Hormone

Nitric oxide (NO)

R

Plasma membrane

GC$_1$

GTP

Cytoplasm

cGMP

GC$_2$ — NO

Activates protein kinase

Appropriate cellular responses

Fig. 26.28 Production of the second messenger cGMP by two routes. In one a membrane receptor, activated by the atrial natriuretic peptide, has a single polypeptide chain with a membrane-spanning sequence and a guanylate cyclase catalytic site on the cytoplasmic face. In the other, nitric oxide directly activates a cytoplasmic guanylate cyclase which is the hormone receptor. (Note that cGMP has a separate function in the visual process, as described on page 442.) GC$_1$, plasma membrane guanylate cyclase; GC$_2$, cytoplasmic guanylate cyclase, activated by nitric oxide (NO); R, receptor for atrial natriuretic peptide.

molecule attachment (Fig. 26.28). The raised cGMP level mediates cell responses by activating specific protein kinases. Hydrolysis of cGMP by a phosphodiesterase reaction analogous to that for cAMP (Fig. 26.17(b)) confers reversibility. An example in which this type of control is the atrial naturetic peptide, produced by endothelial cells, which stimulates the kidney to secrete Na$^+$. The effects of cGMP appear to be more specialized than those of cAMP. They include relaxation of smooth muscle and effects on nerve cells.

Activation of a guanylate cyclase by nitric oxide

There is a second, quite different, guanylate cyclase present in the cytosol (Fig. 26.28). This has a heme molecule as its prosthetic group to which binds a cell-cell signalling molecule of surprising simplicity—**nitric oxide (NO)**. The heme molecule is, in effect, functioning as a sensitive detector of NO and transduces the signal into activation of its attached enzyme. NO is produced from the arginine guanidino group by an enzyme called **nitric oxide synthase**, in cells lining parts of the vascular system. The NO diffuses into the smooth muscle of blood vessels, causing cGMP production which, in turn, causes muscle relaxation and vessel dilatation. Production of NO is also subject to neural control and also is produced in response to a shearing force exerted by blood flow on the endothelial cells lining the vessels, thus resulting in vasodilation and reduction of shearing force. Since NO is oxidized to NO$_2$ and NO$_3$ in seconds, it is a locally acting (paracrine) hormone; being lipid soluble it easily escapes from cells producing it and enters adjacent cells. Trinitroglycerine, a drug long used in the treatment of angina, slowly produces NO thereby relaxing blood vessels, thus reducing the workload of the heart muscle. NO is part of a complex regulatory system with multiple physiological effects. The phosphodiesterase which destroys cGMP is inhibited by the drug sildenafil (Viagra) and thereby potentiates the effect of NO, production of the latter being increased by sexual stimulation. Dilatation of blood vessels causes erection.

Figure 26.29 gives a simplified overview summary of the signalling pathways dealt with on this chapter with emphasis on the protein kinases and second messengers involved.

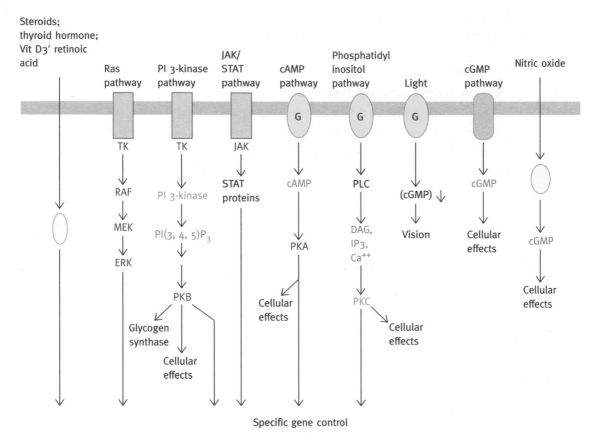

Fig. 26.29 Simplified summary diagram of the signal tranduction pathways. The figure shows only the kinases (red) which are activated by the receptors and second messengers formed (green). The IP$_3$ pathway can be oncogenic but details of the gene control pathway are uncertain. Yellow receptors are intracellular. Receptors shown in red are phosphorylated on binding of the signal. (TK, tyrosine kinase; JAK, Janus kinase). G indicates that the receptor is of the heterotrimeric protein-coupled type. The type of receptor shown in orange is unusual in that the guanylate cyclase which forms the second messenger is part of the receptor itself. Note that the pathways are of extraordinary complexity with multiple parallel pathways. How the the individual pathways all add up to coordinated cellular control is not understood. Note also that the light receptor is an exception in that its effect is to reduce the level of cGMP which is not a second messenger.

FURTHER READING

Signal transduction

Struhl, K. (1998). Fundamentally different logic of gene regulation in eukaryotes and prokaryotes. *Cell*, **98**, 1–4.

A minireview looking in broad terms at the mechanisms of gene transcription control in the two classes.

Hunter, T. (2000). Signalling—2000 and beyond. *Cell*, **100**, 113–27.

A major review which comprehensively discusses all aspects of cell signalling in clearly defined sections. Probably more suitable for the advanced student.

G proteins

Neer, E. J. (1997). Intracellular signalling: turning down G-protein signals. *Current Biology*, **7**, R 31–3.

Reviews the family of regulators of G protein signalling (RGS) which modulate the GTPase activity of the a subunits of G proteins.

Receptors

Pawson, T. (1994). Look at a tyrosine kinase. *Nature*, **372**, 726–7.

A news and views item on the three dimensional structure of the insulin receptor.

Heldin, C.-H. (1995). Dimerisation of cell surface receptors in signal transduction. *Cell*, **80**, 213–23.

Dimerisation or oligomerisation of receptors is the initial response to many growth factors and cytokines. This article reviews the important field.

Pitcher, J. A., Freedman, N. J., and Lefkowitz, R. J. (1998). G protein-coupled receptor kinases. *Ann. Rev. Biochem.*, **67**, 653–92.

A comprehensive review of the densensitisation (downregulation) of receptors. More suitable for advanced students with a special interest in the subject.

Nitric oxide

Snyder, S. H. and Bredt, D. S. (1992). Biological roles of nitric oxide. *Sci. Amer.*, **266**(5), 28–35.

Excellently illustrated article on the roles and mechanism of action of nitric oxide.

Bredt, D. S. and Snyder, S. H. (1994). Nitric oxide: a physiologic messenger molecule. *Ann. Rev. Biochem.*, **63**, 175–95.

Readable general review, but particularly valuable for the survey of the several roles of nitric oxide in the whole body.

Mayer, B. and Hemmens, B. (1997). Biosynthesis and action of nitric oxide in mammalian cells. *Trends Biochem. Sci.*, **22**, 477–81.

A succinct review of this fascinating signalling molecule.

Signalling pathways

Nishizuka, Y. (1994). Protein kinase C and lipid signalling for sustained cellular responses. *FASEB J.*, **9**, 484–96.

Excellent summary of the signalling pathways using inositol phospholipid hydrolysis, and the degradation of various other membrane lipid constituents.

Karin, M. and Hunter, T. (1995). Transcriptional control by protein phosphorylation: signal transmission from the cell surface to the nucleus. *Current Biology*, 5, 747–57.

Well illustrated comparison of the Ras pathway involving translocation of activated protein kinases from the cytoplasm to the nucleus and the JAK kinase pathway involving translocation of activated transcription factors into the nucleus.

Quilliam, L. A., Khosravi-Far, R., Huff, S. Y., and Der, C. J. (1995). Guanine nucleotide exchange factors: activators of the Ras superfamily of proteins. *BioEssays*, **17**, 395–404.

A fairly advanced, but readable review of the control of Ras activity.

Burgering, B. M. T. and Bos, J. L. (1995). Regulation of Ras-mediated signalling: more than one way to skin a cat. *Trends Biochem. Sci.*, **22**, 18–22.

Extensively reviews interaction of the Ras pathway with other signalling routes.

Ihle, J. N. (1996). STATs and MAPKs: obligate or opportunistic partners in signalling. *BioEssays*, **18**, 95–8.

A speculative discussion on the possible cooperation of the Ras and STAT signalling pathways.

Elion, E. A. (1998). Routing MAP kinase cascades. *Science*, **281**, 1625–6.

A 'perspectives' article briefly and clearly summarizing scaffold proteins. See also the two similar following articles (pages 1668 and 1671) by H. J. Schaeffer *et al.* and A. J. Whitmarsh *et al.* respectively.

Hoey, T. and Schindler, U. (1998). STAT structure and function in signalling. *Current Opinion in Genetics and Development*, **8**, 582–7.

Reviews the area with emphasis on structural aspects and includes STAT deficient mouse phenotypes.

Whitmarsh, A. J. and Davis, R. J. (1998). Structural organisation of MAP kinase signalling modules by scaffold proteins in yeast and mammals. *Trends Biochem. Sci.*, **23**, 481–5.

Reviews the role of scaffold proteins in the organization of different MAP kinase pathways.

Evans, D. R. H. and Hemmings, B. A. (1998). What goes up must come down. *Nature*. **394**, 23–4.

A News and Views article dealing with the physical association of MAP kinases and phosphatases and the switch-off of signalling pathways.

Sceffzek, K., Ahmadian, M. R., and Wittinghofer, A. (1998). GTPase-activating proteins: helping hands to complement an active site. *Trends Biochem. Sci.*, **23**, 257–62.

Reviews the GAPs or proteins which regulate the GTPase activities of small GTP-binding proteins involved in signal transduction.

Vanhaesebroek, B., Leevers, S. J., Panoyotou, G., and Waterfield, M. D. (1998). Phosphoinositide 3-kinases: a conserved family of signal transducers. *Trends Biochem. Sci.*, **22**, 267.

Describes several families of PI 3-kinases and their roles in signalling payhways.

Hafen, E. (1998). Kinases and phosphatases—a marriage is consumated. *Science*, **280**, 1212–13.

Deals with the switch-off of MAP kinases in signalling pathways. The 'marriage' is the physical association of kinases and phosphatases.

Krugman, S. and Welch, H. (1998). PI 3-kinase. *Curent Biology*, **8**, R828.

A single-page quick guide giving salient features of the signalling pathways activated by this enzyme.

Alessi, D. R. and Cohen, P. (1998). Mechanism of action and function of protein kinase B. *Current Opinion in Genetics and Development*, **8**, 55–62.

Deals with the activation of the kinase and its role in insulin signalling.

Cohen P. (1999). The Croonian Lecture 1998. Identification of a protein kinase cascade of major importance in insulin signal transduction. *Phil. Trans. Series B, Roy. Soc. Lond.*, **354**, 485–95.

A readable review of this work.

Williams, J. G. (1999). Serpentine receptors and STAT activation: more than one way to twin a STAT. *Trends Biochem. Sci.*, **24**, 333–4.

A concise summary of the way in which membrane receptors activate STAT proteins.

Hemopoietic growth factors

Metcalf, D. (1991). The 1991 Florey Lecture. The colony-stimulating factors: discovery to clinical use. *Phil. Trans. Series B, Roy. Soc., Lond.*, **333**, 147–73.

An in depth review by the discoverer of these important proteins. While there is more research level material than most students would want, it provides an excellent general review of the field, and a fascinating account of the progression from unidentified factors to their widespread clinical use.

Interferons

Ihle, J. N., Witthuhn, B. A., Quelle, F. W., Yamamoto, K., Thierfelder, W. E., Kreider, B., and Silvennoinen, O. (1994). Signalling by the cytokine receptor superfamily: JAKs and STATs. *Trends Biochem. Sci.*, **19**, 222–7.

Comprehensive review on the signalling pathway from cell membrane to nucleus for interferons, interleukins, colony stimulating factors, erythropoietin, and growth hormone.

Stark, G. R., Kerr, I. M., Williams, B. R. G., Silverman, R. H., and Schrieber, R. D. (1998). How cells respond to interferons. *Ann. Rev. Biochem.*, 67, 227–64.

A comprehensive review.

Calcium homeostasis

Bootman, M. D. and Berridge, M. J. (1995). The elemental principles of calcium signalling. *Cell*, **83**, 675–8.

Minireview on the sources of Ca^{2+} for generating signals: Ca^{2+} release from intracellular stores and Ca^{2+} entry across the plasma membrane.

Monteith, G. R. and Roufogalis, B. D. (1995). The plasma membrane calcium pump—a physiological perspective on its regulation. *Cell Calcium*, **18**, 459–70.

The role of the plasma membrane Ca^{2+}/Mg^{2+}-dependent ATPase in cellular signalling and in intracellular calcium homeostasis.

Berridge, M. A., Bootman, M. D., and Lipp, P. (1998). Calcium—a life and death signal. *Nature*, **395**, 645–7.

A News and Views feature which puts into perspective the role of calcium in its universal signalling roles.

PROBLEMS FOR CHAPTER 26

1 How is cAMP production controlled? In particular, describe the role of GTP in the process. What is the relevance of the latter to cholera?

2 How does cAMP exert its effect as second messenger?

3 How does nitric oxide exert its controlling effect?

4 cAMP and cGMP are not the only second messengers. Describe another system.

5 Make a general comparison of the activation mechanism of three types of receptors: (a) the epinephrine receptor activating adenylate cyclase; (b) the EGF receptor of the Ras pathway; and (c) the interferon receptor.

6 What are the classes of signalling molecules between cells in the mammalian body?

7 Lipid-soluble signalling molecules directly enter cells but lipid-insoluble ones do not. Does this mean that the two types act in totally different ways? Explain your answer.

8 If a protein is found to have an SH2 domain, what is its likely function in the cell?

9 In a general way, without detail, compare the salient features of a Ras signalling pathway with a JAK/STAT pathway.

10   G-protein-associated receptors are tremendously versatile in the signals the different receptors respond to. Explain the principle of G protein signalling and why it can be so versatile.

11   Can cGMP be properly regarded as a second messenger in the visual system? What about the nitric oxide signalling system?

12   Most tyrosine kinase-associated receptors dimerize on receipt of a signal. What is the exception to this?

Chapter summary

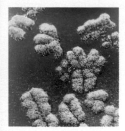

The immune system

The body is constantly under the threat of invasion by pathogenic organisms such as bacteria and viruses. The immune system is the main protection against this. Its absolute necessity is illustrated by the results of an impaired immune system due to genetic defects such as adenosine deaminase deficiency (page 294) or to infection by the HIV (the AIDS virus).

The immune system responds to macromolecules of several types. Proteins are the most important class and we will refer to these throughout this chapter. Carbohydrate attachments to proteins are also often very antigenic, as are some polysaccharides. The immune system does not protect against small foreign molecules such as drugs which enter the body; other chemical protection systems deal with these, as described in Chapter eighteen. A foreign macromolecule in the body is a warning signal that a defensive response may be needed against an invader. People can be allergic to small molecules such as penicillin, the allergy being initiated by an immune response. However, the response is not to penicillin *per se* but rather to penicillin combining with a protein of the body and in doing so making that protein 'look' foreign to the immune system.

You may recall from Chapter five that one of the major problems of the digestive system is how to digest food, the components of which are essentially identical to those of the body, without destroying the body itself by autodigestion. An analogous but much more complex problem is present in immune protection. The body has thousands of proteins differing from foreign proteins only in the minute detail of amino acid sequences and yet the distinction between 'self' proteins and foreign proteins must be made accurately because otherwise the immune system would attack the components of the body in **autoimmune reactions** leading to **autoimmune diseases**. An

example of the latter is the disease **myasthenia gravis** in which an autoimmune reaction destroys the acetylcholine receptors of muscle (page 531), thus preventing nervous stimulation of contraction. Insulin-dependent diabetes is the result of an autoimmune attack on insulin-producing cells of the pancreas. The avoidance of autoimmune attack is of over-riding importance, and much of the elaboration of immune systems is because of this requirement. The protection cannot be absolutely perfect, as evidenced by the existence of autoimmune diseases; there are safeguards which are effective in most cases, but to be certain that there is never a response to any self protein would probably require a system that responded to nothing at all, which obviously would defeat its purpose.

The cells involved in the immune system

All blood cells have their origin in bone marrow stem cells which continually divide (Fig. 27.1), but the principle players of the immune system are **B** cells and **T** cells together with special phagocytic cells known as **antigen-presenting cells (APCs)**. The B and T cells are known as **lymphocytes** (explained below). Although B and T cells both originate in the bone marrow there is a major difference; B cells multiply and undergo their primary maturation in the bone marrow, but those cells committed to becoming T cells migrate to the thymus gland, a small structure behind the breastbone, where they multiply and mature—hence the name T (for thymus) cell.

B cells are responsible for producing antibodies, but only after a very elaborate preparation to be described. To do this they require the cooperation of a class of T cells known as **helper T cells**—they help the B cells to do their job. Two quite separate classes of T cells are generated in the thymus—helper T

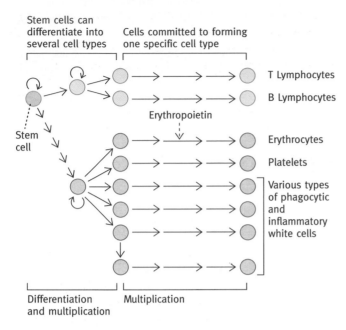

Fig. 27.1 Simplified diagram of hemopoiesis. At each arrow cell multiplication can occur and this may be specifically controlled by various protein cytokines—the colony-stimulating factors, interleukins, and erythropoietin. For example, the latter stimulates erythrocyte production at the step indicated. Differentiation of stem cells into committed cells is likewise controlled. Note that T cells orginate in the bone marrow, then migrate to the thymus gland where they multiply and differentiate.

cells and **cytotoxic or killer T cells**. This leads us to the next important point:

There are two immune protective systems

First, there is the production of **antibodies**. These are soluble proteins which combine with the foreign **antigens**. An antigen is a macromolecule, normally foreign to the body, which leads to anti**body gen**eration. The attachment of the antibody to antigen results in protective events described later. This type of immunity is called **humoral immunity**, for in times past, bodily fluids were referred to as humors, and in this mechanism protection is due to soluble antibodies in body fluids.

The second type of immunity is known as **cell-mediated immunity**. In this, the **cytotoxic T cells** or **killer T cells** (the names are interchangeable) recognize abnormal cells in the body. For instance, the abnormality might result from a virus infection in the cell. The killer cell attaches to the virus-infected cell and destroys it. This is an effective way of stopping a virus infection from spreading, since cell destruction aborts virus replication

inside the cell. Direct contact is made between the killer cell and its target—hence the term cell-mediated immunity. Thus the two mechanisms complement each other—to use the virus infection as the example, the humoral system is part of the defence against virus in the blood or in the mucous membrane before it infects a cell; the cell-mediated immune system destroys the host cell after the latter has become infected and in doing so aborts virus replication in that cell.

Where is the immune system located in the body?

There is no single organ for the immune system. Instead there are vast numbers of separate cells which, *en masse*, would be equivalent to a large organ. They are distributed in spleen, lymph nodes, intestinal and mucosal lymphoid tissue, and about 10% in blood and lymph circulation. Lymph is a 'filtrate' that leaks out of the blood and bathes cells to provide them with nutrients and to remove waste products. It is drained into thin-walled lymphatic vessels which return the lymph to the blood via two main ducts. The lymphocytes are able to squeeze through certain capillary walls and so can migrate from blood to lymph. The lymph is propelled by the movements of the body. On the route of the lymph returning to the blood, the lymphatic vessels expand at places into **lymph nodes**, known as **secondary lymphoid centres** (bone marrow and thymus being the first), in which high concentrations of B and T cells are found. The nodes are where cells can undergo the interactions and the differentiation processes that occur when an antigen response is being mounted. They are strategically placed in the armpit, groin, tonsils, adenoids, and intestine so that lymphocytes and foreign antigens draining from the tissues are likely to encounter nodes. The nodes are in this sense monitoring for infection in the tissues by acting as a filter.

With that introduction we will deal with immunity under two main headings—humoral or antibody-based immune protection and then cell-mediated immunity.

Antibody-based or humoral immunity

Structure of antibodies

Antibodies are **immunoglobulins (Igs)**—the term 'globulin' is from an early protein classification system. There are several different immunoglobulins. **IgG** is the one produced in largest amounts as an immune response develops. We'll look at its structure first. It is a Y-shaped protein made up of two identical **light (L) polypeptide** chains and two identical **heavy (H) chains** held together by disulfide bonds, as shown in Fig. 27.2. The ends

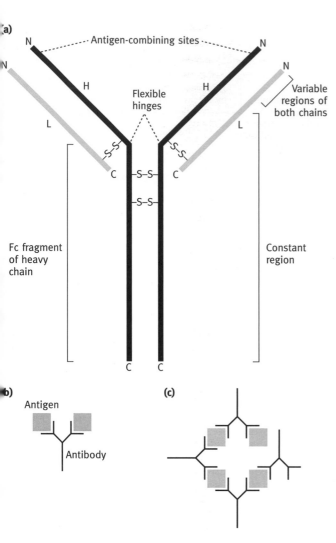

Fig. 27.2 (a) An IgG molecule. IgM and IgA molecules differ in the Fc fraction. but are similar in the variable part. H, heavy chain; L, light chain. N and C refer to the N-terminal and C-terminal ends of the polypeptides, respectively. The Fc fragment is the C-terminal half of the two H chains, bound covalently into a single fragment by —S—S— bonds. The name arises from crystallizable (c) fragment (F)—one of the products of papain hydrolysis of the antibody molecule at the two peptide bonds at the flexible hinges. (b) Cross-linking of antigen with single specific epitopes. (c) Antigens with multiple antigenic determinants form a crosslinked insoluble cluster with antibody.

of the Y arms have variable regions on both the heavy and light chains. It is these variable regions of the two chains that form the antigen-binding site on each of the Y arms. The binding to the antigen is by noncovalent bonds. An antibody thus has two identical antigen-binding sites, which means that it can cross-link antigen molecules. At the fork of the Y there is a flexible 'hinge' region, which increases the cross-linking ability of the molecule Although foreign molecules need to be large to produce an immune response, a given antibody binds to only a small part of the antigen-in the case of a protein antigen only a few amino acids. The specific part of the antigen which is recognized by an antibody is called an **epitope**. Thus a protein antigen provokes the production of a number of different antibodies, each combining with a specific epitope and each produced by a different **clone** of B cells (a clone is a collection of identical cells arising from the multiplication of a single cell).

What are the functions of antibodies?

An IgG antibody molecule is bivalent—the two identical binding sites can each bind to the same epitope on separate antigen molecules. If the antigen has more than two identical epitopes, a single antibody will form a large antibody—antigen complex. The serum of an immunized animal contains multiple antibodies recognizing different epitopes on a given antigen so that a network forms between antigens and antibodies. Such a network activates the **complement system**, which is the name of a group of proteins in blood; antibody—antigen complexes, when assembled on a bacterial cell, attract complement proteins that destroy the bacterial cell. The antibody-coated bacterium or other antigen is engulfed by **phagocytes** which digest the foreign cell.

What are the different classes of antibodies?

There are several classes of immunoglobulins differing in the **constant regions** of their H chains. These differences have nothing to do with the antigen-binding sites but are related to the physiological role of the antibody.

When the body responds to an antigen, the antibody produced first is **IgM, a** multi-subunit, or polymeric, form of antibody with 10 antigen combining sites. It is particularly efficient at binding viruses and bacteria into a network, because of its multiple combining sites, and in activating complement and promoting phagocytosis. More prolonged challenge by the same antigen results in production of **IgG. IgA** is of importance as a first line of defence. It is transported through epithelial cell membranes, by combination with a special secretory polypeptide, into the mucous layer of the intestine and respiratory tract. For example, it prevents the cholera bacillus infecting intestinal cells in humans previously exposed to the disease.

Generation of antibody diversity

The body is potentially exposed to vast numbers of different antigens and it is potentially capable of producing a

corresponding number of different antibodies which combine with these. The specific antibodies differ in the precise amino acid sequences of their binding sites. This is achieved by the L and H chains at their ends being individually variable in their amino acid sequences in different cells so that vast numbers of different combining sites are possible (one specific type per cell). The principle is that the immune system produces B cells differing in their genes coding for antibodies. Each newly developed B cell can produce, in terms of antigen specificity, only a single antibody species, but each cell, as released from the bone marrow, produces a different one. Given the vast numbers involved it is almost certain that, whatever antigen comes along, there will be an antibody which by sheer chance combines with it.

However, human cells contain probably somewhere around 100 000 genes to code for all the proteins of the body, whereas the body has the potential of making about 10^{12} different immunoglobulins, each of course requiring a different gene to code for it. These are needed to cope with the almost limitless number of foreign antigens a human may encounter. Clearly, something special must arrange for the enormous diversity of immunoglobulin genes.

To understand how this generation of diversity is achieved, let us look first at the L chain of immunoglobulin molecules. As outlined, it has two sections—a terminal-variable region which participates in the antigen-binding site and a constant region which is identical in all immunoglobulin molecules. Consider a stem cell in the bone marrow *before* it has become committed to becoming a B cell. At this stage it has not assembled its gene for immunoglobulin L chain; the relevant DNA consists of separate gene sections in the chromosome. There is a section coding for the constant (C) domain of the L chain; there are four separate sections each of which codes for different short peptide sequences used to join the constant to the variable gene section called **joining (J) sections**. And, finally, there are about 300 gene sections (V) each of which codes for an end of an L chain. The 300 **V sections** are all different from one another, as are the four J sections, and they will thus code for different peptide sequences in each case. When the bone marrow stem cell is committed to becoming a B lymphocyte, a rearrangement of these DNA sections occurs so that a functional L chain gene is produced by site-specific or somatic recombination (explained below). In the assembly of the gene, one C section is joined by one of the four J sections to one of the 300 V sections. This results in a new composite gene in each of the B lymphocytes, as shown in Fig. 27.3. *The DNA recombination is a completely random process.* Thus, in the final mRNA coding for an IgG L chain molecule, any one of 300 V sections is joined to any one of the four J sections giving 1200 different combina-

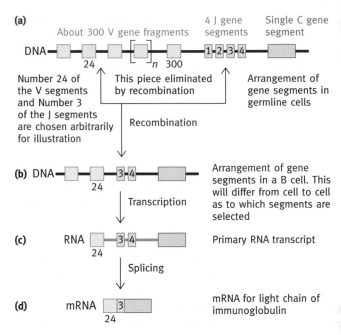

Fig. 27.3 The process of rearrangement leading to a functional L chain immunoglobulin gene: (a) Arrangement of gene segments in the stem cell where the immunoglobulin genes are not being expressed. (b) The randomly chosen V gene (*V24* in this example) is moved next to one of the J genes (*J3* in this example), the intervening DNA being excised. (c) Transcription begins at the *V24* gene segment, the J genes (*J3* and *J4*) also being transcribed. (d) After RNA splicing to remove transcripts of *J4* and introns, those corresponding to *V24, J3,* and *C* now make up the mRNA. Allelic exclusion ensures that only one of the pair of alleles becomes a functional immunoglobulin gene (see text) so that a given cell produces only one immunoglobulin, not two.

tions. The recombination, known as somatic, or site-specific, recombination, is not of the type involved in homologous recombination (page 336) in which two chromosomes exchange sections but involves excision of a piece of DNA and rejoining of the remaining sections. The recombination is brought about by two lymphocyte-specific proteins which cleave the DNA at specific sites, thus excising a section of DNA. The double-stranded breaks so created are end-joined by other proteins in an imprecise way creating more diversity. Yet further diversity is created by a lymphocyte-specific enzyme which adds nucleotides to the cut ends before joining. The number of possible L chain variants is increased to about 3000.

The H chain gene is assembled in a similar random manner from variable sections giving large numbers of H chain gene variants. Since the antigen-binding site is made from the combination of the variable parts of the H and L chains, and there are large numbers of different H and L genes, the number of different binding sites coded by the assembled L and H genes is suffi-

cient to account for the immune system's ability to produce, theoretically, as many as 10^{12} different antibodies. In summary, each developing B cell assembles, at random, genes coding for one antibody with its specific antigen-binding site.

Activation of B cells to produce antibodies

This involves a quite elaborate sequence of events. A preview of the essential concept in each step, without going into detail, may help you keep track of what is happening in the subsequent description:

- In the bone marrow, each B cell produces a limited number of its own particular antibody molecules whose specificity is determined by the random gene assembly process described above. These are not released but are fixed in the membrane with the antigen binding site displayed on the outside.

- In the bone marrow, any B cell that binds an antigen (which will be a self component) is eliminated.

- The survivors, which are likely to be specific only for foreign antigens, are released into the circulation.

- If a released B cell now meets and binds its antigen, it multiplies into a clone of cells. These cells cannot produce secreted antibodies until activated by a helper T cell which itself has been activated by meeting the same antigen (which has to be presented to it by an APC).

- After activation by an activated helper T cell, the B cell multiplies and differentiates into an antibody secreting plasma cell.

We will now describe how these events happen.

Deletion of potentially self-reacting B cells in the bone marrow

The immature B cells formed in the bone marrow have between them the potential to produce antibodies to attack just about every macromolecular component in the body. Any B cell whose antibody, if produced, would be against a self component must be eliminated. This is done during primary maturation in the bone marrow. As they mature there, the B cells are exposed to the self-antigens that circulating B cells are likely to meet.

Each immature B cell in the bone marrow produces about 10^5 molecules of its antibody and displays them on its surface (fixed into the membrane) inviting, as it were, recognition by an anti-

gen. It doesn't release the antibody at this stage of its career; in this role, the antibody is a **receptor** or recognition 'aerial'. If, *during primary maturation* of a B cell in the bone marrow, an antigen comes along which combines with its displayed antibody, it is assumed (as it were) that it is a self-component which must be tolerated (not attacked) by the immune system. *At this stage of development* the B cell dies (or is otherwise functionally eliminated). After this selective killing the surviving B cells should respond only to foreign antigens which should be attacked. They are now released into circulation. *The released population of B cells now has a totally different response to meeting their respective antigens. They are activated instead of being killed*; this is logical, since the antigen now encountered is likely to be foreign.

The theory of clonal selection

When the B cell survivors of the killing process in the bone marrow emerge into the circulation, as stated, they include a vast number of cells each with its own randomly designed antibody receptor exposed so that when a foreign antigen comes along a few of them will by sheer chance combine with it. Of course the number of cells happening to so combine will be small out of the vast population; but if there is a threatening infection, the body has to mount a large-scale response requiring large number of B cells specific for that antigen. The answer to this problem is **clonal selection**, a mechanism proposed by Sir Macfarlane Burnet working in Melbourne for which he was awarded the Nobel prize. The principle is simple. An antigen, by binding, selects the B cells carrying a displayed antibody specific for itself and initiates a train of events which leads to a multiplication of those particular cells. A foreign invader (or other antigen), in short, arranges for its own destruction-which has a rather satisfying poetic justice about it. The displayed antibody is, as stated, an antigen receptor and binding activates signalling pathways that ultimately leads to clonal expansion (increase in the numbers of that particular cell). The principle is illustrated in Fig. 27.4.

What is the mechanism by which B cells are activated to secrete antibody?

Each of the B cells in the resultant clone combines with the specific antigen and internalizes it by endocytosis; it is cleaved into short peptides. On the surface of the B cells are **MHC class II proteins** (encoded by the genes of the **major histocompatibility complex** or locus). These have a groove in their surface in which the peptides produced from antigens are displayed externally on the B cell. The peptides are inserted into newly

B cells in circulation each
displaying a different antibody

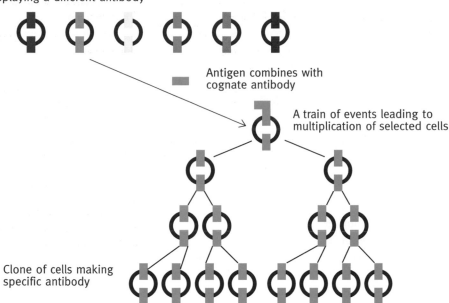

Antigen combines with
cognate antibody

A train of events leading to
multiplication of selected cells

Clone of cells making
specific antibody

Fig. 27.4 The principle of clonal selection. The released population of B cells contains a vast array of cells each of which makes a different antibody which at this stage is displayed as a receptor in the membrane (colored blocks). An antigen binds to its cognate receptor and triggers a train of events leading to multiplication of the selected cell forming a clone. The antigen thus engineers its own destruction by selecting for amplification the clone of cells most suited to attack it. This binding of the antigen is only the initial signal leading to multiplication, which requires a complex set of events to be described. Note that at this stage the antibody is not released but is an integral membrane protein with its binding sites displayed as receptors. See text.

synthesized MHC proteins inside the cell and then displayed on the external surface of the cell membrane as shown in Fig. 27.5.

These B cells are not yet ready to develop into antibody-secreting cells. Helper T cells now come into the picture.

Role of helper T cells

Helper T cells develop in the thymus gland. The T cell has receptors on its surface, somewhat resembling in structure an antibody binding site; these are specific for a single antigen which has to be presented in a special way, as will be described. (Table 27.1 compares the properties of receptors on B and T cells.) There is a mechanism for randomly producing receptors of different specificity (one specificity per cell). Exposure of a T cell to an antigen in the *thymus gland* leads to the death of that cell if it is able to respond to the antigen, the rationale again being that any antigen encountered at that stage will be self (much as happened during the corresponding stage of B cell maturation in the bone marrow). The surviving helper T cells should be specific only for foreign antigens and are released into the circulation.

Activation of helper T cells

The released helper T cells cannot do anything until they are activated by a foreign antigen. For this, a special phagocytic white cell called an antigen-presenting cell (APC) has to engulf

the antigen, hydrolyse it into peptides eight or nine amino acid residues in length and display each in a a membrane-bound MHC class II protein which has a groove for this purpose. When a helper cell combines with this displayed MHC-antigen peptide complex, the APC produces cytokines (Fig. 27.6). These in turn cause the helper T cell to produce a cytokine which acts in an autocrine fashion, stimulating itself to multiply. A clone of *activated* helper T cells results.

Activation of B cells by activated helper T cells

When the activated helper T cell with a receptor specific for a peptide derived from antigen X meets a B cell displaying the same antigen peptide in its MHC proteins, they combine, resulting in cytokine production. These activate signalling pathways that cause the B cell to multiply and differentiate into an antibody-secreting plasma cell (Fig. 27.7).

The interaction of the APC and the helper T cell is facilitated by a CD4 protein on the latter which binds to a constant part of the class II MHC proteins of the former and stabilizes the contact (see Fig. 27.6). The AIDS virus (HIV) uses the CD4 as its receptor to enable it to infect the helper T cell. Helper T cell destruction thus impairs antibody production. (CD proteins are cell surface markers known as clusters of differentiation.) Why is the antibody produced by plasma cells secreted instead of being fixed in the membrane as occurs in the B cells before

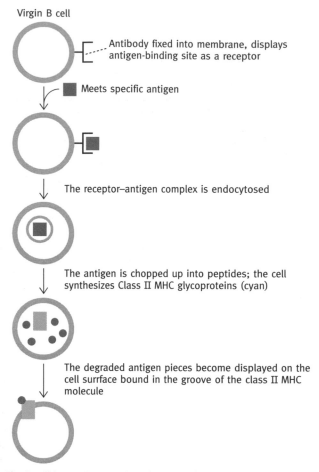

Virgin B cell

Antibody fixed into membrane, displays antigen-binding site as a receptor

Meets specific antigen

The receptor–antigen complex is endocytosed

The antigen is chopped up into peptides; the cell synthesizes Class II MHC glycoproteins (cyan)

The degraded antigen pieces become displayed on the cell surrface bound in the groove of the class II MHC molecule

The B cell is now in an activated state; it does not produce antibodies unless it binds to a helper cell that has been activated by an antigen-presenting cell displaying the same antigen.If this happens, the B cell matures into an antibody-secreting plasma cell and also multiplies due to cytokine release by the helper cell. (This is illustrated in Fig. 27.7.)

Fig. 27.5 Conversion of a virgin B cell to an activated state, but not yet to an antibody-producing plasma cell.

Table 27.1 Comparison of receptors on B and T cells

B Cell receptors	T Cell receptors
Are formed from antibodies displayed in the cell membrane	Special receptor proteins in the membrane
Specific receptors for each cell generated randomly (one cell, one antigen specificity)	Specific receptors for each cell generated randomly (one cell, one antigen specificity)
Bind to antigen molecules in solution. Receptor recognizes the antigen itself	Cannot bind to antigens in solution. Antigen must be internalized by special APC which degrades it and displays peptides in MHC proteins. The receptor recognises the complex of peptide–MHC protein on the APC, not either component alone.

Affinity maturation of antibodies

The B cell(s) were initially selected for clonal expansion by an antigen which combined with the cell's displayed antibody. However, the latter was generated by random variation and the antigen just by chance happened to fit to it. The fit is not likely to be particularly good, so the antibody produced is of low binding efficiency and therefore of low protective effect. There is a process which, during B cell multiplication, causes site-specific mutations that produce yet more variations in the antibody binding sites of different B cells. This is coupled with a selection process of those cells with improved binding affinity for the relevant antigen. Cells with poor binding ability that do not capture an antigen molecule, die. (As the number of antigen molecules falls, owing their removal by phagocytosis, the competition for them increases.) The result is that during an immune response there is a progressive improvement in the antibody quality. The process of **affinity maturation** takes place in special centres (germinal centres) in secondary lymphoid organs.

Memory cells

When B cells are activated and they proliferate, not all of the resultant clone of cells mature into antibody-secreting plasma cells. A proportion become long-lived memory cells which may circulate for years. They are the basis of long-term immunity from a repeat infection. If an appropriate antigen is encountered at a subsequent time, the immunological response is very rapid. T cells are also involved in immunological memory.

activation? The answer lies in differential splicing (page 360) of introns from the immunoglobulin genes. At the 3' end of the immunoglobulin gene are two exons that code for a polypeptide section which fixes the antibody into the membrane. The onset of secretion is caused by a switch in the splicing mechanism which eliminates these two exons from the mRNA. The switch is the result of cytokine signalling. Figure 27.8 summarizes the whole process of antibody production described above.

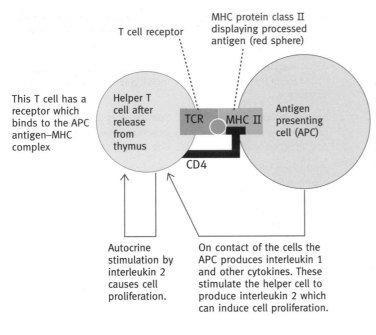

This T cell has a receptor which binds to the APC antigen–MHC complex

Helper T cell after release from thymus

T cell receptor

MHC protein class II displaying processed antigen (red sphere)

TCR   MHC II

CD4

Antigen presenting cell (APC)

This APC has internalized an antigen, processed it into pieces and displayed the latter in its MHC protein, inviting recognition by a helper T cell with the appropriate receptor

Autocrine stimulation by interleukin 2 causes cell proliferation.

On contact of the cells the APC produces interleukin 1 and other cytokines. These stimulate the helper cell to produce interleukin 2 which can induce cell proliferation.

Fig. 27.6 Activation of a helper T cell by an antigen-presenting cell (APC). The autocrine stimulation of the helper T cell allows cell multiplication to occur after separation of the cells. CD4 is a glycoprotein on the T cell that interacts with the MHC II protein of the APC. This interaction is necessary in addition to the binding of the T cell receptor (TCR) with the APC MHC—antigen complex. The CD4 protein is the receptor by which the AIDS virus (HIV) infects the helper T cell.

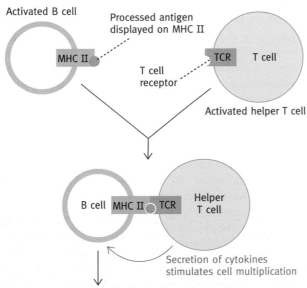

Activated B cell

Processed antigen displayed on MHC II

MHC II

T cell receptor

TCR   T cell

Activated helper T cell

B cell   MHC II   TCR   Helper T cell

Secretion of cytokines stimulates cell multiplication

The B cell, now activated by the helper cell, multiplies and matures into a clone of antibody-secreting B cells known as plasma cells

Fig. 27.7 A helper T cell activating a B cell to become a clone of plasma cells secreting antibody. The cytokines are growth factors that stimulate local cells to divide (paracrine action). On activation of a B cell to become a plasma cell, differential processing of RNA transcripts eliminates the membrane-anchoring section of the antibody.

T cells and cell-mediated immunity

We come now to the second type of immune response involving cytotoxic T cells (killer T cells). B cells and helper T cells are not involved in cell-mediated immunity.

After release from the thymus, the killer T cells must be activated before they can mount an immune response. During their maturation in the thymus, there is the same process of deletion of cells with self-reacting potentiality as happened with helper T cells. The killer T cells, have receptors which recognize antigen peptides on MHC molecules (see Table 27.1) much as do helper T cells. There is a difference between the two in that whereas helper T cells recognize displayed peptides on **class II MHCs**, killer T cells require display on **class I MHCs**. The two classes are very similar but they compartmentalize T cell interactions. All somatic (tissue) cells have class I MHCs. Killer T cells have a CD8 protein which attaches to the MHC to stabilize the association rather than the CD4 site of helper T cells as described earlier. This restricts the attack of cytotoxic cells to cells with MHC I proteins, since the CD8—MHC interaction is essential for effective binding of the cells. B cells are often referred to as CD4 cells and killer cells as CD8 cells.

Activation of cytotoxic T cells

Special APCs are important in this as they were in activation of helper T cells; they display peptides derived from ingested anti-

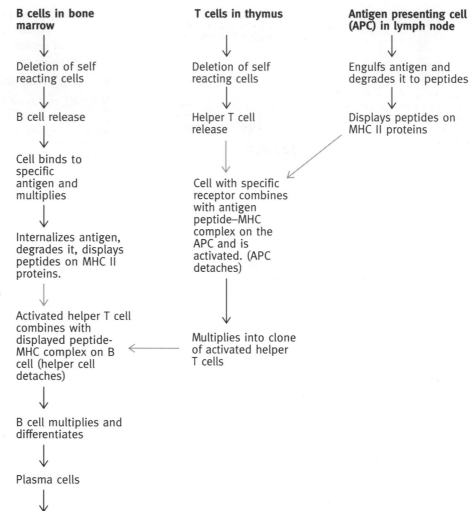

B cells in bone marrow	T cells in thymus	Antigen presenting cell (APC) in lymph node
↓	↓	↓
Deletion of self reacting cells	Deletion of self reacting cells	Engulfs antigen and degrades it to peptides
↓	↓	↓
B cell release	Helper T cell release	Displays peptides on MHC II proteins
↓	↓	
Cell binds to specific antigen and multiplies	Cell with specific receptor combines with antigen peptide–MHC complex on the APC and is activated. (APC detaches)	
↓	↓	
Internalizes antigen, degrades it, displays peptides on MHC II proteins.		
↓	↓	
Activated helper T cell combines with displayed peptide-MHC complex on B cell (helper cell detaches)	Multiplies into clone of activated helper T cells	
↓		
B cell multiplies and differentiates		
↓		
Plasma cells		
↓		
Released antibodies		

Fig. 27.8 Summary scheme of antibody production by B cells. The crucial point to note is that a B cell cannot proceed to release antibodies until it is given the signal to do so by a helper T cell. The helper cell is activated by combining with an APC cell which has engulfed the *same* antigen that was endocytosed by the B cell. The APC displays the antigen peptides on its MHC proteins and this is recognized by the specific helper T cell receptor. The helper T cell thus activated has to see the identical MHC—peptide complex on the B cell. You will notice that in this sense the B cell is acting as an antigen presenting cell. Cytokines are liberated at activation steps and play an essential part in the process (see Fig. 27.6). The red arrows are to draw attention to cell—cell interactions.

gens on their surface. When a killer T cell with a cognate receptor binds to the displayed MHC—peptide complex on the APC, cytokines are released. The T cells multiply so that there is a now clone of activated killer T cells roaming the body looking for trouble.

Mechanism of action of cytotoxic cells

The trouble they are looking for can be illustrated using a viral-infected cell as an example. In the cytoplasm of the cell there are viral proteins synthesized by the cell in response to the infection. Proteasomes in the cell (see page 271) continually 'sample' the cell proteins and break them down into short peptides which become displayed on the outside of the cell, as shown in Fig. 27.9. (The peptides are inserted into the groove of newly synthesized MHC proteins inside the cell and then the complex is transported to the membrane.) The cell is now 'labelled' as containing a foreign protein, and if a killer T cell specific for that MHC—peptide complex sees it, it binds. The cell dies probably by apoptosis or programmed cell death (Fig. 27.9). Proteasomes in the cell break down normal proteins also, so that all cells display peptides form normal proteins in their MHC proteins in addition to abnormal ones. The safeguard lies in the destruction during T cell development in the thymus of killer T cells carrying receptors capable of recognizing 'self'.

To summarize the main steps in cell-mediated immunity:

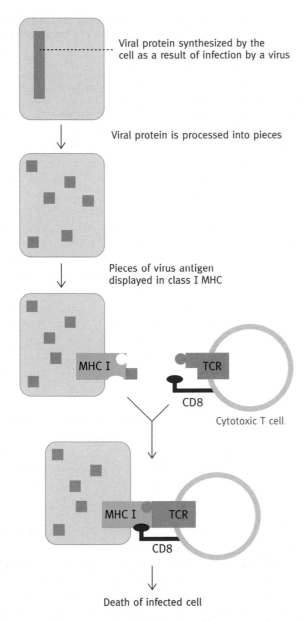

Viral protein synthesized by the cell as a result of infection by a virus

Viral protein is processed into pieces

Pieces of virus antigen displayed in class I MHC

MHC I   TCR

CD8

Cytotoxic T cell

MHC I   TCR

CD8

Death of infected cell

Fig. 27.9 Sequence of events in cell-mediated immune reaction to a foreign antigen synthesized inside cells (for example, as a result of virus infection). A proportion of the clone of cells develops into memory cells. The infected cell may be killed by perforation of its cell membrane due to the release of the protein perforin, or death may be due to apoptosis.

- Cytotoxic cells are produced in the thymus gland; random variations of the cell receptors produces a population capable of recognizing almost any conceivable antigen, each cell being specific for a single antigen.

- Cells which react with self components are deleted.

- Released cells need to be activated; a given cell combines with a special APC displaying an antigen peptide from the cognate foreign antigen (for which the T cell receptor is specific) displayed on class I MHC proteins. Cytotoxic cells have a CD8 protein which is essential for effective binding of these cells to the class I MHCs of other cells, though the specificity of attachment lies in the receptor recognition of the displayed peptide. The APC liberates cytokines which cause the T cell to divide and form a clone of activated cells.

- A virally infected host cell displays peptides from viral protein on its class I MHC complexes.

- Activated killer T cells recognize foreign peptides displayed in this way and destroy the infected cell.

Why does the human immune system so fiercely reject foreign human cells?

At first sight it is quite puzzling why the body should have a mechanism for immune attack on cells transferred from one individual to another. Tissue grafts and organ transplants are recent developments, so there was no evolutionary reason to guard against them. The main cause of rejection of foreign tissue grafts is the MHC molecules on cells. The genes for these are highly polymorphic—there are, in the population, many *alleles* (or *variants*) of genes coding for MHC proteins. It is therefore unlikely that the MHC proteins of one individual are exactly the same as those of the next. Those that are different will therefore be antigenic. The MHC antigens are the same molecules as the class I and II antigen-peptide display proteins discussed above. The killer T cell receptor recognizes the *complex* of MHC molecule plus the antigen peptide, not either component alone. A foreign MHC molecule may be 'seen' by killer T cells as the body's own MHC molecule, complexed to a foreign antigen, and the cell bearing this can be attacked by activated T cells carrying the appropriate receptors. The evolution of such variation in the MHC molecules of the population might be a protection for the species as a whole. Viruses and other pathogens employ devices to outwit the immune system. Suppose that, by chance, a virus evolves so that its processed antigen is not, for some reason, displayed on a host MHC molecule. The cell is not labelled as virally infected and is not attacked. The virus might then

multiply and kill its host. However, the next individual it infects is unlikely to have the same MHC alleles and the chance of the same evasive device being successful again is reduced. The disease is therefore less likely to sweep through the whole population.

Monoclonal antibodies

If an animal is injected (immunized) with an antigen, antibodies appear in the blood against the antigen and from this, an immune serum can be obtained and used for a variety of purposes such as experimentally to detect a particular protein. However, such sera are, in the molecular sense, extremely crude for there are many different antibodies specific for a wide variety of epitopes of different proteins—really to whatever antigens the animal has been exposed to in the preceding period, so that the antibody of interest will be only a very small proportion of the total. It was realized that a completely specific (pure) antibody, in which every immunoglobulin molecule is specific for the antigen of interest, would be a very powerful tool in biochemistry and medicine.

The achievement of this earned Cesar Milstein and Georges Kohler in Cambridge a Nobel prize. A given B cell, as explained, produces one antibody in terms of binding specificity and one only. If a B cell that produces the wanted antibody could be isolated and grown in culture, the antibody produced would be absolutely pure and therefore absolutely specific. It is possible to isolate such a specific B cell, but B cells will not grow in culture for more than a brief period. Only transformed cancerous cells will grow indefinitely in culture.

Pure clones of cancerous B cells are available in tumours called **multiple myelomas**; these are derived from single B cells which multiply uncontrollably and produce large amounts of a pure immunoglobulin and will grow indefinitely in culture. The immunoglobulin it produces will bind to whatever antigen the parent cell was specific for, which could be anything, so its immunoglobulin is of no use. The solution to the problem was to isolate a B cell which produced the wanted specific antibody (but wouldn't grow in culture) and fuse it with a multiple myeloma cancer cell which doesn't produce the wanted antibody but does grow easily. This produces an immortalized B cell which secretes the wanted antibody. The technique, briefly, is to immunize mice with an antigen and a few weeks later to prepare B cells from the spleen which is a secondary lymphoid organ. These are a mixed population of B cells producing different antibodies but including some specific for your injected antigen. They are fused with myeloma cells (this can be done simply by using a common chemical, polyethylene glycol). This gives fused cells called **hybridoma cells** but there will also be large numbers of unwanted cells which have not fused. The latter are eliminated by growing the mixed population in a selective medium which does not allow unfused cells to grow and divide. The hybridomas are collected and inoculated into small wells arranged in rows in a plastic block. By appropriate dilutions it is arranged that only a single hybridoma cell is placed in a small volume of culture fluid in each well. After suitable growth, the wells are tested for the presence of antibody reacting to the antigen of interest. Once identified, the clone can be grown up in unlimited quantities producing the monoclonal antibody also in unlimited amounts. The cell culture can be stored indefinitely in liquid nitrogen and grown up whenever more of the antibody is wanted.

Monoclonal antibodies have become almost the basis of an industry, so powerful are they as a tool. Almost any protein can be detected by developing the appropriate monoclonal antibody. They are used in clinical laboratories for measuring specific human proteins or for HIV and other virus detection; experimentally they can be labelled by fluorescent compounds and used to identify specific protein production for example in cells of embryos. They also have the potential to deliver particular therapeutic agents to their specific target sites.

As a final note, the mammalian immune system is of almost incredible complexity and despite the remarkable advances already made, it is far from completely understood. It is the subject of a massive research effort because of its immense medical importance as well as its basic interest.

FURTHER READING

1993. *Sci. Amer.*, **269**(3), 20–108.

A complete issue of *Scientific American* devoted to all aspects of immunity, superbly illustrated.

Generation of gene diversity

Sekiguchi, A. and Frank, K. (1999). V(D)J recombination. *Current Biology*, **9**, R835.

A one-page quick guide to the essential facts of how the immune system generates gene diversity

Monoclonal antibodies

Yelton, D. E. and Scharff, M. D. (1981). Monoclonal antibodies: a powerful new tool in biology and medicine. *Ann. Rev. Biochem.*, **50**, 657–80.

General review of techniques and applications.

Chien, S. and Silverstein, S. C. (1993). Economic impact of applications of monoclonal antibodies to medicine and biology. *FASEB J.*, 7, 1426–31.

Despite the title, includes a nice summary of how monoclonal antibodies were developed; this is followed by their practical application and monetary value to industries.

Antigen presentation

Engelhard, V. H. (1994). How cells process antigens. *Sci. Amer.*, **271**(2), 44–51.

An account of antigen processing and display of fragments on MHC proteins.

Robertson, M. (1999). Antigen presentation. *Current Biology*, **9**, R829–30.

A 'primer' article describing how antigens are presented on MHC I and MHC II proteins. Includes an account of how viruses sabotage the process for their own ends.

Pathogens and the immune system

Goodenough, U. W. (1991). Deception by pathogens. *American Scientist*, **79**, 344–55.

A fascinating account of how, at the molecular level, bacteria and viruses attempt to escape immunological detection. Summarizes immune mechanisms clearly.

Hemopoeisis and hemopoietic growth factors

Metcalf, D. (1991). The 1991 Florey Lecture. The colony-stimulating factors: discovery to clinical use. *Phil. Trans. Series B, Roy. Soc., Lond.*, **333**, 147–73.

An in-depth review by the discoverer of these important proteins. While there is more research level material than most students would want, it provides an excellent general review of the field, and a fascinating account of the progression from unidentified factors to their widespread clinical use.

Kelso, A. (1993). The immunophysiology of T cell-derived cytokines. *Today's Life Science*, **5**(4), 30–40.

An excellent summary of cytokines in the immune system.

Elimination of self-specific cells

Collins, M. (1991). Death by a thousand cuts. *Current Biology*, **1**, 140–2.

A concise account of immature T cell-progammed cell death in the thymus.

Nowak, M. A. and McMichael, A. J. (1995). How HIV defeats the immune system. *Sci. Amer.*, **273**(2), 42–9.

Presents hypothesis that continuous evolution of the human immunodeficiency virus in the body leads to the immune devastation of AIDS—good graphics of the immune system.

PROBLEMS FOR CHAPTER 27

1 Describe the structure of an IgG molecule.

2 Explain how the body can have genes to code for so many different antibodies.

3   What are the different classes of lymphocyte involved in immune protection and their function?

4   What is an antigen-presenting cell?

5   When a host cell displays, for example, a viral antigen on its surface which class of MHC molecule is it displayed on? Which class of MHC molecule does a cytotoxic T cell recognize?

6   What is meant by the term clonal selection theory?

7   What is the principle of avoidance of autoimmunity, or how does the immune system become self-tolerant?

8   Production of an antibody by a B cell after stimulation (to become a plasma cell) by a helper T cell does not change the antigen-binding site of the antibody, but the antibody has to be released instead of residing in the membrane. How is this achieved?

9   When a B cell (plasma cell) starts to secrete antibody, the latter may have a relatively poor affinity for its antigen but this is rapidly improved. How is this achieved?

10   Which immune systems have the glycoprotein CD4 and which CD8? What are their roles in the immune reaction? What is the relevance of CD4 to the AIDS virus?

Chapter summary

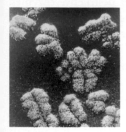

Viruses and viroids

Quite apart from the biological and medical importance of viruses, they are important in biochemical research; they provide simple models for studying complex processes in gene function and also provide important research tools for many purposes, including crucial roles in recombinant DNA technology, described in Chapter 29.

If we ignore the special survival strategies such as the formation of spores in which chemical activities are minimized, living cells are always metabolically active, carrying out functions of ATP production, osmotic work, and synthesis of cellular materials, etc. The cell increases in size and divides. To do this it requires thousands of genes to specify the proteins required.

Viruses are quite different. They are much smaller, an electron microscope being needed to see them rather than the light microscope which is adequate for most bacteria. They have no metabolism—a virus particle or **virion**, as it is called, on its own does absolutely nothing. It generates no energy and catalyses no reactions; it is an inert, organized, complex of molecules that may often be crystallized. Nonetheless, they are reproduced when they infect living cells. Different viruses infect animal and plant cells and also bacteria (the latter ones are known as bacteriophages). Their strategy is to get their genetic material into cells and use the host cell machinery for their own replication.

A virus particle, therefore, is essentially a small amount of nucleic acid surrounded by one or more protective shells. Although the outer shell or shells contain many copies of protein molecules there are only a few different species of these in most viruses. The total number of genes may be as many as about a hundred for a large virus, such as vaccinia, or three or four for the smallest. The protein shell surrounds the genome and is called the **nucleocapsid**. In some viruses there is an additional lipid-bilayer membrane envelope into which are anchored the external viral protein molecules, exposed on the viral surface, that are involved in the first stages of the infective process, as described below.

The life cycle of a virus

A virus must gain entry to the cell that it infects and release its genetic material (RNA or DNA) into the cell. The viral genes must direct the synthesis of the viral-specific enzymes necessary for the reproduction of the virus and also those proteins required for the assembly of new virus particles. Multiple copies of the original genome have to be made, and new virions assembled from the synthesized components. Finally, the progeny virions must escape from the cell.

We will now describe the different stages in the life cycle of viruses. In doing so, we will first deal with different strategies used by different viruses in general terms but, following this, a number of specific viruses will be described in more detail, to illustrate some of the general principles.

A living cell is surrounded by a lipid bilayer membrane that is a barrier to virus penetration. However, there are ways past it. For example, cells have mechanisms for engulfing molecules; receptor-mediated endocytosis in animal cells is one such route (see page 268). Many animal cells are very active in this process and some viruses exploit this to gain entry themselves; they hitch a ride on a normal cellular import mechanism. The first event is that the virion binds to the cell surface, via a **coat protein** complementary to a specific receptor on the host cell surface. The latter requirement is the main restriction on which

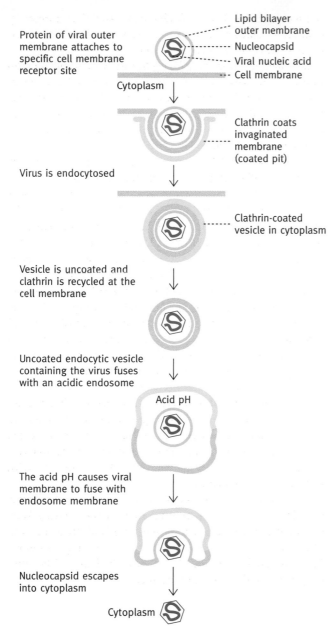

Protein of viral outer membrane attaches to specific cell membrane receptor site

Lipid bilayer outer membrane
Nucleocapsid
Viral nucleic acid
Cell membrane

Cytoplasm

Virus is endocytosed

Clathrin coats invaginated membrane (coated pit)

Clathrin-coated vesicle in cytoplasm

Vesicle is uncoated and clathrin is recycled at the cell membrane

Uncoated endocytic vesicle containing the virus fuses with an acidic endosome

Acid pH

The acid pH causes viral membrane to fuse with endosome membrane

Nucleocapsid escapes into cytoplasm

Cytoplasm

Fig. 28.1 The process of invasion of an animal cell by a membrane-enveloped virus using the receptor-mediated endocytosis route of entry. Receptor-mediated endocytosis is discussed further on page 268.

cells a given virus can infect. After attachment, the receptor-virus complex moves in the membrane to a depression in the latter which has a protein-coat on its internal surface, called **clathrin**, hence the term **coated pit** (see Fig. 28.1). As with normal receptor-mediated endocytosis, the pit invaginates more

and more, until it engulfs the virion into a coated vesicle inside the cell. The clathrin returns to the cell membrane and the vesicle containing the virus fuses with an **endosome**, or cytoplasmic vesicle, leaving the virus inside the latter. In normal (nonviral) endocytosis, the endosome delivers its contents to vesicles from the Golgi sacs, containing an array of destructive enzymes, forming lysosomes in which the engulfed particle is destroyed (Fig. 17.1). However, the endosome has an acidic internal pH due to a proton pump in its membrane. The low pH causes the virus to dissociate from the host cell membrane receptor (to which the virus initially attached). In the case of a membrane-enveloped virus, the lipid membrane envelope now fuses with the endosome membrane and the core of the virus escapes into the cytoplasm. There, it loses its protein coat and the contents are free in the cytoplasm. It is not so obvious how a naked nucleocapsid (that is, without a lipid membrane) virus escapes from the endosome into the cytoplasm.

A second route, available only to certain of the viruses with a lipid membrane coat outside the nucleocapsid, is direct fusion with the external membrane of the host cell. In this, the virus again attaches to the cell surface by interaction of a viral surface protein with a specific protein on the cell surface. The two membranes fuse, a contiguous hole forms in both, and the virion contents enters the cell via this hole. HIV uses this route.

In bacterial viruses, or **bacteriophages**, a different route is followed. The bacterial cell has a rigid cell wall around it. Phage lambda (λ) has a tadpole-like shape (Fig. 28.2). The **head** is a capsule formed by protein molecules in which resides the viral DNA molecule. The phage attaches to the cell wall by its tail fibre and injects the DNA into the bacterial cell, almost like the action of a hypodermic syringe. It may be necessary to remind oneself that this is not a free-living cell, or organism, but a lifeless collection of molecules self-assembled into the phage structure.

Types of genetic material in different viruses

The genetic material of viruses may be double-stranded DNA, single-stranded DNA, double-stranded RNA, or single-stranded RNA.

So far in this book, double-stranded DNA has been taken to be the genetic material of organisms. We have pointed out earlier (page 306) that DNA is chemically more stable than RNA and double-strandedness means that single-strand damage can be corrected (page 330) using the other strand as template for the repair. The DNA replication machinery in the nucleus of cells includes proofreading, which reduces errors by a huge factor (see page 328)—without this, the rate of mutation would be unacceptable high. No such correction machinery has been developed in any RNA synthesis mechanism. Nevertheless, many viruses use RNA as their genetic material. Why then can some

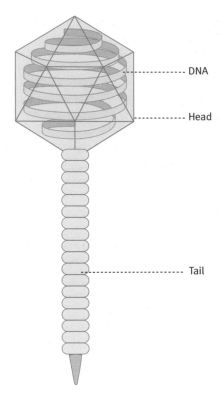

Fig. 28.2 A λ particle. The λ chromosomome—some 50 000 base pairs of DNA—is wrapped around a protein core in the head. After Fig. 1.1 in Ptashne, M. (1986). *A genetic switch*. (2nd edn.) Cell Press and Blackwell Publications, Oxford, with permission from Elsevier Science and M. Ptashne.

viruses get away with having RNA as genetic material? The reasons are probably: (1) that viral genomes are exceedingly small compared with the size of a cellular genome and therefore the chances of deleterious mutations during replication of viral RNA are smaller; (2) although errors in RNA replication are still higher than occur with DNA, giving rise to a rapid mutation rate in RNA viruses some of which will be lethal, this is tolerated because of the rapid multiplication rate and fierce selective pressure, and it becomes an advantage in that rapid mutation helps the virus to escape immunological attack by the animal host.

Different types of viral genetic material demand different biochemical strategies on entering the host cell.

Double-stranded viral DNA is transcribed by host RNA polymerase to produce mRNA in the way with which you are familiar. Vaccinia in this class is unusual in that it carries its own RNA polymerase in its virion, for a reason described later. The minus (template) strand of the double-stranded genome is copied to make mRNA. The mRNA directs the synthesis of virus-specific proteins. In the case of **single-stranded DNA**

viruses, whether (+) or (−), double-stranded DNA is formed by replication using host cell enzymes, from which mRNA is transcribed (see below for strand definition).

Cells have no machinery for replicating viral RNA. **Double-stranded RNA viruses** carry in their structure molecules of an RNA-dependent RNA polymerase. Once in the cell, this enzyme transcribes mRNA from the viral RNA template. This is translated by the host cell machinery into viral-specific proteins necessary for replication of the virus.

Single-stranded RNA genomes in viruses occur in two forms. You will recall (page 345) that, in a cell, one of the two strands of DNA is called the sense (nontemplate) strand, whose base sequence is identical (apart from T becoming U) to the mRNA transcribed from the other (nonsense, or template) strand. In different single-stranded RNA viruses, the RNA strand might be equivalent either to a sense strand or a nonsense strand. These are called the 'plus' (+) and 'minus' (−) strands, respectively. Put in another way, *the RNA of a (+) strand RNA virus is itself a messenger RNA while that of a (−) strand virus is not.* This makes a difference because, if a (+) RNA strand enters the cell, since it *is* an mRNA, the protein synthesizing machinery of the cell can immediately translate it into proteins, including enzymes necessary for viral reproduction. Such a virus requires only its RNA, coding for whatever proteins it needs, for reproduction of the virus to occur after infection. The situation with a (−) strand RNA virus is quite different. It is not an mRNA; the cell cannot translate it; the cell cannot replicate the RNA. An additional RNA-replicating enzyme (see below), carried by the virus, is needed to copy the (−) RNA into (+) RNA, which, as stated, is an mRNA.

Another class of (+) **strand RNA viruses** are the **retroviruses** of which HIV (human immunodeficiency virus), causing AIDS (acquired immune deficiency syndrome) is of major interest. The virion carries an enzyme that converts single-stranded RNA into double-stranded DNA (described later). The latter integrates into the host chromosome.

How are viruses released from cells?

Some viruses are released simply by lysis of the cell; polio is an example. In the case of bacteriophage λ, the rigid inter cell wall would prevent release of new phage particles. However, genes on the phage DNA code for an enzyme, **lysozyme**, which destroys the cell wall, resulting in cell lysis. The gene for this is transcribed only late in the lytic cycle of infection so that the cell is not destroyed before the new phage particles are produced. Despite the fact that bacteriophage λ is essentially just a small piece of DNA, 48 502 nucleotides in length with 63 genes, the latter are subject to an elaborate control system.

The release of animal viruses that are surrounded by a lipid bilayer membrane is more complex. The virus exploits a normal cellular function. You will recall (page ••) that the normal cell produces new cell membrane containing appropriate membrane proteins by inserting the proteins into the rough endoplasmic reticulum. The proteins are glycosylated and transported via Golgi vesicles to the cell membrane; the vesicles fuse with the latter so that we have new cell membrane containing the correct membrane proteins.

Some viruses with lipid membrane envelopes adapt this process to their own ends. The messenger for the envelope proteins codes for leader and anchor sequences (see page 410), and so positions the viral proteins in the rough ER whence they are delivered as described above to the cell membrane. There, the viral membrane proteins accumulate. A new **virus nucleocapsid** (that is, the viral genome plus a shell of capsid proteins) buds off from the membrane, collecting its bilayer envelope containing membrane proteins as it goes. Influenza and the AIDS viruses have this release method of budding off particles (Fig. 28.3).

Replication mechanisms of some selected viruses

The detailed strategies by which viruses cope with replication problems are diverse. Instead of summarizing many of these molecular strategies, which would be a very large topic, we will deal with a few viruses selected because they illustrate some of the general principles given above and because of their intrinsic interest.

Vaccinia

This is the virus causing cowpox and was the one used for small-pox vaccination. It is one of the largest and most complex viruses. It has the type of genome with which you are most familiar—double-stranded DNA. The latter is transcribed into mRNA leading to the synthesis of all of the protein components needed for the entire production of new virions (including enzymes needed for viral DNA replication) by the host ribosomal system.

In the case of most double-stranded DNA viruses, the viral DNA finds its way to the nucleus where host RNA polymerase transcribes the genes to produce mRNA. Vaccinia is unusual in that the entire replication occurs in the cytoplasm. There is no host RNA polymerase in the cytoplasm to transcribe the viral genes and the virus has to carry molecules of its own RNA poly-

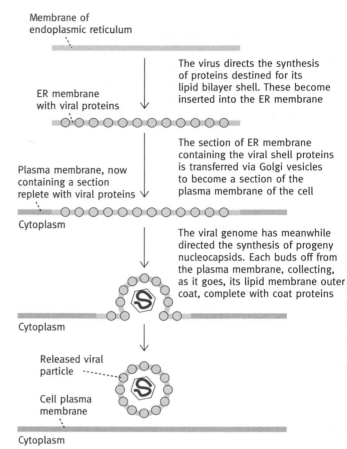

Membrane of endoplasmic reticulum

The virus directs the synthesis of proteins destined for its lipid bilayer shell. These become inserted into the ER membrane

ER membrane with viral proteins

The section of ER membrane containing the viral shell proteins is transferred via Golgi vesicles to become a section of the plasma membrane of the cell

Plasma membrane, now containing a section replete with viral proteins

Cytoplasm

The viral genome has meanwhile directed the synthesis of progeny nucleocapsids. Each buds off from the plasma membrane, collecting, as it goes, its lipid membrane outer coat, complete with coat proteins

Cytoplasm

Released viral particle

Cell plasma membrane

Cytoplasm

Fig. 28.3 Release of viral particles from cells by budding off from the cell surface.

merase in the virion particles, thus permitting mRNA production and the synthesis of all required proteins including, of course, more RNA polymerase.

Poliovirus

This is a **naked virion** (that is, it has only a nucleocapsid shell of coat protein but no membrane). It attaches to specific receptors found only on epithelial cells of humans and other primates. Since its single-strand (+) RNA *is a messenger RNA*, it is translated on entering the cytoplasm producing an **RNA replicase**, which synthesizes RNA using viral RNA as its template. The replicase copies the original (+) strands into (−) strands, which act as templates for more (+) strand synthesis. Poliovirus RNA replicase is unique amongst RNA-synthesizing enzymes in that it cannot, on its own, initiate RNA chains. In this case, the primer is not an oligonucleotide but a protein to which it adds the first nucleotide and then elongates the chain.

Translation of the mRNA occurs via the production of a single large 'polyprotein', which is subsequently cleaved into the separate proteins. This strategy of polyprotein production is, perhaps, surprising for, although many more coat protein molecules are needed for new virion production than are replicase molecules needed for replication, the process produces equal numbers of these.

Influenza virus

This is a (−) strand RNA virus. Its genome is divided into eight sections, each contained in a helical nucleocapsid and the whole surrounded by a lipid membrane into which are inserted molecules of two different glycoproteins (Fig. 28.4). One of these is **hemagglutinin**, present as a surface protein, so called because, if the virus is mixed with red blood cells, it causes the latter to agglutinate. The red blood cell has on its surface a carbohydrate attachment to one of its proteins (**glycophorin A**), terminating in sialic acid, otherwise known as **neuraminic acid** (page 74). When mixed with red blood cells, the hemagglutinin attaches to the neuraminic acid and crosslinks the cells into aggregates. (This is simply an *in vitro* effect.) The virus gains entry to its host cell by attaching to a receptor on the latter which, likewise, terminates in neuraminic acid. It enters via the endocytotic route.

The virion has several hundred hemagglutinin molecules on its surface. It also has an enzyme, **neuraminidase**, as another surface protein, which hydrolyses terminal neuraminic acid groups off glycoproteins. There have been various purposes suggested for this function, for, at first sight, it seems odd that a virus has a surface enzyme to destroy cellular receptors on which it depends for infection. One possibility is that its action liquefies mucins that contain sialic acid (page 45). This would

both permit access to cell surfaces for infection and facilitate escape of new virus particles. Liquefaction of mucins might also help to distribute the virus more efficiently by the sneezing reaction. Finally, it may be that the escaping virus might become anchored to the host cell by neuraminic acid receptors, thus inhibiting its escape. The neuramidinase would guard against this by destroying the receptors. There is evidence that the enzyme is essential for efficient virus multiplication in an infection.

As stated, the virus contains its genome in eight separate (−) RNA strands, each as a helical nucleocapsid (Fig. 28.4). Replication of these requires an RNA replicase, since influenza (−) strand RNA is not a messenger. The virus carries in its nucleocapsids molecules of an RNA replicase that copies the (−) strand into (+) RNA (equals mRNA) and protein synthesis can then produce all the necessary viral proteins including more RNA replicase.

The immune system attack is principally against the hemagglutinin of the influenza virus, which is neutralized in this way, so that an individual experiencing the disease is immune to a second infection by the identical strain of virus. The viral hemagglutinin is constantly changing its amino acid composition as a result of mutations (remember, it is an RNA virus), so that the protection is *gradually* lost as this occurs. This is known as **antigenic drift**. Such mutation of individual amino acids does not produce major epidemics because the different antibody binding sites on the protein are not eliminated by gradual change, but only reduced in number, so that only partial loss of immunity occurs in the population and many infections are mild. However, if a totally new hemagglutinin is present in the virus, a world-wide virulent pandemic can result from this antigenic shift because there is no residual immunological protection against it. Such a situation may occur when two different strains of virus infect a cell. Because of the divided genome, reassortment of the RNA particles can occur when assembly of new virus nucleocapsids takes place. One such pandemic, in 1918, is believed to have resulted from reassortment between a human and bird strains of influenza, and is estimated to have caused 20 million deaths.

The pathogenicity of different strains of influenza viruses is very interesting from the molecular viewpoint. After receptor-mediated endocytosis of the virus, the nucleocapsid has to escape from the endosomal vesicle into the cytoplasm of the infected cell. This requires proteolytic cleavage of hemagglutinin into two fragments following which fusion of the viral envelope with the vesicle membrane occurs leading to the release the nucleocapsid. The cleavage occurs at a linker between the two fragments. Different strains of viruses differ in their susceptibility to cleavage. For some strains the cleavage is restricted to cells of the respiratory system so the infection is localised, but in highly

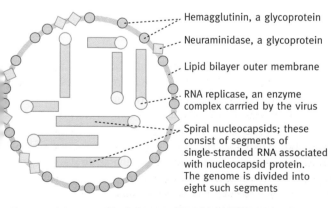

Hemagglutinin, a glycoprotein

Neuraminidase, a glycoprotein

Lipid bilayer outer membrane

RNA replicase, an enzyme complex carrried by the virus

Spiral nucleocapsids; these consist of segments of single-stranded RNA associated with nucleocapsid protein. The genome is divided into eight such segments

Fig. 28.4 Structure of the influenza virion. A layer of matrix protein underlying the outer membrane is not shown.

pathogenic strains it may occur in other cell types so the disease is more widespread in the body. It is speculated that this may have been related to the occurrence of highly pathogenic strains in the past. (See Garten and Klenk reference).

The role of structural determination of proteins in new approaches to disease therapy is illustrated by a current 'rational therapy' approach to influenza. Using crystalline influenza neuraminidase, the precise structure of the neuraminic acid binding site of the enzyme was determined by X-ray crystallography. From this, a synthetic analogue of neuraminic acid was devised that attaches with high affinity to the enzyme and blocks its normal function. It is now used clinically under the name of Relenza. The active site of an enzyme has limited opportunity to mutate and remain functional but this does not expose the active site to immunological attack for it is buried in a cleft not accessible to immune attack. It is, however, accessible to the neuraminic acid analogue. In this particular case, the analogue does not block the hemagglutinin site that binds to the host cell receptor and so would not inhibit infection. It depends on inhibiting the neuraminidase believed to be necessary for new virus release from cells so that the drug limits the spread of the infection.

Retroviruses

As already stated, there is a fourth class of single-stranded RNA viruses of immense current interest—the **retroviruses**. Retroviruses, of which that causing AIDS is one example, are (+) strand RNA viruses. However, they do not, on infection of a cell, follow the molecular strategy outlined above for poliovirus. The retrovirus particle carries within itself a few molecules of an enzyme whose discovery caused initial disbelief, to be followed by the award of a Nobel Prize to its discoverers, Howard Temin and David Baltimore; it is called **reverse transcriptase**. Before its discovery it was, of course, known that DNA could direct RNA synthesis, but the reverse was not known. Viral reverse transcriptase is an amazingly versatile enzyme; when a retrovirus infects a cell, the viral reverse transcriptase copies the (+) RNA into DNA. As with all DNA synthesis, a primer is necessary. Retroviruses, remarkably, use a host cell tRNA molecule (page 382), which hybridizes to the start of the template strand, and it carries the appropriate tRNA in the virion particle. The RNA/DNA hybrid so formed is converted to single-strand DNA by RNA hydrolysis, an enzyme activity also present in the reverse transcriptase molecule. The single-stranded DNA is then copied by the same enzyme to form double-stranded DNA. The viral genome is now in the form of normal duplex DNA which becomes integrated or inserted into the host cell chromosome (Fig. 28.5). The double-stranded DNA copy of the viral genome

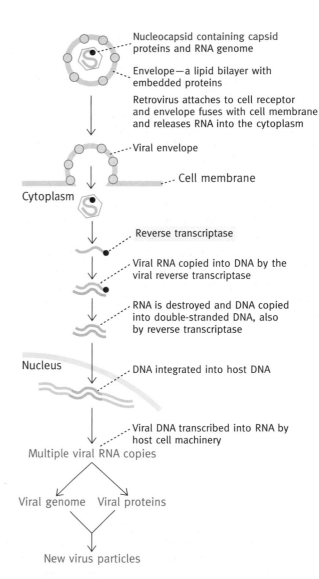

Nucleocapsid containing capsid proteins and RNA genome

Envelope—a lipid bilayer with embedded proteins

Retrovirus attaches to cell receptor and envelope fuses with cell membrane and releases RNA into the cytoplasm

Viral envelope

Cell membrane

Cytoplasm

Reverse transcriptase

Viral RNA copied into DNA by the viral reverse transcriptase

RNA is destroyed and DNA copied into double-stranded DNA, also by reverse transcriptase

Nucleus

DNA integrated into host DNA

Viral DNA transcribed into RNA by host cell machinery

Multiple viral RNA copies

Viral genome   Viral proteins

New virus particles

Fig. 28.5 Replication of a hypothetical retrovirus. See Fig. 28.3 for the release process.

(called **proviral DNA**) has at each end a base sequence called a **long terminal repeat** (LTR) (Fig. 28.6). An integrase enzyme, also carried in the virus, causes integration of the proviral DNA into the host chromosome where it is now, in essence, an extra set of genes carried by the cell. Once in, it is replicated along with host DNA for cell division. For the production of new retrovirus particles, the proviral genes (that is, viral genes in the host chromosomes) are transcribed into (+) RNA transcripts. Some of these become the genome of new retrovirus progeny; some are processed to mRNA to provide the proteins for virus particle assembly.

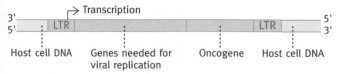

Fig. 28.6 Proviral genome of an oncogenic retrovirus incorporated into a host genome. LTR, Long terminal repeat. The oncogene may code for an abnormal protein involved in cell control or an abnormal amount of such a protein (see Fig. 30.3).

The drug AZT (azidothymidine), whose structure is shown below, is used as therapy against the AIDS virus (HIV); it works by inhibiting the viral reverse transcriptase. For this, it has to be converted into the triphosphate form which is an analogue of dTTP. AZT-triphosphate inhibits reverse transcriptase and also, when incorporated, terminates DNA chain extension because of the lack of a 3′-OH group. Host cell DNA polymerase is inhibited to a smaller extent.

Azidothymidine (AZT)

Retroviruses have the potential to permit gene therapy—to place in a patient's cells that lack a normal essential gene a 'good' gene to remedy the deficiency. The principle is to cripple the reproductive capacity of the retrovirus by gene deletion and replace this with the RNA equivalent of the therapeutic gene, which becomes integrated in DNA form into the host chromosome. The method has the disadvantage that integration occurs randomly in the chromosome, which has potential hazards. Alternative methods for site-specific insertion of genes are being actively developed (see page 488).

Cancer-inducing, or oncogenic, retroviruses

An **oncogenic retrovirus**, when integrated in its DNA form (Fig. 28.6) into the host chromosome, can cause tumour formation (*oncus* equals solid mass or tumour). The first tumour-causing virus discovered was the avian Rous sarcoma virus, which causes tumours in chicken muscle. This is due to a single gene in the retrovirus—an **oncogene**.

Since then, about 30 retroviral oncogenes have been discovered. We described DNA hybridization earlier (page 307). In this technique, a piece of DNA made single-stranded by heating will unerringly find, and hydrogen-bond to, a complementary piece of DNA (that is, with a base sequence complementary in the Watson-Crick base pairing sense) amongst thousands. Techniques exist for detecting the hybridization complexes (see page 481). It was astonishingly discovered that, in each case, a cancer-producing retroviral oncogene had its closely similar counterpart in normal cells. This applies only to the exons of the normal gene since the viral RNA form of the gene lacks introns due to splicing (page 358 and see below). The viral oncogene is clearly derived, in the evolutionary sense, from its cellular counterpart. A single base change may be the only difference between the two. The **viral gene** is given the prefix v and the corresponding **cellular gene**, c. Genes are given curious abbreviations (page 499) so we have names such as c-*myc* and v-*myc*, c-*ras* and v-*ras*, etc. The cellular gene is called a **protooncogene**. It turns out that protooncogenes code for proteins involved in important cellular control mechanisms (described in Chapter 26), abnormalities of which can lead to uncontrolled cell division. Some of them code for abnormal transcription factors (discussed on page 500). It would appear that retroviral oncogenes originated (in the evolutionary sense) through retroviruses accidentally picking up an mRNA molecule derived from a cellular control gene. The oncogenic viral form of the control gene may differ from the normal cellular counterpart due to mutation during, or subsequent to, the 'picking up' process. On reinsertion of this gene (by retroviral infection) into the host chromosome by the mechanism already described, the presence of the oncogene results in either an abnormal control protein, or abnormal amount of such a protein, in turn resulting in abnormal cellular control and cancer. This will become more understandable after you read Chapter 30, which deals with control pathways involving protooncogene protein products.

Bacteriophage lambda

This bacterial virus has been extensively studied and also used in molecular biology and in gene cloning (see Chapter 29 and Fig. 28.2). It is a double-stranded DNA phage but its life cycle is different again. It has an elaborate set of genetic controls; on entering the cell, the DNA has two choices. It can immediately replicate and produce new virus particles to be released by a **phage-coded enzyme, lysozyme**, that disrupts the bacterial cell wall resulting in cell lysis. This is the **virulent** or **lytic route** (Fig. 28.7). The alternative is the **lysogenic route** in which the genome integrates into the *E. coli* chromosome where it can

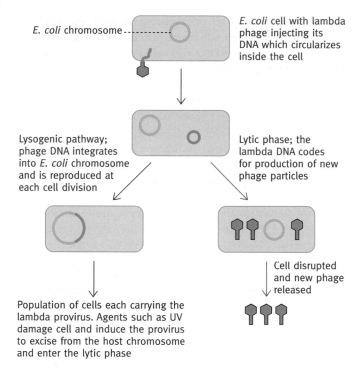

Fig. 28.7 Injection of bacteriophage λ DNA into the host bacterial cell, and the lytic and lysogenic pathways of replication.

remain inert indefinitely being replicated with the host DNA. Thus large populations of otherwise normal *E. coli* cells can result, each carrying the phage DNA. A cell carrying λ DNA in its chromosome is called a lysogenic cell and the incorporated phage DNA a **prophage**. If the DNA of an *E. coli* cell is damaged by UV light, or by a chemical or by ionizing radiation, so-called 'SOS' functions designed to institute repair are activated. However, in a lysogenic cell this results in the prophage being excised from the DNA by a mechanism too complex to describe here. The excised prophage enters the lytic cycle of replication. Bacteriophage λ provides a beautiful example of the self-assembly process. If the component protein and DNA molecules of the virus are mixed together in the test tube, complete functional virus assembles, the DNA being packaged into the head. This constitutes a powerful tool in gene cloning.

What are viroids?

Viroids are the smallest infectious particles known—even smaller and simpler than viruses. They are very small, naked RNA molecules without any protein or other type of coat (see

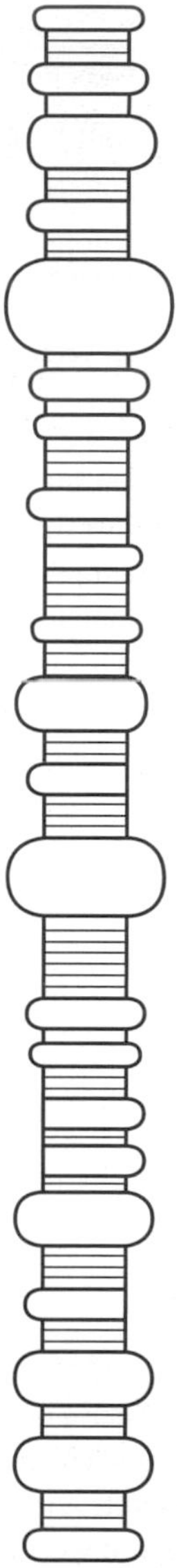

Fig. 28.8 Structure of the coconut cadang-cadang viroid. The loops represent unpaired regions and the cross lines, paired bases. This viroid is the smallest known (246 nucleotides). It almost invariably kills coconut palms 5–10 years after infection.

Fig. 28.8 for an example). Infection of plant cells by them is dependent on mechanical damage of the host plant (to infect plants experimentally the viroid RNA is usually rubbed on the leaves with some abrasive material). The effect of viroid infection varies from some impairment of the health of the plant to certain death of, for example, palm trees. The really remarkable

feature is that no protein coding genes have been identified in viroids. To detect a protein coding gene, you look for an **open reading frame**—that is, a sequence of codons that code for the long stretch of amino acids required for proteins. In viroids, al-though translational initiation sequences are found, the subsequent reading frames are so short (due to hitting stop codons—page 381) that protein production is excluded. It is not known how viroids produce disease.

FUTHER READING

Ptashne, M. (1987). *A genetic switch: gene control and phage λ.* Cell Press and Blackwell Scientific Publications.

Stark, G. R., Kerr, I. M., Williams, B. R. G., Silverman, R. H., and Schrieber, R. D. (1998). How cells respond to interferons. *Ann. Rev. Biochem.*, **67**, 227.

A comprehensive review.

Influenza virus

Hughson, F. M. (1995). Structural characterisation of viral fusion proteins. *Current Biology*, **5**, 265–74.

Deals with infection by influenza virus, and fusion of lipid bilayers. Describes how virus escapes from the endosome, with excellent diagram.

Garten, W. and Klenk, H-D. (1999). Understanding influenza virus pathogenicity. *Trends in Microbiology*, **7**, 99–100.

A brief 'comment' summarizing the evidence that the escape of the virus from the endocytosed vesicle into the cytoplasm is important in determining how widespread and therefore how pathogenic the influenza virus infection is.

McCauley, J. (1999). Relenza. *Current Biology*, **9**, R796.

A one-page quick guide to the new neuraminidase inhibitor drugs against the inflenza virus.

Retroviruses

Katz, R. A. and Skalka, A. M. (1994). The retroviral enzymes. *Ann. Rev. Biochem.*, **63**, 133–73.

Advanced and detailed review of the protease, reverse transcriptase, and integrase enzymes of these viruses, also with HIV inhibitors, and resistance to the latter.

Nowak, M. A. and McMichael, A. J. (1995). How HIV defeats the immune system. *Sci. Amer.*, **273**(2), 42–9.

Describes how continuous evolution of the human immuno-deficiency virus in the body leads to the immune devastation that underlies AIDS—good graphics of the immune system.

Viroids

Symons, R. H. (1989). Self-cleavage of RNA in the replication of small pathogens of plants and animals. *Trends Biochem. Sci.*, **14**, 445–50.

Includes description of the self-cleaving 'hammerhead' RNA structure.

PROBLEMS FOR CHAPTER 28

1   Explain how a virus might gain access to the interior of a eukaryotic cell.

2   Referring only to non-retroviruses, in what way must a (+) single-stranded RNA virus differ from a (−) single-stranded RNA virus?

3   Why does vaccinia carry with it a DNA-dependent RNA polymerase in its virion when the host cell already has this?

4   A retrovirus is a (+) single-stranded RNA virus. In what way does it differ from a non-retroviral (+) single-stranded RNA virus?

5   Explain how a virus with a lipid membrane envelope, such as influenza, acquires this envelope with its assemblage of integral proteins.

6   Influenza epidemics sweep the world regularly. Most are mild but occasionally, as in 1918, a highly lethal one occurs. Why is this so?

7   Why does influenza virus cause red blood cells to agglutinate when the two are simply mixed together?

8   Why is it somewhat surprising that the influenza virus has a surface neuraminidase activity? What might its function be?

9   Describe the 'choice' that bacteriophage λ makes on infecting an *E. coli* cell.

10   (a)   What is a viroid?
      (b)   Given that viroids infect plants and are reproduced, one might expect them to have genes for protein production. Is this the case?

11   Several viruses require for their replication one or more enzymes not present in the host target cell infected. What are the strategies that viruses adopt to cope with this? Explain why viroids cannot use any of these strategies but employ a different one in the replication of their genetic material.

Chapter summary

Gene cloning, recombinant DNA technology, genetic engineering

The terms in the chapter title refer to the same area of molecular biology—they all involve DNA manipulation. DNA manipulation techniques have caused a revolution in biology and biochemistry in particular. In this chapter we deal with the main techniques.

The technologies make it possible to isolate individual genes, to determine their base sequences, to manipulate the latter in any desired way, and to transfer genes from one species to another. The base sequence of the coding region also permits deduction, from the genetic code, of the amino acid sequence of the protein for which it codes (and is often the easiest way to determine the latter). Structural information is a prerequisite for elucidating how a gene functions and what are its control mechanisms. Much of the information in Chapter 23 on the control of gene activity has arisen from studies involving recombinant DNA technology. The isolated gene can be placed in a functional role in many different types of cells to produce the protein it codes for in whatever amount is required. The associated technologies can be used in medical diagnosis of genetic diseases. The general biological applications are almost unlimited.

What were the problems in isolating genes?

A major problem was the gigantic size of the DNA molecules of chromosomes. In a diploid eukaryotic cell there are two copies of a given homozygous gene, one on each chromosome of a pair, amongst scores of thousands of other genes, which, in turn, form only a fraction of the total DNA of a cell. The only thing that distinguishes a gene from all the other DNA is the coded information of its base sequence. The task of isolating a gene resembled not so much finding a needle in a haystack, as finding a particular piece of hay in a haystack. It was not necessarily unduly pessimistic to wonder whether the subject had come up against an impenetrable barrier to further progress. Recombinant DNA technology changed all that.

The first step—cutting the DNA with restriction endonucleases

The first step in isolating a gene is to cut the cellular DNA into defined, suitable pieces, small enough to be handled. This, itself, seemed to be an impossible goal. The only enzymic method of cutting the phosphodiester bonds (page 305) of DNA was with DNases of the type found in the digestive enzymes of pancreatic juice. If DNA is incubated with such an enzyme to give partial hydrolysis, the molecule is fragmented into pieces, but the cutting is completely random.

This point emphasizes the revolutionary importance of the discovery of a different class of **endonucleases** or **DNases** in bacteria. (An endonuclease attacks internal bonds in the molecule, not the terminal bonds as occurs with exonucleases.) The new class of enzymes are usually referred to simply as **restriction enzymes**, the act of cutting with them as restriction, and the DNA so treated as having been restricted. Such enzymes do not cut DNA randomly, but each recognizes a specific short sequence of bases so as to make a cut at a precise point in the sequence. Different bacteria have different restriction enzymes that recognize different base sequences in the DNA and therefore have different cutting sites. A large number of restriction enzymes cutting at specific base sequences are now known.

To illustrate this point, an enzyme from *E. coli* cuts double-stranded DNA at the sequence

GAATTC
CTTAAG.

and that from *Bacillus amyloliquefaciens* cuts at the sequence

GGATCC
CCTAGG.

(There is no point in memorizing such sequences, but note that they have a twofold symmetry—that is, the partner strand read in the 5′ → 3′ direction is identical in sequence.) The enzymes are named after the bacterium (and bacterial strain) of their origin and a Roman numeral where more than one enzyme occurs in that species. Thus the two enzymes mentioned above are called *Eco*RI and *Bam*HI, respectively. *Eco*RI was the first to be isolated from *E. coli* strain R. Other restriction enzymes recognize four-, five-, and eight-base sequences. Restriction enzymes make it possible to cut DNA at precise base sequences with surgical neatness, producing defined fragments.

What is the biological function of restriction enzymes?

Bacterial cells have restriction enzymes to destroy invading foreign DNA. For example, a bacteriophage such as lambda (λ) phage injects its DNA into an *E. coli* cell (page 471). The cell's restriction enzyme, by cutting that DNA at the restriction sites, prevents a successful phage attack; it restricts the ability of foreign DNA to infect the cell. The question arises as to what prevents the enzyme from destroying the cell's own DNA. A hexamer base sequence such as that recognized by *Eco*RI must occur many times in the *E. coli* chromosome—statistically, every 4^6 (4096) base pairs. The cell guards against cutting its own DNA by adding a methyl group to one of the bases in all of the recognition sequences on each new DNA strand as soon as it is synthesized. This does not interfere with base pairing or gene function, but the restriction enzyme no longer recognizes the methylated sequence and the cell's own DNA is therefore immune from attack by that enzyme.

The biological role of restriction enzymes can be illustrated by a brief discussion of lambda phage. Lambda infects *E. coli* cells, but different strains, or isolates, of *E. coli* exist differing in their restriction enzymes. Thus, a λ phage that has replicated in a given strain of *E. coli* will reinfect that same strain with high efficiency, because the phage DNA is protected by methylation as is the *E. coli* host DNA. However, if this phage attempts to infect a different strain of *E. coli*, with a different restriction enzyme, the latter will rapidly attack the phage DNA. Whether, in the latter situation, infection is *ever* successful will be the outcome of a race between the cell's methylation system and the restriction enzyme, for, as soon as the λ DNA enters the cell, it will be subject to both attacks. The protection against infection is very high, for the λ DNA is sufficiently long to incorporate several restriction sites for typical restriction enzymes. When phage DNA, unmethylated in these sites, enters the cell, methylation has to win the race in all cases for a successful infection to result, since a single cut at any one of the sites destroys the infectivity of the phage DNA.

The isolation of specific genes to be used in various ways is of central importance and we will turn now to this technology.

Gene cloning, or how genes are isolated

In this section, we describe two procedures used in recombinant DNA technology. One of these is to isolate a **genomic clone** from DNA; we take human DNA as the example, though the methods apply to DNA from any source. A genomic clone provides a piece of DNA identical in base sequence to the corresponding stretch of DNA in the cell and often is designed to contain a specific gene. The second procedure is the isolation of a human **cDNA**, the 'c' meaning 'complementary'. It is the double-stranded DNA copy of a human mRNA. The genomic DNA and the cDNA differ in that the former contains introns. An important practical aspect of this is that, since introns are not spliced out of RNA transcripts (page 358) of genes in *E. coli*, cloned eukaryotic *genes* cannot direct protein synthesis in that organism. cDNA on the other hand is transcribed into an mRNA-like transcript by *E. coli* (provided that appropriate transcriptional signals are placed on the cDNA). The transcripts will direct protein synthesis in *E. coli*; this also requires that appropriate translational signals have been added to the cDNA. Sometimes the best way to isolate a gene is to first isolate the cDNA related to that gene, which can be used as a hybridization probe (see below). This is particularly true where mRNA preparations rich in one species of mRNA can be obtained, as might be the case for an inducible enzyme. The enriched mRNA preparation may be used as the hybridization probe for cDNA isolation (see below).

Genomic clones are required for studies on gene structure, while cDNA clones are required when, for example, the aim is to produce human insulin or other proteins in *E. coli* cells.

The two procedures, genomic cloning and cDNA cloning, have been selected because they serve to illustrate some general principles of gene cloning. The aim is not to give practical guides on the procedures, but to remove some of the mystery

from gene manipulation. (A current laboratory manual on 'molecular cloning' comprises three volumes each of about 500 pages.)

What does 'cloning' mean?

A **clone** refers to multiple identical copies produced from a single origin. Cloning a gene or a cDNA gives us large numbers of copies of a piece of DNA. This is necessary since many of the techniques used require visualization or quantification of the DNA and this requires many copies of the molecule for detection to be possible.

The essence of cloning is, firstly, that a *single* piece of DNA (inserted into a replicating vector—see below), out of a mixture of thousands, produced, for example, by restricting total DNA, enters a bacterial cell and is replicated in that cell. Secondly, a single bacterial cell will grow into a colony containing large numbers of identical cells—for example, a single *E. coli* cell will produce $\sim10^7$ cells in a single colony when grown on an agar plate overnight. Furthermore, there can exist multiple copies of the vector, each with its piece of inserted DNA, in each cell. Since this will happen to each piece of DNA in the restriction mixture, cloning provides a means of both separating and amplifying each piece of DNA, for it is easy to isolate separate *E. coli* cells by growing them into colonies on a nutritive plate. Provided you can pick out of the thousands of different colonies one that contains the piece of DNA you are interested in, it is possible to produce virtually unlimited numbers of copies of that piece by growing up pure cultures of the selected colony.

Isolation of a genomic clone of a human gene

Preparation of a human gene library

First a cloning vector must be chosen that can accept a piece of human DNA and then be replicated in *E. coli* cells. There are several choices and we will choose λ bacteriophage (described on page 271). The centre portion of the DNA molecule of λ can be replaced with a piece of human DNA, without impairing the ability of the **recombinant phage**, as it is called, to replicate in *E. coli*, as explained below. The λ genes essential for its replication are contained in the arms on either side of the insert. Lambda phage can accept a foreign piece of DNA 15–20 kilobases (kb) long without impairing its replicative ability. This is a convenient size for most cases of genomic cloning.

How are recombinant molecules constructed?

The principle of this lies in 'sticky ends' of DNA pieces. Many restriction enzymes do not make 'straight through' cuts across both strands of a double-stranded DNA, but instead make a 'staggered' cut. *Eco*RI, for example, cuts like this.

—X—X—G ↓ A—A—T—T—C—X—X—
—X—X—C—T—T—A—A ↑ G—X—X—

*Eco*RI ↓

—X—X—G   A—A—T—T—C—X—X—
—X—X—C—T—T—A—A   G—X—X—

Sticky ends

The staggered cut produces cohesive or sticky ends. This is because the 'overhang' sequences automatically will base pair with each other, in this case A with T. If two pieces of DNA, both having identical sticky ends, are mixed, they will join together as indicated in Fig. 29.1. If human DNA and λ phage DNA are restricted so that they have the same sticky ends, and the two sets of pieces are mixed, they will associate to form the recombinant λ DNA molecules described above. (A ligation enzyme is used to seal the nicks in the chain by synthesizing a bond between the 3'-OH and the 5' phosphoryl group. Such a ligase is described on page 328.)

When these are added to the separate protein components of lambda phage, the λ phage assemble and the recombinant phage DNA is automatically packaged into the λ head to produce infective phage, a remarkable illustration of self-assembly. ('Packaging kits' with the necessary phage components are commercially available.)

The packaging process requires that, for any DNA piece to be enclosed in an infective phage, it must have, at its two ends, λ phage DNA and it must be of a correct length. Thus, although the mixture of pieces with sticky ends will associate in all sorts of ways, only the desired type of recombinant molecule illustrated in the text above will be selected. (In preparing the human DNA fragments, steps are taken to ensure that pieces of appropriate size are obtained; we will not go into the details of this.)

The λ phage, now carrying recombinant DNA, are used to infect *E. coli* cells by a procedure in which only a very small proportion of the cells become infected, to maximize the chance that only one phage will enter a given cell. It is essential also that the *E. coli* is a mutant strain lacking its restriction enzyme. The result of all this is that the entire fragmented human genome is collectively contained in a culture of *E. coli* cells, one piece per

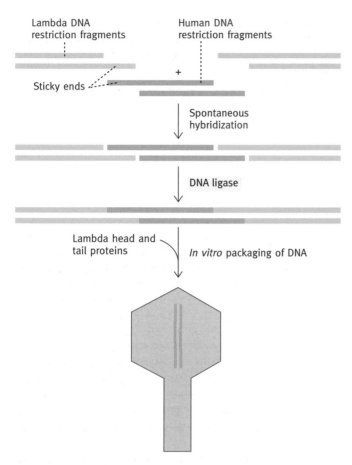

Fig. 29.1 Sequence of steps in producing in λ bacteriophage a recombinant DNA library. Note that, although other associations of DNA pieces will occur due to sticky ends, only those containing a λ left arm and right arm separated by a human DNA fragment of appropriate length (15–20 kb) will be packaged.

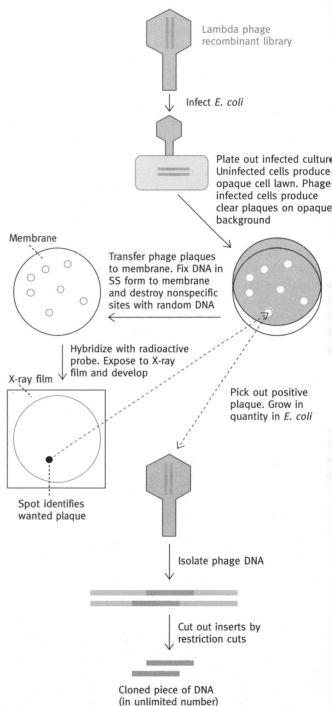

Fig. 29.2 Illustration of gene cloning using bacteriophage lambda phage as the cloning vehicle. See text for explanation.

cell. The collection is known as a **genomic library**. The culture is spread on to a solid growth medium and incubated. The uninfected cells grow to form an opaque 'lawn' over the whole plate, but, wherever there is a phage-infected cell, a clear spot or **plaque** will appear. This is because the single phage multiples in the cell, the progeny are released to infect other, adjacent cells and so on, thus destroying the lawn cells surrounding the originally infected bacterium. (Release of phage involves lysis of the cell.) Each plaque represents multiple copies of phage originating from a single infective particle and each carrying the same copy of a piece of human DNA (Fig. 29.2).

If we can now identify which plaque contains the λ carrying the desired gene, we can grown unlimited numbers of the phage in *E. coli* and isolate its DNA. How is the correct plaque identified?

Screening of the plaques for the desired human gene

To identify the phage carrying the wanted human gene DNA–DNA hybridization is most commonly used (see page 307). For this, a **hybridization probe** is needed, consisting of a short piece of DNA (perhaps 20 nucleotides long) whose base sequence is complementary to a known sequence in the gene in question (see below for how we get this). The method is to transfer phage, from each plaque, to a DNA-absorbent membrane by laying a disc of it on the surface of the culture plate (remembering to mark its orientation so that the position of individual plaques can be identified); phage from each plaque then sticks to it. The membrane is treated with alkali to disrupt the protein–DNA complexes of the phage head and release the DNA and also, importantly, to cause DNA strand separation. The DNA is then fixed to the membrane in single-stranded form by baking or UV light irradiation (which covalently cross-links DNA to the membrane) and the ability of the membrane to fix further DNA is destroyed by flooding with nonspecific DNA. The hybridization probe (made radioactive by enzymic phosphorylation using radioactive ATP) is incubated with the membrane and allowed to hybridize to any complementary DNA on the membrane. Nonhybridized probe is washed away. The probe will stick only to complementary DNA—at the target gene. Exposure of the membrane to X-ray film exposes the silver grains, so that, wherever the probe has hybridized, the developed film will show a black spot due to the radioactivity in the probe (Fig. 29.2). Once the plaque containing the human gene is identified, unlimited numbers of the phage within this plaque can be grown by infecting E. coli, and the human DNA insert released from the phage DNA, in this example, by appropriate restriction.

How is a suitable probe obtained, considering that the gene cannot be base-sequenced until we have isolated it? That is, how can we construct a probe of complementary sequence to identify the presence of the gene in a plaque? If the protein coded for by the gene has been isolated and all, or part, of its amino acid sequence determined, we can work backwards from the genetic code (Table 24.1) and determine what the base sequence of the gene coding for a particular stretch of amino acids in the protein must be. Well, not quite, because of the redundancy of the genetic code (page 382). Where an amino acid has several codons we cannot deduce which one was actually used by the cell to synthesize that protein. (Gene cloners are delighted if they find a stretch of methionines and tryptophans, which have only single codons.) Even where redundancy is unavoidable, there are methods for coping with the ambiguity, such as by synthesizing a mixture of probes. Machines exist that readily chemically synthesize short DNA probes of predetermined base

sequence. A radioactive phosphate group added by an appropriate enzyme completes the probe. If such a probe is not available there are alternative screening possibilities, which we will not go into here.

We will now turn to the isolation of a cDNA clone.

Cloning a human cDNA

Preparation of a human cDNA library

The principle is that mRNA from human cells expressing the gene in question is isolated, copied into double-stranded DNA *in vitro*, and cloned. A significant advantage of this approach is that often a tissue can be selected that expresses the wanted gene in large amounts so that the relevant mRNA is abundant. It is especially favourable if the level of a given protein (and hence its mRNA) can be induced in a tissue, since this results in enrichment of the mRNA specific for the protein.

mRNA is only a small fraction of the total RNA of a cell and, in the case of human cells, has (with a few exceptions) polyA tails (page 370). If the RNA preparation is passed down a column of inert support carrying synthetic oligo-dT ligand (short, single-stranded 'DNA' with only T bases), the ribosomal and transfer RNA run through while mRNA binds due to A=T hydrogen bonding of the tails to the ligand; the mRNA can then be eluted using solutions of low ionic strength.

The isolated mRNA mixture is now copied in the test tube into DNA using the polymerase domain of viral reverse transcriptase (page 470). This gives an RNA–DNA duplex. The RNA is destroyed with NaOH or RNase treatment and the single-stranded DNA converted to double-stranded DNA by an exonuclease-free DNA polymerase I (page 327) derived from a bacteriophage that happens to be particularly suited to the task and is commercially available.

The procedure now is to clone the cDNA molecules. This may be done using λ as the cloning vehicle, as already described for genomic libraries. The cDNA molecules are not sticky ended, but methods are available for enzymically adding appropriate sticky ends to such 'blunt-ended' molecules, as they are called, or alternative procedures for 'blunt-end ligation' are now used.

What can be done with the cloned DNA?

At this point, we have a purified λ phage containing a desired genomic DNA insert or a wanted cDNA insert. We can very easily grown as much as we want of these in E. coli cells, isolate the progeny DNA, and cut out the DNA inserts by restriction.

Subsequent use of the latter, for the various purposes described below, almost always involves the DNA of interest being inserted into a different cloning vector, most frequently a **bacterial plasmid**. Lambda is convenient for the initial preparation of libraries because it can accept large pieces of DNA (around 15 kb) and efficiently infect cells so that a complete genomic 'library', or a complex cDNA library, can be obtained. However, when it comes to studying the cloned piece of genomic DNA or cDNA of interest, λ has the disadvantage of possessing a large amount of its own DNA in the arms (about 30 kb), required for its replication but of no interest in the current context. Plasmids are much smaller and, experimentally, much more convenient to handle. They cannot accept pieces of foreign DNA as large as can λ, but if, for example, the aim is to sequence a genomic clone originally isolated in λ, it is cut up into pieces of appropriate size to be inserted into an engineered sequencing plasmid (described below) and the individual pieces sequenced. cDNA clones may be treated in the same way. However, cDNA clones are commonly small enough to be inserted into **expression vectors**—these include *E. coli* plasmids into which appropriate transcriptional and translational signals have been added. When these are infected into *E. coli* cells, the protein coded for by the cDNA may be produced in quantity. We must now describe what a bacterial plasmid is.

Bacterial plasmids

The *E. coli* cell has a single major circular chromosome carrying the few thousand genes the constitute most of the cell's genetic makeup. In addition, however, there are tiny separate minichromosomes or plasmids in the cytoplasm—circular DNA molecules carrying a handful of genes that usually have a protective role in the cell; typically they carry genes conferring antibiotic resistance on the cell by coding for an enzyme that, for example, destroys the antibiotic. Each plasmid has a replicating origin to provide for duplication of the plasmid in the cell.

Plasmids are not infectious in the sense that bacteriophage are. If, however, *E. coli* cells are treated somewhat unkindly—e.g. exposed to $CaCl_2$ at 0°C and then the temperature suddenly raised to 42°C—they become 'competent' to take up plasmids.

Let us suppose we are dealing with a cloned cDNA molecule isolated as described above. First, a plasmid selected for the purpose in mind is cut with a restriction enzyme and, second, appropriate sticky ends are added to the cDNA molecules by procedures we need not describe. When the cut plasmids and the modified cDNA are mixed and ligated (that is, the nicks are enzymically sealed), **recombinant plasmids** are obtained (see Fig. 29.3). (A procedure that to avoid complexity, we will not describe is available to prevent the cut plasmid merely being re-

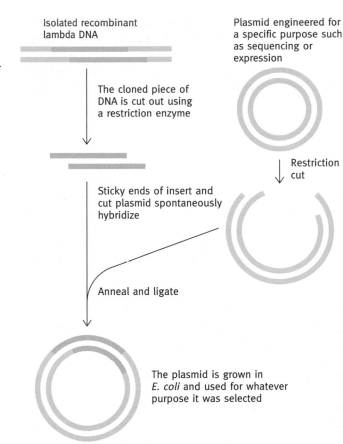

Fig. 29.3 Transferring a λ DNA insert into a plasmid. If it is desired to sequence a DNA insert that has been cloned in λ or, for example, to express a eukaryote cDNA in *E. coli* to produce a specific protein, the selected DNA is transferred to a plasmid engineered for the particular purpose such as sequencing or expression.

circularized to its original form and favours recombinant plasmids being produced.) *E. coli* cells are infected with a low multiplicity of the recombinant plasmids and grown into colonies on agar plates such that each colony arises from a single cell. Because of the low multiplicity of plasmid infection, only a very small proportion of the cells become infected, and the possibility of two plasmids being taken up by one cell is minimized. Since untransformed cells grossly outnumber transformed cells, it is necessary to have a quick method for selecting the latter.

One of the 'engineered' plasmids used for cloning is known as pBR322; this has, inserted into its DNA, genes for ampicillin and tetracycline resistance. Each of these genes has a different restriction site. Suppose that the foreign piece of DNA is inserted into the tetracycline gene; the latter is no longer capable

of directing the synthesis of the protein conferring resistance. A cell carrying such a plasmid will be resistant to ampicillin but sensitive to tetracycline. A cell containing a plasmid without a DNA insert will be resistant to both; a cell without a plasmid will be sensitive to both. This gives a basis for selecting only those cells containing the wanted plasmid.

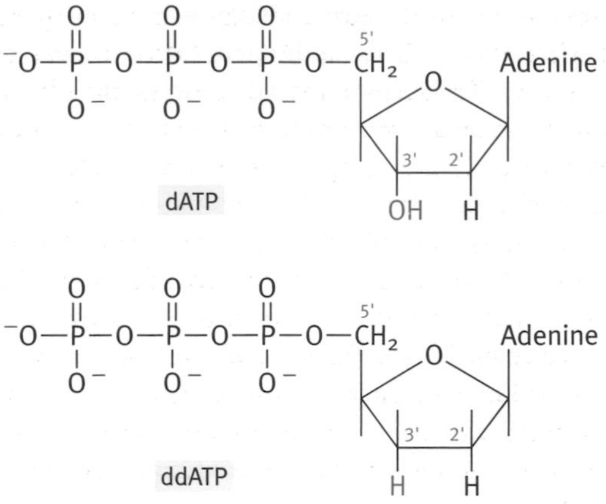

Determination of the base sequence of a cloned piece of DNA

The entire information in a gene lies in its base sequence and the determination of this is of central importance. There are two methods for DNA sequencing as the process is called. One is a direct chemical method—the **Maxam–Gilbert** method. The other is based on enzymic replication of DNA and this is called the **Sanger** or **dideoxy method**; this is the one most often used.

Outline of the dideoxy DNA-sequencing technique

Suppose that you have a cloned piece of DNA and want to sequence it. It is inserted into a specially engineered sequencing plasmid, which is multiplied by infecting *E. coli* cells with it. The plasmids are isolated so that you now have large numbers of recombinant plasmids containing your piece of DNA. The sequencing procedure (described below) requires that your piece of DNA to be sequenced is copied, *in vitro*, by DNA polymerase. This requires (apart from the four deoxynucleoside triphosphates (dNTPs)): (1) that the DNA to be copied is single-stranded; and (2) that a primer is hybridized to the start site because DNA polymerase cannot initiate new chains (page 324). The plasmid is rendered into single-stranded form by treatment with NaOH and heat. The sequencing plasmid is so engineered that at the 3′ end of each strand of your inserted cloned DNA piece is attached a known priming sequence to which a synthetic DNA primer will hybridize. A different primer site is used for the two strands enabling you to sequence one or other of the strands according to which primer you use in the copying process. Often, both strands are sequenced to confirm the results.

Incubation of the single-stranded DNA to be sequenced with primer, exonuclease-free polymerase I, dATP, dGTP, dCTP, and dTTP permits copying of the piece of DNA to proceed. One of the triphosphates (or the primer) is radioactive so that all of the new DNA copies are labelled.

The sequence determination depends on two things. First, an electrophoretic method to separate DNA molecules in which they migrate along a slab of acrylamide gel in an electrical field; the smaller the DNA chain, the faster it moves. Each addition of a nucleotide alters the migration so that chains form separate

Fig. 29.4 The structures of deoxyATP (dATP) and dideoxyATP (ddATP). The absence of the 3′-OH group means that, when a ddNTP is added to a growing DNA chain, the chain is terminated.

bands, each being one nucleotide different in length from the next. They are visualized by autoradiography, which involves exposing the gel to X-ray film sensitive to radioactivity. The second point concerns dideoxy derivatives of nucleoside triphosphates (ddNTPs). DNA polymerase adds a nucleoside to the 3′ OH of a growing DNA chain. Dideoxy NTPs lack the 3′ OH group (Fig. 29.4); they can still be added to a chain via their 5′ P, but the chain is thereby terminated.

It is important to emphasize that, when we talk of a 'piece' of DNA being sequenced, multiple copies of that piece are involved in the experiments. Even a minute amount of DNA contains a large number of individual molecules. (Remember that one mole of any chemical compound contains the astronomic number of 6.03×10^{23} molecules.) Suppose that we have in the copying process all four deoxy NTPs plus a small amount of *one* dideoxy NTP—take dideoxy ATP as an example—every time addition of an A is specified, most of the new chains will have a normal adenine nucleotide added, but a fraction will, by chance, have a dideoxy form of the adenine nucleotide added, thus terminating those particular chains. (The fraction terminated will depend on the relative proportions of dATP and ddATP.) The terminated chains are sufficient in number to be detected as a separate band on an electrophoretic gel. The rest will go on being added to until another A is due to be added and the same will happen again.

How is all this interpreted as a base sequence?

Suppose the piece of DNA being sequenced has a sequence with Ts placed as shown.

3′–X–X–T–X–X–X–X–T–X–X–T–5′

If, during the copying dideoxy ATP is present (along with excess deoxyATP) such that, at the addition of each A, a small proportion of the growing chains are terminated, the copying will produce the following population of chains (attached to the primer)

(a)   5′–X–X–ddA–3′
(b)   5′–X–X–A–X–X–X–X–ddA–3′
(c)   5′–X–X–A–X–X–X–X–A–X–X–ddA–3′.

On a sequencing gel these bands will be seen as the bands in Fig. 29.5 (left-hand column).

If a second incubation contains dideoxy TTP instead of the dideoxy ATP, an analogous set of chains terminating in T will be produced and, similarly, terminating in G and C if dideoxy CTP or GTP, respectively, are present in duplicate incubations. If all four incubations are carried out and the products are run side by side on a gel, the **sequencing ladder** shown in Fig. 29.5 is obtained, from which the base sequence of the piece of DNA can be simply read off. The sequence is read from the bottom of the gel upwards as the sequence of the copy chain (and therefore of the partner to the template strand) rather than of the template. This gives the sequence in the 5′ → 3′ direction since synthesis always proceeds in that direction.

About 200–300 bases can be determined in each sequencing ladder. If overlapping pieces from the original cDNA are obtained by using different restriction enzymes, the entire sequence can be determined. A part of an experimental sequencing ladder is shown in Fig. 29.6.

If the purpose of the sequencing of a cDNA is to determine the amino acid sequence of the protein coded for by the relevant gene, using the genetic code, it is necessary to determine the correct reading frame out of six possible on two strands. The correct frame is the one that does not prematurely run into triplets generating stop codons.

The above procedure is employed in many laboratories. However, sequencing technology has advanced to permit automation. The principle of the new method is unchanged

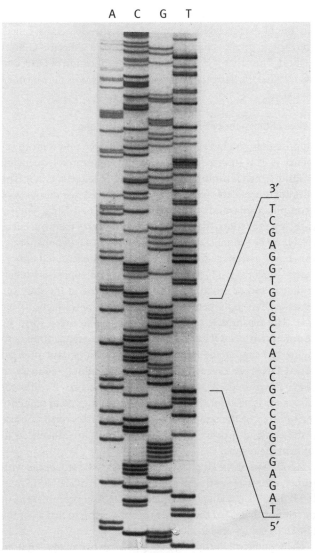

Fig. 29.6 Photograph of an actual sequencing ladder. A, C, G, and T specify which dideoxynucleoside triphosphate was present in each incubation. Kindly provided by Dr Chris Hahn, Department of Biochemistry, University of Adelaide.

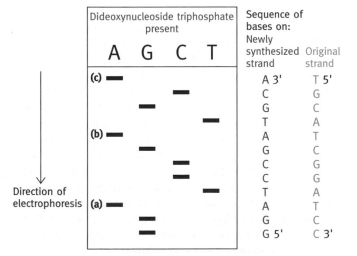

Fig. 29.5 An autoradiograph sequencing gel. The sequence is read from the bottom to the top. **(a)**, **(b)**, and **(c)** are the bands produced in the presence of dideoxy ATP. See text for explanation. A photograph of a real sequencing ladder is shown in Fig. 29.6.

from that devised by Sanger, but fluorescent-labelled dideoxy nucleoside triphosphates are used, the fluorescence being of a different colour for each of the four. A single DNA synthesis incubation is used containing all four dideoxy compounds, plus of course the four deoxynucleoside triphosphates and the products analysed on a single electrophoresis track. The gel is automatically scanned for the different colours, and a computer prints out the base sequence.

DNA databases and genomics

Genomics refers to the computational analysis of the vast amounts of data that is accumulating on genes and genomes (the latter being the collective term for the genetic material of an organism). It is a specialist area used by research workers but it is of such increasing importance to molecular biology, medicine, biotechnology, and indeed to any biological science that we should briefly draw your attention to it.

If a research worker has isolated a gene or other section of DNA and has sequenced it fully or partially, this and other information is usually recorded in one of the international databases; these are generally in the public domain. Software is available which can be used to analyse a sequence in different ways appropriate to the type of questions being asked. We will give only a few illustrative examples. When an unidentified gene or other DNA sequence is isolated and fully or partially sequenced, a search of the databases for matching sequences will reveal whether information on that piece of DNA, or a closely related one, already exists from previous work of other researchers. It is also possible for example to analyse a DNA sequence for the presence of open reading frames or potential splicing patterns. There are many applications available. The automatic sequencing of DNA has advanced in a spectacular way. Smaller genomes have already been completely sequenced including that of the fruit fly, *Drosophila*, which is used so extensively in genetic research and that of the favourite experimental eukaryotic microorganism, yeast. *Arabidopsis* is a plant used extensively for genetic studies; its genome has also been sequenced to the extent of 95% at the time of writing. Databases dedicated to all information of the genomes of particular organisms exist. Yeast is an example.

The most spectacular endeavour is the human genome project. This is a massively funded collaborative international project to determine the nucleotide sequence of the entire human genome (see references in the reading list). It is a gigantic task but is to a large extent completed. If an unidentified gene (or other piece of DNA) is isolated then, for example, determination of a part of its sequence will permit it to be searched for in the human genome database. The complete sequence will then be available and its location in the chromosomes relative to other genes, identifiable. The search might also for example reveal that it is part of a family cluster of related genes; its function or perhaps an association with a disease may have been determined from other work. Depending on the particular case, information of importance to medicine, biotechnology or basic science may be immediately available and speed up the advance of a research project to an almost unimaginable degree.

We have already explained (page 61) that genomics is the partner to **proteomics**, the two being collectively known as **bioinformatics**. As with proteomics use of the databases requires computational skills and expertise in molecular biology. The databases are increasingly powerful research tools in biochemistry and molecular biology. To give a sense reality to the account (or for the interested) the website for one DNA database is given below. It contains about 5 million or more DNA sequences.

Genbank database
http://www.ncbi.nlm.nih.gov/Genbank/GenbankOverview.html

The polymerase chain reaction (PCR) for amplifying a specific DNA segment

We mention this (out of an array of elegant techniques) because it constitutes a major landmark in DNA manipulation. In effect, it is a method of selecting a particular stretch of DNA on a long molecule and amplifying it by making vast numbers of copies. This means that a vanishingly small amount of DNA, perhaps from a few cells, is all that is needed for many purposes.

The principle is as simple as it is brilliant (Fig. 29.7). Take a piece of double-stranded DNA and separate the strands by heating, and copy a selected portion of them into DNA; repeat the process over and over (copy the copies), and this gives an exponential increase in copy numbers. The section to be amplified on a longer piece of DNA is selected by using primers for the DNA synthesis reaction corresponding to the limits of the section to be amplified. Different primers are needed for the synthesis in the left and right direction. To obtain these, of course, one must know the base sequences of the DNA sample to which the primers will hybridize. At the end of each round of synthesis, the DNA duplexes formed by the copying process are separated by heat, the solution cooled to permit further priming, and hence copying—automatic machines are available for achieving this and heat-stable DNA polymerases (from thermophilic bacteria) avoid the necessity of adding more enzyme after each round of copying and strand separation by heating. Cooling in each round is necessary to let the primers bind.

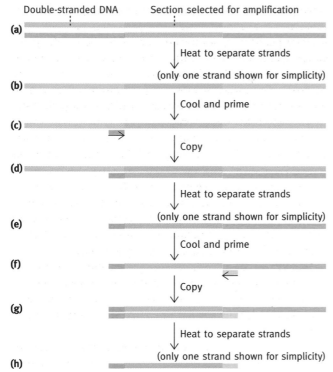

Double-stranded DNA   Section selected for amplification

(a)

↓ Heat to separate strands

(only one strand shown for simplicity)

(b)

↓ Cool and prime

(c)

→

↓ Copy

(d)

↓ Heat to separate strands

(only one strand shown for simplicity)

(e)

↓ Cool and prime

(f)

←

↓ Copy

(g)

↓ Heat to separate strands

(only one strand shown for simplicity)

(h)

This is a copy of the wanted section of DNA; with successive
reaction cycles it will be amplified exponentiially

Fig. 29.7 The basic principle underlying amplification of a DNA section
by the polymerase chain reaction (PCR). **(a)** The green bars represent
the section chosen for amplification by the use of appropriate primers
(→ ←), each complementary to the 3′ end of the section in one strand
to be amplified. **(b)** Heating separates the strands and, from this
point, to keep the diagram to a manageable size, amplification of only
one strand is shown; the process amplifies both strands. **(c)**
Incubations contain four dNTPs, a heat-stable DNA polymerase, and
primers. Cool so that the primer appropriate for this strand binds. **(d)**
DNA is replicated (in this case beyond the end of the desired piece).
(e) Heating separates the strands. **(f)** The new strand is primed. **(g)**
The new strand is replicated. **(h)** The separated wanted section. You
can see that we now have reproduced our selected piece. With about
25 rounds of replication, heating, cooling, priming, and synthesis, the
piece can be amplified a millionfold. Note that, every time priming and
copying occurs, all of the original and new molecules are so primed
and copied so that a detailed diagram of what happens becomes
rather involved. The essential is that an enormous amplification of the
selected piece occurs.

The practical limit of the length of DNA that can be amplified in this way was about two kilobases, but recent techniques
have enabled amplification of up to 35 kb. The DNA copies so
obtained can be used for many purposes. The technique is important in forensic science and in diagnosis of fetal genetic ab-

normalities *in utero* (see below). It is also extensively used in
biochemical research.

Applications of recombinant DNA technology

As stated already, a central role for gene cloning is to determine
base sequences of genes, an essential prerequisite for studying
gene function and control. The DNA base sequence is also a direct 'window' on evolutionary relationships. There are other
uses of practical importance that will be mentioned only briefly
here. These include the following (a far from exhaustive list).

Expression of the gene in bacterial and eukaryote hosts

As already briefly referred to, an isolated cDNA clone or a bacterial gene may be inserted into an 'engineered' expression vector.
This may be a plasmid into which have been inserted the appropriate bacterial DNA promoter signals and translational initiation signals such that the inserted cDNA molecule in each
plasmid is expressed—the cell synthesizes mRNA from the
cDNA and translates this into the protein it codes for. Thus, an
inserted prokaryote gene or eukaryote cDNA (which, remember, has no introns) can be efficiently transcribed and translated
into protein. Expression vectors are available commercially.

Human proteins of therapeutic value, otherwise virtually un-
obtainable, may be produced in unlimited quantities in *E. coli*
or other cells. Human insulin is produced in bacteria and there
is an ever-increasing list of such examples. Apart from making
available, in quantity, human proteins present in the body in
minute amounts, production in *E. coli* eliminates the danger of
a protein prepared from human sources being contaminated
with a human infectious agent. Thus a human growth hormone
is produced via a cDNA. An immunogenic protein of hepatitis B
virus is produced in yeast to be used as a vaccine with no danger
of infective virus being present. In some cases, *E. coli* can be
made to produce a foreign protein as a major percentage of the
total cell protein—so much that it precipitates out as **inclusion
bodies**. There is often a snag, however. Produced like this, the
protein may not fold up properly or only a small proportion
may do so (see page 394). In some cases, the inclusion bodies
have been isolated and the protein partially correctly refolded *in
vitro*. Bacteria do not glycosylate proteins (page 44), but, if appropriate animal cell systems are used as expression hosts, glycosylated proteins can be produced in quantity. Insect cells have
been used for this.

Site-directed mutagenesis

This is a very powerful technique for analysing the functions of
particular amino acid residues in a protein. Consider a case in
which you have an enzyme for which the amino acid sequence is

known, either from direct sequencing or by deduction from the base sequence of its gene or cDNA. You now wish to examine the role of a particular amino acid side chain in the function of that protein. If the bacterial gene or eukaryote cDNA is cloned into an expression vector such as a plasmid, the protein may be produced in *E. coli* in sufficient quantity for study. For example, in the case of an enzyme you could be interested in the rate of its catalytic activity or its specificity. **Site-directed mutagenesis** permits the specific replacement of one or more selected amino acids in the protein. This can be done by replacing the appropriate section of the isolated gene with a synthetic oligonucleotide whose base sequence is such that it codes for the altered amino acid. Methods for such procedures are well established. The altered catalytic activity of the mutant protein can then be studied.

Transgenesis and gene therapy

Genes can be inserted into animals and plants to give different phenotypes. This is referred to a **transgenesis**. It has been found that pieces of DNA injected into cells are incorporated into the cell's DNA where it may be expressed normally. The first example of this was achieved by injecting a cloned gene for growth hormone into the nucleus of a fertilized mouse egg. The egg was then implanted in a mouse to develop. It was found that some cells of the progeny contained multiple copies of the growth hormone gene. As a result it grew to about twice the size of a normal mouse. Moreover, some of the germ line cells contained the extra genes so that it is possible to breed transgenic animals with the extra genes.

This has given rise to hopes that it might permit genetic human diseases to be remedied by **gene therapy** which involves correction of a gene deficiency by inserting a normal gene into cells. In the case of severe **combined immune deficiency syndrome,** children lacking both genes for adenosine deaminase (page 294) in lymphocytes of the immune system fail to produce antibodies correctly and as a consequence suffer from repeated infections. The only known treatment was to isolate the child in a sterile plastic 'bubble'. This disease has been treated successfully in a number of people by inserting a normal gene into bone marrow cells and then returning them to the patient. We have already described (page 470) how retroviruses insert a DNA copy of themselves into host cell DNA and this was employed to introduce the adenosine deaminase gene. The genes needed for replication of the virus were first eliminated by recombinant techniques and then a normal adenosine deaminase gene added. The method has several risks. Insertion of retroviral DNA occurs at random sites on chromosomes and may interrupt and thereby inactivate one or more essential genes. It

could place a proto-oncogene under its control, thereby inappropriately activating it into an oncogene (page 499). Although the viral genes for replication are eliminated, there is always the danger that recombination with other viruses may restore viability to the virus. To mininize these risks, in the case of the adenosine deaminase therapy, the gene insertion was carried out *in vitro* on bone marrow cells obtained from the patient and then returned after transformation. Many difficulties have to be overcome before gene therapy will be a commonly used approach but a lot of work is going on. If bone marrow stem cells (page 488) could be readily obtained, which at present is not the case, then targeted replacement of defective genes by the methods described in the knockout mice section (below) might have the potential of treating gene deficiencies in blood cells. The use of other viruses as gene vectors may also have potential, especially those that would target specific types of cells.

Transgenesis of animals has the potentiality of improving growth rates or of producing wanted human proteins in sheep milk, to give just two examples. It seems that stem cell technology and targeted gene insertion would be potentially useful here.

In plants, foreign genes can be inserted into the chromosomes by using as cloning vector a naturally occurring (but suitably altered) plasmid (the Ti, or tumor-inducing, plasmid) contained in the pathogenic soil bacterium *Agrobacterium tumefaciens*, or the DNA molecules may be literally shot into plant cells, where they become functionally incorporated, with a gun-type instrument. In this way, crop plants are being engineered with specified phenotypic characteristics—for example, to be resistant to herbicides. The purpose of this is to control weeds by blanket spraying with the herbicide—only the resistant crop plant survives.

Knockout mice or gene targeting

Mutants have long played an important role in biochemistry and molecular biology. For decades, prokaryote mutants have provided essential information on the function of genes and of the proteins they code for. The classical principle is to inactivate a gene of interest and then see what are the consequences of this on the **phenotype** of the mutated cell. (Phenotype means the observable characteristics of an organism as opposed to **genotype,** which means the genetic characteristics that determine the phenotype.) It is easier to produce specific genetic mutants of prokaryotes or of lower eukaryotes such as yeast and the fruit fly *Drosophila* than it is to obtain mutants of mammals. With *E. coli*, for example, a culture can be randomly mutated by

chemical or other means and the treated cells grown up in selective media and the wanted mutants isolated. Because of their rapid growth rate, vast numbers of cells can be screened by growing them in selective media so that obtaining mutants of such cells, although not a trivial task, can often be achieved quickly. The problem is obviously much more difficult in mammals whose slow generation times preclude the screening of large numbers. Nonetheless because of the very complexity of animals, information from mutants is all the more important and has great medical relevance, since animal models of human diseases can be constructed. Mice have become the experimental animal of choice and animals in which a specific gene has been functionally eliminated are known as 'knockout mice' or sometimes for greater precision of expression, gene-knockout mice.

Recombinant DNA technology has made the obtaining of specific mutants in mice feasible and it is now a widely used procedure. The development of gene targeting technology in which a specific gene is inactivated has been an important advance over the past decade. In this, a specific gene in the mouse is selected and a targeting vector constructed using the methods of recombinant DNA technology. A common method (and the one we will describe) is to replace the selected gene with a piece of foreign DNA (see below). The basis of the targeting is homologous recombination (described on page 336), so it is essential that at both ends of the section to be incorporated there are regions of homology with the section it is to displace. An homologous recombination event at each end of the DNA to be inserted causes displacement of the targeted section. The technique of gene targeting is only half of the problem, because there follows the question of how you arrange that *all* the cells of the mouse have the mutation, which means that the latter must be present in the germline cells. The solution to this lies in the development of **embryonic stem** (ES) **cell technology**. This promises to a development of great importance, not only in the development of mutant mice. It has major medical potential. We will therefore now spend a little time describing this system.

The embryonic stem (ES) cell system

Stem cells are cells that can divide without limit. There are various groups of them in the body which provide replacement cells in large numbers. Thus there are stem cells for blood cells, sperm cells, skin cells, and epithelial cells. When a stem cell divides, the progeny may terminally differentiate into the appropriate cell type or remain as stem cells. When a cell terminally differentiates into a somatic cell such as a liver cell it has the potentiality to divide to replace wastage but the capacity to replicate is limited to a relatively small number of generations. This is discussed on page 495; the point is mentioned to highlight the

difference between such a cell and a stem cell. **Embryonic stem (ES) cells** are different from other stem cells, and differentiated cells, in a very important way: they give rise every type of cell in the body; they are **totipotent.**

(As an aside, we have to qualify this statement because the famous cloning of **Dolly the sheep** has shown that the nucleus of a differentiated cell which has lost its totipotency and ability to divide for more than a limited number of generations can be restored to totipotency by transfer, with appropriate manipulations, to a denucleated fertlized egg. See reference of the reading list if you would like a clear summary of the work. This has no relevance to the production of knockout mice, to which topic we will now return.)

After fertilization, a mammalian egg divides to form a solid ball of cells, which then develops into a **blastocyst.** This is a mostly hollow sphere of cells containing an inner cell mass (Fig. 29.8). The cells of this inner cell mass are the ES cells. These will give rise to all cells of the mature animal, but at this stage individual cells are not committed to becoming any particular type in the adult. They have another property vital to practical use of these cells: they can be propagated in long-term culture. Under appropriate conditions they do not differentiate, and retain their totipotency. A further vital property, in our present context, is that the cultured cells can be injected back into the cavity of a blastocyst and when the latter is re-introduced into a mouse foster mother it develops into progeny in which the reintroduced cells contribute to all tissues, including germline cells. In any one tissue some of the cells will be derived from the cells of the 'natural' blastocyst and some from the injected ES cells which were obtained from a different blastocyst.

As another aside from the topic of generating mutant mice, ES cell technology promises to be of importance in other directions. By the appropriate use of growth factor (cytokine) signals to promote particular types of differentiation, it may permit the production, in culture, of differentiated cells of any type for use in human therapy or possibly, in the longer term, for growing replacement organs. We will now describe the production of knockout mice by gene targeting.

Gene targeting

The first step is to construct a **targeting vector.** In the example we will use, the target gene is inactivated by replacing it with a **neomycin gene**; the latter has flanking sequences homologous to the targeted gene (Fig. 29.9) so that homologous recombination (page 336) at each end of the replacement section causes the new piece of DNA to replace the targeted gene. The technique is illustrated in Fig. 29.8. We suggest that you closely follow the steps in the figure.

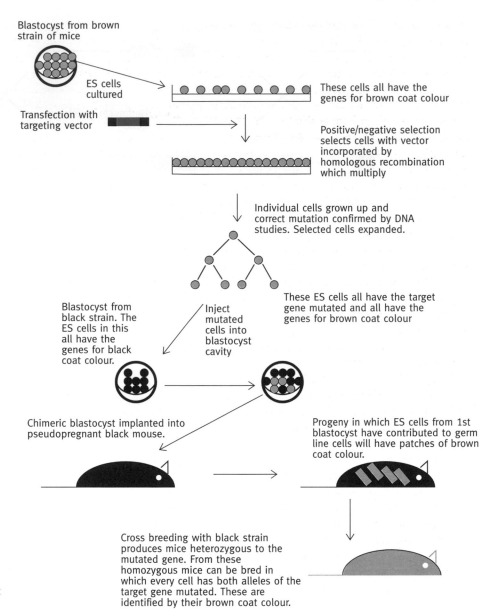

Blastocyst from brown strain of mice

ES cells cultured

These cells all have the genes for brown coat colour

Transfection with targeting vector

Positive/negative selection selects cells with vector incorporated by homologous recombination which multiply

Individual cells grown up and correct mutation confirmed by DNA studies. Selected cells expanded.

Blastocyst from black strain. The ES cells in this all have the genes for black coat colour.

Inject mutated cells into blastocyst cavity

These ES cells all have the target gene mutated and all have the genes for brown coat colour

Chimeric blastocyst implanted into pseudopregnant black mouse.

Progeny in which ES cells from 1st blastocyst have contributed to germ line cells will have patches of brown coat colour.

Cross breeding with black strain produces mice heterozygous to the mutated gene. From these homozygous mice can be bred in which every cell has both alleles of the target gene mutated. These are identified by their brown coat colour.

Fig. 29.8 Procedure for obtaining knockout mice by gene-targeted mutation. The diagram illustrates how coat colour provides a convenient way of picking out wanted progeny. See text for explanation.

- ES cells are obtained from a mouse blastocyst and cultured *in vitro*.

- The cells are then **transfected** by the targeting vector, usually by **electroporation**. This involves giving an electric shock to the cells which transitorily forms holes in the cell membrane and allows the vector to enter. Most frequently, in mouse cells, the introduced DNA integrates into the host chromosomes in a random fashion, which is not what is wanted, but a low percentage (0.5–10% integrate by homologous recombination at the target site. (In the yeast cells most integrations are at the target site.)

- Stably transfected cells are now resistant to the antibiotic due to expression of the neomycin gene so that these cells can be selected by growing the whole sample in the presence of the antibiotic. Cells without the neomycin resistance gene fail to grow. It is necessary to select those cells in which integration has occurred by homologous recombination rather than by random insertion. A rather clever

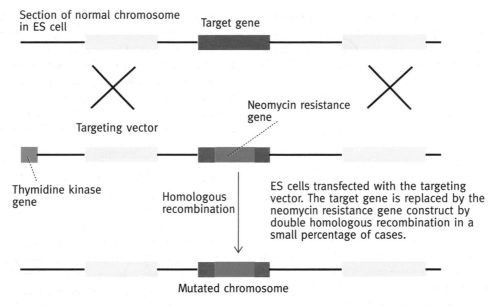

Fig. 29.9 One method of gene targeting by homologous recombination in ES cells. ES cells are obtained from a mouse embryo at the blastocyst stage and cultured. They are then transfected with a targeting vector constructed by recombinant DNA methods. In this, the target gene is replaced by a neomycin resistance gene flanked by sequences homologous to the normal gene. Homologous recombination replaces the target gene with the neomycin resistance gene construct in a small number of cases. Stably transfected cells are cultured and cells which have incorporated the foreign gene by homologous recombination are identified by recombinant DNA methods and cultured. The mutant cells are used to obtain knockout mice as described in the text and illustrated in Fig. 29.8. The yellow rectangles represent normal genes flanking the target gene. The blue (TK) rectangle represents the addition of a viral thymidine kinase gene which is placed outside of the region of homology between the vector and targeted chromosome. This is to permit negative selection of cells in which the vector is incorporated randomly rather than by homologous recombination (see text for explanation).

trick facilitates this. In constructing the targeting vector, a gene for **thymidine kinase**, derived from the herpes simplex virus, is inserted at one end of the construct, well outside the region of homology to the target gene. If the DNA is incorporated randomly, the thymidine kinase gene goes in with it and is expressed; if the incorporation is by homologous recombination the kinase gene is not incorporated (because it is outside the homology region). The relevance of all this is that if the kinase gene is expressed it makes the cell sensitive to **gancyclovir**, since the latter has to be phosphorylated to be toxic to the cell and the viral thymidine kinase does this. Thus, in the **positive–negative selection**, as it is called, only those cells with the neomycin gene but without the thymidine kinase gene will grow.

- The cells which survive this test are separately grown up as clones and individual clones of cells are examined to confirm that the replacement gene has been correctly inserted by homologous recombination. This is done using recombinant DNA techniques such as Southern blotting (page 491) or PCR (page 485).

- Cells in which the targeted gene has been replaced are grown up and injected into the cavity of a mouse blastula.

- This is placed in a pseudopregnant foster mother (pseudopregnancy results from mating with a sterile male). The injected, mutated, cells participate in formation of tissues as the embryo develops so that in each tissue of the progeny mouse, some cells will have originated from the original blastula cell mass and some from the injected transfected cells. In other words, the mouse will be a chimera.

- As a ready guide to what is occurring genetically, coat colour is used. Suppose the original blastula from which the ES cells for manipulation were obtained had a brown coat colour and the recipient blastula into which modified cells were injected came from a black mouse strain, the mouse resulting from implantation of the latter into the foster mother will be a black/brown mixture as shown in the figure. These are not the wanted mutant mice, for only some of the cells are mutated.

- What is needed are mice in which every cell in the body is a null mutant with respect to your target gene. To get these, the chimeric mice are crossed with a black mouse of the strain from which the host blastula was derived. This will give rise to grey progeny, in which all cells have one mutated gene and one normal gene. The latter are heterozygotes derived from germline cells.

- Inbreeding of these results in progeny half of which are brown and half black. The former are homozygous to the mutated gene. In other words, all the cells in the body of the mouse lack the targeted gene. It is a knockout mouse.

Increasingly, knockout mice are being used to test the role of genes in complex biochemical systems and importantly to create animal models of human diseases that are attributed to loss of a specific gene. As an example, a mouse model of the human disease familial hypercholesterolemia has been made by knockout of the gene for the LDL receptor (page 130). However, in some cases, gene knockout has not produced the expected phenotype so that the phenotype of the human disease is not seen in the knockout mice. This is attributed to **redundancy**, which implies is that in an organism as complex as a mammal or even a yeast cell, if one component of a pathway, be it a control or metabolic pathway, is functionally eliminated, the necessary function is carried out or compensated for in an alternative way.

Detection of genetic abnormalities by restriction analysis or Southern blotting

Where the structure of genes is known, hybridization probes can be made to detect gene abnormalities. The amount of DNA required is sufficiently small to permit examination of fetuses so information on potential diseases is available to permit decisions to be made on possible termination of a pregnancy (use of the polymerase chain reaction amplification means that only minute amounts of DNA need be obtained for examination).

For example, in muscular dystrophy, the gene abnormality usually is a deletion of part of the coding DNA and this can be detected by a **Southern blot analysis**. The latter, named after its discoverer, involves cutting DNA with one or more restriction enzymes, running the fragments on an electrophoresis gel, transferring the separated fragments on to a membrane, and probing the membrane with a gene-specific radioactive hybridization probe. This gives a pattern of DNA pieces, visible as bands. Gene abnormalities can show up in a different pattern of bands, because mutations may create a new restriction enzyme site or destroy or delete a pre-existing one. Even though the gel will have vast numbers of different pieces of DNA on it, only those few that hybridize to the specific probe will show up, thus giving a simple pattern of bands.

Even when the gene involved in causing a genetic disease is not precisely localized or isolated, a technique (not described here), known as **restriction fragment length polymorphism (RFLP)** analysis or **genetic linkage analysis,** can be employed to identify which relatives of a person displaying the disease are likely to have the abnormal gene.

A more easily understood use of RFLP analysis is in **DNA fingerprinting**. In this, repetitive human DNA sequences located throughout the genome (page ••) show extreme polymorphisms in restriction sites so that restriction patterns are unique to individuals. This technique is still used in forensic science, but a new technique based on the PCR reaction is replacing it. In this, use is made of the fact that a large number of repeat sequences have been identified in the human genome. These are highly variable from individual to individual in the number of repeats in the sequence. The method is to choose primer sites on either side of the selected repeat sequences and, using PCR, to amplify the latter.

Running the product on an electrophoretic gel gives a band of a size depending on the number of repeats in the amplified segment. By selecting a number of such loci to amplify, a pattern of bands highly characteristic of an individual is obtained.

FURTHER READING

Gene cloning

Jordan, E. and Collins, F. S. (1996). Human genome project: a march of genetic maps. *Nature,* **380,** 111–12.

A short news and views summary of the human genome project, which emphasizes the role of genetic maps in analysing disease genotypes. The article is an excellent guide to future work.

The Human Genome Project. (1999). *Current Biology,* **9,** R908–9.

A summary of the quick guide type summarizing the project in one and a half pages.

The polymerase chain reaction

Saiki, R. K., Gelfand, D. H., Stoffel, S., Scharf, S. J., Higuchi, R., Horn, G. T., Mullis, K. B., and Erlich, H. A. (1988). Primerdirected enzymatic amplification of DNA with a thermostable DNA polymerase. *Science*, **239**, 487–91.

Detailed account of the technology of the PCR reaction.

Arnheim, N. and Erhlich, H. (1992). Polymerase chain reaction strategy. *Ann. Rev. Biochem.*, **61**, 131–56.

Readable account of all aspects of the polymerase chain reaction.

Gene therapy

Morgan, R. A. and Anderson, W. F. (1993). Human gene therapy. *Ann. Rev. Biochem.*, **62**, 191–217.

Deals with the candidate diseases, gene transfer methods, treatment of adenosine deaminase deficiency and concludes with safety and ethical questions.

Capechi, M. R. (1994). Targeted gene replacement. *Sci. Amer*, **270**(3), 34–41.

Describes gene replacement by homologous recombination and transgenesis techniques.

Anderson, W. F. (1995). Gene therapy. *Sci. Amer.*, **273**(3), 96–8B.

An account of the successful insertion of adenosine deaminase into white cells, and their return to the patient to cure the combined immunodeficiency syndrome. Includes a list of 14 diseases being tested in clinical trials of gene therapy.

Gurdon, J. B. and Colman, A. (1999). The future of cloning *Nature*, **402**, 743–6.

A special News and Views article describing how Dolly was cloned by nuclear transplantation including discussion of the potentialities use of stem cell technology for growth of organ transplants the ethical and legal implications and the future of cloning.

Gene targeting

Melton, D. W. (1994). Gene targeting in the mouse. *BioEssays*, **16**, 633–8.

Describes the technology of obtaining knockout mice and discusses its application to study of human diseases.

PROBLEMS FOR CHAPTER 29

1   How does a restriction enzyme differ from pancreatic DNase?

2   *E. coli* RI cuts at a hexamer base sequence; such a sequence must occur many times in *E. coli* DNA. Why doesn't the enzyme destroy the cell's own DNA?

3   What is meant by the term 'sticky ends', as applied to DNA molecules?

4   What is a genomic clone? What is a cDNA clone? How do they differ? Refer to eukaryotes in your answer.

5   What are the steps in making a genomic library in lambda bacteriophage?

6   What are the steps in isolating a particular clone from a genomic library in lambda?

7   What is a dideoxynucleoside triphosphate? What is the precise function of these in the Sanger technique of DNA sequencing?

8   Explain in a few sentences the importance and principle of the polymerase chain reaction, specifying the requirements for it to be used.

9   What is a bacterial plasmid expression vector?

10   What is restriction analysis?

11   What is a stem cell and why are they of great current interest?

12   Describe in conceptual outline how mammalian genes can be targeted to produce specific mutants, generally known as knockout mice.

Chapter summary

Molecular biology of cancer

Cancer involves loss of control of a number of processes; above all, it involves the uncontrolled multiplication of cells. The disease is mainly the result of the normal processes which control cell division being faulty. Rapid cell division occurs normally in stem cells such as those producing blood cells in the bone marrow and germ line cells that produce reproductive cells. Stem cells are capable of multiplying throughout the lifetime of an animal. Each of the daughter cells has the choice of remaining a stem cell or terminally differentiating into a mature, functional cell such as those of skin and liver. The latter for the most cases stop dividing except for the small numbers of divisions needed to replace dead cells, but they retain the ability to divide rapidly if they are given the requisite signals to do so. Wounding of the skin fires off cell replication to heal the wound, but when this is achieved *cell division ceases*. If two-thirds of a rat liver is surgically removed, cell division restores the original size in days but then the *division ceases*. The signalling pathways from growth factors and cytokines control the replication process so that it is appropriate to the needs of the body as a whole.

The cancer cell grows without appropriate signals from other cells or with a diminished requirement for them. It is out of control in the molecular sense, in that cell division occurs when it should not. The difference between normal and cancerous cell division can be illustrated by tissue culture experiments. Mammalian cells can be grown on nutritive medium in Petri dishes if they are supplied with growth factors, usually provided by serum. They multiply and spread out to cover the surface of the plate in a single layer until they are in contact with each other and they then stop growing. Cancer cells are less dependent on growth factors and keep on growing after the plate is completely covered, the cells piling up on one another to form a solid mass, which is essentially is the equivalent of a tumour (Fig. 30.1).

Normal somatic cells can be cultured on nutritive plates for only a relatively small number of generations before some sort of ageing process stops division. In an earlier chapter (page 333) we described how a linear chromosome is shortened after each round of replication, obviously a potentially disastrous situation and one which is now suspected to be the basis of a molecular cell-division counter. This shortening is the inevitable consequence of the mechanism by which DNA is synthesized. It applies to eukaryotic cells whose chromosomes are linear, but not to the circular prokaryotic ones. To cope with this, each eukaryotic chromosome has **telomeres** at its ends. These are extensions of each DNA strand consisting of repeating short units of DNA put on by the enzyme **telomerase** which is a **reverse trancriptase** carrying its own template RNA (see page 334 for a fuller account). The telomeric DNA has no known informational role; it is there to be progressively lost on chromosome replication and, in being sacrificed, protects the 'real' chromosomal DNA carrying the genes. In rapidly dividing stem cells, the telomerase replenishes the lost telomeric DNA so that no matter how many times the chromosome is replicated it is protected from shortening. However, somatic cells do not usually contain telomerase so that once the telomere is reduced to a critical length, functional cell division ceases. It would probably be too simplistic to imply that this is the sole 'safety catch' for limiting cell division, but it is likely to be an important factor and it is speculated that it may contribute to the ageing process in both cells and people. It would seem therefore that, on differentiation, a cell is given its 'ration' of telomeric DNA for so many cell divisions and that is it. There are many experiments supporting this general concept; it has been shown that in skin fibroblasts (described on page 52), telomeres do shorten and, in other experiments, that artificially lengthening telomeres can

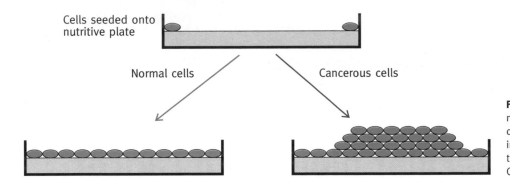

Fig. 30.1 Growth characterisitics of normal and cancerous animal cells in culture. Normal cells show contact inhibition—they stop dividing when they are in contact with each other. Cancer cells do not.

increase the a number of cell divisions a nontransformed cell can undergo. Nonetheless, cancer cells derived from normal somatic cells are immortal; there is no intrinsic limitation on their ability to divide, as is the case with normal cells. It is therefore of great interest that such cells are found to have telomerase. Indeed, conversion of normal human fibroblasts to cancerous cells in culture has also been shown to require telomerase activity.

This does not imply that telomerase activity causes cancer, but that it is a prerequisite for the immortality characteristic of cancer cells. To underline the latter characteristic, a cancer cell line originating from a cervical cancer has been cultivated in countless laboratories for well over half a century and has been through vastly more cell divisions than is possible for a normal somatic animal cell.

There are two broad classes of abnormal cell growth in the body. Some cells grow and form a solid tumour, which simply gets bigger but remains as a single 'lump'; this is a **benign tumour**, so called because it does not spread to adjacent tissues or throughout the body. It may not pose a threat to life other than from the effects of physical compression and is often relatively easy to cure surgically. The other sort are malignant cancer cells; they break off the tumour and extend into adjacent tissues or migrate to lymph nodes and elsewhere and set up further tumours. This spread of a tumour is known as **metastasis.** Such metastasizing cancers are the dangerous ones; the cells have extra properties over and above their immortality. Metastasis involves changes to the cell surface so the cells no longer adhere to each other. Also enzymes are produced which break down basal lamina layers (see page 52) of connective tissue and facilitate the escape and migration of the cells. The tumours become invaded by blood vessels, which supply the solid mass with oxygen and metabolites, the vascularization being stimulated by factors secreted by the tumour itself. A third class of cancers are the **leukaemias** which are 'cancers of the blood'. In this case, large numbers of abnormal white blood cells continue to multiply and circulate without giving rise to, and often at the expense of, normal functional white blood cells.

Returning to the main point, since cancer is, as stated, initially due to uncontrolled cell division it is necessary now to deal with what controls this. We have earlier briefly referred to the eukaryotic cell cycle but now we will look in more detail at its nature and the controls operating on it.

The cell cycle and its control

Overview of the cycle

In a prokaryotic cell such as *E. coli*, DNA synthesis proceeds all the time, given conditions which permit growth. Cell division is not strictly coordinated with rounds of DNA replication so that often a new round commences before the first has been completed. The situation in eukaryotes is very different. Everything is strictly controlled. In terms of what can be observed in the light microscope, at mitosis, chromosomes appear as compact X-shaped structures which align themselves on the midline of the cell, the duplicated chromosomes separate to form two daughter nuclei and then **cytokinesis** or division takes place (see Fig. 33.7). The process takes about 1 hour in a typical animal cell out of the 24 hours needed for a complete cell cycle. Between mitoses, a period known as **interphase**, the chromosomes exist as unravelled threads which pervade the nucleus; the chromosomes are no longer visible in the light microscope. During interphase, the cell synthesizes part of the components needed to grow in size more or less continuously. This, however, does not apply to DNA synthesis, which is confined to **S phase** of perhaps 7 hours' duration out of a 24 hour cycle in mammalian cells (Fig. 30.2(a)). Prior to the S phase there is a G_1 **phase** (G for gap in DNA synthesis) lasting somewhere around 10 hours, although these times vary considerably in different cells. In this, the cell prepares for the S phase and importantly a decision is

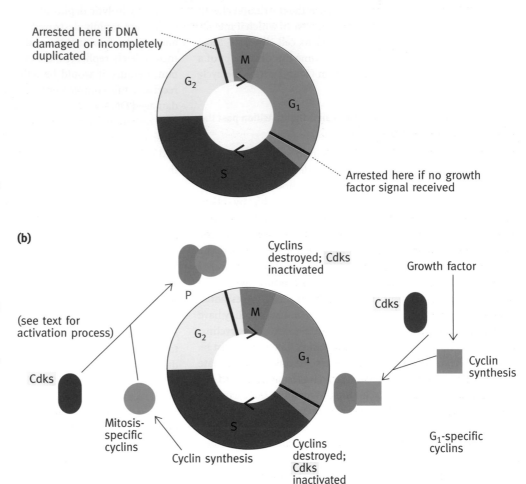

(a)

Arrested here if DNA damaged or incompletely duplicated

Arrested here if no growth factor signal received

(b)

Cyclins destroyed; Cdks inactivated

Growth factor

(see text for activation process)

Cdks

Cyclin synthesis

Mitosis-specific cyclins

Cyclin synthesis

Cyclins destroyed; Cdks inactivated

G_1-specific cyclins

Fig. 30.2 (a) Simplified diagram of cell cycle check points. (b) Simplified diagram of control mechanism. In mammalian cells the cell cycle cannot pass the restriction point in G_1 unless a mitogenic signal is received which activates the synthesis of G_1-specific cyclins. The latter are required to activate kinases (Cdks) necessary for activation of genes involved in the progression into S phase at which stage the cyclins are degraded. Activation of the same kinases by mitosis-specific cyclins is required for progression to the M phase. The different cyclins target the Cdks to different substrates as appropriate for the different phases of the cell cycle. Red represents inactive Cdks and green the activated forms. The dark and light blue shapes represent different cyclins. Note that in the interests of clarity the figure does not show the multiplicity of Cdks or of the activating cyclins involved. The details of these do not affect the essentials of the control scheme.

made whether the cell should proceed to cell division or not. If the decision is yes, the cycle progresses to S phase and the DNA is duplicated. After completion of S phase, the cycle progresses to the second gap phase or G_2 in which the cell prepares for mitosis and once again a decision is made whether to proceed to mitosis. If the answer is yes, mitosis commences but even here if things are not going correctly the process is halted. In Fig. 30.2(a) the checkpoints in G_1 and G_2 are indicated. (The former is known as the **restriction point** but it has no connection with the term restriction as in restriction enzymes, page 477.)

Control of the cell cycle

Let us start at the point at which mitosis and cell division have been completed and the G_1 phase has started. During G_1,

growth occurs and the cell prepares to enter S phase. If the cycle passes the restriction point near the end of G_1, it is committed to proceed all the way to the next checkpoint just before mitosis. As already explained in the chapter on cell signalling, it is vital that animal cells divide only if and when it is appropriate to the body as a whole for them to do so. An individual cell cannot make this decision; it must be given a signal from elsewhere to proceed. This is the role of extracellular signalling molecules, the cytokines and growth factors (page 426) secreted by other cells, so that cell division in the tissue is coordinated properly. Again as explained in the cell signalling chapter, these growth factors perform their role by transmitting signals to the nucleus via signal transduction pathways that result in the control of the activities of specific genes. Thus a mitogenic (growth) factor will activate genes which leads to the synthesis of proteins

necessary for the cell to proceed to S phase. If a mitogenic signal is not received, the cycle is halted at the restriction checkpoint and the cell enters a quiescent G_0 **phase** in which it metabolizes normally but does not proceed to cell division. Most somatic cells remain in G_0 most of the time but on receipt of a mitogenic signal the cell switches from G_0 and enters the cycle again at G_1.

How do growth factors and cytokines cause the transition past the restriction checkpoint in G_1?

It will not by now surprise you that, as in so many cellular controls, protein kinases are central to cell cycle control. We remind you that the overwhelmingly important method of control is by phosphorylation of proteins and reversal of this by phosphatases. (If necessary please refer to pages 215 and 428 which describe this type of control in metabolism and cell signalling respectively). The crucial kinases for us at the moment are the or **cyclin-dependent protein kinases (Cdks)**; there are several different Cdks involved in cell cycle control but we will avoid most of the details of this since it does not affect the essential story. On their own the protein kinases are not active; they have to be activated by combination with proteins known as **cyclins**, particular kinases requiring specific cyclins. The latter must be newly synthesized in G_1 and are synthesized only if the cognate cyclin genes are activated by growth factors or cytokines. If the latter are not present the cycle stops. Why must the cyclins be newly synthesized? As soon as the cell enters S phase, the cyclins relevant to the G_1 phase kinase activity are rapidly destroyed by the ubiquitin–proteasome mechanism already described (page 271) so that in every cell cycle the cyclin proteins are resynthesized and destroyed, an essential basis of cell cycle control. Thus if there is no mitogenic signal there is no cyclin D, the Cdks are not activated, and the cycle cannot proceed into S phase.

Oddly enough, the same Cdks may be required for the progression of the cycle from S phase into mitosis but in this case different cyclins are needed. Nonetheless the kinase/cyclin complexes have different target substrates in the G_1 and G_2 phases. It is not known how different cyclins cause a given Cdk to have different substrate specificities, but one possibility is that the cyclins in associating with the kinase provide, or participate in the formation of, the binding site for the different substrates. The mitotic-related cyclins accumulate in the cell during S and G_2 phases and combine with the relevant Cdks. However, the cyclin—Cdk complex is not immediately active because other enzymes triply phosphorylate the kinase which renders it inactive. Just before mitosis, two of the phosphate groups are removed causing activation of the Cdk. The reason for this convoluted process is probably that it permits build up of the triply phos-phorylated inactivated form and then a very rapid activation by simple hydrolytic dephosphorylation just prior to mitosis. The cyclins are rapidly destroyed after entry into mitosis. There is another checkpoint at the end of G_2. If the DNA is damaged or incompletely replicated, the cycle is halted here. This is important, because it would be lethal if the progeny cells did not receive a full complement of genes and equally disastrous if damaged DNA were to be replicated for it would confirm whatever mutations existed in the damaged section. Halting of the cycle at this point gives time for the replication to be completed and/or for the damage to be repaired. Figure 30.2 summarizes the controls mentioned so far. Other safety checks ensure that cell division does not proceed unless the chromosomes are correctly assembled on the mitotic spindle (see page 541).

What is the molecular nature of cancer?

Cancer always involves multiple genetic changes; that is one very important reason why the incidence of cancer escalates dramatically with age—there is more time for somatic cells to accumulate mutations. For each particular type of cancer, development of the cancerous state involves a progression of changes that differs in the different types and is still incompletely understood. It does not result from a single genetic event but from a series of them which accumulate over long time periods as a clone of cells multiplies into a polyp or tumour. It is well known that mutagens, compounds which damage DNA in some way, are carcinogenic. Painting the skin of mice with such compounds induces skin cancers; to be effective the skin is also painted with a tumour promoter such as a phorbol ester which mimics the second messenger, diacylglycerol (DAG), the activator of protein kinase C (page 440). The difference between a phorbol ester which promotes cancer and DAG, a natural second messenger, is that the former is long lasting whereas DAG is immediately destroyed and so does not give inappropriate prolonged responses. A tumour promoter does not cause cancer but by promoting cell division it increases the chance of the genetic events resulting in cancer

What types of genetic changes are involved in generation of the cancerous state?

There are many different cancers with different specific molecular causes; it will be useful if we deal with the subject in very broad terms giving illustrative examples. Oncogenic changes (those leading to the development of cancer) can be divided into two broad types.

- They can be due to the acquirement or development of **oncogenes,** which give abnormal control signals leading to uncontrolled cell division Their effect is positive; they do something, which actively causes or facilitates the development of cancer. Put colloquially, they are the 'bad' genes.

- The second broad category of genetic changes that contribute to cancer development is due to the loss of functional genes that protect cells against uncontrolled cell division. They are in colloquial terms the 'good' genes. Mutation of these genes does not itself lead to uncontrolled cell growth, but removal of their protective effect means that if oncogenes become activated, cells containing them are now more likely to make the progression to the cancerous state. These protective genes are known as **tumour suppressor genes,** a good many of which are known but the best-studied are the **retinoblastoma genes** and the **p53 genes.** The retinoblastoma genes were first discovered in cases of a rare cancer of the eye but their inactivation is a factor in several cancers. Loss of p53 gene function is known to be a factor in about half of all human cancers.

DNA repair genes are also protective but are not conventionally described as 'tumour suppressor' genes. In Chapter 21 (page 329) we described the mechanism of methyl-directed mismatch repair involving Mut proteins in *E. coli.* It has been found that proteins corresponding to these are present in eukaryotes, mutations of which are associated with a form of heritable bowel cancer in humans. Since mismatched bases (if not corrected) can lead to mutations and possibly to the formation of oncogenic forms of normal genes, it follows that DNA repair systems are protective.

How are oncogenes acquired?

- They may be introduced by viral infection.
- Normal genes involved in control may mutate and become oncogenic.

We have already described the nature of retroviruses and the manner in which they may introduce oncogenes into cells (page 470), but a short recapitulation here will be helpful perhaps.

Retroviruses are RNA viruses. In the evolutionary sense they often have accidentally picked up an mRNA from cells in which they have multiplied. If the RNA picked up happened to be one coding for a cell growth control protein it may have undergone a mutation which rendered it oncogenic when reinserted back into a cell by infection. Presumably such a situation would be selected for, because cell division favours virus replication. When a retrovirus infects a cell, its RNA is transcribed into DNA by reverse transcription and this becomes permanently integrated into the genome of the cell. The important discovery was made that in a normal cell there is a close relative of each retroviral oncogene; these normal counterparts are usually genes for signal transduction control proteins. It appears that the viral oncogenes originated (in the evolutionary sense) from a mRNA of a normal cellular gene coding for such a protein or for a transcription factor. The normal cellular gene is called a **protooncogene.** This name does not imply that there is anything abnormal about it but only that it has the potentiality to become an oncogene. Viral oncogenes are named after the retroviruses carrying them (and printed in italics). Thus, *ras* was found in the **rat sarcoma virus.** The oncogene is identified by the letter 'v' (for virus) and the protooncogene by 'c' (for cytoplasmic). Thus we have v-*ras* and c-*ras* etc. The derived proteins are written as Ras, Raf, etc. (not printed in italics). As an example, the v-*erb* oncogene codes for an abnormal form of the receptor for epidermal growth factor (EGF), described below. DNA viruses are also in some cases potentially oncogenic in humans. (Retroviruses are known to cause cancer in animals but, although no examples are known to do this directly in humans, the retroviral story has greatly helped in our understanding of cancer.)

Other viruses are known to be oncogenic, in some cases in humans, but oncogenes are not solely acquired from viral infection, perhaps not even mainly—protooncogenes can become oncogenes as a result of mutations due to base changes; mutation of the Ras protein so that its GTPase (page 431) activity is absent results in overactivity of the Ras signalling pathway (page 431) is oncogenic in humans. In addition, **chromosomal rearrangements** can result in translocation of genes involved in control of cell division. This can cause oncogene formation in several ways. A protooncogene (normal) may be placed under the control of a different regulatory element (gene promoter) causing overexpression of a particular protein. **Burkitt's lymphoma** is a cancer of human B cells of the immune system. It is due to the c-*myc* gene being put under the control of an immunoglobulin gene promoter due to a chromosomal breakage and rearrangement, resulting in the formation of excessive amounts of the transcription factor Myc. Since Myc controls other genes involved in cell growth, this leads to excessive gene activation leading to the rapid growth of the lymphocytes in an uncontrolled fashion. Alternatively, translocation of a section of DNA may cause gene fusion and result in the production of an oncogenic hybrid protein. Random insertion of retroviral DNA can also have a similar effect since a protooncogene may by chance come under the control of a very active viral promoter again leading to excessive production of a transcriptional factor or other control protein. In addition gene amplification (page 314) may cause overproduction of a control protein.

How do oncogenes cause cancer?

We remind you that cancer occurrence is the result of multiple genetic effects not single ones but now we will look at the molecular basis of oncogenicity.

The central role of signal transduction

Consider the following:

- Cancer involves unregulated cell division.
- Cell division in mammals is controlled by growth factors and cytokines.
- Growth factors and cytokines combine with receptors on cell surfaces and activate them.
- Activated receptors activate several different signalling pathways which result in transcription factors being activated.
- The transcription factors activate genes whose protein products are needed for the control of cell division.

This emphasizes that the reasons for oncogenicity may include one or more of the following:

- Signal transduction pathways leading to formation of active transcription factors may be inappropriately activated.
- There may excessive amounts of particular transcription factors.
- There may be abnormal transcription factors which are inappropriately active.

All of these could lead to inappropriate activation of genes, which lead to cell division.

This list does not cover tumour suppressor gene mutation, which can lead to cancer, but this is in a different category in that it involves a negative effect of the removal of protection, not a positive oncogenic effect. The mechanism by which tumour suppressor genes work will be dealt with shortly.

In the chapter on cell signalling (page 423) we described the known cell signalling pathways. In Fig. 26.29 there is a summary of the various pathways and any one of the genes coding for the protein components of the pathways may in principle become mutated into an oncogene (though, for some reason, not all components have been found to do so). How then can a signalling pathway become uncontrolled? Consider a cascade of events such as:

Growth factor → membrane receptor → A → B → C → D → E → F → transcription factor activation → cell division.

If any of the components of the chain is mutated so that it is in a permanently active state it will be the same as if a growth factor was present even in the absence of the latter. The nucleus is erroneously given the signal to activate genes leading to cell division.

This is illustrated in Fig. 30.3, where we take the growth factor EGF as the example. Binding of the factor to its receptor activates the Ras signal transduction pathway, described on page 430, in which one protein component activates the next until finally transcription factors are activated leading to switch on of specific genes. The essence of the pathway is tight control; the receptor is activated only when EGF is bound to it. Ras has the automatic GTD/GDP switch (page 431) which limits the period

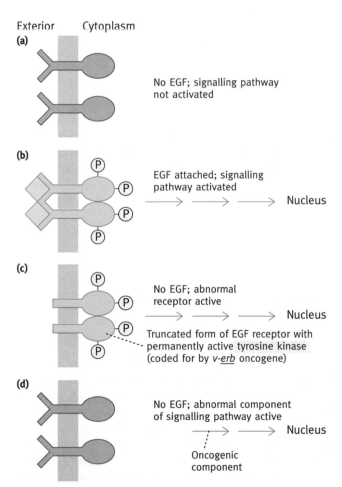

Fig. 30.3 Oncogene products as components of signalling pathways from the membrane to the nucleus—the EGF receptor is used as an example: (a) normal receptor—not activated; (b) normal receptor—activated; (c) permanently active truncated EGF receptor; (d) normal receptor—pathway component permanently active.

of its activation; Ras proteins have in effect to continually ask the receptor whether the EGF signal is still there. If it isn't, the whole pathway is switched off by inactivation of Ras and Raf and phosphatases switch off the MAP kinases (page 432) and activated transcriptional factors. If the receptor is mutated so that it activates Ras in the absence of EGF, it will inappropriately activate the whole pathway. This is what happens with the v-*erb* oncogene. It is a truncated form without the external domain but remains permanently phosphorylated on the tyrosine groups of its cytoplasmic domain. In the case of the Ras protein, the oncogenic form remains in the activated state because its GTPase is defective and therefore the GTP/GDP inactivating switch (page 431) cannot function so that the whole pathway remains activated in the absence of a continued extracellular signal. Many human cancers have an abnormal c-Ras protein; it is also known that a mutated Raf protein which lacks the normal regulatory domain is oncogenic because it continually activates the pathway. This concept is illustrated diagrammatically in Fig. 30.3(d).

Much the same applies to transcription factors. Examples are known in which the latter are in an activated state without any signalling pathway being involved. This results in activation of genes involved in cell division and therefore has a role in the development of the cancerous state. Often, increased amounts of transcription factors or their presence at the 'wrong' time or place is enough to lead to uncontrolled cell division.

What are the roles of tumour suppressor genes?

We have emphasized that oncogenes affect signal transduction pathways. When we come to tumour suppressor genes the emphasis shifts to the cell cycle control and particularly to the restriction point—the checkpoint in the cycle in G_1, which determines whether the cycle may proceed into S phase. At least 12 different tumour suppressor genes are known, mutation of which has been found in particular groups of cancers. The two best-known tumour suppressors are **p53** and the **retinoblastoma** genes, the latter so named because a mutant form of this was first discovered in a rare retinal cancer but has since been found to be present in a number of other cancers. Tumour suppressor genes are a protection against the development of cancer.

The p53 protein is a transcription factor known to have several roles in cellular control but an important one is that it acts as a restraint on the cell cycle progression. It is the *absence* of a normal protective p53 protein that is oncogenic, rather than

mutant versions of the protein promoting cancer. The gene is activated in the presence of damaged DNA, which causes increased amounts of the p53 protein to be synthesized. This prevents the cell from passing from the G_1 phase, across the restriction point, and entering the S phase. It is important that cells with damaged DNA are not allowed to replicate until the DNA is repaired. There is a further safeguard; the p53 gene also has a role in **apoptosis,** or programmed cell death, in situations where DNA damage cannot be adequately repaired. A cell is given a signal to die and a sequence of molecular events is initiated that leads to that result. Thus p53 protein can promote the destruction of cells with badly damaged genomes and this is a protection against the replication of a potentially cancerous cell. As mentioned earlier, defective p53 genes are involved in the development of over half of all human cancers. As is the case with all somatic cell genes there are two p53 genes, one on each chromosome, and most mutations are recessive. If only one gene generates a defective p53 protein there is still functional p53 protein coded for by the other gene, so that mutation of both genes is probably needed to lose the protective effect in most cases, but there is an exception to this. The functional form of the p53 protein is a tetramer; a single abnormal subunit may be able to participate in tetramer formation and interfere with binding of the latter to its control site on DNA. In this event mutation of a single p53 gene causing a faulty p53 protein might interfere with the function of normal p53 protein coded for by the other gene. In this situation a p53 mutation is inherited in a dominant fashion.

Let's look at the mechanism by which the p53 protein acts to protect against cancer (but keeping in mind that its actions are more complex than described here). Consider a normal cell in early G_1 phase. If appropriate growth factors are present, the latter will activate genes leading to the production of the G_1 specific cyclins which activates the Cdks (see above). This activates genes necessary to drive the cell past the restriction point into S phase. In the normal cell the amount of p53 protein is low so that the cycle is able to proceed. If the DNA is damaged however, the p53 gene is activated and more p53 protein is produced which results in the activation of another gene (p21) whose protein product inactivates the G_1-specific Cdk activity. This arrests the cycle at the restriction point, giving time for the DNA to be repaired. If this is done, the p53 gene is no longer activated and the level of p53 protein rapidly returns to its low level and the cycle can continue normally. If the DNA damage is not repaired the cell is signalled to undergo apoptosis, the p53 protein again being involved in this. In the absence of functional p53 genes, the apoptosis signal is not delivered, so a cell with abnormal DNA is allowed to replicate, thus increasing the chance of cancer developing.

The protein coded for by the **retinoblastoma genes** also blocks the cycle at the G_1 checkpoint. It does this by inhibiting a family of transcription factors, which activate genes required for the G_1 phase to pass into S phase. In the normal cycle this inhibitory effect of the retinoblastoma protein is abolished by one of the cyclin-dependent protein kinases phosphorylating it. If an oncogene abnormally activates genes that drive the cycle into S phase without the benefit of cyclin-activated kinase the retinoblastoma protein is not inactivated by phosphorylation and will restrain the cycle thus protecting against the oncogenic effect. If both retinoblastoma genes are mutated so that no functional retinoblastoma protein is produced, the restraining effect on the cell cycle is lost and this contributes to the development of cancer. Retinoblastoma gene defects are involved in certain types of cancers (retinoblastoma, osteosarcoma, and small cell lung cancer).

FURTHER READING

The cell cycle and cancer

Elledge, R. M. and Lee, W.-H. (1995). Life and death by p53. *BioEssays*, **17**, 923–30.

Summarizes the many functions of the p53 gene product, including its role in cancer suppression.

Nigg, E. A. (1995). Cyclin-dependent protein kinases: key regulators of the eukaryotic cell cycle. *BioEssays*, **17**, 471–80.

Describes the role of phosphorylation by cyclin-dependent protein kinases in the regulation of the cell cycle.

Fearon, E. R. (1999). Cancer progression. *Current Biology*, **9**, R873–5. A 'primer' article giving essential information very briefly.

Oncogenes

Cavanee, W. K. and White, R. L. (1995). The genetic basis of cancer. *Sci. Amer.*, **272**(3), 50–7.

Describes the accumulation of genetic defects that can cause normal cells to become cancerous.

Diffley, J. F. X. and Evan, G. (1999). Oncogenes and cell proliferation. Cell cycle, genome integrity and cancer—a millenial view. *Current Opinion in Genetics and Development*, **10**, 13–16.

An informative editorial overview of the whole field. It refers to subsequent reviews in the same issue of the journal also.

Tumour suppressor genes

Brehm, A. and Kouzrides, T. (1999). Retinoblastoma protein meets chromatin. *Trends in Biochem. Sci.*, **24**, 142–5.

Discusses the effect of the tumour suppressor retinoblastoma protein on chromatin remodelling.

P53 tumour suppressor. (1998). *Current Biology*, **8**, R476.

A single page quick guide giving salient facts of this important gene mentioning that at the time of publication there were 11,226 papers on the subject.

Lozano, G. and Elledge, S. J. (2000). P53 sends nucleotides to repair DNA. *Nature*, **404**, 24–5.

A News and Views article describing a new aspect of the tumour suppressor p53, namely that it activates a gene coding for a subunit of ribonucleotide reductase. This produces the doexyribonucleotides required for DNA replication and repair.

DNA mismatch repair

Prolla, T. A. (1998). DNA mismatch repair and cancer. *Current Opinion in Cell Biology*, **10**, 311–16.

Mutations of mismatch repair proteins are associated with the development of tumours in mice and in humans. This paper gives a concise summary of the field.

Telomeres

Johnson, F. B., Marcniak, R. A. and Guarente, L. (1998). Telomeres, the nucleolus and aging. *Current Opinion in Cell Biology*, **10**, 332–8.

Discusses the relationship between telomere lengthening and replicative senescence. Includes discussion of the relationship to human aging.

PROBLEMS FOR CHAPTER 30

1   Explain what a protooncogene is. What are the mechanisms by which a protooncogene can become an oncogene.

2   The restriction point in the G_1 phase of the cell cycle is of major importance Discuss this with reference to the need for a growth factor signal for cell division to proceed.

3   How does the p53 gene protect against the development of cancer?

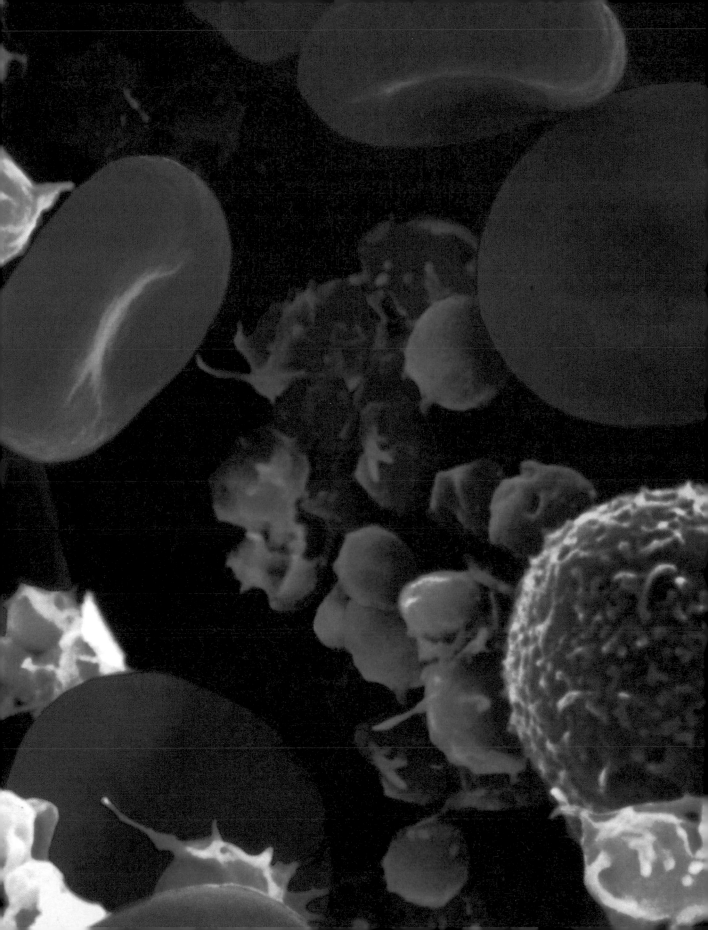

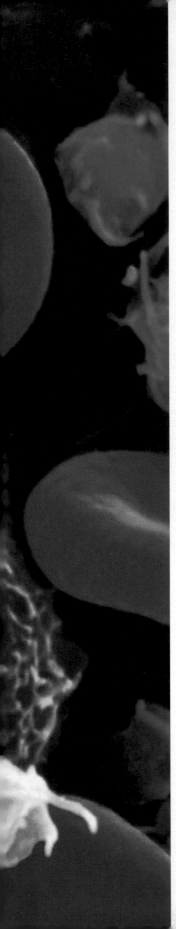

Part 5

Transport of oxygen and CO$_2$

Scanning electron micrograph of human red blood cells, platelets (blue) and a T-lymphocyto (green).

Chapter summary

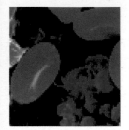

The red blood cell and the role of hemoglobin

The centrally important molecule in the delivery of oxygen from the lungs to the tissues is hemoglobin, which is not just a protein molecule passively capable of combining with oxygen but rather is a sophisticated molecular machine. The red blood cells need to pick up a maximum load of oxygen in the lungs, which implies a high oxygen affinity, and surrender it in the tissue capillaries, which implies a relatively low oxygen affinity. Conversely, it must pick up CO_2 in the tissues and surrender it in the lungs. Because hemoglobin has been studied so intensively, it is the best understood protein and a great deal is understood as to how physiological needs are satisfied.

In this chapter we will discuss several topics all involved with the gas transport system. First, the red cell and its production will be dealt with. Hemoglobin is a protein (globin) combined with heme; the synthesis of heme and its control will therefore be discussed, which, in turn, leads us to iron biochemistry. The control of synthesis of globin will then be discussed. Finally, we will turn to the major topic of how hemoglobin functions.

The red blood cell

The human body produces about 160 million red cells or **erythrocytes** per minute; they circulate for about 110 days and are then destroyed. The mammalian red cell is a flattened biconcave disc lacking a nucleus (Fig. 31.1). The shape is maintained by the submembranous cytoskeleton (see page ••), giving the cell a larger surface area than would a spherical shape. This, together with the smaller diffusion distances involved, facilitates gaseous exchange in and out of the cell. (The total surface area of a human's red cells is about $4000\,\text{m}^2$.)

In the fetus, the liver and spleen produce red cells but in the adult this occurs in the red marrow of flat bones in which **hemopoietic stem cells** continually multiply and produce progenitors of all types of blood cells, as is illustrated in Fig. 27.1. (A stem cell is one that continually reproduces itself, its progeny being capable of differentiating.) Individual cells become irreversibly committed to differentiate into one or other blood cell types (erythrocytes, different white cells, and blood platelets). The production of the various blood cell types is regulated. Different regulatory proteins have been isolated that have the effect of causing multiplication of specific progenitor cells. The protein **erythropoietin**, produced by the kidney medulla, stimulates erythrocyte production, and, since the rate of erythropoietin production is inversely controlled by oxygen tension in the tissue, automatic regulation of red blood cell levels occurs—in oxygen deficiency more hormone and therefore more red cells are produced.

The early cell stages of the erythrocyte are nucleated, but later, in mammals, the nucleus is lost to form the immature red cell known as the **reticulocyte**, which is released into circulation. Reticulocytes contain mRNA and protein synthesis occurs until the cell finally matures into the erythrocyte. When cells are stained, the ribosomes and mRNA result in a reticulum of dark patches—hence the name reticulocyte. The mature cell has no mitochondria and relies on glycolysis for energy production, producing lactate; it has the pentose phosphate pathway to supply NADPH (see page 238).

Heme and its synthesis

The full structure of heme, the complete details of which probably few biochemists carry in their heads, is given in Fig. 31.3 you do not need to learn this detailed structure. It is a ferrous iron complex with protoporphyrin (Fig. 31.2). The latter is a tetrapyrrole, the four substituted pyrroles being linked by methene (=CH—) bridges such that a **conjugated double bond system** exists (that is, you can go right round the molecule via alternating single and double bonds). This gives protoporphyrin and heme their deep red colour.

In heme, the four pyrrole N atoms are bound to Fe^{2+} as shown in Fig. 31.4, leaving two more of the six ligand positions of the Fe^{2+} available for other purposes.

Synthesis of heme

Red blood cells have the vast majority of the body's heme content, so this topic fits well here though heme synthesis occurs in all aerobic cells to supply prosthetic groups for cytochromes and other proteins.

The synthesis of heme appears to be a formidable task but the essentials are surprisingly simple, requiring in animals only two starting reactants, glycine and succinyl-CoA. You have met the latter in the citric acid cycle (page 159). An enzyme, **aminolevulinate synthase**, or **ALA-S**, carries out the reaction shown in Fig. 31.5. 5-Aminolevulinic acid (ALA) is the precursor solely

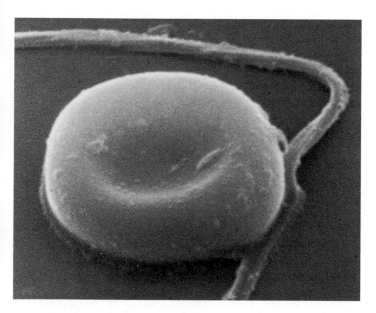

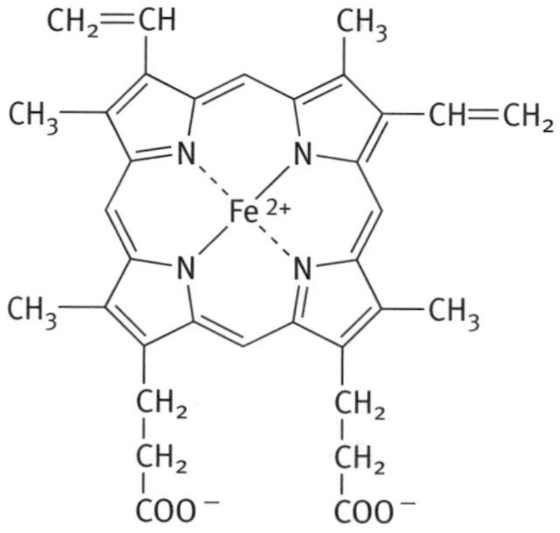

Fig. 31.3 The structure of heme. This form is found in hemoglobin; other hemes differing in their side chains are found in cytochromes. Note that, at one side of the molecule, there are two hydrophilic propionate groups ($-CH_2CH_2COO^-$) while the remaining side groups are all hydrophobic. In myoglobin, heme sits in a cleft of the molecule with the hydrophilic side pointing out towards water and the hydrophobic groups buried into the nonpolar interior of the protein.

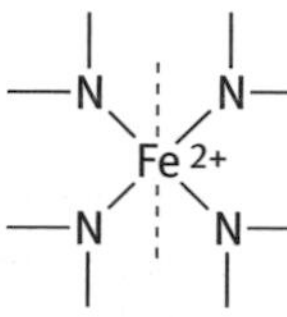

Fig. 31.1 Scanning electron micrograph of an erythrocyte.

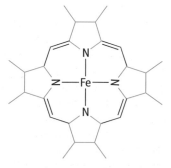

Fig. 31.2 Outline of the heme molecule. (The full structure is given in Fig. 31.3.)

Fig. 31.4 Binding ability of Fe^{2+} in heme. The iron atom in heme can bond to six ligands in total: four bonds to the pyrrole nitrogen atoms as shown and the other two above and below the plane of the page.

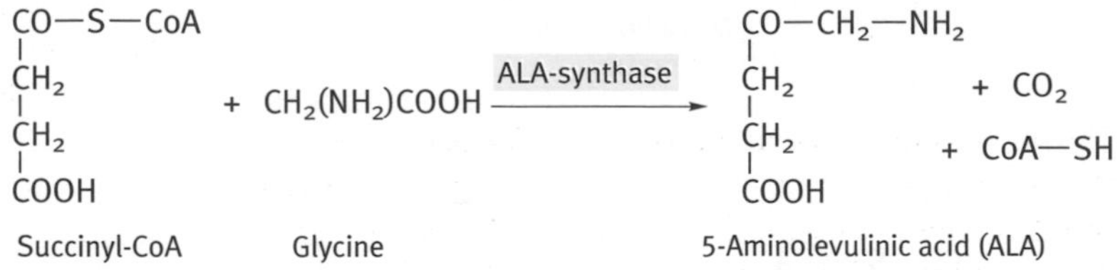

Fig. 31.5 The first step in heme synthesis, catalysed by 5-aminolevulinate synthase (ALA-synthase). For clarity of illustration, nonionized structures are given.

committed to porphyrin synthesis. Two molecules of ALA are used to form the pyrrole, **porphobilinogen** (**PBG**), the two molecules being dehydrated by ALA dehydratase (Fig. 31.6). The remainder of the heme biosynthetic pathway consists of linking four PBG molecules together, modifying the side groups, and chelating an atom of ferrous iron to form heme. The intermediate tetrapyrroles between PBG and heme are the colourless uro- and coproporphyrinogens (methylene bridges) and the red protoporphyrin (methene bridges). The pathway is given in Fig. 31.7; have a quick look at this.

The heme synthesis pathway has a curious feature. The first reaction, synthesis of ALA, occurs inside mitochondria, after which the ALA moves out to the cytoplasm, but the final three steps also occur in the mitochondria. Why this should be so is not clear.

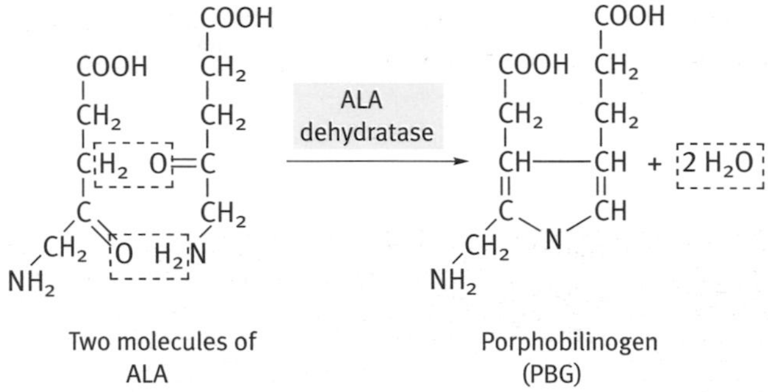

Regulation of heme biosynthesis and iron supply to the erythrocyte

The control of heme biosynthesis is of unusual interest and is intimately tied up with regulation of iron uptake and storage in the cell.

The situation is as follows.

- ALA synthase activity is the rate-limiting step in heme synthesis.
- ALA synthase synthesis and therefore ALA synthase levels are controlled by Fe levels.
- Fe levels are controlled by transferrin receptor levels.
- Fe levels feedback-control the synthesis of transferrin receptor protein.

The rate-limiting step is, as mentioned, the one first committed to heme production—ALA synthesis. The synthesis of the enzyme ALA synthase itself is regulated in erythrocytes at the translational level. mRNA for the enzyme has, at its 5' untranslated end (that is, the translational initiation end), a hairpin loop called an **iron-responsive element** or **IRE**. At low iron levels, an IRE binding protein attaches to the IRE and prevents translation (Fig. 31.8). At higher cellular levels of iron, the latter complexes with an iron–sulfur cluster (page 162) which is

Fig. 31.6 Synthesis of porphobilinogen (PBG)—the ALA-dehydratase step. Nonionized structures are given for simplicity. PBG is a monopyrrole; heme is a tetrapyrrole. The conversion of PBG to heme is shown in Fig. 31.7.

attached to the protein, causing release of the protein, and translation occurs, thus coordinating the rate of heme synthesis with the iron supply. Heme synthesis starts in the nucleated proerythroblast, well before the reticulocyte stage, so that mRNA translation occurs in these cells and continues into the reticulocyte stage.

The question now follows as to what controls the cellular iron level. Iron is transported in the blood plasma as a complex with the protein, **transferrin**, produced in the liver. This is taken up by receptor-mediated endocytosis (page 268). In *non-erythroid cells*, uptake is known to be regulated by control of the synthesis of the receptor protein by cellular iron, which regulates the stability of the transferrin receptor-protein mRNA. The mRNA for the receptor protein has iron-responsive elements at the 3' end of the molecule. In the presence of low cellular levels of iron, the IRE-binding protein binds to the receptor-protein mRNA, which apparently protects the mRNA from degradation. In conditions of high iron levels, the binding protein detaches and removes the protection. Thus, iron causes downregulation or reduction in the number of transferrin receptors which, in turn, reduces iron intake. At low iron levels the reverse applies. It is not established whether this applies to erythroid cells, but it is known that red blood cells have transferrin receptors that increase in number as the cell develops.

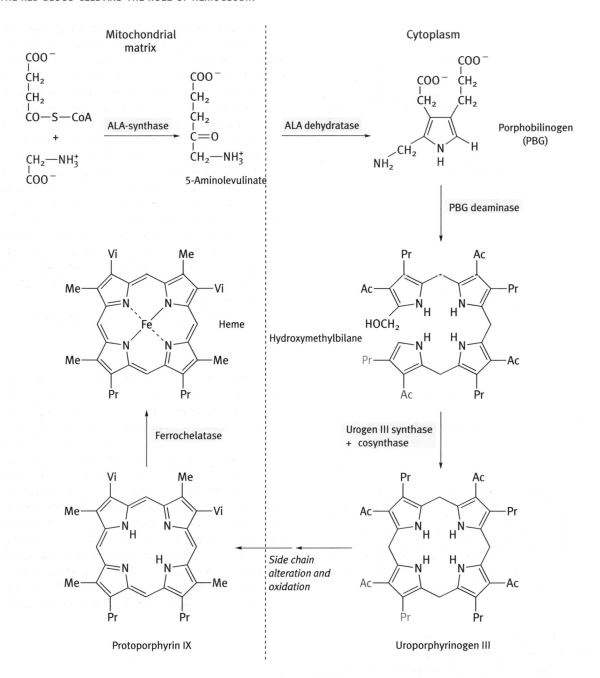

Fig. 31.7 An abbreviated pathway of heme biosynthesis, given for reference purposes.

Iron is stored in red cells as **ferritin**—a complex of the protein, apoferritin, and inorganic iron. (The liver is also an important site for ferritin storage.) It is interesting that apoferritin synthesis is regulated by a mechanism essentially the same as that for ALA synthase—by iron causing detachment of an IRE-binding protein for apoferritin mRNA, permitting synthesis of the protein.

As a slight diversion from blood cells (because this is a convenient place for it), all cells of the body, so far as we know, synthesize heme by the same route; it is needed for cytochromes of the

Situation A: Fe concentration in the cell is high

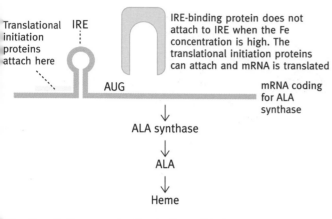

Translational initiation proteins attach here

IRE

IRE-binding protein does not attach to IRE when the Fe concentration is high. The translational initiation proteins can attach and mRNA is translated

AUG

mRNA coding for ALA synthase

↓

ALA synthase

↓

ALA

↓

Heme

Situation B: Fe concentration in the cell is low

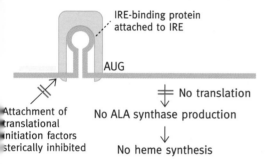

IRE-binding protein attached to IRE

AUG

Attachment of translational initiation factors sterically inhibited

No translation

No ALA synthase production

↓

No heme synthesis

Fig. 31.8 Heme synthesis control by Fe levels in the red blood cell. The regulatory mechanism presumably depends on the ALA synthase having a short half-life in the cell so that, once synthesis of the enzyme stops, heme synthesis stops shortly afterwards. The enzyme is known to be unstable in reticulocyte lysates *in vitro* and presumably is so in red blood cells. The question of how Fe levels in the cell are controlled is dealt with separately. IRE, Iron-responsive element, a stem loop structure in the mRNA. When the IRE-binding protein attaches to the IRE, it probably sterically inhibits the attachment of the translational initiation proteins. Binding of Fe to the IRE-binding protein abolishes its affinity for the IRE.

electron transport chain (page 164) and liver cytochrome P450 (page 280) and other enzymes. Red cells have the greatest need, for the mature cell is packed with hemoglobin; liver is next because of its large cytochrome P450 content. The ALA synthase of erythroid cells is different from that of the 'housekeeping' version (as that in other cells is called). The reaction catalysed is the same, but the control of the ALA-S reaction is different. In liver, synthesis of ALA-S is controlled by heme, which destabilizes its mRNA and also inhibits transport of the enzyme into mitochondria, its functional site. Its mRNA, however, has no iron-responsive elements, as has the erythroid version. Over-

production of hepatic ALA synthase and high blood levels of ALA in acute intermittent porphyria ('mad king' disease) patients correlate with the onset of neurological symptoms in that disease, though the molecular basis of the symptoms is not known. Such patients have a partial block in the heme biosynthetic pathway, in this case at the PBG deaminase step, causing an accumulation of ALA which leaks out into the bloodstream. This also causes an increase in PBG which is excreted in the urine giving it a red colour. In lead poisoning the ALA dehydratase enzyme is inhibited and this causes symptoms with resemblances to those of porphyria. There are seven hereditary porphyria diseases, each associated with a defect in the heme biosynthetic pathway. In three of these the blockage in the heme biosynthetic pathway occurs in erythrocytes rather than the liver. The diseases, as well as a summary of the whole field can be found in the paper by May *et al.* in the reading list. Chlorophyll is also a tetrapyrrole but, interestingly, ALA is synthesized in plants by a quite different reaction.

After that diversion, we'll now return to the red blood cell.

Destruction of heme

Red blood cells are destroyed mainly by **reticuloendothelial cells** of spleen, lymph nodes, bone marrow, and liver. Removal of sialic acid groups from the red cell membrane glycoproteins is a signal that the cell is aged and ready for destruction. The degraded carbohydrate attaches to cell receptors, which leads to endocytosis of the erythrocyte. The enzyme, **heme oxygenase** opens up the tetrapyrrole ring, releasing the iron for re-use and forming biliverdin, a linear tetrapyrrole (Fig. 31.9). Biliverdin is reduced to **bilirubin**. This is water-insoluble but is transported in the blood, attached to serum albumin, to the liver, where it is rendered much more polar by the addition of two glucuronate groups (see page 281) and then excreted, in the bile, into the gut; modification and partial re-absorption of some bile compounds leads to the yellow colour of urine. (Bilirubin has been postulated to function as an antioxidant—see page 283.)

How is globin synthesis regulated?

Hemoglobin consists of heme attached to the protein globin. The two components, **globin** and **heme**, need to be produced in the correct relative amounts if excess of one or the other is to be avoided. A coordinating regulatory link exists. It has been demonstrated in experiments with reticulocyte lysates that, in the absence of heme, a protein kinase phosphorylates one of the factors (eIF2) involved in initiation of protein synthesis

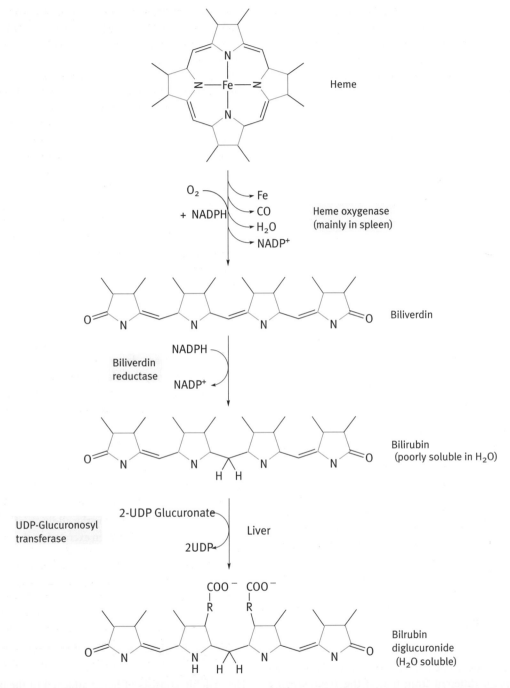

Fig. 31.9 Diagrammatic representation of the heme oxygenase reaction and the reduction of biliverdin. Note that the double bonds of the ring structures are omitted for simplicity. The reaction converts one methene group to carbon monoxide, thus opening up the ring structure. Biliverdin is reduced to bilirubin, conjugated with two glucuronic acid molecules, and excreted. There appears to be some uncertainty as to the stoichiometry of the heme oxygenase reaction; three O_2 molecules are believed to be involved. By analogy with cytochrome P450, it might be assumed that three molecules of NADPH are required to convert three oxygen atoms to H_2O.

(page 392). This has the effect of halting the initiation of translation in the red blood cell and thus prevents globin synthesis. In the presence of heme, the kinase is inactivated and a phosphatase restores the normal activity of the initiation factor. Thus globin synthesis can proceed only when heme is present. This mechanism is found specifically in erythroid cells.

Transport of gases in the blood

The solubility of oxygen in simple aqueous solution is grossly insufficient for tissues to be supplied with oxygen in this manner. A carrier is therefore needed in the blood. Likewise CO_2 cannot be carried from the tissues to the lungs in simple solution at a sufficient rate.

What is the chemical rationale for a heme–protein complex as the oxygen carrier?

Inorganic Fe^{2+}, but not Fe^{3+} ions, are capable of combining with oxygen. However, Fe^{2+} ions are spontaneously, and rapidly, oxidized to Fe^{3+} ions so that inorganic iron, on its own, is no good as an oxygen carrier. The Fe^{2+} in heme can also bind with oxygen, but again this rapidly oxidizes to the Fe^{3+} form of heme, known as **hematin**, so that *free* heme is no good for oxygen-carrying purposes either. For heme to oxidize, two molecules must associate on either side of an oxygen molecule. The binding of heme in a crevice of protein prevents this association so that the Fe^{2+} in the heme of hemoglobin is much more resistant to oxidation than is free heme. It is important to be clear that, while heme in the electron transport cytochromes (page 164) oscillates between the Fe^{3+} and the Fe^{2+} states, hemoglobin functions only in the Fe^{2+} state. Binding of oxygen does not affect this. The iron atom in hemoglobin is bonded to the four pyrrole nitrogens; the fifth and sixth coordination positions are directly above and below the plane of the heme structure; one of these is bonded to a histidine residue of the protein, the other is available for oxygen binding (Fig. 31.13).

Structure of myoglobin and its oxygen binding role

Myoglobin is the red pigment of muscles. A preliminary discussion of this helps the understanding of the more complex hemoglobin.

Myoglobin has the relatively simple job of providing a reservoir of oxygen in the muscle cell. It takes up oxygen from the blood and then gives it up to be used by the mitochondria inside the cells for energy generation. It consists of a single heme

molecule bound to a single protein molecule. The protein is roughly spherical in shape and is made up of a series of α helices (Fig. 31.10(a)) with the heme molecule situated in a crevice between two of these.

Figure 31.11 shows the percentage saturation of myoglobin at increasing oxygen tensions; the result is a hyperbolic curve—the same, in fact, as that observed with an enzyme–substrate concentration curve, described earlier for the 'classical' type of enzyme (see page 25). The affinity of myoglobin for oxygen is high so that it readily extracts oxygen from oxyhemoglobin in the blood (see Fig. 31.11 for the relative affinities of the two carriers). In muscle cells, as the oxygen level falls due to electron transport and its conversion to H_2O, myoglobin gives up its oxygen. It is thus functionally a very simple carrier.

The three-dimensional structure of myoglobin was the first to be determined for a protein and thus is a landmark in biochemistry.

Structure of hemoglobin

Hemoglobin is a tetramer of protein subunits (Fig. 31.10(b)) each of whose folded polypeptide structures closely resembles that of myoglobin and each of which has a similarly located heme molecule capable of binding oxygen. In normal adult hemoglobin, which is the type we will discuss here, the two α subunits are identical, as also are the two β subunits.

Binding of oxygen to hemoglobin

A molecule of hemoglobin binds four molecules of oxygen, one per subunit. When an oxygen-saturation curve is plotted, instead of the hyperbolic curve seen with myoglobin, there is a sigmoid curve that is well to the right of that of myoglobin (Fig. 31.11). Higher oxygen concentrations are required to 50% saturate hemoglobin that in the case for myoglobin, indicating as already stated that the latter has a higher affinity for oxygen.

Hemoglobin needs to pick up as much oxygen as it can in the lungs but surrender it in the tissue capillaries. The sigmoidal oxygenation curve means that it is most steep (that is, surrenders the most oxygen) at oxygen pressures encountered in the capillaries (see Fig. 31.11) but, nonetheless, it is still capable of becoming virtually saturated at the higher oxygen pressures encountered in the lungs.

How is the sigmoidal oxygen saturation curve achieved?

You have met sigmoidal kinetics already when metabolic control was discussed (page 212). It was mentioned there that allosteric enzymes are usually multisubunit proteins that undergo a conformational change on binding of substrate

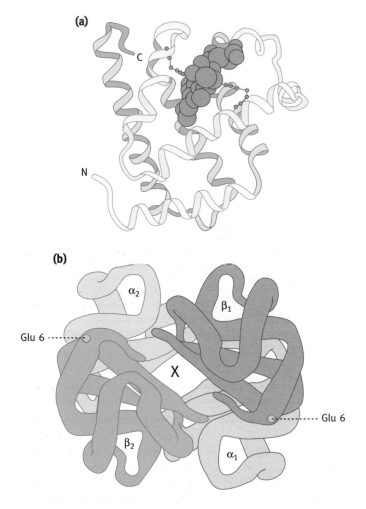

Fig. 31.10 Models of myoglobin and hemoglobin. **(a)** Computer-generated diagram showing the folding of the polypeptide in myoglobin and the positioning of heme in the molecule. **(b)** Model of the subunits and their arrangement in hemoglobin showing the central cavity into which 2:3-bisphosphoglycerate fits in the deoxygenated state (marked X). In sickle cell anemia the two glutamic acid residues at position 6 in the β side chains are mutated to valines, creating hydrophobic patches on the molecule.

molecules. It is this change of structure that results in changes in the affinity of the enzyme for further attachments of its substrate. The same applies in oxygen binding to hemoglobin.

Although the heme molecules in hemoglobin are distant from one another, the initial binding of oxygen to one subunit facilitates the binding of further molecules of oxygen to the other subunits. This is known as a **homotropic positive cooperative effect** (homotropic, because only oxygen is involved). It is this that causes the sigmoidal curve. The effect is that the initial affinity of deoxyhemoglobin for oxygen is 200 times less than that involved in the final addition of oxygen. At the end of the curve, oxygenation becomes limited by availability of binding sites. The number of cooperating sites in the binding of oxygen to hemoglobin can be determined by the **Hill plot** which gives a **Hill coefficient**. This specifies the number of cooperating sites as described on page 213. For hemoglobin the value is 2.8 while for myoglobin with only a single binding site it is 1.

Mechanism of the allosteric change in hemoglobin

When oxygen molecules bind to a hemoglobin molecule, as stated, it makes it easier for the remaining sites to accept oxygen. According to the concerted model (see page 213), hemoglobin exists in two conformational states: one is the 'tense' or T state with *low* oxygen affinity and the other is the 'relaxed' or 'R' state with *high* oxygen affinity, the two being in free equilibrium and the T state predominating in the absence of oxygen. As oxygen binds, it increases the probability of all four subunits of a given molecule being in the R, *high*-affinity, state—it swings the T $\rightleftharpoons$ R equilibrium to the right,

$$T \rightleftharpoons R \xrightarrow{\;O_2\;} RO_2$$

Since, in a study of the oxygenation of hemoglobin, large numbers of individual molecules are involved, as more and more

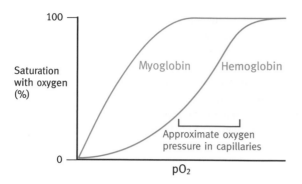

Fig. 31.11 Oxygen–hemoglobin saturation curve. The higher affinity for oxygen of myoglobin as compared to hemoglobin means that myoglobin in the muscles readily accepts oxygen from the blood.

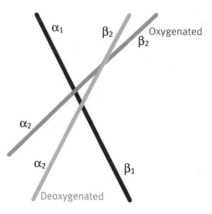

Fig. 31.12 The relative subunit positioning in deoxygenated and oxygenated hemoglobin molecules, the axes of which are represented by straight lines. The α_1, β_1 pair and the α_2, β_2 pair should be regarded as single dimer units. On oxygenation, these dimers rotate and slide, relative to each other, by about 15°. The black and blue lines represent the relative positions of the dimers in the T state. In the oxygenated (R) state the red line shows the rotation of the α_2/β_2 dimer relative to the α_1/β_1 dimer, which is represented as fixed. The change alters the $\alpha_1\beta_2/\alpha_2\beta_1$ contacts between the dimers. A more realistic model of hemoglobin is shown in Fig. 31.10(b).

oxygen binds, more molecules of hemoglobin are in the relaxed state and, therefore, the observed oxygen affinity in the solution increases. This is analogous to the situation with many allosteric enzymes.

By means of X-ray diffraction studies, the conformation of the hemoglobin tetramer has been determined in the oxygenated and deoxygenated states. As already mentioned, hemoglobin has four subunits, two α and two β. Although the α subunits are identical to one another, and the β subunits are also identical to one another, in the tetramer they are distinguishable and are named α_1, α_2, β_1 and β_2. The two α subunits differ slightly in their contacts with the β subunits and vice versa for the β_1 and β_2, something that is evident from the X-ray diffraction studies.

The most useful way to view hemoglobin is to say that the α_1 and β_1 subunits are firmly associated as a dimer and, similarly, that the α_2 and β_2 form a dimer. It is the interaction between the two dimers in the tetramer that undergoes rearrangement in the T $\rightleftharpoons$ R conversion. Figure 31.12 illustrates the relative rotation between the dimers' conversion from the T to R state caused by binding of oxygen. The reality of the change in conformation due to oxygenation is shown by the fact that a crystal of hemoglobin disintegrates on exposure to oxygen.

What causes this allosteric change to the entire hemoglobin molecule when one or more oxygen molecule bind? The movement is initiated by the fact that, on binding of oxygen to heme, the iron atom moves slightly and, in doing so, brings about the T $\rightleftharpoons$ R transformation. Although heme, as usually written, appears as a planar molecule, in the deoxygenated state, the Fe atom lies above the plane of the molecule because it is too big to fit into the tetrapyrrole (Fig. 31.13(a)). In the deoxygenated hemoglobin, the tetrapyrrole itself is not quite flat (Fig. 31.13(a)). The Fe^{2+} atom itself is bonded to a histidine residue in one of the α helices of which the subunit is composed—in fact, the α helix

designated F, the histidine group being identified as F8 (Fig. 31.13(a)).

On binding of oxygen, the iron atom of heme becomes effectively smaller in diameter, due to electronic changes and moves into the plane of the tetrapyrrole, thus flattening the molecule. The protein rearranges itself as a result of the movement of the Fe atom (Fig. 31.13(b)). This tiny movement is amplified because of a lever-like effect of the arrangement of the polypeptide chain of the protein, so that a bigger movement occurs elsewhere in the molecule.

This 'elsewhere' is the point where the α unit of one dimer interacts with the β unit of the other dimer (α_1–β_2/α_2–β_1 interfaces). At each of these locations there is a group of weak bonds between amino acid residues of the two protein subunits involved. In the T state there is one set, in the R state another set of bonds, holding the dimers together. When it moves into the new position in the R state, an alternative set of weak bonds is formed because of close association of different groups on the subunits. The movement initiated by the Fe atom thus causes the two dimers to rearrange from one set of weak bonds to the other set, causing the relative rotation shown in Fig. 31.12.

It should be noted that the structures of the T and R states of hemoglobin are fully proven by X-ray studies on crystals of hemoglobin and of oxyhemoglobin, respectively. However, it is not possible to obtain crystals of intermediate, partially oxygenated hemoglobin and therefore the structure(s) of the

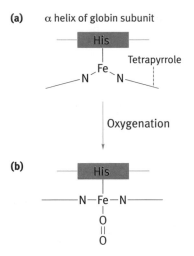

Fig. 31.13 The changes in the heme of hemoglobin upon oxygenation. **(a)** The heme molecule in deoxyhemoglobin with the tetrapyrrole structure strained into a slightly domed shape. **(b)** Attachment of heme in oxyhemoglobin.

latter are not known. The intermediate stages (if any) by which the T and R forms interconvert are therefore not established. On page 213 we described two hypotheses (the concerted and sequential models) both of which could account for the observed changes.

The essential role of 2:3-bisphosphoglycerate (BPG) in hemoglobin function

You are familiar with the glycolytic intermediate 1:3-bisphosphoglycerate (Fig. 9.6) but we have not so far mentioned that **2:3-bisphosphoglycerate (BPG)** is also synthesized in the cell.

1:3-Bisphosphoglycerate   2:3-Bisphosphoglycerate

It plays an important physiological role in oxygen transport by lowering the affinity of hemoglobin for oxygen and thus increases the unloading of oxygen to the tissues; it moves the dissociation curve of oxyhemoglobin to the right. The explanation

is both interesting and simple. The hemoglobin tetramer, looked at from the appropriate viewpoint, has a cavity running through the molecule (see Fig. 31.10(b)). Projecting into this cavity are amino acid side chains with positive charges. The BPG molecule has four negative charges at the pH of the blood and has just the correct size and the configuration to fit into the cavity and make ionic bonds with the positive charges on the protein. This helps to hold hemoglobin in its deoxygenated position; in effect it crosslinks the β units in that position. In the deoxygenated (T) state, the hemoglobin can accommodate a molecule of BPG. However, on oxygenation, because of the conformational change in the protein, the cavity of the R state becomes smaller and is unable to accommodate BPG. If we consider oxyhemoglobin in the capillaries, the ability of BPG to strongly bind to, and stabilize the deoxygenated state favours unloading of the oxygen. In effect, the process is

(a) Hb—oxygen $\rightleftharpoons$ Hb + oxygen;
    (relaxed)   (tense)

(b) Hb + BPG $\rightleftharpoons$ Hb—BPG.
    (tense)   (tense)

Reaction (b) with BPG will tend to pull the equilibrium of reaction (a) over and favour oxygen release.

If blood is stripped of all of its BPG, the hemoglobin remains virtually saturated with oxygen even at oxygen concentrations below that encountered in the tissue capillaries. It therefore would be incapable, in that state, of delivering oxygen to the tissues efficiently. The effect of BPG on oxygen binding by hemoglobin is illustrated in Fig. 31.14.

The higher the concentration of BPG, the more the deoxygenated form is favoured. This constitutes a regulatory system—if oxygen tension in the tissues is low, synthesis of more BPG in the red blood cells (which occurs as a result of regulatory mechanisms we will not go into) favours increased unloading of oxygen. Acclimitization at high altitudes involves the establishment of higher BPG levels in red blood cells. It is to be noted that, while BPG causes greater delivery of oxygen to the tissues, the decreased oxygen affinity has little effect on the degree of oxygenation in the lungs. The normal molar concentration of BPG in the blood is roughly equivalent to that of tetrameric hemoglobin.

There is yet another refinement of the BPG-based regulatory system that illustrates how small changes to proteins can have major physiological effects. For a mother to deliver oxygen to a fetus, it is necessary for the fetal hemoglobin to extract oxygen from the maternal oxyhemoglobin across the placenta. This requires the fetal hemoglobin to have a higher oxygen affinity than that of the maternal carrier. This is achieved by a fetal

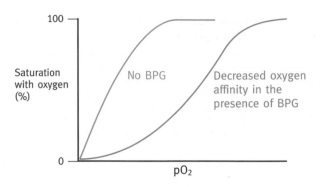

Fig. 31.14 Oxygen saturation curves for hemoglobin, illustrating the effect of 2:3-bisphosphoglycerate (BPG).

hemoglobin subunit (called γ) replacing the adult β chains, each of which lacks one of the positive charges of the β subunit. The missing charges are those that, in adult hemoglobin, line the cavity into which BPG fits; therefore fetal hemoglobin has two fewer ionic groups to bind BPG, the latter therefore being held less tightly. BPG is therefore less efficient in lowering the oxygen affinity, giving fetal hemoglobin a higher O_2 affinity than that of maternal hemoglobin. Thus, the latter readily transfers its oxygen load to the fetus.

Effect of pH on oxygen binding to hemoglobin

Hemoglobin in the deoxygenated state has a higher binding affinity for protons than has oxyhemoglobin. Put in another way, the R form is a stronger acid than is the T (deoxygenated) form, resulting in dissociation of protons from the molecule when oxygen binds, a phenomenon known as the **Bohr effect**,

$$Hb + 4O_2 \rightleftharpoons Hb(O_2)_4 + (H^+)_n. \tag{1}$$

(where n is somewhere around 2; the number depends on a complex set of parameters.)

The protons are released from, for example, histidine side groups of the protein. The pK_a of a dissociable group can be altered somewhat by its immediate chemical environment. The release of protons is caused by the T—R conformational change affecting the ionization (pK_a) of certain histidine groups (page 15).

Role of pH changes in oxygen and CO₂ transport

The Bohr effect, described above, has important physiological repercussions. In the tissues, CO_2 is produced and must be transported to the lungs. It enters the red blood cell where the

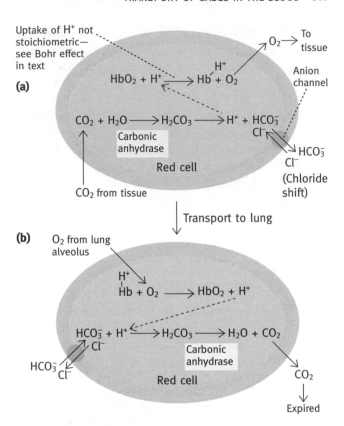

Fig. 31.15 Transport of CO_2 in the blood. (a) Reactions in the tissue capillaries. (b) Reactions in the lungs. The diagrams omit transport of CO_2 as carbamino groups of hemoglobin.

enzyme **carbonic anhydrase** converts it to H_2CO_3 which dissociates into the bicarbonate ion and a proton,

$$CO_2 + H_2O \leftrightarrow H_2CO_3 \rightleftharpoons H^+ + HCO_3^-. \tag{2}$$

The latter will drive the equilibrium shown in equation (1) above, to the left, causing the HbO_2 to unload its oxygen, the effect thus being in harmony with physiological needs.

The HCO_3^- passively moves out via the anion channel (page 83) down the concentration gradient into the serum. The HCO_3^- movement is not accompanied by H^+ movement for there is no channel allowing passage of this across the membrane of the red cell. To electrically balance exit of HCO_3^-, Cl^- moves into the cell via the same anion channel. The dual movement is known as the chloride shift (Fig. 31.15). The HCO_3^- travels in solution in the serum of the venous blood back to the lungs. Here, the proton concentration changes help to achieve the physiologically desirable results too. On oxygenation, release of protons from hemoglobin does two main things. It produces H_2CO_3 from HCO_3^- by a simple equilibrium effect,

$$HCO_3^- + H^+ \rightleftharpoons H_2CO_3,$$

that permits carbonic anhydrase to form CO_2 to be expired because H_2CO_3 (not HCO_3^-) is the substrate of the enzyme

$$H_2CO_3 \rightleftharpoons H_2O + CO_2.$$

The destruction of HCO_3^- in the red blood cell causes HCO_3^- in the serum to enter the cell, down the concentration gradient, and Cl^- exits, that is, a reverse chloride shift occurs in the lungs, resulting in CO_2 expiration.

A small amount of CO_2 is transported in simple solution in the blood, but by far the greatest amount (about 75%) is transported as HCO_3^-. About 10–15% of the CO_2 is carried by the hemoglobin itself. CO_2 chemically and spontaneously reacts with uncharged —NH_2 groups of the globin to form carbamino groups

$$RNH_2 + CO_2 \rightleftharpoons RNHCOOH \rightleftharpoons RNHCOO^- + H^+.$$

The RNH_2 groups available are mainly the terminal amino groups—lysyl and arginine side groups have too high a pK_a (page 15) to be significantly uncharged.

pH buffering in the blood

From the above, it is clear that there are major changes in the hydrogen ion concentration of blood associated with CO_2 and oxygen transport.

When acid is produced in the tissues, the pH of red blood cells cannot be allowed to fall by very much. The buffering power of HCO_3^-, phosphates, and hemoglobin itself is important in maintaining a physiological pH. The Bohr effect, described above, also buffers. When oxygen is unloaded, hemoglobin takes up protons (equation 1, page 517). This carries roughly half of the H^+ ions generated by CO_2 in the tissues, and this again helps to prevent a drop in the pH of the red blood cell to unphysiological levels.

Sickle cell anemia

This disease is usually described in textbooks because it illustrates how a single amino acid change in a protein can have a profound effect, and gave rise to the concept of a molecular disease. In the normal β chains of the human hemoglobin tetramer, amino acid number 6 is glutamic acid whose side group is negatively charged and highly hydrophilic (Fig. 31.10(b)). In the hemoglobin of sickle cell anemia patients, this is replaced by the hydrophobic valine residue. The codons in mRNA for glutamic acid are GAA and GAG; mutation of the central A base to a U is all that is required for this substitution since GUA and GUG are both valine codons. The abnormal hydrophobic patch on the globin subunit caused by valine, by chance, causes a deoxygenated hemoglobin molecule to bind to a particular hydrophobic pocket on another molecule and so on, resulting in the formation of a long multistrand rigid rod. Oxygenated hemoglobin, because of its different conformation, does not permit this. The long deoxyhemoglobin rods distort the normal biconcave disc into sickle-shaped cells that tend to block capillaries and also to break up, causing anemia. If both chromosomes are affected, the disease may be lethal, especially at low oxygen tensions such as occur at high altitude causing unusually high deoxygenation of the hemoglobin.

The disease is prevalent only in geographical areas with a malignant form of malaria (ignoring the effects of immigration) where the high incidence can be explained only by positive selection of the disease genome. The sickled red blood cell is unfavourable for the development of the malarial parasite and thus protects against the parasite. Since malarial infection of normal persons has a higher death rate than does the heterozygous sickle cell state, the latter increases survival of persons with the genetic trait.

FURTHER READING

Heme and its synthesis

Straka, J. G., Rank, J. M., and Bloomer, R. (1990). Porphyria and porphyrin metabolism. *Ann. Rev. Med.*, **41**, 457–69.

The review deals with the defects in the enzymes of heme biosynthesis which lead to porphyrias, and the therapeutic uses of porphyrins.

Chen, J.-J. and London, I. M. (1995). Regulation of protein synthesis by heme-regulated eIF-2α kinase. *Trends Biochem. Sci.*, **20**, 105–8.

A comprehensive review of the control of hemoglobin synthesis by this mechanism.

May, B. K., Dogra, S. C., Sadlon, T. J., Bhasker, C. R., Cox, T. C., and Bottomley, S. S. (1995). Molecular regulation of heme biosynthesis in higher vertebrates. *Progress in Nucleic Acid Research*, **51**, 1–47.
A comprehensive review of all aspects including the hereditary diseases.

Hemoglobin and myohemoglobin

Branden, C. and Tooze, J. (1991). *Introduction to protein structure.* Garland Publishing.
A well illustrated account of all aspects of protein structure, including hemoglobin and myoglobin; very readable.

PROBLEMS FOR CHAPTER 31

1   Describe the first two steps in heme biosynthesis in animals.

2   What is the medical relevance of ALA synthase control in liver?

3   Explain how the red blood cell ALA synthetase activity is coordinated with the availability of iron.

4   Compare the oxygen dissociation curves of myoglobin and hemoglobin.

5   What is the likely mechanism of the sigmoid oxygen-binding curve of hemoglobin?

6   Binding of an oxygen molecule to hemoglobin causes a conformational change in the protein. Describe the mechanism of this.

7   Explain how the fetus is able to oxygenate its hemoglobin from the maternal carrier.

8   Explain the significance of the chloride shift in red blood cells.

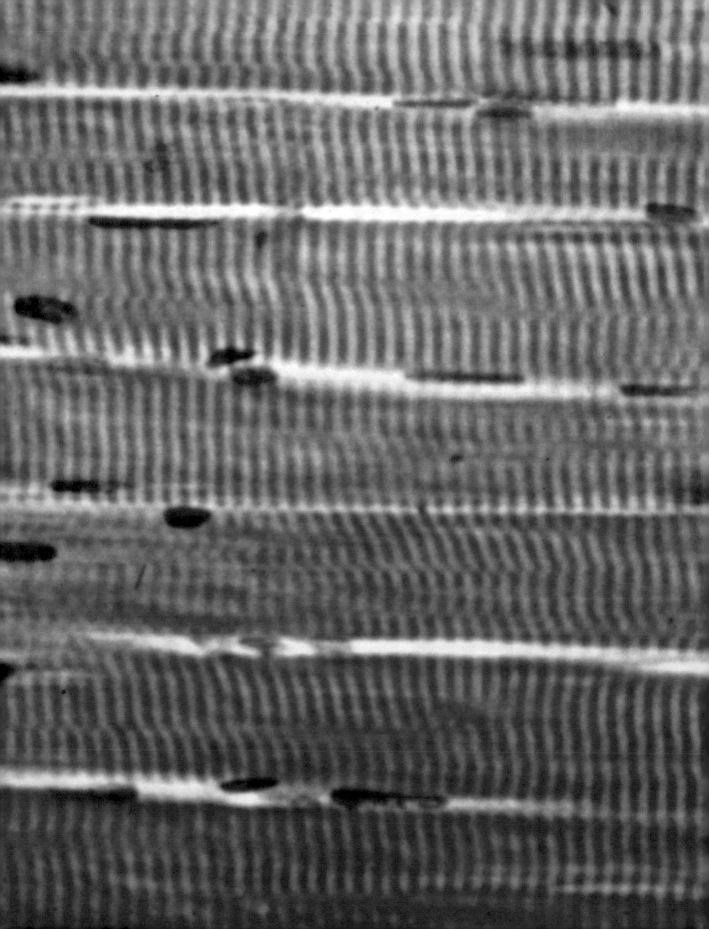

Part 6

Mechanical work by cells

Photomicrograph of myofibrils of striated skeletal muscle.

Chapter summary

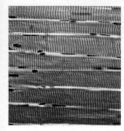

Muscle contraction

Muscles are highly specialized organs in which devices for mechanical movement, present in all cells, whether muscle or nonmuscle, have been developed to the highest and most easily studied state. It might seem logical to deal with the 'molecular motors' present in all cells first but it is probably easier to understand muscle contraction and then move on to movement in nonmuscle cells. This is the arrangement adopted here—we will deal with muscle contraction in this chapter and then with the more generalized movement systems in nonmuscle cells in the following chapter.

A reminder of conformational changes in proteins

The most astonishing thing about muscle contraction is that it occurs at all. The only basis for biological processes are molecules and, for contraction, individual molecules must move in a purposeful fashion. What type of molecular activity is available for this? The only type we know of is conformational change in proteins, minute changes in shape of individual protein molecules that occur as a result of different ligands binding to the surface of proteins.

For movement to occur, whatever exerts force must have something to exert it against. A 'molecular motor' exerting force by conformational change must always have a partner structure to react against—if the motor is fixed, the partner molecule will move; if the latter is fixed, the motor molecule will move. To anticipate the mechanism of muscle contraction, the motor is myosin, which is fixed, and the structure it acts against, and moves, is the actin filament. But more of this later.

Types of muscle cells and their energy supply

The two main classes of muscles are **smooth** and **striated**. Smooth muscle is found in the intestine and blood vessels, which are typically under involuntary nervous control and frequently under control by several hormones. It contracts slowly and can maintain the contraction for extended periods. It is called smooth because it does not have the same striated appearance under the microscope as does the second class of muscle—striated muscle. This is found in skeletal muscle, which is under voluntary nerve control. Striated muscle contracts rapidly. Heart muscle is striated but not identical in structure to skeletal muscle and is under involuntary control.

If we turn now to voluntary striated skeletal muscle there is further specialization of biochemical interest. There are fast twitching fibres and slow twitch fibres. In animals such as lobster and fish, the white striated muscle, consisting of fast twitching fibres with poor blood supply, permits very fast 'escape' reactions dependent on extremely rapid stimulation of ATP production, resulting from the breakdown of glycogen reserves by glycolysis (the mechanism of the almost instant control of this is given earlier—see page 212). Because of the low yield of ATP from glycolysis and lactic acid accumulation (page 138) the white muscle is quickly exhausted. Slow twitch muscle is slower to contract. It is richly supplied with blood and has a high mitochondrial content and a reserve of oxygen bound to myoglobin (page 513). It generates ATP mainly by oxidative phosphorylation. This is much more efficient in terms of ATP yield than is

glycolysis but, in emergency, oxygen supply becomes limiting and it takes time for blood vessels to expand and the heart rate to speed up. It is therefore slower in response than fast twitch muscle but can function much longer without exhaustion. In humans, voluntary striated muscles contain mixtures of both types of fibres, the proportions depending on the precise role of the muscles in the body. Human back muscle, which has to maintain tension for the body posture for extended periods, is mainly slow twitch muscle. The ocular muscles that move the eyeball are good examples of the fast twitch type. In fish, it is the lateral thin stripes of red muscle that cause the contractions involved in ordinary swimming.

In all muscles, the reserve of ATP on which contraction depends is remarkably low—only enough for a very brief period of intensive contraction. This explains a role for a reserve of phosphoryl groups in the form of **creatine phosphate**, whose hydrolysis is associated with a $\Delta G^{0'}$ value of $-43.0\,\mathrm{kJ\,mol^{-1}}$ (cf. $-30.5\,\mathrm{kJ\,mol^{-1}}$ for the hydrolysis of ATP to ADP and P_i). When ATP is converted to ADP and P_i by contraction, the **creatine kinase** reaction regenerates ATP using the phosphoryl group of creatine phosphate.

Contraction event: ATP + H_2O → ADP + P_i.

ATP regeneration event: ADP + creatine—℗ → ATP + creatine.

During muscle recovery, ATP is regenerated by oxidative metabolism and the pool of creatine phosphate is regenerated by the creatine kinase reaction, which is readily reversible in the presence of high levels of ATP and low levels of ADP.

Creatine phosphate has the structure shown below; the compound is of the 'high-energy' type and the phosphoryl group can be reversibly transferred to ADP to form ATP. Because of the reactivity of the phosphoryl group, there is a spontaneous (noncatalysed) formation of creatinine; this has no function and is excreted in the urine at a daily rate essentially proportional to muscle mass. Since creatinine is formed at a relatively constant rate in the body and its only fate is to be excreted in the urine, measurement of its level in the blood is commonly used as an indicator of kidney function. A rise in the blood level indicates impairment of the kidney's ability to excrete it.

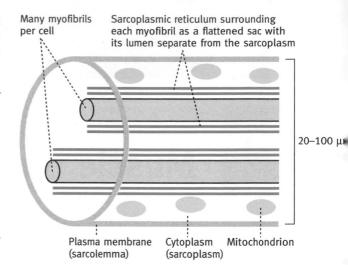

Fig. 32.1 A myofibre (muscle fibre) or muscle cell from striated muscle. Bundles of myofibres make up a muscle.

Structure of skeletal striated muscle

A muscle is composed of multinucleate cells called **myofibres** (Fig. 32.1), which may be very long. The cell membrane (the **sarcolemma**) is excitable (see page 85)—it has nerve endings associated with it, at the neuromuscular junctions, to deliver the nervous signal that triggers contraction. The plasma membrane of the cell surrounds the cytoplasm, multiple nucleii, and many mitochondria. Embedded in this cytoplasm and running lengthwise through it are many long **myofibrils**, each surrounded by a membraneous sac, the **sarcoplasmic reticulum**.

Structure of the myofibril

The myofibril is the structure that does the contracting and, as stated, there are many running the length of the muscle cell. Each myofibril is divided into small segments bounded by **Z discs** (Z for *zwischen* or division). Each small segment is called a **sarcomere** (Fig. 32.2). On contraction, the Z discs are pulled closer together, thus shortening the individual sarcomeres and hence the myofibril. This shortens the entire myofibre, a process that is seen as muscle contraction.

How does the sarcomere shorten?

The Z discs are robust discs of protein at each end of the sarcomere to which are attached 'rods' pointing to the centre of the sarcomere. These rods are thin filaments made up mainly of the protein actin, and are firmly attached by one of their ends to a Z disc. They are essentially inert and serve as 'rachets' on which force can be exerted to drag the Z discs together. In vertebrate

Creatine phosphate

Creatinine

(a)
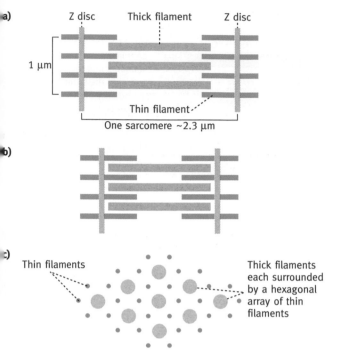

1 μm

Z disc   Thick filament   Z disc

Thin filament

One sarcomere ~2.3 μm

(b)

(c)
Thin filaments

Thick filaments each surrounded by a hexagonal array of thin filaments

Fig. 32.2 Arrangement of thick and thin filaments in a sarcomere. **(a)** Relaxed. The striated appearance of myofibril sections in electron microscopy is caused by the amount of protein the beam traverses; **(b)** contracted; **(c)** arrangement in cross-section. The diagram shows only a few filaments but each sarcomere has a large number, all in lateral register.

(a)

G actin molecules
The arrows indicate the polarity of the molecules

(b)
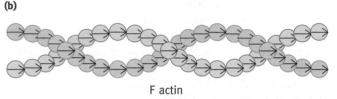

F actin

Fig. 32.3 (a) G (globular) actin. **(b)** F (fibrous) actin made from the polymerization of G actin. The two strands are actually closely apposed forming a rod-like structure, but are presented here and in Fig. 32.8 in more open form for clarity.

(a)
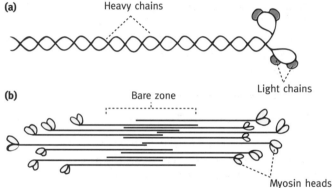

Heavy chains

Light chains

(b)

Bare zone

Myosin heads

Fig. 32.4 (a) A myosin molecule consisting of a dimer structure of two heavy chains, each terminating in a myosin head. The latter has two dissimilar light chains attached to it. **(b)** A thick filament made of about 300 myosin molecules arranged in a bipolar fashion.

muscle, the thin filaments are in hexagonal array on the two discs, and each sarcomere has a number of these arrays. Inside each hexagonal 'cage' formed by the six thin filaments, is a thick filament. The thick filament has finger-like projections that do the actual work of contraction. By a ratchet-like mechanism they 'claw' the thin filaments towards the centre and in doing so pull the Z discs closer together (Fig. 32.2).

That, then, is the overall picture of muscle contraction. To understand how contraction happens we must look at the molecular structures involved.

Structure and action of thick and thin filaments

Thin filaments

Actin is a globular protein molecule (called **G actin**; G for globular) (Fig. 32.3(a)(b)). Each actin molecule has two dissimilar globular parts connected by a narrow waist, giving the molecule a polarity. Under cellular conditions it readily polymerizes into long fibres, always in a definite head to tail polarity. A thin filament consists of two such fibres wound around each other

with a long pitch (and the same polarity), the ends being referred to as (+) and (−).

Thick filaments

We now come to a most remarkable molecule—**myosin**, the protein of the thick filament. Each molecule is a dimer made up of a straight rod section made up of two polypeptides in coiled coil configuration (Fig. 32.4(a))—that is, two α helices (page 39) coiled around each other in helical form. The structure is such that each α helix has regularly spaced hydrophobic residues that form hydrophobic attachments to its partner molecule, the two forming a rigid rod structure. Each chain terminates in a globular head; two other small polypeptide chains, called light myosin chains, are attached to each of the myosin heads.

A thick filament is made up of several hundred myosin molecules arranged in a bipolar fashion as shown in

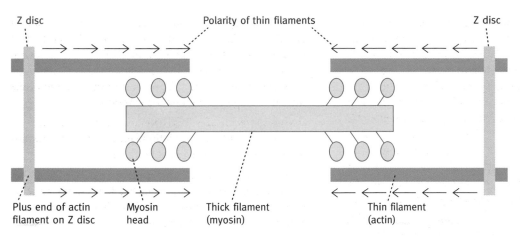

Fig. 32.5 Diagram showing the arrangement of thick and thin filaments in a sarcomere. The arrows in the latter are to show the polarity of the actin filaments. In a contraction event, in effect, the myosin heads track along the actin filaments towards their (+) ends and, in doing so, claw the two discs together. The heads are shown as simple shapes for clarity but their actual structures are described below.

Fig. 32.4(b). The thick filaments are held in a central position relative to the thin filaments by other proteins involved in sarcomere structure. One is **titin**, so named for its huge (titanic) molecular weight. This central positioning is at an optimal distance to allow the myosin heads to contact the actin filaments. The thin filaments are anchored to the Z discs at their (+) ends so that the myosin heads, at both ends of the thick filament, are oriented the same way, relative to the thin filament polarity (Fig. 32.5).

How does the myosin head convert the energy of ATP hydrolysis into mechanical force on the actin filament?

The myosin head is an enzyme capable of hydrolysing ATP to ADP and P_i. It undergoes conformational changes at the expense of ATP hydrolysis, which results in its exerting a force on the actin filament. There is one aspect of this process which is likely to cause confusion unless explained. You are, by now, used to ATP breakdown being used to perform work and it would be natural to assume that the work of muscle contraction occurs when myosin hydrolyses ATP to ADP and P_i. In the ATP-driven processes that you have learned about until now, the hydrolysis has been indirect. In these, first there is a transfer of a phosphoryl group, or an AMP group, to a reactant molecule followed by a second step in which P_i or PP_i is liberated—in other words coupled reactions are involved (page 11) involving the formation of a phosphorylated or adenylated intermediate. In muscle contraction direct hydrolysis of ATP without covalent intermediates takes place, but the actual 'power stroke' in contraction does not occur on ATP hydrolysis itself as you might expect. Hydrolysis of ATP to ADP + P_i results in the myosin head–ADP complex protein taking up a different conformational state; you might think of it as adopting a 'high energy' conformation. It is when the ADP leaves the protein that the

main liberation of free energy occurs so that the 'power step' in muscle contraction correlates, not with ATP hydrolysis *per se*, but with ADP release from the myosin head. The energy transfer is mediated by conformational changes in the protein. (This is very reminiscent of the way synthesis of ATP on mitochondrial ATPase occurs, in which free energy of ATP synthesis from ADP + P_i attached to the protein is very small but the release of the ATP is energy requiring. In both cases the energy transfers occurs via conformational changes in the proteins.) In contraction the conversion of *free* ATP, in solution, to *free* ADP and P_i, in solution, has the expected, large, negative $\Delta G^{0'}$ value so that the overall energetic concept of ATP hydrolysis driving contraction is unaltered.

Mechanism of the conformational changes in the myosin head

Until recently it was believed that the conformational changes in the myosin head involved the entire head swinging on a hinge at its connection to the rod section of the myosin molecule so that the angle of attachment of the head to the actin was believed to swing from perpendicular to oblique. This has however proved to be incorrect. The mechanism of the action has been revealed by X-ray diffraction studies which have revealed the three-dimensional structures of the conformations corresponding to the states preceding and following the power stroke. A myosin head has a compact motor domain which includes the site which attaches to the actin filament. Its orientation to the actin filament never changes; it is always almost perpendicular to it. Attached to this is an extended α helix which, at its distal end, joins to the rod part of the myosin molecule. If necessary look at Fig. 32.4 to remind yourself that the thick filament is a bundle of many myosin molecules held together by aggregation of their rod sections, the end of each emerging from the bundle and carrying a pair of heads. This α

helix is a lever arm; a small conformational movement produced on release of ADP from the compact section is amplified by the swinging lever arm into a much larger movement of the whole head. It is the lever arm moving relative to the actin-binding part which brings about the sliding force of the power stroke (Fig. 32.6(a). Figure 32.6(b) shows models of the myosin head before and after the power stroke. On the right of each model is the site which attaches to the actin. The ATP binding section is known as the **motor domain** and this can cause a conformational change in a domain (known as the **converter**) which causes the lever arm to swing. You will notice that in the models the lever is surrounded by the two **myosin light chains** (not represented in Figs. 32.6(a) and 32.7). These stabilize the lever arm.

Let's now look more closely at the steps involved in this cycle, starting with Fig. 32.7(a) in which the head has just finished its power stroke in a contraction. (You will appreciate that vast numbers of individual heads are involved in any contraction and only one of the two heads of each myosin head is represented in the diagram in the interests of clarity.)

The following sequence of events then occur as depicted in Fig. 32.7.

- In (a) the myosin head is firmly attached to the actin filament *having just completed its power stroke* in which a moving force was applied to the actin filament. The head cannot detach from the actin until a molecule of ATP binds. This resembles the state of muscles in rigor mortis in which muscles are contracted but detachment of the cross-bridges does not occur. (Cross-bridges refer to the binding of the myosin head groups to the actin subunits.)

- In (b) a molecule of ATP has bound causing detachment of the myosin head from the actin filament. This corresponds to the state existing in a relaxed muscle.

- In (c) ATP hydrolysis occurs but the products, ADP and Pi are still attached to the myosin head. A conformational change occurs in the myosin head causing the lever arm to adopt the 'primed' conformation.

- In (d) P_i is released and the head attaches to the actin filament resulting in state (e) in which ADP is still attached.

- In (e) the power stroke now occurs in which ADP is released. This returns the head to our starting point shown in (a). In going from (e) to (a) (the power stroke) the actin filament has been caused to slide thus causing a contractile force.

That the myosin head itself is capable of carrying out the whole process is shown by the *in vitro* experiment in which latex

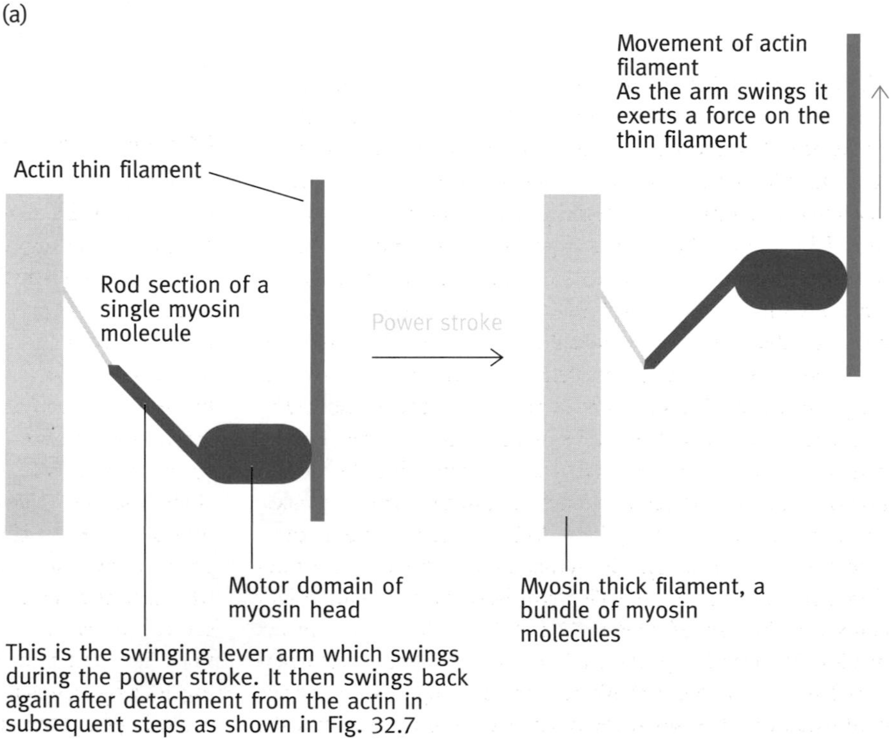

(a)

Movement of actin filament
As the arm swings it exerts a force on the thin filament

Actin thin filament

Rod section of a single myosin molecule

Power stroke

Motor domain of myosin head

Myosin thick filament, a bundle of myosin molecules

This is the swinging lever arm which swings during the power stroke. It then swings back again after detachment from the actin in subsequent steps as shown in Fig. 32.7

Fig. 32.6 (a) Diagrammatic representation of the swinging lever arm mechanism of muscle contraction. The myosin head has a motor domain which binds to the actin filament without changing its angle of attachment as envisaged in the previous model, The motor domain is the site of nucleotide (ATP/ADP) attachment. It is connected to the myosin rod by an α helix which forms the lever arm; this is surrounded by the myosin light chains (not shown). When the power stroke is induced by ADP dissociation from the motor domain (see text) the small resultant conformational change is amplified by the lever arm which swings through an arc of 70°, sufficient to displace the actin filament by about 10 Å. Note that for clarity only a single myosin head is shown; each myosin molecule has two heads.

(b)

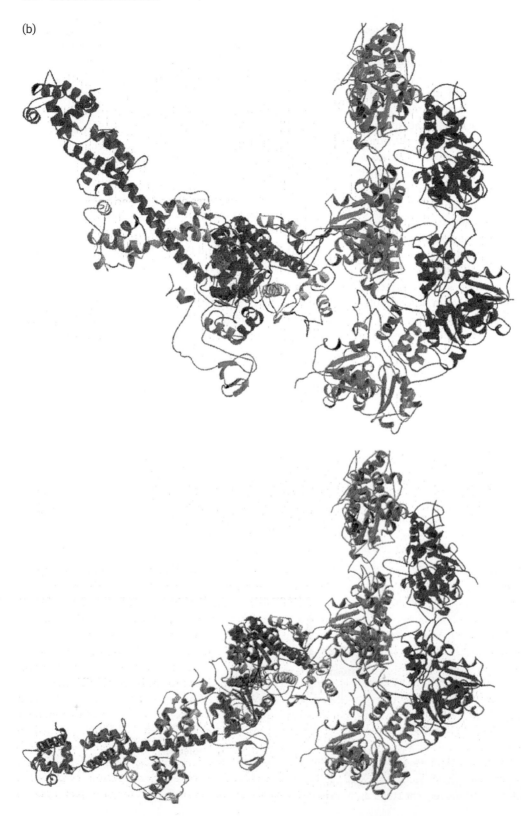

Fig. 32.6 *Continued.* (b) The three-dimensional structure of the myosin head: (upper) before the power stroke with the swinging lever arm in the upright position; (lower) the head immediately after the power stroke.

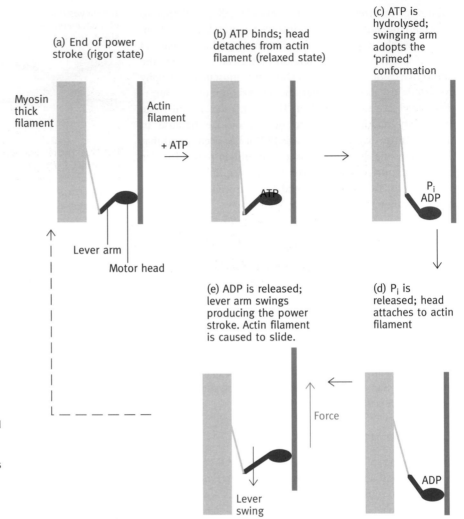

(a) End of power stroke (rigor state)

(b) ATP binds; head detaches from actin filament (relaxed state)

(c) ATP is hydrolysed; swinging arm adopts the 'primed' conformation

Myosin thick filament

Actin filament

+ ATP

Lever arm

Motor head

P_i
ADP

(e) ADP is released; lever arm swings producing the power stroke. Actin filament is caused to slide.

(d) P_i is released; head attaches to actin filament

Force

Lever swing

ADP

Fig. 32.7 Swinging lever arm model of muscle contraction. The pale red bar is the thick filament made of multiple myosin molecules; the thin red line represents an individual myosin molecule; only one of two heads is shown (deep red) for clarity. The green line is the thin actin filament. ATP, P_i, and ADP represent molecules bound to the myosin head. The essence of the model is that the power stroke is associated with ADP release and the force is exerted by the proximal subunit of the myosin head (the lever arm) swinging in a lever-like fashion. See text for individual steps.

beads, coated with isolated myosin heads, literally walk along immobilized actin fibres in the presence of ATP and Ca^{++}—a most dramatic experiment. The determination of the three-dimensional structure of the myosin head has been a major advance in our understanding of how skeletal muscle contracts, but there is still more to learn before muscle contraction is fully understood. The contraction of a muscle involves trillions of myosin heads. It has been pointed out (see reference list) that if billions of heads were exerting the power stroke, and another group of billions were opposing them, it would rip your arm apart. Somehow the movement of the heads is coordinated to avoid this. It is suggested that the heads apparently have a strain sensitivity which can detect what their neighbours are doing and adjust their own activity to produce a coordinated action, though little is known of how this occurs.

Another incompletely understood aspect of muscle contraction relates to **muscular dystrophy**. This term covers a group of related inherited diseases in which there is a progressive muscular weakness and muscle wasting, often leading to death around the age of 20 due to respiratory paralysis. The best-known severe type is **Duchenne muscular dystrophy**. The affected gene in this disease has been cloned; it codes for a very large protein called **dystrophin**. This protein is attached at one end to structural elements of the muscle cell cytoskeleton (the latter described on page 535) and at the other to a transmembrane complex of proteins. The latter on its extracellular domain is linked to proteins of the extracellular matrix which surrounds muscle cells. The linkage of intracellular structural components to the membrane and thence to the strong reinforcing fibres of the extracellular matrix (page 51) is thought to play a part in

stabilizing, or strengthening the muscle cell membrane to withstand the contractile forces involved in muscle contraction. It conceivably might help in transmitting the mechanical force of contraction to the tendon of skeletal muscle via the connective tissue in muscles. In Duchenne muscular dystrophy dystrophin is absent so that the connection of the intracellular cytoskeleton is broken. However, there is still some uncertainty about the nature of the disease because in knockout mice experiments (page 487) it has been found that elimination of a dystrophin-associated protein results in mice having muscular dystrophy even though the linkage between the intracellular cytoskeleton and the extracellular matrix is apparently intact (see reference by Weir).

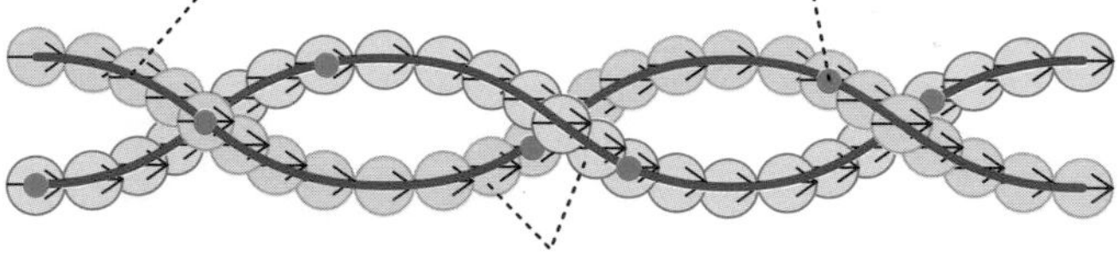

Fig. 32.8 The relationship of the actin and tropomyosin molecules to each other. Each tropomyosin molecule binds to seven actin monomers forming a continuous filament in each of the actin filament grooves. Each tropomyosin molecule also has a troponin complex bound at one end.

How is contraction in voluntary striated muscle controlled?

In skeletal muscles, contraction is initiated by a nerve impulse causing Ca^{2+} ions to be liberated into the myofibril from the sarcoplasmic reticulum and contraction is triggered. The Ca^{2+} ions are rapidly removed from the myofibril (see below), so that, unless more Ca^{2+} is liberated by a continuing nerve impulse, contraction ceases.

How does Ca^{2+} trigger contraction?

Thin filaments are primarily made of actin but they have associated with them additional protein molecules. One of these is **tropomyosin**, consisting of two subunits. This is a fairly short, elongated molecule that lies along the helical groove between the two actin fibres of a thin filament. Each tropomyosin molecule has on it seven actin attachment sites, each one binding to an actin monomer within the groove. The tropomyosin molecule thus is about seven actin monomers long, and the successive molecules overlap one another to form a continuous thread right along the thin filament. A thin filament has two grooves, with a tropomyosin thread in each of them (Fig. 32.8).

Each tropomyosin molecule has attached to it, at one end, an additional complex of three more globular proteins called **troponin**. This is all leading up to the fact that the **troponin–tropomyosin complex** is sensitive to Ca^{2+}, which causes the tropomyosin thread to move slightly, relative to the thin filament. This is the event that permits the myosin head to attach to the actin filament. It was believed that the tropomyosin prevents this attachment to the actin by a simple blocking effect, but other evidence suggests that Ca^{2+} does not affect the binding of myosin to the thin filament. What is known is that, unless Ca^{2+} is present, the **myosin–actin power cycle** does not occur.

In short, Ca^{2+} combines with troponin, causing a conformational change in tropomyosin that somehow activates the myosin–actin power cycle resulting in ATP hydrolysis and contraction. Withdrawl of Ca^{2+} reverses the whole sequence of events and shuts down the contraction event. Which brings us to the mechanism of the control of Ca^{2+} release and withdrawal.

Release and uptake of Ca^{2+} in the muscle

Each myofibril within the muscle cell, is surrounded by a membranous sac—the **sarcoplasmic reticulum**. The membrane is highly specialized in two regards. First, it is rich in a Ca^{2+}/ATPase that pumps Ca^{2+} from the cytosol surrounding the myofibril (sarcoplasm) into the lumen of the reticulum, using ATP hydrolysis energy to drive the process. This continually depletes the myofibril of Ca^{2+} and prevents muscle contraction. Secondly, the sarcoplasmic reticulum membrane has a protein, the **Ca^{2+} channel** or **Ca^{2+} gate**, which is a passive Ca^{2+}-transporting channel; it is normally closed, but, on receipt of a nerve impulse to the muscle cell, it opens and releases Ca^{2+} from the lumen of the reticulum into the myofibril, causing contraction.

A condition known as malignant hyperthermia exists in certain humans and pigs. In this, administration of some anaesthetics causes rigid contraction of the muscles and excess heat production which is potentially fatal. The condition in pigs, and probably in humans, results from a mutation in a Ca^{2+} channel protein causing it to open more readily than normal and it remains open unduly long. The resultant high levels of Ca^{2+} in the

(a)

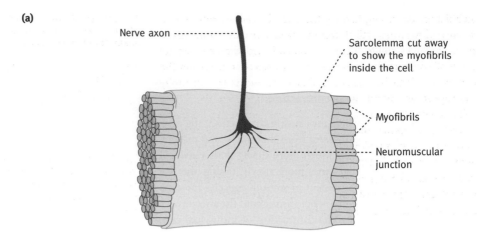

(b)

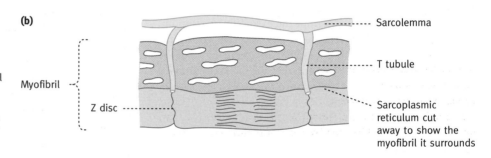

Fig. 32.9 (a) Plasma membrane (sarcolemma) of the myofibre, showing the neuromuscular junction. **(b)** The transverse (T) tubules that carry the plasma membrane depolarization signal to the sarcoplasmic reticulum (SR). It is postulated that the plasma membrane depolarization is directly transmitted to the SR Ca^{2+} channels causing rapid Ca^{2+} release from the SR.

cytosol causes excessive contraction and also stimulates glycogen breakdown.

For a long muscle to contract, all of the sarcomeres in a myofibril and all the myofibres in the muscle need to respond to a motor nerve impulse essentially simultaneously, for otherwise the contraction will be uncoordinated and relatively ineffective. The nerve impulse causes liberation of acetylcholine at the neuromuscular junction; this causes a local depolarization of the plasma membrane, that rapidly propagates throughout the plasma membrane (see page 86). The depolarization causes the voltage-gated Ca^{2+} channels of the sarcoplasmic reticulum to open and release the ion on to the myofibril. In order to ensure that the signal reaches all of the sarcoplasmic reticulum within a cell very rapidly, the plasma membrane is invaginated into transverse (T) tubules that enter the cell at Z discs and make direct contact with the sarcoplasmic reticulum membrane, thus permitting the electrical signal to reach, virtually simultaneously, all of the contractile units controlled by that nerve impulse (Fig. 32.9).

How does smooth muscle differ in structure and control from striated muscle?

Smooth muscle is found in the walls of blood vessels, in the intestine, and in the urinary and reproductive tracts. The long spindle-shaped cells have a single nucleus and associate together to form a muscle in patterns appropriate to their function, such as an annular arrangement in blood vessels and a criss-cross network in the bladder.

The basic principles of contraction are the same as in striated muscle in that myosin molecules exert force on actin filaments, using ATP hydrolysis as the source of energy, and employing a similar 'power cycle'. However, the contractile components are not so highly organized. There are no myofibrils and no repeating sarcomers, the latter being the reason for the absence of a

striated appearance under the microscope. Instead of the sarcomere structure, actin filaments run the length of the whole cell (which is small, compared with a striated muscle cell) and are anchored into the cell membrane.

Control of smooth muscle contraction

Although Ca^{2+} is responsible for causing contraction, the control mechanism is quite different from that in striated muscle. A smooth muscle contracts much more slowly than a striated muscle, typically about 50 times more slowly, taking about 5 seconds. There is no requirement in smooth muscles for contraction to be almost instantaneous throughout the structure as there is in striated muscle; the contraction of individual cells can spread at a more leisurely pace. In smooth muscle cells, a neurological impulse from the autonomic nervous system causes Ca^{2+} gates in the cell membrane to open and allow an in-rush of the ion into the cells from the extracellular medium. There is no sarcoplasmic reticulum. The relatively slow diffusion of Ca^{2+} throughout the cell can be tolerated because of the small distances involved and the slow response requirements. Special junctions between cells allow a neurological signal to spread throughout the muscle.

How does Ca^{2+} control smooth muscle contraction?

In the head of each myosin molecule, as already described, are two small polypeptides known as **myosin light chains**; this is in addition to the heavy myosin chains. In smooth muscle, one of these light chains (the p-light chain) inhibits the binding of the myosin head to the actin fibre, and thus prevents contraction. Ca^{2+} activates a myosin kinase that, with ATP, phosphorylates the p-light chain and abolishes its inhibitory effect, thus triggering contraction.

The Ca^{2+} does not directly activate the kinase; instead it combines with a protein, calmodulin (see page 441), which induces a conformational change in the latter such that it combines with

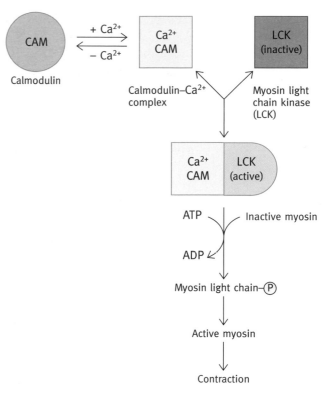

Fig. 32.10 Mechanism of activation of smooth muscle contraction by Ca^{2+}.

the inactive kinase and activates it. When the Ca^{2+} levels falls, the process reverses and a phosphatase dephosphorylates the myosin light chain causing muscle relaxation. The scheme is summarized in Fig. 32.10.

As well as neurological control of smooth muscle contractions, several hormones exert control; prostaglandin (page 196) is one of these; norepinephrine causes contraction of certain blood vessel muscles.

FURTHER READING

Bray, D. (1992). *Cell movements*. Garland Publishing. A comprehensive account of all aspects of movement in cells. Good simple diagrams and also illustrated with electron micrographs.

Muscle contraction

The swinging lever arm hypothesis of muscle contraction. (1997). *Current Biology*, 7, R112–8.

An account of the mechanism of muscle contraction.

Clarke, M. (1998). Sighting of the swinging lever arm of muscle. *Nature*, **395**, 443.

A brief News and Views account of the three-dimensional structure of the myosin motor head.

Cross, R. A. (1999). Molecular motors: walking talking heads. *Current Biology*, **9**, R854–6.

Introduces the concept that myosin heads must be coordinated by strain sensitivity, the latter being the form of molecular talking referred to in the title.

Schliva, M. (1999). Myosin steps backwards. *Nature*, **401**, 431–2.

A News and Views article describing the discovery of a minimyosin that travels towards the (−) end of actin tracks instead of the opposite as has always been found for myosins in the past.

Weir, A. (2000). Muscular dystrophy. *Current Biology*, **10**, R92.

A one-page quick guide giving essentials of the molecular aspects of the disease.

Smooth muscle contraction

Small, J. V. (1995). Structure–function relationships in smooth muscle: the missing links. *BioEssays*, **17**, 785–92.

Discusses the structural organization of the smooth muscle cell and control of contraction.

PROBLEMS FOR CHAPTER 32

1   What are the biochemical differences between fast twitch muscle fibres and slow twitch ones? What are their biological roles?

2   How is contraction of a voluntary striated muscle sarcomere controlled?

3   How is smooth muscle contraction controlled?

4   Can you see any similarities in principle between the mechanisms of ATP synthesis by ATP synthase and its utilization by myosin for contraction?

5   Explain with simple diagrams the mechanism by which myosin causes movement of the actin filament.

Chapter summary

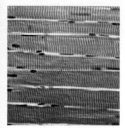

The cytoskeleton, molecular motors, and intracellular transport

An overview

It is easy to form the incorrect impression that mechanical work is not important in nonmuscle cells. A eukaryotic cell such as a liver cell is a veritable hive of intracellular traffic driven by a multitude of motors, carrying loads, running along engineered highways using ATP as fuel; this is true of most eukaryotic cells. It is also easy to envisage an animal cell as a bag of protoplasm without an internal scaffolding. This is also incorrect, for animal cells have a complex internal scaffolding—the **cytoskeleton** - which is involved in movement of cells, movement of structures within them, and in conferring shape. The intracellular scaffolding and motorized traffic are intimately related, which is why we are dealing with them in the same chapter.

It may not be self-evident why these things are needed.

First, shape: an animal cell without a rigid cell wall would be an amorphous blob unless something conferred shape on it. We have already described (page 90) how the erythrocyte maintains its flattened biconcave disc shape by means of a submembranous protein scaffolding that maximixes surface area and minimizes diffusion distances. Most, or all, animal cells have distinct shapes and, for this, internal structures are important. Many cells are constantly changing their shape. This means that the cytoskeleton, which determines shape, must, in most situations, be capable of rapid disassembly and reassembly. This is not true for permanent internal structures such as that of the red cell, or of the intestinal microvilli (Fig. 33.1) but it is true for animal cells in general. Plant cells and bacterial cells with rigid cell walls are in a different category in this respect.

Secondly, movement of animal cells: the most obvious examples are those of the macrophages and other white cells of the body which migrate through tissues by amoeboid-like action. However, motility is a much more widespread property of cells; many migrate over surfaces by a crawling action, and this is important in embryonic development and in normal wound healing. A different type of active contraction is that involved in cell division. When a cell divides, a constriction develops around the equatorial region which progresses to the point of cell separation. Yet another form of movement is due to cilia, whose beating causes movement of surface mucous layers, e.g. in the respiratory passages or of flagellae which sperm use for swimming.

Thirdly, there is internal transport. A eukaryotic cell is a very large object when its dimensions are considered in relation to the low rate of diffusion of solutes. A bacterial cell, being so much smaller, can get away with diffusion of solutes but many eukaryotic cells require elaborate internal transport systems. Membrane-bound vesicles from the endoplasmic reticulum (ER) carry proteins to the Golgi, transport vesicles communicate between the Golgi stacks, and similarly transport vesicles carry proteins to their various cellular destinations (page 405). It is now believed that most or all of these activities are dependent on active transport mechanisms. The latter are most easily seen in exaggerated cases. A nerve axon may be half a metre or more in length (several metres, in giraffes); synthesis of proteins and vesicles occurs in the cell body and some of these have to be transported to the tip of the axon. Diffusion could never do this. A different case is that of the giant algal cell *Nitella* which is so large that solute diffusion would not support the life of the cell; to cope with this, the cell keeps its cytoplasm continually streaming so as to achieve constant mixing. Another example is

that at cell division, chromosomes on the spindle move apart into daughter cells (page 541).

The message of all this is that eukaryotic cells have a most elaborate internal organization known as the **cytoskeleton**, involved in cell shape determination, cell movement, and internal transport. It is less obvious than the contraction of muscles, particularly so because much of it is not visible in cells without appropriate staining and, apart from a few permanent structures, the whole system is dynamic and capable of continuous modification; structures are assembled and disassembled at disconcerting speeds.

There are three distinct intracellular major classes of the cytoskeleton components, two of which are involved in intracellular transport. These are the **actin filaments** and **microtubules**; the third group, **intermediate filaments**, are not involved in transport.

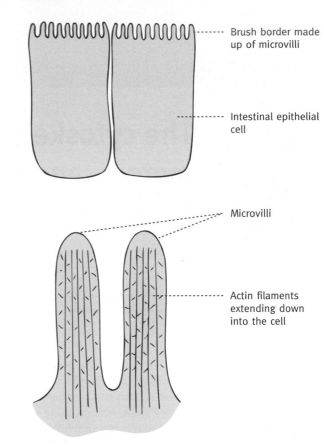

Fig. 33.1 Diagram of actin filaments in microvilli. Note that the filaments have a complex crosslinking and anchoring arrangement with other proteins to form a robust cytoskeletal structure.

The role of actin and myosin in nonmuscle cells

Actin is an abundant protein in most eukaryote cells and in many it is the most abundant single protein. There are slightly different actins in different types of cells, but it is a highly conserved protein and all form actin filaments, as described for muscle cells (see page 525). Myosin also appears to be an almost universal constituent of eukaryotic cells but is present in smaller amounts. The imbalance between actin and myosin in nonmuscle cells correlates with the fact that actin has a structural and transport roles in addition to that in contraction.

Structural role of actin and its involvement in cell movement

A particularly clear example of a structural role for actin is seen in the microvilli of intestinal brush border cells. These are the finger-like projections which enormously increase the absorptive area of the gut lining (Fig. 33.1). Apart from this specialized role, actin filaments pervade all animal cells. They are particularly dense near the cytoplasmic membrane where bundles of them anchored into the membrane form **stress fibres**, involved in maintaining cell shape (Fig. 33.2(a)). The arrangement of actin filaments, their anchoring and cross-linking into networks, depends on an array of other proteins which associate with the actin. The actin filaments in microvilli are permanent structures but, in general, they are disassembled and reassembled rapidly as required, the filaments being reversibly dissociated into actin monomers. When a phagocyte engulfs a particle,

actin filaments are seen to extend into the pseudopodia (these are the footlike extensions the cell puts out in the direction of movement). It is thought that **myosin II**, the 'conventional' muscle-like myosin molecule as found in skeletal muscle, is involved in exerting force on actin filaments anchored to the cell membrane which is involved in the crawling action of cells, though the process is complex. When contracting mechanisms are needed in nonmuscle cells, myosin molecules of the conventional type aggregate when required into small bipolar filaments of about 16 myosin molecules analogous to striated muscle thick filaments. In an action very similar to that described for muscle (page 525); these exert a contracting force on adjacent actin filaments (Fig. 33.3) and exert a pull on the cell membrane to which the filaments are anchored. Control of contraction is associated with phosphorylation of myosin light chains as in smooth muscle (page 532). **Cytokinesis**, or the constriction of a dividing cell into two daughter cells, provides an example of this type of contractile system. During cell division,

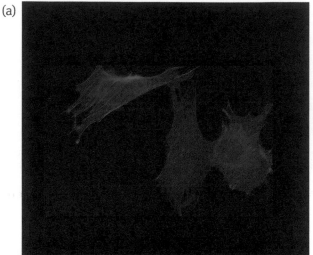

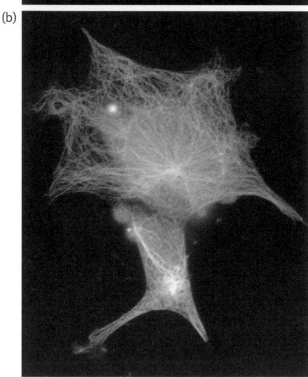

Fig. 33.2 **(a)** Actin filaments in mouse myoblasts visualized with an actin antibody. The filaments pervade the cell and bundles of them (called stress fibres) are attached at focal points to the cell membrane, often at points at which the membrane makes contact with a solid surface. The fibres are rapidly disassembled and assembled. Photograph kindly provided by Dr. P. Gunning, Children's Medical Research Institute, Sydney. **(b)** Microtubules in cytoplasm of a cell radiating from the microtubule organizing centre (MTOC) or centrosome (see text).

an annulus of actin fibres assembles at the equatorial plane where constriction occurs. These actin filaments are postulated to be anchored at the plasma membrane. Overlapping of these filaments (with opposite polarities) presents the opportunity for cytoplasmic myosins to exert a contraction force. The set-up is disbanded as the contraction progresses. The action results in constriction of the cell and ultimately into the separation of two daughter cells.

The role of actin and myosin in intracellular transport of vesicles

In our brief overview, we mentioned that actin filaments act as transport tracks along which ATP-driven molecular motors can move cellular components. The rod-like 'tail' of the muscle type 'conventional' myosin molecule is designed so that, as described, myosin can arrange itself into bipolar (thick) filaments which can cause adjacent actin filaments to slide over one another and thus cause a contraction to occur (page 525), but other types of myosin exist. These have essentially the same myosin heads as those which are the 'motor' components of conventional myosin, but the tails are different. The family of these other types of myosin molecules are called **minimyosins** because instead of the long rod-like tail of muscle myosin, the tail is small. This type of myosin cannot form bipolar bundles but runs along the actin filament, at the expense of ATP hydrolysis, by the same action as described for muscle. The tail of the minimyosin is designed to attach to another structure (the cargo) such as a vesicle membrane so that the minimyosin pulls the vesicle along the actin filament 'track'. As referred to earlier, in the giant alga *Nitella*, minimyosin motors pull the endoplasmic reticulum along stationary actin filaments which circuit the cell. The minimyosin tails attach to the ER which pervades the cytoplasm and, by pulling this along, causes the protoplasm to continually circulate (or stream) in its entirety.

A variety of minimyosins have been identified. They all have the globular actin binding motor head(s) with different tails and presumably bind to specific membranes or possibly other cargoes. They have been found in yeast and vertebrates and in almost all tissues including brain.

We pointed out that in muscle, the myosin heads always track along the actin filament in the (−) → (+) direction. The myosins constitute a large superfamily with 15 subclasses; it was the accepted dogma that all myosins move towards the (+) end of the actin filament. However, very recently a vesicle-transporting class of minimyosins has been discovered which moves in the opposite direction. In the previous chapter we described the swinging lever mechanism of myosin action and this is equally true for the minimyosins. In this mechanism, we remind you,

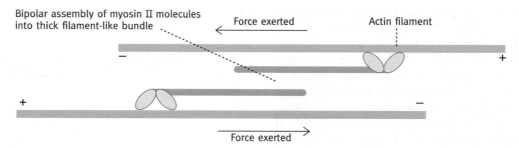

Fig. 33.3 Diagram of the pulling action of myosin II on actin filaments. Note that ATP hydrolysis supplies the energy. The achievement of useful work by this system will depend on appropriate anchoring of the actin filaments to the cell membrane so as to change cell shape. See page 525 for an explanation of the (+) and (−) polarities of the actin filaments Each myosin molecule has a region capable of interacting with other molecules and thus allowing self-assembly of bipolar filaments. The diagram is a cross-sectional representation—the bipolar assemblies are cylindrical. In the arrangement of actin fibres and myosin bundle shown, a contraction results from the relative sliding of the actin filaments.

the energy derived from a molecule of ATP causes a small conformational change in the motor domain of the myosin head; this is amplified in a subdomain called the **converter** which lies at the attachment point of the lever arm and causes the latter to move, thus propelling the head to make a step of a few nanometres along the actin fibre. The minimyosin with the reverse movement has the same general mechanism but at the base of the lever arm there is an extra sectionof about 50 amino acids which causes reversal of the lever arm movement.

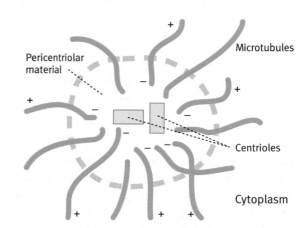

Fig. 33.4 Diagram of microtubules radiating out from the MTOC. The (+) and (−) indicate the polarity of the microtubules. The centrioles are a pair of tube-like structures made of fused microtubules; the radiating microtubules originate in the material surrounding the centrioles. (There is no precise boundary to this.)

Microtubules, cell movement, and intracellular transport

What are microtubules?

Microtubules are made by polymerization of **tubulin** protein subunits. The latter are dimers of α **and** β **tubulin** molecules, the dimers being referred to as tubulin. The tubulin dimer has a definite polarity. Tubulin reversibly polymerises to form a hollow tube bounded by 13 longitudinal rows of subunits. The polymerization of the dimers occurs in a head to tail fashion so that a microtubule has a definite polarity with the ends referred to as plus (+) and minus (−) ends. Microtubules are basically unstable in the cell. They can assemble and disassemble very rapidly. The instability is associated with unprotected ends—where these occur, the microtubules undergo collapse by de-polymerizing into free tubulin subunits. In the cell, the (−) ends are associated with the **microtubule organizing centre (MTOC)**, a structure near the nucleus and containing a pair of small bodies, the **centrioles**, made of fused microtubules

(Fig. 33.4). The cell is pervaded by microtubules radiating out from the MTOC (Fig. 33.2(b)).

What protects the positive (+) ends of microtubules?

As stated, microtubules rapidly collapse unless their ends are protected. The MTOC protects the (−) end. Microtubules grow (and collapse) in the cell from the (+) ends. It is thought that when the microtubule reaches an appropriate target, target proteins 'cap' and protect the microtubule. The arrangement means that microtubules can grow out of the MTOC in completely random directions—those which make contact with an appropriate component of the cell will, on this model, be capped and stabilized while the remainder will collapse. Obviously what

constitutes suitable target proteins and how these are arranged is a separate major question. There is still the problem of what protects a growing microtubule, *before* it reaches a target protein, from collapse. A current model is that tubulin subunits have GTP attached to them; a GTP-tubulin subunit protects the (+) end. The GTP is hydrolysed to GDP and P_i after addition of a subunit to a tubule, but this does not happen until after a slight delay. Thus newly added subunits will be in the GTP form, so temporarily capping the microtubule, which protects the end from collapse. The GTP hydrolysis in this model is acting as a clock; provided the addition of new subunits occurs before the GTP on the last one added is hydrolysed, the microtubule is protected. If not, and the GTP on the exposed end subunit has been hydrolysed, then, on this model, collapse ensues. In cilia and flagella where permanent microtubules occur (see below) covalent modification of the protein occurs after assembly.

Functions of microtubules

As with actin fibres, microtubules are involved in cellular morphology, cell movement, and intracellular transport.

Molecular motors: kinesins and dyneins

Two types of microtubule-associated motors have been identified, kinesin and dynein, both of which are dependent on ATP for their movement. Kinesin travels along a microtubule in the (−) → (+) direction, and dynein in the opposite direction. There are families of kinesin and dynein molecules specialized for different tissue functions. The heads which perform the actual movement along the microtubule are probably a constant motif within each family while the tail structures attach to specific cargoes (Figure 33.5(b)). As an example, vesicles are pulled along microtubules by molecular motors (see below). Kinesin and dynein molecules (supplied with ATP) will walk along microtubule fibres immobilized on a solid, just as myosin heads will walk along immobilized actin fibres.

The best-known kinesin, known as conventional kinesin, is illustrated in Fig. 33.5(a). As with myosin, there are two heads which power the movement along microtubules. The motor heads are about half the size of those of the myosins. The movement differs from that of the myosins in that the two heads indulge in a walking-like action along the microtubule with the heads swivelling past one another at each step. The neck of the molecule is a very flexible section which allows this. It is rather like the action of climbing up a rope hand over hand. At all times at least one hand grips the rope while the other outreaches it. The tail has a pair of light chains which bind to the cargo such as a vesicle. Presumably the multiplicity of different kinesins are

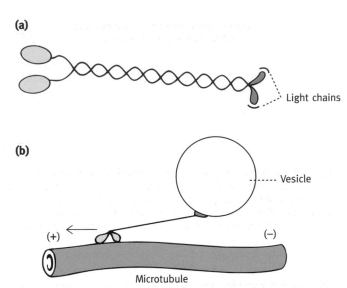

Fig. 33.5 (a) Diagram of the structure of a kinesin molecule. **(b)** Diagram illustrating how a kinesin molecule can transport vesicles along a microtubule. Also see Fig. 33.6 for an electron micrograph of vesicles being transported within a cell.

needed to bind specifically to different cargoes. As stated the kinesin travels towards the (−) end of a microtubule, which means in general toward the periphery of the cell. In a nerve axon they pull vesicles from the cell body onwards along the axon.

Dynein is a larger molecule, again with two motor heads, which travels along microtubules in the direction opposite to that of kinesin. **Axonemal dynein** is confined to cilia and flagellae (see below) but there is also cytoplasmic dynein, found in all cells which have the task of transporting vesicles.

Role of microtubules in cell movement

This is confined to cilia and flagellae. Cells lining the respiratory passages have large numbers of cilia whose beating motion sweep along mucus and its entrapped foreign particles. Sperm propel themselves with flagellae. Cilia and flagella are basically similar, and they have a remarkably elaborate structure. Microtubules run down the length of the organelle, originating in a basal body closely resembling a centriole; these microtubules are permanent structures. Associated with the microtubules are dynein molecules. Using ATP energy, they 'walk' along microtubules (towards the (−) end) causing the latter to slide relative to each other and resulting in a wave motion in the cilium or flagellum.

Role of microtubules in vesicle transport inside the cell

Vesicles produced by the Golgi apparatus (see page 405) are targeted to different destinations and probably guided transport is involved; the multiplicity of different kinesins is presumably needed to carry specific vesicles to their different destinations. It is believed that actin fibres and minimyosins are also involved in some of the vesicle traffic. The transport of vesicles along the nerve axon has already been referred to; microtubules extend down the axon to the tip, with the (+) ends oriented towards the tip. A molecular motor, kinesin, is attached to membrane vesicles at the cell body end and moves along the microtubule in the (+) direction (Figure 33.5(b)). Another motor, cytoplasmic dynein, transports in the reverse direction from the tip to cell body. Another very striking example of vesicle transport along microtubules is shown in the beautiful scanning electron micrograph in Fig. 33.6. Fish and amphibia often have a camouflage mechanism which rapidly changes the colour of the skin. This is achieved by vesicles containing pigment either moving to the periphery of the cells or becoming evenly distributed according to requirements.

Role of microtubules in mitosis

Chromosomes are duplicated during the cell cycle to produce the familiar X-shaped structure shown in Fig. 33.7 in which the two chromosomes are associated together by their centromeric regions. The chromatids, as the new chromosomes are called at this stage, have to be separated so that each of the two daughter cells which will be formed by cell division will receive one copy each. The nuclear membrane disappears and the MTOC divides, one copy each migrating to opposite ends of the cell. From these microtubules grow out to form the **mitotic spindle**. The duplicated chromosomes become arranged in the central plane of the spindle (Fig. 33.7). There are two types of microtubules in the spindle; the first type. called **kinetocore fibres**, attach to the centromeres of each chromatid (see Fig. 33.7). The second type are **polar fibres** which overlap at their positive ends. Two actions are involved in chromosome segregation. Kinesins operating between the overlapping polar fibres are believed to drive the centrosomes to opposite ends of the cell, possibly helped by molecular motors associated with the astral fibres pulling them apart. The separation of the chromosomes is due to shortening of the kinetocore fibres attached to the centromeres. Microtubules cannot contract; the shortening is due to disassembly of the microtubule at the attachment point to the chromosome, a rather remarkable process. The role of microtubule assembly in chromosome segregation is demonstrated by the drug **colchicine** which freezes mitosis at this stage. It binds to tubulin and prevents the microtubule assembly from shortening.

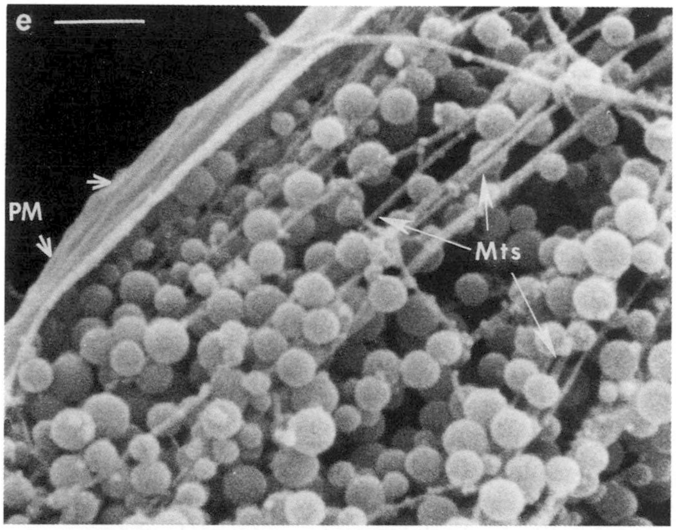

Fig. 33.6 Vesicle transport by microtubules. Scanning electron micrograph of pigment-containing vesicles being transported along microtubules in a chromatophore of a squirrel fish. The change of colour of the cell is effected by the movement of the vesicles to and from the cell centre. Mts, microtubules; PM, plasma membrane. Bar scale = 0.5 μm.

Intermediate filaments

The actin network, described above, has microfilaments 6 nm in diameter; the microtubule filaments are 20 nm in diameter. The third network in eukaryotic cells made of filaments averaging 10 nm in diameter—hence their name of **intermediate filaments (IFs)**.

There are several types of IFs composed of a diverse group of proteins which, nonetheless, have close homology in their structures. Typically there is a core filament about 350 amino acid residues in length with the ends varying in the different types of IF. This diversity of proteins contrasts with the single protein subunits in actin and microtubule filaments.

Different types of IFs occur in different eukaryote cells and are expressed at specific stages of development and differentiation, suggesting that they play important roles. These include keratin IFs in epidermal cells which confer toughness; a number of **lamin proteins** form a network associated with the inner surface of the inner nuclear membrane. They keep the nucleus in position. **Neurofilaments** exist in nerve cells to give necessary mechanical support to the long axons; **desmin filaments** are located in the Z discs of sarcomeres. In short, the role of IFs is to

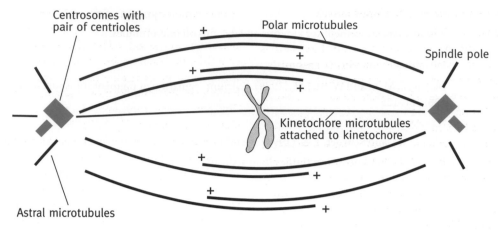

Fig. 33.7 Mitotic spindle at the metaphase state. The microtubules originate at the centrosomes. The duplicated chromosomes (only one duplicate is shown) are arranged at the centre of the spindle attached to the kinetochore fibres at the complex of proteins known as the kinetochore (see Fig. 20.9). The overlapping polar fibres propel the centrosomes apart; it is believed that kinesins in the overlapping regions of the polar fibres cause the separation. The astral microtubules assist in the separation of the spindle poles, (+) directed motors probably being involved. The kinetochore fibres are progressively shortened at the kinetochore attachment site by disassembly.This pulls the duplicated chromosomes apart to opposite ends of the cell in preparation for cell division.

confer mechanical strength to cells. At the extreme end of the spectrum the keratins of hair and fingernails are made from IFs. IFs are not essential for cell growth and division; cells in culture in which IF formation does not occur due to mutation still grow and divide presumably, because they are not subject to the mechanical stresses experienced by cells in functional tissues.

FURTHER READING

Bray, D. (1992). *Cell movements.* Garland Publishing.

A comprehensive readable, account of all aspects of movement in cells. Good simple diagrams and also illustrated with electron micrographs.

Cross, R. A. and Carter, N. J. (2000). Molecular motors. *Current Biology,* **10,** R 177–9.

One of the 'primer' series which gives essential information in a couple of pages.

Sablin, E. P. (2000). Kinesins and microtubules: their structures and motor mechanisms. *Current Opinion in Cell Biology,* **12,** 35–41.

Reviews mainly structural aspects.

Microtubules in cell movement

Byard, E. H. and Lange, B. M. H. (1991). Tubulin and microtubules. *Essays in Biochemistry,* **26,** 13–25.

Excellent general review.

Goldstein, L. S. B. (1993). With apologies to Scheherazade: tails of 1001 kinesin motors. *Ann. Review Genetics,* **27,** 319–51.

An in depth review of these minimotors with useful overviews.

Walker, R. A. and Sheetz, M. P. (1993). Cytoplasmic microtubule-associated motors. *Ann. Rev. Biochem.,* **62,** 429–51.

Detailed review of the minimotors and their roles.

Wallee, R. B. and Sheetz, M. P. (1996). Targeting of motor proteins. *Science*, 271, 1539–44.
Discusses the movements of kinesins and dyneins on microtubules.

Microtubules in vesicle transport

McNiven, M. A. and Ward, J. B. (1988). Calcium regulation of pigment transport in vitro. *J. Cell Biol.*, 106, 111–125.
A research paper but readable.

Allkan, V. J. and Schroer, T. A. (1999). Membrane motors. *Current Opinion in Cell Biology*, 11, 476–82.
Reviews membrane vesicle traffic along actin and microtubule tracks.'

Microtubules in mitosis

Glover, D. M., Gonzalez, C., and Raff, J. W. (1993). The centrosome. *Sci. Amer.*, 268 (6), 32–9.
Very useful review of this relatively little known structure.

PROBLEMS FOR CHAPTER 33

1   Actin is found in nonmuscle cells. What are its roles there?

2   What are microtubules?

3   Microtubules with unprotected ends undergo catastrophic collapse. What protects the ends as they form?

4   What are kinesin and dynein? Illustrate by a diagram.

5   At cell division, chromosomes on the metaphase equatorial plate move apart. Microtubules are attached to the kinetochores and shorten as the chromosomes move apart. Does this mean that microtubules contract? Explain your answer.

6   What are intermediate filaments?

7   What are the functions of intermediate filaments?

Answers to problems

Answers to problems

Chapter 1

1 The amount of ATP that can be synthesized using 5000 kJ of free energy is 5000/55 which equals 90.91 mol. The weight of disodium ATP produced is thus 551×91 g daily, or 50 141 g which is equal to 72% of the man's body weight. The reason why this is possible is that ATP is being continuously recycled to $ADP + P_i$ and back again to ATP.

2 The $\Delta G^{0\prime}$ value refers to standard conditions where ATP, ADP, and P_i are present at 1.0 M concentrations. In the cell the concentrations will be very much lower, the actual ΔG value for ATP synthesis will be different from the $\Delta G^{0\prime}$ value according to the relationship

$$\Delta G = \Delta G^{0\prime} + RT\,2.303\log_{10}\frac{[ADP][P_i]}{[ATP]}.$$

3 ATP and ADP are high-energy phosphoric anhydride compounds, whereas AMP is a low-energy phosphate ester. The factors that make hydrolysis of the former more strongly exergonic include the following.

Release of phosphate relieves the strain caused by the electrostatic repulsion between the negatively charged phosphate groups. The released phosphate ions fly apart. A factor also contributing to the exergonic nature of the hydrolyses is the resonance stabilization of the phosphate ion which exceeds that of the phosphoryl group in ATP. Hydrolysis of AMP causes little increase in resonance stabilization.

4 **(a)** The nonpolar molecule cannot form hydrogen bonds with water molecules so that those of the latter surrounding the benzene molecule are forced into a higher-order arrangement in which they can still hydrogen bond with each other (for the total bonding does not change). This increased order lowers the entropy and increases the energy of the system; insertion of the benzene molecules into the water is therefore opposed and they are forced into a situation with minimal benzene/water interface as spherical globules and then as a separate layer. The effect is known as hydrophobic force.

(b) The polar groups on glucose can form hydrogen bonds with water.

(c) The Na^+ and Cl^- ions become hydrated and the free energy decreases as a result of this exceeds that of the ionic attraction between them. The separation also has a large negative entropy value.

5 The enzyme AMP kinase transfers a phosphoryl group from ATP to AMP by the reaction

$$ATP + AMP \rightarrow 2ADP.$$

Hydrolysis is not involved so that there is no significant $\Delta G^{0\prime}$ in the reaction.

6 In the cell, PP_i will be hydrolysed to $2P_i$, whereas with a completely pure enzyme there will be no inorganic pyrophosphatase present. In the former case, the $\Delta G^{0\prime}$ of the total reaction will be $-32.2 - 33.4 + 10\,\text{kJ mol}^{-1} = -55.6\,\text{kJ mol}^{-1}$. For the completely pure enzyme, the $\Delta G^{0\prime}$ will be $-22.2\,\text{kJ mol}^{-1}$.

7 Ionic bonds, hydrogen bonds, and van der Waals forces with average energies of 20, 12–29, and 4–8 kJ mol^{-1}, respectively. The energies of activation for formation of weak bonds are very low and so can occur without catalysis. Weak bonds in large number can confer definite structures on molecules, but nonetheless can be broken easily, resulting in flexibility of structures.

8 **(a)**

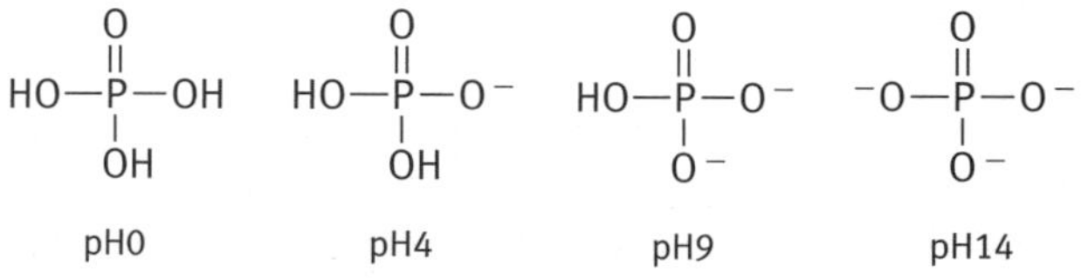

(b)

$$pH = pK_a + \log\frac{[Salt]}{[Acid]}$$

The relevant pK_a is 7.2.

$$pH = 7.2 + \log \frac{[Na_2HPO_4]}{[NaH_2PO_4]}$$
$$= 7.2 + \log 1$$
$$= 7.2.$$

(c) Histidine with a pK_a value of 6.5.

9 The binding of the molecules is via noncovalent bonds. Since these are weak, several are needed to make the binding effective. Noncovalent bonds are very short range and, in the case of hydrogen bonds, highly directional. Put these considerations together and only molecules which exactly fit together can make the required number of bonds. This is the basis of biological specificity. The fact that the bonds are weak (low ΔG) means that they can form spontaneously, without catalytic assistance, and importantly can dissociate to make the associations reversible, which is also essential in many situations. In the case of antibody–antigen binding in which irreversibility is needed for immune protection, very large numbers of bonds are involved. Many structures in the cell, in some cases very elaborate ones, depend on the same principle of specific protein–protein interactions to bring about their self-assemblage.

10 In chemistry, a high-energy bond is one requiring a large amount of energy to break it, which is the reverse of the intended biological concept. Most importantly however, it is incorrect to envisage the energy of a molecule to be located in a particular bond but—rather, it resides in the molecule as a whole. It is the free energy fall in converting ATP to its products that provides the energy for work. A fall in free energy is a property of a reaction, not of the breakage of a particular bond. This does not alter the fact that Lipmann's concept was important in that ATP behaves *as if* the energy is located in the bond. For this reason it can still be found in textbooks.

Chapter 2

1 Very little of significance because the plot will be arbitrary depending on the time period selected for the measurement of activity. This is because, at higher temperatures, destruction of the enzyme is likely to occur and, at any given temperature, the amount of destruction will be proportional to the time the enzyme is exposed to that temperature. If information is sought on the heat stability of the enzyme it is much better to expose it for a standard time at the different temperatures and, after cooling, to measure the activity in each sample at the normal incubation temperature.

2 The $\Delta G^{0\prime}$ value of a reaction determines whether a reaction may proceed, but says nothing about the rate at which it does proceed (if at all). The latter is determined by the energy of activation of the reaction and the rate at which the transition state is formed.

3 This can be due to several factors.
(a) The active site binds the transition state much more firmly than it does the substrate and, in doing so, lowers the activation energy.
(b) It positions molecules in favourable orientations.
(c) It can exert general acid–base catalysis on a reaction.
(d) It may position a metal group which facilitates the reaction.

4 A double reciprocal plot (Lineweaver–Burke plot) is needed. A noncompetitive inhibitor will reduce the V_{max} at infinite substrate concentration (the intercept at the vertical axis) but will not alter the K_m. A competitor inhibitor on the other hand has no effect on the V_{max} at infinite substrate concentration but does change the K_m. This is illustrated in Fig. 2.5.

5 An affinity constant for a substrate binding to an enzyme represents the equilibrium $E + S \rightleftharpoons ES$. A K_m, which is a measure of the concentration of S at which the enzyme is working at half maximum velocity, is derived from a measure of the rate of product formation. This involves the catalytic rate at which the enzyme converts ES to product. For a K_m to represent a true affinity constant, it is necessary that the rate of the reactions $E + S \rightarrow ES$ and the reverse $ES \leftarrow E + S$ are very large compared with the rate of removal of ES by conversion to product so that the latter can be ignored in the substrate binding equilibrium.

6 A transition state binds more tightly to the active centre of an enzyme than does the substrate molecule. This is the fundamental basis of enzyme catalysis. The affinity may be a thousand or more times greater than that of the substrate. Therefore there is considerable interest in the medical potential of stable transition state molecules. They form the basis of 'statin' drugs, which lower cholesterol levels (page 130).

7 It is essential that initial velocities are measured so that product inhibition, substrate depletion, and enzyme denaturation are not significant. This is illustrated in Fig. 3.7.

Chapter 3

1 It is a polypeptide chain composed of large numbers of amino acids linked together by peptide (—CO—NH—) bonds. The structure of such a chain is given on page 37.

2   The polypeptide chain of a protein is folded into a specific, usually compact shape, the precise folding being dependent on noncovalent bonds and covalent disulfide bonds between the amino acid side chains. Heat disrupts the former bonds, causing the polypeptide chains to unravel and become entangled into an insoluble mass, devoid of biological activity. This is known as protein denaturation.

3   Examples of all are given on pages 36 and 37.

4   **(a)** Around 4; **(b)** around 10.5–12.5; **(c)** around 6.0.

5   Aspartic and glutamic acids, lysine, arginine, and histidine. The carboxyl and amino groups of all other amino acids are bound in peptide linkage (excluding the two end ones).

6   Primary, secondary, tertiary, and quaternary. Illustrated in Fig. 3.3.

7   The structures are the α helix, the β-pleated sheet, and the random coil or connecting loop region. The first two are structures that satisfy the hydrogen bonding potential of the polypeptide backbone. The structures are given in Figs. 3.4 and 3.5. The random coil has no consistent structure. It usually is on the exterior of proteins where hydrogen bonding can be satisfied by water.

8   It lies in the four-way desmosine group shown in Fig. 3.10.

9   Collagen fibres are designed for great strength. To achieve this, three polypeptides are associated in a closely packed triple superhelix. The pitch of the supercoil is such that there are three amino acid residues per turn. This arranges that where the polypeptides touch, there is always a glycine residue. The side group of a single hydrogen atom permits close association whereas bigger side groups would prevent this.

10   Proteins are functional only in their native state, which means their correct three-dimensional folded state. This, in most proteins, is due only to noncovalent bonds, which are easily disrupted by mild heat. The polypeptides unfold and become irreversibly tangled up into aggregates. A number of proteins such as insulin have a few disulfide bonds (three in the case of insulin) which, being covalent, are not destroyed by heat. Although noncovalent bonds are still important in insulin structure, the disulfide bonds confer a degree of increased stability. The extreme case is keratin where large numbers of disulfide bonds produce a very stable structure. You can boil hair as long as you like without producing any permanent structural changes.

11   Particularly in the larger proteins, when the three-dimensional structure is determined, separate regions of the protein are often seen (see Fig. 3.7(d)) separated by unstructured polypeptide. The impression is that these are essentially independent globular sections, which could exist on their own. Indeed, in a number of cases, domains have been isolated from multifunctional enzymes and found to have partial catalytic activity on their own. This is sometimes of great practical use in that it permits the preparation of an enzyme fragment with a wanted activity, but is free from other activities of the complete enzyme. It is essential for a protein domain to be formed from a given section of a chain; it can't leave the domain and then double back into it, because then the structure would not be stable on its own if separated from the rest of the protein.

Domains are of very great interest because it is likely that they have played an important role in evolution by a process of domain shuffling. This is described in detail later but for the moment it is often found that domains are coded by sections of genes which are believed to be shuffled about to form new genes. When expressed, these give rise to new functional proteins. Evolution of new enzymes by this means would be much faster than gradually altering proteins by point mutations.

12   Phenylalanine and isoleucine are hydrophobic and are to the maximum possible extent shielded from water in the hydrophobic interior of globular proteins. Aspartic acid and arginine have highly charged side chains at physiological pH and therefore need to be on the outside exposed to water. Inside a protein molecule these groups would have a destabilizing effect because they cannot have their bonding capabilities satisfied in the hydrophobic environment. Formation of bonds releases energy so that only when bonding is maximized do you get the most stable structure.

13   The original Sanger technique was to label the N-terminal amino group of a peptide derived from partial hydrolysis of the protein. The label is coloured or fluorescent for ease of detection and the labelling is stable to acid hydrolysis. The peptide is hydrolysed and the products separated by chromatography. This identifies the first (N-terminal) amino acid residue. Sequencing a protein by this procedure is very laborious, because the exercise has to be performed on many peptides and the complete sequence determined from overlapping peptides. This method would now not be used for anything but very small peptides, if that. The Edman procedure is to label the peptide with a reagent, which under appropriate conditions causes the N-terminal amino acid to detach, still labelled. This exposes the next amino acid residue to be so labelled. The procedure is automated and can be used for sequences up to about 30 residues. The labelled amino acids detached in turn are identified by chromatography. The method is laborious if a complete protein

has to be sequenced and requires the disruption of the protein into oligopeptides which are separately sequenced. Mass spectrometry can determine the sequence of oligopeptides very rapidly and promises to be of increasing importance. It should not be forgotten that often the quickest method of sequencing a protein is to isolate the DNA responsible for coding the amino acid sequence of the protein. The understanding of this statement may have to wait until later sections of the book have been studied.

14 The serine —OH is perfectly positioned next to a histidine group, which readily accepts the proton thus freeing the oxygen atom to make a bond with the target carbon atom.

15 The aspartate —COO⁻ group forms a strong hydrogen bond with the histidine side chain and holds the imidazole ring in the tautomeric form and most favourable orientation to accept the proton from serine (see Fig. 3.11). Amidation of the carboxyl group greatly reduces its hydrogen bonding potential.

16 The charges on the GAGs cause the molecule to adopt an extended conformation and the GAG units permit the formation of very large molecules so that the total volume of water trapped is large.

17 The polypeptide backbone has hydrogen bonding potential which must be satisfied if structures of maximum stability are to be achieved. The backbone cannot avoid crossing the hydrophobic interior of proteins and the hydrophobic side chains of the amino acid residues in these regions offer no possibility of hydrogen bond formation. The solution is for backbone groups to hydrogen bond to other backbone groups and this is what happens in both the α helix and the β-pleated sheet structures.

Chapter 4

1 Two hydrophobic tails rather than a single one.

2 It acts as a fluidity buffer by preventing the polar lipids from associating too closely towards the centre of the bilayer, but also acts as a wedge at the surface layer.

3

```
CH₂—O•CO—R
|
CH—O•CO—R
|        O
|        ||
CH₂—O—P—O⁻
         |
         O⁻
```

4 Lacithin (choline); cephalin (ethanolamine); phosphatidylserine (serine).

5 They put a kink into the tail which prevents close packing of the fatty acyl tails of lipid bilayers and thus make the membrane more fluid.

6 To do so would require stripping the molecules of H_2O molecules, an energetically unfavourable process.

7 It involves a membrane protein forming a hydrophilic channel which permits solutes to traverse the membrane in either direction, according to the direction of the concentration gradient. No energy input is involved. The anion transport of red blood cells is an example; it allows Cl^- and HCO_3^- ions to pass through in either direction.

8 A triacylglycerol is a neutral molecule with no hydrophilic groups. The molecule is therefore water insoluble and to minimize contacts with water the molecules are forced to become spherical globules or form a separate layer. A polar lipid is amphipathic with a charged group at one end and hydrophobic tails at the other. It therefore can assemble into a bilayer structure with hydrophobic tails in contact with each other and shielded from water. The charged group can interact at the surface with water giving the bilayer a stable minimum free energy arrangement of the molecules.

9 It is simply another way of constructing an amphipathic molecule of the same type as a glycerophospholipid—one that can participate in the formation of a lipid bilayer. Sphingosine-based lipids are prevalent in myelinated nerves, but exactly why this particular structure is selected for this is not known nor is there a clear understanding of why such a variety of polar lipids is employed in different membranes.

10 This is done by symport systems. In the case of glucose active transport, the glucose molecule is cotransported into the cell with sodium ions whose concentration is much higher outside than inside. The latter are ejected from the cell by the Na^+/K^+ ATPase that maintains the ion gradient. Thus ATP indirectly supplies the energy for glucose transport.

11 It is a membrane channel formed by proteins, which opens and closes on receipt or cancellation of a signal respectively. A ligand-gated channel opens on binding of a specific molecule. The acetylcholine-gated channel in nerve conduction is one. A voltage-gated channel opens in response to a change in membrane potential. The Na^+ and K^+ voltage-gated channels in neuron axons are examples.

12 The drug 'freezes' the Na^+/K^+ pump in the phosphorylated form when applied to the outside face of the protein in the

presence of K^+. This inactivates the pump and results in the intracellular concentration of Na^+ rising, which lowers the steepness of the gradient of this ion across the cell membrane. This causes an increase in the intracellular Ca^{2+} concentration which increases the strength of the heart muscle contraction. The increase in Ca^{2+} is due to the fact that an antiport system driven by the Na^+ gradient exports calcium from the cell. Lowering the Na^+ gradient reduces the outflow of Ca^{2+}.

Chapter 5

1   Pepsin as pepsinogen; chymotrypsin as chymotrypsinogen; trypsin as trypsinogen; elastase as proelastase; carboxypeptidase as procarboxypepidase.

The proteolytic enzymes are potentially dangerous in that they could attack proteins lining the ducts. There are no components that amylase might attack. Mucins coating the gut cell lining of the intestine protect cells against proteolytic attack.

2   In the stomach the acid pH causes a conformational change in pepsinogen that activates molecules to self-cleave the extra peptide in pepsinogen that inactivates the enzyme. Once some pepsin is formed, it activates more pepsinogen so that an autocatalytic activation cascade occurs. In the small intestine, enteropeptidase, an enzyme produced by gut cells, activates trypsinogen to trypsin. This, in turn, activates the other zymogens.

3   Pancreatitis, or inflammation of the pancreas, due to proteolytic damage of pancreatic cells.

4   Because the gut cells can absorb only monomers (plus monoacylglycerols). Fat, protein, polysaccharides, and disaccharides cannot be absorbed.

5   Milk contains lactose, which must be hydrolysed to glucose and galactose by the enzyme lactose. After infancy, many individuals lose the capacity to produce this. The lactose is not absorbed and is fermented in the large intestine. It attracts water into the gut by its osmotic effect, producing diarrhea.

6   TAG or triacylglycerol (sometimes called triglycerides, but this is chemically incorrect).

```
                / Primary ester
CH₂O · CO—R
|
CHO · CO—R
|
CH₂O · CO—R
                \ Primary ester
```

7   The fat is emulsified by intestinal movement and by the monocylglycerol and free fatty acids produced by initial digestion together with bile salts, so that the lipase can attack the emulsified substrate. The products of digestion (free fatty acids and monoacylglycerol) are carried to the intestinal cells as disc-like mixed micelles with bile salts that have a high carrying capacity for such products. Probably at the cell surface the micelle breaks down.

8   The free fatty acids and monoacylglycerol are resynthesized into TAG. The TAG itself cannot traverse cell membranes but it is incorporated into lipoprotein particles, called chylomicrons. These have a shell of stabilizing phospholipid, some proteins, and a centre of TAG and cholesterol and its ester. The chylomicrons are released by exocytosis into the lymphatics and eventually discharge into the blood as a milky emulsion via the thoracic duct.

Chapter 6

1   The osmotic pressure of glucose precludes this. The osmotic pressure of a solution is related to the number of particles of solute in it. By polymerizing thousands of glucose molecules into a single glycogen molecule, the osmotic pressure is correspondingly reduced.

2   Glycogen stores in the liver are sufficient to last through 24 h starvation only, but there may be sufficient TAG for weeks. TAG is a more concentrated energy source—it is more highly reduced and is not hydrated. If the energy in fat were present as glycogen, the body would have to be much larger.

3   Yes—glucose can be converted to acetyl-CoA.
No—acetyl-CoA cannot be converted into pyruvate.
No—there are no special amino acid storage proteins (excluding milk) in the body.

4   (a)   It is the glucostat of the body. It stores glucose as glycogen in times of plenty; it releases glucose in fasting to keep a constant blood sugar level which the brain needs.
(b)   During prolonged starvation it synthesizes glucose for the same reason; it converts fats to ketone bodies which the brain (and other tissues) can use for part of its energy supplies and so conserve glucose.
(c)   It synthesizes fat and exports it to other tissues.

5   No.

6   They store fat in times of plenty; they release fatty acids in fasting.

7 They are completely dependent on glucose, which they convert to lactate. Since mature red blood cells lack mitochondria, they are unable to produce ATP by any metabolic route except glycolysis.

8 When blood glucose is high, insulin is released. This is the signal for tissues to store food. As the blood glucose levels fall, insulin levels fall and glucagon levels rise. The latter is the signal for the liver to release glucose and for adipose cells to release free fatty acids for the tissues to utilize. The brain is not insulin-dependent for its uptake of glucose, so this can proceed even in starvation, when insulin levels are extremely low.

In addition, epinephrine can override all controls, causing massive turnout of glucose and fatty acids to cope with an emergency requiring the 'fight or flight' reaction.

9 Ketone bodies are produced in excess in uncontrolled type I diabetes. However, in prolonged fasting, when glycogen sources are exhausted, they supply muscles with a ready supply of metabolizable substrate which is preferentially utilized as energy source. This minimizes the utilization of glucose. In starvation, keeping up the blood glucose level assumes top priority because the brain must be supplied with it. Nonetheless during starvation the brain can adapt to use ketone bodies for perhaps half of its energy needs, thus economizing on glucose. Preserving glucose for essential needs in this situation is important. To provide pyruvate for the liver to synthesize glucose, muscle proteins are broken down and it is clearly advantageous to keep muscle wasting to the minimal level consistent with keeping the brain supplied with glucose.

Chapter 7

1 Glucose-1-phosphate + UTP $\rightarrow$ UDP-glucose + PP_i.

$PP_i + H_2O \rightarrow 2P_i$.

UDPG + glycogen$_{(n)}$ $\rightarrow$ UDP + glycogen$_{(n+1)}$.

The hydrolysis of PP_i makes the overall reaction strongly exergonic.

2 The reaction named never occurs in the cell because the required substrate, inorganic pyrophosphate, is immediately destroyed. It is the reverse reaction that occurs in the cell but, for systematic nomenclature reasons, the enzyme is named for the forward reaction.

3 Only the liver (and the less quantitatively important kidney) does so. These alone possess the enzyme glucose-6-phosphatase.

4 Glucose + ATP $\rightarrow$ Glucose-6-phosphate + ADP.

Glucokinase has a much higher K_m than hexokinase. During starvation the liver turns out glucose into the blood, primarily to supply the brain (and red blood cells). The first reaction in the uptake of glucose into brain and liver is phosphorylation of glucose. The lower affinity of glucokinase for glucose, as compared with hexokinase, means that the liver does not compete with brain for blood sugar. It efficiently takes up glucose only when blood sugar levels are high. Since it is not as sensitive to inhibition by glucose-6-phosphate as is hexokinase, it can take up glucose and synthesize glycogen even when cellular levels of glucose-6-phosphate are high.

5 Many glycolipids and glycoproteins contain galactose. Removal of galactose from the diet does not prevent synthesis of these because UDP-glucose can be epimerized to UDP-galactose.

6 In the capillaries, lipoprotein lipase splits free fatty acids from TAG, and these immediately enter adjacent cells.

7 VLDL are lipoproteins, resembling chylomicrons, that carry cholesterol and TAG from the liver to peripheral tissues.

8 Cholesterol moves outwards from the liver to peripheral tissues in VLDL. The VLDL become depleted of TAG and pass through lipoprotein stages known as IDL and finally LDL. The latter are taken up by peripheral tissues. However, some of the IDL and LDL are re-taken up by the liver. The IDL and LDL receive peripheral tissue cholesterol via HDL and this constitutes a reverse flow of cholesterol from tissues to liver.

9 Conversion to bile salts in the liver. Excess cholesterol is a risk factor in causation of heart attacks. Removal of cholesterol is therefore of importance.

10 The enzyme LCAT transfers fatty acyl groups from lecithin to cholesterol.

11 TAG is *not* released; a hormone-sensitive lipase, stimulated by glucagon (or epinephrine in emergency situations), hydrolyses off free fatty acids that are carried in the blood to the tissues attached loosely to serum albumin. By contrast, fat transport from the liver to other tissues occurs via VLDL.

12 In familial hypercholesterolemia the patient lacks LDL receptors and so cannot remove cholesterol-rich LDL from the blood.

13 Adipose cells liberate free fatty acids which are carried attached to serum albumin. This dissociates from the latter and enters cells. Liver exports fat as VLDL. In extrahepatic tissues,

lipoprotein lipases release free fatty acids from the VLDL and these enter the adjacent cells.

Chapter 8

1 Glycolysis, the citric acid cycle, and the electron transport chain; cytoplasm, mitochondrial matrix, and the inner mitochondrial membrane, respectively.

2 Structures on page 136.

$$AH_2 + NAD^+ \rightleftharpoons A + NADH + H^+$$

$$B + NADH + H^+ \rightleftharpoons BH_2 + NAD^+$$

3 FAD is another hydrogen carrier; it is reduced to $FADH_2$. It is not a coenzyme but a prosthetic group attached to enzymes.

4 In aerobic glycolysis, glucose is broken down to pyruvate. The NADH produced is reoxidized by mitochondria. In anaerobic glycolysis, the rate of glycolysis exceeds the capacity of mitochondria to reoxidize NADH. This can occur, for example, in the 'fight or flight' reaction. Since the supply of NAD^+ is limited, if NADH were not reoxidized sufficiently rapidly to NAD^+, glycolysis and its ATP production would cease. An emergency mechanism reoxidizes NADH by reducing pyruvate to lactate.

$$CH_3COCOO^- + NADH + H^+ \rightarrow CH_3CHOHCOO^- + NAD^+$$
Pyruvate   Lactate   Lactate
                      dehydrogenase

5 Structures on page 139.
  -31 as compared with $-20\,kJ\,mol^{-1}$ for the carboxylic ester— that is, the thiol ester is a high-energy compound.
  The vitamin, pantothenic acid, does not participate in the reactions involving CoA, unlike the situation with other coenzymes, where the vitamin is the 'active part' of the molecule.

6 Pyruvate + CoASH + $NAD^+ \rightarrow$ Acetyl-S-CoA + $NADH^+$
+ H^+ + CO_2 $\Delta G^{0'} = -33.5\,kJ\,mol^{-1}$.

7 The acetyl group is fed into the citric acid cycle.

8 They are reoxidized by the electron transport chain producing H_2O and generating ATP in the process (though this is achieved indirectly via a proton gradient).

9 The Nernst equation relates $\Delta G^{0'}$ and $\Delta E_0'$ values as follows. $\Delta G^{0'} = -nF\Delta E_0'$ (F is the Faraday constant = $96.5\,kJ\,V^{-1}\,mol^{-1}$) and $\Delta E_0'$ is the difference between the redox potentials of the electron donor and acceptor.

In the example given $\Delta E_0'$ equals $-1.035\,V(-0.219 - 0.816\,V)$. Therefore,

$$\Delta G^{0'} = -2(96.5\,kJ\,V^{-1}\,mol^{-1})(-1.035\,V) = -193(-1.035)$$
$$= 194.06\,kJ\,mol^{-1}.$$

10 Breakdown of fatty acids to acetyl-CoA.

11 (a) Yes. Glucose is converted to pyruvate and pyruvate dehydrogenase converts this to acetyl-CoA. The latter can be used to synthesize fat.
(b) No. Fatty acids are broken down to acetyl-CoA. To synthesize glucose, pyruvate is needed. The pyruvate dehydrogenase reaction is irreversible. There can, in animals, be no net conversion of acetyl-CoA (and therefore of fatty acids) to glucose.

Chapter 9

1 The splitting of the C_6 molecules into two C_3 molecules occurs by the aldol split catalysed by aldolase. This requires the structure

R
|
C=O
|
R'—C—R''
|
H—C—OH
|
R

The conversion of glucose-6-phosphate to fructose-6-phosphate produces the aldol structure capable of splitting in this way. In straight chain formulae.

CHO   CHOH
| |
CHOH   CO
| |
CHOH   CHOH
| |
CHOH   CHOH
| |
CHOH   CHOH
| |
$CH_2OPO_3^{2-}$ $CH_2OPO_3^{2-}$

Glucose-6-phosphate   Fructose-6-phosphate

2 It occurs when a 'high-energy phosphoryl group' transferable to ADP is generated in covalent attachment to a substrate. Oxidation of glyceraldehyde-3-phosphate by the mechanism given on page 151 is an example.

By contrast, electron transport in mitochondria generates a proton gradient which drives ATP synthesis. No phosphorylated intermediates interpose between $ADP + P_i \rightarrow ATP$.

3 Pyruvate kinase works in the reverse reaction but the nomenclature of kinases always derives from the reaction using ATP. The reaction is irreversible because the product of the reaction, enolpyruvate, spontaneously isomerizes into the keto-form, a reaction with a large negative $\Delta G^{o'}$.

4 3 and 2, respectively. Phosphorolysis of glycogen using P_i generates glucose-1-phosphate, convertible to glucose-6-phosphate. To produce the latter from glucose, a molecule of ATP is consumed.

5 Cytoplasmic NADH itself cannot enter the mitochondrion; instead the electrons must be transported in by one of two alternative shuttles. The malate–aspartate shuttle (Fig. 8.9) reduces mitochondrial NAD^+, but the glycerophosphate shuttle reduces FAD attached to glycerophosphate dehydrogenase of the inner mitochondrial membrane. The redox potential of the former is more negative than that of the latter and the two enter the electron transport chain at different respiratory complexes. The yield of ATP is lower from the glycerophosphate shuttle.

6 Oxidation forms a β-keto acid that readily decarboxylates.

7 The reaction is catalysed by pyruvate carboxylase.

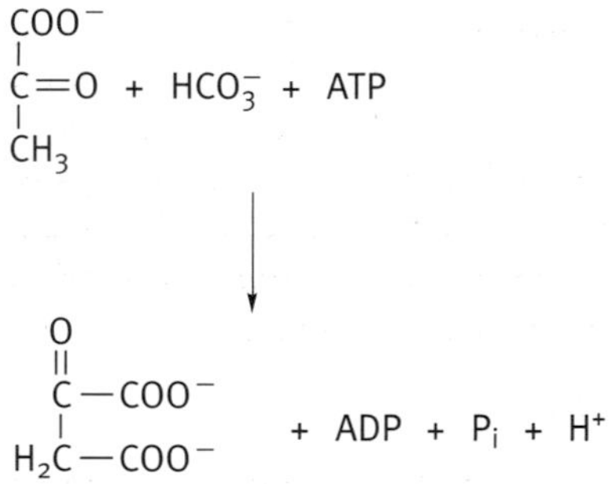

$$
\begin{array}{c}
COO^- \\
| \\
C{=}O \\
| \\
CH_3
\end{array}
+ HCO_3^- + ATP
$$

$$\downarrow$$

$$
\begin{array}{c}
O \\
\| \\
C{-}COO^- \\
| \\
H_2C{-}COO^-
\end{array}
+ ADP + P_i + H^+
$$

8 Biotin, a B group vitamin. Using ATP, it forms a reactive carboxybiotin that can donate a carboxy group to substrates. The reaction is

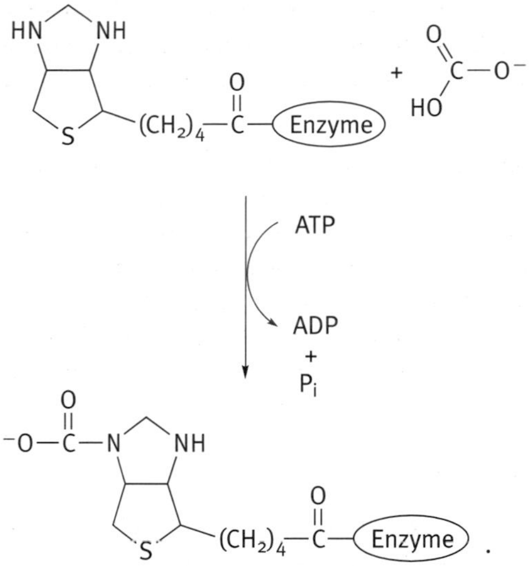

9 This is shown in Fig. 9.19.

10 They are both mobile electron carriers. Ubiquinone connects respiratory complexes I and II with III and cytochrome c connects complexes III and IV. The former exists in the hydrophobic lipid bilayer; the latter in the aqueous phase on the outside of the inner mitochondrial membrane.

11 To pump protons from the mitochondrial matrix to the outside of the inner mitochondrial membrane and so generate a proton gradient and a charge gradient that can be used to drive ATP synthesis.

12 In the eukaryote, NADH generated in the cytoplasm has to have its electrons transferred to the electron transport pathway. If the glycerol-3-phosphate shuttle is employed for this we lose one ATP generated per NADH that has to be so handled. Since there are two NADH molecules generated per glucose glycolysed, this leads to a potential loss of two ATP molecules. In *E. coli* there is no such problem. In addition, in *E. coli* there is no expenditure of the energy used in eukaryotes to exchange ATP and ADP across the mitochondrial membrane.

13 In one turn of the cycle, one acetyl group is disposed of, the products being nine reducing equivalents (effectively nine hydrogen atoms) in the form of three $NADH + H^+$, one $FADH_2$, and two molecules of CO_2. A further proton is needed for generation of CoA–SH making a total of nine hydrogen atoms. The acetyl group supplies three hydrogen atoms and one oxygen atom, leaving a deficit of six hydrogens and three oxygens. These partly come from the input of two water molecules; in addition, if you examine the breakdown of succinyl-CoA, it effectively also splits a further water molecule. This ingenious

thermodynamic device is indirect and easily missed but it is important in appreciating the energetics of the metabolism of the acetyl group in the citric acid cycle. It is seldom referred to in biochemistry texts.

14   The proton flow back into the matrix causes a rotary motion in the F_0 unit in the inner mitochondrial membrane. This drives a shaft inside the F_1 unit, which somehow causes conformation changes in the F_1 subunits. The energy stored in the conformation changes brings about the synthesis of ATP from ADP + P_i but it is important to note that it is the release of ATP which requires the energy. The reaction forming ATP from ADP + P_i at the enzyme surface involves little free energy change (a concept which takes a little getting used to). The diagram in Fig. 9.28 would be suitable in this question.

15   There are three sites capable of synthesizing ATP in each F_1 unit but they progress through different stages referred to as open (O state), the loose binding (of ADP and P_i—the L state) and the tight binding (T state). ATP synthesis occurs at the last stage. Cooperative interdependence means that at any one time only one of the three is in the O state, one in the L state, and one in the T state. The individual sites change simultaneously. Thus the L state can change to the T state only if the preceding O state changes to the L state and so on. Figure 9.26 would be useful here.

Chapter 10

1 (a) From FFA carried by the serum albumin in the blood; these originate from the fat cells.
(b) From chylomicrons—lipoprotein lipase releases FFA and TAG.
(c) From VLDL produced by the liver—released in the same way as (b).

2   The brain, red blood cells (the latter have no mitochondria). See Chapter 6.

3 (a) Activation to convert them to acyl-CoA derivates.
(b) On the outer mitochondrial membrane.
(c) In the mitochondrial matrix.
(d) As carnitine derivatives, as shown in Fig. 10.1

4   The reactions in Fig. 10.2 are analogous to the succinate → fumarate → malate → oxaloacetate reactions in the citric acid cycle, both in the reactions and in the electron acceptors involved.

5   One molecule of palmitic acid produces eight of acetyl-CoA and generates seven $FADH_2$ and seven NADH molecules. NADH oxidation is estimated to produce 2.5, and $FADH_2$ 1.5 molecules of ATP per molecule of NADH and $FADH_2$. If you add to this the yield of ATP from the oxidation of eight molecules of acetyl-CoA (ten per molecule), the total is 108 molecules of ATP (counting the GTP from the citric acid cycle as ATP). From this must be subtracted two used in the activation reaction = a net yield of 106 molecules.

6   After two rounds of β-oxidation, the cis-Δ^3-enoyl-CoA is isomerized to become the $trans$-Δ^2-enoyl-CoA (see page 181 for reactions).

7   In situations of rapid fat release from adipose cells such as occurs in starvation or diabetes, the liver converts acetyl-CoA to ketone bodies that are released into the blood. These are preferentially utilized by muscle, thus conserving glucose; most importantly, the brain can obtain about half its energy needs from ketone bodies.

8   Acetoacetate synthesis occurs in the mitochondrial matrix and cholesterol synthesis in the cytoplasmic compartment, the process occurring on the ER membrane.

Chapter 11

1   Fatty acids are synthesized two carbon atoms at a time, but the donor of these is a three-carbon unit, malonyl-CoA. Acetyl-CoA is converted to the latter by an ATP-dependent carboxylation. The subsequent decarboxylation results in a large negative $\Delta G^{0\prime}$ value. In other words, the point of the carboxylation and decarboxylation is to make the process of adding two carbon atom units to the growing fatty acid chain irreversible.

2   This is given in Fig. 11.1.

3   In eukaryotes, all of the enzyme reactions are organized into a single protein molecule with the enzymic functions catalysed by separate domains. The functional unit is a dimer with the two molecules cooperating as a single entity. In E. coli the different activities are catalysed by separate enzymes. The advantage of the eukaryote situation is that the intermediates are transferred from one active centre to the next. In E. coli the products must diffuse to the next enzyme so that the process is slower.

4   See page 190 for structures. NAD^+ is used in catabolic reactions—it accepts electrons for oxidation and energy generation. $NADP^+$ is involved in the reverse—in reductive syntheses.

The existence of the two is a form of metabolic compartmentation that facilitates independent regulation of the processes.

5 The main sites in the body are liver and adipose cells and the mammary gland during lactation.

6 The acetyl-CoA in the mitochondrion is coverted to citrate; the latter is transported into the cytosol where citrate lyase cleaves citrate to acetyl-CoA and oxaloacetate. This is an ATP-requiring reaction that ensures complete cleavage.

$$\text{Citrate} + \text{ATP} + \text{CoA}-\text{SH} + \text{H}_2 \rightarrow$$
$$\text{acetyl-CoA} + \text{oxaloacetate} + \text{ADP} + \text{P}_i$$

7 The oxaloacetate is reduced to malate by malate dehydrogenase, an NADH-requiring reaction. The malate is oxidized and decarboxylated to pyruvate by the malic enzyme, an $NADP^+$ requiring reaction. This scheme effectively switches reducing equivalents from NADH to NADPH. The pyruvate generated returns to the mitochondrion, as illustrated in Fig. 11.4.

This generates only one NADPH per malonyl-CoA produced, while the reduction steps in fatty acid synthesis require two. The rest is generated by the glucose-6-phosphate dehydrogenase system, described in Chapter 13.

8 The scheme is shown in Fig. 11.5.

9 In the synthesis of glycerol-based phospholipids there are two routes. In one, phosphatidic acid is joined to an alcohol such as ethanolamine (see Fig. 11.6). For this the alcohol is activated; in phospholipid synthesis, the activated molecule is always a CDP-alcohol. For some glycerophospholipid syntheses, the diaclyglycerol component is activated (see Fig. 11.6). Again, this is by formation of the CDP-diaclyglycerol complex. The situation is reminiscent of the use of UDP-glucose whenever an activated glucose moiety is wanted.

10 (a) Eicosanoids have 20 carbon atoms (*eikosi* = 20). They include prostaglandins, thromboxanes, and leukotrienes.
(b) All are related to and synthesized from polyunsaturated fatty acids.
(c) Prostaglandins cause pain, inflammation, and fever. Thromboxanes affect platelet aggregation. Leukotrienes cause smooth muscle contraction and are a factor in asthma, by constricting airways.
(d) Aspirin inhibits cyclooxygenase, an enzyme involved in their synthesis, and thus it can suppress pain and fever, and also inhibit blood clotting.

11 Mevalonic acid is the first metabolite committed solely to cholesterol synthesis. Structural analogue of mevalonic acid have been found to inhibit HMG-CoA reductase, which is the enzyme responsible for mevalonate production. The drugs act in the body as transition-state analogues of HMG-CoA reductase, which they inhibit.

Chapter 12

1 The brain cannot use fatty acids; it must have glucose. So must red blood cells, which have no mitochondria and can generate energy only from glycolysis.

2 The substrate of pyruvate kinase is the enol form of pyruvate, but the keto-enol equilibrium is overwhelmingly to the keto form and hence the enzyme has no substrate. The solution lies in a metabolic route in which two high-energy phosphate groups are expended.

$$\text{Pyruvate} + \text{ATP} + \text{HCO}_3^- \rightarrow \text{Oxaloacetate} + \text{ADP} + \text{P}_i$$
$$\text{(Pyruvate carboxylase)}$$

$$\text{Oxaloacetate} + \text{GTP} + \text{H}_2\text{O} \rightarrow \text{PEP} + \text{GDP} + \text{CO}_2$$
$$\text{(PEP-CK)}$$

3 Fructose-1:6-bisphosphate, producing fructose-6-phosphate and glucose-6-phosphatase.

4 No. Only the liver and kidney produce free glucose.

5 In normal nutritional situations (non-starvation) strenuous muscular activity can generate lactate by anaerobic glycolysis. This travels in the blood to the liver where it is converted back to glucose. Release of glucose into the blood and its uptake by muscle completes the cycle. See Fig. 12.4.

6 Glycerol kinase is required to convert glycerol to glucose by the route shown in Fig. 12.5. Glycerol release occurs in starvation where a prime concern is to produce blood glucose. Since only the liver can do this it makes sense for the glycerol to travel to the liver rather than being metabolized by adipose cells that cannot release blood glucose.

7 Via the glyoxalate cycle shown in Fig. 12.6. The principle is that the two decarboxylating reactions of the citric acid cycle are bypassed.

8 In animals, pyruvate carboxylase synthesizes oxaloacetate (C4) from pyruvate (C3). However, in organisms with the glyoxylate cycle, an extra molecule of acetyl-CoA is converted in the net sense to malate and therefore no topping up reaction is needed. You will appreciate that the citric acid cycle cannot effect a net synthesis of C4 acids.

9 The liver must be supplied with a suitable substrate for gluconeogenesis. Muscle wasting produces free amino acids, which are converted largely to alanine. The alanine migrates to the liver where it is converted to pyruvate.

10 In starvation, the liver must synthesize glucose to supply the brain since glycogen stores are rapidly exhausted. The pyruvate for gluconeogenesis arises from lactate produced by red blood cells and from alanine coming from muscle. Alcohol raises the ratio of reduced to oxidized NAD^+ in the liver; this can impair conversion of lactate to pyruvate because the equilibrium between these is easily shifted by increased NADH levels and also may cause reduction of pyruvate formed from alanine to lactate. Thus the liver may be deprived of pyruvate needed for gluconeogenesis.

Chapter 13

1 One, by allosteric control; two, by covalent modification of the enzyme, the chief mechanism for this being phosphorylation.

2

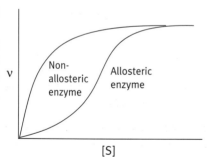

3 Allosteric effectors (with a few exceptions in which v_{max} is changed) work by changing the affinity of an enzyme for its substrate. This requires that the substrate concentrations for such enzymes are sub-saturating, which is usually the case. A positive allosteric effector moves the sigmoid substrate-velocity curve to the left and a negative effector to the right (Fig. 13.4). A sigmoid relationship amplifies the effect of such changes on the reaction velocity and so increases the sensitivity of control. This can be seen in Fig. 13.3.

4 They would have no effect since the effectors change the affinity of the enzyme for its substrate. At saturating concentrations this cannot be used to increase the rate of the reaction.

5 The concerted and sequential mechanisms as illustrated in Figs 13.5 and 13.6.

6 It is that an allosteric effector need have no structural relationship to the substrate of the enzyme. This means that completely distinct metabolic systems can interact in a regulatory fashion.

7 Intrinsic regulation is usually allosteric and can apply to a single cell. It keeps the metabolic pathways in balance. However, it cannot determine the overall direction of metabolism of a cell—whether, for example, it will store glycogen and fat or release these. These are determined by extrinsic controls—hormones, etc.—which direct the activities of cells to be in harmony with the physiological needs of the body.

8 See Fig. 13.20. AMP activates glycogen phosphorylase and phosphofructokinase while ATP inhibits the latter. The salient feature is that high ATP/ADP ratios stop glycolysis while a lower ATP level (resulting in increased AMP) speeds it up. High citrate levels also logically slow the feeding of metabolite via glycolysis into the citric acid cycle. High acetyl-CoA levels may be an indication that oxaloacetate is low—hence activation of pyruvate carboxylase to perform its anaplerotic reaction. At the same time high acetyl-CoA levels indicate that the supply of pyruvate by glycolysis is adequate and inhibition of glycolysis at the PEP step is therefore logical.

9 Direct allosteric controls, controls on a kinase that phosphorylates and inactivates PDH, and controls on protein phosphatase that reverses the phosphorylation (Fig. 13.27).

10 High blood glucose levels stimulate insulin release; low glucose levels stimulate glucagon release.

11 A hormone such as epinephrine is a first messenger; in combining with a cell receptor it causes an increase in a second molecule that exerts metabolic effects. This is a second messenger. For the two hormones mentioned, this is cAMP. It allosterically activates a protein kinase PKA whose activity has diverse metabolic effects.

12 By mobilizing glucose transporters into a functional position in the cell membrane (Fig. 13.13).

13 It causes a cascade of activation, starting with PKA activation as shown in the scheme on page 220.

14 The differences are summarized in Fig. 13.21.

15 Different cells have receptors for different hormones. Thus, cell A has a receptor for hormone X; cAMP in cell A has effects appropriate to hormone X. Cell B does not have a receptor for hormone X but does for hormone Y. cAMP in cell B

elicits responses appropriate to hormone Y but not those appropriate to hormone X.

16 Fructose 2:6-bisphosphate. cAMP decreases its level. See Fig. 13.24.

17 Gluconeogenesis is switched on by glucagon whose second messenger is cAMP. The latter activates a kinase that phosphorylates and inactivates pyruvate kinase. In muscle, epinephrine produces cAMP as second messenger. The aim of epinephrine is to maximize glycolysis so that inactivation of pyruvate kinase here would be inappropriate.

18 cAMP activates a hormone-sensitive lipase that hydrolyses TAG.

19 Such reactions are irreversible in the cell and act as one-way valves in the pathway; they also ensure that the pathway goes to completion. The carboxylation of acetyl-CoA to malonyl-CoA in fatty acid synthesis is an example in which the reaction serves no apparent purpose other than to confer irreversibility due to the subsequent decarboxylation reaction. However, many pathways have to be reversed; an example is glycolysis, which has to operate in the reverse direction for gluconeogenesis. In such situations it is necessary to reciprocally control the different directions or the pathway would be simultaneously breaking down glucose units and synthesizing them. To separately control a pathway in the two directions, different enzymes are required. At the irreversible steps it is necessary to have a different reaction for the reverse direction and this is often a control site. PFK is a typical such step in glycolysis. The reverse step is catalysed by fructose-1:6—bisphosphatase, the two steps being reciprocally controlled.

20 **(a)** Insulin is the signal to store food, in the case of glucose as glycogen. It is therefore logical that insulin should activate glycogen synthase.
(b) The default state is that glycogen synthase is inactivated by GSK3. There are several kinases phosphorylating the synthase in different positions but insulin activation involves removal of the —P groups added by GSK3 and these are the sites mainly involved in synthase control. To activate the synthase, the GSK3 has therefore to be inhibited and, in addition, the protein phosphatase which dephosphorylates the synthase and thereby activates it, is activated by insulin. PKB is the enzyme which inactivates GSK3 by phosphorylating the latter. PKB is activated by a signal pathway activated by binding of insulin to its cell membrane receptor. Figure 13.18(b) summarizes this complicated sequence.

21 It has an important role in the control of phosphorylase. If there is plenty of glucose around there is no point in breaking down glycogen. Glucose binds to phosphorylase a (the active phosphorylated form) and induces a conformational change which makes the enzyme more susceptible to attack by the protein phosphatase 1. This converts the phosphorylase a into the relatively inactive 'b' form.

22 Malonyl-CoA, the first metabolite committed solely to fat synthesis, inhibits the conversion of fatty acyl-CoAs to the carnitine derivative. The latter is essential for transporting the acyl-CoAs into mitochondria where fatty acid oxidation takes place.

23 In the presence of an allosteric activator of the enzyme. The sigmoid response is due to increasing numbers of subunits in the total enzyme present being in the high affinity form. If sufficient activator is present so that most of the subunits are in the high affinity form activated at low substrate concentration, the enzyme tends towards a Michaelis-Menten type of hyperbolic response.

24 The strongly charged phosphate group is very effective at producing a conformation change in a protein, which in most cases is responsible for altering the activity of the enzyme.

25 This is a broadly acting control which shuts down ATP-consuming synthesis reactions and activates catabolic ATP-generating pathways. The rationale is that AMP is a sensitive signal of reduced phosphorylation charge—that ATP supplies are low. There is only a relatively small amount of ATP in a cell and life depends on its very rapid regeneration. ATP shortage would be much more dangerous than shutting down anabolic reactions temporarily. The AMPK achieves this.

Chapter 14

1 It supplies ribose-5-phosphate for nucleotide synthesis; it can supply NADPH for fat synthesis; it provides for the metabolism of pentose sugars.

2 Glucose-6-phosphate is converted to ribose-5-phosphate + CO_2 and $NADP^+$ is reduced. The reactions are given in Fig. 14.1.

3 Transaldolase and transketolase are the main ones whose reactions are given in Fig. 14.2. (Other enzymes of the glycolytic pathway may also participate).

4 Part of the ribose-5-phosphate is converted to xylulose-5-phosphate, a ketose sugar (since transaldolase and transketolase must have a ketose sugar as donor). The following manipulations now occur.

1. $2C_5 \rightarrow C_3 + C_7$ (transketolase)

2. $C_7 + C_3 \rightarrow C_4 + C_6$ (transaldolase)

3. $C_5 + C_4 \rightarrow C_3 + C_6$ (transketolase)

The $2C_5$ in reaction 1 are ribose-5-phosphate and xylulose-5-phosphate; the final C_3 compound is glyceraldehyde-3-phosphate, convertible to glucose-6-phosphate with loss of P_i. The net effect is that six molecules of ribose-5-phosphate are converted to five molecules of glucose-6-phosphate + P_i. Thus the cell can produce NADPH with no net increase in ribose-5-phosphate.

5   The oxidative part of the pathway converts glucose-6-phosphate to ribose-5-phosphate and CO_2. If you consider that six molecules of the former, and the six molecules of the latter are recycled to 5 glucose-6-phosphate, then, on a balance sheet, six molecules of glucose-6-phosphate have produced six molecules of CO_2 which, on paper, looks like the oxidation of a glucose molecule. It isn't; one molecule of glucose-6-phosphate is not converted to six molecules of CO_2.

6   NADPH is required to reduce glutathione, a molecule necessary for the protection of the red blood cell. Patients lacking the enzyme are sensitive to the anti-malarial drug, pamaquine, resulting in a hemolytic anemia.

Chapter 15

1   Light reactions involve the splitting of water using light energy, and the reduction of $NADP^+$ to NADPH. Dark reactions mean the utilization of that NADPH to reduce CO_2 and water to carbohydrate. The term 'dark' implies that light is not essential, not that it occurs only in the dark. In fact, dark reactions will occur maximally in bright sunlight.

2   Photosystems contain many chlorophyll molecules. When excited by a photon, one of the electrons in the latter is excited to a higher energy level. Resonance energy transfer permits this excitation to jump from one molecule to another until it becomes trapped in special reaction centre molecules. The excitation is insufficient for resonance energy transfer but sufficient for an electron to enter the electron transport chain shown in Fig. 15.4.

3   (a) Electrons passing along the carriers of Photosystem II result in ATP synthesis by the chemiosmotic mechanism as they traverse the cytochrome *bf* complex (see Fig. 15.6).

(b) When all of the $NADP^+$ is reduced, electrons passing along the carriers of photosystem I are diverted to the cytochrome *bf* complex and so generate more ATP (see Fig. 15.7).

4   It is the chlorophyll $P680^+$, that is, the reaction centre pigment of PSII that has been excited by resonance energy transfer from antenna chlorophyll molecules to donate an electron to pheophytin, the first component of the PSII electron transport chain. $P680^+$ is an electron short and has a strong tendency to accept an electron, that is, it is a strong oxidizing agent.

5   Thylakoids are formed by invagination of the inner chloroplast membrane (cf. inner mitochondrial membrane), which explains the apparent opposite orientation of proton pumping.

6   The enzyme Rubisco (ribulose-1:5-bisphosphate carboxylase) splits ribulose-1:5-bisphosphate into two molecules of 3-phosphoglycerate, a molecule of CO_2 being fixed in the process (See reaction on page 257.)

7   As in Fig. 15.10.

8   If CO_2 is fixed first into 3-phosphoglycerate by Rubisco, the plant is known as a C_3 plant. These are subject to full competition between oxygen and CO_2 in the Rubisco reaction. However, in high temperature—high sunlight areas, in C_4 plants CO_2 is first fixed into oxaloacetate by pyruvate-P_i dikinase and PEP carboxylase (see Fig. 15.11). This is reduced to malate which is transported into the underlying bundle sheath cell where the Calvin cycle occurs. Decarboxylation of the malate occurs so that the ratio of CO_2/O_2 in the cell is greatly increased—the concentration of CO_2 in bundle sheath cells may be raised 10–60-fold by this means. The whole scheme is given in Fig. 15.11. Variations in detail of the scheme given occur in different C_4 plants but all are devoted to the same basic strategy.

9   In animals this cannot happen directly—pyruvate kinase cannot form PEP from pyruvate. However, the enzyme in plants is pyruvate-P_i dikinase in which two phosphoryl groups of ATP are used, making the reaction thermodynamically feasible.

Chapter 16

1   Oxidation results in the formation of a Schiff base, hydrolysable by water.

$$CHNH_2 \xrightarrow{-2H} C{=}NH \xrightarrow{H_2O} C{=}O + NH_3$$

2 Glutamic acid.

3 Transdeamination is the most prevalent mechanism. The amino group of many amino acids are transferred to α-ketoglutarate forming glutamate. The latter is deaminated by glutamate dehydrogenase. As an example,

1. Alanine + α-ketoglutarate $\rightarrow$ pyruvate + glutamate;

2. Glutamate + NAD^+ + H_2O $\rightarrow$ α-ketoglutarate + NADH + NH_4^+

Net reaction: Alanine + NAD^+ + H_2O $\rightarrow$ pyruvate + NADH + NH_4^+.

4 Pyridoxal phosphate (Fig. 16.2). The mechanism of transamination is given in Fig. 16.3.

5 By removal of H_2O and H_2S, respectively, as illustrated in Fig. 16.4.

6 A glucogenic amino acid is one that, after deamination, can give rise to pyruvate (or phosphoenolpyruvate). This may be indirect—any acid of the citric acid cycle is glucogenic. A ketogenic amino acid is one that cannot give rise to the above but rather gives rise to acetyl-CoA. Only leucine and lysine are purely ketogenic but some, such as phenylalanine, are mixed. Ketogenic amino acids produce ketone bodies only in circumstances appropriate to this such as starvation. Otherwise the acetyl-CoA is oxidised normally.

7 Phenylalanine is not normally transaminated; it is converted to tyrosine and then metabolized (see Fig. 16.5). If the phenylalanine conversion to tyrosine is defective, phenylalanine does transaminate, producing phenylpyruvate, which causes irreparable brain damage to babies and early death.

8 To supply the reducing equivalent for the formation of H_2O from one atom of the oxygen molecule used in the hydroxylation reaction.

9 By formation of 5-adenosylmethionine ('SAM'); this has a sulfonium ion structure conferring a strong leaving tendency on the methyl group. The formation of SAM is illustrated in Fig. 16.6.

10 This is shown in Fig. 16.8.

11 Both situations call for high rates of deamination of amino acids (in starvation muscle proteins break down to allow glucose synthesis). The amino nitrogen must be converted to urea.

12 (a) Ammonia is converted to glutamine, which is transported to the liver and hydrolysed.

Glutamate + ammonia $\xrightarrow{\text{ATP} \quad \text{ADP} + P_i}$ Glutamine

Glutamine + H_2O $\longrightarrow$ Glutamate + ammonia .

(b) Amino nitrogen is transported from muscle as alanine. The alanine cycle is shown in Fig. 16.9.

Chapter 17

1 To destroy unwanted molecules and structures imported into the cell by endocytosis and to destroy components of the cell destined for destruction.

2 They are vesicles produced by the Golgi sacs containing an array of hydrolytic enzymes. The enzymes are acid hydrolases with an optimum pH of 4.5–5.0. Proton pumps in the membrane maintain this pH inside the vesicle.

They fuse with endosomes to form the lysosomes.

3 A large variety of genetically determined lysosomal storage disorders exist in which the absence of a specific hydrolytic enzyme causes the lysosomes to become overloaded with material normally disposed of.

4 In Pompe s disease, a fatal genetically determined one, there is a lack of a lysosomal α-1:4-glycosidase that normally degrades glycogen. The lysosomes become overloaded with glycogen. However, it is not clear why disposal of glycogen by this means should be needed. It is not part of the usual accounts of glycogen metabolism.

5 These are membrane-bounded vesicles in the cytosol that contain oxidases. These enzymes oxidize a variety of substrates, using oxygen, and generate H_2O_2.

$R'H_2 + O_2 \rightarrow R' + H_2O_2$.

Catalase destroys the hydrogen peroxide.

Substrates include those not metabolized elsewhere, for example, very long chain fatty acids are shortened. Oxidation of the cholesterol side chain to form bile salts is believed to occur here. The essential role of peroxisomes is underlined by a fatal genetic disease in which some tissues lack peroxisomes.

6 Proteasomes are protein organelles consisting of a central core made of rings of protein subunits with caps at both ends. Inside the cavity of the central core are proteolytic enzymes which hydrolyse proteins into peptides and amino acids. The caps act as the gateway for entry of proteins destined for destruction. The proteasomes selectively destroy individual protein molecules, which are targeted into the cavity. The vital role is selective destruction of proteins such as regulatory proteins in cell cycle control. They also play a vital part in the immune system by producing peptides, for example from viruses to be displayed of the outside of cells inviting attack by killer T cells of

the immune system. The importance of proteasomes is underlined by the fact that they have been highly conserved throughout evolution. Mutations causing nonfunctional proteasomes in yeast are lethal. In very recent times it has been realized that protein breakdown in cells especially by proteasomes has very great fundamental importance, so the field has become one of the most intensively studied—a revolution in the image of the field.

7 The entry ticket is the attachment of the small protein ubiquitin to target proteins. Polyubiquinated proteins are allowed to enter the proteasome cavity for destruction. There is still must to learn about what determines the selection of proteins for ubiquitination but certain N-terminal sequences are known to be a factor.

Chapter 18

1 The initial stimulus for blood clotting is, in quantitative terms, very small. To obtain a sufficiently rapid response amplification is needed. Cascades are the biological method of amplification. The final enzyme activation is that of prothrombin to thrombin, an active proteolytic enzyme.

2 Fibrinogen monomer proteins are prevented from spontaneous polymerization by negatively charged fibrinopeptides (Fig. 18.2) that cause mutual repulsion. Removal of these by thrombin permits the association of fibrin monomers, illustrated in Fig. 18.3.

3 Covalent crosslinks are formed between monomers in the polymer by an enzymic transamidation process between glutamine and lysine side chains,

$$-CONH_2 + H_3N^+ - \rightarrow -CO-NH- + NH_4^+.$$

4 In the conversion of prothrombin to thrombin, a glutamic acid side chain is carboxylated; the carboxyglutamate binds Ca^{2+}. Vitamin K is a cofactor in the carboxylation reaction.

$$\begin{array}{ccc} | & & | \\ CH_2 & & CH_2 \\ | & + CO_2 \longrightarrow & | \\ CH_2 & & HC-COO^- \\ | & & | \\ COO^- & & C-COO^- \end{array}$$

Other conversions in the proteolytic cascade involve this conversion.

5 It is involved in the conversion of hydrophobic xenobiotics to soluble ones. A typical reaction is

$$AH + O_2 + NADPH + H^+ \rightarrow AOH + H_2O + NAD^+.$$

6 To reduce one oxygen atom to H_2. It is known as a mixed function oxygenase.

7 The glucuronyl group is donated to an —OH on the foreign molecule, rendering it much more polar (see Fig. 18.5).

8 It involves an ATP-driven membrane transport system that removes a variety of compounds from cells. It transports steroids in steroid-secreting adrenal cortical cells, but all transported compounds (including anticancer drugs used in chemotherapy) are lipid-soluble amphipathic compounds, indicating a more general role. It has recently been shown to transport the cholesterol to the outside of the cells to be removed by HDL.

9 Neutrophils secrete elastase in the mucous lining of lung tissue that destroys elastin, the elastic structural material of lungs. This, unchecked, converts the minute alveoli into much larger structures with a greatly reduced surface area for gas exchange. α_1-Antitrypsin in the blood diffuses into the lung and keeps elastase in check and thus prevents damage. Smoking has two effects: (a) it inactivates α_1-antitrypsin by converting a crucial methionine side chain to a sulfoxide (S → S=O); (2) by irritating the lungs it attracts neutrophils resulting in more elastase being liberated.

10 It is an oxygen molecule that has acquired an extra electron.

$$O_2 + e^- \rightarrow O_2^-$$

Superoxide is extremely reactive, causing damage.

11 Most eukaryotic cells have superoxide dismutase and catalase

$$2O_2^- + 2H^+ \rightarrow H_2O_2 + O_2 \text{ (dismutase)};$$

$$2H_2O_2 \rightarrow 2H_2O + O_2 \text{ (catalase)}.$$

Secondly, there are antioxidants such as ascorbic acid and vitamin E. Superoxide is dangerous because it attacks a molecule and generates another free radical, causing a self-perpetuating chain of reactions. Antioxidants are quenching agents. When attacked by superoxide, the free radical produced by these is insufficiently reactive to perpetuate the chain of reactions.

Chapter 19

1 PRPP or 5'-phosphoribosyl-1-pyrophosphate is the universal agent. Its formation from ribose-5-phosphate and ATP and its mode of action are given on page 290.

2 Tetrahydrofolate (FH_4).

Formyl-FH$_4$ =

3 Serine. Serine hydroxymethylase transfers —CH$_2$ OH to FH$_4$ leaving glycine and forming N^5,N^{10}-methylene FH$_4$. This is oxidized to N^5,N^{10}-methylene FH$_4$ by an NADP$^+$ requiring reaction. Hydrolysis of this produces formyl-FH$_4$. The reactions are shown on page 293.

4 It ribotidizes guanine and hypoxanthine in the purine salvage pathway.

5 HGPRT is missing so that purine salvage cannot occur. However, the brain has the direct pathway and it may be that overproduction of purine nucleotides *de novo* occurs due to increased levels of PRPP (because it is not used by salvage). In such patients, excess uric acid is formed but prevention of this by allopurinol does not relieve the neurological symptoms. Nor do gout patients suffer the latter, so these symptoms are unexplained.

6 It is converted to alloxanthine which inhibits xanthine oxidase. The conversion is due to xanthine oxidase itself—referred to as suicide inhibition.

7 As in Fig. 19.8.

8 No. Thymidylate synthetase transfers a C$_1$ group from methylene FH$_4$ do dUMP and at the same time reduces it to a methyl group. The H atoms for this are taken from FH$_4$ so that FH$_2$ is a product as shown on page 297.

9 It inhibits reduction of FH$_2$, generated by thymidylate synthase, and so prevents dTMP production essential for cell multiplication. FH$_4$ is essential for the thymidylate synthase reaction.

10 Vitamin B$_{12}$ is required for the methylation of homocystene to methionine, the methyl donor being methyl FH$_4$. Lack of B$_{12}$ causes FH$_4$ to be 'trapped' as the methyl compound and unavailable for the other folate-dependent reactions.

Chapter 20

1 This is shown on page 305.

2 Genetic material needs to be as chemically stable as possible. DNA is more stable than RNA. This is because, as shown on page 305, the 2'-OH of RNA can make a nucleophilic attack on the phosphodiester bond making RNA less chemically stable than DNA.

3 The phosphodiester bond is longer than the thickness of a base; a straight chain structure of DNA would leave hydrophobic faces of the bases exposed to water. Because of hydrophobic forces the bases are collapsed together by sloping the phosphodiester bond as shown in Fig. 20.3(a).

4 B DNA; right-handed; 10 base pairs.

5 One chain runs 5' → 3' in one direction and the other 5' → 3' in the other direction. Thus in a linear DNA molecule each end has a 5' end of one chain and a 3' end of another chain.

6 A 5' → 3' direction means that you are moving from a terminal 5'-OH group to a 3'-OH group of the polynucleotide chain.

7 5' CATAGCCG 3'

  3' GTATCGGC 5'

Watson-Crick base pairing explains the complementary sequence. By convention a single sequence is written with the 5' end to the left.

8 (a) The DNA of a eukaryote chromosome is wound around an octamer of histone proteins, two turns per nucleosome occupying 146 base pairs. Successive nucleosomes are connected by linked DNA.
(b) When chromatin is digested with DNase, the 146 base-pair sections of DNA are protected; this observation led to the discovery of nucleosomes (see Fig. 20.7).

9 Repetitive DNA; an Alu sequence is a few hundred bases long but repeated almost exactly hundreds of thousands of times; scattered throughout the human chromosome.

Chapter 21

1 A section of DNA whose replication is initiated at a single origin of replication.

2 Ahead of the replicative fork positive supercoils occur.

3 In *E. coli*, gyrase, a topoisomerase II, introduces negative supercoils. In eukaryotes, a topoisomerase I relaxes positive supercoils.

4 As shown in Figs 21.6 and 21.7.

5 In winding DNA around a nucleosome, a local negative supercoil is introduced, but, since no bonds are broken, there cannot be a net negative supercoiling. The local negative supercoil is compensated for by a local positive supercoiling elsewhere. This is relaxed by topoisomerase I, leaving a net negative supercoiling in the DNA.

6 dATP, dGTP, dCTP, and dTTP.

7 Cytosine readily deaminates to uracil; if uracil were a normal constituent of DNA it would be impossible to recognize and correct the mutation since A–U pairing is the same as A–T pairing.

8 (a) No. A primer is required. DNA polymerase cannot initiate new chains.
(b) Synthesis proceeds in the 5′ → 3′ direction; but this is meant that the new chain is elongated in the 5′ → 3′ direction, new nucleotides being added to the free 3′OH of the preceding nucleotide. It does not refer to the template strand, which runs antiparallel.

9 Hydrolysis of inorganic pyrophosphate; base pairing of the nucleotides.

10 Polymerase I is the enzyme that processes Okazaki fragments into a continuous lagging strand. It has a 5′–3′ exonuclease and a 3′–5′ exonuclease, and is a DNA polymerase with low processivity. Its activities are summarized in Fig. 21.16.

11 Correct base pairing of the incoming nucleotide triphosphate with the template nucleotide is the primary essential. The enzyme also has a proofreading activity. It has a 3′ → 5′ exonuclease activity that removes the last incorporated nucleotide if this is improperly base paired.

12 The methyl-directed mismatch repair system is described in Fig. 21.19. There is evidence that proteins similar to *E. coli* proteins exist in humans and that, where these are deficient, there is an increased risk of cancer.

13 Thymine dimers are formed when DNA is subjected to UV irradiation. Two adjacent thymine bases become covalently linked. Repair can either be direct by a light-dependent repair system that disrupts the bonds and restores the separate bases; or it can be subject to excision repair (Fig. 21.20).

14 As shown in Fig. 21.23, removal of the 3′ Okazaki fragment primer leaves an unreplicated section.

15 The answer lies in telomeric DNA synthesized as described in Fig. 21.24.

16 Processivity is the ability of a DNA polymerase to duplicate the coding strand of DNA for the very long stretches involved in DNA replication without falling off. The sliding clamps achieve this. These are annular proteins that are placed around the DNA to be copied, at the site of an RNA primer; the polymerase attaches to the clamp and so is prevented from falling off. DNA polymerase I has low processivity because its job is to remove the RNA primer from Okazaki fragments. If it had high processivity it would remove the newly synthesized DNA of the fragments and replace them unnecessarily. It is best that soon after it has reached the DNA part, it detaches to allow the ligase to heal the nick.

Chapter 22

1 RNA polymerase uses ATP, CTP, GTP, and UTP as substrates (cf. dATP, dCTP, dGTP, and dTTP). Most importantly, RNA polymerase can initiate new chains; DNA polymerase requires a primer. RNA synthesis never includes proofreading.

2 This is illustrated in Fig. 22.4. In more detail the promoter has a −10 (Pribnow) box and a −35 box.

3 RNA polymerase attaches nonspecifically to the DNA, but, when joined by a sigma factor molecule, it binds firmly to a promoter. The Pribnow box and the −35 box position orientate the polymerase correctly. The enzyme synthesizes a few phosphodiester bonds and then the sigma protein flies off and the polymerase progresses down the gene transcribing mRNA.

4 One is by the G—C stem loop structure shown in Fig. 22.7. In this, internal G≡C base pairing in the mRNA prevents bonding of the mRNA to the template DNA strand. Following this, a string of uracil nucleotides further weakens the attachment (only two hydrogen bonds in a U═A pair) facilitating detachment. There is still some uncertainty as to why this is a reliable terminator. The second method depends on the Rho factor, a helicase that unwinds the mRNA—DNA hybrid. At the termination site, the polymerase pauses (possibly due to a G≡C rich region in the DNA) and the Rho factor catches up and detaches the mRNA.

5 The precise base sequence of the −10 and −35 boxes, their distance apart, and the bases in the +1 to +10 region.

6 This is covered in Fig. 22.10.

7 Helix-turn-helix proteins, leucine zipper proteins, zinc fingers proteins, and homeodomain proteins.

Chapter 23

1 The primary transcript has introns that must be spliced out; the 5′ end is capped; the mechanism of termination is unknown; the mRNA is, with few exceptions, polyadenylated at the 3′ end.

2 The mechanism is shown in Fig. 22.3. It is based on a consensus sequence that determines where splicing occurs and on a transesterification reaction involving little change in free energy. Split genes may have facilitated evolution by promoting exon shuffling. Differential splicing can also result in a single gene giving rise to different proteins.

3 The eukaryotic polymerase III does not attach to the DNA directly but rather attaches to a protein complex that assembles on the DNA (Fig. 23.12). In addition, any number of transcriptional factors may be associated with the complex (Fig. 23.13).

4 In the default condition, eukaryotic genes are shut down. This is due to nucleosomes blocking the promoter sites. Chromatin remodelling unblocks the site by removal of the nucleosomes. This is done by the enzyme histone acetyltransferase (HAT) which acetylates the tails of the histones that constitute the nucleosome octamer. This has the effect of removing the positive charge on the histone lysine side chains. Somehow this results in chromosome remodelling.

5 In insect salivary glands, the formation of 'chromosome puffs' at the time of localized gene transcription can be seen in the microscope. Secondly it is known that genes at the time of activation become hypersensitive to added DNase, indicating that the enzyme has increased access to the DNA. This has been demonstrated for the hemoglobin gene among others.

6 The enzyme histone deacetylase is believed to reverse the activation process started by the histone acetyl-CoA transferase.

7 In the case of HAT, the enzyme is part of the coactivator. The latter binds to the activating transcription factor(s) and brought into proximity to the nucleosomes on the promoter selected by the transcription factors. Negative control may be effected in essentially the same way, a repressing transcription factor binding to the promoter and attracting the deacetylase.

8 The 'introns early' view is that they have always been present in modern genes. The latter are postulated to have been formed from primitive mini genes which fused together, the introns being the fused nontranslated regions. On this view, uninterrupted prokaryotic genes are due to all excess DNA having been discarded in the interests of rapid cell division.

The alternative view is that introns are late additions to prokaryotic genes which are regarded as primitive, and the introns may almost be regarded as parasitic DNA. It is generally accepted that introns may have been important in evolution by facilitating exon shuffling. The argument on intron origins still rages.

Chapter 24

1 If only 20 codons were used, there would be 44 not coding for any amino acid. Any mutation in a gene coding region would then be highly likely to inactivate the gene by prematurely introducing a stop codon. Instead, by using 61 codons, a base-change would either cause no change or would substitute a different amino acid. Because of the arrangement of the genetic code, many of these substitutions would be conservative and cause minimal change to the change to the protein structure.

2 The answer lies in the wobble mechanism described on page 383.

3 The convention is that RNA is shown with the 5′ end to the left. When mRNA is shown in this manner, a tRNA anticodon is base-paired in an antiparallel fashion. The tRNA molecule is therefore shown with the 5′ end to the right.

4 In the case of some amino acids, at the stage of attaching the amino acid to tRNA. In all cases at the stage of elongation; a pause in GTP hydrolysis on the EF-Tu gives time for unpaired aminoacyl tRNAs to leave the ribosome.

5 In general GTP hydrolysis to GDP and P_i is believed to result in conformational changes in the relevant proteins. GTP is involved in assemblage of the *E. coli* initiation complex. It is involved in the delivery of aminoacyl-tRNAs to the ribosome by EF-Tu. It is involved in the translocation step.

6 Figure 24.10 is necessary for this explanation. The straddling has the advantage that in moving from one site to the other the tRNA is never completely detached. It also means that the peptidyl group does not have to physically move relative to the ribosome. The two possible mechanisms are shown in the figure. Model II would explain why ribosomes have two subunits.

7 Figure 24.12 explains this. The scanning mechanism of selecting the start AUG means that there can be only one start site. As illustrated in Fig. 24.8, prokaryote initiation permits multiple start sites.

8 They bind to nascent polypeptides and prevent premature, improper folding associations. Appropriate release of chaperones facilitates correct folding though the process is far from fully understood. Other roles are summarized on page 395.

9 The prion diseases described on page 396. In these, the faulty protein is due to different folding of the identical polypeptide. They are typified by 'mad cow' disease.

10 (a) Yes, you can deduce the amino acid sequence of the protein that an mRNA will code for because each codon unambiguously specifies an amino acid.
(b) No, you cannot deduce the base sequence of an mRNA from the amino acid sequence of the protein it codes for. This is because of the degenerate nature of the genetic code or redundancy. Since several codons may code for a given amino acid it is impossible to deduce which of the alternatives was actually used by the cell in translating the messenger.

11 Synthesis of an aminoacyl tRNA requires the expenditure of two phosphoryl groups. This is the conventional utilization of ATP to promote chemical synthesis by liberating inorganic pyrophosphate which is then hydrolysed to P_i. GTP hydrolysis to GDP and P_i is involved in two steps. The first is to allow the release of the EFTU from the amino acyl tRNA on the ribosome. The hydrolysis occurs only after a delay believed to be necessary for checking that the correct aminoacyl tRNA is in the acceptor site. Another GTP is hydrolysed during the translocation step. Thus for each peptide bond formed in the elongation process four phosphoryl groups are consumed equal to a standard free energy change of 117.2 kJ.

12 If one or two bases were deleted the reading frame would be shifted so that improper codons would be read thereafter. The result of this would be meaningless polypeptide after the first 100 amino acids. (Redundancy of the code might produce a few correct amino acid residues.) If the new reading frame encountered a termination codon, the polypeptide would be prematurely terminated. If all three bases were deleted, the only effect would be to delete one amino acid residue and it is possible that the protein would be functional, depending on the effect of the deletion on the structure of the protein.

13 The peptidyl transferase has been shown to be active even after extraction of proteins from the ribosome. It is in fact not an enzyme but a ribozyme. Also, deletion of a single adenine base from one of the ribosomal RNAs by ricin inactivates the ribosome.

14 Chaperones are involved in assisting the correct folding of newly synthesized proteins. Heat shock proteins are produced in response to heat shock. The two seem quite unrelated but the connection is that a newly synthesized protein, before it folds, is essentially denatured or partly so. Heat shock also produces denatured proteins, which are unfolded proteins or partly unfolded. Therefore the problem of folding them correctly is the same and it is not surprising that chaperones are produced on heat shock nor that the workers who observed the production of such proteins named them heat shock proteins. 'Chaperone' is a collective term; individual types are called Hsp 60 and Hsp 70, etc.

15 Transcription is the synthesis of mRNA coded for by the gene; translation is the synthesis of polypeptide coded for by the mRNA. The terms derive from the fact that transcription implies copying in the same language (DNA and RNA are both polynucleotides.) Translation implies copying in a different language (mRNA is a polynucleotide, proteins are polypeptides).

16 A codon is a triplet of bases in an mRNA. Out of a total of 64, 61 are used to specify amino acids. If only 20 were used for this purpose there would be 44 codons that did not code for any amino acid. Thus a lot of mutations which now have minimal effects would produce stop codons and destroy the function of the gene. In short, redundancy (multiple codons for single amino acids) confers genetic buffering by minimizing the potential effect of point mutations.

17 It may create a stop codon and prematurely terminate translation. It may have no effect if it simply creates a codon, which because of redundancy, codes for the same amino acid. It may substitute a different amino acid in the protein. This can have effects ranging from negligible (if a very similar amino acid is substituted at a noncritical point) to completely destroying the activity of the protein if it affects a critical residue, for example at the active site of an enzyme or if it prevents normal folding of the polypeptide.

Chapter 25

1 In contranslational transport, the polypeptide is transferred through the target membrane as it is synthesized. This occurs in the transport of proteins into the ER. Posttranslational transport is where the protein is fully synthesized in the cytoplasm, released, and then transported through its target

membrane. Examples are mitochondrial proteins, nuclear proteins, and probably peroxisomal proteins.

2 (It will probably be necessary to refer to Fig. 25.6 to follow this explanation). In almost every case, GTP hydrolysis occurs to produce a conformational change in a protein. Commonly the hydrolysis occurs only after a slight delay, this essentially being a timing device to permit other happenings to occur. An example of this is in the transport of proteins through the ER. After the SRP in its GDP-bound form joins the docking protein on the ER, it has to lose its grip on the signal peptide so that the latter can insert into the translocon. To do this the GDP is exchanged for GTP; this reduces the affinity of the SRP for the signal peptide and increases its affinity for the docking protein. Since the SRP now has to leave the latter and return to the cytoplasm, it hydrolyses the GTP back to GDP which reduces its affinity for the docking protein. However, in its GDT form it would also bind to the signal peptide again if the latter was not safely esconced in the translocon. This is the point of the delay in hydrolysing the GTP—it allows time for the signal peptide to enter the channel.

3 (It will probably be necessary to refer to Fig. 25.15 to follow this explanation). The Ran GDP/GTP exchange enzyme exists only inside the nucleus. The function of Ran-GTP is to effect the release of the cargo from the importin–cargo complex arriving from the cytoplasm. The Ran-GTP–importin complex migrates back to the cytoplasm. It is now necessary for the GTP to be hydrolysed to allow the Ran-GDP to dissociate and release the importin. The Ran-GTPase however must be activated by GAP, the GTPase activating protein. This is found only in the cytoplasm, not in the nucleus (the Ran-GDP itself must return to the nucleus). The released importin picks up another cargo and carries it into the nucleus to start the whole cycle again.

Chapter 26

1 The production is described in Fig. 26.18. GTP hydrolysis has a timing function. In cholera the GTPase is inactivated so that cAMP production remains activated once stimulated.

2 It allosterically activates protein kinase A (PKA). This can have metabolic effects depending on phosphorylation of key enzymes but, in addition, there are genes with cAMP-responsive elements (CREs). Inactive transcriptional factors can be activated by phosphorylation by PKA. Therefore cAMP can have extensive gene control effects.

3 It activates a cytoplasmic guanylate cyclase that produces cGMP. The latter is a second messenger in some systems.

4 The phosphatidylinositol cascade involves the receptor-mediated activation of membrane-bound phospholipase C. This releases inositol trisphosphate (IP$_3$) and diacylglycerol (DAG). The former increases cytoplasmic Ca^{2+}; the latter activates a protein kinase PKC. Thus, IP$_3$ and DAG are second messengers. The scheme is shown in Fig. 26.21.

5 **(a)** Involves an allosteric change of the cytoplasmic domain on binding of the hormone.
(b) Involves receptor dimerization and activation of self-phosphorylation of tyrosine residues by an intrinsic kinase.
(c) Involves receptor dimerization and association with a separate tyrosine kinase that phosphorylates the receptor (see Figs 26.7 and 26.15).

6 Neurotransmitters; homones, cytokines, and growth factors; vitamin D$_3$ and retinoic acid; nitric oxide is in a class of its own.

7 In both cases the signal combines with a receptor protein and this triggers a response which in terms of gene control usually causes a protein to enter the nucleus and effect gene control. In the case of a lipid-soluble signal such as glucocorticoid it combines with a cytoplasmic receptor protein which enters the nucleus and functions as a transcription factor. In the case of a water-soluble signal such as EGF, it combines with a membrane receptor, and this results ultimately in a kinase entering the nucleus and activating a transcription factor. The two systems are fundamentally similar, the differences being in the detail. (In some cases the lipid signal receptor resides in the nucleus, but the same general principle applies.)

8 It is likely to be a component of a signalling pathway which specifically binds to an activated membrane receptor of the tyrosine kinase type.

9 Both have tyrosine kinase-associated receptors. The Ras associated receptors are themselves tyrosine kinases, whereas the JAK/STAT pathways depend on cytoplasmic kinases associating with the receptors. The Ras pathway is activated by many hormones whereas the other pathway is best known for its association with cytokine signals such as interferon. The Ras receptor activates a long cascade of proteins terminating in a kinase that migrates into the nucleus and activates transcription factors. The JAK/STAT receptors attract the JAK kinases which phosphorylate tyrosine groups of the receptor. The STAT proteins are attracted from the cytoplasm to bind to these and are themselves phosphorylated by the double-headed JAK kinases. The

phosphorylated STAT proteins migrate into the nucleus where they activate genes. Thus the latter pathway is a much more direct system from receptor to gene whereas the Ras pathway is very long. This is probably to allow amplification of the signal. Less certainly, it may also give more opportunity for cross-talk between Ras and other pathways and provide opportunities for extra controls.

10 All G proteins are heterotrimeric proteins associated with membrane receptors. The latter do not become phosphorylated but on the binding of the cognate ligand undergo a conformational change. This in turn causes a change in the conformation of the attached G protein such that the GDP attached to the α subunit is exchanged for GTP. This GTP-subunit detaches from its partner β and γ subunits and migrates to a target enzyme located on the membrane. In the case of a typical G protein, the one associated with the β-adrenergic receptor, it activates adenylate cyclase which produces a second messenger cAMP. The activation is terminated by the slow intrinsic GTPase activity of the α subunit which acts as a timing device. The α subunit in its GDP form migrates back to the receptor to reform the heterotrimeric G protein. If the receptor is still activated, the process is repeated. The versatility is due to the fact that different receptors have different G proteins whose activated α subunit activates or inhibits different enzymes involved in production of second messengers. Thus stimulation of receptors by epinephrine may activate or inhibit cAMP production depending on the nature of the particular G protein subunit associated with the receptor; in the phosphatidylinositol system, the G protein subunit activates PLC as one example of the different target enzyme specificities that exist. In this case the second messengers are DAG and PIP_3. (Note that other GTPase proteins exist, but the term G protein is restricted to the type described above.)

11 Not really; cGMP is continually produced in the rod cells to keep the cation channels open. The effect of light is to cause the activation of the enzyme which destroys cGMP. In the case of the nitric oxide signalling system cGMP is a second messenger for the signal activates either a membrane receptor or an intracellular receptor to produce the cGMP.

12 The insulin receptor does not dimerize but it structurally resembles a covalently permanently dimerized receptor of the EGF type.

Chapter 27

1 As in Fig. 27.3.

2 The principle is that each B cell assembles its own Ig genes randomly from a selection as described in Fig. 27.4.

3 B cells produce antibodies (after activation and maturation into plasma cells). Helper T cells are (in most cases) required for B cells to do this. Cytotoxic or killer T cells bind to host cells displaying a foreign antigen and kill them by perforating their membrane or inducing apoptosis.

4 It is a type of phagocyte that engulfs a foreign antigen, processes it into pieces, and displays it in combination with its MHC molecules. If an inactive helper T cell combines with the antigen–MHC complex it is activated (see Fig. 27.4).

5 In both cases MHC class I.

6 Each B cell produces a different antibody. For any given antigen there will be very few B cells specific for it, but, when the antigen binds to the displayed antibody, the B cell proliferates into a clone. Thus the antigen automatically selects which B cells are to proliferate.

7 During *primary* maturation of B cells in the bone marrow or of T cells in the thymus, if an antigen binds, the cell is eliminated, the principle being that such antigens will be 'self'. After release from the bone marrow or thymus, the binding to an antigen activates the cells.

8 Differential splicing of introns; at the 3′ end of the immunoglobulin gene is an exon that codes for an anchoring polypeptide sequence. At the onset of secretion a switch in the splicing eliminates this.

9 The onset of secretion is accompanied by rapid somatic mutation in the cells, which modifies the variable site. Since binding of the antigen causes cell proliferation this constitutes a selection mechanism for those cells producing as 'better' antibody. This is known as affinity maturation.

10 CD4 is present on helper T cells. The protein binds to a constant protein component of MHC class II molecules and thus confines interactions to B cells, helper T cells, and antigen-presenting cells. Cytotoxic T cells have CD8 which binds to MHC class I molecules; this restricts the interaction of the killer T cells to host cells. The CD4 protein is the receptor by which the AIDS virus infects helper T cells.

Chapter 28

1 By exploiting the normal receptor-mediated endocytosis mechanism as illustrated in Fig. 28.1.

2 The latter must carry into the cell an RNA replicase; the RNA of the (+) virus is an mRNA which codes for an RNA replicase.

3 The vaccinia replicates in the cytoplasm; the host polymerase is in the nucleus.

4 The retroviral RNA is copied into double-stranded DNA which integrates into the host chromosome as shown in Fig. 28.5.

5 The mechanism is explained in Fig. 28.3. It collects its membrane plus proteins as it buds off from the plasma membrane.

6 Antibody protection against influenza is directed at the hemagglutinin protein. This is continually mutated but, because there are many epitopes on the molecule, the immune protection loss is partial and gradual. This antigenic drift leaves people susceptible to infection but residual protection means that it is mild. If by a recombination between different strains a totally new hemagglutinin is present (antigenic shift), a lethal pandemic can result. The pathogenicity is also related to the ability of the nucleocapsid to escape into the cytoplasm (see page 469).

7 The virus has, on its surface, proteins that specifically bind to neuraminic acid (sialic acid). The red blood cell glycophorin terminates in this acid. On mixing, the multiple binding sites on the virus crosslink the red blood cells.

8 The virus binds to its host cell by a glycoprotein terminating in neuraminic acid. Why should the virus have an enzyme for destroying the very receptor which it needs to gain entry to the cell? The answer may be (a) to facilitate release of new virus particles that might adhere to the cell glycoproteins; or (b) to liquefy mucins that have neuraminic acid in their structure and so facilitate the virus reaching the cell surface. Mucin liquefaction might also help spread the virus through sneezing.

9 It may enter the lytic or lysogenic phase as illustrated in Fig. 28.7.

1 (a) Small, infectious, naked RNA molecules without any ~~ein~~ or other type of coat.

(b) No. There are no sufficiently long coding stretches of bases to permit protein synthesis.

11 The virus may contain in its capsid the required enzyme which it carries into the infected cell. The influenza virus contains its own RNA replicase; the vaccinia virion carries its own RNA polymerase to transcribe its genes because it replicates in the cytoplasm. Other DNA viruses which enter the nucleus use host RNA polymerase which is present in the nucleus (but not in the cytoplasm). Any other enzymes that may be required are coded for by the virus and synthesized by the host protein-synthesizing system. A viroid is a naked piece of RNA and therefore cannot carry any enzymes. Since there are no known reading frames in viroid RNA capable of coding for proteins, it cannot induce the plant cell to synthesize any required enzymes. Viroids and other small circular RNA pathogens replicate by the rolling circle method producing concatenated RNA genomes which have to be separated. There is no known enzyme capable of doing this, so the strategy is to self-cleave by ribozyme action.

Chapter 29

1 Pancreatic DNase randomly cuts DNA. A restriction enzyme cuts only at specific short sequences of bases.

2 The adenine bases in all of the relevant hexamer sequences of *E. coli* strain R DNA are methylated so that the enzyme does not recognize it. Invading foreign DNA will not be so protected.

3 It refers to ends resulting from a staggered cut made by many restriction enzymes.

The overhanging ends will automatically base pair and 'stick' together.

4 A genomic clone refers to a cloned section of DNA identical to the sequence of DNA in a chromosome. A cDNA clone means complementary DNA. It refers to cloned DNA identical to the mRNA for a gene. The cDNA lacks introns; genomic clones have them.

5 These are outlined in Fig. 29.1.

6 These are outlined in Fig. 29.2.

7 The nucleotide lacks a 3′-OH and therefore, when added by DNA polymerase to a growing DNA chain, terminates that chain. The application of this to sequencing is described on page 484.

8 A specific section of DNA on a chromosome can be amplified logarithmically. It involves copying the section of DNA and copying the copies *ad infinitum*. Its importance is that a minute amount of DNA, too small for any studies, can be amplified at will. The essential, apart from enzymes and substrates, is that you must have primers for copying the section in both directions and this means knowing the base sequences at either end of the piece to be amplified.

9 It is an engineered plasmid with a convenient insertion site for, say, a cDNA clone for a specific protein, but which also contains appropriate bacterial DNA transcriptional signals (a promoter) and translational signals. The DNA is transcribed and the mRNA translated in a bacterial cell. It can be used to produce large quantities of specific proteins.

10 It is a technique for detecting gene abnormalities. If DNA is cut by a pattern of restriction enzymes and the resultant pieces separated on an electrophoretic gel, a pattern of bands can be visualized by hybridization methods. Mutations may produce or remove restriction sites so that the pattern may be altered.

11 Stem cells divide and the progeny cells can either proceed to develop into mature terminally differentiated cells or divide into more stem cells. This can occur indefinitely, unlike somatic cells which have a limited potential to divide. Thus a stem cell population keeps up a continual supply of replacement cells. Some stem cells are committed to become a certain class of cells; bone marrow stem cells differentiate into many types of blood cells. Embryonic stem cells found in blastocysts are totipotent; they can give rise to any type of the cell in the body. They are of especial interest because they can be cultivated *in vitro* and retain this totipotency. Apart from their use in producing knock-out mice, they have the potential to be of enormous medical importance in that they might be developed in controlled *in vitro* conditions into replacement cells for the human body. Production of nerve cells to repair neurological damage is but one possible example.

12 (The procedure is complex and you will probably have to follow Figs. 29.8 and 29.9 to understand this answer.) The principle is to inactivate a specific gene by homologous recombination. For this a targeting vector is constructed so that the normal gene is replaced by an inactivated one. The principle of the method is to use embryonic stem cells obtained from a mouse blastocyst. These are mutated by the vector and cells with the targeted gene correctly replaced are grown up and injected into the blastocyst of another mouse. The blastocyst is implanted into a mouse to develop. The offspring contains cells derived from both the mutated cells and the recipient blastocystcells. By appropriate cross-breeding, mice which are homozygous to the mutated gene can be obtained. In the procedure, the targeted stem cells are taken from a blastocyst from a mouse of a coat colour which is different from that from which the recipient blastocyst is taken. This enables the process to be followed by coat colour of the progeny.

Chapter 30

1 A protooncogene is a normal gene which can become converted into an oncogene. It is a gene coding for a protein which is involved in cellular control processes most commonly those which affect cell replication. They were discovered when it was found that oncogenes of retroviruses, known to be cancer producing, have almost identical counterparts in normal cells. Conversion of a protooncogene to an oncogene usually involves the creation of excessive signalling—either the gene is overexpressed or the gene product (the signalling molecule) has inappropriately prolonged life. The change may be caused by a single base mutation or by the gene being placed under the control of an excessively active promoter. In the case of Burkitt's lymphoma chromosomal rearrangements places a protooncogene under the control of an immunoglobulin gene promoter; or a very active retroviral promoter may be inserted at a point in a chromosome so as to control the expression of the protooncogene. Alternatively a mutation may produce an excessively long-lived transcription factor. In the case of the Ras oncogenes, mutations eliminate the GTPase activity so that the signalling pathway cannot be switched off.

2 In G_1, a growth factor signal is necessary to activate genes that code for the G_1-specific cyclins needed to activate the kinases which push the cell cycle through the restriction point. In the absence of growth factor signal the cycle is arrested in the G_0 phase awaiting the arrival of a signal.

3 The p53 gene is normally expressed at a low level but this increases if the DNA is damaged. The resultant high levels of p53 protein activate genes that cause the inhibition of the cyclin-dependent kinases which are needed for the cell to proceed through the restriction point in G1. The delay allows time for the DNA to be repaired. When this is done, the p53 expression is no longer stimulated, the p53 protein diminishes and the

cell cycle proceeds. If the DNA is not repaired the p53 gene can also induce apotosis of the cell.

Chapter 31

1 5-Aminolevulinate synthase (ALA) carries out the reaction shown in Fig. 31.4 and ALA dehydratase produces the pyrrole, porphobilinogen, by the reaction in Fig. 31.6.

2 Increased activity in liver is associated with acute intermittent porphyria (the 'mad king' disease) though the connection between ALA production and the neurological effect is not understood.

3 It is based on an iron-dependent inhibition of mRNA translation as shown in Fig. 31.8.

4 As shown in Fig. 31.11, myoglobin has the higher affinity and the curve is hyperbolic as compared with the sigmoid curve of hemoglobin.

5 According to the concerted model, the hemoglobin exists in a low-affinity state and a high-affinity state. Binding of oxygen displaces the equilibrium between the two to increase the amount in the high-affinity state (see page 514).

6 On binding of oxygen to the heme iron, Fe moves into the plane of the porphyrin ring. This requires the slightly domed tetrapyrrole to flatten. The Fe is attached to the F8 histidine of the protein and the molecule rearranges itself. This allows the tetrapyrrole to flatten. The movement affects subunit interactions causing the tetramer to change its conformation from the T to the R state (see Fig. 31.13).

7 Adult hemoglobin in the deoxygenated state has a cavity that binds 2:3-bisphosphoglycerate (BPG) by positive charges on the protein side chains. Binding of BPG is possible only in the deoxygenated state and therefore binding increases unloading of oxygen—it reduces affinity of the hemoglobin for oxygen. Fetal hemoglobin has a γ subunit instead of the adult β one. It lacks one of the BPG charge-binding groups. BPG therefore binds to fetal hemoglobin less tightly, thus raising its affinity for oxygen above that of the maternal hemoglobin.

8 It is necessary for the major transport of CO_2 as HCO_3^- as illustrated in Fig. 31.15.

Chapter 32

1 Fast twitch fibres rely mainly on glycolysis for immediate contraction since this can be rapidly activated. However, they quickly exhaust. The white muscle of fish exemplify this—this muscle is for escape reactions. Fast twitch fibres exist in humans in the eye muscles. Slow twitch fibres obtain their ATP from oxidative metabolism and can function for much longer without exhaustion but the energy supply cannot be increased as rapidly as in slow fibres. In man, muscles have both types. Back muscles, which maintain body posture for long periods, are mainly of the slow twitch type.

2 Acetylcholine stimulation of the muscle receptor leads to liberation of Ca^{2+} from the sarcoplasmic reticulum into the myofibril. On the actin filaments are tropomyosin molecules and, in turn, these have a Ca^{2+}-sensitive troponin complex attached to them. When Ca^{2+} binds, a conformational change occurs; it was believed that the tropomyosin blocked the attachment of myosin heads to actin. This is now questioned, A Ca^{2+}/ATPase returns the Ca^{2+} to the sarcoplasmic reticulum and terminates the contraction.

3 In smooth muscle myosin heads there is a p-light chain that inhibits the binding of the myosin head to the actin fibre, preventing contraction. Neurological stimulation opens channels allowing Ca^{2+} to enter the cell. The Ca^{2+} combines with calmodulin regulatory protein, which activates a myosin light chain kinase. Phosphorylation of the light chain abolishes its inhibitory effect and contraction occurs.

4 In one case ATP is synthesized from ADP + P_i. The synthesis on the enzyme surface involves little or no free energy change, but energy is required for release of ATP. This is mediated by a conformation change in the proteins of the ATP synthase head. In the case of myosin, ATP is split into ADP + P_i but on the enzyme surface there is little free energy change involved. It is on the release of the ADP that the power stroke of contraction occurs. This is mediated by a conformation change in the myosin head. In neither case is there a covalent intermediate in the synthesis or breakdown of the ATP.

5 It used to be thought that the myosin head swung at its point of attachment to the coiled-coil rod of myosin so that the angle of the head to the actin filament changed. This is now known to be incorrect. The angle of attachment of the head to the actin filament does not change but an α helix of the head known as the lever arm swings to cause the power stroke. The diagram in Fig. 32.6(a) provides an illustration of this.

Chapter 33

1 Contraction occurs in such cells. Actin filaments anchored in the cell membrane provide the means for small bundles of myosin molecules to exert a contractile force. A second role is

for actin filaments to form a transport track along which special minimyosin molecules move. The latter have a myosin head but the rod-like structure is replaced by a short tail to which vesicles may be attached.

2 Hollow tubes formed by the polymerization of tubulin protein subunits. They have a definite polarity with (+) and (−) ends.

3 The microtubule organizing centre protects the (−) end. The (+) ends of *growing* tubules are protected by a tubulin–GTP cap. The GTP is slowly hydrolysed, removing the protection. Unless a new tubulin–GTP molecule is added before this, the microtubule collapses. When the microtubule reaches a 'target' structure it is then protected.

4 They are molecular motors that move along microtubule tracks and can pull a load with them. Kinesin and dynein travel in opposite directions (in terms of microtubule polarity) on the microtubule track.

5 No, they cannot contract. It is believed that the shortening is due to depolymerization but precisely what causes chromosome movement is not certain.

6 Filaments 10 nm in diameter which are intermediate in this respect between microtubule filaments (20 nm) and actin filaments (6 nm).

7 In specialized cases, they form the structural basis of hair. They may be associated with conferring toughness on structures such as neurofilaments and the Z discs of sarcomeres. However, mutant cultured cells survive without them.

Figure acknowledgements

Fig. 3.7 (a)–(d). Redrawn from figs 2.10a, 2.15b, 4.1a, and 4.4, respectively, in Branden, C. and Tooze, J. (1991), *Introduction to protein structure*, Garland Publishing, New York. Figs 2.10a, 4.1a and 4.4 by permission of Taylor and Francis, Inc. and C. Branden; Fig. 2.15b, structure determined by Cotton, Hazer, Richardson, and colleagues, and reproduced with permission.

Fig. 3.8 Redrawn from fig. 2 in Jentoft, N. (1990), *Trends Biochem. Sci.* **15**, 293, with permission of Elsevier Science.

Fig. 3.10 Slightly modified from fig. 1 of Dodson, G. and Wlodawer, A. (1998). *Trends Biochem Sci.*, **23**, 347, with permission from Elsevier Science.

Fig. 3.24 Photograph courtesy of Dr Anne Chapman-Smith, Department of Molecular Biosciences, University of Adelaide.

Fig. 3.25 The spectrum is taken from fig. 5 of Wabnitz, P. A., Bowie, J. H., and Tyler, M. J. (1999), *Rapid Comm. Mass Spectrom.*, **13**, 2498–2, reproduced with permission of John Wiley & Sons Limited, and J. Bowie.

Fig. 7.16 Modified from fig. 1 of Osborne, T. F. (1997), *Current Biology*, **7**, R172, with permission from Elsevier Science.

Fig. 9.27 Reproduced from fig. 1a of Noji, H., Yasuda, R., Yoshida, M., and Kinosita Jr, K. (1997), *Nature*, **386**, 299–302, with permission from Macmillan Magazines Ltd., and M. Yoshida.

Fig. 9.28 Modified from fig. 1 of Rastogi, V. K. and Girvin, M. E. (1999), *Nature*, **402**, 263–8, with permission from Macmillan Magazines Ltd., and M. Girvin.

Fig. 9.30 Reproduced from fig. 2(c) of Abrahams, J. P., Leslie, A. G. W., Lutter, R., and Walker, J. E. (1994), *Nature*, **370**, 621–8, with permission from Macmillan Magazines Ltd., and J. Walker.

Fig. 13.13 Modified from fig. 7 in Karnieli, E. *et al.* (1981), *J. Biol. Chem.*, **256**, 4772, with permission of E. Karnieli.

Fig. 17.3 Reproduced from fig. 5(b) of Baumeister, W., Walz, J., Zühl, F., and Seemüller, E. (1998), *Cell*, **92**, 367–80, with permission from Elsevier Science and W. Baumeister.

Fig. 20.3 (a) and (b) adapted from figs 2.4a and 3.1 respectively in Calladine, C. R. and Drew, H. W. (1994), *Understanding DNA*, Academic Press, with permission of Academic Press Ltd.

Fig. 20.4 Reproduced from fig. 2.1 in Ptashne, M. (1987), *A genetic switch*, Cell Press and Blackwell Scientific Publications, Oxford, with permission of Elsevier Science and M. Ptashne.

Fig. 20.5 Reproduced from fig. 1.8 in Lewin, B. (2000), *Genes VII*, Cell Press and Oxford University Press.

Fig. 20.8 Fig. 28.19 in Lewin, B. (1994), *Genes V.* Oxford University Press, Oxford. (The electron micrograph was provided to Lewin by Barbara Hamkalo.)

Fig. 21.6 Fig. 2.24 in Singer, M. and Berg, P. (1991), *Genes and genomes.* University Science Books, Sausalito, California.

Fig. 21.7 Fig. 2.26a in Singer, M. and Berg, P. (1991), *Genes and genomes.* University Science Books, Sausalito, California.

Fig. 21.15 Fig. 3a from Krishna, T. S. R., Kong, X-P., Gary, S., Burgers, P. M. and Kuriyan, J. (1994), *Cell*, **79**, 1233. Copyright by Elsevier Science. Photograph kindly provided by Dr John Kuriyan.

Fig. 21.16 Based on fig. 1 of Kelman, Z. and O'Donnell, M. (1995), *Ann. Rev. Biochem.*, **64**, 171, with permission of Annual Reviews.

Fig. 21.18 Fig. 3 from Echols, H. and Goodman, M. F. (1991), *Ann. Rev. Biochem.*, **60**, 490; reproduced with copyright permission of Annual Reviews Inc. Original from Kennard, O. (1987), *Nucleic Acids and Molecular Biology*, **1**, 25.

Fig. 21.25 After fig. 34.1 from Lewin, B. (1994), *Genes V*, Oxford University Press, Oxford.

Fig. 22.11 (b). Fig. 3a in Ohlendorf, D. H., Anderson, U. F. and Matthews, B. W. (1983), *J. Molec. Evolution*, **19**, 109. Reproduced with copyright permission of Springer-Verlag.

Fig. 23.6 Reproduced from fig. 29.10 in Lewin, B. (1994), *Genes V*, Oxford University Press, Oxford.

Fig. 23.15 Adapted from fig. 4 in Ross, J. (1995), *Microbiol. Rev.*, **59**, 423.

Fig. 23.16 Fig. 2 from Ross, J. (1995), *Microbiol. Rev.*, **59**, 423.

Fig. 23.21 Based on fig. 1 of Haseloff, J. and Gerlach, W. L. (1988), *Nature*, **334**, 585.

Fig. 24.4 Fig. 9.5 from Lewin, B. (1994), *Genes V*, Oxford University Press, Oxford.

Fig. 24.9 Adapted from fig. 4 of Noller, H. F. (1991), *Ann. Rev. Biochem.*, **60**, 191; modified with copyright permission of Annual Reviews, Inc.

Fig. 24.10 From fig. 1 of Noller, H. F. (1991), *Ann. Rev. Biochem.*, **60**, 193. Reproduced with copyright permission of Annual Reviews, Inc.

Fig. 24.14 From fig. 1 of Sigler, R. B. *et al.* (1998), *Ann. Rev. Biochem.*, **67**, 581–608, with permission from Annual Reviews.

Fig. 25.15 Based on fig. 2 of Mattaj, I. W. and Englmeier, L., *Ann. Rev. Biochem.* (1998), **67**, 265, with permission from Annual Reviews and I. Mattaj.

Fig. 25.16 Based on fig. 2 of Mattaj, I. W. and Englmeier, L., *Ann. Rev. Biochem.* (1998), **67**, 265, with permission from Annual Reviews and I. Mattaj.

Fig. 25.17 Modified from fig. 1 of Jans, D. A. (1998), *Australian Biochemist*, **29**, 5–10, with permission of the Australian Society for Biochemistry and Molecular Biology Inc. and D. Jans.

Fig. 26.16 After fig. 2 in Dohlman, H. G., Caron, M. G., and Lefkowitz, R. J. (1987), *Biochemistry*, **26**, 2660, with copyright permission of the American Chemical Society.

Fig. 27.1 After fig. 1 in Metcalf, D. (1991), *Phil. Trans. Roy. Soc., Lond.* **B333**, 147. Reproduced (modified) by permission of the author and the Royal Society.

Fig. 28.2 After fig. 1.1 in Ptashne, M. (1986), *A genetic switch* (2nd edn), Cell Press and Blackwell Publications, Oxford, with permission of Elsevier Science and M. Ptashne.

Fig. 28.8 Kindly provided by Professor R. H. Symons, Department of Plant Science, Waite Institute, University of Adelaide.

Fig. 31.1 Kindly provided by Professor W. G. Breed, Department of Anatomy, University of Adelaide.

Fig. 32.6 From figs 7 and 8 in Greeves, M. A. and Holmes, K. C. (1999), *Ann. Rev. Biochem.*, **68**, 687–728, reproduced with permission from Annual Reviews and M. Greeves.

Fig. 33.2 After fig. 2.13 in Lewin, B. (1994), *Genes V*, Oxford University Press, Oxford.

Fig. 33.6 After fig. 1e in McNiver, M. A. and Ward, J. B. (1998), *J. Cell Biol.*, **106**, 111, with permission from the Rockefeller University Press.

Index